Optimization Models

Emphasizing practical understanding over the technicalities of specific algorithms, this elegant textbook is an accessible introduction to the field of optimization, focusing on powerful and reliable convex optimization techniques. Students and practitioners will learn how to recognize, simplify, model and solve optimization problems – and apply these basic principles to their own projects.

A clear and self-contained introduction to linear algebra, accompanied by relevant real-world examples, demonstrates core mathematical concepts in a way that is easy to follow, and helps students to understand their practical relevance.

Requiring only a basic understanding of geometry, calculus, probability and statistics, and striking a careful balance between accessibility and mathematical rigor, it enables students to quickly understand the material, without being overwhelmed by complex mathematics.

Accompanied by numerous end-of-chapter problems, an online solutions manual for instructors, and examples from a diverse range of fields including engineering, data science, economics, finance, and management, this is the perfect introduction to optimization for both undergraduate and graduate students.

Giuseppe C. Calafiore is an Associate Professor at Dipartimento di Automatica e Informatica, Politecnico di Torino, and a Research Fellow of the Institute of Electronics, Computer and Telecommunication Engineering, National Research Council of Italy.

Laurent El Ghaoui is a Professor in the Department of Electrical Engineering and Computer Science, and the Department of Industrial Engineering and Operations Research, at the University of California, Berkeley.

Optimization Models

Giuseppe C. Calafiore
Politecnico di Torino

Laurent El Ghaoui
University of California, Berkeley

CAMBRIDGE
UNIVERSITY PRESS

CAMBRIDGE
UNIVERSITY PRESS

University Printing House, Cambridge CB2 8BS, United Kingdom

One Liberty Plaza, 20th Floor, New York, NY 10006, USA

477 Williamstown Road, Port Melbourne, VIC 3207, Australia

314-321, 3rd Floor, Plot 3, Splendor Forum, Jasola District Centre, New Delhi - 110025, India

79 Anson Road, #06-04/06, Singapore 079906

Cambridge University Press is part of the University of Cambridge.

It furthers the University's mission by disseminating knowledge in the pursuit of education, learning and research at the highest international levels of excellence.

www.cambridge.org
Information on this title: www.cambridge.org/9781107050877

First published 2014
4th printing 2019

A catalogue record for this publication is available from the British Library

ISBN 978-1-107-05087-7 Hardback

Internal design based on tufte-latex.googlecode.com

Additional resources for this publication at www.cambridge.org/optimizationmodels

Cambridge University Press has no responsibility for the persistence or accuracy of URLs for external or third-party internet websites referred to in this publication, and does not guarantee that any content on such websites is, or will remain, accurate or appropriate.

Dedicated to my parents, and to Charlotte.
G. C.

Dedicated to Louis, Alexandre and Camille.
L. El G.

Contents

Preface

OPTIMIZATION REFERS TO a branch of applied mathematics concerned with the minimization or maximization of a certain function, possibly under constraints. The birth of the field can perhaps be traced back to an astronomy problem solved by the young Gauss. It matured later with advances in physics, notably mechanics, where natural phenomena were described as the result of the minimization of certain "energy" functions. Optimization has evolved towards the study and application of algorithms to solve mathematical problems on computers.

Today, the field is at the intersection of many disciplines, ranging from statistics, to dynamical systems and control, complexity theory, and algorithms. It is applied to a widening array of contexts, including machine learning and information retrieval, engineering design, economics, finance, and management. With the advent of massive data sets, optimization is now viewed as a crucial component of the nascent field of data science.

In the last two decades, there has been a renewed interest in the field of optimization and its applications. One of the most exciting developments involves a special kind of optimization, convex optimization. Convex models provide a reliable, practical platform on which to build the development of reliable problem-solving software. With the help of user-friendly software packages, modelers can now quickly develop extremely efficient code to solve a very rich library of convex problems. We can now address convex problems with almost the same ease as we solve a linear system of equations of similar size. Enlarging the scope of tractable problems allows us in turn to develop more efficient methods for difficult, non-convex problems.

These developments parallel those that have paved the success of numerical linear algebra. After a series of ground-breaking works on computer algorithms in the late 80s, user-friendly platforms such as Matlab or R, and more recently Python, appeared, and allowed generations of users to quickly develop code to solve numerical problems. Today, only a few experts worry about the actual algorithms and techniques for solving numerically linear systems with a few thousands of variables and equations; the rest of us take the solution, and the algorithms underlying it, for granted.

Optimization, more precisely, convex optimization, is at a similar stage now. For these reasons, most of the students in engineering, economics, and science in general, will probably find it useful in their professional life to acquire the ability to recognize, simplify, model, and solve problems arising in their own endeavors, while only few of them will actually need to work on the details of numerical algorithms. With this view in mind, we titled our book *Optimization Models*, to highlight the fact that we focus on the "art" of understanding the nature of practical problems and of modeling them into solvable optimization paradigms (often, by discovering the "hidden convexity" structure in the problem), rather than on the technical details of an ever-growing multitude of specific numerical optimization algorithms. For completeness, we do provide two chapters, one covering basic linear algebra algorithms, and another one extensively dealing with selected optimization algorithms; these chapters, however, can be skipped without hampering the understanding of the other parts of this book.

Several textbooks have appeared in recent years, in response to the growing needs of the scientific community in the area of convex optimization. Most of these textbooks are graduate-level, and indeed contain a good wealth of sophisticated material. Our treatment includes the following distinguishing elements.

- The book can be used both in undergraduate courses on linear algebra and optimization, and in graduate-level introductory courses on convex modeling and optimization.

- The book focuses on *modeling* practical problems in a suitable optimization format, rather than on *algorithms* for solving mathematical optimization problems; algorithms are circumscribed to two chapters, one devoted to basic matrix computations, and the other to convex optimization.

- About a third of the book is devoted to a self-contained treatment of the essential topic of linear algebra and its applications.

- The book includes many real-world examples, and several chapters devoted to practical applications.

- We do not emphasize general non-convex models, but we do illustrate how convex models can be helpful in solving some specific non-convex ones.

We have chosen to start the book with a first part on linear algebra, with two motivations in mind. One is that linear algebra is perhaps the most important building block of convex optimization. A good command of linear algebra and matrix theory is essential for understanding convexity, manipulating convex models, and developing algorithms for convex optimization.

A second motivation is to respond to a perceived gap in the offering in linear algebra at the undergraduate level. Many, if not most, linear algebra textbooks focus on abstract concepts and algorithms, and devote relatively little space to real-life practical examples. These books often leave the students with a good understanding of concepts and problems of linear algebra, but with an incomplete and limited view about *where* and *why* these problems arise. In our experience, few undergraduate students, for instance, are aware that linear algebra forms the backbone of the most widely used machine learning algorithms to date, such as the PageRank algorithm, used by Google's web-search engine.

Another common difficulty is that, in line with the history of the field, most textbooks devote a lot of space to eigenvalues of general matrices and Jordan forms, which do have many relevant applications, for example in the solutions of ordinary differential systems. However, the central concept of singular value is often relegated to the final chapters, if presented at all. As a result, the classical treatment of linear algebra leaves out concepts that are crucial for understanding linear algebra as a building block of practical optimization, which is the focus of this textbook.

Our treatment of linear algebra is, however, necessarily partial, and biased towards models that are instrumental for optimization. Hence, the linear algebra part of this book is not a substitute for a reference textbook on theoretical or numerical linear algebra.

In our joint treatment of linear algebra and optimization, we emphasize tractable models over algorithms, contextual important applications over toy examples. We hope to convey the idea that, in terms of reliability, a certain class of optimization problems should be considered on the same level as linear algebra problems: reliable models that can be confidently used without too much worry about the inner workings.

In writing this book, we strove to strike a balance between mathematical rigor and accessibility of the material. We favored "operative" definitions over abstract or too general mathematical ones, and practical relevance of the results over exhaustiveness. Most proofs of technical statements are detailed in the text, although some results are provided without proof, when the proof itself was deemed not to be particularly instructive, or too involved and distracting from the context.

Prerequisites for this book are kept at a minimum: the material can be essentially accessed with a basic understanding of geometry and calculus (functions, derivatives, sets, etc.), and an elementary knowledge of probability and statistics (about, e.g., probability distributions, expected values, etc.). Some exposure to engineering or economics may help one to better appreciate the applicative parts in the book.

Book outline

The book starts out with an overview and preliminary introduction to optimization models in Chapter 1, exposing some formalism, specific models, contextual examples, and a brief history of the optimization field. The book is then divided into three parts, as seen from Table 1.

Part I is on linear algebra, Part II on optimization models, and Part III discusses selected applications.

Table 1 Book outline.

	1	Introduction
I Linear algebra	2	Vectors
	3	Matrices
	4	Symmetric matrices
	5	Singular value decomposition
	6	Linear equations and least squares
	7	Matrix algorithms
II Convex optimization	8	Convexity
	9	Linear, quadratic, and geometric models
	10	Second-order cone and robust models
	11	Semidefinite models
	12	Introduction to algorithms
III Applications	13	Learning from data
	14	Computational finance
	15	Control problems
	16	Engineering design

The first part on linear algebra starts with an introduction, in Chapter 2, to basic concepts such as vectors, scalar products, projections, and so on. Chapter 3 discusses matrices and their basic properties, also introducing the important concept of factorization. A fuller story on factorization is given in the next two chapters. Symmetric matrices and their special properties are treated in Chapter 4, while Chapter 5 discusses the singular value decomposition of general matrices, and its applications. We then describe how these tools can be used for solving linear equations, and related least-squares problems, in Chapter 6. We close the linear algebra part in Chapter 7, with a short overview of some classical algorithms. Our presentation in Part I seeks to emphasize the optimization aspects that underpin many linear algebra concepts; for example, projections and the solution of systems of linear equations are interpreted as a basic optimization problem and, similarly, eigenvalues of symmetric matrices result from a "variational" (that is, optimization-based) characterization.

The second part contains a core section of the book, dealing with optimization models. Chapter 8 introduces the basic concepts of convex functions, convex sets, and convex problems, and also focuses on some theoretical aspects, such as duality theory. We then proceed with three chapters devoted to specific convex models, from linear, quadratic, and geometric programming (Chapter 9), to second-order cone (Chapter 10) and semidefinte programming (Chapter 11). Part II closes in Chapter 12, with a detailed description of a selection of important algorithms, including first-order and coordinate descent methods, which are relevant in large-scale optimization contexts.

A third part describes a few relevant applications of optimization. We included machine learning, quantitative finance, control design, as well as a variety of examples arising in general engineering design.

How this book can be used for teaching

This book can be used as a resource in different kinds of courses.

For a senior-level undergraduate course on *linear algebra and applications*, the instructor can focus exclusively on the first part of this textbook. Some parts of Chapter 13 include relevant applications of linear algebra to machine learning, especially the section on principal component analysis.

For a senior-level undergraduate or beginner graduate-level course on *introduction to optimization*, the second part would become the central component. We recommend to begin with a refresher on basic linear algebra; in our experience, linear algebra is more difficult to teach than convex optimization, and is seldom fully mastered by students. For such a course, we would exclude the chapters on algorithms, both Chapter 7, which is on linear algebra algorithms, and Chapter 12, on optimization ones. We would also limit the scope of Chapter 8, in particular, exclude the material on duality in Section 8.5. For a graduate-level course on *convex optimization*, the main material would be the second part again. The instructor may choose to emphasize the material on duality, and Chapter 12, on algorithms. The applications part can serve as a template for project reports.

Bibliographical references and sources

By choice, we have been possibly incomplete in our bibliographical references, opting to not overwhelm the reader, especially in the light of the large span of material covered in this book. With today's online resources, interested readers can easily find relevant material. Our only claim is that we strove to provide the appropriate search terms. We hope that the community of researchers who have contributed to this fascinating field will find solace in the fact that the success of an idea can perhaps be measured by a lack of proper references.

In writing this book, however, we have been inspired by, and we are indebted to, the work of many authors and instructors. We have drawn in particular from the largely influential textbooks listed on the side.[1] We also give credit to the excellent course material of the courses EE364a, EE364b (S. Boyd), EE365 (S. Lall) at Stanford University, and of EE236a, EE236b, EE236c (L. Vandenberghe) at UCLA, as well as the slides that S. Sra developed for the course EE 227A in 2012 at UC Berkeley.

[1] S. Boyd and L. Vandenberghe, *Convex Optimization*, Cambridge University Press, 2004.
D. P. Bertsekas, *Nonlinear Optimization*, Athena Scientific, 1999.
D. P. Bertsekas (with A. Nedic, A. Ozdaglar), *Convex Analysis and Optimization*, Athena Scientific, 2003.
Yu. Nesterov, *Introductory Lectures on Convex Optimization: A Basic Course*, Springer, 2004.
A. Ben-Tal and A. Nemirovski, *Lectures on Modern Convex Optimization*, SIAM, 2001.
J. Borwein and A. Lewis, *Convex Analysis and Nonlinear Optimization: Theory and Examples*, Springer, 2006.

Acknowledgments

In the last 20 years, we witnessed many exciting developments in both theory and applications of optimization. The prime stimulus for writing this book came to us from the thriving scientific community involved in optimization research, whose members gave us, directly or indirectly, motivation and inspiration. While it would be impossible to mention all of them, we wish to give special thanks to our colleagues Dimitris Bertsimas, Stephen Boyd, Emmanuel Candès, Constantin Caramanis, Vu Duong, Michael Jordan, Jitendra Malik, Arkadi Nemirovksi, Yuri Nesterov, Jorge Nocedal, Kannan Ramchandran, Anant Sahai, Suvrit Sra, Marc Teboulle, Lieven Vandenberghe, and Jean Walrand, for their support, and constructive discussions over the years. We are also thankful to the anonymous reviewers of our initial draft, who encouraged us to proceed. Special thanks go to Daniel Lyons, who reviewed our final draft and helped improve our presentation.

Our gratitude also goes to Phil Meyler and his team at Cambridge University Press, and especially to Elizabeth Horne for her technical support.

This book has been typeset in Latex, using a variant of Edward Tufte's book style.

1
Introduction

OPTIMIZATION IS A TECHNOLOGY that can be used to devise effective *decisions* or *predictions* in a variety of contexts, ranging from production planning to engineering design and finance, to mention just a few. In simplified terms, the process for reaching the decision starts with a phase of construction of a suitable mathematical *model* for a concrete problem, followed by a phase where the model is *solved* by means of suitable numerical algorithms. An optimization model typically requires the specification of a quantitative *objective criterion* of goodness for our decision, which we wish to maximize (or, alternatively, a criterion of cost, which we wish to minimize), as well as the specification of *constraints*, representing the physical limits of our decision actions, budgets on resources, design requirements that need be met, etc. An optimal design is one which gives the best possible objective value, while satisfying all problem constraints.

In this chapter, we provide an overview of the main concepts and building blocks of an optimization problem, along with a brief historical perspective of the field. Many concepts in this chapter are introduced without formal definition; more rigorous formalizations are provided in the subsequent chapters.

1.1 *Motivating examples*

We next describe a few simple but practical examples where optimization problems arise naturally. Many other more sophisticated examples and applications will be discussed throughout the book.

1.1.1 *Oil production management*

An oil refinery produces two products: jet fuel and gasoline. The profit for the refinery is $0.10 per barrel for jet fuel and $0.20 per barrel for gasoline. Only 10,000 barrels of crude oil are available for processing. In addition, the following conditions must be met.

1. The refinery has a government contract to produce at least 1,000 barrels of jet fuel, and a private contract to produce at least 2,000 barrels of gasoline.

2. Both products are shipped in trucks, and the delivery capacity of the truck fleet is 180,000 barrel-miles.

3. The jet fuel is delivered to an airfield 10 miles from the refinery, while the gasoline is transported 30 miles to the distributor.

How much of each product should be produced for maximum profit?

Let us formalize the problem mathematically. We let x_1, x_2 represent, respectively, the quantity of jet fuel and the quantity of gasoline produced, in barrels. Then, the profit for the refinery is described by function $g_0(x_1, x_2) = 0.1x_1 + 0.2x_2$. Clearly, the refinery interest is to maximize its profit g_0. However, constraints need to be met, which are expressed as

$$
\begin{aligned}
x_1 + x_2 &\leq 10,000 &&\text{(limit on available crude barrels)} \\
x_1 &\geq 1,000 &&\text{(minimum jet fuel)} \\
x_2 &\geq 2,000 &&\text{(minimum gasoline)} \\
10x_1 + 30x_2 &\leq 180,000 &&\text{(fleet capacity).}
\end{aligned}
$$

Therefore, this production problem can be formulated mathematically as the problem of finding x_1, x_2 such that $g_0(x_1, x_2)$ is maximized, subject to the above constraints.

1.1.2 Prediction of technology progress

Table 1.1 reports the number N of transistors in 13 microprocessors as a function of the year of their introduction.

If one observes a plot of the logarithm of N_i versus the year y_i (Figure 1.1), one sees an approximately linear trend. Given these data, we want to determine the "best" line that approximates the data. Such a line quantifies the trend of technology progress, and may be used to estimate the number of transistors in a microchip in the future. To model this problem mathematically, we let the approximating line be described by the equation

$$z = x_1 y + x_2, \tag{1.1}$$

where y is the year, z represents the logarithm of N, and x_1, x_2 are the unknown parameters of the line (x_1 is the slope, and x_2 is the

year: y_i	no. transistors: N_i
1971	2250
1972	2500
1974	5000
1978	29000
1982	120000
1985	275000
1989	1180000
1993	3100000
1997	7500000
1999	24000000
2000	42000000
2002	220000000
2003	410000000

Table 1.1 Number of transistors in a microprocessor at different years.

intercept of the line with the vertical axis). Next, we need to agree on a criterion for measuring the level of *misfit* between the approximating line and the data. A commonly employed criterion is one which measures the sum of squared deviations of the observed data from the line. That is, at a given year y_i, Eq. (1.1) predicts $x_1 y_i + x_2$ transistors, while the observed number of transistors is $z_i = \log N_i$, hence the squared error at year y_i is $(x_1 y_i + x_2 - z_i)^2$, and the accumulated error over the 13 observed years is

$$f_0(x_1, x_2) = \sum_{i=1}^{13} (x_1 y_i + x_2 - z_i)^2.$$

The best approximating line is thus obtained by finding the values of parameters x_1, x_2 that minimize the function f_0.

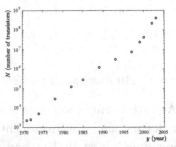

Figure 1.1 Semi-logarithmic plot of the number of transistors in a microprocessor at different years.

1.1.3 An aggregator-based power distribution model

In the electricity market, an *aggregator* is a marketer or public agency that combines the loads of multiple end-use customers in facilitating the sale and purchase of electric energy, transmission, and other services on behalf of these customers. In simplified terms, the aggregator buys wholesale c units of power (say, Megawatt) from large power distribution utilities, and resells this power to a group of n business or industrial customers. The i-th customer, $i = 1, \dots, n$, communicates to the aggregator its ideal level of power supply, say c_i Megawatt. Also, the customer *dislikes* to receive more power than its ideal level (since the excess power has to be paid for), as well as it dislikes to receive less power that its ideal level (since then the customer's business may be jeopardized). Hence, the customer communicates to the aggregator its own model of dissatisfaction, which we assume to be of the following form

$$d_i(x_i) = \alpha_i (x_i - c_i)^2, \quad i = 1, \dots, n,$$

where x_i is the power allotted by the aggregator to the i-th customer, and $\alpha_i > 0$ is a given, customer-specific, parameter. The aggregator problem is then to find the power allocations x_i, $i = 1, \dots, n$, so as to minimize the average customer dissatisfaction, while guaranteeing that the whole power c is sold, and that no single customer incurs a level of dissatisfaction greater than a contract level $\bar{d}$.

The aggregator problem is thus to minimize the average level of customer dissatisfaction

$$f_0(x_1, \dots, x_n) = \frac{1}{n} \sum_{i=1}^{n} d_i(x_i) = \frac{1}{n} \sum_{i=1}^{n} \alpha_i (x_i - c_i)^2,$$

while satisfying the following constraints:

$$\sum_{i=1}^{n} x_i = c, \quad \text{(all aggregator power must be sold)}$$

$$x_i \geq 0, \ i = 1, \ldots, n, \quad \text{(supplied power cannot be negative)}$$

$$\alpha_i(x_i - c_i)^2 \leq \bar{d}, \ i = 1, \ldots, n, \quad \text{(dissatisfaction cannot exceed } \bar{d} \text{)}.$$

1.1.4 An investment problem

An investment fund wants to invest (all or in part) a total capital of c dollars among n investment opportunities. The cost for the i-th investment is w_i dollars, and the investor expects a profit p_i from this investment. Further, at most b_i items of cost w_i and profit p_i are available on the market ($b_i \leq c/w_i$). The fund manager wants to know how many items of each type to buy in order to maximize his/her expected profit.

This problem can be modeled by introducing decision variables $x_i, \ i = 1, \ldots, n$, representing the (integer) number of units of each investment type to be bought. The expected profit is then expressed by the function

$$f_0(x_1, \ldots, x_n) = \sum_{i=1}^{n} p_i x_i.$$

The constraints are instead

$$\sum_{i=1}^{n} w_i x_i \leq c, \quad \text{(limit on capital to be invested)}$$

$$x_i \in \{0, 1, \ldots, b_i\}, \quad i = 1, \ldots, n \quad \text{(limit on availability of items)}.$$

The investor goal is thus to determine $x_1, \ldots, x_n$ so as to maximize the profit f_0 while satisfying the above constraints. The described problem is known in the literature as the *knapsack* problem.

Remark 1.1 *A warning on limits of optimization models.* Many, if not all, real-world decision problems and engineering design problems *can*, in principle, be expressed mathematically in the form of an optimization problem. However, we warn the reader that having a problem expressed as an optimization model does not necessarily mean that the problem can then be *solved* in practice. The problem described in Section 1.1.4, for instance, belongs to a category of problems that are "hard" to solve, while the examples described in the previous sections are "tractable," that is, easy to solve numerically. We discuss these issues in more detail in Section 1.2.4. Discerning between hard and tractable problem formulations is one of the key abilities that we strive to teach in this book.

1.2 Optimization problems

1.2.1 Definition

A standard form of optimization. We shall mainly deal with optimization problems[1] that can be written in the following standard form:

$$p^* = \min_x \quad f_0(x) \tag{1.2}$$
$$\text{subject to:} \quad f_i(x) \leq 0, \quad i = 1, \ldots, m,$$

where

- vector[2] $x \in \mathbb{R}^n$ is the *decision variable*;
- $f_0 : \mathbb{R}^n \to \mathbb{R}$ is the *objective* function,[3] or *cost*;
- $f_i : \mathbb{R}^n \to \mathbb{R}$, $i = 1, \ldots, m$, represent the *constraints*;
- p^* is the *optimal value*.

In the above, the term "subject to" is sometimes replaced with the shorthand "s.t.:," or simply by colon notation ":".

[2] A vector x of dimension n is simply a collection of real numbers $x_1, x_2, \ldots, x_n$. We denote by $\mathbb{R}^n$ the space of all possible vectors of dimension n.

[3] A function f describes an operation that takes a vector $x \in \mathbb{R}^n$ as an input, and assigns a real number, denoted $f(x)$, as a corresponding output value. The notation $f : \mathbb{R}^n \to \mathbb{R}$ allows us to define the input space precisely.

Example 1.1 (*An optimization problem in two variables*) Consider the problem

$$\min_x 0.9x_1^2 - 0.4x_1x_2 + 0.6x_2^2 - 6.4x_1 - 0.8x_2 : \ -1 \leq x_1 \leq 2, \ 0 \leq x_2 \leq 3.$$

The problem can be put in the standard form (1.2), where:

- the decision variable is $x = (x_1, x_2) \in \mathbb{R}^2$;
- the objective function $f_0 : \mathbb{R}^2 \to \mathbb{R}$, takes values

$$f_0(x) = 0.9x_1^2 - 0.4x_1x_2 - 0.6x_2^2 - 6.4x_1 - 0.8x_2;$$

- the constraint functions $f_i : \mathbb{R}^n \to \mathbb{R}$, $i = 1, 2, 3, 4$ take values

$$f_1(x) = -x_1 - 1, f_2(x) = x_1 - 2, f_3(x) = -x_2, f_4(x) = x_2 - 3.$$

Problems with equality constraints. Sometimes the problem may present explicit *equality* constraints, along with inequality ones, that is

$$p^* = \min_x \quad f_0(x)$$
$$\text{s.t.:} \quad f_i(x) \leq 0, \quad i = 1, \ldots, m,$$
$$h_i(x) = 0, \quad i = 1, \ldots, p,$$

where the h_is are given functions. Formally, however, we may reduce the above problem to a standard form with inequality constraints only, by representing each equality constraint via a pair of inequalities. That is, we represent $h_i(x) = 0$ as $h_i(x) \leq 0$ and $h_i(x) \geq 0$.

Problems with set constraints.   Sometimes, the constraints of the problem are described abstractly via a set-membership condition of the form $x \in \mathcal{X}$, for some subset $\mathcal{X}$ of $\mathbb{R}^n$. The corresponding notation is

$$p^* = \min_{x \in \mathcal{X}} \ f_0(x),$$

or, equivalently,

$$p^* = \min_{x} \ f_0(x)$$
$$\text{s.t.:} \quad x \in \mathcal{X}.$$

Problems in maximization form.   Some optimization problems come in the form of maximization (instead of minimization) of an objective function, i.e.,

$$p^* = \max_{x \in \mathcal{X}} \ g_0(x). \tag{1.3}$$

Such problems, however, can be readily recast in standard minimization form by observing that, for any g_0, it holds that

$$\max_{x \in \mathcal{X}} g_0(x) = - \min_{x \in \mathcal{X}} -g_0(x).$$

Therefore, problem (1.3) in maximization form can be reformulated as one in minimization form as

$$-p^* = \min_{x \in \mathcal{X}} \ f_0(x),$$

where $f_0 = -g_0$.

Feasible set.   The *feasible set*[4] of problem (1.2) is defined as

$$\mathcal{X} = \{x \in \mathbb{R}^n \ \text{s.t.:} \ f_i(x) \leq 0, \ i = 1, \ldots, m\}.$$

A point x is said to be *feasible* for problem (1.2) if it belongs to the feasible set $\mathcal{X}$, that is, if it satisfies the constraints. The feasible set may be empty, if the constraints cannot be satisfied simultaneously. In this case the problem is said to be *infeasible*. We take the convention that the optimal value is $p^* = +\infty$ for infeasible minimization problems, while $p^* = -\infty$ for infeasible maximization problems.

1.2.2   *What is a solution?*

In an optimization problem, we are usually interested in computing the optimal value p^* of the objective function, possibly together with a corresponding *minimizer*, which is a vector that achieves the optimal value, and satisfies the constraints. We say that the problem is attained if there is such a vector.[5]

[4] In the optimization problem of Example 1.1, the feasible set is the "box" in $\mathbb{R}^2$, described by $-1 \leq x_1 \leq 2$, $0 \leq x_2 \leq 3$.

[5] In the optimization problem of Example 1.1, the optimal value $p^* = -10.2667$ is attained by the optimal solution $x_1^* = 2$, $x_2^* = 1.3333$.

Feasibility problems. Sometimes an objective function is not provided. This means that we are just interested in finding a feasible point, or determining that the problem is infeasible. By convention, we set f_0 to be a constant in that case, to reflect the fact that we are indifferent to the choice of a point x, as long as it is feasible. For problems in the standard form (1.2), solving a feasibility problem is equivalent to finding a point that solves the system of inequalities $f_i(x) \leq 0$, $i = 1, \ldots, m$.

Optimal set. The *optimal set*, or *set of solutions*, of problem (1.2) is defined as the set of feasible points for which the objective function achieves the optimal value:

$$\mathcal{X}_{\text{opt}} = \{x \in \mathbb{R}^n \text{ s.t.: } f_0(x) = p^*, \ f_i(x) \leq 0, \ i = 1, \ldots, m\}.$$

A standard notation for the optimal set is via the arg min notation:

$$\mathcal{X}_{\text{opt}} = \arg\min_{x \in \mathcal{X}} f_0(x).$$

A point x is said to be *optimal* if it belongs to the optimal set, see Figure 1.2.

When is the optimal set empty? Optimal points may not exist, and the optimal set may be empty. This can be for two reasons. One is that the problem is infeasible, i.e., $\mathcal{X}$ itself is empty (there is no point that satisfies the constraints). Another, more subtle, situation arises when $\mathcal{X}$ is nonempty, but the optimal value is only reached in the limit. For example, the problem

$$p^* = \min_x \ e^{-x}$$

has no optimal points, since the optimal value $p^* = 0$ is only reached in the limit, for $x \to +\infty$. Another example arises when the constraints include strict inequalities, for example with the problem

$$p^* = \min_x \ x \ \text{ s.t.: } \ 0 < x \leq 1. \tag{1.4}$$

In this case, $p^* = 0$ but this optimal value is not attained by any x that satisfies the constraints. Rigorously, the notation "inf" should be used instead of "min" (or, "sup" instead of "max") in situations when one doesn't know *a priori* if optimal points are attained. However, in this book we do not dwell too much on such subtleties, and use the min and max notations, unless the more rigorous use of inf and sup is important in the specific context. For similar reasons, we only consider problems with non-strict inequalities. Strict inequalities can be

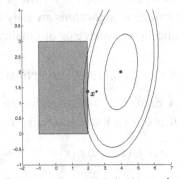

Figure 1.2 A toy optimization problem, with lines showing the points with constant value of the objective function. The optimal set is the singleton $\mathcal{X}_{\text{opt}} = \{x^*\}$.

safely replaced by non-strict ones, whenever the objective and constraint functions are continuous. For example, replacing the strict inequality by a non-strict one in (1.4) leads to a problem with the same optimal value $p^* = 0$, which is now attained at a well-defined optimal solution $x^* = 0$.

Sub-optimality. We say that a point x is ϵ-suboptimal for problem (1.2) if it is feasible, and satisfies

$$p^* \leq f_0(x) \leq p^* + \epsilon.$$

In other words, x is ϵ-close to achieving the best value p^*. Usually, numerical algorithms are only able to compute suboptimal solutions, and never reach true optimality.

1.2.3 Local vs. global optimal points

A point z is *locally optimal* for problem (1.2) if there exists a value $R > 0$ such that z is optimal for problem

$$\min_x f_0(x) \text{ s.t.: } f_i(x) \leq 0, \ i = 1, \ldots, m, \ |x_i - z_i| \leq R, \ i = 1, \ldots, n.$$

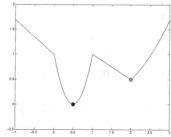

In other words, a local minimizer x minimizes f_0, but only for nearby points on the feasible set. The value of the objective function at that point is *not* necessarily the (global) optimal value of the problem. Locally optimal points might be of no practical interest to the user.

The term *globally optimal* (or optimal, for short) is used to distinguish points in the optimal set $\mathcal{X}_{opt}$ from local optima. The existence of local optima is a challenge in general optimization, since most algorithms tend to be trapped in local minima, if these exist, thus failing to produce the desired global optimal solution.

Figure 1.3 Local (gray) vs. global (black) minima. The optimal set is the singleton $\mathcal{X}_{opt} = \{0.5\}$. The point $x = 2$ is a local minimum.

1.2.4 Tractable vs. non-tractable problems

Not all optimization problems are created equal. Some problem classes, such as finding a solution to a finite set of linear equalities or inequalities, can be solved numerically in an efficient and reliable way. On the contrary, for some other classes of problems, no reliable efficient solution algorithm is known.

Without entering a discussion on the *computational complexity* of optimization problems, we shall here refer to as "tractable" all those optimization models for which a globally optimal solution can be found numerically in a reliable way (i.e., always, in any problem instance), with a computational effort that grows gracefully with the *size* of the problem (informally, the size of the problem is measured by the number of decision variables and/or constraints in the model). Other problems are known to be "hard," and yet for other problems the computational complexity is unknown.

The examples presented in the previous sections all belong to problem classes that are tractable, with the exception of the problem in Section 1.1.4. The focus of this book is on tractable models, and a key message is that models that can be formulated in the form of linear algebra problems, or in *convex*[6] form, are typically tractable. Further, if a convex model has some special structure,[7] then solutions can typically be found using existing and very reliable numerical solvers, such as CVX, Yalmip, etc.

It is also important to remark that tractability is often *not* a property of the problem itself, but a property of our formulation and modeling of the problem. A problem that may seem hard under a certain formulation may well become tractable if we put some more effort and intelligence in the modeling phase. Just to make an example, the raw data in Section 1.1.2 could not be fit by a simple linear model. However, a logarithmic transformation in the data allowed a good fit by a linear model.

One of the goals of this book is to provide the reader with some glimpse into the "art" of manipulating problems so as to model them in a tractable form. Clearly, this is not always possible: some problems are just hard, no matter how much effort we put in trying to manipulate them. One example is the *knapsack* problem, of which the investment problem described in Section 1.1.4 is an instance (actually, most optimization problems in which the variable is constrained to be integer valued are computationally hard). However, even for intrinsically hard problems, for which *exact* solutions may be unaffordable, we may often find useful tractable models that provide us with readily computable *approximate*, or relaxed, solutions.

[6] See Chapter 8.

[7] See Section 1.3, Chapter 9, and subsequent chapters.

1.2.5 Problem transformations

The optimization formalism in (1.2) is extremely flexible and allows for many transformations, which may help to cast a given problem in a tractable formulation. For example, the optimization problem

$$\min_x \sqrt{(x_1 + 1)^2 + (x_2 - 2)^2} \ \ \text{s.t.:} \ \ x_1 \geq 0$$

has the same optimal set as

$$\min_x ((x_1 + 1)^2 + (x_2 - 2)^2) \ \ \text{s.t.:} \ \ x_1 \geq 0.$$

The advantage here is that the objective is now differentiable. In other situations, it may be useful to change variables. For example, the problem

$$\max_{x} \ x_1 x_2^3 x_3 \ \text{s.t.:} \ x_i \geq 0, \ i = 1,2,3, \ x_1 x_2 \leq 2, \ x_2^2 x_3 \leq 1$$

can be equivalently written, after taking the log of the objective, in terms of the new variables $z_i = \log x_i$, $i = 1,2,3$, as

$$\max_{z} \ z_1 + 3z_2 + z_3 \ \text{s.t.:} \ z_1 + z_2 \leq \log 2, \ 2z_2 + z_3 \leq 0.$$

The advantage is that now the objective and constraint functions are all linear. Problem transformations are treated in more detail in Section 8.3.4.

1.3   Important classes of optimization problems

In this section, we give a brief overview of some standard optimization models, which are then treated in detail in subsequents parts of this book.

1.3.1   Least squares and linear equations

A linear least-squares problem is expressed in the form

$$\min_{x} \ \sum_{i=1}^{m} \left(\sum_{j=1}^{n} A_{ij} x_j - b_i \right)^2, \tag{1.5}$$

where A_{ij}, b_i, $1 \leq i \leq m$, $1 \leq j \leq n$, are given numbers, and $x \in \mathbb{R}^n$ is the variable. Least-squares problems arise in many situations, for example in statistical estimation problems such as linear regression.[8]

An important application of least squares arises when solving a set of linear equations. Assume we want to find a vector $x \in \mathbb{R}^n$ such that

$$\sum_{j=1}^{n} A_{ij} x_j = b_i, \ i = 1, \ldots, m.$$

Such problems can be cast as least-squares problems of the form (1.5). A solution to the corresponding set of equations is found if the optimal value of (1.5) is zero; otherwise, an optimal solution of (1.5) provides an approximate solution to the system of linear equations. We discuss least-squares problems and linear equations extensively in Chapter 6.

[8] The example in Section 1.1.2 is an illustration of linear regression.

1.3.2 Low-rank approximations and maximum variance

The problem of *rank-one approximation* of a given matrix (a rectangular array of numbers A_{ij}, $1 \le i \le m$, $1 \le j \le n$) takes the form

$$\min_{x \in \mathbb{R}^n, z \in \mathbb{R}^m} \sum_{i=1}^{m} \left(\sum_{j=1}^{n} A_{ij} - z_i x_j \right)^2 .$$

The above problem can be interpreted as a variant of the least-squares problem (1.5), where the functions inside the squared terms are non-linear, due to the presence of products between variables $z_i x_j$. A small value of the objective means that the numbers A_{ij} can be well approximated by $z_i x_j$. Hence, the "rows" $(A_{i1}, \dots, A_{in})$, $i = 1, \dots, m$, are all scaled version of the same vector $(x_1, \dots, x_n)$, with scalings given by the elements in $(z_1, \dots, z_m)$.

This problem arises in a host of applications, as illustrated in Chapters 4 and 5, and it constitutes the building block of a technology known as the singular value decomposition (SVD).

A related problem is the so-called maximum-variance problem:

$$\max_x \sum_{i=1}^{m} \left(\sum_{j=1}^{n} A_{ij} x_j \right)^2 \quad \text{s.t.:} \quad \sum_{i=1}^{n} x_i^2 = 1.$$

The above can be used, for example, to find a line that best fits a set of points in a high-dimensional space, and it is a building block for a data dimensionality reduction technique known as principal component analysis, as detailed in Chapter 13.

1.3.3 Linear and quadratic programming

A linear programming (LP) problem has the form

$$\min_x \sum_{j=1}^{n} c_j x_j \quad \text{s.t.:} \quad \sum_{j=1}^{n} A_{ij} x_j \le b_i, \quad i = 1, \dots, m,$$

where c_j, b_i and A_{ij}, $1 \le i \le m$, $1 \le j \le n$, are given real numbers. This problem is a special case of the general problem (1.2), in which the functions f_i, $i = 0, \dots, m$, are all affine (that is, linear plus a constant term). The LP model is perhaps the most widely used model in optimization.

Quadratic programming problems (QPs for short) are an extension of linear programming, which involve a sum-of-squares function in the objective. The linear program above is modified to

$$\min_x \sum_{i=1}^{r} \left(\sum_{j=1}^{n} C_{ij} x_j \right)^2 + \sum_{j=1}^{n} c_j x_j \quad \text{s.t.:} \quad \sum_{j=1}^{n} A_{ij} x_j \le b_i, \quad i = 1, \dots, m,$$

where the numbers C_{ij}, $1 \leq i \leq r$, $1 \leq j \leq n$, are given. QPs can be thought of as a generalization of both least-squares and linear programming problems. They are popular in many areas, such as finance, where the linear term in the objective refers to the expected negative return on an investment, and the squared term corresponds to the risk (or variance of the return). LP and QP models are discussed in Chapter 9.

1.3.4 Convex optimization

Convex optimization problems are problems of the form (1.2), where the objective and constraint functions have the special property of convexity. Roughly speaking, a convex function has a "bowl-shaped" graph, as exemplified in Figure 1.4. Convexity and general convex problems are covered in Chapter 8.

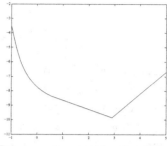

Figure 1.4 A convex function has a "bowl-shaped" graph.

Not all convex problems are easy to solve, but many of them are indeed computationally tractable. One key feature of convex problems is that all local minima are actually global, see Figure 1.5 for an example.

The least-squares, LP, and (convex) QP models are examples of tractable convex optimization problems. This is also true for other specific optimization models we treat in this book, such as the geometric programming (GP) model discussed in Chapter 9, the second-order cone programming (SOCP) model covered in Chapter 10, and the semidefinite programming (SDP) model covered in Chapter 11.

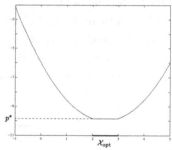

Figure 1.5 For a convex function, any local minimum is global. In this example, the minimizer is not unique, and the optimal set is the interval $\mathcal{X}_{opt} = [2,3]$. Every point in the interval achieves the global minimum value $p^* = -9.84$.

1.3.5 Combinatorial optimization

In combinatorial optimization, variables are Boolean (0 or 1), or more generally, integers, reflecting discrete choices to be made. The knapsack problem, described in Section 1.1.4, is an example of an integer programming problem, and so is the Sudoku problem shown in Figure 1.6. Many practical problems actually involve a mix of integer and real-valued variables. Such problems are referred to as mixed-integer programs (MIPs).

Combinatorial problems and, more generally, MIPs, belong to a class of problems known to be computationally hard, in general. Although we sometimes discuss the use of convex optimization to find approximate solutions to such problems, this book does not cover combinatorial optimization in any depth.

	9						3	
8			5		7			2
			3		2			
	1	2				5	6	
	3	4				7	2	
			7		8			
9			1		3			4
	6						1	

Figure 1.6 The Sudoku problem, as it is the case for many other popular puzzles, can be formulated as a feasibility problem with integer variables.

1.3.6   Non-convex optimization

Non-convex optimization corresponds to problems where one or more of the objective or constraint functions in the standard form (1.2) does not have the property of convexity.

In general, such problems are very hard to solve. In fact, this class comprises combinatorial optimization: if a variable x_i is required to be Boolean (that is, $x_i \in \{0,1\}$), we can model this as a pair of constraints $x_i^2 - x_i \leq 0$, $x_i - x_i^2 \leq 0$, the second of which involves a non-convex function. One of the reasons for which general non-convex problems are hard to solve is that they may present *local minima*, as illustrated in Figure 1.3. This is in contrast with convex problems, which do not suffer from this issue.

It should, however, be noted that not every non-convex optimization problem is hard to solve. The maximum variance and low-rank approximation problems discussed in Section 1.3.2, for example, are non-convex problems that can be reliably solved using special algorithms from linear algebra.

Example 1.2 (*Protein folding*) The *protein folding* problem amounts to predicting the three-dimensional structure of a protein, based on the sequence of the amino-acids that constitutes it. The amino-acids interact with each other (for example, they may be electrically charged). Such a problem is difficult to address experimentally, which calls for computer-aided methods.

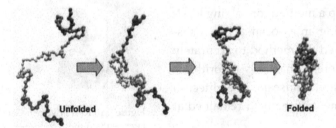

Unfolded   **Folded**

Figure 1.7 Protein folding problem.

In recent years, some researchers have proposed to express the problem as an optimization problem, involving the minimization of a potential energy function, which is usually a sum of terms reflecting the interactions between pairs of amino-acids. The overall problem can be modeled as a nonlinear optimization problem.

Unfortunately, protein folding problems remain challenging. One of the reasons is the very large size of the problem (number of variables and constraints). Another difficulty comes from the fact that the potential energy function (which the actual protein is minimizing) is not exactly known. Finally, the fact that the energy function is usually not convex

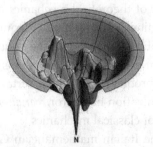

Figure 1.8 Graph of energy function involved in protein folding models.

may lead algorithms to discover "spurious" (that is, wrong) molecular conformations, corresponding to local minima of the potential function. Figure 1.8 is a three-dimensional rendition of the level sets of a protein's energy function.

1.4 History

1.4.1 Early stages: birth of linear algebra

The roots of optimization, as a field concerned with algorithms for solving numerical problems, can perhaps be traced back to the earliest known appearance of a system of linear equations in ancient China. Indeed, the art termed *fangcheng* (often translated as "rectangular arrays") was used as early as 300 BC to solve practical problems which amounted to linear systems. Algorithms identical to Gauss elimination for solving such systems appear in Chapter 8 of the treatise *Nine Chapters on the Mathematical Art*, dated around 100 CE.

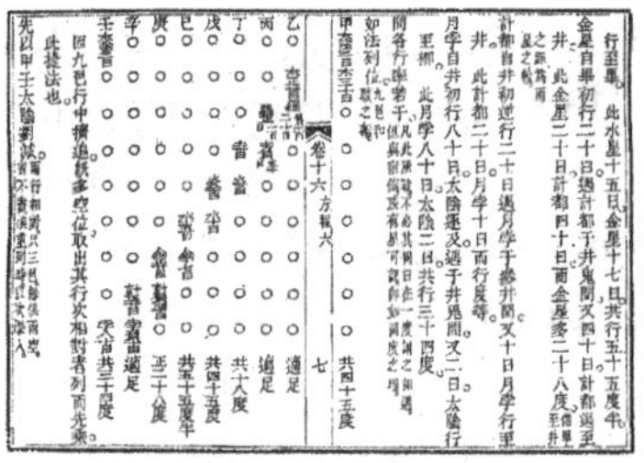

Figure 1.9 pictures a 9×9 matrix found in the treatise, as printed in the 1700s (with a reversed convention for the column's order). It is believed that many of the early Chinese results in linear algebra gradually found their way to Europe.

Figure 1.9 Early Chinese linear algebra text.

1.4.2 Optimization as a theoretical tool

In the 1800s, Gauss (Figure 1.10) built on early results (and his own contributions) in linear algebra to develop a method for solving least-squares problems, which relied on solving an associated linear system (the famous *normal equations*). He used the method to accurately predict the trajectory of the planetoid Ceres. This early algorithmic result was an exception in the optimization landscape in eighteenth century Europe, as most of the development of the field remained at a theoretical level.

The notion of optimization problems was crucial to the development of theoretical mechanics and physics between the seventeenth and nineteenth centuries. Around 1750, Maupertuis introduced (and later Euler formalized) the *principle of least action*, according to which the motion of natural systems could be described as a minimization problem involving a certain cost function called "energy." This optimization-based (or, *variational*) approach is indeed the foundation of classical mechanics.

The Italian mathematician Giuseppe Lodovico (Luigi) Lagrangia (Figure 1.11), also known as Lagrange, was a key player in this development, and his name is associated with the notion of duality,

Figure 1.10 Karl Friedrich Gauss (1777—1855).

Figure 1.11 Giuseppe Lodovico (Luigi) Lagrangia (1736–1813).

which is central in optimization. While optimization *theory* played a central role in physics, it was only with the birth of computers that it could start making its mark in practical applications, and venture into fields other than physics.

1.4.3   Advent of numerical linear algebra

With computers becoming available in the late 40s, the field of numerical linear algebra was ready to take off, motivated in no small part by the cold war effort. Early contributors include Von Neumann, Wilkinson, Householder, and Givens.

Early on, it was understood that a key challenge was to handle the numerical errors that were inevitably propagated by algorithms. This led to an intense research activity into the so-called stability of algorithms, and associated perturbation theory.[9] In that context, researchers recognized the numerical difficulties associated with certain concepts inherited from some nineteenth century physics problems, such as the eigenvalue decomposition of general square matrices. More recent decompositions, such as the singular value decomposition, were recognized as playing a central role in many applications.[10]

Optimization played a key role in the development of linear algebra. First, as an important source of applications and challenges; for example, the simplex algorithm for solving linear programming problems, which we discuss below, involves linear equations as the key step. Second, optimization has been used as a model of computation: for example, finding the solution to linear equations can be formulated as a least-squares problem, and analyzed as such.

In the 70s, practical linear algebra was becoming inextricably linked to software. Efficient packages written in FORTRAN, such as LINPACK and LAPACK, embodied the progress on the algorithms and became available in the 80s. These packages were later exported into parallel programming environments, to be used on super-computers. A key development came in the form of scientific computing platforms, such as Matlab, Scilab, Octave, R, etc. Such platforms hid the FORTRAN packages developed earlier behind a user-friendly interface, and made it very easy to, say, solve linear equations, using a coding notation which is very close to the natural mathematical one. In a way, linear algebra became a commodity technology, which can be called upon by users without any knowledge of the underlying algorithms.

A recent development can be added to the long list of success stories associated with applied linear algebra. The PageRank algo-

[9] See, e.g., N. J. Higham, *Accuracy and Stability of Numerical Algorithms*, SIAM, 2002.

[10] See the classical reference textbook: G. H. Golub and C. F. Van Loan, *Matrix Computations*, IV ed, John Hopkins University Press, 2012.

rithm,[11] which is used by a famous search engine to rank web pages, relies on the power iteration algorithm for solving a special type of eigenvalue problem.

[11] This algorithm is discussed in Section 3.5.

Most of the current research effort in the field of numerical linear algebra involves the solution of extremely large problems. Two research directions are prevalent. One involves solving linear algebra problems on distributed platforms; here, the earlier work on parallel algorithms[12] is revisited in the light of cloud computing, with a strong emphasis on the bottleneck of data communication. Another important effort involves sub-sampling algorithms, where the input data is partially loaded into memory in a random fashion.

[12] See, e.g., Bertsekas and Tsitsiklis, *Parallel and Distributed Computation: Numerical Methods*, Athena Scientific, 1997.

1.4.4 *Advent of linear and quadratic programming*

The LP model was introduced by George Dantzig in the 40s, in the context of logistical problems arising in military operations.

George Dantzig, working in the 1940s on Pentagon-related logistical problems, started investigating the numerical solution to linear inequalities. Extending the scope of linear algebra (linear equalities) to inequalities seemed useful, and his efforts led to the famous *simplex algorithm* for solving such systems. Another important early contributor to the field of linear programming was the Soviet Russian mathematician Leonid Kantorovich.

QPs are popular in many areas, such as finance, where the linear term in the objective refers to the expected negative return on an investment, and the squared term corresponds to the risk (or variance of the return). This model was introduced in the 50s by H. Markowitz (who was then a colleague of Dantzig at the RAND Corporation), to model investment problems. Markowitz won the Nobel prize in Economics in 1990, mainly for this work.

In the 60s–70s, a lot of attention was devoted to nonlinear optimization problems. Methods to find local minima were proposed. In the meantime, researchers recognized that these methods could fail to find global minima, or even to converge. Hence the notion formed that, while linear optimization was numerically tractable, nonlinear optimization was not, in general. This had concrete practical consequences: linear programming solvers could be reliably used for day-to-day operations (for example, for airline crew management), but nonlinear solvers needed an expert to baby-sit them.

In the field of mathematics, the 60s saw the development of convex analysis, which would later serve as an important theoretical basis for progress in optimization.

1.4.5   Advent of convex programming

Most of the research in optimization in the United States in the 60s–80s focused on nonlinear optimization algorithms, and contextual applications. The availability of large computers made that research possible and practical.

In the Soviet Union at that time, the focus was more towards optimization theory, perhaps due to more restricted access to computing resources. Since nonlinear problems are hard, Soviet researchers went back to the linear programming model, and asked the following (at that point theoretical) question: what makes linear programs easy? Is it really linearity of the objective and constraint functions, or some other, more general, structure? Are there classes of problems out there that are nonlinear but still easy to solve?

In the late 80s, two researchers in the former Soviet Union, Yurii Nesterov and Arkadi Nemirovski, discovered that a key property that makes an optimization problem "easy" is not linearity, but actually *convexity*. Their result is not only theoretical but also algorithmic, as they introduced so-called *interior-point methods* for solving convex problems efficiently.[13] Roughly speaking, convex problems are easy (and that includes linear programming problems); non-convex ones are hard. Of course, this statement needs to be qualified. Not *all* convex problems are easy, but a (reasonably large) subset of them is. Conversely, some non-convex problems are actually easy to solve (for example some path planning problems can be solved in linear time), but they constitute some sort of "exception."

[13] Yu. Nesterov and A. Nemirovski, *Interior-point Polynomial Algorithms in Convex Programming*, SIAM, 1994.

Since the seminal work of Nesterov and Nemirovski, convex optimization has emerged as a powerful tool that generalizes linear algebra and linear programming: it has similar characteristics of reliability (it always converges to the global minimum) and tractability (it does so in reasonable time).

1.4.6   Present

In present times there is a very strong interest in applying optimization techniques in a variety of fields, ranging from engineering design, statistics and machine learning, to finance and structural mechanics. As with linear algebra, recent interfaces to convex optimization solvers, such as CVX[14] or YALMIP[15], now make it extremely easy to prototype models for moderately-sized problems.

[14] cvxr.com/cvx/
[15] users.isy.liu.se/johanl/yalmip/

In research, motivated by the advent of very large datasets, a strong effort is currently being made towards enabling solution of extremely large-scale convex problems arising in machine learning, image processing, and so on. In that context, the initial focus of the 90s on interior-point methods has been replaced with a revisitation and development of earlier algorithms (mainly, the so-called "first-order" algorithms, developed in the 50s), which involve very cheap iterations.

I
Linear algebra models

2

Vectors and functions

> *Le contraire du simple n'est pas*
> *le complexe, mais le faux.*
>
> André Comte-Sponville

A VECTOR IS A COLLECTION of numbers, arranged in a column or
a row, which can be thought of as the coordinates of a point in n-
dimensional space. Equipping vectors with sum and scalar multiplication allows us to define notions such as independence, span,
subspaces, and dimension. Further, the *scalar product* introduces a
notion of the angle between two vectors, and induces the concept of
length, or norm. Via the scalar product, we can also view a vector as
a linear function. We can compute the projection of a vector onto a
line defined by another vector, onto a plane, or more generally onto
a subspace. Projections can be viewed as a first elementary optimization problem (finding the point in a given set at minimum distance
from a given point), and they constitute a basic ingredient in many
processing and visualization techniques for high-dimensional data.

2.1 Vector basics

2.1.1 Vectors as collections of numbers

Vectors are a way to represent and manipulate a single collection of
numbers. A vector x can thus be defined as a collection of elements
$x_1, x_2, \ldots, x_n$, arranged in a column or in a row. We usually write

vectors in column format:

$$x = \begin{bmatrix} x_1 \\ x_2 \\ \vdots \\ x_n \end{bmatrix}.$$

Element x_i is said to be the i-th component (or the i-th element, or entry) of vector x, and the number n of components is usually referred to as the *dimension* of x.

When the components of x are real numbers, i.e. $x_i \in \mathbb{R}$, then x is a real vector of dimension n, which we indicate with the notation $x \in \mathbb{R}^n$. We shall seldom need *complex* vectors, which are collections of complex numbers $x_i \in \mathbb{C}$, $i = 1, \ldots, n$. We denote the set of such vectors by $\mathbb{C}^n$.

To transform a column-vector x to row format and vice versa, we define an operation called *transpose*, denoted by a superscript $^\top$:

$$x^\top = \begin{bmatrix} x_1 & x_2 & \cdots & x_n \end{bmatrix}; \quad x^{\top\top} = x.$$

Sometimes, we use the notation $x = (x_1, \ldots, x_n)$ to denote a vector, if we are not interested in specifying whether the vector is in column or in row format. For a column vector $x \in \mathbb{C}^n$, we use the notation $x^\star$ to denote the transpose-conjugate, that is the row vector with elements set to the conjugate of those of x.

A vector x in $\mathbb{R}^n$ can be viewed as a point in that space, where the Cartesian coordinates are the components x_i; see Figure 2.1 for an example in dimension 3.

For example, the position of a ship at sea with respect to a given reference frame, at some instant in time, can be described by a two-dimensional vector $x = (x_1, x_2)$, where x_1, x_2 are the coordinates of the center of mass of the ship. Similarly, the position of an aircraft can be described by a three-dimensional vector $x = (x_1, x_2, x_3)$, where x_1, x_2, x_3 are the coordinates of the center of mass of the aircraft in a given reference frame.

Note that vectors need not be only two- or three-dimensional. For instance, one can represent as a vector the coordinates, at a given instant of time, of a whole swarm of m robots, each one having coordinates $x^{(i)} = (x_1^{(i)}, x_2^{(i)})$, $i = 1, \ldots, m$. The swarm positions are therefore described by the vector

$$x = (x_1^{(1)}, x_2^{(1)}, x_1^{(2)}, x_2^{(2)}, \ldots, x_1^{(m)}, x_2^{(m)})$$

of dimension $2m$; see Figure 2.2.

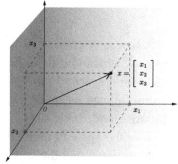

Figure 2.1 Cartesian representation of a vector in $\mathbb{R}^3$.

Figure 2.2 The position of m robots in a swarm can be represented by a $2m$-dimensional vector $x = (x_1^{(1)}, x_2^{(1)}, x_1^{(2)}, x_2^{(2)}, \ldots, x_1^{(m)}, x_2^{(m)})$, where $x^{(i)} = (x_1^{(i)}, x_2^{(i)})$, $i = 1, \ldots, m$, are the coordinates of each robot in a given fixed reference frame.

Example 2.1 (*Bag-of-words representations of text*) Consider the following text:

> "A (real) vector is just a collection of real numbers, referred to as the components (or, elements) of the vector; $\mathbb{R}^n$ denotes the set of all vectors with n elements. If $x \in \mathbb{R}^n$ denotes a vector, we use subscripts to denote elements, so that x_i is the i-th component in x. Vectors are arranged in a column, or a row. If x is a column vector, $x^\top$ denotes the corresponding row vector, and vice versa."

The row vector $c = [5, 3, 3, 4]$ contains the number of times each word in the list $V = \{vector, elements, of, the\}$ appears in the above paragraph. Dividing each entry in c by the total number of occurrences of words in the list (15, in this example), we obtain a vector $x = [1/3, 1/5, 1/5, 4/15]$ of relative word frequencies. Vectors can be thus used to provide a frequency-based representation of text documents; this representation is often referred to as the bag-of-words representation. In practice, the ordered list V contains an entire or restricted *dictionary* of words. A given document d may then be represented as the vector $x(d)$ that contains as elements a score, such as the relative frequency, of each word in the dictionary (there are many possible choices for the score function). Of course, the representation is not faithful, as it ignores the order of appearance of words; hence, the term "bag-of-words" associated with such representations.

Example 2.2 (*Temperatures at different airports*) Assume we record the temperatures at four different airports at a given time, and obtain the data in Table 2.1.

We can view the triplet of temperatures as a point in a three-dimensional space. Each axis corresponds to temperatures at a specific location. The vector representation is still legible if we have more than one triplet of temperatures, e.g., if we want to trace a curve of temperature as a function of time. The vector representation cannot, however, be visualized graphically in more than three dimensions, that is, if we have more than three airports involved.

Table 2.1 Airport temperature data.

Airport	Temp. (°F)
SFO	55
ORD	32
JFK	43

Example 2.3 (*Time series*) A *time series* represents the evolution in (discrete) time of a physical or economical quantity, such as the amount of solar radiation or the amount of rainfall (e.g., expressed in millimeters) at a given geographical spot, or the price of a given stock at the closing of the market. If $x(k)$, $k = 1, \ldots, T$, describes the numerical value of the quantity of interest at time k (say, k indexes discrete intervals of time, like minutes, days, months, or years), then the whole time series, over the time horizon from 1 to T, can be represented as a T-dimensional vector x containing all the values of $x(k)$, for $k = 1$ to $k = T$, that is

$$x = [x(1)\, x(2)\, \cdots\, x(T)]^\top \in \mathbb{R}^T.$$

Figure 2.3 shows for instance the time series of the adjusted close price of the Dow Jones Industrial Average Index, over a 66 trading day period from April 19, 2012 to July 20, 2012. This time series can be viewed as a vector x in a space of dimension $T = 66$.

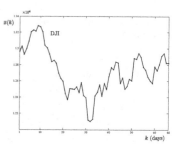

Figure 2.3 The DJI time series from April 19 to July 20, 2012.

2.1.2 Vector spaces

Seeing vectors as collections of numbers, or as points, is just the beginning of the story. In fact, a much richer understanding of vectors comes from their correspondence with linear functions. To understand this, we first examine how we can define some basic operations between vectors, and how to generate *vector spaces* from a collection of vectors.

2.1.2.1 Sum and scalar multiplication of vectors. The operations of sum, difference, and scalar multiplication are defined in an obvious way for vectors: for any two vectors $v^{(1)}, v^{(2)}$ having equal number of elements, we have that the sum $v^{(1)} + v^{(2)}$ is simply a vector having as components the sum of the corresponding components of the addends, and the same holds for the difference; see Figure 2.4.

Similarly, if v is a vector and α is a scalar (i.e., a real or complex number), then αv is obtained by multiplying each component of v by α. If $\alpha = 0$, then αv is the *zero vector*, or *origin*, that is, a vector in which all elements are zero. The zero vector is simply denoted by 0, or sometimes with 0_n, when we want to highlight the fact that it is a zero vector of dimension n.

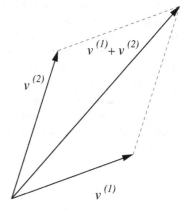

Figure 2.4 The sum of two vectors $v^{(1)} = [v_1^{(1)} \cdots v_n^{(1)}]$, $v^{(2)} = [v_1^{(2)} \cdots v_n^{(2)}]$ is the vector $v = v^{(1)} + v^{(2)}$ having components $[v_1^{(1)} + v_1^{(2)} \cdots v_n^{(1)} + v_n^{(2)}]$.

2.1.2.2 Vector spaces. From a slightly more abstract perspective, a *vector space*, $\mathcal{X}$, is obtained by equipping vectors with the operations of addition and multiplication by a scalar. A simple example of a vector space is $\mathcal{X} = \mathbb{R}^n$, the space of n-tuples of real numbers. A less obvious example is the set of single-variable polynomials of a given degree.

Example 2.4 (*Vector representation of polynomials*) The set of real polynomials of degree at most $n - 1$, $n \geq 1$, is

$$P_{n-1} = \{p : p(t) = a_{n-1}t^{n-1} + a_{n-2}t^{n-2} + \cdots + a_1 t + a_0, \ t \in \mathbb{R}\},$$

where $a_0, \ldots, a_{n-1} \in \mathbb{R}$ are the coefficients of the polynomial. Any polynomial $p \in P_{n-1}$ is uniquely identified by a vector $v \in \mathbb{R}^n$ containing its coefficients $v = [a_{n-1} \ldots a_0]^\top$ and, conversely, each vector $v \in \mathbb{R}^n$ uniquely defines a polynomial $p \in P_{n-1}$. Moreover, the operations of multiplication of a polynomial by a scalar and sum of two polynomials correspond respectively to the operations of multiplication and sum of the corresponding vector representations of the polynomials. In mathematical terminology, we say that that P_{n-1} is a vector space *isomorphic* to the standard vector space $\mathbb{R}^n$.

2.1.2.3 Subspaces and span. A nonempty subset $\mathcal{V}$ of a vector space $\mathcal{X}$ is called a *subspace* of $\mathcal{X}$ if, for any scalars α, β,

$$x, y \in \mathcal{V} \ \Rightarrow \ \alpha x + \beta y \in \mathcal{V}.$$

In other words, $\mathcal{V}$ is "closed" under addition and scalar multiplication.

Note that a subspace always contains the zero element. A *linear combination* of a set of vectors $S = \{x^{(1)}, \ldots, x^{(m)}\}$ in a vector space $\mathcal{X}$ is a vector of the form $\alpha_1 x^{(1)} + \cdots + \alpha_m x^{(m)}$, where $\alpha_1, \ldots, \alpha_m$ are given scalars. The set of all possible linear combinations of the vectors in $S = \{x^{(1)}, \ldots, x^{(m)}\}$ forms a subspace, which is called the subspace generated by S, or the *span* of S, denoted by $\mathrm{span}(S)$.

In $\mathbb{R}^n$, the subspace generated by a singleton $S = \{x^{(1)}\}$ is a line passing through the origin; see Figure 2.5.

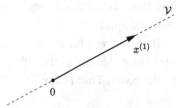

Figure 2.5 Line generated by scaling of a vector $x^{(1)}$.

The subspace generated by two non-collinear (i.e., such that one is not just a scalar multiple of the other) vectors $S = \{x^{(1)}, \ x^{(2)}\}$ is the plane passing through points $0, x^{(1)}, x^{(2)}$; see Figure 2.6 and Figure 2.7.

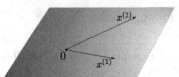

Figure 2.6 Plane generated by linear combinations of two vectors $x^{(1)}$, $x^{(2)}$.

More generally, the subspace generated by S is a *flat* passing through the origin.

2.1.2.4 Direct sum. Given two subspaces $\mathcal{X}, \mathcal{Y}$ in $\mathbb{R}^n$, the direct sum of $\mathcal{X}, \mathcal{Y}$, which we denote by $\mathcal{X} \oplus \mathcal{Y}$, is the set of vectors of the form $x + y$, with $x \in \mathcal{X}, y \in \mathcal{Y}$. It is readily checked that $\mathcal{X} \oplus \mathcal{Y}$ is itself a subspace.

2.1.2.5 Independence, bases, and dimensions. A collection $x^{(1)}, \ldots, x^{(m)}$ of vectors in a vector space $\mathcal{X}$ is said to be *linearly independent* if no vector in the collection can be expressed as a linear combination of the others. This is the same as the condition

$$\sum_{i=1}^{m} \alpha_i x^{(i)} = 0 \Longrightarrow \alpha = 0.$$

Later in this book we will see numerically efficient ways to check the independence of a collection of vectors.

Given a set $S = \{x^{(1)}, \ldots, x^{(m)}\}$ of m elements from a vector space $\mathcal{X}$, consider the subspace $\mathcal{S} = \mathrm{span}(S)$ generated by S, that is the set of vectors that can be obtained by taking all possible linear combinations of the elements in S. Suppose now that one element in

S, say the last one $x^{(m)}$, can itself be written as a linear combination of the remaining elements. Then, it is not difficult to see that we could remove $x^{(m)}$ from S and still obtain the same span, that is[1] $\mathrm{span}(S) = \mathrm{span}(S \setminus x^{(m)})$. Suppose then that there is another element in $\{S \setminus x^{(m)}\}$, say $x^{(m-1)}$, that can be be written as a linear combination of elements in $\{S \setminus x^{(m)}, x^{(m-1)}\}$. Then again this last set has the same span as S. We can go on in this way until we remain with a collection of vectors, say $B = \{x^{(1)}, \ldots, x^{(d)}\}$, $d \leq m$, such that $\mathrm{span}(B) = \mathrm{span}(S)$, and no element in this collection can be written as a linear combination of the other elements in the collection (i.e., the elements are linearly independent). Such an "irreducible" set is called a *basis* for $\mathrm{span}(S)$, and the number d of elements in the basis is called the *dimension* of $\mathrm{span}(S)$. A subspace can have many different bases (actually, infinitely many), but the number of elements in any basis is fixed and equal to the dimension of the subspace (d, in our example).

If we have a basis $\{x^{(1)}, \ldots, x^{(d)}\}$ for a subspace $\mathcal{S}$, then we can write any element in the subspace as a linear combination of elements in the basis. That is, any $x \in \mathcal{S}$ can be written as

$$x = \sum_{i=1}^{d} \alpha_i x^{(i)}, \tag{2.1}$$

for appropriate scalars α_i.

[1] We use the notation $A \setminus B$ to denote the difference of two sets, that is the set of elements in set A that do not belong to set B.

Example 2.5 (*Bases*) The following three vectors constitute a basis for $\mathbb{R}^3$:

$$x^{(1)} = \begin{bmatrix} 1 \\ 1 \\ 1 \end{bmatrix}, \ x^{(2)} = \begin{bmatrix} 1 \\ 2 \\ 0 \end{bmatrix}, \ x^{(3)} = \begin{bmatrix} 1 \\ 3 \\ 1 \end{bmatrix}.$$

Given, for instance, the vector $x = [1, 2, 3]^\top$, we can express it as a linear combination of the basis vectors as in (2.1):

$$\begin{bmatrix} 1 \\ 2 \\ 3 \end{bmatrix} = \alpha_1 \begin{bmatrix} 1 \\ 1 \\ 1 \end{bmatrix} + \alpha_2 \begin{bmatrix} 1 \\ 2 \\ 0 \end{bmatrix} + \alpha_3 \begin{bmatrix} 1 \\ 3 \\ 1 \end{bmatrix}.$$

Finding the suitable values for the α_i coefficients typically requires the solution of a system of linear equations (see Chapter 6). In the present case, the reader may simply verify that the correct values for the coefficients are $\alpha_1 = 1.5, \alpha_2 = -2, \alpha_3 = 1.5$.

There are, however, infinitely many other bases for $\mathbb{R}^3$. A special one is the so-called *standard basis* for $\mathbb{R}^3$, which is given by

$$x^{(1)} = \begin{bmatrix} 1 \\ 0 \\ 0 \end{bmatrix}, \ x^{(2)} = \begin{bmatrix} 0 \\ 1 \\ 0 \end{bmatrix}, \ x^{(3)} = \begin{bmatrix} 0 \\ 0 \\ 1 \end{bmatrix}.$$

More generally, the standard basis for $\mathbb{R}^n$ is

$$x^{(1)} = e_1, \; x^{(2)} = e_2, \; \ldots, \; x^{(n)} = e_n,$$

where e_i denotes a vector in $\mathbb{R}^n$ whose entries are all zero, except for the i-th entry, which is equal to one; see Figure 2.7.

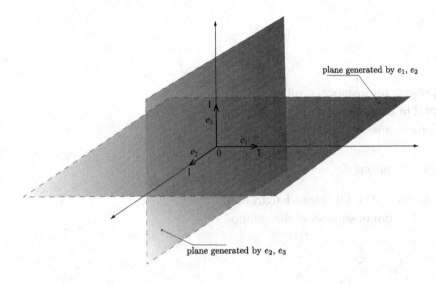

plane generated by e_1, e_2

plane generated by e_2, e_3

Figure 2.7 Standard basis of $\mathbb{R}^3$ and planes generated by linear combinations of (e_1, e_2) and of (e_2, e_3).

In all the rest of this book we deal with *finite dimensional* vector spaces, that is with spaces having a basis of finite cardinality. There are some vector spaces that are infinite dimensional (one such example is the space of polynomials with unspecified degree). However, a rigorous treatment of such vector spaces requires tools that are out of the scope of the present exposition. From now on, any time we mention a vector space, we tacitly assume that this vector space is *finite dimensional*.

2.1.2.6 Affine sets. A concept related to the one of subspaces is that of *affine sets*, which are defined as a translation of subspaces. Namely, an affine set is a set of the form[2]

$$\mathcal{A} = \{ x \in \mathcal{X} : x = v + x^{(0)}, v \in \mathcal{V} \},$$

where $x^{(0)}$ is a given point and $\mathcal{V}$ is a given subspace of $\mathcal{X}$. Subspaces are just affine spaces containing the origin. Geometrically, an affine set is a flat passing through $x^{(0)}$. The dimension of an affine set $\mathcal{A}$ is defined as the dimension of its generating subspace $\mathcal{V}$. For example, if $\mathcal{V}$ is a one-dimensional subspace generated by a vector $x^{(1)}$, then $\mathcal{A}$ is a one-dimensional affine set parallel to $\mathcal{V}$ and passing through $x^{(0)}$, which we refer to as a *line*, see Figure 2.8.

[2] We shall sometimes use the shorthand notation $\mathcal{A} = x^{(0)} + \mathcal{V}$ to denote an affine set, and we shall refer to $\mathcal{V}$ as the subspace generating $\mathcal{A}$.

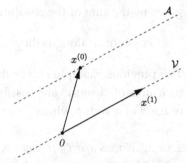

Figure 2.8 A line is a one-dimensional affine set.

A line can hence be described by means of two elements: a point $x^{(0)}$ belonging to the line, and a vector $u \in \mathcal{X}$ describing the direction of the line in space. Then, the line through x_0 along direction u is the set

$$L = \{x \in \mathcal{X} : x = x_0 + v, \ v \in \text{span}(u)\},$$

where in this case $\text{span}(u) = \{\lambda u : \lambda \in \mathbb{R}\}$.

2.2 Norms and inner products

As we have seen, vectors may represent, for instance, positions of objects in space. It is therefore natural to introduce a notion of *distance* between vectors, or of the *length* of a vector.

2.2.1 Euclidean length and general ℓ_p norms

2.2.1.1 The concept of length and distance. The Euclidean length of a vector $x \in \mathbb{R}^n$ is the square-root of the sum of squares of the components of x, that is

$$\text{Euclidean length of } x \doteq \sqrt{x_1^2 + x_2^2 + \cdots + x_n^2}.$$

This formula is an obvious extension to the multidimensional case of the Pythagoras theorem in $\mathbb{R}^2$; see Figure 2.9.

The Euclidean length represents the actual distance to be "travelled" to reach point x from the origin 0, along the most direct way (the straight line passing through 0 and x). It may, however, be useful to have a slightly more general notion of length and distance in a vector space, besides the Euclidean one. Suppose for instance that for going from 0 to x we cannot move along the direct route, but we have to follow some path along an orthogonal grid, as exemplified in Figure 2.10. This is the situation experienced, for example, by a driver who needs to move along a network of orthogonal streets to reach its destination. In this case, the shortest distance from 0 to x is given by the sum of the absolute values of the components of x:

$$\text{Length of } x \text{ (along orthogonal grid)} = |x_1| + |x_2| + \cdots + |x_n|.$$

The previous example shows that in vector spaces several different measures of "length" are possible. This leads to the general concept of *norm* of a vector, which generalizes the idea of Euclidean length.

2.2.1.2 Norms and ℓ_p norms. A *norm* on a vector space $\mathcal{X}$ is a real-valued function with special properties that maps any element $x \in \mathcal{X}$ into a real number $\|x\|$.

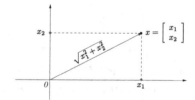

Figure 2.9 The Euclidean length of a vector in $\mathbb{R}^2$ is computed by means of the Pythagoras theorem.

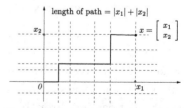

Figure 2.10 Length of path from 0 to x along an orthogonal grid is $|x_1| + |x_2|$.

Definition 2.1 *A function from $\mathcal{X}$ to $\mathbb{R}$ is a* norm, *if*

$\|x\| \geq 0 \; \forall x \in \mathcal{X}$, *and* $\|x\| = 0$ *if and only if $x = 0$;*

$\|x + y\| \leq \|x\| + \|y\|$, *for any $x, y \in \mathcal{X}$ (triangle inequality);*

$\|\alpha x\| = |\alpha| \|x\|$, *for any scalar α and any $x \in \mathcal{X}$.*

Examples of norms on the vector space $\mathcal{X} = \mathbb{R}^n$ are the so-called ℓ_p norms, defined as

$$\|x\|_p \doteq \left(\sum_{k=1}^n |x_k|^p \right)^{1/p}, \quad 1 \leq p < \infty.$$

In particular, for $p = 2$ we obtain the standard Euclidean length

$$\|x\|_2 \doteq \sqrt{\sum_{k=1}^n x_k^2},$$

and for $p = 1$ we obtain the sum-of-absolute-values length

$$\|x\|_1 \doteq \sum_{k=1}^n |x_k|.$$

The limit case $p = \infty$ defines the ℓ_∞ norm (max absolute value norm, or Chebyshev norm)[3]

$$\|x\|_\infty \doteq \max_{k=1,\ldots,n} |x_k|.$$

In some applications, we may encounter functions of a vector $x \in \mathbb{R}^n$ that are not formally norms but still encode some measure of "size" of a vector. A prime example is the *cardinality* function, which we shall discuss in Section 9.5.1. The cardinality of a vector x is defined as the number of nonzero elements in x:

$$\text{card}(x) \doteq \sum_{k=1}^n \mathbb{I}(x_k \neq 0), \quad \text{where} \quad \mathbb{I}(x_k \neq 0) \doteq \begin{cases} 1 & \text{if } x_k \neq 0 \\ 0 & \text{otherwise.} \end{cases}$$

The cardinality of a vector x is often called the ℓ_0 norm and denoted by $\|x\|_0$, although this is not a norm in the proper sense,[4] since it does not satisfy the third property in Definition 2.1.

2.2.1.3 Norm balls. The set of all vectors with ℓ_p norm less than or equal to one,

$$\mathcal{B}_p = \{ x \in \mathbb{R}^n : \|x\|_p \leq 1 \},$$

is called the unit ℓ_p norm ball. The shape of this set completely characterizes the norm. Depending on the value of p, the sets $\mathcal{B}_p$ have a different geometrical shape. Figure 2.11 shows the shapes of the $\mathcal{B}_2$, $\mathcal{B}_1$, and $\mathcal{B}_\infty$ balls in $\mathbb{R}^2$.

[3] If, for instance, x is a vector representation of a time-series (see Example 2.3), then $\|x\|_\infty$ returns its peak amplitude.

[4] This slight abuse of terminology is justified by the fact that

$$\text{card}(x) = \|x\|_0 = \lim_{p \to 0} \left(\sum_{k=1}^n |x_k|^p \right)^{1/p}.$$

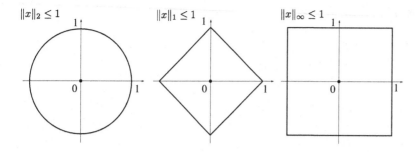

$\|x\|_2 \leq 1$ $\|x\|_1 \leq 1$ $\|x\|_\infty \leq 1$

Figure 2.11 Norm balls in $\mathbb{R}^2$.

We observe, for instance, that the ℓ_2 norm does not "favor" any direction in space, i.e., it is rotationally invariant, meaning that a vector of fixed length that rotates arbitrarily will maintain the same ℓ_2 norm. On the contrary, the same vector will have different ℓ_1 norms, reaching its smallest value when aligned to the coordinate axes.

2.2.2 Inner product, angle, orthogonality

2.2.2.1 Inner product. A fundamental operation that can be defined between two vectors is the *inner product*.

Definition 2.2 *An* inner product *on a (real) vector space $\mathcal{X}$ is a real-valued function which maps any pair of elements $x, y \in \mathcal{X}$ into a scalar denoted by $\langle x, y \rangle$. The inner product satisfies the following axioms: for any $x, y, z \in \mathcal{X}$ and scalar α*

$$\langle x, x \rangle \geq 0;$$
$$\langle x, x \rangle = 0 \text{ if and only if } x = 0;$$
$$\langle x + y, z \rangle = \langle x, z \rangle + \langle y, z \rangle;$$
$$\langle \alpha x, y \rangle = \alpha \langle x, y \rangle;$$
$$\langle x, y \rangle = \langle y, x \rangle.$$

A vector space equipped with an inner product is called an *inner product space*.

The *standard inner product*[5] defined in $\mathbb{R}^n$ is the "row-column" product of two vectors

$$\langle x, y \rangle = x^\top y = \sum_{k=1}^{n} x_k y_k. \tag{2.2}$$

However, other inner products can be defined in $\mathbb{R}^n$, see, for example, Exercise 2.4. Moreover, the inner product remains well defined also on vector spaces different from $\mathbb{R}^n$, such as, for instance, the space of matrices, defined in Chapter 3.

[5] The standard inner product is also often referred to as the *scalar* product (since it returns a scalar value), or as the *dot* product (since it is sometimes denoted by $x \cdot y$).

In an inner product space, the function $\sqrt{\langle x, x \rangle}$ is a norm, which will often be denoted simply by $\|x\|$, with no subscript. For example, for $\mathbb{R}^n$ equipped with the standard inner product, we have

$$\|x\| = \sqrt{\langle x, x \rangle} = \|x\|_2.$$

The few examples below further illustrate the usefulness of the concept of the standard inner product between vectors.

Example 2.6 (*Rate of return of a financial portfolio*) The rate of return r (or return) of a single financial asset over a given period (say, a year, or a day) is the interest obtained at the end of the period by investing in it. In other words, if, at the beginning of the period, we invest a sum S in the asset, we will earn $S_{\text{end}} = (1 + r)S$ at the end of the period. That is,

$$r = \frac{S_{\text{end}} - S}{S}.$$

Whenever the rate of return is small ($r \ll 1$), the following approximation is considered:

$$r = \frac{S_{\text{end}}}{S} - 1 \approx \log \frac{S_{\text{end}}}{S},$$

with the latter quantity being known as the log-return. For n assets, we can define a vector $r \in \mathbb{R}^n$ such that the i-th component of r is the rate of return of the i-th asset. Assume that at the beginning of the period we invest a total sum S over all assets, by allocating a fraction x_i of S in the i-th asset. Here, $x \in \mathbb{R}^n$ represents the portfolio "mix," and it is a non-negative vector whose components sum to one. At the end of the period the total value of our portfolio would be

$$S_{\text{end}} = \sum_{i=1}^{n} (1 + r_i) x_i S,$$

hence the rate of return of the portfolio is the relative increase in wealth:

$$\frac{S_{\text{end}} - S}{S} = \sum_{i=1}^{n} (1 + r_i) x_i - 1 = \sum_{i=1}^{n} x_i - 1 + \sum_{i=1}^{n} r_i x_i = r^\top x.$$

The rate of return is thus the standard inner product between the vectors of individual returns r and of the portfolio allocation weights x. Note that, in practice, rates of return are never known exactly in advance, and they can be negative (although, by construction, they are never less than -1).

Example 2.7 (*The arithmetic mean, the weighted mean, and the expected value*) The arithmetic mean (or, average) of given numbers $x_1, \ldots, x_n$, is defined as

$$\hat{x} = \frac{1}{n}(x_1 + \cdots + x_n).$$

The arithmetic mean can be interpreted as a scalar product:

$$\hat{x} = p^\top x,$$

where $x = [x_1, \ldots, x_n]^\top$ is the vector containing the numbers (samples), and p is a vector of weights to be assigned to each of the samples. In the specific case of arithmetic mean, each sample has equal weight $1/n$, hence $p = \frac{1}{n}\mathbf{1}$, where $\mathbf{1}$ is a vector of all ones.

More generally, for any weight vector $p \in \mathbb{R}^n$ such that $p_i \geq 0$ for every i, and $p_1 + \cdots + p_n = 1$, we can define the corresponding weighted average of the elements of x as $p^\top x$. The interpretation of p is in terms of a discrete probability distribution of a random variable X, which takes the value x_i with probability p_i, $i = 1, \ldots, n$. The weighted average is then simply the expected value (or, mean) of X, under the discrete probability distribution p. The expected value is often denoted by $\mathbb{E}_p\{X\}$, or simply $\mathbb{E}\{X\}$, if the distribution p is clear from context.

2.2.2.2 Angle between vectors. The standard inner product on $\mathbb{R}^n$ is related to the notion of angle between two vectors. If two nonzero vectors x, y are visualized as two points in Cartesian space, one can consider the triangle constituted by x, y and the origin 0; see Figure 2.12. Let θ be the angle at 0 between the $0x$ and $0y$ sides of the triangle, and let $z = x - y$. Applying the Pythagoras theorem to the triangle with vertices yxx', we have that

$$
\begin{aligned}
\|z\|_2^2 &= (\|y\|_2 \sin\theta)^2 + (\|x\|_2 - \|y\|_2 \cos\theta)^2 \\
&= \|x\|_2^2 + \|y\|_2^2 - 2\|x\|_2\|y\|_2 \cos\theta.
\end{aligned}
$$

But,

$$
\|z\|_2^2 = \|x - y\|_2^2 = (x - y)^\top(x - y) = x^\top x + y^\top y - 2x^\top y,
$$

which, compared with the preceding equation, yields

$$
x^\top y = \|x\|_2\|y\|_2 \cos\theta.
$$

The angle between x and y is therefore defined via the relation

$$
\cos\theta = \frac{x^\top y}{\|x\|_2\|y\|_2}. \tag{2.3}
$$

When $x^\top y = 0$, the angle between x and y is $\theta = \pm 90°$, i.e., x, y are *orthogonal*. When the angle θ is $0°$, or $\pm 180°$, then x is aligned with y, that is $y = \alpha x$, for some scalar α, i.e., x and y are *parallel*. In this situation $|x^\top y|$ achieves its maximum value $|\alpha|\|x\|_2^2$.

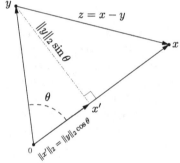

Figure 2.12 Angle θ between vectors x, y: $\cos\theta = \frac{x^\top y}{\|x\|\|y\|}$.

Example 2.8 (*Comparing text*) We can use the word-frequency vector representation of text introduced in Example 2.1, for comparing text documents. In this context, similarity between two documents may be measured by means of the angle θ between the two frequency vectors representing the documents, the documents being maximally "different" when the corresponding frequency vectors are orthogonal. As an example, consider the following headlines from the web edition of the New York Times on Dec. 7, 2010:

(a) Suit Over Targeted Killing in Terror Case Is Dismissed. A federal judge on Tuesday dismissed a lawsuit that sought to block the United States from attempting to kill an American citizen, Anwar Al-Awlaki, who has been accused of aiding Al Qaeda.

(b) In Tax Deal With G.O.P., a Portent for the Next 2 Years. President Obama made clear that he was willing to alienate his liberal base in the interest of compromise. Tax Deal suggests new path for Obama. President Obama agreed to a tentative deal to extend the Bush tax cuts, part of a package to keep jobless aid and cut payroll taxes.

(c) Obama Urges China to Check North Koreans. In a frank discussion, President Obama urged China's president to put the North Korean government on a tighter leash after a series of provocations.

(d) Top Test Scores From Shanghai Stun Educators. With China's debut in international standardized testing, Shanghai students have surprised experts by outscoring counterparts in dozens of other countries.

The text has been first simplified (e.g., plurals removed, verbs converted to present tense, etc.) and then compared against the dictionary $V = \{$aid, kill, deal, president, tax, china$\}$. The frequency vectors are

$$x^{(a)} = \begin{bmatrix} \frac{1}{3} \\ \frac{2}{3} \\ 0 \\ 0 \\ 0 \\ 0 \end{bmatrix}, \ x^{(b)} = \begin{bmatrix} \frac{1}{10} \\ 0 \\ \frac{3}{10} \\ \frac{1}{5} \\ \frac{2}{5} \\ 0 \end{bmatrix}, \ x^{(c)} = \begin{bmatrix} 0 \\ 0 \\ 0 \\ \frac{1}{2} \\ 0 \\ \frac{1}{2} \end{bmatrix}, \ x^{(d)} = \begin{bmatrix} 0 \\ 0 \\ 0 \\ 0 \\ 0 \\ 1 \end{bmatrix}.$$

Table 2.2 displays the $\cos\theta$ between pairs of vectors representing the text. A high value of $\cos\theta$ suggests a high correlation between the two texts, while a $\cos\theta$ near zero means that the two texts are nearly orthogonal (uncorrelated).

Table 2.2 Cosine of angle θ between texts.

$\cos\theta$	$x^{(a)}$	$x^{(b)}$	$x^{(c)}$	$x^{(d)}$
$x^{(a)}$	1	0.0816	0	0
$x^{(b)}$	*	1	0.2582	0
$x^{(c)}$	*	*	1	0.7071
$x^{(d)}$	*	*	*	1

2.2.2.3 *Cauchy–Schwartz inequality and its generalization.* Since $|\cos\theta| \leq 1$, it follows from equation (2.3) that

$$|x^{\top}y| \leq \|x\|_2 \|y\|_2,$$

and this inequality is known as the Cauchy–Schwartz inequality. A generalization of this inequality involves general ℓ_p norms and is known as the Hölder inequality: for any vectors $x, y \in \mathbb{R}^n$ and for any $p, q \geq 1$ such that $1/p + 1/q = 1$, it holds that[6]

$$|x^{\top}y| \leq \sum_{k=1}^{n} |x_k y_k| \leq \|x\|_p \|y\|_q. \tag{2.4}$$

[6] See Exercise 2.7.

2.2.2.4   Maximization of inner product over norm balls. Given a nonzero vector $y \in \mathbb{R}^n$, consider the problem of finding some vector $x \in \mathcal{B}_p$ (the unit ball in ℓ_p norm) that maximizes the inner product $x^\top y$: that is, solve

$$\max_{\|x\|_p \leq 1} \ x^\top y.$$

For $p = 2$ the solution is readily obtained from Eq. (2.3): x should be aligned (parallel) to y, so as to form a zero angle with it, and have the largest possible norm, that is, a norm equal to one. Therefore the unique solution is

$$x_2^* = \frac{y}{\|y\|_2},$$

hence $\max_{\|x\|_2 \leq 1} x^\top y = \|y\|_2$.

Consider next the case with $p = \infty$: since $x^\top y = \sum_{i=1}^n x_i y_i$, where each element x_i is such that $|x_i| \leq 1$, then the maximum in the sum is achieved by setting $x_i = \mathrm{sgn}(y_i)$,[7] so that $x_i y_i = |y_i|$. Hence,

$$x_\infty^* = \mathrm{sgn}(y),$$

and $\max_{\|x\|_\infty \leq 1} x^\top y = \sum_{i=1}^n |y_i| = \|y\|_1$. The optimal solution may not be unique, since corresponding to any $y_i = 0$ any value $x_i \in [-1, 1]$ could be selected without modifying the optimal objective.

Finally, we consider the case with $p = 1$: the inner product $x^\top y = \sum_{i=1}^n x_i y_i$ can now be interpreted as a weighted average of the y_is, where the x_is are the weights, whose absolute values must sum up to one. The maximum of the weighted average is achieved by first finding the y_i having the largest absolute value, that is by finding one index m such that $|y_i| \leq |y_m|$ for all $i = 1, \dots, n$, and then setting

$$[x_1^*]_i = \begin{cases} \mathrm{sgn}(y_i) & \text{if } i = m \\ 0 & \text{otherwise} \end{cases}, \quad i = 1, \dots, n.$$

We thus have $\max_{\|x\|_1 \leq 1} x^\top y = \max_i |y_i| = \|y\|_\infty$. Again, the optimal solution may not be unique since in the case when vector y has several entries with identical maximum absolute value then m can be chosen to be any of the indices corresponding to these entries.

[7] sgn denotes the sign function, which, by definition, takes values $\mathrm{sgn}(x) = 1$, if $x > 0$, $\mathrm{sgn}(x) = -1$, if $x < 0$, and $\mathrm{sgn}(x) = 0$, if $x = 0$.

Example 2.9 (*Production margin*) Consider a production process involving two raw materials r_1, r_2 and one finished product. The unit cost for the raw materials is subject to variability, and it is given by

$$c_i = \bar{c}_i + \alpha_i x_i, \quad i = 1, 2,$$

where, for $i = 1, 2$, $\bar{c}_i$ is the nominal unit cost of material r_i, $\alpha_i \geq 0$ is the cost spread, and $|x_i| \leq 1$ is an unknown term accounting for cost uncertainty. Production of one unit of the finished product requires a fixed amount m_1 of raw material r_1 and a fixed amount m_2 of raw material r_2. Each finished product can be sold on the market at a price p which is not precisely known in advance. We assume that

$$p = \bar{p} + \beta x_3,$$

where $\bar{p}$ is the nominal selling price for one unit of the finished product, $\beta \geq 0$ is the price spread, and $|x_3| \leq 1$ is an unknown term accounting for price uncertainty. The production margin (income minus cost) for each unit of finished product is thus given by

$$\begin{aligned} \text{margin} &= p - c_1 m_1 - c_2 m_2 \\ &= \bar{p} + \beta x_3 - \bar{c}_1 m_1 - \alpha_1 x_1 m_1 - \bar{c}_2 m_2 - \alpha_2 x_2 m_2 \\ &= \text{nom\_margin} + x^\top y, \end{aligned}$$

where we defined $\text{nom\_margin} \doteq \bar{p} - \bar{c}_1 m_1 - \bar{c}_2 m_2$, and

$$x^\top = [x_1, x_2, x_3], \quad y = [-\alpha_1 m_1, -\alpha_2 m_2, \beta]^\top.$$

We then see that the production margin is given by a constant term reflecting the nominal material costs and sale price, plus a variable term of the form $x^\top y$, with uncertainty vector x such that $\|x\|_\infty \leq 1$. Our problem is to determine the maximum and the minimum production margin under the given uncertainty. Clearly, the margin lies in an interval centered at the nominal margin nom_margin, of half-length

$$\max_{\|x\|_\infty \leq 1} x^\top y = \|y\|_1 = \alpha_1 m_1 + \alpha_2 m_2 + \beta.$$

2.2.3 Orthogonality and orthogonal complements

2.2.3.1 Orthogonal vectors. Generalizing the concept of orthogonality to generic inner product spaces, we say that two vectors x, y in an inner product space $\mathcal{X}$ are *orthogonal* if $\langle x, y \rangle = 0$. Orthogonality of two vectors $x, y \in \mathcal{X}$ is symbolized by $x \perp y$.

Nonzero vectors $x^{(1)}, \ldots, x^{(d)}$ are said to be mutually orthogonal if $\langle x^{(i)}, x^{(j)} \rangle = 0$ whenever $i \neq j$. In words, each vector is orthogonal to all other vectors in the collection. The following proposition holds (the converse of this proposition is instead false, in general).

Proposition 2.1 *Mutually orthogonal vectors are linearly independent.*

Proof Suppose, for the purpose of contradiction, that $x^{(1)}, \ldots, x^{(d)}$ are orthogonal but linearly dependent vectors. This would mean that there exist $\alpha_1, \ldots, \alpha_d$, not all identically zero, such that

$$\sum_{i=1}^{d} \alpha_i x^{(i)} = 0.$$

But, taking the inner product of both sides of this equation with $x^{(j)}$, $j = 1, \ldots, d$, we would get

$$\langle \sum_{i=1}^{d} \alpha_i x^{(i)}, x^{(j)} \rangle = 0, \quad j = 1, \ldots, d.$$

Since

$$\langle \sum_{i=1}^{d} \alpha_i x^{(i)}, x^{(j)} \rangle = \sum_{i=1}^{d} \langle \alpha_i x^{(i)}, x^{(j)} \rangle = \alpha_j \|x^{(j)}\|^2 = 0,$$

it would follow that $\alpha_i = 0$ for all $j = 1, \ldots, d$, which contradicts the hypothesis. □

2.2.3.2 *Orthonormal vectors.* A collection of vectors $S = \{x^{(1)}, \ldots, x^{(d)}\}$ is said to be *orthonormal* if, for $i, j = 1, \ldots, d$,

$$\langle x^{(i)}, x^{(j)} \rangle = \begin{cases} 0 & \text{if } i \neq j, \\ 1 & \text{if } i = j. \end{cases}$$

In words, S is orthonormal if every element has unit norm, and all elements are orthogonal to each other. A collection of orthonormal vectors S forms an *orthonormal basis* for the span of S.

2.2.3.3 *Orthogonal complement.* A vector $x \in \mathcal{X}$ is orthogonal to a subset S of an inner product space $\mathcal{X}$ if $x \perp s$ for all $s \in S$. The set of vectors in $\mathcal{X}$ that are orthogonal to S is called the *orthogonal complement* of S, and it is denoted by $S^{\perp}$; see Figure 2.13.

2.2.3.4 *Direct sum and orthogonal decomposition.* A vector space $\mathcal{X}$ is said to be the *direct sum* of two subspaces A, B if any element $x \in \mathcal{X}$ can be written in a unique way as $x = a + b$, with $a \in A$ and $b \in B$; this situation is symbolized by the notation $X = A \oplus B$. The following theorem holds.

Theorem 2.1 (Orthogonal decomposition) *If S is a subspace of an inner-product space $\mathcal{X}$, then any vector $x \in \mathcal{X}$ can be written in a unique way as the sum of one element in S and one in the orthogonal complement $S^{\perp}$ (see Figure 2.14), that is*

$$\mathcal{X} = S \oplus S^{\perp} \quad \text{for any subspace } S \subseteq \mathcal{X}.$$

Proof We first observe that $S \cap S^{\perp} = 0$, since if $v \in S \cap S^{\perp}$, then $\langle v, v \rangle = \|v\|^2 = 0$, which implies that $v = 0$. Next, we denote $\mathcal{W} = S + S^{\perp}$ (the space of vectors obtained by summing elements from S and elements from $S^{\perp}$). We can choose an orthonormal basis of $\mathcal{W}$

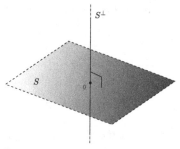

Figure 2.13 Example of a two-dimensional subspace S in $\mathbb{R}^3$ and its orthogonal complement $S^{\perp}$.

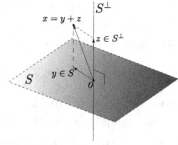

Figure 2.14 Any vector can be written in a unique way as the sum of an element in a subspace S and one in its orthogonal complement $S^{\perp}$.

and extend it to an orthonormal basis[8] of $\mathcal{X}$. Thus, if $\mathcal{W} \neq \mathcal{X}$, there is an element z in the basis of $\mathcal{X}$ which is orthogonal to $\mathcal{W}$. Since $\mathcal{S} \subseteq \mathcal{W}$, z is orthogonal to $\mathcal{S}$ as well, which means that z belongs to $\mathcal{S}^\perp$. The latter is a subspace of $\mathcal{W}$, therefore z is in $\mathcal{W}$, and we arrive at a contradiction. Thus, we proved that $\mathcal{S} + \mathcal{S}^\perp = \mathcal{X}$, that is each element $x \in \mathcal{X}$ can be written as the sum of one element from $x_s \in \mathcal{S}$ and one element from $y \in \mathcal{S}^\perp$, i.e., $x = x_s + y$. It remains to be proved that such a decomposition is unique. Suppose it is not, for the purpose of contradiction. Then, there would exist $x_{s_1}, x_{s_2} \in \mathcal{S}$, and $y_1, y_2 \in \mathcal{S}^\perp$, $x_{s_1} \neq x_{s_2}$, $y_1 \neq y_2$, such that $x = x_{s_1} + y_1$ and $x = x_{s_2} + y_2$. But then, taking the difference of these last two expressions we would have

$$0 \neq x_{s_1} - x_{s_2} = y_2 - y_1,$$

where the left-hand side belongs to $\mathcal{S}$ and the right-hand side belongs to $\mathcal{S}^\perp$, which is impossible since $\mathcal{S} \cap \mathcal{S}^\perp = 0$. $\square$

The following proposition summarizes some fundamental properties of inner product spaces.

Proposition 2.2 *Let x, z be any two elements of a (finite dimensional) inner product space $\mathcal{X}$, let $\|x\| = \sqrt{\langle x, x \rangle}$, and let α be a scalar. Then:*

1. $|\langle x, z \rangle| \leq \|x\| \|z\|$, *and equality holds iff $x = \alpha z$, or $z = 0$ (Cauchy–Schwartz);*

2. $\|x + z\|^2 + \|x - z\|^2 = 2\|x\|^2 + 2\|z\|^2$ *(parallelogram law);*

3. *if $x \perp z$, then $\|x + z\|^2 = \|x\|^2 + \|z\|^2$ (Pythagoras theorem);*

4. *for any subspace $S \subseteq \mathcal{X}$ it holds that $\mathcal{X} = S \oplus S^\perp$;*

5. *for any subspace $S \subseteq \mathcal{X}$ it holds that $\dim \mathcal{X} = \dim S + \dim S^\perp$.*

2.3   Projections onto subspaces

The idea of projection is central in optimization, and it corresponds to the problem of finding a point on a given set that is closest (in norm) to a given point. Formally, given a vector x in an inner product space $\mathcal{X}$ (say, e.g., $\mathcal{X} = \mathbb{R}^n$) and a closed set[9] $\mathcal{S} \subseteq \mathcal{X}$, the projection of x onto $\mathcal{S}$, denoted by $\Pi_S(x)$, is defined as the point in $\mathcal{S}$ at minimal distance from x:

$$\Pi_S(x) = \arg \min_{y \in S} \|y - x\|,$$

where the norm used here is the norm induced by the inner product, that is $\|y - x\| = \sqrt{\langle y - x, y - x \rangle}$. This simply reduces to the Euclidean norm, when using the standard inner product (see Eq. (2.2)), in which case the projection is called the *Euclidean projection*.

[8] We discuss in Section 2.3.3 how it is always possible to construct an orthonormal basis for a subspace, starting from any given basis for that subspace.

[9] A set is closed if it contains its boundary. For instance, the set of points $x \in \mathbb{R}^2$ such that $|x_1| \leq 1$, $|x_2| \leq 1$ is closed, whereas the set characterized by $|x_1| \leq 1$, $|x_2| < 1$ is not. See Section 8.1.1 for further details.

In this section we focus in particular on the case when $\mathcal{S}$ is a subspace.

2.3.1   Projection onto a one-dimensional subspace

To introduce the concept of projections, we begin by studying a one-dimensional case. Given a point (vector) $x \in \mathcal{X}$ and a nonzero vector $v \in \mathcal{X}$, where $\mathcal{X}$ is an inner product space, the projection $\Pi_{\mathcal{S}_v}(x)$ of x onto the subspace generated by v (i.e., the one-dimensional subspace of vectors $\mathcal{S}_v = \{\lambda v, \lambda \in \mathbb{R}\}$) is the vector belonging to $\mathcal{S}_v$ at minimum distance (in the sense of the norm induced by the inner product) from x. In formulas, we seek

$$\Pi_{\mathcal{S}_v}(x) = \arg \min_{y \in \mathcal{S}_v} \|y - x\|.$$

We next show that the projection is characterized by the fact that the difference vector $(x - \Pi_{\mathcal{S}_v}(x))$ is orthogonal to v. To see this fact, let x_v be a point in $\mathcal{S}_v$ such that $(x - x_v) \perp v$, and consider an arbitrary point $y \in \mathcal{S}_v$. By the Pythagoras theorem, we have that (see Figure 2.15)

$$\|y - x\|^2 = \|(y - x_v) - (x - x_v)\|^2 = \|y - x_v\|^2 + \|x - x_v\|^2.$$

Since the first quantity in the above expression is always nonnegative, it follows that the minimum over y is obtained by choosing $y = x_v$, which proves that x_v is the projection we sought. To find a formula for x_v, we start from the orthogonality condition

$$(x - x_v) \perp v \iff \langle x - x_v, v \rangle = 0.$$

Then, we exploit the fact that x_v is by definition a scalar multiple of v, that is $x_v = \alpha v$ for some $\alpha \in \mathbb{R}$, and we solve for α, obtaining

$$x_v = \alpha v, \quad \alpha = \frac{\langle v, x \rangle}{\|v\|^2}.$$

Vector x_v is usually called the *component* of x along the direction v, see Figure 2.15. If v has unit norm, then this component is simply given by $x_v = \langle v, x \rangle v$.

2.3.2   Projection onto an arbitrary subspace

We now extend the previous result to the case when $\mathcal{S}$ is an arbitrary subspace (i.e., not necessarily one-dimensional). This is stated in the following key theorem, which is illustrated in Figure 2.16.

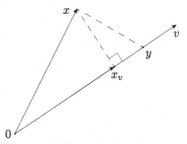

Figure 2.15 The projection of x onto the subspace $\mathcal{S}_v$ generated by a single vector v is a point $x_v \in \mathcal{S}_v$ such that the difference $x - x_v$ is orthogonal to v. x_v is also called the *component* of x along the direction v.

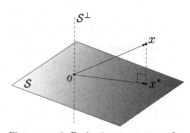

Figure 2.16 Projection onto a subspace.

Theorem 2.2 (projection theorem) *Let $\mathcal{X}$ be an inner product space, let x be a given element in $\mathcal{X}$, and let S be a subspace of $\mathcal{X}$. Then, there exists a unique vector $x^* \in S$ which is the solution to the problem*

$$\min_{y \in S} \|y - x\|.$$

Moreover, a necessary and sufficient condition for x^* being the optimal solution for this problem is that

$$x^* \in S, \quad (x - x^*) \perp S.$$

Proof Let $S^\perp$ be the orthogonal subspace of S, then, by virtue of Theorem 2.1, any vector $x \in \mathcal{X}$ can be written in a unique way as

$$x = u + z, \ u \in S, \ z \in S^\perp.$$

Hence, for any vector y,

$$\|y - x\|^2 \quad = \quad \|(y - u) - z\|^2 = \|y - u\|^2 + \|z\|^2 - 2\langle y - u, z\rangle.$$

The last inner product term in the sum is zero, since $z \in S^\perp$ is orthogonal to all vectors in S. Therefore

$$\|y - x\|^2 \quad = \quad \|y - u\|^2 + \|z\|^2,$$

from which it follows that the unique minimizer of the distance $\|y - x\|$ is $x^* = y = u$. Finally, with this choice, $y - x = z \in S^\perp$, which concludes the proof. □

A simple generalization of Theorem 2.2 considers the problem of projecting a point x onto an affine set $\mathcal{A}$, see Figure 2.17. This is formally stated in the next corollary.

Corollary 2.1 (Projection on affine set) *Let $\mathcal{X}$ be an inner product space, let x be a given element in $\mathcal{X}$, and let $\mathcal{A} = x^{(0)} + S$ be the affine set obtained by translating a given subspace S by a given vector $x^{(0)}$. Then, there exists a unique vector $x^* \in \mathcal{A}$ which is the solution to the problem*

$$\min_{y \in \mathcal{A}} \|y - x\|.$$

Moreover, a necessary and sufficient condition for x^* to be the optimal solution for this problem is that

$$x^* \in \mathcal{A}, \quad (x - x^*) \perp S.$$

Proof We reduce the problem to that of projecting a point onto a subspace. Considering that any point $y \in \mathcal{A}$ can be written as $y = z + x^{(0)}$, for some $z \in S$, our problem becomes

$$\min_{y \in \mathcal{A}} \|y - x\| = \min_{z \in S} \|z + x^{(0)} - x\|.$$

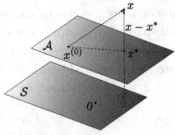

Figure 2.17 Projection on affine set.

The latter problem thus amounts to projecting the point $(x - x^{(0)})$ onto the subspace $\mathcal{S}$. By the projection theorem, the optimality conditions for this problem are $z^* \in \mathcal{S}$ and $z^* - (x - x^{(0)}) \perp \mathcal{S}$. In terms of our original variable, the optimum $x^* = z^* + x^{(0)}$ is characterized by

$$x^* \in \mathcal{A}, \quad (x^* - x) \perp \mathcal{S},$$

which concludes the proof.　　　　　　　　　　　　□

2.3.2.1 Euclidean projection of a point onto a line. Let $p \in \mathbb{R}^n$ be a given point. We want to compute the Euclidean projection p^* of p onto a line $L = \{x_0 + \text{span}(u)\}$, $\|u\|_2 = 1$, as defined in Section 2.1.2.6; see Figure 2.18.

The Euclidean projection is a point in L at minimum Euclidean distance from p, that is

$$p^* = \arg\min_{x \in L} \|x - p\|_2.$$

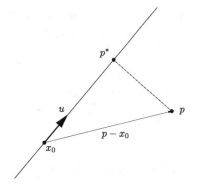

Figure 2.18 Projection of point p onto the line $L = \{x_0 + v,\ v \in \text{span}(u)\}$.

Since any point $x \in L$ can be written as $x = x_0 + v$, for some $v \in \text{span}(u)$, the above problem is equivalent to finding a value v^* for v, such that

$$v^* = \arg\min_{v \in \text{span}(u)} \|v - (p - x_0)\|_2.$$

We recognize here the classical situation addressed by the projection theorem: for $z = p - x_0$, we need to find the projection of z onto the subspace $\mathcal{S} = \text{span}(u)$. The solution must hence satisfy the orthogonality condition $(z - v^*) \perp u$, i.e., $\langle (z - v^*), u \rangle = 0$ where the inner product to be used here is the standard one. Recalling that $v^* = \lambda^* u$ and $u^\top u = \|u\|_2^2 = 1$, we hence have

$$u^\top z - u^\top v^* = 0 \Leftrightarrow u^\top z - \lambda^* = 0 \Leftrightarrow \lambda^* = u^\top z = u^\top (p - x_0).$$

The optimal point p^* is thus given by

$$p^* = x_0 + v^* = x_0 + \lambda^* u = x_0 + u^\top (p - x_0) u,$$

and the squared distance from p to the line is

$$\|p - p^*\|_2^2 = \|p - x_0\|_2^2 - \lambda^{*2} = \|p - x_0\|_2^2 - (u^\top (p - x_0))^2.$$

2.3.2.2 Euclidean projection of a point onto an hyperplane. A hyperplane is an affine set defined as follows

$$H = \{z \in \mathbb{R}^n : a^\top z = b\},$$

where $a \neq 0$ is called a *normal direction* of the hyperplane, since for any two vectors $z_1, z_2 \in H$ it holds that $(z_1 - z_2) \perp a$, see Section 2.4.4.

Given a point $p \in \mathbb{R}^n$ we want to determine the Euclidean projection p^* of p onto H. The projection theorem requires $p - p^*$ to be orthogonal to H. Since a is a direction orthogonal to H, the condition $(p - p^*) \perp H$ is equivalent to saying that

$$p - p^* = \alpha a$$

for some $\alpha \in \mathbb{R}$. To find α, consider that $p^* \in H$, thus $a^\top p^* = b$, and multiply the previous equation on the left by $a^\top$, obtaining

$$a^\top p - b = \alpha \|a\|_2^2,$$

whereby

$$\alpha = \frac{a^\top p - b}{\|a\|_2^2},$$

and

$$p^* = p - \frac{a^\top p - b}{\|a\|_2^2} a. \tag{2.5}$$

The distance from p to H is

$$\|p - p^*\|_2 = |\alpha| \cdot \|a\|_2 = \frac{|a^\top p - b|}{\|a\|_2}. \tag{2.6}$$

2.3.2.3 *Projection on a vector span.* Suppose we have a basis for a subspace $\mathcal{S} \subseteq \mathcal{X}$, that is

$$\mathcal{S} = \mathrm{span}(x^{(1)}, \ldots, x^{(d)}).$$

Given a vector $x \in \mathcal{X}$, the projection theorem readily tells us that the unique projection x^* of x onto $\mathcal{S}$ is characterized by the orthogonality condition $(x - x^*) \perp \mathcal{S}$. Since $x^* \in \mathcal{S}$, we can write x^* as some (unknown) linear combination of the elements in the basis of $\mathcal{S}$, that is

$$x^* = \sum_{i=1}^{d} \alpha_i x^{(i)}. \tag{2.7}$$

Then $(x - x^*) \perp \mathcal{S} \Leftrightarrow \langle x - x^*, x^{(k)} \rangle = 0$, $k = 1, \ldots, d$, and these conditions boil down to the following system of d linear equations[10] in d unknowns (the α_is):

[10] Linear equations are studied in detail in Chapter 6.

$$\sum_{i=1}^{d} \alpha_i \langle x^{(k)}, x^{(i)} \rangle = \langle x^{(k)}, x \rangle, \quad k = 1, \ldots, d. \tag{2.8}$$

Solving this system of linear equations provides the coefficients α, and hence the desired x^*.

Projection onto the span of orthonormal vectors. If we have an orthonormal basis for a subspace $S = \text{span}(S)$, then it is immediate to obtain the projection x^* of x onto that subspace. This is due to the fact that, in this case, the system of equations in (2.8) immediately gives the coefficients $\alpha_k = \langle x^{(k)}, x \rangle$, $i = 1, \ldots, d$. Therefore, from Eq. (2.7) we have that

$$x^* = \sum_{i=1}^{d} \langle x^{(i)}, x \rangle x^{(i)}. \tag{2.9}$$

We next illustrate a standard procedure to construct an orthonormal basis for $\text{span}(S)$.

2.3.3 The Gram–Schmidt procedure

Given a basis $S = \{x^{(1)}, \ldots, x^{(d)}\}$ (i.e., a collection of linearly independent elements) for a subspace $S = \text{span}(S)$, we wish to construct an orthonormal basis for the same subspace. The Gram–Schmidt procedure does so by working recursively as follows. Choose any element from S, say $x^{(1)}$, and let

$$\zeta^{(1)} = x^{(1)}, \quad z^{(1)} = \frac{\zeta^{(1)}}{\|\zeta^{(1)}\|}.$$

Note that $\zeta^{(1)}$ is nonzero, for otherwise $x^{(1)}$ would be a linear combination (with zero coefficients) of the remaining vectors, which is ruled out by the independence of the elements in S (since S is assumed to be a basis), hence the division by $\|\zeta^{(1)}\|$ is well defined.

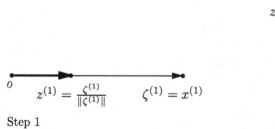

Step 1

Step 2

Figure 2.19 First two steps of the Gram–Schmidt procedure.

Observe now that if we project $x^{(2)}$ onto $\text{span}(z^{(1)})$, obtaining $\tilde{x}^{(2)} = \langle x^{(2)}, z^{(1)} \rangle z^{(1)}$ (see Eq. (2.9)), then, by definition of projection, the residual $x^{(2)} - \langle x^{(2)}, z^{(1)} \rangle z^{(1)}$ is orthogonal to $\text{span}(z^{(1)})$: we thus obtained our second element in the orthonormal basis:

$$\zeta^{(2)} = x^{(2)} - \langle x^{(2)}, z^{(1)} \rangle z^{(1)}, \quad z^{(2)} = \frac{\zeta^{(2)}}{\|\zeta^{(2)}\|}.$$

Observe again that $\zeta^{(2)}$ is nonzero, since otherwise $x^{(2)}$ would be proportional to $z^{(1)}$ and hence also to $x^{(1)}$, which is not allowed by the independence assumption. The first two iterations of the Gram–Schmidt procedure are illustrated in Figure 2.19.

Next, we take the projection of $x^{(3)}$ onto the subspace generated by $\{z^{(1)}, z^{(2)}\}$. Since $z^{(1)}, z^{(2)}$ are orthonormal, this projection is readily computed using Eq. (2.9) as $\tilde{x}^{(3)} = \langle x^{(3)}, z^{(1)} \rangle z^{(1)} + \langle x^{(3)}, z^{(2)} \rangle z^{(2)}$. Considering the residual

$$\zeta^{(3)} = x^{(3)} - \langle x^{(3)}, z^{(1)} \rangle z^{(1)} - \langle x^{(3)}, z^{(2)} \rangle z^{(2)}$$

and normalizing[11] $\zeta^{(3)}$, we obtain the third element of the orthonormal basis (see Figure 2.20): $z^{(3)} = \dfrac{\zeta^{(3)}}{\|\zeta^{(3)}\|}$.

[11] Note again that the independence assumption prevents $\zeta^{(3)}$ from being zero.

Figure 2.20 Third step of the Gram–Schmidt procedure.

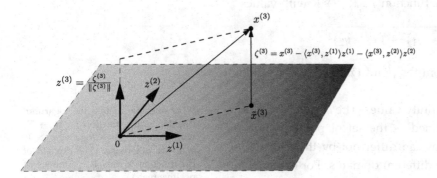

The process is iterated, yielding at the generic iteration k:

$$\zeta^{(k)} = x^{(k)} - \sum_{i=1}^{k-1} \langle x^{(k)}, z^{(i)} \rangle z^{(i)},$$

$$z^{(k)} = \frac{\zeta^{(k)}}{\|\zeta^{(k)}\|}.$$

The collection $\{z^{(1)}, \ldots, z^{(d)}\}$ is, by construction, an orthonormal basis for $\mathcal{S} = \operatorname{span}(S)$. The Gram–Schmidt procedure is a simple way to obtain an orthogonal basis, although it is not the most numerically reliable. In Section 7.3.1 we shall discuss a simple modification of the procedure that has better numerical properties.[12] The Gram–Schmidt procedure is strictly related to the so-called QR (orthogonal-triangular) factorization of a matrix, and it is a key ingredient in the solution of least-squares problems and linear equations (more on these matters in Chapter 6).

[12] See also Section 7.3.2 for a variant of the method that does not require the original vectors to be independent.

2.4   Functions

Besides the standard operations of sum and scalar multiplication, other operations can be defined on vectors, and this leads to the notion of functions, a basic object in optimization problems. We also show how the concepts of linear and affine functions are closely related to inner products.

2.4.1   Functions and maps

A *function* takes a vector argument in $\mathbb{R}^n$, and returns a unique value in $\mathbb{R}$. We use the notation

$$f : \mathbb{R}^n \to \mathbb{R},$$

to refer to a function with "input" space $\mathbb{R}^n$. The "output" space for functions is $\mathbb{R}$. For example, the function $f : \mathbb{R}^2 \to \mathbb{R}$ with values

$$f(x) = \sqrt{(x_1 - y_1)^2 + (x_2 - y_2)^2}$$

gives the Euclidean distance from the point (x_1, x_2) to a given point (y_1, y_2).

We allow functions to take infinity values. The *domain*[13] of a function f, denoted dom f, is defined as the set of points where the function is finite. Two functions can differ not by their formal expression, but because they have different domains. For example, the functions f, g defined as

$$f(x) = \begin{cases} 1/x & \text{if } x \neq 0 \\ +\infty & \text{otherwise,} \end{cases} \quad , \quad g(x) = \begin{cases} 1/x & \text{if } x > 0 \\ +\infty & \text{otherwise,} \end{cases}$$

have the same formal expression inside their respective domains. However, they are not the same function, since their domain is different.

We usually reserve the term *map* to refer to vector-valued functions That is, maps are functions that return a vector of values. We use the notation

$$f : \mathbb{R}^n \to \mathbb{R}^m,$$

to refer to a map with input space $\mathbb{R}^n$ and output space $\mathbb{R}^m$. The *components* of the map f are the (scalar-valued) functions $f_i, i = 1, \ldots, m$.

2.4.2   Sets related to functions

Consider a function $f : \mathbb{R}^n \to \mathbb{R}$. We define a number of sets relevant to f. The *graph* and the *epigraph* of the function f are both subsets of $\mathbb{R}^{n+1}$; see Figure 2.21. The *graph* of f is the set of input/output pairs that f can attain, that is:

$$\text{graph } f = \left\{ (x, f(x)) \in \mathbb{R}^{n+1} \ : \ x \in \mathbb{R}^n \right\}.$$

[13] For example, define the logarithm function as the function $f : \mathbb{R} \to \mathbb{R}$, with values $f(x) = \log x$ if $x > 0$, and $-\infty$ otherwise. The domain of the function is thus $\mathbb{R}_{++}$ (the set of positive reals).

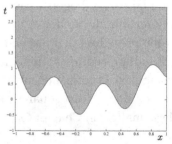

Figure 2.21 The graph of the function is the set of input/output pairs, shown as a solid line. The epigraph corresponds to points on and above the graph, in light grey.

The *epigraph*, denoted by epi f, describes the set of input/output pairs that f can achieve, as well as "anything above":

$$\text{epi}\, f = \left\{ (x,t) \in \mathbb{R}^{n+1} \; : \; x \in \mathbb{R}^n, \; t \geq f(x) \right\}.$$

Level and sublevel sets correspond to the notion of the contour of the function f. Both depend on some scalar value t, and are subsets of $\mathbb{R}^n$. A *level set* (or *contour* line) is simply the set of points that achieve exactly some value for the function f. For $t \in \mathbb{R}$, the t-level set of the function f is defined as

$$C_f(t) = \{x \in \mathbb{R}^n : \; f(x) = t\}.$$

A related notion is that of *sublevel* set. The t-sublevel set of f is the set of points that achieve at most a certain value for f:

$$L_f(t) = \{x \in \mathbb{R}^n : \; f(x) \leq t\}.$$

See Figure 2.22 for an example.

2.4.3   Linear and affine functions

Linear functions are functions that preserve scaling and addition of the input argument. A function $f : \mathbb{R}^n \to \mathbb{R}$ is *linear* if and only if

$$\forall x \in \mathbb{R}^n \text{ and } \alpha \in \mathbb{R}, f(\alpha x) = \alpha f(x);$$
$$\forall x_1, x_2 \in \mathbb{R}^n, f(x_1 + x_2) = f(x_1) + f(x_2).$$

A function f is *affine* if and only if the function $\tilde{f}(x) = f(x) - f(0)$ is linear (affine = linear + constant).

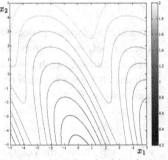

Figure 2.22 Level and sublevel sets of a function $f : \mathbb{R}^2 \to \mathbb{R}$, with domain $\mathbb{R}^2$ itself, and values on the domain given by

$$f(x) = \ln \left(e^{\sin(x_1 + 0.3x_2 - 0.1)} + e^{0.2x_2 + 0.7} \right).$$

Example 2.10 Consider the functions $f_1, f_2, f_3 : \mathbb{R}^2 \to \mathbb{R}$ defined below:

$$
\begin{aligned}
f_1(x) &= 3.2x_1 + 2x_2, \\
f_2(x) &= 3.2x_1 + 2x_2 + 0.15, \\
f_3(x) &= 0.001x_2^2 + 2.3x_1 + 0.3x_2.
\end{aligned}
$$

The function f_1 is linear; f_2 is affine; f_3 is neither linear nor affine (f_3 is a quadratic function).

Linear or affine functions can be conveniently defined by means of the standard inner product. Indeed, a function $f : \mathbb{R}^n \to \mathbb{R}$ is affine if and only if it can be expressed as

$$f(x) = a^\top x + b,$$

for some unique pair (a, b), with a in $\mathbb{R}^n$ and $b \in \mathbb{R}$. The function is linear if and only if $b = 0$. Vector $a \in \mathbb{R}^n$ can thus be viewed as a (linear) map from the "input" space $\mathbb{R}^n$ to the "output" space $\mathbb{R}$. More generally, any element a of a vector space $\mathcal{X}$ defines a linear functional $f_a : \mathcal{X} \to \mathbb{R}$, such that $f_a(z) = \langle a, z \rangle$. For any affine function f, we can obtain a and b as follows: $b = f(0)$, and $a_i = f(e_i) - b$, $i = 1, \ldots, n$. We leave it to the reader to prove that the identity $f(x) = a^\top x + b$ then holds for any x.

Example 2.11 (*Linear functions and power laws*) Sometimes, a nonlinear function can be "made" linear (or affine) via an appropriate change of variables. An example of this approach is given by the so-called power laws in physics.

Consider a physical process which has inputs $x_j > 0$, $j = 1, \ldots, n$, and a scalar output y. Inputs and output are physical, positive quantities, such as volume, height, or temperature. In many cases, we can (at least empirically) describe such physical processes by power laws, which are nonlinear models of the form

$$y = \alpha x_1^{a_1} \ldots x_n^{a_n},$$

where $\alpha > 0$, and the coefficients a_j, $j = 1, \ldots, n$ are real numbers. We find power laws, for example, in the relationship between area, volume, and size of basic geometric objects; in the Coulomb law in electrostatics; in birth and survival rates of (say) bacteria as functions of concentrations of chemicals; in heat flows and losses in pipes as functions of the pipe geometry; in analog circuit properties as functions of circuit parameters; etc. The relationship $x \to y$ is not linear nor affine, but if we introduce the new variables

$$\tilde{y} = \log y, \quad \tilde{x}_j = \log x_j, \quad j = 1, \ldots, n,$$

then the above equation becomes an affine one:

$$\tilde{y} = \log \alpha + \sum_{j=1}^{n} a_j \log x_j = a^\top \tilde{x} + b, \quad \text{where } b = \log \alpha.$$

2.4.4   Hyperplanes and half-spaces

2.4.4.1   Hyperplanes. As defined in Section 2.3.2.2, a hyperplane is a set described by a single scalar product equality. Precisely, a hyperplane in $\mathbb{R}^n$ is a set of the form

$$H = \left\{ x \in \mathbb{R}^n : a^\top x = b \right\}, \tag{2.10}$$

where $a \in \mathbb{R}^n$, $a \neq 0$, and $b \in \mathbb{R}$ are given. Equivalently, we can think of hyperplanes as the level sets of linear functions, see Figure 2.23.

When $b = 0$, the hyperplane is simply the set of points that are orthogonal to a (i.e., H is a $(n-1)$-dimensional subspace). This is

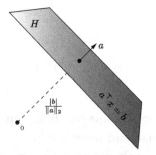

Figure 2.23 Representation of a hyperplane as the level set of a linear function: $H = \{x : a^\top x = b\}$.

evident since the condition $a^\top x = 0$ means that x has to be orthogonal to vector a, which in turn means that it lies in the orthogonal complement of $\mathrm{span}(a)$, which is a subspace of dimension $n - 1$.

When $b \neq 0$, then the hyperplane is a translation, along the direction a, of that previously mentioned subspace. Vector a is the *normal* direction to the hyperplane; see Figure 2.23. The b term is related to the distance of the hyperplane from the origin. Indeed, for computing this distance we can project the origin onto the hyperplane (which is an affine set), obtaining (using Eq. (2.6) with $p = 0$)

$$\mathrm{dist}(H, 0) = \frac{|b|}{\|a\|_2}.$$

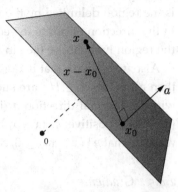

Figure 2.24 Representation of a hyperplane as the set of vectors orthogonal to a given direction a: $H = \{x : a^\top(x - x_0) = 0\}$.

If $\|a\|_2 = 1$, then $|b|$ is precisely the distance of the hyperplane from the origin.

If $x_0 \in H$, then for any other element $x \in H$, we have $a^\top x = a^\top x_0 = b$. Hence, the hyperplane can be equivalently characterized as the set of vectors x such that $x - x_0$ is orthogonal to a:

$$H = \left\{ x \in \mathbb{R}^n : a^\top(x - x_0) = 0 \right\},$$

see Figure 2.24.

2.4.4.2 *Equivalent representations of hyperplanes.* We have seen that hyperplanes are affine sets of dimension $n - 1$ that generalize the usual notion of a plane in $\mathbb{R}^3$. In fact, any affine set of dimension $n - 1$ is a hyperplane of the form (2.10), for some $a \in \mathbb{R}^n$ and $b \in \mathbb{R}$. We next prove that the following two representation of a hyperplane are in fact equivalent:

$$
\begin{aligned}
H &= \{x \in \mathbb{R}^n : a^\top x = b\}, \ a \in \mathbb{R}^n, b \in \mathbb{R} & (2.11) \\
&= x_0 + \mathrm{span}(u_1, \dots, u_{n-1}), & (2.12)
\end{aligned}
$$

for some linearly independent vectors $u_1, \dots, u_{n-1} \in \mathbb{R}^n$ and some vector $x_0 \in \mathbb{R}^n$. Indeed, if H is given as in (2.11), then we take any $x_0 \in H$ and it holds that $a^\top(x - x_0) = 0$ for any $x \in H$. Hence, we choose $\{u_1, \dots, u_{n-1}\}$ to be a basis[14] for $\mathrm{span}\{a\}^\perp$ and immediately obtain the representation in (2.12). Conversely, starting from the representation in (2.12), we obtain (2.11) by choosing a to be a vector orthogonal to $\{u_1, \dots, u_{n-1}\}$, and $b = a^\top x_0$.

[14] Numerically, a basis can be computed via the singular value decomposition (SVD), discussed in Chapter 5. Alternatively, one can use a variant of the Gram–Schmidt procedure, see Section 7.3.2.

2.4.4.3 *Half-spaces.* A hyperplane H separates the whole space into two regions:

$$H_- \doteq \left\{ x : a^\top x \leq b \right\}, \quad H_{++} \doteq \left\{ x : a^\top x > b \right\}.$$

These regions are called half-spaces (H_- is a closed half-space, H_{++} is an open half-space). As shown in Figure 2.25, the half-space H_- is the region delimited by the hyperplane $H = \{a^\top x = b\}$ and lying in the direction opposite to vector a. Similarly, the half-space H_{++} is the region lying above (i.e., in the direction of a) the hyperplane.

Also, we remark that if x_0 is any point lying in the hyperplane H, then all points $x \in H_+$ are such that $(x - x_0)$ forms an acute angle with the normal direction a (i.e., the inner product between a and $(x - x_0)$ is positive: $a^\top (x - x_0) > 0$). Similarly, for all points $x \in H_-$, we have that $a^\top (x - x_0) \le 0$, see Figure 2.26.

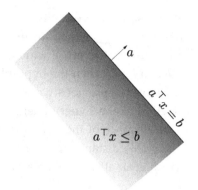

Figure 2.25 Half-space.

2.4.5 Gradients

The gradient of a function $f : \mathbb{R}^n \to \mathbb{R}$ at a point x where f is differentiable, denoted by $\nabla f(x)$, is a column vector of first derivatives of f with respect to $x_1, \ldots, x_n$:

$$\nabla f(x) = \left[\begin{array}{ccc} \frac{\partial f(x)}{\partial x_1} & \cdots & \frac{\partial f(x)}{\partial x_n} \end{array} \right]^\top.$$

When $n = 1$ (there is only one input variable), the gradient is simply the derivative.

An affine function $f : \mathbb{R}^n \to \mathbb{R}$, represented as $f(x) = a^\top x + b$, has a very simple gradient: $\nabla f(x) = a$. The (a, b) terms of f thus have the following interpretation: $b = f(0)$ is the constant term, referred to as the bias, or intercept (it corresponds to the point where the graph of f crosses the vertical axis); the terms a_j, $j = 1, \ldots, n$, which correspond to the components of the gradient of f, give the coefficients of influence of x_j on f.

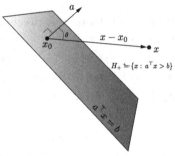

Figure 2.26 Geometry of half-spaces.

Example 2.12 (*Gradient of a nonlinear function*) The function $f : \mathbb{R}^2 \to \mathbb{R}$, with values for $x = [x_1\, x_2]^\top$:

$$f(x) = \sin x_1 + 2x_1 x_2 + x_2^2$$

has partial derivatives

$$\frac{\partial f}{\partial x_1}(x) = \cos x_1 + 2x_2,$$

$$\frac{\partial f}{\partial x_2}(x) = 2x_1 + 2x_2,$$

hence the gradient at x is the vector $\nabla f(x) = [\cos x_1 + 2x_2,\ 2x_1 + 2x_2]^\top$.

Example 2.13 (*Gradient of the distance function*) The distance function from a point $p \in \mathbb{R}^n$ to another point $x \in \mathbb{R}^n$ is defined as

$$\rho(x) = \|x - p\|_2 = \sqrt{\sum_{i=1}^n (x_i - p_i)^2}.$$

The function is differentiable at all $x \neq p$, and we have that

$$\nabla \rho(x) = \frac{1}{\|x - p\|_2}(x - p).$$

Example 2.14 (*Gradient of the log-sum-exp function*) The log-sum-exp function[15] $\mathrm{lse} : \mathbb{R}^n \to \mathbb{R}$ is defined as

$$\mathrm{lse}(x) \doteq \ln\left(\sum_{i=1}^{n} e^{x_i}\right).$$

The gradient at x of this function is

$$\nabla \mathrm{lse}(x) = \frac{z}{Z},$$

where $z = [e^{x_1} \cdots e^{x_n}]^\top$, and $Z = \sum_{i=1}^{n} z_i$.

[15] This function appears in the objective of an important class of learning problems called logistic regression, as discussed in Section 13.3.5. It also appears in the objective and constraints of the so-called geometric programming models, discussed in Section 9.7.2.

2.4.5.1 Chain rule for the gradient. Suppose $f : \mathbb{R}^m \to \mathbb{R}$ is a differentiable function in n variables $z = (z_1, \ldots, z_m)$, and each z_i is a differentiable function of n variables $x = (x_1, \ldots, x_n)$: $z_i = g_i(x)$, $i = 1, \ldots, m$ (which we write compactly as $z = g(x)$, where $g : \mathbb{R}^n \to \mathbb{R}^m$). Then, the composite function $\varphi : \mathbb{R}^n \to \mathbb{R}$, with values $\varphi(x) = f(g(x))$, has a gradient $\nabla \varphi(x) \in \mathbb{R}^n$ whose j-th component is

$$[\nabla \varphi(x)]_j = \left[\frac{\partial g_1(x)}{\partial x_j} \quad \cdots \quad \frac{\partial g_m(x)}{\partial x_j} \right] \nabla f(g(x)), \quad j = 1, \ldots, n.$$

As a relevant example, when the functions g_i are affine,

$$z_i = g_i(x) \doteq a_i^\top x + b_i, \quad a_i \in \mathbb{R}^n, b_i \in \mathbb{R}, i = 1, \ldots, m,$$

then

$$[\nabla \varphi(x)]_j = [a_{1j} \cdots a_{mj}] \nabla f(g(x)), \quad j = 1, \ldots, n,$$

where a_{ij} denotes the j-th element of vector a_i, $i = 1, \ldots, m$, $j = 1, \ldots, n$.

2.4.5.2 Affine approximation of nonlinear functions. A nonlinear function $f : \mathbb{R}^n \to \mathbb{R}$ can be approximated locally via an affine function, using a first-order Taylor series expansion, see Figure 2.27.

Specifically, if f is differentiable at point x_0, then for all points x in a neighborhood of x_0, we have that

$$f(x) = f(x_0) + \nabla f(x_0)^\top (x - x_0) + \epsilon(x),$$

where the error term $\epsilon(x)$ goes to zero faster than first order, as $x \to x_0$, that is

$$\lim_{x \to x_0} \frac{\epsilon(x)}{\|x - x_0\|_2} = 0.$$

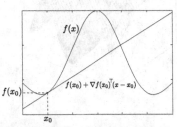

Figure 2.27 Affine approximation of $f(x)$ in the neighborhood of a given point x_0.

In practice, this means that for x sufficiently close to x_0, we can write the approximation

$$f(x) \simeq f(x_0) + \nabla f(x_0)^\top (x - x_0).$$

2.4.5.3 *Geometric interpretation of the gradient.* The gradient of a function can be nicely interpreted in the context of the level sets defined in Section 2.4.2. Indeed, geometrically, the gradient of f at a point x_0 is a vector $\nabla f(x_0)$ perpendicular to the contour line of f at level $\alpha = f(x_0)$, pointing from x_0 outwards to the α-sublevel set (that is, it points towards higher values of the function).

Consider for example the function

$$f(x) = \text{lse}(g(x)), \ \ g(x) = [\sin(x_1 + 0.3x_2), \ 0.2x_2]^\top, \tag{2.13}$$

where lse is the log-sum-exp function defined in Example 2.14. The graph of this function is shown in the left panel of Figure 2.28, some of its contour lines are shown in the center panel of the figure. The right panel in Figure 2.28 also shows a detail of the contour lines of (2.13), together with arrows representing the gradient vectors at some grid points.

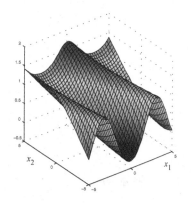
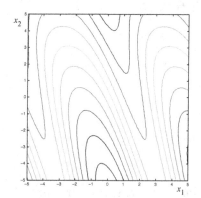
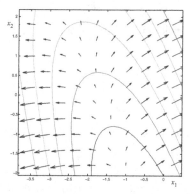

Figure 2.28 Graph of the function in (2.13) (left), its contour lines (center), and gradient vectors (arrows) at some grid points.

The gradient $\nabla f(x_0)$ also represents the direction along which the function has the maximum rate of increase (steepest ascent direction). Indeed, let v be a unit direction vector (i.e., $\|v\|_2 = 1$), let $\epsilon \geq 0$, and consider moving away at distance ϵ from x_0 along direction v, that is, consider a point $x = x_0 + \epsilon v$. We have that

$$f(x_0 + \epsilon v) \simeq f(x_0) + \epsilon \nabla f(x_0)^\top v, \text{ for } \epsilon \to 0,$$

or, equivalently,

$$\lim_{\epsilon \to 0} \frac{f(x_0 + \epsilon v) - f(x_0)}{\epsilon} = \nabla f(x_0)^\top v.$$

Looking at this formula we see that whenever $\epsilon > 0$ and v is such that $\nabla f(x_0)^\top v > 0$, then f is increasing along the direction v, for small ϵ. Indeed, the inner product $\nabla f(x_0)^\top v$ measures the rate of variation of f at x_0, along direction v, and it is usually referred to as the *directional derivative* of f along v. The rate of variation is thus zero, if v is orthogonal to $\nabla f(x_0)$: along such a direction the function value remains constant (to first order), that is, this direction is tangent to the contour line of f at x_0. On the contrary, the rate of variation is maximal when v is parallel to $\nabla f(x_0)$, hence along the normal direction to the contour line at x_0; see Figure 2.29.

2.4.6 Application to visualization of high-dimensional data

Vectors with dimension higher than three cannot be visualized graphically. However, we can try to gain insight into high-dimensional data by looking at projections of the data onto lower-dimensional affine sets, such as lines (one-dimensional), planes (two-dimensional), or three-dimensional subspaces. Each "view" corresponds to a particular projection, that is, a particular one-, two- or three-dimensional subspace on which we choose to project the data.

Projecting data on a line.   A one-dimensional subspace of $\mathbb{R}^m$ is simply a line passing through the origin. Such a line is described by means of a unit-norm vector $u \in \mathbb{R}^m$ defining the direction of the line in space. For each point $x \in \mathbb{R}^m$, the Euclidean projection of x onto the subspace span(u) is readily obtained from the projection theorem as

$$x^* = (u^\top x)u.$$

The projected datum x^* is still an m-dimensional vector, so it is still impossible to visualize it, if $m > 3$. The point here, however, is that we are not interested in x^* itself, but in the component of x along u, that is in the value of $u^\top x$, which is a scalar.

If we have a batch of n points $x^{(i)} \in \mathbb{R}^m$, $i = 1, \ldots, n$, we can visualize these points along the direction u as points along a line:

$$y^{(i)} = u^\top x^{(i)}, \quad y^{(i)} \in \mathbb{R}, \quad i = 1, \ldots, n.$$

Also, it is sometimes useful to offset the scalar data so as to center them in an appropriate way, e.g., by setting the average of the data to zero.

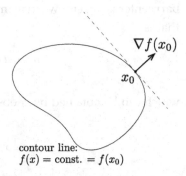

Figure 2.29 The gradient $\nabla f(x_0)$ is normal to the contour line of f at x_0, and defines the direction of maximum increase rate.

Scoring. In effect, we are defining an affine function $f : \mathbb{R}^m \to \mathbb{R}$ mapping each point $x \in \mathbb{R}^m$ into a scalar value representing a sort of "score" of point x along the direction u:

$$f(x) = u^\top x + v,$$

where v is an offset. If we want to center the data so that their barycenter is at zero, we may impose that $f(x^{(1)}) + \cdots + f(x^{(n)}) = 0$, that is

$$nv + u^\top \sum_{i=1}^{n} x^{(i)} = 0,$$

which can be obtained by choosing the offset

$$v = -u^\top \hat{x}, \quad \hat{x} = \frac{1}{n} \sum_{i=1}^{n} x^{(i)},$$

being $\hat{x}$ the average of the data points. The centered projection map can now be expressed as $f(x) = u^\top (x - \hat{x})$.

Example 2.15 (*Visualizing US Senate voting data*) We consider a data set representing the votes of US Senators in the period 2004–2006. This dataset is a collection of n vectors $x^{(j)} \in \mathbb{R}^m$, $j = 1, \ldots, n$, with $m = 645$ being the number of bills voted, and $n = 100$ the number of Senators. Thus, $x^{(j)}$ contains all the votes of Senator j, and the i-th component of $x^{(j)}$, $x_i(j)$, contains the vote of Senator j on bill i. Each vote is encoded as a binary number, with the convention that $x_i(j) = 1$ if the vote is in favor of the bill, and $x_i(j) = 0$ if it is against. The vector $\hat{x}$ can now be interpreted as the average vote across Senators. A particular projection (that is, a direction in $u \in \mathbb{R}^m$, the "bill space") corresponds to assigning a "score" to each Senator, thus allowing us to represent each Senator as a single scalar value on a line. Since we centered our data, the average score across Senators is zero.

As a tentative direction for projecting the data we choose the direction that corresponds to the "average bill." That is, we choose the direction u to be the parallel to the vector of ones in $\mathbb{R}^m$, scaled appropriately so that its Euclidean norm is one. The scores obtained with this all-ones direction are shown in Figure 2.30. This figure shows the values of the projections of the Senators' votes $x^{(j)} - \hat{x}$ (that is, with average across Senators removed) on a normalized "average bill" direction. This projection reveals clearly the party affiliation of many senators. The interpretation is that the behavior of senators on an "average bill" almost fully determines her or his party affiliation. We do observe that the direction does not perfectly predict the party affiliation. In Chapter 5, we will see methods to determine better directions.

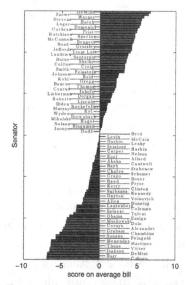

Figure 2.30 US Senators' scores on "average bill." Darker shade of gray indicates Republican Senators.

2.5   Exercises

Exercise 2.1 (Subspaces and dimensions) Consider the set $\mathcal{S}$ of points such that
$$x_1 + 2x_2 + 3x_3 = 0, \quad 3x_1 + 2x_2 + x_3 = 0.$$

Show that $\mathcal{S}$ is a subspace. Determine its dimension, and find a basis for it.

Exercise 2.2 (Affine sets and projections) Consider the set in $\mathbb{R}^3$ defined by the equation
$$\mathcal{P} = \left\{ x \in \mathbb{R}^3 \; : \; x_1 + 2x_2 + 3x_3 = 1 \right\}.$$

1. Show that the set $\mathcal{P}$ is an affine set of dimension 2. To this end, express it as $x^{(0)} + \text{span}(x^{(1)}, x^{(2)})$, where $x^{(0)} \in \mathcal{P}$, and $x^{(1)}, x^{(2)}$ are linearly independent vectors.

2. Find the minimum Euclidean distance from 0 to the set $\mathcal{P}$, and a point that achieves the minimum distance.

Exercise 2.3 (Angles, lines, and projections)

1. Find the projection z of the vector $x = (2, 1)$ on the line that passes through $x_0 = (1, 2)$ and with direction given by vector $u = (1, 1)$.

2. Determine the angle between the following two vectors:
$$x = \begin{bmatrix} 1 \\ 2 \\ 3 \end{bmatrix}, \quad y = \begin{bmatrix} 3 \\ 2 \\ 1 \end{bmatrix}.$$

   Are these vectors linearly independent?

Exercise 2.4 (Inner product) Let $x, y \in \mathbb{R}^n$. Under which condition on $\alpha \in \mathbb{R}^n$ does the function
$$f(x, y) = \sum_{k=1}^{n} \alpha_k x_k y_k$$

define an inner product on $\mathbb{R}^n$?

Exercise 2.5 (Orthogonality) Let $x, y \in \mathbb{R}^n$ be two unit-norm vectors, that is, such that $\|x\|_2 = \|y\|_2 = 1$. Show that the vectors $x - y$ and $x + y$ are orthogonal. Use this to find an orthogonal basis for the subspace spanned by x and y.

Exercise 2.6 (Norm inequalities)

1. Show that the following inequalities hold for any vector x:

$$\frac{1}{\sqrt{n}}\|x\|_2 \le \|x\|_\infty \le \|x\|_2 \le \|x\|_1 \le \sqrt{n}\|x\|_2 \le n\|x\|_\infty.$$

Hint: use the Cauchy–Schwartz inequality.

2. Show that for any nonzero vector x,

$$\mathrm{card}(x) \ge \frac{\|x\|_1^2}{\|x\|_2^2},$$

where $\mathrm{card}(x)$ is the *cardinality* of the vector x, defined as the number of nonzero elements in x. Find vectors x for which the lower bound is attained.

Exercise 2.7 (Hölder inequality) Prove Hölder's inequality (2.4). *Hint:* consider the normalized vectors $u = x/\|x\|_p$, $v = y/\|y\|_q$, and observe that

$$|x^\top y| = \|x\|_p\|y\|_q \cdot |u^\top v| \le \|x\|_p\|y\|_q \sum_k |u_k v_k|.$$

Then, apply Young's inequality (see Example 8.10) to the products $|u_k v_k| = |u_k||v_k|$.

Exercise 2.8 (Bound on a polynomial's derivative) In this exercise, you derive a bound on the largest absolute value of the derivative of a polynomial of a given order, in terms of the size of the coefficients.[16] For $w \in \mathbb{R}^{k+1}$, we define the polynomial p_w, with values

$$p_w(x) \doteq w_1 + w_2 x + \cdots + w_{k+1}x^k.$$

Show that, for any $p \ge 1$

$$\forall x \in [-1, 1] : \left| \frac{dp_w(x)}{dx} \right| \le C(k, p)\|v\|_p,$$

where $v = (w_2, \ldots, w_{k+1}) \in \mathbb{R}^k$, and

$$C(k, p) = \begin{cases} k & p = 1, \\ k^{3/2} & p = 2, \\ \frac{k(k+1)}{2} & p = \infty. \end{cases}$$

Hint: you may use Hölder's inequality (2.4) or the results from Exercise 2.6.

[16] See the discussion on regularization in Section 13.2.3 for an application of this result.

3
Matrices

A MATRIX IS A COLLECTION of numbers, arranged in columns and rows in a tabular format. Suitably defining operations such as sum, product, and norms on matrices, we can treat matrices as elements of a vector space. A key perspective in this chapter is the interpretation of a matrix as defining a *linear map* between an input and an output space. This leads to the introduction of concepts such as range, rank, nullspace, eigenvalues, and eigenvectors, that permit a complete analysis of (finite dimensional) linear maps. Matrices are a ubiquitous tool in engineering for organizing and manipulating data. They constitute the fundamental building block of numerical computation methods.

3.1 Matrix basics

3.1.1 Matrices as arrays of numbers

Matrices are rectangular arrays of numbers. We shall mainly deal with matrices whose elements are real (or sometimes complex) numbers, that is with arrays of the form

$$A = \begin{bmatrix} a_{11} & a_{12} & \cdots & a_{1n} \\ a_{21} & a_{22} & \cdots & a_{2n} \\ \vdots & \vdots & \ddots & \vdots \\ a_{m1} & a_{m2} & \cdots & a_{mn} \end{bmatrix}.$$

This matrix has m *rows* and n *columns*. In the case of real elements, we say that $A \in \mathbb{R}^{m,n}$, or $A \in \mathbb{C}^{m,n}$ in the case of complex elements. The i-th row of A is the (row) vector $[a_{i1} \cdots a_{in}]$; the j-th column of A is the (column) vector $[a_{1j} \cdots a_{mj}]^{\top}$. The transposition operation[1]

[1] In case of a complex A, we denote by $A^{\star}$ the Hermitian-conjugate of a complex matrix, obtained by taking the transpose of the matrix, with the conjugate values of the elements of A.

works on matrices by exchanging rows and columns, that is

$$[A^\top]_{ij} = [A]_{ji},$$

where the notation $[A]_{ij}$ (or sometimes simply A_{ij}) refers to the element of A positioned in row i and column j. The *zero matrix* in $\mathbb{R}^{m,n}$ is denoted by $0_{m,n}$, or simply by 0, when dimensions are obvious from the context.

The operations of multiplication by a scalar and of sum (of matrices of the same size) are defined in the obvious way (i.e., for multiplying by a scalar one multiplies every entry in the matrix by the scalar, and for summing two matrices with the same shape one sums corresponding elements in the same position). With these operations defined, we can see $\mathbb{R}^{m,n}$ as a vector space.[2]

[2] The terminology can be confusing at times: an element of a *vector space* need not be only a "vector" intended as a column of elements. Matrices indeed constitute elements of a vector space. We have also already seen that, for example, polynomials of degree at most n are elements of a vector space too.

Example 3.1 (*Images*) A gray-scale image is represented as a matrix of numerical values, where each entry in the matrix contains the value of intensity of the corresponding pixel in the image (a "double" type value in $[0,1]$, where 0 is for black and 1 for white; or an "integer" type value, between 0 and 255). Figure 3.1 shows a gray-scale image with 400 horizontal pixels and 400 vertical pixels.

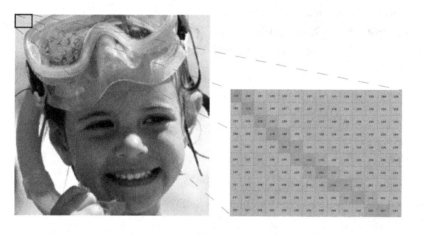

Figure 3.1 400×400 pixels gray-scale image (left) and intensity values for a rectangular detail in the upper-left position of the image (right).

3.1.2 *Matrix products*

Two matrices can be multiplied if conformably sized, i.e., if $A \in \mathbb{R}^{m,n}$ and $B \in \mathbb{R}^{n,p}$, then the matrix product $AB \in \mathbb{R}^{m,p}$ is defined as a matrix whose (i,j)-th entry is

$$[AB]_{ij} = \sum_{k=1}^{n} A_{ik} B_{kj}. \tag{3.1}$$

The matrix product is non-commutative, meaning that, in general, $AB \neq BA$. For example

$$\begin{bmatrix} 1 & 2 \\ 3 & 4 \end{bmatrix} \begin{bmatrix} 0 & 1 \\ 1 & 1 \end{bmatrix} = \begin{bmatrix} 2 & 3 \\ 4 & 7 \end{bmatrix}$$

$$\begin{bmatrix} 0 & 1 \\ 1 & 1 \end{bmatrix} \begin{bmatrix} 1 & 2 \\ 3 & 4 \end{bmatrix} = \begin{bmatrix} 3 & 4 \\ 4 & 6 \end{bmatrix}.$$

The $n \times n$ *identity matrix* (often denoted by I_n, or simply I, depending on context), is a matrix with all zero elements, except for the elements on the diagonal (that is, the elements with row index equal to the column index), which are equal to one. This matrix satisfies $AI_n = A$ for every matrix A with n columns, and $I_n B = B$ for every matrix B with n rows.

A matrix $A \in \mathbb{R}^{m,n}$ can also be seen as a collection of columns, each column being a vector, or as a collection of rows, each row being a (transposed) vector. We shall write correspondingly

$$A = \begin{bmatrix} a_1 & a_2 & \cdots & a_n \end{bmatrix}, \text{ or } A = \begin{bmatrix} \alpha_1^\top \\ \alpha_2^\top \\ \vdots \\ \alpha_m^\top \end{bmatrix},$$

where $a_1, \ldots, a_n \in \mathbb{R}^m$ denote the columns of A, and $\alpha_1^\top, \ldots, \alpha_m^\top \in \mathbb{R}^n$ denote the rows of A.

If the columns of B are given by the vectors $b_i \in \mathbb{R}^n$, $i = 1, \ldots, p$, so that $B = [b_1 \cdots b_p]$, then AB can be written as

$$AB = A \begin{bmatrix} b_1 & \ldots & b_p \end{bmatrix} = \begin{bmatrix} Ab_1 & \ldots & Ab_p \end{bmatrix}.$$

In other words, AB results from transforming each column b_i of B into Ab_i. The matrix–matrix product can also be interpreted as an operation on the rows of A. Indeed, if A is given by its rows $\alpha_i^\top$, $i = 1, \ldots, m$, then AB is the matrix obtained by transforming each one of these rows into $\alpha_i^\top B$, $i = 1, \ldots, m$:

$$AB = \begin{bmatrix} \alpha_1^\top \\ \vdots \\ \alpha_m^\top \end{bmatrix} B = \begin{bmatrix} \alpha_1^\top B \\ \vdots \\ \alpha_m^\top B \end{bmatrix}.$$

Finally, the product AB can be given the interpretation as the sum of so-called *dyadic* matrices (matrices of rank one, see Section 3.4.7) of the form $a_i \beta_i^\top$, where $\beta_i^\top$ denote the rows of B:

$$AB = \sum_{i=1}^{n} a_i \beta_i^\top, \quad A \in \mathbb{R}^{m,n}, B \in \mathbb{R}^{n,p}.$$

3.1.2.1 Matrix–vector product. Rule (3.1) for matrix multiplication also works when $A \in \mathbb{R}^{m,n}$ is a matrix and $b \in \mathbb{R}^n$ is a vector. In this case, the matrix–vector multiplication rule becomes (using the column representation of A)

$$Ab = \sum_{k=1}^{n} a_k b_k, \quad A \in \mathbb{R}^{m,n}, b \in \mathbb{R}^n.$$

That is, Ab is a vector in $\mathbb{R}^m$ obtained by forming a linear combination of the columns of A, using the elements in b as coefficients. Similarly, we can multiply matrix $A \in \mathbb{R}^{m,n}$ on the left by (the transpose of) vector $c \in \mathbb{R}^m$ as follows:

$$c^\top A = \sum_{k=1}^{m} c_k \alpha_k^\top, \quad A \in \mathbb{R}^{m,n}, c \in \mathbb{R}^m.$$

That is, $c^\top A$ is a vector in $\mathbb{R}^{1,m}$ obtained by forming a linear combination of the rows α_k of A, using the elements in c as coefficients.

Example 3.2 (*Incidence matrix and network flows*) A network can be represented as a graph of m nodes connected by n directed arcs. Here, we assume that arcs are ordered pairs of nodes, with at most one arc joining any two nodes; we also assume that there are no self-loops (arcs from a node to itself). We can fully describe such a kind of network via the so-called (directed) arc–node incidence matrix, which is an $m \times n$ matrix defined as follows:

$$A_{ij} = \begin{cases} 1 & \text{if arc } j \text{ starts at node } i \\ -1 & \text{if arc } j \text{ ends at node } i \\ 0 & \text{otherwise.} \end{cases} \quad , \; 1 \le i \le m, \; 1 \le j \le n. \quad (3.2)$$

Figure 3.2 shows an example of a network with $m = 6$ nodes and $n = 8$ arcs. The (directed) arc–node incidence matrix for this network is

$$A = \begin{bmatrix} 1 & 1 & 0 & 0 & 0 & 0 & 0 & -1 \\ -1 & 0 & 1 & 0 & 0 & 0 & 0 & 1 \\ 0 & -1 & -1 & -1 & 1 & 1 & 0 & 0 \\ 0 & 0 & 0 & 1 & 0 & 0 & -1 & 0 \\ 0 & 0 & 0 & 0 & 0 & -1 & 1 & 0 \\ 0 & 0 & 0 & 0 & -1 & 0 & 0 & 0 \end{bmatrix}.$$

We describe a flow (of goods, traffic, charge, information, etc.) across the network as a vector $x \in \mathbb{R}^n$, where the j-th component of x denotes the amount flowing through arc j. By convention, we use positive values when the flow is in the direction of the arc, and negative ones in the opposite case. The total flow leaving a given node i is then

$$\sum_{j=1}^{n} A_{ij} x_j = [Ax]_i,$$

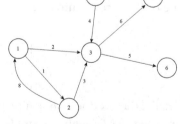

Figure 3.2 Example of a directed graph.

where $[Ax]_i$ denotes the i-th component of vector Ax. Next, we define the external supply as a vector $b \in \mathbb{R}^m$, with negative b_i representing an external demand at node i, and positive b_i a supply. We assume that the total supply equals the total demand, which means that $\mathbf{1}^\top b = 0$. The flows x must satisfy the flow-balance equations, which represent a "conservation of mass"-type constraint at each node (the total in-flow at node i, plus supply/demand, must equal the total out-flow from node i). These constraints are represented by the vector equality $Ax = b$.

3.1.2.2 *Product and transposition.* For any two conformably sized matrices A, B, it holds that

$$(AB)^\top = B^\top A^\top,$$

hence, for a generic chain of products $A_1 A_2 \cdots A_p$ it holds that

$$(A_1 A_2 \cdots A_p)^\top = A_p^\top \cdots A_2^\top A_1^\top.$$

3.1.3 *Block matrix products*

Matrix algebra generalizes to blocks, provided block sizes are consistent. To illustrate this, consider the matrix–vector product between a $m \times n$ matrix A and a n-vector x, where A, x are partitioned in blocks, as follows:

$$A = \begin{bmatrix} A_1 & A_2 \end{bmatrix}, \quad x = \begin{bmatrix} x_1 \\ x_2 \end{bmatrix},$$

where A_i is $m \times n_i$, $x_i \in \mathbb{R}^{n_i}$, $i = 1, 2$, $n_1 + n_2 = n$. Then

$$Ax = A_1 x_1 + A_2 x_2.$$

Symbolically, it works as if we formed the inner product between the "row vector" $[A_1, A_2]$ and the column vector $\begin{bmatrix} x_1 \\ x_2 \end{bmatrix}$. Likewise, if an $n \times p$ matrix B is partitioned into two blocks B_i, each of size n_i, $i = 1, 2$, with $n_1 + n_2 = n$, then

$$AB = \begin{bmatrix} A_1 & A_2 \end{bmatrix} \begin{bmatrix} B_1 \\ B_2 \end{bmatrix} = A_1 B_1 + A_2 B_2.$$

Again, symbolically, we apply the same rules as for the scalar product, except that now the result is a matrix.

Finally, we discuss the so-called outer products. Consider the case for example when A is a $m \times n$ matrix partitioned row-wise into two blocks A_1, A_2, and B is a $n \times p$ matrix that is partitioned column-wise into two blocks B_1, B_2:

$$A = \begin{bmatrix} A_1 \\ A_2 \end{bmatrix}, \quad B = \begin{bmatrix} B_1 & B_2 \end{bmatrix}.$$

Then, the product $C = AB$ can be expressed in terms of the blocks, as follows:

$$C = AB = \begin{bmatrix} A_1 \\ A_2 \end{bmatrix} \begin{bmatrix} B_1 & B_2 \end{bmatrix} = \begin{bmatrix} A_1 B_1 & A_1 B_2 \\ A_2 B_1 & A_2 B_2 \end{bmatrix}.$$

In the special case when A is a column vector and B is a row vector, that is, we have

$$A = \begin{bmatrix} a_1 \\ a_2 \\ \vdots \\ a_m \end{bmatrix}, \quad B = \begin{bmatrix} b_1 & b_2 & \cdots & b_p \end{bmatrix}$$

then

$$AB = \begin{bmatrix} a_1 b_1 & \cdots & a_1 b_p \\ a_2 b_1 & \cdots & a_2 b_p \\ \vdots & \ddots & \vdots \\ a_m b_1 & \cdots & a_m b_p \end{bmatrix}.$$

3.1.4 Matrix space and inner product

The vector space $\mathbb{R}^{m,n}$ can be endowed with a standard inner product: for $A, B \in \mathbb{R}^{m,n}$, we define

$$\langle A, B \rangle = \text{trace } A^\top B,$$

where $\text{trace}(X)$ is the *trace* of (square) matrix X, defined as the sum of the diagonal elements of X. This inner product induces the so-called Frobenius norm

$$\sqrt{\langle A, A \rangle} = \sqrt{\text{trace } AA^\top} = \|A\|_\text{F} \doteq \sqrt{\sum_{ij} a_{ij}^2}.$$

Our choice is consistent with the one made for vectors. In fact, the inner product above represents the scalar product between two vectors obtained from the matrices A, B, by stacking all the columns on top of each other; thus, the Frobenius norm is the Euclidean norm of the vectorized form of the matrix.

The trace operator is a linear operator, and it has several important properties. In particular, the trace of a square matrix is equal to that of its transpose and, for any two matrices $A \in \mathbb{R}^{m,n}$, $B \in \mathbb{R}^{n,m}$, it holds that

$$\text{trace } AB = \text{trace } BA.$$

3.2 Matrices as linear maps

3.2.1 Matrices, linear and affine maps

We can interpret matrices as linear maps (vector-valued functions), or "operators," acting from an "input" space to an "output" space. We recall that a map $f : \mathcal{X} \to \mathcal{Y}$ is linear if any points x and z in $\mathcal{X}$ and any scalars λ, μ satisfy $f(\lambda x + \mu z) = \lambda f(x) + \mu f(z)$. Any linear map $f : \mathbb{R}^n \to \mathbb{R}^m$ can be represented by a matrix $A \in \mathbb{R}^{m,n}$, mapping input vectors $x \in \mathbb{R}^n$ to output vectors $y \in \mathbb{R}^m$ (see Figure 3.3):

$$y = Ax.$$

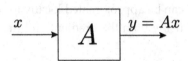

Figure 3.3 Linear map defined by a matrix A.

Affine maps are simply linear functions plus a constant term, thus any affine map $f : \mathbb{R}^n \to \mathbb{R}^m$ can be represented as

$$f(x) = Ax + b,$$

for some $A \in \mathbb{R}^{m,n}$, $b \in \mathbb{R}^m$.

Example 3.3 A linear map that scales each component x_i of vector x by some scalar factor α_i, $i = 1, \ldots, n$, is described by a *diagonal* matrix

$$A = \mathrm{diag}\,(\alpha_1, \ldots, \alpha_n) = \begin{bmatrix} \alpha_1 & 0 & \cdots & 0 \\ 0 & \alpha_2 & 0 & \cdots \\ \vdots & \cdots & \ddots & \vdots \\ 0 & \cdots & 0 & \alpha_n \end{bmatrix}.$$

For such a diagonal A, we thus have

$$y = Ax \quad \Leftrightarrow \quad y_i = \alpha_i x_i, \; i = 1, \ldots, n.$$

3.2.2 Approximation of nonlinear functions

A nonlinear map $f : \mathbb{R}^n \to \mathbb{R}^m$ can be approximated by an affine map, in the neighborhood of a given point x_0 (at which f is differentiable), as

$$f(x) = f(x_0) + J_f(x_0)(x - x_0) + o(\|x - x_0\|),$$

where $o(\|x - x_0\|)$ are terms that go to zero faster than first order for $x \to x_0$, and where $J_f(x_0)$ is the Jacobian of f at x_0, defined as

$$J_f(x_0) \doteq \begin{bmatrix} \frac{\partial f_1}{\partial x_1} & \cdots & \frac{\partial f_1}{\partial x_n} \\ \vdots & \ddots & \vdots \\ \frac{\partial f_m}{\partial x_1} & \cdots & \frac{\partial f_m}{\partial x_n} \end{bmatrix}_{x=x_0}.$$

Thus, for x "near" x_0, the variation $\delta_f(x) \doteq f(x) - f(x_0)$ can be approximately described, to first order, by a linear map defined by the Jacobian matrix, i.e.,

$$\delta_f(x) \simeq J_f(x_0)\delta_x, \quad \delta_x \doteq x - x_0.$$

Also, a scalar-valued function[3] $f : \mathbb{R}^n \to \mathbb{R}$, twice differentiable at x_0, can be approximated locally to second-order around a given point x_0, using both the gradient and the second-derivatives matrix (Hessian):

[3] For a scalar-valued function the Jacobian coincides with the transpose of the gradient vector.

$$f \simeq f(x_0) + \nabla f(x_0)^\top (x - x_0) + \frac{1}{2}(x - x_0)^\top \nabla^2 f(x_0)(x - x_0),$$

where $\nabla^2 f(x_0)$ is the Hessian matrix at x_0, defined as

$$\nabla^2 f(x_0) \doteq \begin{bmatrix} \frac{\partial^2 f}{x_1^2} & \cdots & \frac{\partial^2 f}{\partial x_1 \partial x_n} \\ \vdots & \ddots & \vdots \\ \frac{\partial^2 f}{\partial x_n \partial x_1} & \cdots & \frac{\partial^2 f}{x_n^2} \end{bmatrix}_{x=x_0}.$$

In this case, f is approximated locally via a *quadratic* function defined via the Hessian matrix $\nabla^2 f(x_0)$.

3.2.3 *Range, rank, and nullspace*

Range and rank. Consider an $m \times n$ matrix A, and denote by a_i, $i = 1, \ldots, n$, its i-th column, so that $A = [a_1 \ldots a_n]$. The set of vectors y obtained as a linear combination of the a_is are of the form $y = Ax$ for some vector $x \in \mathbb{R}^n$. This set is commonly known as the *range* of A, and is denoted $\mathcal{R}(A)$:

$$\mathcal{R}(A) = \{Ax \,:\, x \in \mathbb{R}^n\}.$$

By construction, the range is a subspace. The dimension of $\mathcal{R}(A)$ is called the *rank* of A and denoted by $\text{rank}(A)$; by definition the rank represents the number of linearly independent columns of A. It can be shown[4] that the rank is also equal to the number of linearly independent rows of A; that is, the rank of A is the same as that of its transpose $A^\top$. As a consequence, we always have the bounds $0 \leq \text{rank}(A) \leq \min(m, n)$.

[4] This result, as well as some other results that are reported in this chapter without proof, can be found in any classical textbook on linear algebra and matrix analysis, such as, for instance:

G. Strang, *Introduction to Linear Algebra*, Wellesley-Cambridge Press, 2009;

R. A. Horn, C. R. Johnson, *Matrix Analysis*, Cambridge University Press, 1990;

C. D. Meyer, *Matrix Analysis and Applied Linear Algebra*, SIAM, 2001.

Nullspace. The nullspace of the matrix A is the set of vectors in the input space that are mapped to zero, and is denoted by $\mathcal{N}(A)$:

$$\mathcal{N}(A) = \{x \in \mathbb{R}^n \,:\, Ax = 0\}.$$

This set is again a subspace.

3.2.4   The fundamental theorem of linear algebra

The so-called *fundamental theorem* of linear algebra is a result that establishes a key connection between the nullspace of a matrix and the range of its transpose. We start by observing that any vector in the range of $A^\top$ is orthogonal to any vector in the nullspace of A, that is, for any $x \in \mathcal{R}(A^\top)$, and any $z \in \mathcal{N}(A)$, it holds that $x^\top z = 0$. This fact can be easily proved by observing that every $x \in \mathcal{R}(A^\top)$ is, by definition, a linear combination of the rows of A, that is, it can be written as $x = A^\top y$ for some $y \in \mathbb{R}^m$. Hence,

$$x^\top z = (A^\top y)^\top z = y^\top A z = 0, \quad \forall z \in \mathcal{N}(A).$$

Thus, $\mathcal{R}(A^\top)$ and $\mathcal{N}(A)$ are mutually orthogonal subspaces, i.e., $\mathcal{N}(A) \perp \mathcal{R}(A^\top)$, or, equivalently, $\mathcal{N}(A) = \mathcal{R}(A^\top)^\perp$. We recall from Section 2.2.3.4 that the direct sum of a subspace and its orthogonal complement equals the whole space, thus,

$$\mathbb{R}^n = \mathcal{N}(A) \oplus \mathcal{N}(A)^\perp = \mathcal{N}(A) \oplus \mathcal{R}(A^\top).$$

With a similar reasoning, we also argue that

$$\begin{aligned}
\mathcal{R}(A)^\perp &= \{y \in \mathbb{R}^m : y^\top z = 0, \forall z \in \mathcal{R}(A)\} \\
&= \{y \in \mathbb{R}^m : y^\top A x = 0, \forall x \in \mathbb{R}^n\} = \mathcal{N}(A^\top),
\end{aligned}$$

hence we see that

$$\mathcal{R}(A) \perp \mathcal{N}(A^\top),$$

and, therefore, the output space $\mathbb{R}^m$ is decomposed as

$$\mathbb{R}^m = \mathcal{R}(A) \oplus \mathcal{R}(A)^\perp = \mathcal{R}(A) \oplus \mathcal{N}(A^\top).$$

The previous findings are summarized in the following theorem.

Theorem 3.1 (Fundamental theorem of linear algebra) *For any given matrix* $A \in \mathbb{R}^{m,n}$, *it holds that* $\mathcal{N}(A) \perp \mathcal{R}(A^\top)$ *and* $\mathcal{R}(A) \perp \mathcal{N}(A^\top)$, *hence*

$$\begin{aligned}
\mathcal{N}(A) \oplus \mathcal{R}(A^\top) &= \mathbb{R}^n, \\
\mathcal{R}(A) \oplus \mathcal{N}(A^\top) &= \mathbb{R}^m,
\end{aligned}$$

and

$$\begin{aligned}
\dim \mathcal{N}(A) + \mathrm{rank}(A) &= n, & (3.3) \\
\dim \mathcal{N}(A^\top) + \mathrm{rank}(A) &= m. & (3.4)
\end{aligned}$$

Consequently, we can decompose any vector $x \in \mathbb{R}^n$ *as the sum of two vectors orthogonal to each other, one in the range of* $A^\top$, *and the other in the nullspace of* A:

$$x = A^\top \xi + z, \quad z \in \mathcal{N}(A).$$

Similarly, we can decompose any vector $w \in \mathbb{R}^m$ as the sum of two vectors orthogonal to each other, one in the range of A, and the other in the nullspace of $A^\top$:

$$w = A\varphi + \zeta, \quad \zeta \in \mathcal{N}(A^\top).$$

A geometrical intuition for the theorem is provided in Figure 3.4.

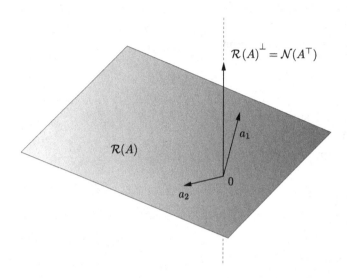

Figure 3.4 Illustration of the fundamental theorem of linear algebra in $\mathbb{R}^3$. Here, $A = [a_1 \; a_2]$. Any vector in $\mathbb{R}^3$ can be written as the sum of two orthogonal vectors, one in the range of A, the other in the nullspace of $A^\top$.

3.3   Determinants, eigenvalues, and eigenvectors

As we have seen, any matrix A represents a linear map. If the matrix is square, i.e. $A \in \mathbb{R}^{n,n}$, then it represents a linear map from $\mathbb{R}^n$ into itself. In this section, we briefly discuss how certain simple geometric shapes, such as lines and cubes in $\mathbb{R}^n$, are mapped by the transformation $y = Ax$, and use these geometric interpretations to introduce the concepts of determinants, eigenvalues, and eigenvectors of a square matrix.

3.3.1   Action of a matrix along lines

We start by asking how does a linear map A act on lines through the origin (one-dimensional subspaces). Consider a nonzero vector $u \in \mathbb{R}^n$ and the line passing from the origin through u, that is the set $L_u = \{x = \alpha u, \, \alpha \in \mathbb{R}\}$. When A is applied to a vector $x \in \mathbb{R}^n$ belonging to L_u, it transforms it into an output vector $y \in \mathbb{R}^n$:

$$y = Ax = \alpha Au.$$

We shall next show that the effect of A on any point $x \in L_u$ is to rotate the point by a fixed angle θ_u, and then to shrink/amplify the length of x by a fixed amount γ_u. Note that the angle of rotation θ_u and the length gain γ_u are constant and fixed for all points along the line L_u. To see this fact, consider the original length of x, as measured by the Euclidean norm: $\|x\|_2 = |\alpha| \|u\|_2$, then

$$\|y\|_2 = \|Ax\|_2 = |\alpha| \|Au\|_2 = \frac{\|Au\|_2}{\|u\|_2} |\alpha| \|u\|_2 = \gamma_u \|x\|_2,$$

where we set $\gamma_u = \|Au\|_2 / \|u\|_2$ to represent the *gain* in the direction u. Similarly, for the angle between x and y, we have

$$\cos \theta_u = \frac{y^\top x}{\|x\|_2 \|y\|_2} = \frac{x^\top A^\top x}{\|x\|_2 \|y\|_2} = \frac{u^\top A^\top u}{\gamma_u \|u\|_2^2},$$

which again depends only on the line direction u, and not on the actual point along the line. The next example helps in visualizing the concept via a simple numerical experiment.

Example 3.4 (*Action of a square matrix on lines*) Consider the following 2×2 matrix

$$A = \begin{bmatrix} 1.2 & 0.4 \\ 0.6 & 1 \end{bmatrix}.$$

Figure 3.5 shows how points x along an input direction u are mapped to points $y = Ax$ along an output direction which forms an angle θ_u with u. Also, as $\|x\|_2$ is kept constant and the direction u sweeps all possible directions, x moves along a circle, and the picture shows the corresponding locus for y (which, incidentally, turns out to be an ellipse). Three loci are displayed in Figure 3.5, corresponding to $\|x\|_2 = 1, 1.3, 2$ respectively.

Figure 3.5 Graphical illustration in $\mathbb{R}^2$ of how points along lines and circles are mapped by the linear transformation $y = Ax$.

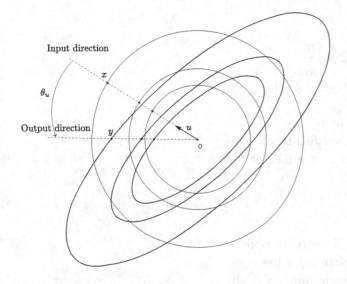

Interestingly, one may discover by numerical experiments that there are in this example two input directions $u^{(1)}, u^{(2)}$ that are angle invariant under the map defined by A. By angle-invariant direction we here mean a direction such that when the input point x is along this direction, then the output point y is also along the same direction. In other words, the angle θ_u is zero (or $\pm 180°$) for these special input directions; see Figure 3.6.

In the current example, these invariant directions are described by the vectors

$$u^{(1)} = \frac{\sqrt{2}}{2} \begin{bmatrix} 1 \\ 1 \end{bmatrix}, \quad u^{(2)} = \frac{2}{\sqrt{13}} \begin{bmatrix} 1 \\ -1.5 \end{bmatrix}.$$

The action of A on points x lying along an invariant direction u is indeed very simple: A acts as a scalar multiplication along these lines, that is $Ax = \lambda x$, for some scalar λ.

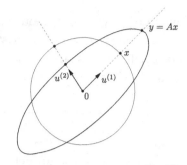

Figure 3.6 Input directions $u^{(1)}, u^{(2)}$ that are angle-invariant under A.

3.3.2 Determinants and the transformation of the unit cube

Consider a 2×2 matrix

$$A = \begin{bmatrix} a_{11} & a_{12} \\ a_{21} & a_{22} \end{bmatrix}.$$

The *determinant* of this matrix is a real number, defined as follows:

$$\det A \doteq a_{11}a_{22} - a_{21}a_{12}.$$

Suppose we apply the linear map $y = Ax$ to the four vectors defining the vertices of the unit square in $\mathbb{R}^2$, that is to the points $x^{(1)} = [0\,0]^\top$, $x^{(2)} = [1\,0]^\top$, $x^{(3)} = [0\,1]^\top$, $x^{(4)} = [1\,1]^\top$. The transformed points

$$y^{(1)} = Ax^{(1)} = \begin{bmatrix} 0 \\ 0 \end{bmatrix}, \quad y^{(2)} = Ax^{(2)} = \begin{bmatrix} a_{11} \\ a_{21} \end{bmatrix},$$

$$y^{(3)} = Ax^{(3)} = \begin{bmatrix} a_{12} \\ a_{22} \end{bmatrix}, \quad y^{(4)} = Ax^{(4)} = \begin{bmatrix} a_{11} + a_{12} \\ a_{21} + a_{22} \end{bmatrix}$$

form the vertices of a parallelogram, see Figure 3.7.

The area of the unit square is one. It can be verified by elementary geometry that the area of the transformed square (i.e., of the parallelogram) is instead equal to

$$\text{area} = |\det A|.$$

The (absolute value of the) determinant of a 2×2 matrix thus gives the factor by which the area (two-dimensional measure) of the input unit square is increased or decreased when passing through A. In dimension larger than two, the determinant can be uniquely defined as the only real-valued function of an $n \times n$ matrix such that:

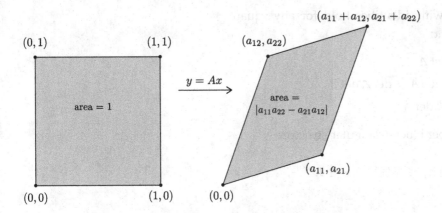

Figure 3.7 Linear mapping of the unit square.

(1) switching two rows or two columns of the matrix changes the sign of the function;

(2) the function is linear in each row (or column) of the matrix;

(3) the function is equal to one for the identity matrix.

The determinant of a generic matrix $A \in \mathbb{R}^{n,n}$ can be computed by defining $\det a = a$ for a scalar a, and then applying the following inductive formula (Laplace's determinant expansion):

$$\det(A) = \sum_{j=1}^{n}(-1)^{i+j}a_{ij}\det A_{(i,j)},$$

where i is any row, chosen at will (e.g., one may choose $i = 1$), and $A_{(i,j)}$ denotes an $(n-1) \times (n-1)$ submatrix of A obtained by eliminating row i and column j from A.

It can be proved that, in generic dimension n, the absolute value of the determinant of A still describes the volume (n-dimensional measure) of the parallelotope obtained by transforming the unit hypercube through A. An interesting situation arises when the volume of the transformed cube is zero, that is, when $\det A = 0$.[5] In the 2×2 example, this happens whenever $a_{11}a_{22} = a_{21}a_{12}$, i.e., when one of the rows (or one of the columns) is a multiple of the other. In such a case, the columns (and the rows) are not linearly independent, and the matrix has a non-trivial null space. This means that there exist directions in the input space along which all input vectors are mapped to zero by A. The same concept extends to generic dimension n, whence it can be proved that

[5] A square matrix A for which $\det A = 0$ is said to be *singular*.

$A \in \mathbb{R}^{n,n}$ is singular $\Leftrightarrow$ $\det A = 0$ $\Leftrightarrow$ $\mathcal{N}(A)$ is not equal to $\{0\}$.

We finally recall that the following identities hold for any square matrices $A, B \in \mathbb{R}^{n,n}$ and scalar α:

$$
\begin{aligned}
\det A &= \det A^\top \\
\det AB &= \det BA = \det A \det B \\
\det \alpha A &= \alpha^n \det A.
\end{aligned}
$$

Moreover, for a matrix with upper block-triangular structure

$$
X = \begin{bmatrix} X_{11} & X_{12} \\ 0 & X_{22} \end{bmatrix}, \quad X_{11} \in \mathbb{R}^{n_1,n_1}, X_{22} \in \mathbb{R}^{n_2,n_2},
$$

it holds that

$$
\det X = \det X_{11} \det X_{22},
$$

an analogous result holds for lower block-triangular matrices.

3.3.3 Matrix inverses

If $A \in \mathbb{R}^{n,n}$ is nonsingular (i.e., $\det A \neq 0$), then we define the *inverse matrix* A^{-1} as the unique $n \times n$ matrix such that

$$
AA^{-1} = A^{-1}A = I_n.
$$

If A, B are square and nonsingular, then it holds for the inverse of the product that

$$
(AB)^{-1} = B^{-1}A^{-1}.
$$

Also, if A is square and nonsingular, then

$$
(A^\top)^{-1} = (A^{-1})^\top,
$$

that is, the order of transposition and inversion is exchangeable. It holds for the determinant of a square and nonsingular matrix A that

$$
\det A = \det A^\top = \frac{1}{\det A^{-1}}.
$$

Non-square, or square-but-singular, matrices do not possess a regular inverse. However, for a generic matrix $A \in \mathbb{R}^{m,n}$, a generalized inverse (or, *pseudoinverse*) can be defined. In particular, if $m \geq n$, then A^{li} is said to be a *left inverse* of A, if

$$
A^{\mathrm{li}}A = I_n.
$$

Similarly, if $n \geq m$, then A^{ri} is said to be a *right inverse* of A, if

$$
AA^{\mathrm{ri}} = I_m.
$$

In general, matrix A^{pi} is a pseudoinverse of A, if $AA^{\mathrm{pi}}A = A$. Left/right inverses and pseudoinverses are further discussed in Section 5.2.3.

3.3.4 Similar matrices

Two matrices $A, B \in \mathbb{R}^{n,n}$ are said to be *similar* if there exists a non-singular matrix $P \in \mathbb{R}^{n,n}$ such that

$$B = P^{-1}AP.$$

Similar matrices are related to different representations of the same linear map, under a change of basis in the underlying space. Consider the linear map

$$y = Ax$$

mapping $\mathbb{R}^n$ into itself. Since $P \in \mathbb{R}^{n,n}$ is nonsingular, its columns are linearly independent, hence they represent a basis for $\mathbb{R}^n$. Vectors x and y can thus be represented in this basis as linear combinations of the columns of P, that is there exist vectors $\tilde{x}, \tilde{y}$ such that

$$x = P\tilde{x}, \quad y = P\tilde{y}.$$

Writing the relation $y = Ax$, substituting the representations of x, y in the new basis, we obtain

$$P\tilde{y} = AP\tilde{x} \quad \Rightarrow \quad \tilde{y} = P^{-1}AP\tilde{x} = B\tilde{x},$$

that is, matrix $B = P^{-1}AP$ represents the linear map $y = Ax$, in the new basis defined by the columns of P.

3.3.5 Eigenvectors and eigenvalues

Eigenvalues and characteristic polynomial. We are now ready to give a formal definition for eigenvectors and eigenvalues. We use the same concepts introduced when studying the action of A along lines, only we now allow for a slightly more general perspective by viewing A as a linear map from $\mathbb{C}^n$ (the space of complex vectors with n components) into itself. Eigenvectors are simply directions in $\mathbb{C}^n$ that are angle-invariant under A. More precisely, we say that $\lambda \in \mathbb{C}$ is an *eigenvalue* of matrix $A \in \mathbb{R}^{n,n}$, and $u \in \mathbb{C}^n$ is a corresponding *eigenvector*, if it holds that

$$Au = \lambda u, \quad u \neq 0,$$

or, equivalently,

$$(\lambda I_n - A)u = 0, \quad u \neq 0.$$

This latter equation shows that in order for (λ, u) to be an eigenvalue/eigenvector pair it must happen that: (a) λ makes matrix $\lambda I_n - A$ singular (so that it possesses a non-trivial nullspace), and (b) u lies in the nullspace of $\lambda I_n - A$. Since $\lambda I_n - A$ is singular if and only if its determinant is zero, eigenvalues can be easily characterized as those real or complex numbers that satisfy the equation[6]

$$\det(\lambda I_n - A) = 0.$$

In particular, $p(\lambda) \doteq \det(\lambda I_n - A)$ is a polynomial of degree n in λ, known as the *characteristic polynomial* of A.

Multiplicities and eigenspaces. The eigenvalues of $A \in \mathbb{R}^{n,n}$ are thus the roots of the characteristic polynomial. Some of these eigenvalues can indeed be "repeated" roots of the characteristic polynomial, hence their multiplicity can be larger than one. Also, some eigenvalues can be complex, with nonzero imaginary part, in which case they appear in complex conjugate pairs.[7] The following theorem holds.

Theorem 3.2 (Fundamental theorem of algebra) *Any matrix $A \in \mathbb{R}^{n,n}$ has n eigenvalues λ_i, $i = 1, \ldots, n$, counting multiplicities.*

We call *distinct* eigenvalues the eigenvalues of A not counting multiplicities; i.e., we put in the set of distinct eigenvalues only one representative per each group of repeated eigenvalues with identical value. Each distinct eigenvalue λ_i, $i = 1, \ldots, k$, has an associated *algebraic multiplicity* $\mu_i \geq 1$, defined as the number of times the eigenvalue is repeated as a root of the characteristic polynomial. We thus have $\sum_{i=1}^{k} \mu_i = n$.

To each distinct eigenvalue λ_i, $i = 1, \ldots, k$, there corresponds a whole subspace $\phi_i \doteq \mathcal{N}(\lambda_i I_n - A)$ of eigenvectors associated with this eigenvalue, called the *eigenspace*. Eigenvectors belonging to different eigenspaces are linearly independent, as formalized next.

Theorem 3.3 *Let λ_i, $i = 1, \ldots, k \leq n$ be the distinct eigenvalues of $A \in \mathbb{R}^{n,n}$. Let $\phi_i = \mathcal{N}(\lambda_i I_n - A)$, and let $u^{(i)}$ be any nonzero vectors such that $u^{(i)} \in \phi_i$, $i = 1, \ldots, k$. Then, the $u^{(i)}$s are linearly independent.*

Proof Suppose initially that $u^{(i)} \in \phi_j$ for $j \neq i$. This would mean that $Au^{(i)} = \lambda_j u^{(i)} = \lambda_i u^{(i)}$, hence $\lambda_j = \lambda_i$, which is impossible since the λs are distinct. We then conclude that $j \neq i$ implies $u_i \notin \phi_j$.

Suppose now, again for the purpose of contradiction, that there exist an $u^{(i)}$ (say without loss of generality, the first one, $u^{(1)}$) which is a linear combination of the other eigenvectors:

$$u^{(1)} = \sum_{i=2}^{k} \alpha_i u^{(i)}. \tag{3.5}$$

[6] Note incidentally that A and $A^\top$ have the same eigenvalues, since the determinant of a matrix and that of its transpose coincide.

[7] This is only true for matrices with real elements, i.e., for $A \in \mathbb{R}^{n,n}$.

Then we have the two identities:

$$\lambda_1 u^{(1)} = \sum_{i=2}^{k} \alpha_i \lambda_1 u^{(i)},$$

$$\lambda_1 u^{(1)} = Au^{(1)} = \sum_{i=2}^{k} \alpha_i Au^{(i)} = \sum_{i=2}^{k} \alpha_i \lambda_i u^{(i)},$$

and, subtracting these two equations, we obtain

$$\sum_{i=2}^{k} \alpha_i (\lambda_i - \lambda_1) u^{(i)} = 0,$$

where $\lambda_i - \lambda_1 \neq 0$, since the eigenvalues are distinct by hypothesis. This would mean that $u^{(2)}, \ldots, u^{(k)}$ are linearly dependent, hence at least one of these vectors, say without loss of generality $u^{(2)}$, can be written as a linear combination of the other vectors $u^{(3)}, \ldots, u^{(k)}$. At this point, by repeating the initial reasoning, we would conclude that also $u^{(3)}, \ldots, u^{(k)}$ are linearly dependent. Proceeding in this way, we would eventually arrive at the conclusion that $u^{(k-1)}, u^{(k)}$ are linearly dependent, which would mean in particular that $u^{(k-1)} \in \phi_k$. However, this is impossible, by virtue of the initial statement in our proof. We thus conclude that (3.5) was contradicted, so the proposition stands proved. $\qquad \square$

Block-triangular decomposition. Thanks to eigenvalues and eigenvectors, a square matrix can be shown to be similar to a block-triangular matrix, that is, a matrix of the form[8]

$$\begin{bmatrix} A_{11} & A_{12} & \cdots & A_{1p} \\ 0 & A_{22} & & A_{2p} \\ \vdots & \ddots & \ddots & \vdots \\ 0 & \cdots & 0 & A_{pp} \end{bmatrix},$$

[8] See Section 3.4.8 for properties of block-triangular matrices.

where the matrices A_{ii}, $i = 1, \ldots, p$ are square.

Let ν_i be the dimension of ϕ_i and let $U^{(i)} = [u_1^{(i)} \cdots u_{\nu_i}^{(i)}]$ be a matrix containing by columns a basis of ϕ_i. Note that, without loss of generality, this matrix can be chosen to have orthonormal columns. Indeed, take any basis of ϕ_i and apply the Gram–Schmidt procedure (see Section 2.3.3) to this basis to obtain an orthonormal basis spanning the same subspace. With this choice, $U^{(i)\perp}U^{(i)} = I_{\nu_i}$. Let further $Q^{(i)}$ be a $n \times (n - \nu_i)$ matrix with orthonormal columns spanning the subspace orthogonal to $\mathcal{R}(U^{(i)})$. The following corollary holds.

Corollary 3.1 *Any matrix $A \in \mathbb{R}^{n,n}$ is similar to a block-triangular matrix having a block $\lambda_i I_{\nu_i}$ on the diagonal, where λ_i is a distinct eigenvalue of A, and ν_i is the dimension of the associated eigenspace.*

Proof The compound matrix $P_i \doteq [U^{(i)} \, Q^{(i)}]$ is an orthogonal matrix (the columns of P_i form an orthonormal basis spanning the whole space $\mathbb{C}^n$, see Section 3.4.6), hence it is invertible, and $P_i^{-1} = P_i^\top$, see Section 3.4.6. Then, since $AU^{(i)} = \lambda_i U^{(i)}$, we have that

$$U^{(i)\top} A U^{(i)} = \lambda_i U^{(i)\top} U^{(i)} = \lambda_i I_{\nu_i},$$

and $Q^{(i)\top} A U^{(i)} = \lambda_i Q^{(i)\top} U^{(i)} = 0$. Therefore

$$P_i^{-1} A P_i = P_i^\top A P_i = \begin{bmatrix} \lambda_i I_{\nu_i} & U^{(i)\top} A Q^{(i)} \\ 0 & Q^{(i)\top} A U^{(i)} \end{bmatrix}, \tag{3.6}$$

which proves the claim. □

Since similar matrices have the same set of eigenvalues (counting multiplicities),[9] and since the set of eigenvalues of a block-triangular matrix is the union of the eigenvalues of the diagonal blocks, we can also conclude from Eq. (3.6) that it must always be $\nu_i \leq \mu_i$ (if $\nu_i > \mu_i$, then the block-triangular form in (3.6) would imply that A has at least ν_i identical eigenvalues at λ_i, so the algebraic multiplicity of λ_i would be $\mu_i = \nu_i$, which is a contradiction).

3.3.6 Diagonalizable matrices

A direct consequence of Theorem 3.3 is that, under certain hypotheses, $A \in \mathbb{R}^{n,n}$ is similar to a diagonal matrix, i.e., in current terminology, A is *diagonalizable*.[10] This is stated in the following theorem.

Theorem 3.4 *Let λ_i, $i = 1, \ldots, k \leq n$, be the distinct eigenvalues of $A \in \mathbb{R}^{n,n}$, let μ_i, $i = 1, \ldots, k$, denote the corresponding algebraic multiplicities, and let $\phi_i = \mathcal{N}(\lambda_i I_n - A)$. Let further $U^{(i)} = [u_1^{(i)} \cdots u_{\nu_i}^{(i)}]$ be a matrix containing by columns a basis of ϕ_i, being $\nu_i \doteq \dim \phi_i$. Then, it holds that $\nu_i \leq \mu_i$ and, if $\nu_i = \mu_i$, $i = 1, \ldots, k$, then*

$$U = [U^{(1)} \cdots U^{(k)}]$$

is invertible, and

$$A = U \Lambda U^{-1}, \tag{3.7}$$

where

$$\Lambda = \begin{bmatrix} \lambda_1 I_{\mu_1} & 0 & \cdots & 0 \\ 0 & \lambda_2 I_{\mu_2} & \cdots & 0 \\ \vdots & \vdots & \ddots & \vdots \\ 0 & \cdots & 0 & \lambda_k I_{\mu_k} \end{bmatrix}.$$

Proof The fact that $\nu_i \leq \mu_i$ has already been proved below Eq. (3.6). Let then $\nu_i = \mu_i$. Vectors $u_1^{(i)}, \ldots, u_{\nu_i}^{(i)}$ are linearly independent, since

[9] This fact is readily proved by constructing the characteristic polynomial of a matrix $B = P^{-1}AP$, since

$$\det(\lambda I - B) = \det(\lambda I - P^{-1}AP)$$
$$= \det(P^{-1}(\lambda I - A)P)$$
$$= \det(P^{-1}) \det(\lambda I - A) \det(P)$$
$$= \det(\lambda I - A).$$

[10] Not all matrices are diagonalizable. For example, $A = \begin{bmatrix} 1 & 1 \\ 0 & 1 \end{bmatrix}$ is not diagonalizable. However, it can be proved that for any given square matrix, there always exist an arbitrarily small additive perturbation that makes it diagonalizable; i.e., diagonalizable matrices form a *dense* subset of $\mathbb{R}^{n,n}$.

they are, by definition, a basis of Φ_i. Moreover, from Theorem 3.3, $u_{j_1}^{(1)}, \ldots, u_{j_k}^{(k)}$ are linearly independent, for any $j_i \in \{1, \ldots, v_i\}$, $i = 1, \ldots, k$. This implies that the whole collection $\{u_j^{(i)}\}_{i=1,\ldots,k;j=1,\ldots,v_i}$ is linearly independent. Now, since $v_i = \mu_i$ for all i, then $\sum_{i=1}^k v_i = \sum_{i=1}^k \mu_i = n$, hence matrix U is full rank, and thus invertible.

For each $i = 1, \ldots, k$, we have that $Au_j^{(i)} = \lambda_i u_j^{(i)}$, $j = 1, \ldots, \mu_i$, and this can be rewritten in compact matrix form as

$$AU^{(i)} = \lambda_i U^{(i)}, \ i = 1, \ldots, k,$$

which is also rewritten as

$$AU = U\Lambda,$$

whereby the statement (3.7) follows by multiplying both sides by U^{-1} on the right. □

Example 3.5 (*Eigenvectors and the Google PageRank*) The effectiveness of Google's search engine largely relies on its PageRank (so named after Google's founder Larry Page) algorithm, which quantitatively ranks the importance of each page on the web, allowing Google to thereby present to the user the more important (and typically most relevant and helpful) pages first.

If the web of interest is composed of n pages, each labelled with integer k, $k = 1, \ldots, n$, we can model this web as a directed graph, where pages are the nodes of the graph, and a directed edge exists pointing from node k_1 to node k_2 if the web page k_1 contains a link to k_2. Let x_k, $k = 1, \ldots, n$, denote the importance *score* of page k. A simple initial idea would be to assign the score to any page k according to the number of other web pages that link to the considered page (backlinks).

In the web depicted in Figure 3.8, for example, the scores would be $x_1 = 2$, $x_2 = 1$, $x_3 = 3$, $x_4 = 2$, so that page $k = 3$ appears to be the most relevant page, whereas page $k = 2$ is the least important. In this approach, a page's score can be interpreted as the number of "votes" that a page has received from other pages, where each incoming link is a vote. However, the web is not *that* democratic, since the relevance of a page typically depends on the relevance, or "authority" of the pages pointing to it. In other words, your page relevance should be higher if your page is pointed directly by Yahoo.com rather then, say, by Nobody.com. Votes should therefore be weighted, rather than merely counted, and the weight should be related to the score of the pointing page itself. The actual scoring count goes then as follows: each page j has a score x_j and n_j outgoing links; as an assumption, we do not allow links from a page to itself, and we do not allow for dangling pages, that is pages with no outgoing links, therefore $n_j > 0$ for all j. The score x_j represents the total voting power of node j, which is to be evenly subdivided among the n_j outgoing links; each outgoing link thus carries x_j/n_j units of vote. Let

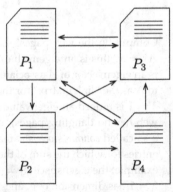

Figure 3.8 A small web.

B_k denote the set of labels of the pages that point to page k, i.e., B_k is the set of backlinks for page k. Then, the score of page k is computed as

$$x_k = \sum_{j \in B_k} \frac{x_j}{n_j}, \quad k = 1, \ldots, n.$$

Note that this approach is less direct than the initial pure counting one, since now scores are defined in an apparently circular way, i.e., page k's score is defined as a function of other pages' scores, which in turn depend on the score of page k, etc.

We now apply this approach to the web in Figure 3.8. We have $n_1 = 3$, $n_2 = 2$, $n_3 = 1$, $n_4 = 2$, hence

$$
\begin{aligned}
x_1 &= x_3 + \frac{1}{2}x_4, \\
x_2 &= \frac{1}{3}x_1, \\
x_3 &= \frac{1}{3}x_1 + \frac{1}{2}x_2 + \frac{1}{2}x_4, \\
x_4 &= \frac{1}{3}x_1 + \frac{1}{2}x_2.
\end{aligned}
$$

We can write this system of equations in compact form exploiting the matrix–vector product rule, as follows:

$$
x = Ax, \quad A = \begin{bmatrix} 0 & 0 & 1 & \frac{1}{2} \\ \frac{1}{3} & 0 & 0 & 0 \\ \frac{1}{3} & \frac{1}{2} & 0 & \frac{1}{2} \\ \frac{1}{3} & \frac{1}{2} & 0 & 0 \end{bmatrix}, \quad x = \begin{bmatrix} x_1 \\ x_2 \\ x_3 \\ x_4 \end{bmatrix}.
$$

Computing the web pages' scores thus amounts to finding x such that $Ax = x$: this is an eigenvalue/eigenvector problem and, in particular, x is an eigenvector of A associated with the eigenvalue $\lambda = 1$. We will refer to A as the "link matrix" for the given web. It can actually be proved that $\lambda = 1$ is indeed an eigenvalue of A, for any link matrix A (provided the web has no dangling pages), due to the fact that A is, by construction, a so-called *column stochastic* matrix,[11] that is a matrix with non-negative entries for which the sum of the elements over each column is one. In this example, the eigenspace $\phi_1 = \mathcal{N}(I_n - A)$ associated with the eigenvalue $\lambda = 1$ has dimension one, and it is given by

[11] See Exercise 3.11.

$$
\phi_1 = \mathcal{N}(I_n - A) = \mathrm{span}\left(\begin{bmatrix} 12 \\ 4 \\ 9 \\ 6 \end{bmatrix} \right),
$$

hence the solution to $Ax = x$ is any vector x in ϕ_1. Usually, the solution x is chosen so that the entries are normalized by summing up to one. In this case,

$$
x = \frac{1}{31} \begin{bmatrix} 12 \\ 4 \\ 9 \\ 6 \end{bmatrix} = \begin{bmatrix} 0.3871 \\ 0.1290 \\ 0.2903 \\ 0.1935 \end{bmatrix}.
$$

Page 1 thus appears to be the most relevant, according to the PageRank scoring.

Note that the method discussed so far can lead to ambiguities in certain cases where the eigenspace ϕ_1 has dimension larger than one. In such cases, in fact, there are multiple eigenvectors corresponding to the eigenvalue $\lambda = 1$, therefore the ranking of pages is not uniquely defined. To overcome this difficulty, a modification of the basic approach is used in Google. Specifically, one considers, instead of A, the modified matrix

$$\tilde{A} = (1 - \mu)A + \mu E,$$

where $\mu \in [0,1]$ and E is an $n \times n$ matrix whose entries are all equal to $1/n$. A typical choice for μ is $\mu = 0.15$. The modified link matrix $\tilde{A}$ has an eigenvalue at $\lambda = 1$, and one can prove that the corresponding eigenspace has always dimension one, hence the page ranking vector is unique, up to scaling.

The challenge in the real-world application resides, of course, in the huge size of the eigenvector problem that one is faced with. According to Google, the PageRank problem tallies up to about two billion variables, and it is solved about once a week for the whole World-Wide Web.

3.4 Matrices with special structure and properties

We here briefly review some important classes of matrices with special structure and properties.

3.4.1 Square matrices

A matrix $A \in \mathbb{R}^{m,n}$ is said to be *square* if it has as many columns as rows, that is if $m = n$.

3.4.2 Sparse matrices

Informally, a matrix $A \in \mathbb{R}^{m,n}$ is said to be *sparse* if most of its elements are zero. Several improvements in computational efficiency can be obtained when dealing with sparse matrices. For instance, a sparse matrix can be stored in memory by storing only its $p \ll mn$ nonzero elements. Also, operations such as addition and multiplication can be performed efficiently by dealing only with the nonzero elements of the matrices.[12]

[12] See Exercise 7.1.

3.4.3 Symmetric matrices

Symmetric matrices are square matrices that satisfy $a_{ij} = a_{ji}$ for every pair $i, j = 1, \ldots, n$. More compactly, A is symmetric if $A = A^\top$. A symmetric $n \times n$ matrix is defined by the entries on and above the main diagonal, the entries below the diagonal being a symmetric copy of those above the diagonal. The number of "free" entries of a symmetric matrix is therefore

$$n + (n-1) + \cdots + 1 = \frac{n(n+1)}{2}.$$

Symmetric matrices play an important role in optimization and are further discussed in Chapter 4.

Figure 3.9 shows an "image" of a sparse symmetric matrix, where nonzero elements are represented by gray levels, and zero elements in white.

3.4.4 Diagonal matrices

Diagonal matrices are square matrices with $a_{ij} = 0$ whenever $i \neq j$. A diagonal $n \times n$ matrix can be denoted by $A = \text{diag}(a)$, where a is an n-dimensional vector containing the diagonal elements of A. We usually write

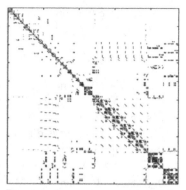

Figure 3.9 A sparse symmetric matrix.

$$A = \text{diag}(a_1, \ldots, a_n) = \begin{bmatrix} a_1 & & \\ & \ddots & \\ & & a_n \end{bmatrix},$$

where by convention the zero elements outside the diagonal are not written. It can be readily verified that the eigenvalues of a diagonal matrix are simply the elements on the diagonal. Further, $\det A = a_1 a_2 \cdots a_n$, hence a diagonal matrix is nonsingular if and only if $a_i \neq 0, i = 1, \ldots, n$. The inverse of a nonsingular diagonal matrix is simply

$$A^{-1} = \begin{bmatrix} \frac{1}{a_1} & & \\ & \ddots & \\ & & \frac{1}{a_n} \end{bmatrix}.$$

3.4.5 Triangular matrices

Triangular matrices are square matrices in which all elements either above or below the diagonal are zero. In particular, an upper-triangular matrix A is such that $a_{ij} = 0$ whenever $i > j$:

$$A = \begin{bmatrix} a_{11} & \cdots & a_{1n} \\ & \ddots & \vdots \\ & & a_{nn} \end{bmatrix} \text{ upper-triangular matrix,}$$

and a lower-triangular matrix is such that $a_{ij} = 0$ whenever $i < j$:

$$A = \begin{bmatrix} a_{11} & & \\ \vdots & \ddots & \\ a_{n1} & \cdots & a_{nn} \end{bmatrix} \quad \text{lower-triangular matrix.}$$

Similarly to diagonal matrices, the eigenvalues of a triangular matrix are the elements on the diagonal, and $\det A = a_{11}a_{22} \cdots a_{nn}$. The product of two upper (resp. lower) triangular matrices is still upper (resp. lower) triangular. The inverse of a nonsingular upper (resp. lower) triangular matrices is still upper (resp. lower) triangular.

3.4.6 Orthogonal matrices

Orthogonal matrices are square matrices, such that the columns form an orthonormal basis of $\mathbb{R}^n$. If $U = [u_1 \cdots u_n]$ is an orthogonal matrix, then

$$u_i^\top u_j = \begin{cases} 1 & \text{if } i = j, \\ 0 & \text{otherwise.} \end{cases}$$

Thus, $U^\top U = UU^\top = I_n$. Orthogonal matrices preserve length and angles. Indeed, for every vector x,

$$\|Ux\|_2^2 = (Ux)^\top (Ux) = x^\top U^\top Ux = x^\top x = \|x\|_2^2.$$

Thus, the underlying linear map $x \to Ux$ preserves the length (measured in Euclidean norm). In addition, angles are also preserved by orthogonal maps: if x, y are two vectors with unit norm, then the angle θ between them satisfies $\cos\theta = x^\top y$, while the angle θ' between the rotated vectors $x' = Ux$, $y' = Uy$, satisfies $\cos\theta' = (x')^\top y'$. Since $(Ux)^\top (Uy) = x^\top U^\top Uy = x^\top y$, we obtain that the angles are the same. The converse statement is also true: any square matrix that preserves lengths and angles is orthogonal. Further, pre- and post-multiplication of a matrix by an orthogonal matrix does not change the Frobenius norm (nor the ℓ_2-induced norm formally defined later in Section 3.6.3), that is

$$\|UAV\|_\mathrm{F} = \|A\|_\mathrm{F}, \quad \text{for } U, V \text{ orthogonal.}$$

Figure 3.10 shows an "image" of an orthogonal matrix.

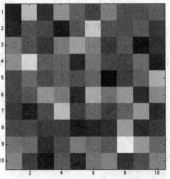

Figure 3.10 Image of an orthogonal matrix. Orthogonality is not apparent in the image.

Example 3.6 The matrix

$$U = \frac{1}{\sqrt{2}} \begin{bmatrix} 1 & 1 \\ 1 & -1 \end{bmatrix}$$

is orthogonal. The vector $x = [2 \; 1]^\top$ is transformed by the orthogonal matrix above into

$$Ux = \frac{1}{\sqrt{2}} \begin{bmatrix} 1 \\ 3 \end{bmatrix}.$$

Thus, U corresponds to a rotation of angle $45°$ counter-clockwise. More generally, the map defined by the orthogonal matrix

$$U(\theta) = \begin{bmatrix} \cos\theta & -\sin\theta \\ \sin\theta & \cos\theta \end{bmatrix}$$

represents a counter-clockwise rotation of angle θ.

3.4.7 Dyads

A matrix $A \in \mathbb{R}^{m,n}$ is a *dyad* if it is of the form $A = uv^\top$, for some vectors $u \in \mathbb{R}^m$ and $v \in \mathbb{R}^n$. If u, v have the same dimension, then the dyad $A = uv^\top$ is square.

A dyad acts on an input vector $x \in \mathbb{R}^n$ as follows:

$$Ax = (uv^\top)x = (v^\top x)u.$$

The elements A_{ij} are of the form $u_i v_j$; thus each row (resp. column) is a scaled version of the others, with "scalings" given by vector u (resp. v). In terms of the associated linear map $x \to Ax$, for a dyad $A = uv^\top$, the output always points in the same direction u, no matter what the input x is. The output is thus always a simple scaled version of u. The amount of scaling depends on the vector v, via the linear function $x \to v^\top x$.

If u and v are nonzero, the dyad $uv^\top$ has rank one, since its range is the line generated by u. A square dyad ($m = n$) has only one nonzero eigenvalue in $\lambda = v^\top u$, with corresponding eigenvector u.

We can always normalize the dyad, by assuming that both u, v are of unit (Euclidean) norm, and using a factor to capture their scale. That is, any dyad can be written in normalized form:

$$A = uv^\top = (\|u\|_2 \cdot \|v\|_2) \frac{u}{\|u\|_2} \frac{v^\top}{\|v\|_2} = \sigma \tilde{u}\tilde{v}^\top,$$

where $\sigma > 0$, and $\|\tilde{u}\|_2 = \|\tilde{v}\|_2 = 1$. Figure 3.11 shows an "image" of a dyad.

Figure 3.11 Image of a dyad. Perhaps the reader can feel that rows and columns are scaled versions of each other.

Example 3.7 (*Single factor model of financial price data*) Consider an $m \times T$ data matrix A which contains the log-returns (see Example 2.6) of m assets over T time periods (say, days). A single-factor model for these data is one based on the assumption that the matrix is a dyad: $A = uv^\top$, where $v \in \mathbb{R}^T$, and $u \in \mathbb{R}^m$. According to the single-factor model, the entire market behaves as follows. At any time t, $1 \leq t \leq T$, the log-return of asset i, $1 \leq i \leq m$, is of the form $[A]_{it} = u_i v_t$. The vectors u and v have the following interpretation.

- For any asset, the rate of change in log-returns between two time instants $t_1 \leq t_2$ is given by the ratio v_{t_2}/v_{t_1}, independent of the asset. Hence, v gives the time profile for all the assets: every asset shows the same time profile, up to a scaling given by u.

- Likewise, for any time t, the ratio between the log-returns of two assets i and j at time t is given by u_i/u_j, independent of t. Hence u gives the asset profile for all the time periods. Each time shows the same asset profile, up to a scaling given by v.

While single-factor models may seem crude, they often offer a reasonable amount of information. It turns out that with many financial market data, a good single factor model involves a time profile v equal to the log-returns of the average of all the assets, or some weighted average (such as the SP 500 index). With this model, all assets follow the profile of the entire market.

3.4.8 Block-structured matrices

Any matrix $A \in \mathbb{R}^{m,n}$ can be partitioned into blocks, or submatrices, of compatible dimensions:

$$A = \begin{bmatrix} A_{11} & A_{12} \\ A_{21} & A_{22} \end{bmatrix}. \tag{3.8}$$

When A is square, and $A_{12} = 0$, $A_{21} = 0$ (here by 0 we mean a matrix block of all zeros of suitable dimension), then A is said to be *block diagonal*:

$$A = \begin{bmatrix} A_{11} & 0 \\ 0 & A_{22} \end{bmatrix}.$$

The set $\lambda(A)$ of eigenvalues of A is the union of the sets of eigenvalues of A_{11} and of A_{22}:

$$A \text{ block diagonal} \quad \Rightarrow \quad \lambda(A) = \lambda(A_{11}) \cup \lambda(A_{22}).$$

Also, a block-diagonal matrix is invertible if and only if its diagonal blocks are invertible, and it holds that

$$\begin{bmatrix} A_{11} & 0 \\ 0 & A_{22} \end{bmatrix}^{-1} = \begin{bmatrix} A_{11}^{-1} & 0 \\ 0 & A_{22}^{-1} \end{bmatrix}.$$

A square, block-partitioned matrix A of the form (3.8) is said to be *block upper triangular*, if $A_{21} = 0$, and *block lower triangular*, if $A_{12} = 0$:

$$A = \begin{bmatrix} A_{11} & 0 \\ A_{21} & A_{22} \end{bmatrix} \quad \text{block lower triangular,}$$

$$A = \begin{bmatrix} A_{11} & A_{12} \\ 0 & A_{22} \end{bmatrix} \quad \text{block upper triangular.}$$

Also for block-triangular matrices it holds that the eigenvalues of A are the union of the eigenvalues of the diagonal blocks, that is

A block (upper or lower) triangular $\Rightarrow \lambda(A) = \lambda(A_{11}) \cup \lambda(A_{22})$.

The inverse of a nonsingular block-triangular matrix can be expressed as follows:

$$\begin{bmatrix} A_{11} & 0 \\ A_{21} & A_{22} \end{bmatrix}^{-1} = \begin{bmatrix} A_{11}^{-1} & 0 \\ -A_{22}^{-1}A_{21}A_1^{-1} & A_{22}^{-1} \end{bmatrix},$$

$$\begin{bmatrix} A_{11} & A_{12} \\ 0 & A_{22} \end{bmatrix}^{-1} = \begin{bmatrix} A_{11}^{-1} & -A_{11}^{-1}A_{12}A_{22}^{-1} \\ 0 & A_{22}^{-1} \end{bmatrix}.$$

Both these formulas can be proved by checking directly that $AA^{-1} = I$ and $A^{-1}A = I$, which are the properties defining unequivocally the inverse of a matrix. Two equivalent formulas also exist for the inverse of a nonsingular full block matrix (3.8). Let

$$S_1 \doteq A_{11} - A_{12}A_{22}^{-1}A_{21}, \quad S_2 \doteq A_{22} - A_{21}A_{11}^{-1}A_{12},$$

then

$$\begin{bmatrix} A_{11} & A_{12} \\ A_{21} & A_{22} \end{bmatrix}^{-1} = \begin{bmatrix} S_1^{-1} & -A_{11}^{-1}A_{12}S_2^{-1} \\ -A_{22}^{-1}A_{21}S_1^{-1} & S_2^{-1} \end{bmatrix}$$

$$= \begin{bmatrix} S_1^{-1} & -S_1^{-1}A_{12}A_{22}^{-1} \\ -S_2^{-1}A_{21}A_{11}^{-1} & S_2^{-1} \end{bmatrix}.$$

Further equivalent expressions can be obtained by expanding the inverses of the S_1 and S_2 blocks using a handy matrix identity, known as the *matrix inversion lemma*, or Woodbury formula:

$$\left(A_{11} - A_{12}A_{22}^{-1}A_{21} \right)^{-1} =$$
$$A_{11}^{-1} + A_{11}^{-1}A_{12}(A_{22} - A_{21}A_{11}^{-1}A_{12})^{-1}A_{21}A_{11}^{-1}. \qquad (3.9)$$

3.4.9 *Rank-one perturbations*

A special case of Eq. (3.9) arises when A_{12} and A_{21} are vectors, that is, when a rank-one matrix (i.e., a dyad) is added to A_{11}. Specifically, for $A_{12} = u \in \mathbb{R}^n$, $A_{21}^\top = v \in \mathbb{R}^n$, and $A_{22} = -1$, the above formula becomes

$$(A_{11} + uv^\top)^{-1} = A_{11}^{-1} - \frac{A_{11}^{-1} uv^\top A_{11}^{-1}}{1 + v^\top A_{11}^{-1} u}. \tag{3.10}$$

This formula permits us to easily compute the inverse of a *rank-one perturbation* of A_{11}, based on the inverse of A_{11} itself.

A further interesting property is that a rank-one perturbation cannot alter the rank of a matrix by more than one unit. This fact actually holds for generic (i.e., possibly rectangular) matrices, as stated next.

Lemma 3.1 (Rank of rank-one perturbation) *Let $A \in \mathbb{R}^{m,n}$ and $q \in \mathbb{R}^m$, $p \in \mathbb{R}^n$. Then*

$$|\operatorname{rank}(A) - \operatorname{rank}(A + qp^\top)| \leq 1.$$

Proof We next show that $\operatorname{rank}(A) \leq \operatorname{rank}(A + qp^\top) + 1$; the symmetric condition $\operatorname{rank}(A + qp^\top) \leq \operatorname{rank}(A) + 1$ can be proved via an identical argument, exchanging the role of the A and $A + qp^\top$ matrices. Since the rank coincides with the dimension of the range subspace of a matrix, what we need to prove is that

$$\dim \mathcal{R}(A) \leq \dim \mathcal{R}(A + qp^\top) + 1.$$

Recalling that, by the fundamental theorem of linear algebra, $\dim \mathcal{R}(A) + \dim \mathcal{N}(A^\top) = m$, the previous condition is also equivalent to

$$\dim \mathcal{N}(A^\top + pq^\top) \leq \dim \mathcal{N}(A^\top) + 1. \tag{3.11}$$

We prove (3.11) by contradiction: let $\nu \doteq \dim \mathcal{N}(A^\top)$, and suppose, for the purpose of contradiction, that

$$\dim \mathcal{N}(A^\top + pq^\top) > \nu + 1.$$

Then, there would exist $\nu + 2$ linear independent vectors $v_1, \ldots, v_{\nu+2}$, all belonging to the nullspace of $A^\top + pq^\top$, that is $(A^\top + pq^\top)v_i = 0$, $i = 1, \ldots, \nu + 2$, which implies that

$$A^\top v_i = -\alpha_i p, \ \alpha_i \doteq q^\top v_i, \quad i = 1, \ldots, \nu + 2.$$

Now, at least one of the scalars α_i must be nonzero, for otherwise $A^\top v_i = 0$ for $i = 1, \ldots, \nu + 2$, which would contradict the fact that $\dim \mathcal{N}(A^\top) = \nu$, and the result would be proved immediately. Assume then, without loss of generality, that $\alpha_1 \neq 0$, and define vectors $w_i = v_{i+1} - (\alpha_{i+1}/\alpha_1)v_1$, $i = 1, \ldots, \nu + 1$. It can then be checked directly that

$$A^\top w_i = A^\top v_{i+1} - \frac{\alpha_{i+1}}{\alpha_1} A^\top v_1 = -\alpha_{i+1}p + \alpha_{i+1}p = 0, \quad i = 1, \ldots, \nu + 1.$$

We would then have $\nu + 1$ linearly independent vectors w_i belonging to the nullspace of $A^\top$, which is a contradiction, since $\dim \mathcal{N}(A^\top) = \nu$. $\square$

3.5   *Matrix factorizations*

A substantial part of theoretical and numerical linear algebra is dedicated to the problem of matrix factorizations. That is, given a matrix
$A \in \mathbb{R}^{m,n}$, write this matrix as the product of two or more matrices
with special structure. Usually, once a matrix is suitably factorized,
several quantities of interest become readily accessible, and subsequent computations are greatly simplified. For example, it is known
that any square matrix A can be written as the product of an orthogonal matrix and a triangular matrix, that is $A = QR$, where Q is
orthogonal and R is upper triangular. Once such a factorization is
obtained we can, for instance, immediately evaluate the rank of A,
which is just the number of nonzero elements on the diagonal of R.
Also, we can readily solve for the unknown x a system of linear equations of the type $Ax = b$, as further discussed in Section 6.4.4.1. In
terms of the linear map defined by a matrix A, a factorization can be
interpreted as a decomposition of the map into a series of successive
stages, see, e.g., Figure 3.12.

We next briefly describe some of the most used matrix factorizations.

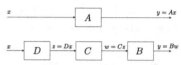

Figure 3.12 Given a matrix factorization $A = BCD$, the linear map $y = Ax$
is interpreted as the series connection
of three stages.

3.5.1   *Orthogonal-triangular decomposition (QR)*

Any square $A \in \mathbb{R}^{n,n}$ can be decomposed as

$$A = QR,$$

where Q is an orthogonal matrix, and R is an upper triangular matrix.
If A is nonsingular, then the factors Q, R are uniquely defined, if the
diagonal elements in R are imposed to be positive.

If $A \in \mathbb{R}^{m,n}$ is rectangular, with $m \geq n$, a similar decomposition
holds:

$$A = Q \begin{bmatrix} R_1 \\ 0_{m-n,n} \end{bmatrix},$$

where $Q \in \mathbb{R}^{m,m}$ is orthogonal, and $R_1 \in \mathbb{R}^{n,n}$ is upper triangular, see Figure 3.13. The QR decomposition is closely related to the
Gram–Schmidt orthonormalization procedure, and it is useful in the
numerical solution of linear equations and least-squares problems,
see Sections 7.3 and 6.4.4.1.

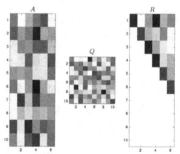

Figure 3.13 Image of a 10×6 matrix,
and its QR decomposition.

3.5.2   *Singular value decomposition (SVD)*

Any nonzero $A \in \mathbb{R}^{m,n}$ can be decomposed as

$$A = U\tilde{\Sigma}V^{\top},$$

where $V \in \mathbb{R}^{n,n}$ and $U \in \mathbb{R}^{m,m}$ are orthogonal matrices, and

$$\bar{\Sigma} = \begin{bmatrix} \Sigma & 0_{r,n-r} \\ 0_{m-r,r} & 0_{m-r,n-r} \end{bmatrix}, \quad \Sigma = \mathrm{diag}\,(\sigma_1, \ldots, \sigma_r),$$

where r is the rank of A, and the scalars $\sigma_i > 0$, $i = 1, \ldots, r$, are called the *singular values* of A. The first r columns $u_1, \ldots, u_r$ of U (resp. $v_1, \ldots, v_r$ of V) are called the left (resp. right) singular vectors, and satisfy

$$Av_i = \sigma_i u_i, \quad A^\top u_i = \sigma_i v_i, \quad i = 1, \ldots, r.$$

The SVD is a fundamental factorization in numerical linear algebra, as it exposes all relevant information about the linear map described by A, such as the range, nullspace, and rank. SVD is discussed in depth in Section 5.1. Figure 3.14 shows a pictorial representation of the SVD of a 4×8 matrix.

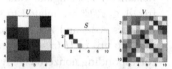

Figure 3.14 Image of the SVD of a 4×8 matrix.

3.5.3 Eigenvalue decomposition for diagonalizable matrices

A square, diagonalizable,[13] matrix $A \in \mathbb{R}^{n,n}$ can be decomposed as

$$A = U \Lambda U^{-1},$$

[13] Diagonalizable matrices are introduced in Section 3.3.6.

where $U \in \mathbb{C}^{n,n}$ is an invertible matrix containing by columns the eigenvectors of A, and Λ is a diagonal matrix containing the eigenvalues $\lambda_1, \ldots, \lambda_n$ of A in the diagonal. For generic matrices these eigenvalues are real or complex numbers, with complex ones coming in complex-conjugate pairs (see Section 3.3.6). The columns $u_1, \ldots, u_n$ of U are called the eigenvectors of A, and satisfy

$$Au_i = \lambda_i u_i, \quad i = 1, \ldots, n.$$

Indeed, these relations can be compactly written as $AU = \Lambda U$, which is equivalent to $A = U \Lambda U^{-1}$.

3.5.4 Spectral decomposition for symmetric matrices

Any *symmetric* matrix $A \in \mathbb{R}^{n,n}$ can be factored as

$$A = U \Lambda U^\top,$$

where $U \in \mathbb{R}^{n,n}$ is an orthogonal matrix, and Λ is a diagonal matrix containing the eigenvalues $\lambda_1, \ldots, \lambda_n$ of A in the diagonal. All these eigenvalues are real numbers, for symmetric matrices. Thus, symmetric matrices are diagonalizable, their eigenvalues are always real, and corresponding eigenvectors can be chosen also to be real and to form an orthonormal basis. The columns $u_1, \ldots, u_n$ of U are indeed the eigenvectors of A, and satisfy

$$Au_i = \lambda_i u_i, \quad i = 1, \ldots, n.$$

This factorization is known as the *spectral* decomposition for symmetric matrices, and it is further discussed in Section 4.2.

Remark 3.1 The singular values and singular vectors of a *rectangular* matrix $A \in \mathbb{R}^{m,n}$ are related to spectral decompositions of the symmetric matrices $AA^\top$ and $A^\top A$. Precisely, if $A = U\tilde{\Sigma}V^\top$ is an SVD of A, then the columns of U (resp. V) are eigenvectors of $AA^\top$ (resp. $A^\top A$), with corresponding nonzero eigenvalues $\sigma_i^2, i = 1, \ldots, r$.

3.6 Matrix norms

3.6.1 Definition

A function $f : \mathbb{R}^{m,n} \to \mathbb{R}$ is a matrix norm if, analogously to the vector case, it satisfies three standard axioms. Specifically, for all $A, B \in \mathbb{R}^{m,n}$ and all $\alpha \in \mathbb{R}$:

- $f(A) \geq 0$, and $f(A) = 0$ if and only if $A = 0$;

- $f(\alpha A) = |\alpha| f(A)$;

- $f(A + B) \leq f(A) + f(B)$.

Many of the popular matrix norms also satisfy a fourth condition called *sub-multiplicativity*: for any conformably sized matrices A, B

$$f(AB) \leq f(A)f(B).$$

Among the most frequently encountered matrix norms, we find the Frobenius norm:

$$\|A\|_{\mathrm{F}} \doteq \sqrt{\operatorname{trace} AA^\top},$$

and with $p = 1, 2, \infty$, the ℓ_p-induced, or operator, matrix norms

$$\|A\|_p \doteq \max_{\|u\|_p \neq 0} \frac{\|Au\|_p}{\|u\|_p}.$$

3.6.2 Frobenius norm

The Frobenius norm $\|A\|_{\mathrm{F}}$ is nothing but the standard Euclidean (ℓ_2) vector norm applied to the vector formed by all elements of $A \in \mathbb{R}^{m,n}$:

$$\|A\|_{\mathrm{F}} = \sqrt{\operatorname{trace} AA^\top} = \sqrt{\sum_{i=1}^{m} \sum_{j=1}^{n} |a_{ij}|^2}.$$

The Frobenius norm also has an interpretation in terms of the eigenvalues of the symmetric matrix $AA^\top$ (we recall that the trace of a square matrix represents the sum of the elements on the diagonal, as well as the sum of the eigenvalues of that matrix):

$$\|A\|_{\mathrm{F}} = \sqrt{\operatorname{trace} AA^\top} = \sqrt{\sum_{i=1}^{m} \lambda_i(AA^\top)}.$$

Let $a_1^\top, \ldots, a_m^\top$ denote the rows of $A \in \mathbb{R}^{m,n}$, then we have that

$$\|A\|_{\mathrm{F}}^2 = \sum_{i=1}^{m} \|a_i^\top\|_2^2,$$

therefore, for any $x \in \mathbb{R}^n$, it holds that

$$\|Ax\|_2 \leq \|A\|_{\mathrm{F}} \|x\|_2. \tag{3.12}$$

This is a consequence of the Cauchy–Schwartz inequality applied to $|a_i^\top x|$:

$$\|Ax\|_2^2 = \sum_{i=1}^{m} |a_i^\top x|^2 \leq \sum_{i=1}^{m} \|a_i^\top\|_2^2 \|x\|_2^2 = \|A\|_{\mathrm{F}}^2 \|x\|_2^2.$$

Inequality (3.12) also implies that the Frobenius norm is indeed submultiplicative, that is, for any $B \in \mathbb{R}^{n,p}$, it holds that

$$\|AB\|_{\mathrm{F}} \leq \|A\|_{\mathrm{F}} \|B\|_{\mathrm{F}}.$$

To see this fact, let $b_1, \ldots, b_p$ denote the columns of B, then $AB = [Ab_1 \cdots Ab_p]$, and

$$\|AB\|_{\mathrm{F}}^2 = \sum_{j=1}^{p} \|Ab_j\|_2^2 \leq \sum_{j=1}^{p} \|A\|_{\mathrm{F}}^2 \|b_j\|_2^2 = \|A\|_{\mathrm{F}}^2 \|B\|_{\mathrm{F}}^2.$$

The Frobenius norm of A can be interpreted in terms of the linear map associated with A: $u \to y = Au$, as pictured Figure 3.15. Specifically, it provides a measure of the output's variance, when inputs are random; see Exercise 3.9 for details.

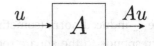

Figure 3.15 A matrix as an operator. Matrix norms measure "typical" output sizes.

3.6.3 Operator norms

While the Frobenius norm measures the response to random inputs, the so-called operator norms give a characterization of the *maximum* input–output *gain* of the linear map $u \to y = Au$. Choosing to measure both inputs and outputs in terms of a given ℓ_p norm, with typical values $p = 1, 2, \infty$, leads to the definition

$$\|A\|_p \doteq \max_{u \neq 0} \frac{\|Au\|_p}{\|u\|_p} = \max_{\|u\|=1} \|Au\|_p,$$

where the last equality follows from the fact that we can divide both terms in the fraction by the (nonzero) norm of u.

By definition, for every u, $\|Au\|_p \leq \|A\|_p\|u\|_p$. From this property follows that any operator norm is sub-multiplicative, that is, for any two conformably sized matrices A, B, it holds that

$$\|AB\|_p \leq \|A\|_p\|B\|_p.$$

This fact is easily seen by considering the product AB as the series connection of the two operators B, A:

$$\|Bu\|_p \leq \|B\|_p\|u\|_p, \quad \|ABu\|_p \leq \|A\|_p\|Bu\|_p \leq \|A\|_p\|B\|_p\|u\|_p,$$

see Figure 3.16. For the typical values of $p = 1, 2, \infty$, we have the following results (proofs of the first two cases are left to the reader as an exercise).

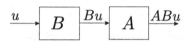

Figure 3.16 Submultiplicativity of operator norms.

- The ℓ_1-induced norm corresponds to the largest absolute column sum:

$$\|A\|_1 = \max_{\|u\|_1=1} \|Au\|_1 = \max_{j=1,\dots,n} \sum_{i=1}^{m} |a_{ij}|.$$

- The ℓ_∞-induced norm corresponds to the largest absolute row sum:

$$\|A\|_\infty = \max_{\|u\|_\infty=1} \|Au\|_\infty = \max_{i=1,\dots,m} \sum_{j=1}^{n} |a_{ij}|.$$

- The ℓ_2-induced norm (sometimes referred to as the *spectral* norm) corresponds to the square-root of the largest eigenvalue $\lambda_{\max}$ of $A^\top A$:

$$\|A\|_2 = \max_{\|u\|_2=1} \|Au\|_2 = \sqrt{\lambda_{\max}(A^\top A)}.$$

The latter identity follows from the variational characterization of the eigenvalues of a symmetric matrix, see Section 4.3.1.

Remark 3.2 It is possible to define other matrix norms, for example operator norms where different norms are used to measure input and output sizes. Some of these norms may be hard to compute. For example, the norm

$$\max_u \|Au\|_2 \; : \; \|u\|_\infty \leq 1,$$

is hard to compute exactly, although good approximations are available.

3.6.3.1 *Spectral radius.* The spectral radius $\rho(A)$ of a matrix $A \in \mathbb{R}^{n,n}$ is defined as the maximum modulus of the eigenvalues of A, that is

$$\rho(A) \doteq \max_{i=1,\dots,n} |\lambda_i(A)|.$$

Clearly, $\rho(A) \geq 0$ for all A, and $A = 0$ implies $\rho(A) = 0$. However, the converse is not true, since $\rho(A) = 0$ does not imply[14] necessarily that $A = 0$, hence $\rho(A)$ is not a matrix norm. However, for any induced matrix norm $\| \cdot \|_p$, it holds that

$$\rho(A) \leq \|A\|_p.$$

To prove this fact, let $\lambda_i, v_i \neq 0$ be an eigenvalue/eigenvector pair for A, then

$$\|A\|_p \|v_i\|_p \geq \|Av_i\|_p = \|\lambda_i v_i\|_p = |\lambda_i| \|v_i\|_p,$$

where the first inequality follows from the definition of induced matrix norm, hence

$$|\lambda_i| \leq \|A\|_p, \quad \text{for all } i = 1, \dots, n,$$

which proves the claim.

It follows, in particular, that $\rho(A) \leq \min(\|A\|_1, \|A\|_\infty)$, that is $\rho(A)$ is no larger than the maximum row or column sum of $|A|$ (the matrix whose entries are the absolute values of the entries in A).

3.7 Matrix functions

We have already encountered several scalar functions of matrix argument. For instance, given a matrix $X \in \mathbb{R}^{n,n}$, the determinant $\det X$, the trace of X, and any norm $\|X\|$ are examples of functions $f : \mathbb{R}^{n,n} \to \mathbb{R}$. Here we briefly discuss functions that return a *matrix* argument, that is, functions $f : \mathbb{R}^{n,n} \to \mathbb{R}^{n,n}$. One example of such a function is the matrix inverse, defined on the domain of nonsingular matrices, that given $X \in \mathbb{R}^{n,n}$ nonsingular, returns a matrix X^{-1} such that $XX^{-1} = X^{-1}X = I_n$.

3.7.1 Matrix powers and polynomials

The integer power function

$$f(X) = X^k, \quad k = 0, 1, \dots$$

can be quite naturally defined via the matrix product, by observing that $X^k = XX \cdots X$ (k times; we take the convention that $X^0 = I_n$).

[14] Take for instance $A = \begin{bmatrix} 0 & 1 \\ 0 & 0 \end{bmatrix}$. Matrices such that $\rho(A) = 0$ are called *nilpotent*.

Similarly, negative integer power functions can be defined over non-singular matrices as integer powers of the inverse:

$$f(X) = X^{-k} = (X^{-1})^k, \quad k = 0, 1, \ldots$$

A polynomial matrix function of degree $m \geq 0$ can hence be naturally defined as

$$p(X) = a_m X^m + a_{m-1} X^{m-1} + \cdots + a_1 X + a_0 I_n,$$

where a_i, $i = 0, 1, \ldots, m$, are the scalar coefficients of the polynomial. A first interesting result for matrix polynomials is the following one.

Lemma 3.2 (Eigenvalues and eigenvectors of a matrix polynomial)
Let $X \in \mathbb{R}^{n,n}$ and let λ, u be an eigenvalue/eigenvector pair for X (that is, $Xu = \lambda u$, and let

$$p(X) = a_m X^m + a_{m-1} X^{m-1} + \cdots + a_1 X + a_0 I_n.$$

Then, it holds that

$$p(X)u = p(\lambda)u,$$

where $p(\lambda) = a_m \lambda^m + a_{m-1} \lambda^{m-1} + \cdots + a_1 \lambda + a_0$. That is, if λ, u is an eigenvalue/eigenvector pair for X, then $p(\lambda), u$ is an eigenvalue/eigenvector pair for the polynomial matrix $p(X)$.

Proof The proof is immediate, by observing that $Xu = \lambda u$ implies that

$$X^2 u = X(Xu) = X(\lambda u) = \lambda^2 u, \quad X^3 u = X(X^2 u) = X(\lambda^2 u) = \lambda^3 u, \ldots$$

hence,

$$\begin{aligned}
p(X)u &= (a_m X^m + \cdots + a_1 X + a_0 I_n)u = (a_m \lambda^m + \cdots + a_1 \lambda + a_0)u \\
&= p(\lambda)u.
\end{aligned}$$

$\square$

A simple consequence of Lemma 3.2 is a result known as the *eigenvalue shift* rule: if $\lambda_i(A)$, $i = 1, \ldots, n$, denote the eigenvalues of a matrix $A \in \mathbb{R}^{n,n}$, then

$$\lambda_i(A + \mu I_n) = \lambda_i(A) + \mu, \quad i = 1, \ldots, n. \tag{3.13}$$

For matrices X that admit a diagonal factorization, a polynomial of such a matrix argument can be expressed according to the same type of factorization, as detailed in the next lemma.

Lemma 3.3 (Diagonal factorization of a matrix polynomial) *Let $X \in \mathbb{R}^{n,n}$ admit a diagonal factorization of the form (3.7)*

$$X = U\Lambda U^{-1},$$

where Λ is a diagonal matrix containing the eigenvalues of X, and U is a matrix containing by columns the corresponding eigenvectors. Let $p(t)$, $t \in \mathbb{R}$, be a polynomial

$$p(t) = a_m t^m + a_{m-1} t^{m-1} + \cdots + a_1 t + a_0.$$

Then

$$p(X) = U p(\Lambda) U^{-1},$$

where

$$p(\Lambda) = \mathrm{diag}\left(p(\lambda_1), \ldots, p(\lambda_n)\right).$$

Proof If $X = U\Lambda U^{-1}$, then $X^2 = XX = U\Lambda U^{-1} U\Lambda U^{-1} = U\Lambda^2 U^{-1}$, $X^3 = X^2 X = U\Lambda^3 U^{-1}$, etc., hence for any $k = 1, 2, \ldots$,

$$X^k = U\Lambda^k U^{-1}.$$

Therefore,

$$
\begin{aligned}
p(X) &= a_m X^m + a_{m-1} X^{m-1} + \cdots + a_1 X + a_0 I_n \\
&= U(a_m \Lambda^m + a_{m-1} \Lambda^{m-1} + \cdots + a_1 \Lambda + a_0 I_n) U^{-1} \\
&= U p(\Lambda) U^{-1}.
\end{aligned}
$$

$\square$

3.7.1.1 Convergence of matrix powers. A topic of interest in many applications (such as numerical algorithms, linear dynamical systems, Markov chains, etc.) is the convergence of the matrix powers X^k for $k \to \infty$. If X is diagonalizable, then our previous analysis states that

$$X^k = U\Lambda^k U^{-1} = \sum_{i=1}^{n} \lambda_i^k u_i v_i^\top,$$

where u_i is the i-th column of U and $v_i^\top$ is the i-th row of U^{-1}. We can draw some interesting conclusions from this expression. First, if $|\lambda_i| < 1$ for all i (that is, if $\rho(X) < 1$), then each term in the sum tends to zero as $k \to \infty$, hence $X^k \to 0$. Conversely, if $X^k \to 0$, then it must hold that $\rho(X) < 1$, for otherwise there would exist a λ_i with $|\lambda_i| \geq 1$ and the corresponding term in the above sum will either remain bounded in norm (if $|\lambda_i| = 1$), or grow indefinitely in norm, thus $X^k \to 0$ could not happen. Also, suppose $|\lambda_i| < 1$ for all i except for one (say, without loss of generality, the first one), for which we have $\lambda_1 = 1$. Then, it is easy to see that in such case X^k converges to a constant matrix: $X^k \to U_1 V_1^\top$, where U_1 contains by columns the eigenvectors associated with the eigenvalue $\lambda_1 = 1$, and V_1 contains the corresponding columns in $V = U^{-1}$.

The above analysis, however, is limited by the assumption of diagonalizability of X. The following theorem states a general result on convergence of matrix powers.[15]

[15] For a proof, see, for instance, Chapter 7 in C. D. Meyer, *Matrix Analysis and Applied Linear Algebra*, SIAM, 2001.

Theorem 3.5 (Convergence of matrix powers) *Let* $X \in \mathbb{R}^{n,n}$. *Then:*

1. $\lim_{k \to \infty} X^k = 0$ *if and only if* $\rho(X) < 1$;

2. $\sum_{k=0}^{\infty} X^k$ *converges if and only if* $\rho(X) < 1$ *(in which case the limit of the series is* $(I - X)^{-1}$*);*

3. $\lim_{k \to \infty} X^k = \bar{X} \neq 0$ *if and only if* $|\lambda_i| < 1$ *for all except one (possibly repeated) eigenvalue* $\lambda = 1$, *whose corresponding eigenspace has dimension equal to the algebraic multiplicity of* $\lambda = 1$. *Also,* $\bar{X} = U_1 V_1^\top$, *where* U_1 *contains by columns a basis for the eigenspace of* X *associated with* $\lambda = 1$, *and* V_1 *is such that* $V_1^\top U_1 = I$, *and* $V_1^\top u_i = 0$ *for all eigenvectors* u_i *of* X *associated with eigenvalues* $\lambda_i \neq 1$. *Further, if* X^k *is convergent, then* $\bar{X}(I - X) = (I - X)\bar{X} = 0$, *and conversely if* $\lim_{k \to \infty} X^k (I - X) = 0$ *then* X^k *is convergent.*

3.7.2   *Non-polynomial matrix functions*

Let $f : \mathbb{R} \to \mathbb{R}$ be an *analytic* function, that is, a function which is locally representable by a power series

$$f(t) = \sum_{k=0}^{\infty} a_k t^k$$

which is convergent for all t such that $|t| \leq R$, $R > 0$. If $\rho(X) < R$ (where $\rho(X)$ is the spectral radius of X), then the value of the matrix function $f(X)$ can be defined as the sum of the convergent series

$$f(X) = \sum_{k=0}^{\infty} a_k X^k.$$

Moreover, if X is diagonalizable, then $X = U\Lambda U^{-1}$, and

$$
\begin{aligned}
f(X) &= \sum_{k=0}^{\infty} a_k X^k = U \left(\sum_{k=0}^{\infty} a_k \Lambda^k \right) U^{-1} \\
&= U\mathrm{diag}\left(f(\lambda_1), \ldots, f(\lambda_n) \right) U^{-1} \\
&= U f(\Lambda) U^{-1}. \qquad\qquad (3.14)
\end{aligned}
$$

Equation (3.14) states in particular that the *spectrum* (i.e., the set of eigenvalues) of $f(A)$ is the image of the spectrum of A under the mapping f. This fact is known as the *spectral mapping theorem*.

Example 3.8 A notable example of application of Eq. (3.14) is given by the *matrix exponential*: the function $f(t) = e^t$ has a power series representation which is globally convergent

$$e^t = \sum_{k=0}^{\infty} \frac{1}{k!} t^k,$$

hence, for any diagonalizable $X \in \mathbb{R}^{n,n}$, we have

$$e^X \doteq \sum_{k=0}^{\infty} \frac{1}{k!} X^k = U \text{diag} \left(e^{\lambda_1}, \ldots, e^{\lambda_n} \right) U^{-1}.$$

Another example is given by the geometric series

$$f(t) = (1-t)^{-1} = \sum_{k=0}^{\infty} t^k, \quad \text{for } |t| < 1 = R,$$

from which we obtain that

$$f(X) = (I - X)^{-1} = \sum_{k=0}^{\infty} X^k, \quad \text{for } \rho(X) < 1.$$

More generally, for $\sigma \neq 0$, we have that

$$f(t) = (t - \sigma)^{-1} = -\frac{1}{\sigma} \sum_{k=0}^{\infty} \left(\frac{t}{\sigma} \right)^k, \quad \text{for } |t| < |\sigma| = R,$$

hence

$$f(X) = (X - \sigma I)^{-1} = -\frac{1}{\sigma} \sum_{k=0}^{\infty} \left(\frac{X}{\sigma} \right)^k, \quad \text{for } \rho(X) < |\sigma|,$$

and $X = U \Lambda U^{-1}$ implies that

$$(X - \sigma I)^{-1} = U(\Lambda - \sigma I)^{-1} U^{-1}. \tag{3.15}$$

3.8 Exercises

Exercise 3.1 (Derivatives of composite functions)

1. Let $f : \mathbb{R}^m \to \mathbb{R}^k$ and $g : \mathbb{R}^n \to \mathbb{R}^m$ be two maps. Let $h : \mathbb{R}^n \to \mathbb{R}^k$ be the composite map $h = f \circ g$, with values $h(x) = f(g(x))$ for $x \in \mathbb{R}^n$. Show that the derivatives of h can be expressed via a matrix–matrix product, as $J_h(x) = J_f(g(x)) \cdot J_g(x)$, where $J_h(x)$ is the Jacobian matrix of h at x, i.e., the matrix whose (i,j) element is $\frac{\partial h_i(x)}{\partial x_j}$.

2. Let g be an affine map of the form $g(x) = Ax + b$, for $A \in \mathbb{R}^{m,n}$, $b \in \mathbb{R}^m$. Show that the Jacobian of $h(x) = f(g(x))$ is

$$J_h(x) = J_f(g(x)) \cdot A.$$

3. Let g be an affine map as in the previous point, let $f : \mathbb{R}^n \to \mathbb{R}$ (a scalar-valued function), and let $h(x) = f(g(x))$. Show that

$$\begin{aligned} \nabla_x h(x) &= A^\top \nabla_g f(g(x)), \\ \nabla_x^2 h(x) &= A^\top \nabla_g^2 f(g(x)) A. \end{aligned}$$

Exercise 3.2 (Permutation matrices) A matrix $P \in \mathbb{R}^{n,n}$ is a permutation matrix if its columns are a permutation of the columns of the $n \times n$ identity matrix.

1. For an $n \times n$ matrix A, we consider the products PA and AP. Describe in simple terms what these matrices look like with respect to the original matrix A.

2. Show that P is orthogonal.

Exercise 3.3 (Linear maps) Let $f : \mathbb{R}^n \to \mathbb{R}^m$ be a linear map. Show how to compute the (unique) matrix A such that $f(x) = Ax$ for every $x \in \mathbb{R}^n$, in terms of the values of f at appropriate vectors, which you will determine.

Exercise 3.4 (Linear dynamical systems) Linear dynamical systems are a common way to (approximately) model the behavior of physical phenomena, via recurrence equations of the form[16]

$$x(t+1) = Ax(t) + Bu(t), \quad y(t) = Cx(t), \quad t = 0, 1, 2, \ldots,$$

where t is the (discrete) time, $x(t) \in \mathbb{R}^n$ describes the state of the system at time t, $u(t) \in \mathbb{R}^p$ is the input vector, and $y(t) \in \mathbb{R}^m$ is the output vector. Here, matrices A, B, C, are given.

[16] Such models are the focus of Chapter 15.

1. Assuming that the system has initial condition $x(0) = 0$, express the output vector at time T as a linear function of $u(0), \ldots, u(T-1)$; that is, determine a matrix H such that $y(T) = HU(T)$, where

$$U(T) \doteq \begin{bmatrix} u(0) \\ \vdots \\ u(T-1) \end{bmatrix}$$

contains all the inputs up to and including at time $T - 1$.

2. What is the interpretation of the range of H?

Exercise 3.5 (Nullspace inclusions and range) Let $A, B \in \mathbb{R}^{m,n}$ be two matrices. Show that the fact that the nullspace of B is contained in that of A implies that the range of $B^\top$ contains that of $A^\top$.

Exercise 3.6 (Rank and nullspace) Consider the image in Figure 3.17, a gray-scale rendering of a painting by Mondrian (1872–1944). We build a 256×256 matrix A of pixels based on this image by ignoring grey zones, assigning $+1$ to horizontal or vertical black lines, $+2$ at the intersections, and zero elsewhere. The horizontal lines occur at row indices 100, 200, and 230, and the vertical ones at column indices 50, 230.

1. What is the nullspace of the matrix?

2. What is its rank?

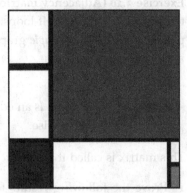

Figure 3.17 A gray-scale rendering of a painting by Mondrian.

Exercise 3.7 (Range and nullspace of $A^\top A$) Prove that, for any matrix $A \in \mathbb{R}^{m,n}$, it holds that

$$
\begin{aligned}
\mathcal{N}(A^\top A) &= \mathcal{N}(A), \\
\mathcal{R}(A^\top A) &= \mathcal{R}(A^\top).
\end{aligned}
\tag{3.16}
$$

Hint: use the fundamental theorem of linear algebra.

Exercise 3.8 (Cayley–Hamilton theorem) Let $A \in \mathbb{R}^{n,n}$ and let

$$
p(\lambda) \doteq \det(\lambda I_n - A) = \lambda^n + c_{n-1}\lambda^{n-1} + \cdots + c_1 \lambda + c_0
$$

be the characteristic polynomial of A.

1. Assume A is diagonalizable. Prove that A annihilates its own characteristic polynomial, that is

$$
p(A) = A^n + c_{n-1}A^{n-1} + \cdots + c_1 A + c_0 I_n = 0.
$$

Hint: use Lemma 3.3.

2. Prove that $p(A) = 0$ holds in general, i.e., also for non-diagonalizable square matrices. *Hint:* use the facts that polynomials are continuous functions, and that diagonalizable matrices are dense in $\mathbb{R}^{n,n}$, i.e., for any $\epsilon > 0$ there exist $\Delta \in \mathbb{R}^{n,n}$ with $\|\Delta\|_F \leq \epsilon$ such that $A + \Delta$ is diagonalizable.

Exercise 3.9 (Frobenius norm and random inputs) Let $A \in \mathbb{R}^{m,n}$ be a matrix. Assume that $u \in \mathbb{R}^n$ is a vector-valued random variable, with zero mean and covariance matrix I_n. That is, $\mathbb{E}\{u\} = 0$, and $\mathbb{E}\{uu^\top\} = I_n$.

1. What is the covariance matrix of the output, $y = Au$?

2. Define the total output variance as $\mathbb{E}\{\|y - \hat{y}\|_2^2\}$, where $\hat{y} = \mathbb{E}\{y\}$ is the output's expected value. Compute the total output variance and comment.

Exercise 3.10 (Adjacency matrices and graphs) For a given undirected graph G with no self-loops and at most one edge between any pair of nodes (i.e., a *simple* graph), as in Figure 3.18, we associate a $n \times n$ matrix A, such that

$$A_{ij} = \begin{cases} 1 & \text{if there is an edge between node } i \text{ and node } j, \\ 0 & \text{otherwise.} \end{cases}$$

This matrix is called the *adjacency* matrix of the graph.[17]

1. Prove the following result: for positive integer k, the matrix A^k has an interesting interpretation: the entry in row i and column j gives the number of *walks* of length k (i.e., a collection of k edges) leading from vertex i to vertex j. *Hint:* prove this by induction on k, and look at the matrix–matrix product $A^{k-1}A$.

2. A *triangle* in a graph is defined as a subgraph composed of three vertices, where each vertex is reachable from each other vertex (i.e., a triangle forms a complete subgraph of order 3). In the graph of Figure 3.18, for example, nodes $\{1, 2, 4\}$ form a triangle. Show that the number of triangles in G is equal to the trace of A^3 divided by 6. *Hint:* For each node in a triangle in an undirected graph, there are two walks of length 3 leading from the node to itself, one corresponding to a clockwise walk, and the other to a counter-clockwise walk.

Exercise 3.11 (Non-negative and positive matrices) A matrix $A \in \mathbb{R}^{n,n}$ is said to be *non-negative* (resp. *positive*) if $a_{ij} \geq 0$ (resp. $a_{ij} > 0$) for all $i, j = 1, \ldots, n$. The notation $A \geq 0$ (resp. $A > 0$) is used to denote non-negative (resp. positive) matrices.

A non-negative matrix is said to be column (resp. row) *stochastic*, if the sum of the elements along each column (resp. row) is equal to one, that is if $\mathbf{1}^\top A = \mathbf{1}^\top$ (resp. $A\mathbf{1} = \mathbf{1}$). Similarly, a vector $x \in \mathbb{R}^n$ is said to be non-negative if $x \geq 0$ (element-wise), and it is said to be a *probability vector*, if it is non-negative and $\mathbf{1}^\top x = 1$. The set of probability vectors in $\mathbb{R}^n$ is thus the set $S = \{x \in \mathbb{R}^n : x \geq 0, \mathbf{1}^\top x = 1\}$, which is called the *probability simplex*. The following points you are requested to prove are part of a body of results known as the Perron–Frobenius theory of non-negative matrices.

1. Prove that a non-negative matrix A maps non-negative vectors into non-negative vectors (i.e., that $Ax \geq 0$ whenever $x \geq 0$), and that a column stochastic matrix $A \geq 0$ maps probability vectors into probability vectors.

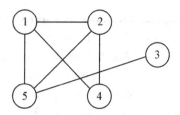

Figure 3.18 An undirected graph with $n = 5$ vertices.

[17] The graph in Figure 3.18 has adjacency matrix

$$A = \begin{bmatrix} 0 & 1 & 0 & 1 & 1 \\ 1 & 0 & 0 & 1 & 1 \\ 0 & 0 & 0 & 0 & 1 \\ 1 & 1 & 0 & 0 & 0 \\ 1 & 1 & 1 & 0 & 0 \end{bmatrix}.$$

2. Prove that if $A > 0$, then its spectral radius $\rho(A)$ is positive. *Hint:* use the Cayley–Hamilton theorem.

3. Show that it holds for any matrix A and vector x that

$$|Ax| \leq |A||x|,$$

where $|A|$ (resp. $|x|$) denotes the matrix (resp. vector) of moduli of the entries of A (resp. x). Then, show that if $A > 0$ and λ_i, v_i is an eigenvalue/eigenvector pair for A, then

$$|\lambda_i||v_i| \leq A|v_i|.$$

4. Prove that if $A > 0$ then $\rho(A)$ is actually an eigenvalue of A (i.e., A has a positive real eigenvalue $\lambda = \rho(A)$, and all other eigenvalues of A have modulus no larger than this "dominant" eigenvalue), and that there exist a corresponding eigenvector $v > 0$. Further, the dominant eigenvalue is simple (i.e., it has unit algebraic multiplicity), but you are not requested to prove this latter fact.

   Hint: For proving this claim you may use the following fixed-point theorem due to Brouwer: *if S is a compact and convex set*[18] *in $\mathbb{R}^n$, and $f : S \to S$ is a continuous map, then there exist an $x \in S$ such that $f(x) = x$.* Apply this result to the continuous map $f(x) \doteq \frac{Ax}{1^\top Ax}$, with S being the probability simplex (which is indeed convex and compact).

   [18] See Section 8.1 for definitions of compact and convex sets.

5. Prove that if $A > 0$ and it is column or row stochastic, then its dominant eigenvalue is $\lambda = 1$.

4

Symmetric matrices

THIS CHAPTER IS DEVOTED TO symmetric matrices and to their special properties. A fundamental result, the spectral theorem, shows that we can decompose any symmetric matrix as a three-term product of matrices, involving an orthogonal matrix and a real diagonal matrix. The theorem has a direct implication for quadratic functions, as it allows us to decompose any quadratic function having no linear or constant terms into a weighted sum of squared linear functions involving vectors that are mutually orthogonal.

The spectral theorem also allows us to determine when a quadratic function enjoys an important property called *convexity*. Next, we provide a characterization of the eigenvalues of symmetric matrices in terms of optimal values of certain quadratic optimization problems (variational characterization). We shall further discuss a special class of symmetric matrices, called positive semidefinite matrices, that play a relevant role in optimization models. Finally, we show that many important properties of a matrix $A \in \mathbb{R}^{m,n}$, such as the range, the nullspace, the Frobenius, and the spectral norms, can be studied by analyzing the related symmetric matrices $A^\top A$ and $AA^\top$. This observation naturally leads to the topic of singular value decomposition (SVD), that will be treated in Chapter 5.

4.1 Basics

4.1.1 Definitions and examples

A square matrix $A \in \mathbb{R}^{n,n}$ is *symmetric* if it is equal to its transpose: $A = A^\top$, that is: $A_{ij} = A_{ji}, 1 \le i, j \le n$. Elements above the diagonal in a symmetric matrix are thus identical to corresponding elements below the diagonal. Symmetric matrices are ubiquitous in engineering applications. They arise, for instance, in the description of graphs with undirected weighted edges between the nodes, in geometric distance arrays (between, say, cities), in defining the Hessian of a nonlinear function, in describing the covariances of random vectors, etc. The following is an example of a 3×3 symmetric matrix:

$$A = \begin{bmatrix} 4 & 3/2 & 2 \\ 3/2 & 2 & 5/2 \\ 2 & 5/2 & 2 \end{bmatrix}.$$

The set of symmetric $n \times n$ matrices is a subspace of $\mathbb{R}^{n,n}$, and it is denoted by $\mathbb{S}^n$.

Example 4.1 (*Edge weight and Laplacian matrices of a graph*) Consider an (undirected) graph consisting of $m = 6$ nodes connected by $n = 9$ arcs (or edges), as in Figure 4.1.

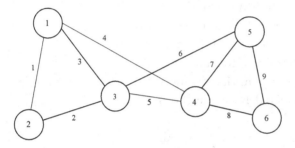

Figure 4.1 An undirected graph with $m = 6$ nodes and $n = 9$ edges.

If we assume that each undirected edge between nodes i and j is assigned a weight $w_{ij} = w_{ji}$, $1 \leq i, j \leq m$, we obtain a symmetric matrix $W \in \mathbb{S}^m$ of edge weights. In the example in Figure 4.1, assuming all weights are equal to one, we would obtain

$$W = \begin{bmatrix} 0 & 1 & 1 & 1 & 0 & 0 \\ 1 & 0 & 1 & 0 & 0 & 0 \\ 1 & 1 & 0 & 1 & 1 & 0 \\ 1 & 0 & 1 & 0 & 1 & 1 \\ 0 & 0 & 1 & 1 & 0 & 1 \\ 0 & 0 & 0 & 1 & 1 & 0 \end{bmatrix}.$$

Also, the *Laplacian* matrix of a graph is defined as an $m \times m$ symmetric matrix

$$[L]_{ij} = \begin{cases} \text{number of arcs incident to node } i & \text{if } i = j, \\ -1 & \text{if there is an arc joining node } i \text{ to } j, \\ 0 & \text{otherwise.} \end{cases}$$

Several key properties of a graph are related to the Laplacian matrix. If the graph has no self-loops and only one edge between any pair of nodes, then the Laplacian matrix is related to the (directed) node-arc incidence matrix A of any orientation of the graph,[1] as

[1] By an *orientation* of an undirected graph we indicate a directed graph obtained from an undirected graph by choosing some orientation of the edges.

$$L = AA^\top \in S^m.$$

For the example in Figure 4.1, we have

$$L = \begin{bmatrix} 3 & -1 & -1 & -1 & 0 & 0 \\ -1 & 2 & -1 & 0 & 0 & 0 \\ -1 & -1 & 4 & -1 & -1 & 0 \\ -1 & 0 & -1 & 4 & -1 & -1 \\ 0 & 0 & -1 & -1 & 3 & -1 \\ 0 & 0 & 0 & -1 & -1 & 2 \end{bmatrix}.$$

Example 4.2 (*Sample covariance matrix*) Given m points $x^{(1)}, \ldots, x^{(m)}$ in $\mathbb{R}^n$, we define the *sample covariance matrix* to be the $n \times n$ symmetric matrix

$$\Sigma \doteq \frac{1}{m} \sum_{i=1}^{m} (x^{(i)} - \hat{x})(x^{(i)} - \hat{x})^\top,$$

where $\hat{x} \in \mathbb{R}^n$ is the sample average of the points:

$$\hat{x} \doteq \frac{1}{m} \sum_{i=1}^{m} x^{(i)}.$$

The covariance matrix Σ is obviously a symmetric matrix. This matrix arises when computing the sample variance of the scalar products $s_i \doteq w^\top x^{(i)}$, $i = 1, \ldots, m$, where $w \in \mathbb{R}^n$ is a given vector. Indeed, the sample average of vector s is

$$\hat{s} = \frac{1}{m}(s_1 + \cdots + s_m) = w^\top \hat{x},$$

while the sample variance is

$$\sigma^2 = \sum_{i=1}^{m} (w^\top x^{(i)} - \hat{s})^2 = \sum_{i=1}^{m} (w^\top (x^{(i)} - \hat{x}))^2 = w^\top \Sigma w.$$

Example 4.3 (*Portfolio variance*) For n financial assets, we can define a vector $r \in \mathbb{R}^n$ whose components r_k are the rate of returns of the k-th asset, $k = 1, \ldots, n$; see Example 2.6. Assume now that we have observed m samples of historical returns $r^{(i)}$, $i = 1, \ldots, m$. The sample average over that history of return is $\hat{r} = (1/m)(r^{(1)} + \cdots + r^{(m)})$, and the sample covariance matrix has (i,j) component given by

$$\Sigma_{ij} = \frac{1}{m} \sum_{t=1}^{m} (r_i^{(t)} - \hat{r}_i)(r_j^{(t)} - \hat{r}_j), \ \ 1 \le i, \ j \le n.$$

If $w \in \mathbb{R}^n$ represents a portfolio "mix," that is $w_k \ge 0$ is the fraction of the total wealth invested in asset k, then the return of such a portfolio is given by $\rho = r^\top w$. The sample average of the portfolio return is $\hat{r}^\top w$, while the sample variance is given by $w^\top \Sigma w$.

Example 4.4 (*Hessian matrix of a function*) The Hessian of a twice differentiable function $f : \mathbb{R}^n \to \mathbb{R}$ at a point $x \in \text{dom } f$ is the matrix containing the second derivatives of the function at that point. That is, the Hessian is the matrix with elements given by

$$H_{ij} = \frac{\partial^2 f(x)}{\partial x_i \partial x_j}, \quad 1 \leq i, \; j \leq n.$$

The Hessian of f at x is often denoted by $\nabla^2 f(x)$. Since the second derivative is independent of the order in which derivatives are taken, it follows that $H_{ij} = H_{ji}$ for every pair (i,j), thus the Hessian is always a symmetric matrix.

Hessian of a quadratic function. Consider the quadratic function (a polynomial function is said to be quadratic if the maximum degree of its monomials is equal to two)

$$q(x) = x_1^2 + 2x_1 x_2 + 3x_2^2 + 4x_1 + 5x_2 + 6.$$

The Hessian of q at x is given by

$$H = \left[\frac{\partial^2 q(x)}{\partial x_i \partial x_j} \right]_{1 \leq i,j \leq 2} = \left[\begin{array}{cc} \frac{\partial^2 q}{\partial x_1^2} & \frac{\partial^2 q}{\partial x_1 \partial x_2} \\ \frac{\partial^2 q}{\partial x_2 \partial x_1} & \frac{\partial^2 q}{\partial x_2^2} \end{array} \right] = \left[\begin{array}{cc} 2 & 2 \\ 2 & 6 \end{array} \right].$$

For quadratic functions, the Hessian is a constant matrix, that is, it does not depend on the point x at which it is evaluated. The monomials in $q(x)$ of degree two can also be written compactly as

$$x_1^2 + 2x_1 x_2 + 3x_2^2 = \frac{1}{2} x^\top H x.$$

Therefore, the quadratic function can be written as the sum of a quadratic term involving the Hessian, and an affine term:

$$q(x) = \frac{1}{2} x^\top H x + c^\top x + d, \quad c^\top = [4\ 5], \; d = 6.$$

Hessian of the log-sum-exp function. Consider the log-sum-exp function $\text{lse} : \mathbb{R}^n \to \mathbb{R}$, taking values

$$\text{lse}(x) = \ln \sum_{i=1}^n e^{x_i}.$$

The Hessian of this function can be determined as follows. First, we determine the gradient at a point x, as done in Example 2.14:

$$\nabla \text{lse}(x) = \frac{1}{Z} z,$$

where $z \doteq [e^{x_1} \cdots e^{x_n}]$, and $Z \doteq \sum_{i=1}^n z_i$. Then, the Hessian at a point x is obtained by taking derivatives of each component of the gradient. If $g_i(x)$ is the i-th component of the gradient, that is,

$$g_i(x) = \frac{\partial f(x)}{\partial x_i} = \frac{e^{x_i}}{\sum_{i=1}^n e^{x_i}} = \frac{z_i}{Z},$$

then

$$\frac{\partial g_i(x)}{\partial x_i} = \frac{z_i}{Z} - \frac{z_i^2}{Z^2},$$

and, for $j \neq i$:

$$\frac{\partial g_i(x)}{\partial x_j} = -\frac{z_i z_j}{Z^2}.$$

More compactly:

$$\nabla^2 \mathrm{lse}(x) = \frac{1}{Z^2} \left(Z \operatorname{diag}(z) - zz^\top \right).$$

Example 4.5 (*Projections and the Gram matrix*) Suppose we are given d linearly independent vectors $x^{(1)}, \ldots, x^{(d)}$ in $\mathbb{R}^n$, and a vector $x \in \mathbb{R}^n$. In Section 2.3.2.3 we considered the problem of computing the projection x^* of x onto the subspace spanned by $x^{(1)}, \ldots, x^{(d)}$. Such a projection can be computed as

$$x^* = X\alpha, \quad X = [x^{(1)} \cdots x^{(d)}],$$

where $\alpha \in \mathbb{R}^d$ is a vector of coefficients that must satisfy the so-called Gram system of linear equations (2.8)

$$\begin{bmatrix} x^{(1)\top} x^{(1)} & \cdots & x^{(1)\top} x^{(d)} \\ \vdots & \ddots & \vdots \\ x^{(d)\top} x^{(1)} & \cdots & x^{(d)\top} x^{(d)} \end{bmatrix} \begin{bmatrix} \alpha_1 \\ \vdots \\ \alpha_d \end{bmatrix} = \begin{bmatrix} x^{(1)\top} x \\ \vdots \\ x^{(d)\top} x \end{bmatrix}.$$

The right-hand term in these equations is a vector containing the components of x along the directions $x^{(1)}, \ldots, x^{(d)}$, and the coefficient matrix appearing in the left-hand side of these equation is a symmetric matrix called the *Gram matrix*: $G = X^\top X \in \mathbb{S}^n$.

4.1.2 Quadratic functions

A quadratic function $q : \mathbb{R}^n \to \mathbb{R}$ is a second-order multivariate polynomial in x, that is a function containing a linear combination of all possible monomials of degree at most two. Such a function can hence be written as

$$q(x) = \sum_{i=1}^n \sum_{j=1}^n a_{ij} x_i x_j + \sum_{i=1}^n c_i x_i + d,$$

where a_{ij} are the coefficients of the monomials of degree two $x_i x_j$, c_i are the coefficients of the monomials of degree one x_i, and d is the zero-degree (constant) term. The above expression has a more compact representation in matrix format as

$$q(x) = x^\top A x + c^\top x + d,$$

where $A \in \mathbb{R}^{n,n}$ is a matrix containing in row i and column j the coefficient a_{ij}, and c is the vector of c_i coefficients. Notice that, since

$x^\top Ax$ is a scalar it is equal to its transpose, hence it holds that $x^\top Ax = x^\top A^\top x$, thus

$$x^\top Ax = \frac{1}{2}x^\top (A + A^\top)x,$$

where $H = A + A^\top$ is a symmetric matrix. A generic quadratic function can thus be represented as

$$q(x) = \frac{1}{2}x^\top Hx + c^\top x + d = \frac{1}{2}\begin{bmatrix} x \\ 1 \end{bmatrix}^\top \begin{bmatrix} H & c \\ c^\top & 2d \end{bmatrix}\begin{bmatrix} x \\ 1 \end{bmatrix}, \quad (4.1)$$

where $H \in \mathbb{S}^n$. A *quadratic form* is a quadratic function with no linear and constant terms, that is $c = 0, d = 0$:

$$q(x) = \frac{1}{2}x^\top Hx, \quad H \in \mathbb{S}^n.$$

Note that the Hessian of a quadratic function $q(x)$ is constant: $\nabla^2 q(x) = H$.

A generic, twice differentiable function $f : \mathbb{R}^n \to \mathbb{R}$ can be locally approximated in the neighborhood of a point x_0 via a quadratic function, by means of Taylor series expansion, see Section 3.2.2:

$$\begin{aligned} f(x) &\simeq q(x) \\ &= f(x_0) + \nabla f(x_0)^\top (x - x_0) + \frac{1}{2}(x - x_0)^\top \nabla^2 f(x_0)(x - x_0), \end{aligned}$$

where $|f(x) - q(x)|$ goes to zero faster than second order as $x \to x_0$. Here, the approximating quadratic function has the standard format (4.1), with $H = \nabla^2 f(x_0)$, $c = \nabla f(x_0) - \nabla^2 f(x_0)x_0$, $d = f(x_0) - \nabla f(x_0)^\top x_0 + \frac{1}{2}x_0^\top \nabla^2 f(x_0)x_0$.

Two special cases: diagonal matrices and dyads. Let $a = [a_1 \cdots a_n]^\top$, then a diagonal matrix

$$A = \mathrm{diag}\,(a) = \begin{bmatrix} a_1 & 0 & \cdots & 0 \\ 0 & a_2 & \cdots & 0 \\ \vdots & \vdots & \ddots & \vdots \\ 0 & \cdots & 0 & a_n \end{bmatrix}.$$

is a special case of a symmetric matrix. The quadratic form associated with $\mathrm{diag}\,(a)$ is

$$q(x) = x\,\mathrm{diag}\,(a)\,x = \sum_{i=1}^{n} a_i x_i^2,$$

that is, $q(x)$ is a linear combination of pure squares x_i^2 (i.e., no cross product terms of type $x_i x_j$ appear in the sum).

Another important class of symmetric matrices is that formed by symmetric *dyads*, that is by vector products of the form

$$A = aa^\top = \begin{bmatrix} a_1^2 & a_1 a_2 & \cdots & a_1 a_n \\ a_2 a_1 & a_2^2 & \cdots & a_2 a_n \\ \vdots & \vdots & \ddots & \vdots \\ a_n a_1 & \cdots & \cdots & a_n^2 \end{bmatrix}.$$

Dyads are matrices of rank one, and the quadratic form associated with a dyad has the form

$$q(x) = x^\top aa^\top x = (a^\top x)^2,$$

that is, it is the square of a linear form in x. It follows that the quadratic form associated with a dyad is always non-negative: $q(x) \geq 0$, for all x.

4.2 The spectral theorem

4.2.1 Eigenvalue decomposition of symmetric matrices

We recall the definition of eigenvalues and eigenvectors of a square matrix from Section 3.3. Let A be an $n \times n$ matrix. A scalar λ is said to be an eigenvalue of A if there exists a nonzero vector u such that

$$Au = \lambda u.$$

The vector u is then referred to as an eigenvector associated with the eigenvalue λ. The eigenvector u is said to be normalized if $\|u\|_2 = 1$. In this case, we have[2] $u^\star Au = \lambda u^\star u = \lambda$.

The interpretation of u is that it defines a direction along which the linear map defined by A behaves just like scalar multiplication. The amount of scaling is given by λ.

The eigenvalues of A are the roots of the characteristic polynomial

$$p_A(\lambda) = \det(\lambda I - A).$$

That is, the eigenvalues λ_i, $i = 1, \ldots, n$, are the roots (counting multiplicities) of the polynomial equation of degree n: $p_A(\lambda) = 0$. For a generic matrix $A \in \mathbb{R}^{n,n}$, the eigenvalues λ_i can thus be real and/or complex (in complex conjugate pairs), and likewise the corresponding eigenvectors can be real or complex. However, the situation is different in the case of symmetric matrices: symmetric matrices have real eigenvalues and eigenvectors and, moreover, for each distinct eigenvalue λ_i, the dimension of the eigenspace $\phi_i = \mathcal{N}(\lambda_i I_n - A)$ coincides with the algebraic multiplicity of that eigenvalue. This is summarized in the following key theorem.

[2] Recall that the superscript $\star$ denotes the Hermitian conjugate of a vector/matrix, obtained by transposing and taking the conjugate; if the vector/matrix is real, then the Hermitian conjugate simply coincides with the transpose.

Theorem 4.1 (Eigendecomposition of a symmetric matrix) *Let $A \in \mathbb{R}^{n,n}$ be symmetric, let λ_i, $i = 1, \ldots, k \leq n$, be the distinct eigenvalues of A. Let further μ_i, $i = 1, \ldots, k$, denote the algebraic multiplicity of λ_i (the multiplicity of λ_i as a root of the characteristic polynomial), and let $\phi_i = \mathcal{N}(\lambda_i I_n - A)$. Then, for all $i = 1, \ldots, k$:*

1. $\lambda_i \in \mathbb{R}$;

2. $\phi_i \perp \phi_j$, $i \neq j$;

3. $\dim \phi_i = \mu_i$.

Proof

Part 1. Let λ, u be any eigenvalue/eigenvector pair for A. Then

$$Au = \lambda u$$

and, by taking the Hermitian conjugate of both sides,

$$u^\star A^\star = \lambda^\star u^\star.$$

Multiplying the first of the previous equations on the left by $u^\star$ and the second on the right by u, we have

$$u^\star A u = \lambda u^\star u, \quad u^\star A^\star u = \lambda^\star u^\star u.$$

Since $u^\star u = \|u\|_2^2 \neq 0$, recalling that A real implies that $A^\star = A^\top$, and subtracting the two equalities in the previous expression it follows that

$$u^\star (A - A^\top) u = (\lambda - \lambda^\star) \|u\|_2.$$

Now, $A - A^\top = 0$, since A is symmetric, hence it must be that

$$\lambda - \lambda^\star = 0,$$

which implies that λ must be a real number. Notice that also the associated eigenvector u can always be chosen to be real. Indeed, if a complex u satisfies $Au = \lambda u$, with A, λ real, then we also have that $\text{Re}(Au) = A\text{Re}(u) = \lambda \text{Re}(u)$, which means that $\text{Re}(u)$ is an eigenvector of A associated with λ.

Part 2. Let $v_i \in \phi_i$, $v_j \in \phi_j$, $i \neq j$. Since $Av_i = \lambda_i v_i$, $Av_j = \lambda_j v_j$, we have

$$v_j^\top A v_i = \lambda_i v_j^\top v_i$$

and

$$v_j^\top A v_i = v_i^\top A^\top v_j = v_i^\top A v_j = \lambda_j v_i^\top v_j = \lambda_j v_j^\top v_i.$$

Thus, subtracting these two equalities, we get

$$(\lambda_i - \lambda_j) v_j^\top v_i = 0.$$

Since $\lambda_i \neq \lambda_j$ by hypothesis, it must be that $v_j^\top v_i = 0$, i.e., v_j, v_i are orthogonal.

Part 3. Let λ be an eigenvalue of A, let $\mu \geq 1$ be its algebraic multiplicity, and let ν be the dimension of $\phi = \mathcal{N}(\lambda I_n - A)$. We know that, in general, $\nu \leq \mu$, that is, the geometric multiplicity (i.e., the dimension of the eigenspace) is no larger than the algebraic multiplicity, see Section 3.3.5.

We next prove that for symmetric matrices it actually holds that $\nu = \mu$, by constructing an orthonormal basis for ϕ composed of μ elements.

To this end, we state a preliminary result. *Let B be a symmetric $m \times m$ matrix, and let λ be an eigenvalue of B. Then, there exists an orthogonal matrix $U = [u\ Q] \in \mathbb{R}^{m,m}$, $Q \in \mathbb{R}^{m,m-1}$, such that $Bu = \lambda u$ and*

$$U^\top B U = \begin{bmatrix} \lambda & 0 \\ 0 & B_1 \end{bmatrix}, \quad B_1 = Q^\top B Q \in \mathbb{S}^{m-1}. \tag{4.2}$$

To see this fact, let u be any unit-norm eigenvector of B associated with λ, i.e., $Bu = \lambda u$, $\|u\|_2 = 1$. Proceeding similarly to the reasoning done before Eq. (3.6), we can take Q to be a matrix containing by columns an orthonormal basis of the orthogonal complement of $\mathcal{R}(u)$, so that $U = [u\ Q]$ is by construction an orthogonal matrix: $U^\top U = I_m$. Equation (4.2) then follows from the fact that $Bu = \lambda u$, $u^\top B = \lambda u^\top$, and that $u^\top Q = Q^\top u = 0$.

We now first apply this result to $A \in \mathbb{S}^n$: since $\mu \geq 1$, there exists an orthogonal matrix $U_1 = [u_1\ Q_1] \in \mathbb{R}^{n,n}$, such that $Au_1 = \lambda u_1$, and

$$U_1^\top A U_1 = \begin{bmatrix} \lambda & 0 \\ 0 & A_1 \end{bmatrix}, \quad A_1 = Q_1^\top A Q_1 \in \mathbb{S}^{n-1}.$$

Now, if $\mu = 1$ we have finished the proof, since we found a subspace of ϕ of dimension one (the subspace is $\mathcal{R}(u_1)$). If instead $\mu > 1$ then, due to the block diagonal structure of $U_1^\top A U_1$ and to the fact that this matrix is similar to A, we conclude that λ is an eigenvalue of A_1 of multiplicity $\mu - 1$. We hence apply the same reasoning to the symmetric matrix $A_1 \in \mathbb{S}^{n-1}$: there exist an orthogonal matrix $U_2 = [\tilde{u}_2\ Q_2] \in \mathbb{R}^{n-1,n-1}$ such that $A_1 \tilde{u}_2 = \lambda \tilde{u}_2$, $\|\tilde{u}_2\|_2 = 1$, and

$$U_2^\top A_1 U_2 = \begin{bmatrix} \lambda & 0 \\ 0 & A_2 \end{bmatrix}, \quad A_2 = Q_2^\top A Q_2 \in \mathbb{S}^{n-2}.$$

We next show that the vector

$$u_2 = U_1 \begin{bmatrix} 0 \\ \tilde{u}_2 \end{bmatrix}$$

is a unit-norm eigenvector of A, and it is orthogonal to u_1. Indeed,

$$
\begin{aligned}
Au_2 &= U_1 \begin{bmatrix} \lambda & 0 \\ 0 & A_1 \end{bmatrix} U_1^\top U_1 \begin{bmatrix} 0 \\ \tilde{u}_2 \end{bmatrix} = U_1 \begin{bmatrix} 0 \\ A_1 \tilde{u}_2 \end{bmatrix} \\
&= U_1 \begin{bmatrix} 0 \\ \lambda \tilde{u}_2 \end{bmatrix} = \lambda u_2.
\end{aligned}
$$

Moreover,

$$
\|u_2\|^2 = u_2^\top u_2 = \begin{bmatrix} 0 \\ \tilde{u}_2 \end{bmatrix}^\top U_1^\top U_1 \begin{bmatrix} 0 \\ \tilde{u}_2 \end{bmatrix} = \|\tilde{u}_2\|^2 = 1,
$$

and

$$
u_1^\top u_2 = u_1^\top [u_1 \; Q_1] \begin{bmatrix} 0 \\ \tilde{u}_2 \end{bmatrix} = 0,
$$

hence u_2 is orthogonal to u_1. If $\mu = 2$, then the proof is finished, since we have found an orthonormal basis of dimension two for ϕ (such a basis is u_1, u_2). If otherwise $\mu > 2$, we iterate the same reasoning on matrix A_2 and find an eigenvector u_3 orthogonal to u_1, u_2, etc. We can continue in this way until we reach the actual value of μ, and at this point we exit the procedure with an orthonormal basis of ϕ composed of exactly μ vectors.   □

4.2.2   The spectral theorem

Putting together Theorem 4.1 and Theorem 3.4 it is easy to conclude that any symmetric matrix is orthogonally similar to a diagonal matrix. This is stated in the following so-called *spectral theorem* for symmetric matrices.

Theorem 4.2 (Spectral theorem) *Let $A \in \mathbb{R}^{n,n}$ be symmetric, let $\lambda_i \in \mathbb{R}$, $i = 1, \ldots, n$, be the eigenvalues of A (counting multiplicities). Then, there exists a set of orthonormal vectors $u_i \in \mathbb{R}^n$, $i = 1, \ldots, n$, such that $Au_i = \lambda_i u_i$. Equivalently, there exists an orthogonal matrix $U = [u_1 \cdots u_n]$ (i.e., $UU^\top = U^\top U = I_n$) such that*

$$
A = U \Lambda U^\top = \sum_{i=1}^{n} \lambda_i u_i u_i^\top, \quad \Lambda = \mathrm{diag}\,(\lambda_1, \ldots, \lambda_n).
$$

The spectral theorem also shows that any symmetric matrix can be decomposed as a weighted sum of simple rank-one matrices (dyads) of the form $u_i u_i^\top$, where the weights are given by the eigenvalues λ_i.

Example 4.6 (*Eigenvalue decomposition of a symmetric 2×2 matrix*) We give a practical numerical example of the spectral theorem. Let

$$
A = \begin{bmatrix} 3/2 & -1/2 \\ -1/2 & 3/2 \end{bmatrix}.
$$

To determine the eigenvalues, we solve the characteristic equation:

$$0 = \det(\lambda I - A) = (\lambda - 3/2)^2 - (1/4) = (\lambda - 1)(\lambda - 2),$$

hence the eigenvalues are $\lambda_1 = 1$, $\lambda_2 = 2$. For each eigenvalue λ_i, we look for a corresponding unit-norm vector u_i such that $Au_i = \lambda u_i$. For λ_1, we obtain the equation in u_1

$$0 = (A - \lambda_1)u_1 = \begin{bmatrix} 1/2 & -1/2 \\ -1/2 & 1/2 \end{bmatrix} u_1,$$

which leads (after normalization) to the eigenvector $u_1 = (1/\sqrt{2})[1 \ 1]^\top$. Similarly, for λ_2 we obtain the eigenvector $u_2 = (1/\sqrt{2})[1 \ -1]^\top$. Hence, A admits the factorization

$$A = \frac{1}{\sqrt{2}} \begin{bmatrix} 1 & 1 \\ 1 & -1 \end{bmatrix}^\top \begin{bmatrix} 1 & 0 \\ 0 & 2 \end{bmatrix} \frac{1}{\sqrt{2}} \begin{bmatrix} 1 & 1 \\ 1 & -1 \end{bmatrix}.$$

4.3 Spectral decomposition and optimization

In this section, we illustrate how the spectral decomposition of symmetric matrices can be used to solve very specific types of optimization problems, namely those that involve the maximization or minimization of quadratic forms[3] over the Euclidean ball.

[3] Recall that a quadratic function is a quadratic form if it has quadratic terms but no linear or constant terms.

4.3.1 Variational characterization of eigenvalues

We begin with expressing the eigenvalues of a symmetric matrix as optimal values of certain optimization problems. Since the eigenvalues of $A \in S^n$ are real, we can arrange them in decreasing order:[4]

$$\lambda_{\max}(A) = \lambda_1(A) \geq \lambda_2(A) \geq \cdots \geq \lambda_n(A) = \lambda_{\min}(A).$$

[4] We shall maintain throughout the book this ordering convention for the eigenvalues of symmetric matrices.

The extreme eigenvalues can be related to the minimum and the maximum attained by the quadratic form induced by A over the unit Euclidean sphere. For $x \neq 0$ the ratio

$$\frac{x^\top A x}{x^\top x}$$

is called a *Rayleigh quotient*. The following theorem holds.

Theorem 4.3 (Rayleigh quotients) *Given $A \in S^n$, it holds that*

$$\lambda_{\min}(A) \leq \frac{x^\top A x}{x^\top x} \leq \lambda_{\max}(A), \quad \forall x \neq 0.$$

Moreover,

$$\lambda_{\max}(A) = \max_{x:\ \|x\|_2 = 1} x^\top A x,$$

$$\lambda_{\min}(A) = \min_{x:\ \|x\|_2 = 1} x^\top A x,$$

and the maximum and minimum are attained for $x = u_1$ and for $x = u_n$, respectively, where u_1 (resp. u_n) is the unit-norm eigenvector of A associated with its largest (resp. smallest) eigenvalue of A.

Proof The proof is based on the spectral theorem for symmetric matrices and on the invariance of the Euclidean norm under orthogonal transformations. Let $A = U\Lambda U^\top$ be the spectral factorization of A, where Λ contains the ordered eigenvalues on the diagonal, and U is orthogonal. Defining $\bar{x} \doteq U^\top x$, we have

$$
\begin{aligned}
x^\top A x &= x^\top U \Lambda U^\top x = \bar{x}^\top \Lambda \bar{x} \\
&= \sum_{i=1}^{n} \lambda_i \bar{x}_i^2.
\end{aligned}
$$

Now,

$$
\lambda_{\min} \sum_{i=1}^{n} \bar{x}_i^2 \leq \sum_{i=1}^{n} \lambda_i \bar{x}_i^2 \leq \lambda_{\max} \sum_{i=1}^{n} \bar{x}_i^2,
$$

that is, considering that $\sum_{i=1}^{n} \bar{x}_i^2 = \|\bar{x}\|_2^2 = \|U^\top x\|_2^2 = \|x\|_2^2$,

$$
\lambda_{\min} \|x\|_2^2 \leq x^\top A x \leq \lambda_{\max} \|x\|_2^2,
$$

from which the first claim follows. Moreover, it is easy to check that the upper and the lower bound in the above inequalities are actually attained for $x = u_1$ (the first column of U) and $x = u_n$ (the last column of U), respectively, thus concluding the proof. □

4.3.2 Minimax principle

Theorem 4.3 is actually a particular case of a more general principle called the *minimax* principle for eigenvalues of symmetric matrices. Let us first state the following result.

Theorem 4.4 (Poincaré inequality) *Let $A \in \mathbb{S}^n$ and let $\mathcal{V}$ be any k-dimensional subspace of $\mathbb{R}^n$, $1 \leq k \leq n$. Then, there exist vectors $x, y \in \mathcal{V}$, with $\|x\|_2 = \|y\|_2 = 1$, such that*

$$
x^\top A x \leq \lambda_k(A), \quad y^\top A y \geq \lambda_{n-k+1}(A).
$$

Proof Let $A = U\Lambda U^\top$ be the spectral factorization of A, and denote by $\mathcal{Q} = \mathcal{R}(U_k)$ the subspace spanned by the columns of $U_k = [u_k \cdots u_n]$. Since $\mathcal{Q}$ has dimension $n - k + 1$ and $\mathcal{V}$ has dimension k, the intersection $\mathcal{V} \cap \mathcal{Q}$ must be nonempty (since otherwise the direct sum $\mathcal{Q} \oplus \mathcal{V}$ would have a dimension larger than n, the dimension of the embedding space). Take then a unit-norm vector $x \in \mathcal{V} \cap \mathcal{Q}$.

Then $x = U_k \xi$, for some ξ with $\|\xi\|_2 = 1$, hence

$$
\begin{aligned}
x^\top A x &= \xi^\top U_k^\top U \Lambda U^\top U_k \xi = \sum_{i=k}^{n} \lambda_i(A) \xi_i^2 \\
&\leq \lambda_k(A) \sum_{i=k}^{n} \xi_i^2 = \lambda_k(A),
\end{aligned}
$$

which proves the first statement. The second one can be proved analogously, by applying the same reasoning to $-A$. $\qquad \square$

From the Poincaré inequality follows the minimax principle stated next, also known as *variational characterization* of the eigenvalues.

Corollary 4.1 (Minimax principle) *Let $A \in S^n$ and let $\mathcal{V}$ denote a subspace of $\mathbb{R}^n$. Then, for $k \in \{1, \ldots, n\}$ it holds that*

$$
\begin{aligned}
\lambda_k(A) &= \max_{\dim \mathcal{V} = k} \ \min_{x \in \mathcal{V}, \|x\|_2 = 1} x^\top A x \\
&= \min_{\dim \mathcal{V} = n-k+1} \ \max_{x \in \mathcal{V}, \|x\|_2 = 1} x^\top A x.
\end{aligned}
$$

Proof From the Poincaré inequality, if $\mathcal{V}$ is any k-dimensional subspace of $\mathbb{R}^n$ then $\min_{x \in \mathcal{V}, \|x\|_2 = 1} x^\top A x \leq \lambda_k(A)$. In particular, if we take $\mathcal{V}$ to be the span of $\{u_1, \ldots, u_k\}$ then equality is achieved, which proves the first statement. The second statement follows by applying the first one to matrix $-A$. $\qquad \square$

Example 4.7 (*Matrix gain*) Given a matrix $A \in \mathbb{R}^{m,n}$, let us consider the linear function associated with A, which maps input vectors $x \in \mathbb{R}^n$ to output vectors $y \in \mathbb{R}^m$:
$$
y = Ax.
$$

Given a vector norm, the matrix *gain*, or operator norm, is defined as the maximum value of the ratio $\|Ax\|/\|x\|$ between the size (norm) of the output and that of the input, see Section 3.6. In particular, the gain with respect to the Euclidean norm is defined as

$$
\|A\|_2 = \max_{x \neq 0} \frac{\|Ax\|_2}{\|x\|_2},
$$

and it is often referred to as the *spectral* norm of A. The square of the input–output ratio in the Euclidean norm is

$$
\frac{\|Ax\|_2^2}{\|x\|_2^2} = \frac{x^\top (A^\top A) x}{x^\top x}.
$$

In view of Theorem 4.3, we see that this quantity is upper and lower bounded by the maximum and the minimum eigenvalue of the symmetric matrix $A^\top A \in S^n$, respectively:

$$
\lambda_{\min}(A^\top A) \leq \frac{\|Ax\|_2^2}{\|x\|_2^2} \leq \lambda_{\max}(A^\top A)
$$

(notice incidentally that all eigenvalues of $A^\top A$, $\lambda_i(A^\top A)$, $i = 1, \ldots, n$ are non-negative, since $A^\top A$ is a positive-semidefinite matrix, as discussed next in Section 4.4). We also know from Theorem 4.3 that the upper and lower bounds are actually attained when x is equal to an eigenvector of $A^\top A$ corresponding respectively to the maximum and the minimum eigenvalues of $A^\top A$. Therefore,

$$\|A\|_2 = \max_{x \neq 0} \frac{\|Ax\|_2}{\|x\|_2} = \sqrt{\lambda_{\max}(A^\top A)}, \qquad (4.3)$$

where this maximum gain is obtained for x along the direction of eigenvector u_1 of $A^\top A$, and

$$\min_{x \neq 0} \frac{\|Ax\|_2}{\|x\|_2} = \sqrt{\lambda_{\min}(A^\top A)},$$

where this minimum gain is obtained for x along the direction of eigenvector u_n of $A^\top A$.

One important consequence of the minimax property is the following result comparing the ordered eigenvalues of A, B with those of $A + B$.

Corollary 4.2 *Let $A, B \in \mathbb{S}^n$. Then, for each $k = 1, \ldots, n$, we have*

$$\lambda_k(A) + \lambda_{\min}(B) \leq \lambda_k(A + B) \leq \lambda_k(A) + \lambda_{\max}(B). \qquad (4.4)$$

Proof From Corollary 4.1 we have that

$$\begin{aligned}
\lambda_k(A + B) &= \min_{\dim \mathcal{V} = n-k+1} \max_{x \in \mathcal{V}, \|x\|_2 = 1} (x^\top A x + x^\top B x) \\
&\geq \min_{\dim \mathcal{V} = n-k+1} \max_{x \in \mathcal{V}, \|x\|_2 = 1} x^\top A x + \lambda_{\min}(B) \\
&= \lambda_k(A) + \lambda_{\min}(B),
\end{aligned}$$

which proves the left-hand side inequality in (4.4); the right-hand side inequality follows from an analogous reasoning. □

A special case of Corollary 4.2 arises when a symmetric matrix $A \in \mathbb{S}^n$ is perturbed by adding to it a rank-one matrix $B = qq^\top$. Since $\lambda_{\max}(qq^\top) = \|q\|_2^2$, and $\lambda_{\min}(qq^\top) = 0$, we immediately obtain from (4.4) that

$$\lambda_k(A) \leq \lambda_k(A + qq^\top) \leq \lambda_k(A) + \|q\|_2^2, \quad k = 1, \ldots, n.$$

4.4 Positive semidefinite matrices

4.4.1 Definition

A symmetric matrix $A \in \mathbb{S}^n$ is said to be *positive semidefinite* (PSD) if the associated quadratic form is non-negative, i.e.,

$$x^\top A x \geq 0, \quad \forall x \in \mathbb{R}^n.$$

If, moreover,

$$x^\top A x > 0, \quad \forall 0 \neq x \in \mathbb{R}^n,$$

then A is said to be *positive definite*. To denote a symmetric positive semidefinite (resp. positive definite) matrix, we use the notation $A \succeq 0$ (resp. $A \succ 0$). We say that A is negative semidefinite, written $A \preceq 0$, if $-A \succeq 0$, and likewise A is negative definite, written $A \prec 0$, if $-A \succ 0$. It can immediately be seen that a positive semidefinite matrix is actually positive definite if and only if it is invertible.

Example 4.8 (*Sample covariance matrix*) With the notation used in Example 4.2, we observe that any sample covariance matrix Σ is by definition positive semidefinite. This is due to the fact that for any $u \in \mathbb{R}^n$, the quantity $u^\top \Sigma u$ is nothing but the sample variance of the scalar products $u^\top x^{(i)}, i = 1, \ldots, m$, hence it is a non-negative number.[5]

[5] As seen in Section 4.4.4, the converse is also true: any PSD matrix can be written as a sample covariance of certain data points.

The set of real positive semidefinite matrices in $\mathbb{R}^{n,n}$ is denoted by

$$\mathbf{S}^n_+ = \{ A \in \mathbf{S}^n : A \succeq 0 \}.$$

Similarly, $\mathbf{S}^n_{++}$ denotes the set of positive definite matrices in $\mathbb{R}^{n,n}$.

Remark 4.1 *Principal submatrices of a PSD matrix.* A simple observation is the following one: let $\mathcal{I} = \{ i_1, \ldots, i_m \}$ be a subset of the indices $\{ 1, \ldots, n \}$, and denote by $A_\mathcal{I}$ a submatrix obtained from $A \in \mathbb{R}^{n,n}$ by taking the rows and columns with indices in $\mathcal{I}$ (this is called a *principal submatrix* of A of dimension $m \times m$). Then,

$$A \succeq 0 \quad \Rightarrow \quad A_\mathcal{I} \succeq 0, \ \forall \mathcal{I} \tag{4.5}$$

and, similarly, $A \succ 0$ implies that $A_\mathcal{I} \succ 0$. This fact is easily seen from the definition of PSD matrices, since forming the product $x_\mathcal{I}^\top A_\mathcal{I} x_\mathcal{I}$ is the same as forming the product $x^\top A x$ with a vector x whose entries x_i are nonzero only if $i \in \mathcal{I}$. One consequence of this fact, for instance, is that $A \succeq 0$ implies that the diagonal elements $a_{ii} \geq 0, i = 1, \ldots, n$, and, likewise, $A \succ 0$ implies that $a_{ii} > 0, i = 1, \ldots, n$.

4.4.2 Eigenvalues of PSD matrices

We here maintain the notation that, for $A \in \mathbf{S}^n$, the eigenvalues are arranged in increasing order as $\lambda_1(A) \geq \cdots \geq \lambda_n(A)$. The following facts hold:

$$A \succeq 0 \quad \Leftrightarrow \quad \lambda_i(A) \geq 0, \ i = 1, \ldots, n,$$
$$A \succ 0 \quad \Leftrightarrow \quad \lambda_i(A) > 0, \ i = 1, \ldots, n.$$

To prove the first of these statements (the second is proved analogously), let $A = U\Lambda U^\top$ be the spectral factorization of A, then

$$x^\top A x = x^\top U\Lambda U^\top x = z\Lambda z = \sum_{i=1}^n \lambda_i(A)z_i^2,$$

where $z \doteq Ux$. Now,

$$x^\top A x \geq 0 \,\forall x \in \mathbb{R}^n \quad \Leftrightarrow \quad z\Lambda z \geq 0 \,\forall z \in \mathbb{R}^n,$$

and the latter condition is clearly equivalent to $\lambda_i(A) \geq 0$, $i = 1,\ldots,n$.

The eigenvalues of a matrix $A \in \mathbb{S}^m$ cannot decrease if a PSD matrix B is added to A. Indeed, if $B \succeq 0$, then $\lambda_{\min}(B) \succeq 0$, and it follows immediately from (4.4) that

$$B \succeq 0 \quad \Rightarrow \quad \lambda_k(A+B) \geq \lambda_k(A), \quad k = 1,\ldots,n. \qquad (4.6)$$

4.4.3   Congruence transformations

The following theorem characterizes the definiteness of a matrix when it is pre- and post-multiplied by another matrix.

Theorem 4.5 *Let $A \in \mathbb{S}^n$, $B \in \mathbb{R}^{n,m}$, and consider the product*

$$C = B^\top A B \in \mathbb{S}^m. \qquad (4.7)$$

1. *If $A \succeq 0$, then $C \succeq 0$;*

2. *if $A \succ 0$, then $C \succ 0$ if and only if $\operatorname{rank} B = m$;*

3. *if B is square and invertible, then $A \succ 0$ (resp. $A \succeq 0$) if and only if $C \succ 0$ (resp. $C \succeq 0$).*

Proof For fact 1, we have that, for all $x \in \mathbb{R}^m$,

$$x^\top C x = x^\top B^\top A B x = z^\top A z \geq 0,$$

with $z = Bx$, hence $C \succeq 0$ as desired. For point 2, observe that, since $A \succ 0$, then $C \succ 0$ if and only if $Bx \neq 0$ for all $x \neq 0$, i.e., if and only if $\dim \mathcal{N}(B) = 0$. By the fundamental theorem of linear algebra, $\dim \mathcal{N}(B) + \operatorname{rank}(B) = m$, from which the statement follows. For point 3, the direct implication follows from point 2. Conversely, let $C \succ 0$, and suppose for purpose of contradiction that $A \not\succ 0$. Then there would exist $z \neq 0$ such that $z^\top A z \leq 0$. Since B is invertible, let $x = B^{-1}z$, then

$$x^\top C x = x^\top B^\top A B x = z^\top A z \leq 0,$$

which is impossible, since $C \succ 0$. An analogous argument shows that $C \succeq 0$ implies $A \succeq 0$, thus concluding the proof. □

When B is square and invertible, then (4.7) defines a so-called *congruence* transformation, and A, C are said to be congruent. The *inertia* of a symmetric matrix $A \in S^n$, $\text{In}(A) = (\text{npos}(A), \text{nneg}(A), \text{nzero}(A))$, is defined as a triple of non-negative integers representing respectively the number of positive, negative, and zero eigenvalues of A (counting multiplicities). It can be proved that two matrices $A \in S^n$, $C \in S^n$ have the same inertia if and only if they are congruent.

The following corollary follows from Theorem 4.5, by simply observing that the identity matrix is positive definite.

Corollary 4.3 *For any matrix $A \in \mathbb{R}^{m,n}$ it holds that:*

1. $A^\top A \succeq 0$, *and* $AA^\top \succeq 0$;

2. $A^\top A \succ 0$ *if and only if A is full-column rank, i.e., $\text{rank } A = n$;*

3. $AA^\top \succ 0$ *if and only if A is full-row rank, i.e., $\text{rank } A = m$.*

An interesting fact is that, under a certain positivity assumption on their linear combination, two symmetric matrices can be simultaneously "diagonalized" via an appropriate congruence transformation, as stated in the next theorem.

Theorem 4.6 (Joint diagonalization by congruence) *Let $A_1, A_2 \in S^n$ be such that*

$$A = \alpha_1 A_1 + \alpha_2 A_2 \succ 0$$

for some scalars α_1, α_2. Then, there exists a nonsingular matrix $B \in \mathbb{R}^{n,n}$ such that both $B^\top A_1 B$ and $B^\top A_2 B$ are diagonal.

Proof Assume without loss of generality that $\alpha_2 > 0$. Since $A \succ 0$, A is congruent to the identity matrix, that is there exists some nonsingular B_1 such that $B_1^\top A B_1 = I_n$. Since $B_1^\top A_1 B_1$ is symmetric, there exists an orthogonal matrix W such that $W^\top B_1^\top A_1 B_1 W = D$, where D is diagonal. Taking $B = B_1 W$ we have that

$$B^\top A B = W^\top B_1^\top A B_1 W = W^\top I_n W = I$$

and

$$B^\top A_1 B = D,$$

therefore, since $A_2 = (A - \alpha_1 A_1)/\alpha_2$, matrix $B^\top A_2 B$ is also diagonal, which concludes the proof. □

The following fact is also readily established from Theorem 4.6.

Corollary 4.4 *Let $A \succ 0$ and $C \in S^n$. Then there exists a nonsingular matrix B such that $B^\top C B$ is diagonal and $B^\top A B = I_n$.*

4.4.4 Matrix square-root and Cholesky decomposition

Let $A \in \mathbb{S}^n$. Then

$$A \succeq 0 \quad \Leftrightarrow \quad \exists\, B \succeq 0 : A = B^2, \tag{4.8}$$

$$A \succ 0 \quad \Leftrightarrow \quad \exists\, B \succ 0 : A = B^2. \tag{4.9}$$

Indeed, any $A \succeq 0$ admits the spectral factorization $A = U\Lambda U^\top$, with U orthogonal and $\Lambda = \mathrm{diag}\,(\lambda_1, \ldots, \lambda_n)$, $\lambda_i \geq 0$, $i = 1, \ldots, n$. Defining $\Lambda^{1/2} = \mathrm{diag}\,\left(\sqrt{\lambda_1}, \ldots, \sqrt{\lambda_1}\right)$ and $B = U\Lambda^{1/2}U^\top$, we have that

$$B^2 = U\Lambda^{1/2}U^\top U\Lambda^{1/2}U^\top = U\Lambda U^\top = A.$$

Conversely, if for some symmetric B it holds that $A = B^\top B = B^2$, then $A \succeq 0$ follows from Corollary 4.3, and this proves (4.8). The proof of (4.9) is analogous. Further, it can be proved (not done here) that the matrix B in (4.8), (4.9) is *unique*. This matrix is called the *matrix square-root* of A: $B = A^{1/2}$.

Repeating the previous reasoning with $B = \Lambda^{1/2}U^\top$, we can also conclude that

$$A \succeq 0 \quad \Leftrightarrow \quad \exists\, B : A = B^\top B,$$

$$A \succ 0 \quad \Leftrightarrow \quad \exists\, B \text{ nonsingular} : A = B^\top B. \tag{4.10}$$

In particular, Eq. (4.10) states that A is positive definite if and only if it is congruent to the identity.

Notice further that every square matrix B has a QR factorization: $B = QR$, where Q is orthogonal and R is an upper-triangular matrix with the same rank as B (see Section 7.3). Then, for any $A \succeq 0$ we have that

$$A = B^\top B = R^\top Q^\top Q R = R^\top R,$$

that is, any PSD matrix can be factorized as $R^\top R$ where R is upper triangular. Further, R can be chosen to have non-negative diagonal entries. If $A \succ 0$, then these diagonal entries are positive. In this case, the factorization is unique and it is called the *Cholesky decomposition* of A.

Using the matrix square-root, we can prove the following result, relating the eigenvalues of B and AB, for B symmetric and $A \succ 0$.

Corollary 4.5 *Let $A, B \in \mathbb{S}^n$, with $A \succ 0$. Then, the matrix AB is diagonalizable, has purely real eigenvalues, and it has the same inertia as B.*

Proof Let $A^{1/2} \succ 0$ be the matrix square-root of A. Then,

$$A^{-1/2}ABA^{1/2} = A^{1/2}BA^{1/2}.$$

The matrix $A^{-1/2}ABA^{1/2}$ on the left in this equation is *similar* to AB, hence it has the same eigenvalues as AB. Since the matrix on the right is symmetric, it is diagonalizable and its eigenvalues are real; therefore, also AB is diagonalizable and has purely real eigenvalues. Further, the matrix on the right is congruent to B, hence it has the same inertia as B. Thus, AB has the same inertia as B.   □

4.4.5 *Positive-definite matrices and ellipsoids*

Positive-definite matrices are intimately related to geometrical objects called *ellipsoids*, which are further discussed in Section 9.2.2. A full-dimensional, bounded ellipsoid with center in the origin can indeed be defined as the set

$$\mathcal{E} = \{x \in \mathbb{R}^n : x^\top P^{-1} x \leq 1\},$$

where $P \succ 0$. The eigenvalues and eigenvectors of P define the orientation and shape of the ellipsoid: the eigenvectors u_i of P define the directions of the semi-axes of the ellipsoid, while their lengths are given by $\sqrt{\lambda_i}$, where $\lambda_i > 0$, $i = 1, \ldots, n$, are the eigenvalues of P, see Figure 4.2.

Since $P \succ 0$ is equivalent to $P^{-1} \succ 0$, by the Cholesky decomposition (4.10) there exists a nonsingular A such that $P^{-1} = A^\top A$. Hence, the previous definition of ellipsoid $\mathcal{E}$, which involved the product $x^\top P^{-1} x = x^\top A^\top A x = \|Ax\|_2^2$, is also equivalent to the following one:

$$\mathcal{E} = \{x \in \mathbb{R}^n : \|Ax\|_2 \leq 1\}.$$

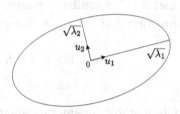

Figure 4.2 A two-dimensional ellipsoid.

4.4.6 *The PSD cone and partial order*

The set of positive semidefinite matrices $\mathbb{S}_+^n$ is a *convex cone*, as defined later in Section 8.1. First, it is a *convex set*, since it satisfies the defining property of convex sets (see Section 8.1), that is for any two matrices $A_1, A_2 \in \mathbb{S}_+^n$ and any $\theta \in [0, 1]$, it holds that

$$x^\top (\theta A_1 + (1 - \theta) A_2) x = \theta x^\top A_1 x + (1 - \theta) x^\top A_2 x \geq 0, \ \forall x,$$

hence $\theta A_1 + (1 - \theta) A_2 \in \mathbb{S}_+^n$. Moreover, for any $A \succeq 0$ and any $\alpha \geq 0$, we have that $\alpha A \succeq 0$, which says that $\mathbb{S}_+^n$ is a *cone*. The relation "$\succeq$" defines a partial order on the cone of PSD matrices. That is, we say that $A \succeq B$ if $A - B \succeq 0$ and, similarly, $A \succ B$ if $A - B \succ 0$. This is a *partial* order, since not every two symmetric matrices may be put in a $\preceq$ or $\succeq$ relation. Take for example

$$A = \begin{bmatrix} 2 & 1 \\ 1 & 1 \end{bmatrix}, \quad B = \begin{bmatrix} 1 & 1 \\ 1 & 1 \end{bmatrix}, \quad C = \begin{bmatrix} 1 & 1 \\ 1 & 2 \end{bmatrix}.$$

Then, one may check that $A \succeq B$, $B \preceq C$, but neither $A \preceq C$ nor $A \succeq C$.

Theorem 4.7 *Let $A \succ 0$ and $B \succeq 0$, and denote by $\rho(\cdot)$ the spectral radius of a matrix (that is, the maximum modulus of the eigenvalues of a matrix). Then*

$$A \succeq B \quad \Leftrightarrow \quad \rho(BA^{-1}) \leq 1, \tag{4.11}$$

$$A \succ B \quad \Leftrightarrow \quad \rho(BA^{-1}) < 1. \tag{4.12}$$

Proof We prove (4.11), the reasoning for proving (4.12) being equivalent. By Corollary 4.4, there exists a nonsingular matrix M such that $A = MIM^\top$ and $B = MDM^\top$, with $D = \mathrm{diag}\,(d_1,\ldots,d_n)$, $d_i \geq 0$, $i = 1,\ldots,n$. Then, $A - B \succeq 0$ if and only if $M(I - D)M^\top \succeq 0$ which, by Theorem 4.5, is satisfied if and only if $I - D \succeq 0$, i.e., for $d_i \leq 1$, $i = 1,\ldots,n$. Since $BA^{-1} = MDM^\top M^{-\top}M^{-1} = MDM^{-1}$, we see that BA^{-1} is similar to matrix D, hence the eigenvalues of BA^{-1} are precisely d_i, $i = 1,\ldots,n$. Since the d_is are non-negative, we conclude that

$$A - B \succeq 0 \; \Leftrightarrow \; d_i \leq 1 \,\forall i \; \Leftrightarrow \; \rho(BA^{-1}) \leq 1.$$

$\square$

Observe that for any two square matrices X, Y the products XY and YX have the same eigenvalues, hence $\rho(XY) = \rho(YX)$. Therefore for $A \succ 0$, $B \succ 0$ by Theorem 4.7 we have that

$$A \succeq B \; \Leftrightarrow \; \rho(BA^{-1}) = \rho(A^{-1}B) \leq 1 \; \Leftrightarrow \; B^{-1} \succeq A^{-1}.$$

More generally, the relation $A \succeq B$ induces a corresponding relation among the ordered eigenvalues of A and B, and likewise on monotone functions of the eigenvalues, such as the determinant and the trace. Indeed, for any $A, B \in \mathbb{S}^n$, it follows directly from Eq. (4.6) that $A - B \succeq 0$ implies that

$$\lambda_k(A) = \lambda_k(B + (A - B)) \geq \lambda_k(B), \quad k = 1,\ldots,n, \tag{4.13}$$

(notice, however, that the converse is *not* true, i.e. $\lambda_k(A) \geq \lambda_k(B)$, $i = 1,\ldots,n$, does *not* imply $A - B \succeq 0$). Therefore, by (4.13), $A - B \succeq 0$ also implies that

$$\det A \;=\; \prod_{k=1,\ldots,n} \lambda_k(A) \geq \prod_{k=1,\ldots,n} \lambda_k(B) = \det B,$$

$$\mathrm{trace}\, A \;=\; \sum_{k=1,\ldots,n} \lambda_k(A) \geq \sum_{k=1,\ldots,n} \lambda_k(B) = \mathrm{trace}\, B.$$

The following result is related to an important matrix equation arising in systems and control theory, called the Lyapunov equation.

Theorem 4.8 (Symmetric sum) *Let $A \succ 0$ and $B \in \mathbb{S}^n$, and consider the symmetric sum*

$$S = AB + BA.$$

Then, $S \succeq 0$ (resp., $S \succ 0$) implies that $B \succeq 0$ (resp., $B \succ 0$).

Proof Since $B \in \mathbb{S}^n$ it admits the spectral factorization $B = U\Lambda U^\top$ where U is orthogonal and Λ is diagonal. Then, by Theorem 4.5, $S \succeq 0$ if and only if $U^\top S U \succeq 0$, that is if and only if

$$(U^\top A U)\Lambda + \Lambda(U^\top A U) \succeq 0.$$

This implies that the diagonal elements $[U^\top S U]_{ii} \geq 0$, $i = 1, \ldots, n$, that is

$$2\alpha_i \lambda_i(B) \geq 0, \ i = 1, \ldots, n, \text{ where } \alpha_i \doteq u_i^\top A u_i > 0,$$

being u_i the i-th column of U. The latter condition clearly implies that $\lambda_i(B) \geq 0$, $i = 1, \ldots, n$, i.e., that $B \succeq 0$. An analogous reasoning similarly shows that $S \succ 0$ implies $B \succ 0$. □

Example 4.9 Taking the matrix square-root preserves the PSD ordering. In particular, if $A \succ 0$, $B \succeq 0$, then

$$A \succ B \quad \Rightarrow \quad A^{1/2} \succ B^{1/2}. \tag{4.14}$$

To see this fact consider the identity

$$2(A - B) = (A^{1/2} + B^{1/2})(A^{1/2} - B^{1/2}) + (A^{1/2} - B^{1/2})(A^{1/2} + B^{1/2}).$$

Since $A \succ 0$, $B \succeq 0$, then $A^{1/2} \succ 0$, $B^{1/2} \succeq 0$, hence $A^{1/2} + B^{1/2} \succ 0$. Thus, by applying Theorem 4.8 to the above sum, we have that $A - B \succ 0$ implies that $A^{1/2} - B^{1/2} \succ 0$, which is the claim.

Notice that a converse of (4.14) does not hold in general. Take for example

$$A = \begin{bmatrix} 2 & 1 \\ 1 & 1 \end{bmatrix}, \quad B = \begin{bmatrix} 1.2 & 1 \\ 1 & 0.9 \end{bmatrix}.$$

Then, $A \succ 0$, $B \succ 0$, $A \succ B$, but $A^2 \nsucc B^2$.

4.4.7 Schur complements

Let $A \in \mathbb{S}^n$, $B \in \mathbb{S}^m$, and consider the block-diagonal matrix

$$M = \begin{bmatrix} A & 0_{n,m} \\ 0_{m,n} & B \end{bmatrix}.$$

Then, it is easy to verify that

$$M \succeq 0 \ (\text{resp.}, M \succ 0) \quad \Leftrightarrow \quad A \succeq 0, B \succeq 0 \ (\text{resp.}, A \succ 0, B \succ 0).$$

We now state an important result on positive definiteness of block matrices that are not necessarily block diagonal.

Theorem 4.9 (Schur complements) *Let $A \in \mathbb{S}^n$, $B \in \mathbb{S}^m$, $X \in \mathbb{R}^{n,m}$, with $B \succ 0$. Consider the symmetric block matrix*

$$M = \begin{bmatrix} A & X \\ X^\top & B \end{bmatrix},$$

and define the so-called Schur complement matrix of A in M

$$S \doteq A - XB^{-1}X^\top.$$

Then,

$$M \succeq 0 \ (\text{resp.}, M \succ 0) \quad \Leftrightarrow \quad S \succeq 0 \ (\text{resp.}, S \succ 0).$$

Proof Define the block matrix

$$C = \begin{bmatrix} I_n & 0_{n,m} \\ -B^{-1}X^\top & I_m \end{bmatrix}.$$

This matrix is square lower triangular and has all nonzero diagonal entries, hence it is nonsingular. Consider then the congruence transformation on M

$$C^\top M C = \begin{bmatrix} S & 0_{n,m} \\ 0_{m,n} & B \end{bmatrix}.$$

From Theorem 4.5 we have that $M \succeq 0$ (resp., $M \succ 0$) if and only if $C^\top M C \succeq 0$ (resp., $C^\top M C \succ 0$). But $C^\top M C$ is block diagonal and $B \succ 0$ by assumption, therefore we conclude that $M \succeq 0$ (resp., $M \succ 0$) if and only if $S \succeq 0$ (resp., $S \succ 0$). $\square$

4.5   Exercises

Exercise 4.1 (Eigenvectors of a symmetric 2×2 matrix) Let $p, q \in \mathbb{R}^n$ be two linearly independent vectors, with unit norm ($\|p\|_2 = \|q\|_2 = 1$). Define the symmetric matrix $A \doteq pq^\top + qp^\top$. In your derivations, it may be useful to use the notation $c \doteq p^\top q$.

1. Show that $p + q$ and $p - q$ are eigenvectors of A, and determine the corresponding eigenvalues.

2. Determine the nullspace and rank of A.

3. Find an eigenvalue decomposition of A, in terms of p, q. *Hint:* use the previous two parts.

4. What is the answer to the previous part if p, q are not normalized?

Exercise 4.2 (Quadratic constraints) For each of the following cases, determine the shape of the region generated by the quadratic constraint $x^\top A x \leq 1$.

1. $A = \begin{bmatrix} 2 & 1 \\ 1 & 2 \end{bmatrix}$.

2. $A = \begin{bmatrix} 1 & -1 \\ -1 & 1 \end{bmatrix}$.

3. $A = \begin{bmatrix} -1 & 0 \\ 0 & -1 \end{bmatrix}$.

Hint: use the eigenvalue decomposition of A, and discuss depending on the sign of the eigenvalues.

Exercise 4.3 (Drawing an ellipsoid)

1. How would you efficiently draw an ellipsoid in $\mathbb{R}^2$, if the ellipsoid is described by a quadratic inequality of the form

$$\mathcal{E} = \left\{ x^\top A x + 2b^\top x + c \leq 0 \right\},$$

where A is 2×2 and symmetric, positive definite, $b \in \mathbb{R}^2$, and $c \in \mathbb{R}$? Describe your method as precisely as possible.

2. Draw the ellipsoid

$$\mathcal{E} = \left\{ 4x_1^2 + 2x_2^2 + 3x_1 x_2 + 4x_1 + 5x_2 + 3 \leq 1 \right\}.$$

Exercise 4.4 (Interpretation of covariance matrix) As in Example 4.2, we are given m points $x^{(1)}, \ldots, x^{(m)}$ in $\mathbb{R}^n$, and denote by Σ the sample covariance matrix:

$$\Sigma \doteq \frac{1}{m} \sum_{i=1}^{m} (x^{(i)} - \hat{x})(x^{(i)} - \hat{x})^\top,$$

where $\hat{x} \in \mathbb{R}^n$ is the sample average of the points:

$$\hat{x} \doteq \frac{1}{m} \sum_{i=1}^{m} x^{(i)}.$$

We assume that the average and variance of the data projected along a given direction do not change with the direction. In this exercise we will show that the sample covariance matrix is then proportional to the identity.

We formalize this as follows. To a given normalized direction $w \in \mathbb{R}^n$, $\|w\|_2 = 1$, we associate the line with direction w passing through the origin, $\mathcal{L}(w) = \{tw : t \in \mathbb{R}\}$. We then consider the projection of the points $x^{(i)}$, $i = 1, \ldots, m$, on the line $\mathcal{L}(w)$, and look at the associated coordinates of the points on the line. These *projected values* are given by

$$t_i(w) \doteq \arg\min_t \|tw - x^{(i)}\|_2, \quad i = 1, \ldots, m.$$

We assume that for any w, the sample average $\hat{t}(w)$ of the projected values $t_i(w)$, $i = 1, \ldots, m$, and their sample variance $\sigma^2(w)$, are both constant, independent of the direction w. Denote by $\hat{t}$ and σ^2 the (constant) sample average and variance. Justify your answer to the following questions as carefully as you can.

1. Show that $t_i(w) = w^\top x^{(i)}$, $i = 1, \ldots, m$.

2. Show that the sample average $\hat{x}$ of the data points is zero.

3. Show that the sample covariance matrix Σ of the data points is of the form $\sigma^2 I_n$. *Hint:* the largest eigenvalue $\lambda_{\max}$ of the matrix Σ can be written as: $\lambda_{\max} = \max_w \{w^\top \Sigma w : w^\top w = 1\}$, and a similar expression holds for the smallest eigenvalue.

Exercise 4.5 (Connected graphs and the Laplacian) We are given a graph as a set of vertices in $V = \{1, \ldots, n\}$, with an edge joining any pair of vertices in a set $E \subseteq V \times V$. We assume that the graph is undirected (without arrows), meaning that $(i, j) \in E$ implies $(j, i) \in E$. As in Section 4.1, we define the Laplacian matrix by

$$L_{ij} = \begin{cases} -1 & \text{if } (i,j) \in E, \\ d(i) & \text{if } i = j, \\ 0 & \text{otherwise.} \end{cases}$$

Here, $d(i)$ is the number of edges adjacent to vertex i. For example, $d(4) = 3$ and $d(6) = 1$ for the graph in Figure 4.3.

1. Form the Laplacian for the graph shown in Figure 4.3.

2. Turning to a generic graph, show that the Laplacian L is symmetric.

3. Show that L is positive-semidefinite, proving the following identity, valid for any $u \in \mathbb{R}^n$:

$$u^\top L u = q(u) \doteq \frac{1}{2} \sum_{(i,j) \in E} (u_i - u_j)^2.$$

Hint: find the values $q(k)$, $q(e_k \pm e_l)$, for two unit vectors e_k, e_l such that $(k, l) \in E$.

4. Show that 0 is always an eigenvalue of L, and exhibit an eigenvector. *Hint:* consider a matrix square-root[6] of L.

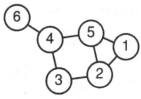

Figure 4.3 Example of an undirected graph.

[6] See Section 4.4.4.

5. The graph is said to be connected if there is a path joining any pair of vertices. Show that if the graph is connected, then the zero eigenvalue is simple, that is, the dimension of the nullspace of L is 1. *Hint:* prove that if $u^\top L u = 0$, then $u_i = u_j$ for every pair $(i, j) \in E$.

Exercise 4.6 (Component-wise product and PSD matrices) Let $A, B \in S^n$ be two symmetric matrices. Define the component-wise product of A, B, by a matrix $C \in S^n$ with elements $C_{ij} = A_{ij} B_{ij}$, $1 \le i, j \le n$. Show that C is positive semidefinite, provided both A, B are. *Hint:* prove the result when A is rank-one, and extend to the general case via the eigenvalue decomposition of A.

Exercise 4.7 (A bound on the eigenvalues of a product) Let $A, B \in S^n$ be such that $A \succ 0$, $B \succ 0$.

1. Show that all eigenvalues of BA are real and positive (despite the fact that BA is not symmetric, in general).

2. Let $A \succ 0$, and let $B^{-1} \doteq \mathrm{diag}\left(\|a_1^\top\|_1, \ldots, \|a_n^\top\|_1\right)$, where $a_i^\top$, $i = 1, \ldots, n$, are the rows of A. Prove that

$$0 < \lambda_i(BA) \le 1, \quad \forall i = 1, \ldots, n.$$

3. With all terms defined as in the previous point, prove that

$$\rho(I - \alpha BA) < 1, \quad \forall \alpha \in (0, 2).$$

Exercise 4.8 (Hadamard's inequality) Let $A \in S^n$ be positive semidefinite. Prove that

$$\det A \le \prod_{i=1}^n a_{ii}.$$

Hint: Distinguish the cases $\det A = 0$ and $\det A \ne 0$. In the latter case, consider the normalized matrix $\tilde{A} \doteq DAD$, where $D = \mathrm{diag}\left(a_{11}^{-1/2}, \ldots, a_{nn}^{-1/2}\right)$, and use the geometric–arithmetic mean inequality (see Example 8.9).

Exercise 4.9 (A lower bound on the rank) Let $A \in S^n_+$ be a symmetric, positive semidefinite matrix.

1. Show that the trace, trace A, and the Frobenius norm, $\|A\|_F$, depend only on its eigenvalues, and express both in terms of the vector of eigenvalues.

2. Show that

$$(\mathrm{trace}\, A)^2 \le \mathrm{rank}(A) \|A\|_F^2.$$

3. Identify classes of matrices for which the corresponding lower bound on the rank is attained.

Exercise 4.10 (A result related to Gaussian distributions) Let $\Sigma \in S_{++}^n$ be a symmetric, positive definite matrix. Show that

$$\int_{\mathbb{R}^n} e^{-\frac{1}{2}x^\top \Sigma^{-1}x}dx = (2\pi)^{n/2}\sqrt{\det\Sigma}.$$

You may assume known that the result holds true when $n = 1$. The above shows that the function $p : \mathbb{R}^n \to \mathbb{R}$ with (non-negative) values

$$p(x) = \frac{1}{(2\pi)^{n/2} \cdot \sqrt{\det\Sigma}}e^{-\frac{1}{2}x^\top\Sigma^{-1}x}$$

integrates to one over the whole space. In fact, it is the density function of a probability distribution called the multivariate Gaussian (or normal) distribution, with zero mean and covariance matrix Σ. *Hint:* you may use the fact that for any integrable function f, and invertible $n \times n$ matrix P, we have

$$\int_{x\in\mathbb{R}^n} f(x)dx = |\det P| \cdot \int_{z\in\mathbb{R}^n} f(Pz)dz.$$

5
Singular value decomposition

THIS CHAPTER is devoted to the singular value decomposition (SVD) of general rectangular matrices, and its applications. The singular value decomposition provides a full insight into the structure of linear maps, and gives an effective computational tool for solving a wealth of linear algebra problems. In optimization, SVD applications include some problems that are *convex*, in the sense that will be discussed in Chapter 8, as well as some non-convex problems that might seem very hard to solve at first glance, such as those involving rank minimization (Section 5.3.1), or optimization over rotation matrices (Section 5.3.3).

5.1   Singular value decomposition

5.1.1   Preliminaries

The singular value decomposition (SVD) of a matrix provides a three-term factorization which is similar to the spectral factorization, but holds for any, possibly non-symmetric and rectangular, matrix $A \in \mathbb{R}^{m,n}$. The SVD allows us to fully describe the linear map associated with A via the matrix–vector product $y = Ax$ as a three-step process: first the input vector x undergoes an orthogonal transformation (rotation or reflection); then a non-negative scaling is performed on the entries of the rotated input vector, and possibly dimensions are added to or removed from this vector in order to match the dimension of the output space. Finally, another orthogonal transformation is performed in the output space. In formulas, we shall see that any matrix $A \in \mathbb{R}^{m,n}$ can be factored as

$$A = U\tilde{\Sigma}V^{\top},$$

where $V \in \mathbb{R}^{n,n}$ and $U \in \mathbb{R}^{m,m}$ are orthogonal matrices (describing the mentioned rotations/reflections in input and output space, respectively), and

$$\tilde{\Sigma} = \begin{bmatrix} \Sigma & 0_{r,n-r} \\ 0_{m-r,r} & 0_{m-r,n-r} \end{bmatrix}, \quad \Sigma = \text{diag}\,(\sigma_1, \dots, \sigma_r) \succ 0, \qquad (5.1)$$

where r is the rank of A, and the scalars $\sigma_i > 0$, $i = 1, \dots, r$, represent the scaling factors on the rotated input vector, see Figure 5.1.

Figure 5.1 Input–output map $y = Ax$, with $A = U\tilde{\Sigma}V^\top$.

Most of the relevant characteristics of A can be derived from its SVD. For instance, as we shall see next, if we know the SVD of a matrix A, then we also know the rank of A, the spectral norm (maximum gain) of A, and the condition number of A. Further, we can readily obtain orthonormal bases for the range and for the nullspace of A; we can solve systems of linear equations involving A as the coefficient matrix (see Section 6.4.2), and analyze the effect of errors in those equations; we can determine least-squares solutions to overdetermined systems of linear equations, or minimum-norm solutions to underdetermined systems.

The SVD is also of fundamental importance in many applications. For example, the SVD arises in several nonlinear (and non-convex) optimization problems, it is a key tool for data compression, and it is employed in statistics for factor analysis and principal component analysis (PCA), where it can be used to reduce the dimensionality of high-dimensional data sets, by "explaining" the variance in the data in terms of a few factors, as seen in Section 5.3.2 and in Chapter 13.

5.1.2   The SVD theorem

We here state the main SVD theorem and then provide a schematic proof for it.

Theorem 5.1 (SVD decomposition) *Any matrix $A \in \mathbb{R}^{m,n}$ can be factored as*

$$A = U\tilde{\Sigma}V^\top \qquad (5.2)$$

where $V \in \mathbb{R}^{n,n}$ and $U \in \mathbb{R}^{m,m}$ are orthogonal matrices (i.e., $U^\top U = I_m$, $V^\top V = I_n$), and $\tilde{\Sigma} \in \mathbb{R}^{m,n}$ is a matrix having the first $r \doteq \text{rank}\,A$

diagonal entries $(\sigma_1, \ldots, \sigma_r)$ *positive and decreasing in magnitude, and all other entries zero (see Eq. (5.1)).*

Proof Consider the matrix $A^\top A \in \mathbb{S}^n$. This matrix is symmetric and positive semidefinite, and it admits the spectral factorization

$$A^\top A = V \Lambda_n V^\top \qquad (5.3)$$

where $V \in \mathbb{R}^{n,n}$ is orthogonal (i.e., $V^\top V = I_n$) and Λ_n is diagonal, containing the eigenvalues $\lambda_i = \lambda_i(A^\top A) \geq 0,\, i = 1, \ldots, n$, on the diagonal, arranged in decreasing order. Since $r = \operatorname{rank} A = \operatorname{rank} A^\top A$, the first r of these eigenvalues are strictly positive. Notice that $AA^\top$ and $A^\top A$ have the same nonzero eigenvalues, hence the same rank r. We thus define

$$\sigma_i \doteq \sqrt{\lambda_i(A^\top A)} = \sqrt{\lambda_i(AA^\top)} > 0,\ i = 1, \ldots, r,$$

Now, denote by $v_1, \ldots, v_r$ the first r columns of V, i.e., the eigenvectors of $A^\top A$ associated with $\lambda_1, \ldots, \lambda_r$. By definition,

$$A^\top A v_i = \lambda_i v_i, \quad i = 1, \ldots, r,$$

hence, multiplying both sides by A,

$$(AA^\top) A v_i = \lambda_i A v_i, \quad i = 1, \ldots, r,$$

which means that $Av_i,\, i = 1, \ldots, r$, are eigenvectors of $AA^\top$. These eigenvectors are mutually orthogonal, since

$$v_i^\top A^\top A v_j = \lambda_j v_i^\top v_j = \begin{cases} \lambda_i & \text{if } i = j \\ 0 & \text{otherwise.} \end{cases}$$

Therefore, the normalized vectors

$$u_i = \frac{Av_i}{\sqrt{\lambda_i}} = \frac{Av_i}{\sigma_i}, \quad i = 1, \ldots, r$$

forms an orthonormal set of r eigenvectors of $AA^\top$ associated with the nonzero eigenvalues $\lambda_1, \ldots, \lambda_r$. Then, for $i, j = 1, \ldots, r$,

$$\begin{aligned} u_i^\top A v_j &= \frac{1}{\sigma_i} v_i^\top A^\top A v_j = \frac{\lambda_j}{\sigma_i} v_i^\top v_j \\ &= \begin{cases} \sigma_i & \text{if } i = j \\ 0 & \text{otherwise.} \end{cases} \end{aligned}$$

Rewritten in matrix format, the previous relation yields

$$\begin{bmatrix} u_1^\top \\ \vdots \\ u_r^\top \end{bmatrix} A \begin{bmatrix} v_1 & \cdots & v_r \end{bmatrix} = \operatorname{diag}(\sigma_1, \ldots, \sigma_r) \doteq \Sigma. \qquad (5.4)$$

This is already the SVD in its "compact" form. We next derive the "full"-version SVD. Notice that, by definition,

$$A^\top A v_i = 0, \quad \text{for } i = r+1, \dots, n,$$

and this implies that

$$A v_i = 0, \quad \text{for } i = r+1, \dots, n.$$

To verify this latter statement, suppose by contradiction that $A^\top A v_i = 0$ and $A v_i \neq 0$. Then $A v_i \in \mathcal{N}(A^\top) \equiv \mathcal{R}(A)^\perp$, which is impossible, since clearly $A v_i \in \mathcal{R}(A)$. Then, we can find orthonormal vectors $u_{r+1}, \dots, u_m$ such that $u_1, \dots, u_r, u_{r+1}, \dots, u_m$ is an orthonormal basis for $\mathbb{R}^m$, and

$$u_i^\top A v_j = 0, \quad \text{for } i = 1, \dots, m; \ j = r+1, \dots, n.$$

Therefore, completing (5.4), we obtain

$$\begin{bmatrix} u_1^\top \\ \vdots \\ u_m^\top \end{bmatrix} A \begin{bmatrix} v_1 & \cdots & v_n \end{bmatrix} = \begin{bmatrix} \Sigma & 0_{r,n-r} \\ 0_{m-r,r} & 0_{m-r,n-r} \end{bmatrix} \doteq \tilde{\Sigma}.$$

Defining the orthogonal matrix $U = [u_1 \cdots u_m]$, the latter expression is rewritten as $U^\top A V = \tilde{\Sigma}$ which, multiplied on the left by U and on the right by $V^\top$ finally yields the full SVD factorization

$$A = U \tilde{\Sigma} V^\top.$$

$\square$

The following corollary can be easily derived from Theorem 5.1.

Corollary 5.1 (Compact-form SVD) *Any matrix $A \in \mathbb{R}^{m,n}$ can be expressed as*

$$A = \sum_{i=1}^{r} \sigma_i u_i v_i^\top = U_r \Sigma V_r^\top$$

where $r = \operatorname{rank} A$, $U_r = [u_1 \cdots u_r]$ is such that $U_r^\top U_r = I_r$, $V_r = [v_1 \cdots v_r]$ is such that $V_r^\top V_r = I_r$, and $\sigma_1 \geq \sigma_2 \geq \cdots \geq \sigma_r > 0$. The positive numbers σ_i are called the singular values of A, vectors u_i are called the left singular vectors of A, and v_i the right singular vectors. These quantities satisfy

$$A v_i = \sigma_i u_i, \quad u_i^\top A = \sigma_i v_i, \quad i = 1, \dots, r.$$

Moreover, $\sigma_i^2 = \lambda_i(A A^\top) = \lambda_i(A^\top A)$, $i = 1, \dots, r$, and u_i, v_i are the eigenvectors of $A^\top A$ and of $A A^\top$, respectively.

5.2   Matrix properties via SVD

In this section we review several properties of a matrix $A \in \mathbb{R}^{m,n}$ that can be derived directly from its SVD in full form

$$A = U\tilde{\Sigma}V^\top,$$

or compact form

$$A = U_r \Sigma V_r^\top.$$

5.2.1   Rank, nullspace, and range

The rank r of A is the cardinality of the nonzero singular values, that is the number of nonzero entries on the diagonal of $\tilde{\Sigma}$. Also, since in practice the diagonal elements in $\tilde{\Sigma}$ may be very small but not exactly zero (e.g., due to numerical errors), the SVD makes it possible to define a more reliable *numerical rank*, defined as the largest k such that $\sigma_k > \epsilon\sigma_1$, for a given tolerance $\epsilon \geq 0$.

Since $r = \text{rank}\,A$, by the fundamental theorem of linear algebra the dimension of the nullspace of A is $\dim \mathcal{N}(A) = n - r$. An orthonormal basis spanning $\mathcal{N}(A)$ is given by the last $n - r$ columns of V, i.e.

$$\mathcal{N}(A) = \mathcal{R}(V_{nr}), \quad V_{nr} \doteq [v_{r+1} \cdots v_n].$$

Indeed, $v_{r+1}, \ldots, v_n$ form an orthonormal set of vectors (they are columns of an orthogonal matrix). Moreover, for any vector $\xi = V_{nr}z$ in the range of V_{nr}, we have

$$A\xi \;=\; U_r \Sigma V_r^\top \xi = U_r \Sigma V_r^\top V_{nr}z = 0,$$

since $V_r^\top V_{nr} = 0$ due to the fact that $V = [V_r\, V_{nr}]$ is orthogonal. This shows that the columns of V_{nr} provide an orthonormal basis for the nullspace of A.

Similarly, an orthonormal basis spanning the range of A is given by the first r columns of U, i.e.

$$\mathcal{R}(A) = \mathcal{R}(U_r), \quad U_r \doteq [u_1 \cdots u_r].$$

To see this fact notice first that, since $\Sigma V_r^\top \in \mathbb{R}^{r,n}$, $r \leq n$, is full row rank, then as x spans the whole $\mathbb{R}^n$ space, $z = \Sigma V_r^\top x$ spans the whole $\mathbb{R}^r$ space, whence

$$
\begin{aligned}
\mathcal{R}(A) \;&=\; \{y : y = Ax,\, x \in \mathbb{R}^n\} = \{y : y = U_r \Sigma V_r^\top x,\, x \in \mathbb{R}^n\} \\
&=\; \{y : y = U_r z,\, z \in \mathbb{R}^r\} \equiv \mathcal{R}(U_r).
\end{aligned}
$$

Example 5.1 Consider the $m \times n$ matrix ($m = 4$, $n = 5$)

$$A = \begin{bmatrix} 1 & 0 & 0 & 0 & 2 \\ 0 & 0 & 3 & 0 & 0 \\ 0 & 0 & 0 & 0 & 0 \\ 0 & 4 & 0 & 0 & 0 \end{bmatrix}.$$

A singular value decomposition of this matrix is given by $A = U\tilde{\Sigma}V^\top$, with

$$U = \begin{bmatrix} 0 & 0 & 1 & 0 \\ 0 & 1 & 0 & 0 \\ 0 & 0 & 0 & -1 \\ 1 & 0 & 0 & 0 \end{bmatrix}, \quad V^\top = \begin{bmatrix} 0 & 1 & 0 & 0 & 0 \\ 0 & 0 & 1 & 0 & 0 \\ \sqrt{0.2} & 0 & 0 & 0 & \sqrt{0.8} \\ 0 & 0 & 0 & 1 & 0 \\ -\sqrt{0.8} & 0 & 0 & 0 & \sqrt{0.2} \end{bmatrix},$$

and

$$\tilde{\Sigma} = \begin{bmatrix} 4 & 0 & 0 & 0 & 0 \\ 0 & 3 & 0 & 0 & 0 \\ 0 & 0 & \sqrt{5} & 0 & 0 \\ 0 & 0 & 0 & 0 & 0 \end{bmatrix}.$$

Notice that $\tilde{\Sigma}$ has a 3×3 nonzero diagonal block

$$\Sigma = \text{diag}\,(\sigma_1, \sigma_2, \sigma_3)$$

$\sigma_1 = 4$, $\sigma_2 = 3$, $\sigma_3 = \sqrt{5}$. The rank of A (which is the number of nonzero elements on the diagonal of $\tilde{\Sigma}$) is thus $r = 3 \le \min(m, n)$. We can also check that $V^\top V = VV^\top = I_5$, and $UU^\top = U^\top U = I_4$. An orthonormal basis for the range of A is given by the first three columns of U:

$$\mathcal{R}(A) = \mathcal{R}\left(\begin{bmatrix} 0 & 0 & 1 \\ 0 & 1 & 0 \\ 0 & 0 & 0 \\ 1 & 0 & 0 \end{bmatrix} \right).$$

Similarly, an orthonormal basis for the nullspace of A is given by the last $n - r = 2$ columns of V:

$$\mathcal{N}(A) = \mathcal{R}\left(\begin{bmatrix} 0 & -\sqrt{0.8} \\ 0 & 0 \\ 0 & 0 \\ 1 & 0 \\ 0 & \sqrt{0.2} \end{bmatrix} \right).$$

5.2.2   Matrix norms

The squared Frobenius matrix norm of a matrix $A \in \mathbb{R}^{m,n}$ can be defined as

$$\|A\|_{\mathrm{F}}^2 = \text{trace}\, A^\top A = \sum_{i=1}^{n} \lambda_i(A^\top A) = \sum_{i=1}^{n} \sigma_i^2,$$

where σ_i are the singular values of A. Hence the squared Frobenius norm is nothing but the sum of the squares of the singular values.

The squared spectral matrix norm $\|A\|_2^2$ is equal to the maximum eigenvalue of $A^\top A$, see Eq. (4.3), therefore

$$\|A\|_2^2 = \sigma_1^2,$$

that is, the spectral norm of A coincides with the maximum singular value of A.

Also, the so-called *nuclear* norm[1] of a matrix A is defined in terms of its singular values:

[1] The norm is also called trace norm, or Ky Fan norm, after Ky Fan (1914–2010).

$$\|A\|_* = \sum_{i=1}^{r} \sigma_i, \quad r = \operatorname{rank} A.$$

appears in several problems related to low-rank matrix completion or rank minimization problems[2], due to the fact that $\|A\|_*$ is the largest possible convex lower bound on rank A, over the set of matrices with spectral norm bounded by unity.

[2] See Section 11.4.1.4.

5.2.2.1 *Condition number.* The *condition number* of an invertible matrix $A \in \mathbb{R}^{n,n}$ is defined as the ratio between the largest and the smallest singular value:

$$\kappa(A) = \frac{\sigma_1}{\sigma_n} = \|A\|_2 \cdot \|A^{-1}\|_2. \tag{5.5}$$

This number provides a quantitative measure of how close A is to being singular (the larger $\kappa(A)$ is, the more close to singular A is). The condition number also provides a measure of the sensitivity of the solution of a system of linear equations to changes in the equation coefficients, see, e.g., Section 6.5.

5.2.3 *Matrix pseudoinverse*

Given any matrix $A \in \mathbb{R}^{m,n}$, let $r = \operatorname{rank} A$, and let $A = U\tilde{\Sigma}V^\top$ be its SVD, where $\tilde{\Sigma}$ has the structure (5.1), with Σ being the diagonal matrix containing the positive singular values. The so-called Moore–Penrose pseudoinverse (or generalized inverse) of A is defined as

$$A^\dagger = V\tilde{\Sigma}^\dagger U^\top \in \mathbb{R}^{n,m} \tag{5.6}$$

where

$$\tilde{\Sigma}^\dagger = \begin{bmatrix} \Sigma^{-1} & 0_{r,m-r} \\ 0_{n-r,r} & 0_{n-r,m-r} \end{bmatrix}, \quad \Sigma^{-1} = \operatorname{diag}\left(\frac{1}{\sigma_1}, \ldots, \frac{1}{\sigma_r}\right) \succ 0.$$

Due to the zero blocks in $\tilde{\Sigma}$, Eq. (5.6) can be also written in a compact form that involves only the first r columns of V and U:

$$A^\dagger = V_r \Sigma^{-1} U_r^\top. \tag{5.7}$$

Notice that, according to these definitions

$$\tilde{\Sigma}\tilde{\Sigma}^\dagger = \begin{bmatrix} I_r & 0_{r,m-r} \\ 0_{m-r,r} & 0_{m-r,m-r} \end{bmatrix}, \quad \tilde{\Sigma}^\dagger\tilde{\Sigma} = \begin{bmatrix} I_r & 0_{r,n-r} \\ 0_{n-r,r} & 0_{n-r,n-r} \end{bmatrix},$$

whereby the following properties hold for the pseudoinverse[3]:

$$\begin{aligned}
AA^\dagger &= U_r U_r^\top \\
A^\dagger A &= V_r V_r^\top \\
AA^\dagger A &= A \\
A^\dagger A A^\dagger &= A^\dagger.
\end{aligned} \tag{5.8}$$

[3] Notice that any matrix $A^\dagger$ satisfying $AA^\dagger A = A$ is a legitimate pseudoinverse of A. The Moore–Penrose pseudoinverse is just one of the possibly many pseudoinverses of A. In this book, however, the symbol $A^\dagger$ typically denotes the Moore–Penrose pseudoinverse, unless specified otherwise.

The following three special cases are particularly interesting.

1. If A is square and nonsingular, then $A^\dagger = A^{-1}$.

2. If $A \in \mathbb{R}^{m,n}$ is full column rank, that is $r = n \leq m$, then

$$A^\dagger A = V_r V_r^\top = VV^\top = I_n,$$

that is, $A^\dagger$ is in this case a *left inverse* of A (i.e., a matrix that, when multiplied by A on the left, yields the identity: $A^\dagger A = I_n$). Notice that in this case $A^\top A$ is invertible and, from (5.3), we have that

$$\begin{aligned}
(A^\top A)^{-1} A^\top &= (V\Sigma^{-2}V^\top)V\tilde{\Sigma}^\top U^\top = V\Sigma^{-2}\Sigma U_r^\top \\
&= V\Sigma^{-1}U_r^\top = A^\dagger.
\end{aligned}$$

Any possible left inverse of A can be expressed as

$$A^{\text{li}} = A^\dagger + Q^\top, \tag{5.9}$$

where Q is some matrix such that $A^\top Q = 0$ (i.e., the columns of Q belong to the nullspace of $A^\top$).

To summarize, in the full-column rank case the pseudoinverse is a left inverse of A, and it has an explicit expression in terms of A:

$$A \in \mathbb{R}^{m,n},\, r = \text{rank}\, A = n \leq m \Rightarrow A^\dagger A = I_n,\, A^\dagger = (A^\top A)^{-1}A^\top.$$

3. If $A \in \mathbb{R}^{m,n}$ is full row rank, that is $r = m \leq n$, then

$$AA^\dagger = U_r U_r^\top = UU^\top = I_m,$$

that is, $A^\dagger$ is in this case a *right inverse* of A (i.e., a matrix that, when multiplied by A on the right, yields the identity: $AA^\dagger = I_m$).

Notice that in this case $AA^\top$ is invertible and, from (7.3), we have that

$$
\begin{aligned}
A^\top (AA^\top)^{-1} &= V\tilde{\Sigma}^\top U^\top (U\Sigma^2 U^\top)^{-1} = V\tilde{\Sigma}^\top U^\top U\Sigma^{-2}U^\top \\
&= V_r\Sigma^{-1}U^\top = A^\dagger.
\end{aligned}
$$

Any possible right inverse of A can be expressed as

$$
A^{\mathrm{ri}} = A^\dagger + Q,
$$

where Q is some matrix such that $AQ = 0$ (i.e., the columns of Q belong to the nullspace of A).

To summarize, in the full-row rank case the pseudoinverse is a right inverse of A, and it has an explicit expression in terms of A:

$$
A \in \mathbb{R}^{m,n},\ r = \operatorname{rank} A = m \le n \ \Rightarrow\ AA^\dagger = I_m,\ A^\dagger = A^\top (AA^\top)^{-1}.
$$

5.2.4 Orthogonal projectors

We have seen that any matrix $A \in \mathbb{R}^{m,n}$ defines a linear map $y = Ax$ between the input space $\mathbb{R}^n$ and the output space $\mathbb{R}^m$. Moreover, by the fundamental theorem of linear algebra, the input and output spaces are decomposed into orthogonal components as follows:

$$
\begin{aligned}
\mathbb{R}^n &= \mathcal{N}(A) \oplus \mathcal{N}(A)^\perp = \mathcal{N}(A) \oplus \mathcal{R}(A^\top) \\
\mathbb{R}^m &= \mathcal{R}(A) \oplus \mathcal{R}(A)^\perp = \mathcal{R}(A) \oplus \mathcal{N}(A^\top).
\end{aligned}
$$

As previously discussed, the SVD $A = U\tilde{\Sigma}V^\top$ provides orthonormal bases for all four of these subspaces: with the usual notation

$$
U = [U_r\ U_{nr}], \quad V = [V_r\ V_{nr}],
$$

where $U_r,\ V_r$ contain the first $r = \operatorname{rank} A$ columns of U and V, respectively, we have that

$$
\begin{aligned}
\mathcal{N}(A) &= \mathcal{R}(V_{nr}), \quad \mathcal{N}(A)^\perp \equiv \mathcal{R}(A^\top) = \mathcal{R}(V_r), \quad &(5.10) \\
\mathcal{R}(A) &= \mathcal{R}(U_r), \quad \mathcal{R}(A)^\perp \equiv \mathcal{N}(A^\top) = \mathcal{R}(U_{nr}). \quad &(5.11)
\end{aligned}
$$

We next discuss how to compute the projections of a vector $x \in \mathbb{R}^n$ onto $\mathcal{N}(A)$, $\mathcal{N}(A)^\perp$, and the projections of a vector $y \in \mathbb{R}^m$ onto $\mathcal{R}(A)$, $\mathcal{R}(A)^\perp$.

First, we recall that given a vector $x \in \mathbb{R}^n$ and d linearly independent vectors $b_1, \ldots, b_d \in \mathbb{R}^n$, the orthogonal projection of x onto the subspace span by $\{b_1, \ldots, b_d\}$ is the vector

$$
x^* = B\alpha,
$$

where $B = [b_1 \cdots b_d]$, and $\alpha \in \mathbb{R}^d$ should solve the Gram systems of linear equations

$$B^\top B\alpha = B^\top x,$$

see Section 2.3 and Example 4.5. Notice in particular that if the basis vectors in B are actually orthonormal, then it holds that $B^\top B = I_d$, hence the Gram system has an immediate solution $\alpha = B^\top x$, and the projection is simply computed as $x^* = BB^\top x$.

Returning to our case of interest, let $x \in \mathbb{R}^n$ be given, and suppose we want to compute the projection of x onto $\mathcal{N}(A)$. Since an orthonormal basis for $\mathcal{N}(A)$ is given by the columns of V_{nr}, by the previous reasoning we have immediately that

$$[x]_{\mathcal{N}(A)} = (V_{nr}V_{nr}^\top)x,$$

where we used the notation $[x]_S$ to denote the projection of a vector onto the subspace S. Now, observe that

$$I_n = VV^\top = V_rV_r^\top + V_{nr}V_{nr}^\top,$$

hence, using (5.8),

$$P_{\mathcal{N}(A)} = V_{nr}V_{nr}^\top = I_n - V_rV_r^\top = I_n - A^\dagger A.$$

Matrix $P_{\mathcal{N}(A)}$ is called an *orthogonal projector* onto the subspace $\mathcal{N}(A)$. In the special case when A is full row rank, then $A^\dagger = A^\top(AA^\top)^{-1}$, and

$$P_{\mathcal{N}(A)} = I_n - A^\top(AA^\top)^{-1}A, \quad \text{if } A \text{ is full row rank.}$$

Via an analogous reasoning we obtain that the projection of $x \in \mathbb{R}^n$ onto $\mathcal{N}(A)^\perp \equiv \mathcal{R}(A^\top)$ is given by

$$[x]_{\mathcal{N}(A)^\perp} = (V_rV_r^\top)x = P_{\mathcal{N}(A)^\perp}x, \quad P_{\mathcal{N}(A)^\perp} = A^\dagger A,$$

and specifically

$$P_{\mathcal{N}(A)^\perp} = A^\top(AA^\top)^{-1}A, \quad \text{if } A \text{ is full row rank.}$$

Similarly, for $y \in \mathbb{R}^m$, we have that

$$[y]_{\mathcal{R}(A)} = (U_rU_r^\top)y = P_{\mathcal{R}(A)}y, \quad P_{\mathcal{R}(A)} = AA^\dagger,$$

with

$$P_{\mathcal{R}(A)} = A(A^\top A)^{-1}A^\top, \quad \text{if } A \text{ is full column rank,}$$

and finally

$$[y]_{\mathcal{R}(A)^\perp} = (U_{nr}U_{nr}^\top)y = P_{\mathcal{R}(A)^\perp}y, \quad P_{\mathcal{R}(A)^\perp} = I_m - AA^\dagger, \quad (5.12)$$

with

$$P_{\mathcal{R}(A)^\perp} = I_m - A(A^\top A)^{-1}A^\top, \quad \text{if } A \text{ is full column rank.}$$

5.2.4.1   *Projections on subspaces.* We consider the problem of computing the projection of a given vector $y \in \mathbb{R}^m$ onto a subspace span of a given set of vectors $S = \mathrm{span}(a^{(1)}, \ldots, a^{(n)}) \subseteq \mathbb{R}^m$, as already discussed in Section 2.3 and in Section 5.2. Clearly, S coincides with the range space of the matrix having these vectors as columns, $A \doteq [a^{(1)} \cdots a^{(n)}]$, hence the problem we face is

$$\min_{z \in \mathcal{R}(A)} \|z - y\|_2. \tag{5.13}$$

If $r \doteq \dim S = \mathrm{rank}(A)$, then the compact SVD of A is $A = U_r \Sigma V_r^\top$, and the unique minimum norm solution of (5.13) is given by the projection theorem as

$$z^* = [y]_S = P_{\mathcal{R}(A)} y = A A^\dagger y = (U_r U_r^\top) y,$$

where $P_{\mathcal{R}(A)}$ is the orthogonal projector onto $\mathcal{R}(A)$. [4] Notice that the projection z^* is a linear function of y, and that the matrix defining this projection is provided by the U_r factor of the SVD of A.

Similarly, suppose we want to find the projection of y onto $S^\perp$, the orthogonal complement of S. Since $S^\perp = \mathcal{N}(A^\top)$, the problem is written as

$$\min_{z \in \mathcal{N}(A^\top)} \|z - y\|_2$$

and the solution is given in (5.12) as

$$z^* = [y]_{S^\perp} = (I_m - A A^\dagger) y.$$

[4] Since $z \in \mathcal{R}(A)$ means that $z = Ax$ for some $x \in \mathbb{R}^n$, the problem (5.13) can also be rewritten equivalently as

$$\min_{x \in \mathbb{R}^n} \|Ax - y\|_2,$$

which is know as the least-squares problem (thoroughly covered in Chapter 6). Since we found that $z^* = A A^\dagger y$, it follows that the optimal solution of the LS problem above is $x^* = A^\dagger y$.

5.3   SVD and optimization

In this section, we illustrate how certain optimization problems can be conveniently solved via SVD. Further applications of SVD in optimization are given in Chapter 6.

5.3.1   Low-rank matrix approximations

Let $A \in \mathbb{R}^{m,n}$ be a given matrix, with $\mathrm{rank}(A) = r > 0$. We here consider the problem of approximating A with a matrix of lower rank. In particular, we consider the following rank-constrained approximation problem

$$\min_{A_k \in \mathbb{R}^{m,n}} \|A - A_k\|_F^2 \tag{5.14}$$
$$\text{s.t.:} \quad \mathrm{rank}(A_k) = k,$$

where $1 \le k \le r$ is given. Let

$$A = U \tilde{\Sigma} V^\top = \sum_{i=1}^r \sigma_i u_i v_i^\top$$

be an SVD of A. We next show that an optimal solution of problem (5.14) can be simply obtained by truncating the previous summation to the k-th term, that is

$$A_k = \sum_{i=1}^{k} \sigma_i u_i v_i^\top. \tag{5.15}$$

Remark 5.1 *Matrix of movie scores.* Assume that A contains scores of user ratings of movies,[5] where A_{ij} contains the score of the i-th user in the j-th movie. Thus, the i-th row gives the score profile of user i. Then, the rank-one approximation $A \approx \sigma_1 u_1 v_1^\top$ corresponds to a model where all the movies are rated to a single profile (given by v_1), and users differ from each other only by a scalar multiple as given by u_1: $A_{ij} \approx \sigma u_{1,i} v_{1,j}$.

[5] See, e.g., the Netflix problem described in Section 11.4.1.4.

To prove the low-rank approximation result above, observe that the Frobenius norm is *unitarily invariant*, meaning that $\|Y\|_F = \|QYR\|_F$ for all $Y \in \mathbb{R}^{m,n}$ and any orthogonal matrices $Q \in \mathbb{R}^{m,m}$, $R \in \mathbb{R}^{n,n}$. Therefore,

$$\|A - A_k\|_F^2 = \|U^\top (A - A_k)V\|_F^2 = \|\tilde{\Sigma} - Z\|_F^2,$$

where $Z = U^\top A_k V$. With this change of variables, problem (5.14) reads

$$\min_{Z \in \mathbb{R}^{m,n}} \left\| \begin{bmatrix} \operatorname{diag}(\sigma_1, \ldots, \sigma_r) & 0_{r,n-r} \\ 0_{m-r,r} & 0_{m-r,n-r} \end{bmatrix} - Z \right\|_F^2 \tag{5.16}$$
$$\text{s.t.:} \quad \operatorname{rank}(Z) = k.$$

Notice then that Z can be assumed to be diagonal, since considering nonzero off-diagonal elements in Z only worsens the Frobenius norm objective in this problem. Therefore, the objective (5.16) becomes

$$f_0 = \|\operatorname{diag}(\sigma_1, \ldots, \sigma_r) - \operatorname{diag}(z_1, \ldots, z_r)\|_F^2 = \sum_{i=1}^{r} (\sigma_i - z_i)^2.$$

Since the constraint $\operatorname{rank}(Z) = k$ requires that exactly k of the diagonal entries z_i are nonzero, the best choice is to set $z_i = \sigma_i$, for $i = 1, \ldots, k$, and $z_i = 0$, for $i > k$. In this way, the z_i, $i = 1, \ldots, k$, "neutralize" the largest singular values of A, so that the residual terms in the objective only contain the $r - k$ smallest singular values, that is, an optimal solution is

$$Z^* = \begin{bmatrix} \operatorname{diag}(\sigma_1, \ldots, \sigma_k, 0, \ldots, 0) & 0_{r,n-r} \\ 0_{m-r,r} & 0_{m-r,n-r} \end{bmatrix},$$

and the optimal objective is

$$f_0^* = \sum_{i=k+1}^{r} \sigma_i^2.$$

The optimal solution for the original problem (5.14) can then be recovered from the change of variables $Z = U^\top A_k V$, giving

$$A_k = U Z^* V^\top = \sum_{i=1}^{k} \sigma_i u_i v_i^\top,$$

which coincides indeed with (5.15), as desired. Following the very same reasoning, we can actually prove that the solution in (5.15) is optimal not only for the Frobenius norm objective, but also for the spectral (maximum singular value) matrix norm. That is, A_k is optimal also for the following problem (hint: also the spectral norm is unitarily invariant):

$$\min_{A_k \in \mathbb{R}^{m,n}} \quad \|A - A_k\|_2^2$$
$$\text{s.t.:} \quad \text{rank}(A_k) = k.$$

The ratio

$$\eta_k = \frac{\|A_k\|_F^2}{\|A\|_F^2} = \frac{\sigma_1^2 + \cdots + \sigma_k^2}{\sigma_1^2 + \cdots + \sigma_r^2} \tag{5.17}$$

indicates what fraction of the total *variance* (Frobenius norm) in A is explained by the rank k approximation of A. A plot of η_k as a function of k may give useful indications on a good rank level k at which to approximate A, see, e.g., Example 5.2 for an application in image compression. Clearly, η_k is related to the relative norm approximation error

$$e_k = \frac{\|A - A_k\|_F^2}{\|A\|_F^2} = \frac{\sigma_{k+1}^2 + \cdots + \sigma_r^2}{\sigma_1^2 + \cdots + \sigma_r^2} = 1 - \eta_k.$$

Remark 5.2 *Minimum "distance" to rank deficiency.* Suppose $A \in \mathbb{R}^{m,n}$, $m \geq n$ is full rank, i.e., $\text{rank}(A) = n$. We ask what is a minimal perturbation δA of A that makes $A + \delta A$ rank deficient. The Frobenius norm (or the spectral norm) of the minimal perturbation δA measures the "distance" of A from rank deficiency. Formally, we need to solve

$$\min_{\delta A \in \mathbb{R}^{m,n}} \quad \|\delta A\|_F^2$$
$$\text{s.t.:} \quad \text{rank}(A + \delta A) = n - 1.$$

This problem is equivalent to (5.14), for $\delta A = A_k - A$. The optimal solution is thus readily obtained as

$$\delta A^* = A_k - A,$$

where $A = \sum_{i=1}^{n} \sigma_i u_i v_i^\top$ is a compact SVD of A, and $A_k \doteq \sum_{i=1}^{n-1} \sigma_i u_i v_i^\top$. Therefore, we have

$$\delta A^* = -\sigma_n u_n v_n^\top.$$

This result shows that the minimal perturbation that leads to rank deficiency is a rank-one matrix, and that the distance to rank deficiency is $\|\delta A^*\|_F = \|\delta A^*\|_2 = \sigma_n$.

Example 5.2 (*SVD-based image compression*) Figure 5.2 shows a gray-scale image, which is represented by a 266×400 matrix A of integers corresponding to the gray levels of the pixels. This matrix is full rank, i.e., $\text{rank}(A) = 266$.

We computed the SVD of matrix A representing the image, and then plotted the ratio η_k in (5.17), for k from 1 to 266, see Figure 5.3.

We see from this curve, for instance, that $k = 9$ already captures 96% of the image variance (the relative approximation error is $e_k \simeq 0.04$); $k = 23$ corresponds to $\eta_k = 0.98$; $k = 49$ corresponds to $\eta_k = 0.99$, and $k = 154$ corresponds to $\eta_k = 0.999$.

Figure 5.4 shows visually a comparison of these approximations: the image is already intelligible at coarse approximation $k = 9$; for $k = 49$ we have a reasonably good compressed image, whilst at $k = 154$ the image is barely distinguishable from the original. To better understand the usefulness of the approximations, suppose we need to transmit the image over a communication channel. If transmitting the original image required sending $N_1 = 266 \times 600 = 106,400$ numerical values, transmitting the $k = 49$ approximation requires sending k right singular vectors, k left singular vectors, and k singular values, that is $N_2 = 266 \times 49 + 49 + 400 \times 49 = 32,683$ numerical values. The relative compression would then be

$$\frac{N_1 - N_2}{N_1} \times 100 = 225\,\%.$$

Figure 5.2 A 266×400 gray-scale image.

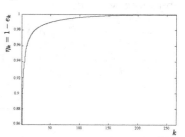

Figure 5.3 Complement of the approximation error e_k as a function of rank k.

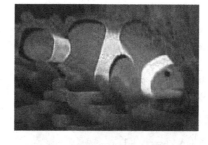

Figure 5.4 Rank k approximations of the original image in Figure 5.2, for $k = 9$ (top left), $k = 23$ (top right), $k = 49$ (bottom left), and $k = 154$ (bottom right).

5.3.2   *Principal component analysis*

Principal component analysis (PCA) is a technique of *unsupervised learning* (see Section 13.5 for further discussion on learning from data), widely used to "discover" the most important, or informative, directions in a data set, that is the directions along which the data varies the most.

5.3.2.1   *Basic idea.*
To make the intuition clear, consider for example the two-dimensional data cloud in Figure 5.5: it is apparent that there exists a direction (at about 45 degrees from the horizontal axis) along which almost all the variation of the data is contained. In contrast, the direction at about 135 degrees contains very little variation of the data. This means that, in this example, the important phenomena underlying the data are essentially uni-dimensional along the 45 degrees line. The important direction was easy to spot in this two-dimensional example. However, graphical intuition does not help when analyzing data in dimension $n > 3$, which is where principal component analysis (PCA) comes in handy.

Let $x_i \in \mathbb{R}^n$, $i = 1, \ldots, m$, be the given data points one wishes to analyze, denote by $\bar{x} = \frac{1}{m} \sum_{i=1}^{m} x_i$ the average of the data points, and let $\tilde{X}$ be the $n \times m$ matrix containing the centered data points:

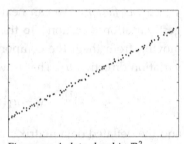

Figure 5.5 A data cloud in $\mathbb{R}^2$.

$$\tilde{X} = [\tilde{x}_1 \cdots \tilde{x}_m], \quad \tilde{x}_i \doteq x_i - \bar{x}, \ i = 1, \ldots, m.$$

We look for a normalized direction in data space, $z \in \mathbb{R}^n$, $\|z\|_2 = 1$, such that the variance of the projections of the centered data points on the line determined by z is maximal. Our choice of the Euclidean norm in the normalization of z is made because it does not favor any particular direction.

The components of the centered data along direction z are given by (see, e.g., Section 2.3.1)

$$\alpha_i = \tilde{x}_i^\top z, \quad i = 1, \ldots, m.$$

Notice that $\alpha_i z$ are the projections of $\tilde{x}_i$ along the span of z. The mean-square variation of the data along direction z is thus given by

$$\frac{1}{m} \sum_{i=1}^{m} \alpha_i^2 = \sum_{i=1}^{m} z^\top \tilde{x}_i \tilde{x}_i^\top z = z^\top \tilde{X} \tilde{X}^\top z.$$

The direction z along which the data has the largest variation can thus be found as the solution to the following optimization problem:

$$\max_{z \in \mathbb{R}^n} \quad z^\top (\tilde{X} \tilde{X}^\top) z \qquad (5.18)$$
$$\text{s.t.:} \quad \|z\|_2 = 1.$$

Let us now solve this problem via the SVD of $\tilde{X}$: let

$$\tilde{X} = U_r \Sigma V_r^\top = \sum_{i=1}^{r} \sigma_i u_i v_i^\top$$

be a compact SVD of $\tilde{X}$, where $r = \text{rank}(\tilde{X})$. Then

$$H \doteq \tilde{X}\tilde{X}^\top = U_r \Sigma^2 U_r^\top$$

is a spectral factorization of H. From Theorem 4.3 we have that the optimal solution of this problem is given by the column u_1 of U_r corresponding to the largest eigenvalue of H, which is σ_1^2. The direction of largest data variation is thus readily found as $z = u_1$, and the mean-square variation along this direction is proportional to σ_1^2.

5.3.2.2 *Deflation.* We can next proceed to determine a second-largest variation direction. To this end, we first *deflate* the data by removing from them the component along the already found largest variation direction u_1. That is, we consider the deflated data points

$$\tilde{x}_i^{(1)} = \tilde{x}_i - u_1(u_1^\top \tilde{x}_i), \quad i = 1, \dots, m,$$

and the deflated data matrix

$$\tilde{X}^{(1)} = [\tilde{x}_1^{(1)} \cdots \tilde{x}_m^{(1)}] = (I_n - u_1 u_1^\top)\tilde{X}.$$

We can readily obtain the compact SVD for the deflated matrix $\tilde{X}^{(1)}$ from the SVD of $\tilde{X} = \sum_{i=1}^{r} \sigma_i u_i v_i^\top$. Namely:

$$
\begin{aligned}
\tilde{X}^{(1)} &= \sum_{i=1}^{r} \sigma_i u_i v_i^\top - \sum_{i=1}^{r} \sigma_i u_1 u_1^\top u_i v_i^\top \\
&= \sum_{i=1}^{r} \sigma_i u_i v_i^\top - \sigma_1 u_1 v_1^\top \\
&= \sum_{i=2}^{r} \sigma_i u_i v_i^\top,
\end{aligned}
$$

where the first line follows from the fact that u_i, $i = 1, \dots, m$, form an orthonormal set, thus $u_1^\top u_i$ is zero whenever $i \neq 1$, and it is equal to one when $i = 1$. Notice that the summation in the dyadic expansion of $\tilde{D}^{(1)}$ now starts from $i = 2$.

The second-largest direction of variation z thus solves the following optimization problem

$$
\begin{aligned}
\min_{z \in \mathbb{R}^n} \quad & z^\top (\tilde{X}^{(1)} \tilde{X}^{(1)\top}) z \\
\text{s.t.:} \quad & \|z\|_2 = 1,
\end{aligned}
$$

and the solution, again from Theorem 4.3, is $z = u_2$, that is the direction of the left singular vector corresponding to the largest singular value of $\tilde{X}^{(1)}$, which is indeed σ_2.

We can actually iterate this deflation process until we find all r principal directions. The above reasoning shows that these principal directions are nothing but the left singular vectors $u_1, \ldots, u_r$ of the centered data matrix $\tilde{X}$, and the mean-square data variations along these directions are proportional to the corresponding squared singular values $\sigma_1^2, \ldots, \sigma_r^2$.

Remark 5.3 *Explained variance.* We have seen that the eigenvectors $u_1, \ldots, u_r$ of $\tilde{X}\tilde{X}^\top$ (which coincide with the left singular vectors of $\tilde{X}$) provide us with principal directions corresponding to decreasing mean-square variation in the data. The mean-square variation is also sometimes referred to as *variance in the data*. The variance along direction u_1 is σ_1^2; the variance of the deflated data along u_2 is σ_2^2, etc. Therefore, the *total variance* in the data is $\sigma_1^2 + \cdots + \sigma_r^2$, which coincides with the trace of the Gram data matrix $H = \tilde{X}\tilde{X}^\top$:

$$
\begin{aligned}
\mathrm{trace}(\tilde{X}\tilde{X}^\top) &= \mathrm{trace}(U_r \Sigma^2 U_r^\top) = \mathrm{trace}(\Sigma^2 U_r^\top U_r) = \mathrm{trace}(\Sigma^2) \\
&= \sigma_1^2 + \cdots + \sigma_r^2.
\end{aligned}
$$

If we project the centered data onto the span of the first $k \le r$ principal directions, we obtain a projected data matrix

$$
\tilde{X}_k = U_k^\top \tilde{X},
$$

where U_k contains the columns $u_1, \ldots, u_k$. The data variance contained in this projection (i.e., the variance *explained* by the first k principal directions) is

$$
\begin{aligned}
\mathrm{trace}(\tilde{X}_k \tilde{X}_k^\top) &= \mathrm{trace}(U_k^\top \tilde{X}\tilde{X}^\top U_k) = \mathrm{trace}\left([I_k\ 0]\Sigma^2 [I_k\ 0]^\top\right) \\
&= \sigma_1^2 + \cdots + \sigma_k^2.
\end{aligned}
$$

Hence, we can define the ratio between the variance "explained" by the projected data and the total variance, as

$$
\eta_k = \frac{\sigma_1^2 + \cdots + \sigma_k^2}{\sigma_1^2 + \cdots + \sigma_r^2}. \tag{5.19}
$$

If this ratio is high, we can say that much of the variation in the data can be observed on the projected k-dimensional subspace.

Example 5.3 (*PCA of market data*) As a numerical example, we considered data consisting in the returns of six financial indices: (1) the MSCI US index, (2) the MSCI EUR index, (3) the MSCI JAP index, (4) the MSCI PACIFIC index, the (5) MSCI BOT liquidity index, and the (6) MSCI WORLD index. We used monthly return data, from Feb. 26, 1993 to Feb. 28, 2007, for a total of 169 data points, as shown in Figure 5.6.

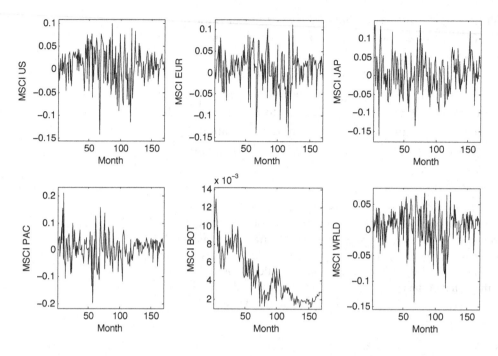

Figure 5.6 Monthly returns of six financial indices.

The data matrix X has thus $m = 169$ data points in dimension $n = 6$. Centering the data, and performing the SVD on the centered data matrix $\tilde{X}$, we obtain the principal axes u_i, and the corresponding singular values:

$$
U = \begin{bmatrix}
-0.4143 & 0.2287 & -0.3865 & -0.658 & 0.0379 & -0.4385 \\
-0.4671 & 0.1714 & -0.3621 & 0.7428 & 0.0172 & -0.2632 \\
-0.4075 & -0.9057 & 0.0690 & -0.0431 & 0.0020 & -0.0832 \\
-0.5199 & 0.2986 & 0.7995 & -0.0173 & 0.0056 & -0.0315 \\
-0.0019 & 0.0057 & 0.0005 & -0.0053 & -0.9972 & -0.0739 \\
-0.4169 & 0.0937 & -0.2746 & -0.1146 & -0.0612 & 0.8515
\end{bmatrix}
$$

$$
\sigma = \begin{bmatrix} 1.0765 & 0.5363 & 0.4459 & 0.2519 & 0.0354 & 0.0114 \end{bmatrix}.
$$

Computing the ratios η_k in (5.19), we have

$$
\eta \times 100 = [67.77 \quad 84.58 \quad 96.21 \quad 99.92 \quad 99.99 \quad 100].
$$

From this we deduce, for instance, that over 96% of the variability in the returns of these six assets can be explained in terms of only three implicit "factors" (say, $z = [z_1 \ z_2 \ z_3]^\top$). In statistical terms, this means that each realization of the return vector $x \in \mathbb{R}^6$ can be expressed (up to a 96% "approximation") as

$$
x = \bar{x} + U_3 z,
$$

where z is a zero-mean vector of random factors, and $U_3 = [u_1 \ u_2 \ u_3]$ is the *factor loading* matrix, composed of the first three principal directions of the data.

5.3.2.3   Computing the PCA. In practice, one is interested in computing just a few principal directions, hence the full SVD is not required. The power iteration algorithm, described in Section 12.5.3, allows us to address PCA problems when only a few directions are sought, even for large data sets.

5.3.2.4   Link with low-rank matrix approximation. The PCA problem is closely linked with the low-rank approximation problem examined in Section 5.3.1. Precisely, the vector z that is optimal for problem (5.18) is nothing else than an eigenvector of the matrix $\tilde{X}\tilde{X}^\top$ that corresponds to the largest eigenvalue of that matrix. As such, it is a left singular vector corresponding to the largest eigenvalue of the centered data matrix $\tilde{X}$ (see Corollary 5.1).

In effect, we can perform a low-rank approximation to the centered data matrix, and the resulting low-rank matrix provides the principal components. For example, if we solve the rank-one approximation problem

$$\min_{u,v,\sigma} \|\tilde{X} - \sigma u v^\top\|_\mathrm{F} \; : \; \|u\|_2 = \|v\|_2 = 1, \; \sigma \geq 0,$$

then at optimum $u = u_1$ is the principal direction with the largest variance.[6] The k-rank approximation to an $n \times m$ data matrix X:

$$X \approx \sum_{i=1}^{k} \sigma_i u_i v_i^\top$$

allows one to interpret it as a sum of k different "factors," where each factor is a dyad of the form $pq^\top$.

For instance, linking back to Example 5.3, if our data matrix X contains the returns of different assets over a time period, with X_{ij} being the one-period return of asset i in period j, then a rank-one approximation $X \approx pq^\top$, so that $X_{ij} = p_i q_j$ for every pair (i, j), is such that vector q could be interpreted as the return of a "typical" asset, while vector p contains some positive or negative factors that are specific to each asset. A general low-rank approximation can be interpreted as an effort to represent data as a (small) linear combination of typical profiles, each asset assigning its own scaling to each profile.

[6] See Section 12.5.3

5.3.3   Procrustean transformation problems

The problem of approximately superimposing two ordered groups of three-dimensional points by means of a rigid displacement (rotation and translation) is a classical one in robotics, manufacturing, and computer vision, where it is encountered under various names such as the absolute orientation, pose estimation, procrustean transformation problem, or the matching problem. In this section, we illustrate the matching problem for data in generic dimension n, and highlight its connection with the SVD. Let

$$A = [a_1 \cdots a_m] \in \mathbb{R}^{n,m}, \quad B = [b_1 \cdots b_m] \in \mathbb{R}^{n,m}$$

be given matrices containing by columns two ordered sets of points. The matching problem considered here amounts to determining a rigid rotation and a translation of the data set B that brings this set to approximately match the A set. In a manufacturing context, for instance, the data points in A are interpreted as "template" points of a machined part (e.g, from computer-assisted design, or CAD), while the data in B are points physically measured on the actual machined part, and one is interested in knowing whether the actual part is in tolerance, by bringing it to match the template by a suitable displacement. Formally, the rotation is represented by means of an orthogonal matrix $R \in \mathbb{R}^{n,n}$, and the translation by a vector $t \in \mathbb{R}^n$. The displaced points are described by the matrix

$$B_d = RB + t\mathbf{1}^\top.$$

The problem is to minimize the matching error, as measured by the squared Frobenius norm of $A - B_d$:

$$\begin{aligned} \min_{R,t} \quad & \|A - (RB + t\mathbf{1}^\top)\|_F^2 \\ \text{s.t.:} \quad & RR^\top = I_n. \end{aligned} \tag{5.20}$$

The following theorem holds.

Theorem 5.2 *Given $A, B \in \mathbb{R}^{n,m}$, let*

$$\begin{aligned} P &= I_m - \frac{1}{m}\mathbf{1}\mathbf{1}^\top, \\ \tilde{A} &= AP, \\ \tilde{B} &= BP \end{aligned}$$

and let

$$\tilde{B}\tilde{A}^\top = U\tilde{\Sigma}V^\top$$

be an SVD for $\tilde{B}\tilde{A}^\top$. Then, an optimal solution for problem (5.20) is given by

$$\begin{aligned} R^* &= VU^\top, \\ t^* &= \frac{1}{m}(A - R^*B)\mathbf{1}. \end{aligned}$$

Proof For any fixed R, the objective (5.20) as a function of t is

$$
\begin{aligned}
f_0(R,t) &= \|A - RB - t\mathbf{1}^\top\|_{\mathrm{F}}^2 \\
&= \|A - RB\|_{\mathrm{F}}^2 - 2\operatorname{trace}(A - RB)\mathbf{1}t^\top + mt^\top t.
\end{aligned}
$$

This can be minimized with respect to t by setting the gradient with respect to t to zero:

$$
\nabla_t f_0 = -2^\top(A - RB)\mathbf{1} + 2mt = 0,
$$

which yields the optimal translation vector as a function of R:

$$
t = \frac{1}{m}(A - RB)\mathbf{1}.
$$

Substituting this t back into the objective we obtain

$$
f_0(R) = \|\tilde{A} - R\tilde{B}\|_{\mathrm{F}}^2.
$$

The minimum of $f_0(R)$ over orthogonal matrices can be determined as follows: recall that for orthogonal R it holds that $\|R\tilde{B}\|_{\mathrm{F}} = \|\tilde{B}\|_{\mathrm{F}}$, and let $U\tilde{\Sigma}V^\top$ be the singular value factorization of $\tilde{B}\tilde{A}^\top$. Then, we write

$$
\begin{aligned}
\|\tilde{A} - R\tilde{B}\|_{\mathrm{F}}^2 &= \|\tilde{A}\|_{\mathrm{F}}^2 + \|R\tilde{B}\|_{\mathrm{F}}^2 - 2\operatorname{trace}R\tilde{B}\tilde{A}^\top \\
&= \|\tilde{A}\|_{\mathrm{F}}^2 + \|\tilde{B}\|_{\mathrm{F}}^2 - 2\operatorname{trace}T\tilde{\Sigma} \\
&= \|\tilde{A}\|_{\mathrm{F}}^2 + \|\tilde{B}\|_{\mathrm{F}}^2 - 2\sum_{i=1}^{r} T_{ii}\sigma_i, \quad\quad (5.21)
\end{aligned}
$$

where $T \doteq V^\top RU$ is an orthogonal matrix, and $r = \operatorname{rank}(\tilde{B}\tilde{A}^\top)$. Clearly, (5.21) is minimized if $\sum_{i=1}^{r} T_{ii}\sigma_i$ is maximized. Since orthogonality of T imposes $|T_{ii}| \le 1$, the maximum is achieved by choosing $T_{ii} = 1$, i.e. $T = I_n$, which results in an optimal orthogonal matrix $R^* = VU^\top$. Notice finally that real orthogonal matrices may represent reflections as well as proper rotations. Proper rotations are characterized by the conditions that $\det R = +1$, whereas for reflections we have $\det R = -1$. If in the problem we insist on having proper rotation matrices, and it turns out that $\det R^* = -1$, then we may renormalize the last column of U by multiplying it by -1. $\square$

Example 5.4 We show a simple example consisting of $n = 4$ points in $\mathbb{R}^3$, as shown in Figure 5.7, with

$$
A = \begin{bmatrix} 1 & 1 & -1 & -1 \\ 1 & -1 & -1 & 1 \\ 0 & 0 & 0 & 0 \end{bmatrix},
$$

$$
B = \begin{bmatrix} 3.0000 & 4.4142 & 3.0000 & 1.6858 \\ 2.5142 & 0.9000 & -0.5142 & 1.0000 \\ 0 & 0 & 0 & 0 \end{bmatrix}.
$$

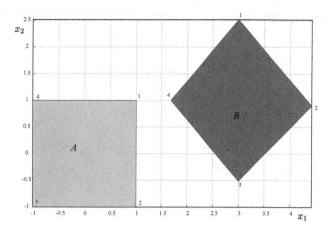

Figure 5.7 Matching two-dimensional data sets.

We want to determine a displacement that superimposes B on A, minimizing the matching cost (5.20). Applying Theorem 5.2, we compute the SVD

$$\tilde{B}\tilde{A}^\top = \begin{bmatrix} 2.7284 & -2.7284 & 0 \\ 2.9284 & 3.1284 & 0 \\ 0 & 0 & 0 \end{bmatrix} = U\tilde{\Sigma}V^\top,$$

where

$$U = \begin{bmatrix} -0.1516 & 0.9884 & 0 \\ 0.9884 & 0.1516 & 0 \\ 0 & 0 & 1 \end{bmatrix},$$

$$\tilde{\Sigma} = \begin{bmatrix} 4.2949 & 0 & 0 \\ 0 & 3.8477 & 0 \\ 0 & 0 & 0 \end{bmatrix},$$

$$V^\top = \begin{bmatrix} 0.5776 & 0.8163 & 0 \\ 0.8163 & -0.5776 & 0 \\ 0 & 0 & 1 \end{bmatrix}.$$

Hence

$$R^* = VU^\top = \begin{bmatrix} 0.7193 & 0.6947 & 0 \\ -0.6947 & 0.7193 & 0 \\ 0 & 0 & 1.0000 \end{bmatrix},$$

$$t^* = \frac{1}{m}(A - R^*B)\mathbf{1} = \begin{bmatrix} -2.8532 \\ 1.4002 \\ 0 \end{bmatrix},$$

with optimal matching error

$$\|A - R^*B - t^*\mathbf{1}^\top\|_F = 0.1804.$$

Matrix R^* corresponds to a rotation of $-44.0048°$ around an axis pointing outwards to the x_1, x_2 plane.

5.4 Exercises

Exercise 5.1 (SVD of an orthogonal matrix) Consider the matrix

$$A = \frac{1}{3} \begin{bmatrix} -1 & 2 & 2 \\ 2 & -1 & 2 \\ 2 & 2 & -1 \end{bmatrix}.$$

1. Show that A is orthogonal.

2. Find a singular value decomposition of A.

Exercise 5.2 (SVD of a matrix with orthogonal columns) Assume a matrix $A = [a_1, \dots, a_m]$ has columns $a_i \in \mathbb{R}^n$, $i = 1, \dots, m$ that are orthogonal to each other: $a_i^\top a_j = 0$ for $1 \le i \ne j \le n$. Find an SVD for A, in terms of the a_is. Be as explicit as you can.

Exercise 5.3 (Singular values of augmented matrix) Let $A \in \mathbb{R}^{n,m}$, with $n \ge m$, have singular values $\sigma_1, \dots, \sigma_m$.

1. Show that the singular values of the $(n + m) \times m$ matrix

$$\tilde{A} \doteq \begin{bmatrix} A \\ I_m \end{bmatrix}$$

are $\tilde{\sigma}_i = \sqrt{1 + \sigma_i^2}$, $i = 1, \dots, m$.

2. Find an SVD of the matrix $\tilde{A}$.

Exercise 5.4 (SVD of score matrix) An exam with m questions is given to n students. The instructor collects all the grades in a $n \times m$ matrix G, with G_{ij} the grade obtained by student i on question j. We would like to assign a difficulty score to each question, based on the available data.

1. Assume that the grade matrix G is well approximated by a rank-one matrix $sq^\top$, with $s \in \mathbb{R}^n$ and $q \in \mathbb{R}^m$ (you may assume that both s, q have non-negative components). Explain how to use the approximation to assign a difficulty level to each question. What is the interpretation of vector s?

2. How would you compute a rank-one approximation to G? State precisely your answer in terms of the SVD of G.

Exercise 5.5 (Latent semantic indexing) Latent semantic indexing is an SVD-based technique that can be used to discover text documents similar to each other. Assume that we are given a set of m documents $D_1, \dots, D_m$. Using a "bag-of-words" technique described in

Example 2.1, we can represent each document D_j by an n-vector d_j, where n is the total number of distinct words appearing in the whole set of documents. In this exercise, we assume that the vectors d_j are constructed as follows: $d_j(i) = 1$ if word i appears in document D_j, and 0 otherwise. We refer to the $n \times m$ matrix $M = [d_1, \ldots, d_m]$ as the "raw" term-by-document matrix. We will also use a normalized[7] version of that matrix: $\tilde{M} = [\tilde{d}_1, \ldots, \tilde{d}_m]$, where $\tilde{d}_j = d_j / \|d_j\|_2$, $j = 1, \ldots, m$.

Assume we are given another document, referred to as the "query document," which is not part of the collection. We describe that query document as an n-dimensional vector q, with zeros everywhere, except a 1 at indices corresponding to the terms that appear in the query. We seek to retrieve documents that are "most similar" to the query, in some sense. We denote by $\tilde{q}$ the normalized vector $\tilde{q} = q / \|q\|_2$.

1. A first approach is to select the documents that contain the largest number of terms in common with the query document. Explain how to implement this approach, based on a certain matrix–vector product, which you will determine.

2. Another approach is to find the closest document by selecting the index j such that $\|q - d_j\|_2$ is the smallest. This approach can introduce some biases, if for example the query document is much shorter than the other documents. Hence a measure of similarity based on the normalized vectors, $\|\tilde{q} - \tilde{d}_j\|_2$, has been proposed, under the name of "cosine similarity". Justify the use of this name for that method, and provide a formulation based on a certain matrix–vector product, which you will determine.

3. Assume that the normalized matrix $\tilde{M}$ has an SVD $\tilde{M} = U\Sigma V^\top$, with Σ an $n \times m$ matrix containing the singular values, and the unitary matrices $U = [u_1, \ldots, u_n]$, $V = [v_1, \ldots, v_m]$ of size $n \times n$, $m \times m$ respectively. What could be an interpretation of the vectors $u_l, v_l, l = 1, \ldots, r$? *Hint:* discuss the case when r is very small, and the vectors $u_l, v_l, l = 1, \ldots, r$, are sparse.

4. With real-life text collections, it is often observed that M is effectively close to a low-rank matrix. Assume that a optimal rank-k approximation ($k \ll \min(n, m)$) of $\tilde{M}$, $\tilde{M}_k$, is known. In the latent semantic indexing approach[8] to document similarity, the idea is to first project the documents and the query onto the subspace generated by the singular vectors $u_1, \ldots, u_k$, and then apply the cosine similarity approach to the projected vectors. Find an expression for the measure of similarity.

[7] In practice, other numerical representation of text documents can be used. For example we may use the relative frequencies of words in each document, instead of the ℓ_2-norm normalization employed here.

[8] In practice, it is often observed that this method produces better results than cosine similarity in the original space, as in part 2.

Exercise 5.6 (Fitting a hyperplane to data) We are given m data points $d_1, \ldots, d_m \in \mathbb{R}^n$, and we seek a hyperplane

$$\mathcal{H}(c,b) \doteq \{x \in \mathbb{R}^n : c^\top x = b\},$$

where $c \in \mathbb{R}^n$, $c \neq 0$, and $b \in \mathbb{R}$, that best "fits" the given points, in the sense of a minimum sum of squared distances criterion, see Figure 5.8.

Formally, we need to solve the optimization problem

$$\min_{c,b} \; \sum_{i=1}^{m} \text{dist}^2(d_i, \mathcal{H}(c,b)) \; : \; \|c\|_2 = 1,$$

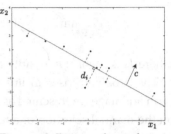

Figure 5.8 Fitting a hyperplane to data.

where $\text{dist}(d, \mathcal{H})$ is the Euclidean distance from a point d to $\mathcal{H}$. Here the constraint on c is imposed without loss of generality, in a way that does not favor a particular direction in space.

1. Show that the distance from a given point $d \in \mathbb{R}^n$ to $\mathcal{H}$ is given by

$$\text{dist}(d, \mathcal{H}(c,b)) = |c^\top d - b|.$$

2. Show that the problem can be expressed as

$$\min_{b,c \,:\, \|c\|_2=1} \; f_0(b,c),$$

where f_0 is a certain quadratic function, which you will determine.

3. Show that the problem can be reduced to

$$\min_{c} \; c^\top(\tilde{D}\tilde{D}^\top)c$$
$$\text{s.t.:} \quad \|c\|_2 = 1,$$

where $\tilde{D}$ is the matrix of centered data points: the i-th column of $\tilde{D}$ is $d_i - \bar{d}$, where $\bar{d} \doteq (1/m)\sum_{i=1}^{m} d_i$ is the average of the data points. *Hint:* you can exploit the fact that at optimum, the partial derivative of the objective function with respect to b must be zero, a fact justified in Section 8.4.1.

4. Explain how to find the hyperplane via SVD.

Exercise 5.7 (Image deformation) A rigid transformation is a mapping from $\mathbb{R}^n$ to $\mathbb{R}^n$ that is the composition of a translation and a rotation. Mathematically, we can express a rigid transformation ϕ as $\phi(x) = Rx + r$, where R is an $n \times n$ orthogonal transformation and $r \in \mathbb{R}^n$ a vector.

We are given a set of pairs of points (x_i, y_i) in $\mathbb{R}^n$, $i = 1, \ldots, m$, and wish to find a rigid transformation that best matches them. We can write the problem as

$$\min_{R \in \mathbb{R}^{n,n}, r \in \mathbb{R}^n} \sum_{i=1}^m \|Rx_i + r - y_i\|_2^2 \; : \; R^\top R = I_n, \qquad (5.22)$$

where I_n is the $n \times n$ identity matrix.

The problem arises in image processing, to provide ways to deform an image (represented as a set of two-dimensional points) based on the manual selection of a few points and their transformed counterparts.

Figure 5.9 Image deformation via rigid transformation. The image on the left is the original image, and that on the right is the deformed image. Dots indicate points for which the deformation is chosen by the user.

1. Assume that R is fixed in problem (5.22). Express an optimal r as a function of R.

2. Show that the corresponding optimal value (now a function of R only) can be written as the original objective function, with $r = 0$ and x_i, y_i replaced with their centered counterparts,

$$\bar{x}_i = x_i - \hat{x}, \;\; \hat{x} = \frac{1}{m} \sum_{j=1}^m x_j, \;\; \bar{y}_i = y_i - \hat{y}, \;\; \hat{y} = \frac{1}{m} \sum_{j=1}^m y_j.$$

3. Show that the problem can be written as

$$\min_R \|RX - Y\|_{\mathrm{F}} \; : \; R^\top R = I_n,$$

for appropriate matrices X, Y, which you will determine. *Hint:* explain why you can square the objective; then expand.

4. Show that the problem can be further written as

$$\max_R \operatorname{trace} RZ \; : \; R^\top R = I_n,$$

for an appropriate $n \times n$ matrix Z, which you will determine.

5. Show that $R = VU^\top$ is optimal, where $Z = USV^\top$ is the SVD of Z. *Hint:* reduce the problem to the case when Z is diagonal, and use without proof the fact that when Z is diagonal, I_n is optimal for the problem.

6. Show the result you used in the previous question: assume Z is diagonal, and show that $R = I_n$ is optimal for the problem above. *Hint:* show that $R^\top R = I_n$ implies $|R_{ii}| \leq 1$, $i = 1, \ldots, n$, and using that fact, prove that the optimal value is less than or equal to trace Z.

7. How woud you apply this technique to make Mona Lisa smile more? *Hint:* in Figure 5.9, the two-dimensional points x_i are given (as dots) on the left panel, while the corresponding points y_i are shown on the left panel. These points are manually selected. The problem is to find how to transform all the other points in the original image.

6

Linear equations and least squares

WE INTRODUCE HERE LINEAR EQUATIONS and a standard formalism for their representation as an element-wise vector equality of the form $Ax = y$, where $x \in \mathbb{R}^n$ is the unknown variable, $A \in \mathbb{R}^{m,n}$ is the coefficient matrix, and $y \in \mathbb{R}^m$ is the known term vector. Linear equations constitute a fundamental building block of numerical linear algebra, and their solution is usually a key part in many optimization algorithms. Actually, the problem of finding a solution to a set of linear equations $Ax = y$ can also be interpreted as an optimization problem, that of minimizing $\|Ax - y\|_2$ with respect to x. We shall characterize the set of solutions of linear equations, and then discuss approximate solution approaches, which are useful when no exact solution exists. This leads to the introduction of the least-squares problem, and its variants. Numerical sensitivity issues and solution techniques are also discussed, together with relations with matrix factorizations such as the QR factorization and the SVD.

6.1   Motivation and examples

Linear equations describe the most basic form of relationship among variables in an engineering problem. Systems of linear equations are ubiquitous in all branches of science: they appear for instance in elastic mechanical systems, relating forces to displacements, in resistive electrical networks, relating voltages to currents, in curve fitting, in many geometrical problems such as triangulation, trilateration, and localization from relative position measurements, in discrete-time dynamical systems relating input and output signals, etc. Linear equations form the core of linear algebra, and often arise as constraints in optimization problems. They are also an important building block of optimization methods, since many optimization algorithms rely on solution of a set of linear equations as a key step in the algorithm's iterations. We next provide a few illustrative examples of linear equations.

Example 6.1 (*An elementary* 3×2 *system*) The following is an example of a system of three equations in two unknowns:

$$
\begin{aligned}
x_1 + 4.5x_2 &= 1, \\
2x_1 + 1.2x_2 &= -3.2, \\
-0.1x_1 + 8.2x_2 &= 1.5.
\end{aligned}
$$

This system can be written in vector format as $Ax = y$, where A is a 3×2 matrix, and y is a 3-vector:

$$
A = \begin{bmatrix} 1 & 4.5 \\ 2 & 1.2 \\ -0.1 & 8.2 \end{bmatrix}, \quad y = \begin{bmatrix} 1 \\ -3.2 \\ 1.5 \end{bmatrix}.
$$

A solution to the linear equations is a vector $x \in \mathbb{R}^2$ that satisfies the equations. In the present example, it can be readily verified by hand calculations that the equations have no solution, i.e., the system is *infeasible*.

Example 6.2 (*Trilateration*) Trilateration is a method for determining the position of a point, given the distances to known control points (anchors). Trilateration can be applied to many different areas such as geographic mapping, seismology, navigation (e.g., GPS systems), etc.

In Figure 6.1, the coordinates of the three anchor points $a_1, a_2, a_3 \in \mathbb{R}^2$ are known, and the distances from point $x = [x_1\ x_2]^\top$ to the anchors are measured as d_1, d_2, d_3. The unknown coordinates of x are related to the distance measurements by three *nonlinear* equations

$$
\|x - a_1\|_2^2 = d_1^2, \quad \|x - a_2\|_2^2 = d_2^2, \quad \|x - a_3\|_2^2 = d_3^2.
$$

However, by subtracting the second and the third equation from the first one, we obtain a system of two *linear* equations in the variable x

$$
\begin{aligned}
2(a_2 - a_1)^\top x &= d_1^2 - d_2^2 + \|a_2\|_2^2 - \|a_1\|_2^2, \\
2(a_3 - a_1)^\top x &= d_1^2 - d_3^2 + \|a_3\|_2^2 - \|a_1\|_2^2,
\end{aligned}
$$

such that each solution to the original nonlinear system is also a solution to this linear system[1] of two equations in two unknowns (the desired coordinates). This system is put in the standard vector format $Ax = y$, with

$$
A = \begin{bmatrix} 2(a_2 - a_1)^\top \\ 2(a_3 - a_1)^\top \end{bmatrix}, \quad y = \begin{bmatrix} d_1^2 - d_2^2 + \|a_2\|_2^2 - \|a_1\|_2^2 \\ d_1^2 - d_3^2 + \|a_3\|_2^2 - \|a_1\|_2^2 \end{bmatrix}.
$$

The matrix A is invertible whenever vectors $a_2 - a_1$ and $a_3 - a_1$ are not parallel, i.e., whenever the three centers are not collinear. We shall see that, when A is invertible, the linear system has a unique solution. Under

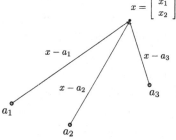

Figure 6.1 A trilateration problem on a plane (view from above). At point x, we measure the distances from three beacons a_1, a_2, a_3, in order to determine the coordinates of x.

[1] The converse statement, however, may not hold. That is, not every solution to the linear system is necessarily a solution to the original system of nonlinear equations. For a given solution x^* of the linear system, we have to check a-posteriori whether $\|x^* - a_3\|_2^2 = d_3^2$ also holds, to ensure that x^* is also a solution to the original system of nonlinear equations.

such a hypothesis, if the solution of the linear system also satisfies $\|x - a_3\|_2^2 = d_3^2$, then we obtained the (unique) solution of the original system of nonlinear equations; if it does not, then we conclude that the original system has no solution (the measurements are inconsistent).

Example 6.3 (*Force/torque generation*) Consider a rigid body moving in a horizontal plane, equipped with n thrusters as shown in Figure 6.2. Each thruster i is placed at coordinates (x_i, y_i) with respect to the center of mass, and can impress a force of intensity f_i to the rigid body, along its direction of action θ_i. Suppose we want to impress to the body an overall resultant force $f = [f_x \; f_y]^\top$ and resultant torque τ: determine the intensities f_i, $i = 1, \ldots, n$, such that the desired overall resultant force and torque are attained. Notice that the resultant force along the x and y axes is given by $\sum_{i=1}^n f_i \cos \theta_i$ and $\sum_{i=1}^n f_i \sin \theta_i$, respectively, while the resultant torque is $\sum_{i=1}^n f_i (y_i \cos \theta_i - x_i \sin \theta_i)$.

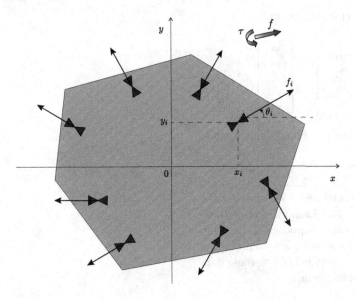

Figure 6.2 Determine thrust intensities f_i, $i = 1, \ldots, n$, so that the resultant force f and torque τ are attained.

In order to match the desired force and torque, the thruster intensities should therefore satisfy the following system of three linear equations in the n unknowns $f_1, \ldots, f_n$:

$$
\begin{aligned}
f_1 \cos \theta_1 + \cdots + f_n \cos \theta_n &= f_x, \\
f_1 \sin \theta_1 + \cdots + f_n \sin \theta_n &= f_y, \\
f_1 \alpha_1 + \cdots + f_n \alpha_n &= \tau,
\end{aligned}
$$

where we defined the coefficients $\alpha_i \doteq y_i \cos \theta_i - x_i \sin \theta_i$, $i = 1, \ldots, n$. This system of linear equations can be written in more compact vector notation as

$$
\begin{bmatrix}
\cos \theta_1 & \cdots & \cos \theta_n \\
\sin \theta_1 & \cdots & \sin \theta_n \\
\alpha_1 & \cdots & \alpha_n
\end{bmatrix}
\begin{bmatrix}
f_1 \\
\vdots \\
f_n
\end{bmatrix}
=
\begin{bmatrix}
f_x \\
f_y \\
\tau
\end{bmatrix}.
$$

Example 6.4 (*Polynomial interpolation*) Consider the problem of interpolating a given set of points (x_i, y_i), $i = 1, \ldots, m$, with a polynomial of degree $n - 1$:

$$p(x) = a_{n-1}x^{n-1} + \cdots + a_1 x + a_0.$$

Clearly, the polynomial interpolates the i-th point if and only if $p(x_i) = y_i$, and each of such conditions is a linear equation in the polynomial coefficients a_j, $j = 0, \ldots, n - 1$. An interpolating polynomial is hence found if the following system of linear equations in the a_j variables has a solution:

$$\begin{aligned}
a_0 + x_1 a_1 + \cdots + a_{n-1} x_1^{n-1} &= y_1, \\
a_0 + x_2 a_1 + \cdots + a_{n-1} x_2^{n-1} &= y_2, \\
&\vdots \quad \vdots \quad \vdots \\
a_0 + x_m a_1 + \cdots + a_{n-1} x_m^{n-1} &= y_m.
\end{aligned}$$

This system can be rewritten in compact vector notation as

$$\begin{bmatrix} 1 & x_1 & x_1^2 & \cdots & x_1^{n-1} \\ 1 & x_2 & x_2^2 & \cdots & x_2^{n-1} \\ \vdots & & \cdots & \cdots & \vdots \\ 1 & x_m & x_m^2 & \cdots & x_m^{n-1} \end{bmatrix} \begin{bmatrix} a_0 \\ a_1 \\ a_2 \\ \vdots \\ a_{n-1} \end{bmatrix} = \begin{bmatrix} y_1 \\ y_2 \\ \vdots \\ y_m \end{bmatrix},$$

where the matrix of coefficients on the left has a so-called Vandermonde structure.

Example 6.5 (*Fitting a power law to experimental data*) We consider the problem of constructing a power-law model (see Example 2.11) that explains a batch of experimentally observed data. Suppose we are given experimental data with input vectors $x^{(i)} > 0$ and associated outputs $y_i > 0$, $i = 1, \ldots, m$, and that we have an *a priori* belief that these data may come from a power-law form model of the form

$$y = \alpha x_1^{a_1} \cdots x_n^{a_n}.$$

Here, the variables of the problem are $\alpha > 0$, and the vector $a \in \mathbb{R}^n$. Taking logarithms, we have that each observation produces a linear equation in the a_i variables:

$$\tilde{y}_i = a^\top \tilde{x}^{(i)} + b, \quad i = 1, \ldots, m, \tag{6.1}$$

where we defined

$$b = \log \alpha, \quad \tilde{x}_i = \log x_i, \quad \tilde{y}_i = \log y_i.$$

These equations form a system of linear equations that can be written in compact matrix form as follows:

$$\begin{bmatrix} \tilde{y}_1 \\ \vdots \\ \tilde{y}_m \end{bmatrix} = \begin{bmatrix} \tilde{x}^{(1)\top} & 1 \\ \vdots & \vdots \\ \tilde{x}^{(m)\top} & 1 \end{bmatrix} \begin{bmatrix} a \\ b \end{bmatrix}.$$

In practice, one cannot expect that the experimental data are perfectly explained by equation (6.1). It is much more realistic to assume that the power model explains the observations up to a certain residual error, that is

$$\tilde{y}_i = a^\top \tilde{x}^{(i)} + b + r_i, \quad i = 1, \ldots, m,$$

where $r = [r_1 \cdots r_m]^\top$ is the vector of residuals, accounting for the mismatch between model and reality. In this situation, it is very reasonable to seek a model (i.e., an (a, b) vector) that makes the mismatch minimal in some sense. If the Euclidean norm of r is chosen as the mismatch criterion, then the best fit model can be found by solving the following optimization problem:

$$\min_z \|Xz - \tilde{y}\|_2,$$

where $z = [a^\top, b]^\top \in \mathbb{R}^{n+1}$, and $X \in \mathbb{R}^{(n+1),m}$ is a matrix whose i-th column is $[\tilde{x}^{(1)\top}, 1]^\top$. This problem belongs to the class of so-called least-squares problems, which are discussed in Section 6.3.1.

Example 6.6 (*CAT scan imaging*) Tomography means reconstruction of an image from its sections. The word comes from the Greek "tomos" (slice) and "graph" (description). The problem arises in many fields, ranging from astronomy to medical imaging. Computerized axial tomography (CAT) is a medical imaging method that processes large amounts of two-dimensional X-ray images in order to produce a three-dimensional image. The goal is to picture, for example, the tissue density of the different parts of the brain, in order to detect anomalies, such as brain tumors.

Typically, the X-ray images represent "slices" of the part of the body that is examined. Those slices are indirectly obtained via axial measurements of X-ray attenuation, as explained below. Thus, in CAT for medical imaging, one uses axial (line) measurements to get two-dimensional images (slices), and from those slices one may proceed to digitally reconstruct a three-dimensional view. Here, we focus on the process that produces a single two-dimensional image from axial measurements. Figure 6.3 shows a collection of slices of a human brain obtained by CAT scan. The pictures offer an image of the density of tissue in the various parts of the brain. Each slice is actually a reconstructed image obtained by a tomography technique explained below.

From 1D to 2D: axial tomography. In CAT-based medical imaging, a number of X-rays are sent through the tissues to be examined along different directions, and their intensity after they have traversed the tissues is captured by a receiver sensor. For each direction, we record the attenuation of the X-ray, by comparing the intensity of the X-ray at the source to the intensity after the X-ray has traversed the tissues, at the receiver's end, see Figure 6.4.

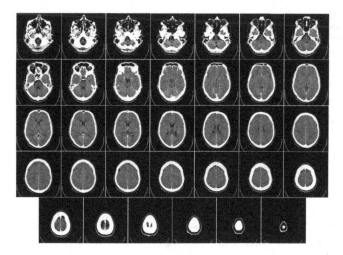

Figure 6.3 CAT scan slices of a human brain (Source: Wikipedia).

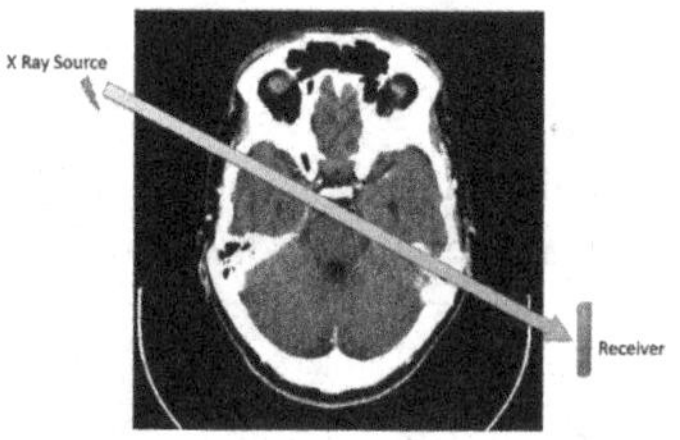

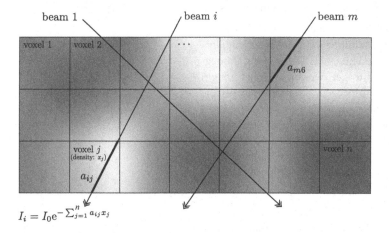

Figure 6.4 X-ray traversing tissues (Source: Wikipedia).

Similarly to the Beer–Lambert law of optics, it turns out that, to a reasonable degree of approximation, the log-ratio of the intensities at the source and at the receiver is linear in the densities of the tissues traversed. To formalize this idea, consider a discretized version of a rectangular slice of a tissue, divided into a number n of volume elements (called *voxels*), see Figure 6.5, each having unknown density x_j, $j = 1, \ldots, n$.

Figure 6.5 Beams traversing a section of tissue divided into n voxels.

beam 1   beam i beam m

voxel 1   voxel 2 $\cdots$

a_{m6}

voxel j
(density: x_j)

a_{ij}

voxel n

$$I_i = I_0 e^{-\sum_{j=1}^n a_{ij} x_j}$$

A (typically large) number m of beams of intensity I_0 at the source travel across the tissue: the i-th beam, $i = 1, \ldots, m$, has a path of length a_{ij} through voxel j, $j = 1, \ldots, n$. The log-attenuation of the i-th beam intensity due to the j-th voxel is proportional to the density of the voxel x_j times the length of the path, that is $a_{ij} x_j$. The total log-attenuation for beam i is therefore given by the sum of the log-attenuations:

$$y_i = \log \frac{I_0}{I_i} = \sum_{j=1}^n a_{ij} x_j, \quad i = 1, \ldots, m,$$

where I_i is the intensity of the i-th beam at the receiver end. As a simplified example, consider a square section containing four voxels traversed by four beams, as shown in Figure 6.6.

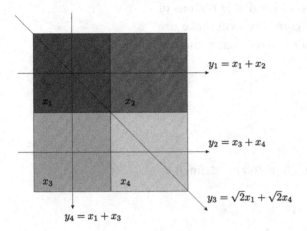

$$y_1 = x_1 + x_2$$

$$y_2 = x_3 + x_4$$

$$y_3 = \sqrt{2}x_1 + \sqrt{2}x_4$$

$$y_4 = x_1 + x_3$$

Figure 6.6 A simplified 4 voxel example.

The vector of unknown densities $x = [x_1 \cdots x_4]^\top$ is linearly related to the vector of observed log-intensity ratios $y = [y_1 \cdots y_4]^\top$ via the system of linear equations

$$\begin{bmatrix} y_1 \\ y_2 \\ y_3 \\ y_4 \end{bmatrix} = \begin{bmatrix} 1 & 1 & 0 & 0 \\ 0 & 0 & 1 & 1 \\ \sqrt{2} & 0 & 0 & \sqrt{2} \\ 1 & 0 & 1 & 0 \end{bmatrix} \begin{bmatrix} x_1 \\ x_2 \\ x_3 \\ x_4 \end{bmatrix}.$$

Recovering the densities x_j from the y_i measurements thus amounts to solving a system of linear equations of the form $y = Ax$, where $A \in \mathbb{R}^{m,n}$. Note that depending on the number n of voxels used, and on the number m of measurements, the matrix A can be quite large. In general, the matrix is "fat," in the sense that it has (many) more columns than rows ($n \gg m$). Thus, the system of equations resulting from CAT scan problems is usually *underdetermined* (has more unknowns than equations).

As shown in the previous examples, generic linear equations can be expressed in vector format as

$$Ax = y, \tag{6.2}$$

where $x \in \mathbb{R}^n$ is the vector of unknowns, $y \in \mathbb{R}^m$ is a given vector, and $A \in \mathbb{R}^{m,n}$ is a matrix containing the coefficients of the linear equations. The examples motivate us to address the problems of solving linear equations. They also raise the issues of existence of a solution (does a solution x exist) and unicity (if a solution exists, is it unique?). We next discuss some fundamental properties of linear equations, focusing on issues of existence, uniqueness, and characterization of all possible solutions. We anticipate that, depending on the size and properties of A and y, system (6.2) can have no solution, or a unique solution, or a whole infinity of possible solutions. In the latter case, the set of solutions actually forms a *subspace* of $\mathbb{R}^n$; in the first case (no solution), we shall introduce suitable notions of *approximate solution*. Our analysis of linear equations will make use of much of the definitions and facts related to vector spaces that we introduced in Chapter 2 and Chapter 3.

6.2   The set of solutions of linear equations

6.2.1   Definition and properties

The solution set of the system of linear equations (6.2) is defined as as

$$S \doteq \{x \in \mathbb{R}^n : Ax = y\}.$$

Let $a_1, \ldots, a_n \in \mathbb{R}^m$ denote the columns of A, i.e. $A = [a_1 \cdots a_n]$, and notice that the product Ax is nothing but a linear combination of the columns of A, with coefficients given by x:

$$Ax = x_1 a_1 + \cdots + x_n a_n.$$

We recall that, by definition, the range of a matrix is the subspace generated by its columns, therefore, no matter what the value of the x coefficients is, the vector Ax always lies in $\mathcal{R}(A)$. It then follows that whenever $y \notin \mathcal{R}(A)$ equations (6.2) do not admit a solution (i.e., they are *infeasible*), hence the solution set S is empty. Equivalently, system (6.2) admits a solution if and only if $y \in \mathcal{R}(A)$, that is if and only if y is a linear combination of the columns of A. This condition can be checked via the rank test[2]

$$\text{rank}([A\ y]) = \text{rank}(A). \tag{6.3}$$

Suppose next that condition (6.3) is satisfied, hence a solution $\bar{x}$ exists such that $y = A\bar{x}$. We next show that the solution set is an affine set: notice that another solution $x \neq \bar{x}$ for the system exists if and only if

$$A(x - \bar{x}) = 0,$$

hence $x - \bar{x}$ must lie in the nullspace of A, $\mathcal{N}(A)$. All possible solutions for the system must therefore have the form $x = \bar{x} + z$, for $z \in \mathcal{N}(A)$. That is, the solution set S is the affine set given by a translation of the nullspace of A: $S = \{x = \bar{x} + z : z \in \mathcal{N}(A)\}$. It also follows from this fact that the solution $\bar{x}$ is *unique* if and only if $\mathcal{N}(A) = \{0\}$. We now recap our findings in the following fundamental proposition.

[2] Clearly, it always holds that $\mathcal{R}(A) \subseteq \mathcal{R}([A\,y])$. Thus, the rank test expresses the condition that $\dim \mathcal{R}(A) = \dim \mathcal{R}([A\,y])$, implying that $\mathcal{R}(A) = \mathcal{R}([A\,y])$.

Proposition 6.1 (The solution set of linear equations) *The linear equation*

$$Ax = y, \quad A \in \mathbb{R}^{m,n}$$

admits a solution if and only if $\text{rank}([A\ y]) = \text{rank}(A)$. *When this existence condition is satisfied, the set of all solutions is the affine set*

$$S = \{x = \bar{x} + z : z \in \mathcal{N}(A)\},$$

where $\bar{x}$ is any vector such that $A\bar{x} = y$. In particular, the system has a unique solution if (6.3) is satisfied, and $\mathcal{N}(A) = \{0\}$.

6.2.2 *Underdetermined, overdetermined, and square systems*

We briefly discuss three typical situations that may arise in systems of linear equations, namely when there are more unknowns than equations (underdetermined), when there are more equations than unknowns (overdetermined), and when there are as many equations as unknowns. These three cases are discussed under the hypothesis that A is full rank. The following theorem holds for full rank matrices (see also an equivalent result previously stated in Corollary 4.3).

Theorem 6.1 *The following two statements hold:*

1. *$A \in \mathbb{R}^{m,n}$ is full column rank (i.e., $\text{rank}(A) = n$) if and only if $A^\top A$ is invertible;*

2. *$A \in \mathbb{R}^{m,n}$ is full row rank (i.e., $\text{rank}(A) = m$) if and only if $AA^\top$ is invertible.*

Proof Consider the first point. If $A^\top A$ is not invertible, then there exists $x \neq 0$ such that $A^\top A x = 0$. Then $x^\top A^\top A x = \|Ax\|_2^2 = 0$, hence $Ax = 0$. Hence A is not full column rank. Conversely, if $A^\top A$ is invertible, then $A^\top A x \neq 0$ for every $x \neq 0$, which implies that $Ax \neq 0$ for every nonzero x, as desired. The proof for the second point in the theorem follows similar lines. □

6.2.2.1 Overdetermined systems. The system $Ax = y$ is said to be *overdetermined* when it has more equations than unknowns, i.e., when matrix A has more rows then columns ("skinny" matrix): $m > n$. Assume that A is full column rank, that is $\text{rank}(A) = n$. By (3.3) it follows that $\dim \mathcal{N}(A) = 0$, hence the system has either one or no solution at all. Indeed, the most common case for overdetermined systems is that $y \notin \mathcal{R}(A)$, so that no solution exists. In this case, it is often useful to introduce a notion of approximate solution, that is a solution that renders minimal some suitable measure of the mismatch between Ax and y, as further discussed in Section 6.3.1.

6.2.2.2 Underdetermined systems. The system $Ax = y$ is said to be *underdetermined* if it has more unknowns than equations, i.e., when matrix A has more columns than rows ("wide" matrix): $n > m$. Assume that A is full row rank, that is $\mathrm{rank}(A) = m$, and then $\mathcal{R}(A) = \mathbb{R}^m$. Recall from (3.3) that

$$\mathrm{rank}(A) + \dim \mathcal{N}(A) = n,$$

hence $\dim \mathcal{N}(A) = n - m > 0$. The system of linear equations is therefore solvable with infinite possible solutions, and the set of solutions has "dimension" $n - m$. Among all possible solutions, it is often of interest to single out one specific solution having minimum norm: this issue is discussed in detail in Section 6.3.2.

6.2.2.3 Square systems. The system $Ax = y$ is said to be *square* when the number of equations is equal to the number of unknowns, i.e. when matrix A is square: $m = n$. If a square matrix is full rank, then it is invertible, and the inverse A^{-1} is unique and has the property that $A^{-1}A = I$. In the case of square full rank A the solution of the linear system is thus unique and it is formally written as

$$x = A^{-1}y.$$

Note that the solution x is, however, rarely computed by actually determining A^{-1} and multiplying it by y; see instead Section 7.2 for numerical methods for computing the solution of nonsingular systems of linear equations.

6.3 Least-squares and minimum-norm solutions

6.3.1 Approximate solutions: least squares

When $y \notin \mathcal{R}(A)$, the system of linear equations is infeasible: there is no x such that $Ax = y$. This situation happens frequently in the case of overdetermined systems of equations. In such cases it may, however, make sense to determine an "approximate solution" to the system, that is a solution that renders the *residual* vector $r \doteq Ax - y$ as "small" as possible. A natural way of measuring the size of the residual is by the use of a norm: we thus wish to determine x such that the norm of the residual is minimized. In this section we discuss in particular the most common case where the norm selected for measuring the residual is the standard Euclidean norm, whence the problem becomes

$$\min_{x} \quad \|Ax - y\|_2. \tag{6.4}$$

Since the function z^2 is monotone increasing for $z \geq 0$, the previous problem is also equivalent to minimizing the square of the Euclidean norm:

$$\min_{x} \quad \|Ax - y\|_2^2, \tag{6.5}$$

and from this latter formulation derives the name of *least-squares* (LS) solution of the linear equations, that is a solution that minimizes the sum of the squares of the equation residuals:

$$\|Ax - y\|_2^2 = \sum_{i=1}^{m} (a_i^\top x - y_i)^2,$$

where $a_i^\top$ denotes the i-th row of A.

Problem (6.4) has an interesting geometric interpretation: since the vector Ax lies in $\mathcal{R}(A)$, the problem amounts to determining a point $\tilde{y} = Ax$ in $\mathcal{R}(A)$ at *minimum distance* from y, see also Section 5.2.4.1. The projection theorem (Theorem 2.2) then tells us that this point is indeed the orthogonal projection of y onto the subspace $\mathcal{R}(A)$, see Figure 6.7.

We can thus apply Theorem 2.2 to find an explicit solution to problem (6.5), as formalized in the following proposition.

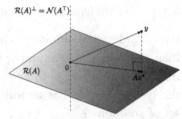

Figure 6.7 Projection of y onto $\mathcal{R}(A)$.

Proposition 6.2 (LS approximate solution of linear equations) *Let* $A \in \mathbb{R}^{m,n}, y \in \mathbb{R}^m$. *The LS problem*

$$\min_{x} \|Ax - y\|_2$$

always admits (at least) a solution. Moreover, any solution $x^* \in \mathbb{R}^n$ *of (6.5) is a solution of the following system of linear equations (the* normal *equations)*

$$A^\top A x^* = A^\top y, \tag{6.6}$$

and vice versa. Further, if A is full column rank (i.e., $\text{rank}(A) = n$), then the solution to (6.5) is unique, and it is given by

$$x^* = (A^\top A)^{-1} A^\top y. \tag{6.7}$$

Proof Given any $y \in \mathbb{R}^m$, by Theorem 2.2 there exists a unique point $\tilde{y} \in \mathcal{R}(A)$ at minimal distance from y, and this point is such that $(y - \tilde{y}) \in \mathcal{R}(A)^\perp \equiv \mathcal{N}(A^\top)$, that is

$$A^\top (y - \tilde{y}) = 0.$$

Since $\tilde{y} \in \mathcal{R}(A)$, there certainly exists an x such that $\tilde{y} = Ax$, which proves that (6.5) admits a solution. Then, substituting $\tilde{y} = Ax$ in the previous orthogonality condition, we have

$$A^\top A x = A^\top y,$$

which shows the equivalence between the LS problem (6.5) and (6.6). Finally, if A is full column rank then, by Theorem 6.1, $A^\top A$ is invertible, hence the unique solution of (6.6) is given by (6.7). $\square$

Remark 6.1 *Normal equations and optimality.* The normal equations are nothing else than the optimality conditions for the optimization problem

$$\min_x f(x),$$

where $f(x) = \|Ax - y\|_2^2$. As we will see in Section 8.4, when the function is differentiable, convex, and the problem has no constraints, optimal points are characterized by the conditions $\nabla f(x) = 0$. In our case, the gradient of f at a point x is easily seen to be $\nabla f(x) = A^\top(Ax - y)$.

6.3.2   *The underdetermined case: minimum-norm solution*

We consider next the case when the matrix A has more columns than rows: $m < n$. Assuming that A is full row rank, we have that $\dim \mathcal{N}(A) = n - m > 0$, hence it follows from Proposition 6.1 that the system $y = Ax$ has an infinite number of solutions and that the set of solutions is $\mathcal{S}_{\bar{x}} = \{x : x = \bar{x} + z, z \in \mathcal{N}(A)\}$, where $\bar{x}$ is any vector such that $A\bar{x} = y$. (We are interested in singling out from the set of solutions $\mathcal{S}_{\bar{x}}$ the one solution x^* with minimal Euclidean norm. That is, we want to solve the problem

$$\min_{x : Ax = y} \|x\|_2,$$

which is equivalent to $\min_{x \in \mathcal{S}_{\bar{x}}} \|x\|_2$. Corollary 2.1 can be directly applied to the case at hand: the (unique) solution x^* must be orthogonal to $\mathcal{N}(A)$ or, equivalently, $x^* \in \mathcal{R}(A^\top)$, which means that $x^* = A^\top \xi$, for some suitable ξ. Since x^* must solve the system of equations, it must be $Ax^* = y$, i.e., $AA^\top \xi = y$. Since A is full row rank, $AA^\top$ is invertible and the unique ξ that solves the previous equation is $\xi = (AA^\top)^{-1}y$. This finally gives us the unique minimum-norm solution of the system:

$$x^* = A^\top(AA^\top)^{-1}y. \tag{6.8}$$

The previous discussion constitutes a proof for the following proposition.

Proposition 6.3 (Minimum norm solution) *Let $A \in \mathbb{R}^{m,n}$, $m \leq n$, be full rank and let $y \in \mathbb{R}^m$. Among the solutions of the system of linear equations $Ax = y$ there exists a unique one having minimal Euclidean norm. This solution is given by (6.8).*

6.3.3   LS and the pseudoinverse

For $A \in \mathbb{R}^{m,n}$, $y \in \mathbb{R}^m$, consider the LS problem

$$\min_x \|Ax - y\|_2. \tag{6.9}$$

Under the hypothesis that a solution to the linear equations $Ax = y$ exists, any solution of these equations is also a minimizer of (6.9) and, vice versa, any minimizer of (6.9) is a solution of the linear equations. Considering problem (6.9) is therefore, in some sense, "more general" that considering the linear equations $Ax = y$, since (6.9) has a solution even when the linear equations do not, and it has the same solution set as $Ax = y$, when this set is nonempty. Notice further that (6.9) has multiple (infinitely many) solutions, whenever A has a non-trivial nullspace. Indeed, all solutions to (6.9) are the solutions of the normal equations (6.6), and these equations have multiple solutions if and only if $\mathcal{N}(A^\top A) = \mathcal{N}(A)$ is non-trivial.

Among all possible solutions to the normal equations $A^\top Ax = A^\top y$ we are now interested in finding the unique minimum norm one (note that, due to (3.16), these equations always admit at least one solution). From Corollary 2.1, we have that the unique minimum norm solution x^* must be orthogonal to $\mathcal{N}(A)$ or, which is the same, must belong to $\mathcal{R}(A^\top)$. Therefore, x^* is uniquely determined by the following two conditions: (a) it must belong to $\mathcal{R}(A^\top)$, and (b) it must satisfy the normal equations (6.6). We claim that such a solution is simply expressed in terms of the Moore–Penrose pseudoinverse as follows:

$$x^* = A^\dagger y. \tag{6.10}$$

This fact is readily proved as follows. Let $A = U_r \Sigma V_r^\top$ be a compact SVD of A. Then, the Moore–Penrose pseudoinverse is expressed in (5.7) as $A^\dagger = V_r \Sigma^{-1} U_r^\top$, hence it follows that $x^* = A^\dagger y = V_r \Sigma^{-1} U_r^\top y = V_r \xi$, thus $x^* \in \mathcal{R}(V_r)$, but from (5.10) we have $\mathcal{R}(V_r) = \mathcal{R}(A^\top)$, hence condition (a) is satisfied by x^* in (6.10). Furthermore,

$$
\begin{aligned}
A^\top A x^* &= A^\top A A^\dagger y = V_r \Sigma U_r^\top U_r \Sigma V_r^\top V_r \Sigma^{-1} U_r^\top y \\
&= V_r \Sigma^2 V_r^\top V_r \Sigma^{-1} U_r^\top y = V_r \Sigma U_r^\top y \\
&= A^\top y,
\end{aligned}
$$

which shows that also condition (b) is satisfied, hence $x^* = A^\dagger y$ provides the unique minimum norm solution to the LS problem (6.9). This is summarized in the following corollary.

Corollary 6.1 (Set of solutions of LS problem) *The set of optimal solutions of the LS problem*

$$p^* = \min_x \|Ax - y\|_2$$

can be expressed as

$$\mathcal{X}_{\text{opt}} = A^\dagger y + \mathcal{N}(A),$$

where $A^\dagger y$ is the minimum-norm point in the optimal set. The optimal value p^* is the norm of the projection of y onto orthogonal complement of $\mathcal{R}(A)$: for $x^* \in \mathcal{X}_{\text{opt}}$,

$$p^* = \|y - Ax^*\|_2 = \|(I_m - AA^\dagger)y\|_2 = \|P_{\mathcal{R}(A)^\perp}y\|_2,$$

where the matrix $P_{\mathcal{R}(A)^\perp}$ is the projector onto $\mathcal{R}(A)^\perp$, defined in (5.12). If A is full column rank, then the solution is unique, and equal to

$$x^* = A^\dagger y = (A^\top A)^{-1} A^\top y.$$

6.3.4 Interpretations of the LS problem

The LS problem (6.4) can be given a variety of different (but of course related) interpretations, depending on the application context. Some of these interpretations are briefly summarized in the next paragraphs, where the first two items have already been diffusely discussed in the previous sections.

6.3.4.1 Approximate solution of linear equations. Given a system of linear equations $Ax = y$ which is possibly infeasible (i.e., may have no exact solution), we relax the requirement, and ask for a solution x that *approximately* solves the system, i.e., such that $Ax \simeq y$. In the LS method, the approximate solution is such that the equation residual vector $r = Ax - y$ has minimal Euclidean norm.

6.3.4.2 Projection onto $\mathcal{R}(A)$. Given a point $y \in \mathbb{R}^m$, the LS problem seeks a coefficient vector x such that y is approximated in the best possible way (according to the Euclidean norm criterion) by a linear combination of the columns $a_1, \ldots, a_n$ of A. An LS solution x^* gives the optimal coefficients for this linear combination, such that

$$y^* = Ax^* = x_1^* a_1 + \cdots + x_n^* a_n$$

is the projection of y onto the subspace spanned by the columns of A.

6.3.4.3 Linear regression. Denoting by $a_i^\top$, $i = 1, \ldots, m$, the rows of A, the LS problem (6.4) can be rewritten as

$$\min_x \sum_{i=1}^m (a_i^\top x - y_i)^2,$$

that is, given "output" points y_i, and "input" points a_i, $i = 1,\ldots,m$, we are trying to approximate the output points with a linear function $f(a_i) = a_i^\top x$ of the input points, where x here is the parameter defining the linear function.

A classical example in two dimensions is the fitting of a straight line through experimental or measured data. Given scalar output observations $y_i \in \mathbb{R}$, and input observations $\xi_i \in \mathbb{R}$, $i = 1,\ldots,m$, we seek an affine function

$$f(\xi) = x_1\xi + x_2 = a^\top x, \quad a = \begin{bmatrix} \xi \\ 1 \end{bmatrix}, \quad x = \begin{bmatrix} x_1 \\ x_2 \end{bmatrix}$$

(x_1 is the slope of the line, x_2 is the intercept with the vertical axis) approximating the output in the LS sense:

$$\min_x \sum_{i=1}^m (x_1\xi_i + x_2 - y_i)^2 = \min_x \sum_{i=1}^m (a_i^\top x - y_i)^2.$$

Figure 6.8 shows an example where the data points ξ_i represent the market price of a given item, and y_i represents the average number of customers who buy the item at price ξ_i. The straight line in the figure represents the linear model obtained from the observed data via an LS fit. This model shows how customers react to variations in the price of a given item and can be used, for instance, to predict the value of the average number of customers buying the item at new price tags.

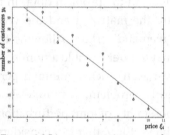

Figure 6.8 Linear regression example.

Another example of a linear regression fit comes from a popular model used for prediction of time series, called the *auto-regressive model*. This model assumes that the value of a discrete time signal y_t is a linear combination of a certain number of past values of the signal itself:

$$y_t = x_1y_{t-1} + \cdots + x_ny_{t-n}, \quad t = 1,\ldots,m,$$

where x_i are constant coefficients, and n is the "memory length" of the model. The interpretation of such a model is that the next output is a linear function of the past. Elaborate variants of auto-regressive models are widely used for prediction of time series arising in finance and economics. If we want to approximately fit an auto-regressive model to an observed signal, we collect observations of the actual signal $\{y_t\}_{1-n \le t \le m}$, with $m \ge n$, and seek parameters x such that the total squared error of fit is minimized:

$$\min_x \sum_{t=1}^m (y_t - x_1y_{t-1} - \cdots - x_ny_{t-n})^2.$$

This problem is readily expressed as an LS problem, with appropriate data A, y.

6.3.4.4   Minimal perturbation to feasibility. Suppose the linear equations $Ax = y$ are infeasible, that is no $x \in \mathbb{R}^n$ satisfies the equations. We may then consider the following *perturbed* version of the problem

$$Ax = y + \delta y,$$

where $\delta y \in \mathbb{R}^m$ is a perturbation on the right-hand side of the equations, and ask what is the smallest (in the Euclidean norm sense) perturbation δy that renders the equations feasible. Clearly, since $\delta y = Ax - y$, the answer is again given by the LS solution x^*, which renders $\|\delta y\|_2$ minimal, that is $\delta y^* = Ax^* - y$. This interpretation raises the important issue of the presence of uncertainty, or perturbations, in the problem data. In the linear equations $Ax = y$ the "data" are the matrix A and the vector y, and these are considered to be given and certain. Allowing for possible perturbations in the y term may render feasible a nominally unfeasible problem and, as we have just seen, such a minimal perturbation is given via the solution of an LS problem. A more elaborate perturbation model considers the possibility of joint perturbations on both the coefficient matrix A and the vector y, that is it considers the perturbed equations

$$(A + \delta A)x = y + \delta y,$$

and seeks the minimal perturbation matrix $\Delta = [\delta A \; \delta y]$ such that the equations become feasible. When the size of the perturbation matrix is measured by the spectral norm, $\|\Delta\|_2$, then determining such a minimal perturbation is known as a *total least-squares* (TLS) problem, see further discussion in Section 6.7.5.

6.3.4.5   Best linear unbiased estimator (BLUE). A further important interpretation of the LS problem arises in the context of statistical estimation. Suppose one assumes a linear statistical model between an unknown deterministic vector of parameters $x \in \mathbb{R}^n$ and its "measurements" $y \in \mathbb{R}^m$:

$$y = Ax + z, \tag{6.11}$$

where z is a vector of random errors, and $A \in \mathbb{R}^{m,n}$ is assumed to be full rank, with $m \geq n$. The meaning of equation (6.11) is that each measurement reading y_i is equal to a linear function $a_i^\top x$ of the unknown parameter x, plus a random noise term z_i. We assume that z has zero mean and unit covariance matrix, that is

$$\mathbb{E}\{z\} = 0, \quad \text{var}\{z\} = \mathbb{E}\{zz^\top\} = I_m.$$

According to this model, the readings vector y is, *a priori*, a random vector with

$$\mathbb{E}\{y\} = Ax, \quad \text{var}\{y\} = \mathbb{E}\{(y - \mathbb{E}\{y\})(y - \mathbb{E}\{y\})^\top\} = I_m.$$

A *linear estimator* $\hat{x}$ of the unknown parameter x is defined as a linear function of y:

$$\hat{x} = Ky, \qquad (6.12)$$

where $K \in \mathbb{R}^{n,m}$ is some to-be-determined *gain* of the estimator. Notice that the estimator $\hat{x}$ is itself a random vector, with

$$\mathbb{E}\{\hat{x}\} = K\mathbb{E}\{y\} = KAx.$$

An estimator is said to be *unbiased* if its expectation coincides with the unknown parameter to be estimated, that is if $\mathbb{E}\{\hat{x}\} = x$. We see from the previous equation that in order for $\hat{x}$ to be an unbiased estimator, for any x, we must have

$$KAx = x,$$

that is, $KA = I_n$, which means that K must be a left inverse of A. According to (5.9), any left inverse of A can be written as follows:

$$K = A^\dagger + Q = (A^\top A)^{-1}A^\top + Q,$$

where $Q \in \mathbb{R}^{n,m}$ is any matrix such that $QA = 0$. A BLUE estimator is an unbiased linear estimator of the form (6.12) that has the minimal possible covariance matrix. Letting

$$\hat{x} = (A^\dagger + Q)y = (A^\dagger + Q)(Ax + z) = x + (A^\dagger + Q)z,$$

the covariance of $\hat{x}$ is

$$\begin{aligned} \mathrm{var}\{\hat{x}\} &= \mathbb{E}\{(\hat{x} - x)(\hat{x} - x)^\top\} = \mathbb{E}\{(A^\dagger + Q)zz^\top(A^\dagger + Q)^\top\} \\ &= (A^\dagger + Q)\mathbb{E}\{zz^\top\}(A^\dagger + Q)^\top = (A^\dagger + Q)(A^\dagger + Q)^\top \\ &= A^\dagger A^{\dagger\top} + QQ^\top, \end{aligned}$$

where the last passage follows from the fact that $A^\dagger Q^\top = 0$. Since $QQ^\top \succeq 0$, it follows from (4.6) that

$$\mathrm{var}\{\hat{x}\} = A^\dagger A^{\dagger\top} + QQ^\top \succeq A^\dagger A^{\dagger\top}, \quad \forall Q,$$

hence the minimal covariance matrix is achieved by taking $Q = 0$, that is with the estimator

$$\hat{x} = Ky, \quad K = A^\dagger = (A^\top A)^{-1}A^\top.$$

The BLUE estimator for the linear model (6.11) therefore coincides with the solution of the LS problem (6.4).

6.3.5   Recursive least squares

In the context of the parameter estimation interpretation of the LS problem discussed in the previous section, we seek to estimate an unknown parameter $x \in \mathbb{R}^n$ from a series of $m \geq n$ noisy linear measurements y_i:

$$y_i = a_i^\top x + z_i, \quad i = 1, \ldots, m,$$

where z_i are independent and identically distributed (iid), zero-mean and unit variance random noise terms, and $a_i^\top$ are the rows of matrix $A \in \mathbb{R}^{m,n}$. In such a context, it makes sense to ask the following question: suppose we observed $m > k \geq n$ measurements and that we solved the estimation problem, determining the optimal estimate $x^{(k)}$ based on the k available measurements. Then, a new measurement y_{k+1} arrives. Can we avoid re-solving the whole problem again from scratch, and instead find a simpler way to *update* the previous estimate by incorporating the new information? The answer to this question is positive, and we next derive a well-known *recursive* solution for the LS estimation problem.

Let $A_k \in \mathbb{R}^{k,n}$ be the matrix containing the first k rows $a_1^\top, \ldots, a_k^\top$, and let $y^{(k)} \in \mathbb{R}^k$ be the vector containing the first k measurements: $y^{(k)} = [y_1 \cdots y_k]^\top$. Let further the new measurement be

$$y_{k+1} = a_{k+1}^\top x + z_{k+1},$$

and let

$$A_{k+1} = \begin{bmatrix} A_k \\ a_{k+1}^\top \end{bmatrix}, \quad y^{(k+1)} = \begin{bmatrix} y^{(k)} \\ y_{k+1} \end{bmatrix}, \quad H_k = A_k^\top A_k \succ 0,$$

where we assume that A_k has full rank n. The optimal estimate based on the first k measurements is, from (6.7),

$$x^{(k)} = H_k^{-1} A_k^\top y^{(k)},$$

whereas the optimal estimate based also on the additional $(k+1)$-th measurement is

$$x^{(k+1)} = H_{k+1}^{-1} A_{k+1}^\top y^{(k+1)} = H_{k+1}^{-1}(A_k^\top y^{(k)} + a_{k+1} y_{k+1}), \quad (6.13)$$

where

$$H_{k+1} = A_{k+1}^\top A_{k+1} = H_k + a_{k+1} a_{k+1}^\top.$$

Now, using the rank-one perturbation formula (3.10) for the inverse of H_{k+1}, we have that

$$H_{k+1}^{-1} = H_k^{-1} - \frac{1}{\gamma_{k+1}} H_k^{-1} a_{k+1} a_{k+1}^\top H_k^{-1}, \quad (6.14)$$

$$\gamma_{k+1} \doteq 1 + a_{k+1}^\top H_k^{-1} a_{k+1},$$

which, substituted in (6.13), gives

$$
\begin{aligned}
x^{(k+1)} &= \left(I - \frac{1}{\gamma_{k+1}} H_k^{-1} a_{k+1} a_{k+1}^\top\right) H_k^{-1}(A_k^\top y^{(k)} + a_{k+1} y_{k+1}) \\
&= \left(I - \frac{1}{\gamma_{k+1}} H_k^{-1} a_{k+1} a_{k+1}^\top\right) x^{(k)} + \frac{1}{\gamma_{k+1}} H_k^{-1} a_{k+1} y_{k+1} \\
&= x^{(k)} + \frac{(y_{k+1} - a_{k+1}^\top x^{(k)})}{\gamma_{k+1}} H_k^{-1} a_{k+1}.
\end{aligned}
\tag{6.15}
$$

Formulas (6.14) and (6.15) provide the desired recursion for updating the current LS solution when a new measurement is available. An advantage of this formulation is that (6.15) permits us to compute the new solution $x^{(k+1)}$ in a number of operations (scalar multiplications and additions) of the order of n^2, whereas the direct formula (6.13) would require a number of operations of order of n^3 (if the inverse of H_{k+1} were to be computed from scratch). This approach can be used recursively, as further measurements are collected: one starts at some k_0 with H_{k_0} (invertible) and $x^{(k_0)}$. Then, for each $k \geq k_0$, we update the inverse matrix H_k^{-1} according to (6.14) and the estimate $x^{(k)}$ according to (6.13), and keep iterating as long as new measurements are available.

6.4   Solving systems of linear equations and LS problems

We first discuss techniques for solving a square and nonsingular system of equations of the form

$$
Ax = y, \quad A \in \mathbb{R}^{n,n}, \; A \text{ nonsingular.}
\tag{6.16}
$$

6.4.1   Direct methods

If $A \in \mathbb{R}^{n,n}$ has a special structure, such as being an upper (resp., lower) triangular matrix, then the algorithms of *backward substitution* (resp., *forward substitution*), described in Section 7.2.2, can be directly applied to solve (6.16). If A is not triangular, then the method of *Gaussian elimination*, described in Section 7.2.3, applies a sequence of elementary operations that reduce the system to upper triangular form. Then, backward substitution can be applied to this transformed system in triangular form. A possible drawback of these methods is that they work simultaneously on the coefficient matrix A and on the right-hand side term y, hence the whole process has to be redone if one needs to solve the system for several different right-hand sides.

6.4.2 Factorization-based methods

Another common approach for solving (6.16) is the so-called *factor-solve* method. With this approach, the coefficient matrix A is first factored into the product of matrices having a particular structure (such as orthogonal, diagonal, or triangular), and then the solution is found by solving a sequence of simpler systems of equations, where the special structure of the factor matrices can be exploited. Some of these factorization-based methods are described next. An advantage of factorization methods is that, once the factorization is computed, it can be used to solve systems for many different values of the right-hand side y.

6.4.2.1 Via SVD.
If the SVD of $A \in \mathbb{R}^{n,n}$ is available, we can readily solve (6.16) as follows. Let $A = U\Sigma V^\top$, where $U, V \in \mathbb{R}^{n,n}$ are orthogonal, and Σ is diagonal and nonsingular. Then, we write the system $Ax = y$ as a sequence of systems (see Figure 6.9)

Figure 6.9 Factor-solve method with SVD factorization.

$$Uw = y, \quad \Sigma z = w, \quad V^\top x = z,$$

which are readily solved sequentially as

$$w = U^\top y, \quad z = \Sigma^{-1}w, \quad x = Vz.$$

6.4.2.2 Via QR factorization.
We show in Section 7.3 that any nonsingular matrix $A \in \mathbb{R}^{n,n}$ can be factored as $A = QR$, where $Q \in \mathbb{R}^{n,n}$ is orthogonal, and R is upper triangular with positive diagonal entries. Then, the linear equations $Ax = y$ can be solved by first multiplying both sides on the left by $Q^\top$, obtaining

$$Q^\top Ax = Rx = \tilde{y}, \quad \tilde{y} = Q^\top y,$$

and then solving the triangular system $Rx = \tilde{y}$ by backward substitution. This factor-solve process is represented graphically in Figure 6.10.

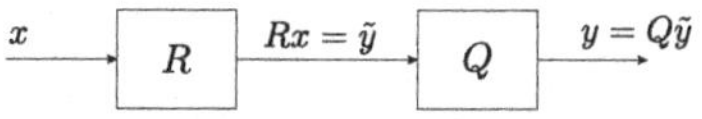

Figure 6.10 Factor-solve method with QR factorization: first solve for $\tilde{y}$ the system $Q\tilde{y} = y$ (with Q orthogonal), then solve for x the system $Rx = \tilde{y}$ (with R triangular).

6.4.3 SVD method for non-square systems

Consider the linear equations

$$Ax = y,$$

where $A \in \mathbb{R}^{m,n}$, and $y \in \mathbb{R}^m$. We can completely describe the set of solutions via SVD, as follows. Let $A = U\tilde{\Sigma}V^\top$ be an SVD of A, and pre-multiply the linear equation by the inverse of U, $U^\top$; then we express the equation in terms of the "rotated" vector $\tilde{x} = V^\top x$ as

$$\tilde{\Sigma}\tilde{x} = \tilde{y},$$

where $\tilde{y} = U^\top y$ is the "rotated" right-hand side of the equation. Due to the simple form of $\tilde{\Sigma}$ in (5.1), the above becomes

$$\sigma_i \tilde{x}_i = \tilde{y}_i, \ i = 1,\dots,r; \quad 0 = \tilde{y}_i, \ i = r+1,\dots,m. \tag{6.17}$$

Two cases can occur.

1. If the last $m - r$ components of $\tilde{y}$ are not zero, then the second set of conditions in (6.17) are not satisfied, hence the system is infeasible, and the solution set is empty. This occurs when y is not in the range of A.

2. If y is in the range of A, then the second set of conditions in (6.17) hold, and we can solve for $\tilde{x}$ with the first set of conditions, obtaining

$$\tilde{x}_i = \frac{\tilde{y}_i}{\sigma_i}, \ i = 1,\dots,r.$$

The last $n - r$ components of $\tilde{x}$ are free. This corresponds to elements in the nullspace of A. If A is full column rank (its nullspace is reduced to $\{0\}$), then there is a unique solution. Once vector $\tilde{x}$ is obtained, the actual unknown x can then be recovered as $x = V\tilde{x}$.

6.4.4 Solving LS problems

Given $A \in \mathbb{R}^{m,n}$ and $y \in \mathbb{R}^m$, we here discuss solution of the LS problem

$$\min_x \|Ax - y\|_2.$$

All solutions of the LS problem are solutions of the system of normal equations (see Proposition 6.2)

$$A^\top A x = A^\top y. \tag{6.18}$$

Therefore, LS solutions can be obtained either by using Gaussian elimination and backward substitution to the normal equations, or by applying a factor-solve method to the normal equations.

6.4.4.1 Using the QR factorization.
Given $A \in \mathbb{R}^{m,n}$ and $y \in \mathbb{R}^m$, with $m \geq n$, $\text{rank}(A) = n$, Theorem 7.1 guarantees that we may write $A = QR$, where $R \in \mathbb{R}^{n,n}$ is upper triangular, and $Q \in \mathbb{R}^{m,n}$ has orthonormal columns. Then $A^\top A = R^\top Q^\top QR = R^\top R$, since $Q^\top Q = I_n$. Therefore, the normal equations become

$$R^\top R x = R^\top Q^\top y,$$

and, multiplying both sides on the left by $R^{-\top}$ (the inverse of $R^\top$), we obtain an equivalent system in upper-triangular form

$$Rx = Q^\top y$$

which can be solved by backward substitution. The numerical cost of solving the LS problem using QR can thus be evaluated as follows: we need $\sim 2mn^2$ operations to compute the QR factorization of A, plus $2nm$ operations for forming $Q^\top y$, and then n^2 operations for applying backward substitution, thus overall still $\sim 2mn^2$ operations.

6.4.4.2 Using the Cholesky factorization. Another possibility, when $m \geq n$, rank$(A) = n$, is to use the Cholesky factorization of $M = A^\top A$ for solving the normal equations (6.18). With the Cholesky factorization, a symmetric positive definite matrix $M \in \mathbb{S}^n$ is factored as $M = LL^\top$, where $L \in \mathbb{R}^{n,n}$ is nonsingular and lower triangular. This factorization requires $\sim n^3/3$ operations. The normal equations then become

$$LL^\top x = b, \quad b = A^\top y.$$

These can be solved by first finding z such that

$$Lz = b,$$

which can be done by forward substitution (n^2 operations), and then determining x such that

$$L^\top x = z,$$

which can be done via backward substitution (n^2 operations). This method requires $\sim mn^2$ (for computing the product $M = A^\top A$), plus $\sim n^3/3$ (for the Cholesky factorization), plus $2mn$ for computing b, plus $2n^2$ operations for solving the two auxiliary triangular systems. The overall complexity is thus $\sim mn^2 + n^3/3$. This complexity figure is lower than the $\sim 2mn^2$ complexity from the QR approach; the Cholesky method is indeed twice as fast as QR, if $m \gg n$. However, the Cholesky method is very sensitive to roundoff errors in the finite precision computations, so the QR method is preferable for dense and medium sized A matrices. When A is very large and sparse, one can exploit specialized algorithms for sparse Cholesky factorization, yielding a complexity figure much smaller than $\sim mn^2 + n^3/3$. The Cholesky method may thus be preferable in these cases.

6.4.4.3 Using the SVD. Yet another possibility is to solve the normal equations (6.18) via SVD. If $A = U_r \Sigma V_r^\top$ is a compact SVD of A, then the unique minimum norm solution to the normal equations is given by (see (6.10))

$$x = A^\dagger y = V_r \Sigma^{-1} U_r^\top y.$$

6.5   Sensitivity of solutions

In this section, we analyze the effect of small perturbations in the data on the solution of square and nonsingular linear equations. The following results also apply to the normal equations, hence to the LS approximate solution of linear equations.

6.5.1   Sensitivity to perturbations in the input

Let x be the solution of a linear system $Ax = y$, with A square and nonsingular, and $y \neq 0$. Assume that we change y slightly by adding to it a small perturbation term δy, and call $x + \delta x$ the solution of the perturbed system:

$$A(x + \delta x) = y + \delta y. \tag{6.19}$$

Our key question is: if δy is "small," will δx also be small or not? We see from (6.19), and from the fact that $Ax = y$, that the perturbation δx is itself a solution of a linear system

$$A\delta x = \delta y,$$

and, since A is assumed to be invertible, we can formally write

$$\delta x = A^{-1}\delta y.$$

Taking the Euclidean norm of both sides of this equation yields

$$\|\delta x\|_2 = \|A^{-1}\delta y\|_2 \leq \|A^{-1}\|_2\|\delta y\|_2, \tag{6.20}$$

where $\|A^{-1}\|_2$ is the spectral (maximum singular value) norm of A^{-1}. Similarly, from $Ax = y$ it follows that $\|y\|_2 = \|Ax\|_2 \leq \|A\|_2\|x\|_2$, hence

$$\|x\|_2^{-1} \leq \frac{\|A\|_2}{\|y\|_2}. \tag{6.21}$$

Multiplying (6.20) and (6.21), we get

$$\frac{\|\delta x\|_2}{\|x\|_2} \leq \|A^{-1}\|_2\|A\|_2\frac{\|\delta y\|_2}{\|y\|_2}.$$

This result is what we were looking for, since it relates a relative variation on the "input term" y to the relative variation of the "output" x. The quantity

$$\kappa(A) = \|A^{-1}\|_2\|A\|_2, \quad 1 \leq \kappa(A) \leq \infty$$

is the condition number of matrix A, see Eq. (5.5). Large $\kappa(A)$ means that perturbations on y are greatly amplified on x, i.e. the system is very sensitive to variations in the input data. If A is singular, then $\kappa(A) = \infty$. Very large $\kappa(A)$ indicates that A is close to being numerically singular; we say in this case that A is *ill conditioned*. We summarize our findings in the following lemma.

Lemma 6.1 (Sensitivity to input perturbations) *Let A be square and nonsingular, and let $x, \delta x$ be such that*

$$
\begin{aligned}
Ax &= y \\
A(x + \delta x) &= y + \delta y.
\end{aligned}
$$

Then it holds that

$$
\frac{\|\delta x\|_2}{\|x\|_2} \leq \kappa(A) \frac{\|\delta y\|_2}{\|y\|_2},
$$

where $\kappa(A) = \|A^{-1}\|_2 \|A\|_2$ is the condition number of A.

6.5.2 Sensitivity to perturbations in the coefficient matrix

We next consider the effect on x of perturbations on the A matrix. Let $Ax = y$ and let δA be a perturbation such that

$$
(A + \delta A)(x + \delta x) = y, \quad \text{for some } \delta x.
$$

Then we see that

$$
A\delta x = -\delta A(x + \delta x),
$$

hence $\delta x = -A^{-1}\delta A(x + \delta x)$. Therefore

$$
\|\delta x\|_2 = \|A^{-1}\delta A(x + \delta x)\|_2 \leq \|A^{-1}\|_2 \|\delta A\|_2 \|x + \delta x\|_2,
$$

and

$$
\frac{\|\delta x\|_2}{\|x + \delta x\|_2} \leq \|A^{-1}\|_2 \|A\|_2 \frac{\|\delta A\|_2}{\|A\|_2}.
$$

We see again that the relative effect on x of a small perturbation $\frac{\|\delta A\|_2}{\|A\|_2} \ll 1$ is small only if the condition number is not too large, i.e. if it is not too far away from one, $\kappa(A) \simeq 1$. This is summarized in the next lemma.

Lemma 6.2 (Sensitivity to perturbations in the coefficients matrix)
Let A be square and nonsingular, and let $x, \delta A, \delta x$ be such that

$$
\begin{aligned}
Ax &= y, \\
(A + \delta A)(x + \delta x) &= y.
\end{aligned}
$$

Then it holds that

$$
\frac{\|\delta x\|_2}{\|x + \delta x\|_2} \leq \kappa(A) \frac{\|\delta A\|_2}{\|A\|_2}.
$$

6.5.3 Sensitivity to joint perturbations on A, y

We finally consider the effect on x of simultaneous perturbations on A and y. Let $Ax = y$, and let δA, δy be perturbations such that

$$(A + \delta A)(x + \delta x) = y + \delta y, \quad \text{for some } \delta x.$$

Then, $A\delta x = \delta y - \delta A(x + \delta x)$, hence $\delta x = A^{-1}\delta y - A^{-1}\delta A(x + \delta x)$. Therefore

$$
\begin{aligned}
\|\delta x\|_2 &= \|A^{-1}\delta y - A^{-1}\delta A(x + \delta x)\|_2 \\
&\leq \|A^{-1}\delta y\|_2 + \|A^{-1}\delta A(x + \delta x)\|_2 \\
&\leq \|A^{-1}\|_2\|\delta y\|_2 + \|A^{-1}\|_2\|\delta A\|_2\|x + \delta x\|_2,
\end{aligned}
$$

and, dividing by $\|x + \delta x\|_2$,

$$\frac{\|\delta x\|_2}{\|x + \delta x\|_2} \leq \|A^{-1}\|_2 \frac{\|\delta y\|_2}{\|y\|_2} \frac{\|y\|_2}{\|x + \delta x\|_2} + \kappa(A)\frac{\|\delta A\|_2}{\|A\|_2}.$$

But $\|y\|_2 = \|Ax\|_2 \leq \|A\|_2\|x\|_2$, hence

$$\frac{\|\delta x\|_2}{\|x + \delta x\|_2} \leq \kappa(A) \frac{\|\delta y\|_2}{\|y\|_2} \frac{\|x\|_2}{\|x + \delta x\|_2} + \kappa(A)\frac{\|\delta A\|_2}{\|A\|_2}.$$

Next, we use $\|x\|_2 = \|x + \delta x - \delta x\|_2 \leq \|x + \delta x\|_2 + \|\delta x\|_2$, to write

$$\frac{\|\delta x\|_2}{\|x + \delta x\|_2} \leq \kappa(A) \frac{\|\delta y\|_2}{\|y\|_2} \left(1 + \frac{\|\delta x\|_2}{\|x + \delta x\|_2}\right) + \kappa(A)\frac{\|\delta A\|_2}{\|A\|_2},$$

from which we obtain

$$\frac{\|\delta x\|_2}{\|x + \delta x\|_2} \leq \kappa(A) \frac{\|\delta y\|_2}{\|y\|_2} \left(1 + \frac{\|\delta x\|_2}{\|x + \delta x\|_2}\right) + \kappa(A)\frac{\|\delta A\|_2}{\|A\|_2},$$

and hence

$$\frac{\|\delta x\|_2}{\|x + \delta x\|_2} \leq \frac{\kappa(A)}{1 - \kappa(A)\frac{\|\delta y\|_2}{\|y\|_2}} \left(\frac{\|\delta y\|_2}{\|y\|_2} + \frac{\|\delta A\|_2}{\|A\|_2}\right).$$

The "amplification factor" of the perturbations is upper bounded by $\frac{\kappa(A)}{1 - \kappa(A)\frac{\|\delta y\|_2}{\|y\|_2}}$. Therefore, this bound is smaller than some given γ, if

$$\kappa(A) \leq \frac{\gamma}{1 + \gamma\frac{\|\delta y\|_2}{\|y\|_2}}.$$

We thus see that the effect of joint perturbations is still controlled by the condition number of A, as formalized next.

Lemma 6.3 (Sensitivity to joint perturbations in A, y) *Let A be square and nonsingular, let $\gamma > 1$ be given, and let x, δy, δA, δx be such that*

$$
\begin{aligned}
Ax &= y, \\
(A + \delta A)(x + \delta x) &= y + \delta y.
\end{aligned}
$$

Then

$$
\kappa(A) \le \frac{\gamma}{1 + \gamma \frac{\|\delta y\|_2}{\|y\|_2}},
$$

which implies

$$
\frac{\|\delta x\|_2}{\|x + \delta x\|_2} \le \gamma \left(\frac{\|\delta y\|_2}{\|y\|_2} + \frac{\|\delta A\|_2}{\|A\|_2} \right).
$$

6.5.4   Sensitivity of the LS solution

We have seen in Section 6.5 that the sensitivity of the solution of a system of linear equations to perturbations in the y term or in the A matrix is proportional to the condition number of A. We now apply these results to study the sensitivity of the LS solution. Since any solution of the LS problem (6.5) is a solution of the normal equations (6.6), we immediately have that the sensitivity of an LS solution is dictated by the condition number of the matrix $A^\top A$. Suppose $\mathrm{rank}(A) = n$, and let $A = U_r \Sigma V_r^\top$ be a compact SVD for A. Then, $A^\top A = V_r \Sigma^2 V_r^\top$, and we have that

$$
\kappa(A^\top A) = \frac{\sigma_{\max}(A^\top A)}{\sigma_{\min}(A^\top A)} = \kappa^2(A).
$$

Also, since the LS solution is a linear function of the y term, $x^* = A^\dagger y$, we see that if the y term is perturbed to $y + \delta y$, then the LS solution is perturbed to $x^* + \delta x$, where δx must satisfy

$$
\delta x = A^\dagger \delta y.
$$

If the perturbation δy is bounded in norm as $\|\delta y\|_2 \le 1$, then the set of corresponding perturbations δx is the ellipsoid (see Section 4.4.5 and Lemma 6.4)

$$
\mathcal{E}_{\delta x} = \{ \delta x = A^\dagger \delta y, \ \|\delta y\|_2 \le 1 \}.
$$

This ellipsoid has its axes aligned to the directions v_i, and semi-axis lengths given by σ_i^{-1}. When $\mathrm{rank}(A) = n$, then $A^\dagger = (A^\top A)^{-1} A^\top$, and the ellipsoid has the explicit representation

$$
\mathcal{E}_{\delta x} = \{ \delta x : \ \delta x^\top (A^\top A) \delta x \le 1 \}.
$$

6.6 *Direct and inverse mapping of a unit ball*

Here we focus on the linear map

$$y = Ax, \quad A \in \mathbb{R}^{m,n}, \tag{6.22}$$

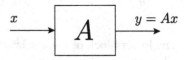

Figure 6.11 Input–output map $y = Ax$.

where $x \in \mathbb{R}^n$ is the input vector, and $y \in \mathbb{R}^m$ is the output, see Figure 6.11, and we consider two problems that we call the direct and the inverse (or estimation) problem.

6.6.1 *Direct problem*

In the direct problem, we assume that the input x lies in a unit Euclidean ball centered at zero, and we ask where the output y is. That is, we let

$$x \in \mathcal{B}^n, \quad \mathcal{B}^n = \{z \in \mathbb{R}^n : \|z\|_2 \le 1\}$$

and we want to characterize the image of this set under the linear mapping (6.22), that is the set

$$\mathcal{E}_y = \{y : y = Ax, \, x \in \mathcal{B}^n\}. \tag{6.23}$$

Let $A = U\tilde{\Sigma}V^\top$ be a full SVD of A, let $r = \operatorname{rank} A$, and let Σ be the $r \times r$ diagonal matrix containing the nonzero singular values of A. Notice first that if the input x is in the direction of one of the left singular vectors v_i, then the output y is either zero (if $i > r$), or it is along the direction of the corresponding right singular vector u_i, scaled by a factor σ_i. Indeed, denoting by e_i a vector of all zeros except for a one in position i, we have that

$$Av_i = U\tilde{\Sigma}V^\top v_i = U\tilde{\Sigma}e_i = \begin{cases} 0 & \text{if } i > r, \\ \sigma_i u_i & \text{otherwise.} \end{cases}$$

Notice also that the unit ball $\mathcal{B}^n$ is invariant under a linear map defined via an orthogonal matrix, that is

$$\{y : y = Qx, \, x \in \mathcal{B}^n\} = \mathcal{B}^n, \text{ for any orthogonal matrix } Q \in \mathbb{R}^{n,n}.$$

This is due to the fact that $QQ^\top = Q^\top Q = I_n$, since $\|y\|_2^2 = y^\top y = x^\top Q^\top Qx = \|x\|_2^2$. It follows that we may equivalently define the image set $\mathcal{E}_y$ in (6.23) by assuming $x = Vz, \, z \in \mathcal{B}^n$, hence

$$\mathcal{E}_y = \{y : y = U\tilde{\Sigma}z, \, z \in \mathcal{B}^n\}.$$

Now, defining $U = [U_r \ U_{nr}]$ and $z^\top = [z_1^\top \ z_2^\top]$, where U_r contains the first r columns of U, U_{nr} contains the last $n - r$ columns of U, and

z_1, z_2 are the first r components of z and the last $n - r$ components of z, respectively, recalling the block structure of $\tilde{\Sigma}$ in (5.1), we have

$$y = U\tilde{\Sigma}z \quad \Leftrightarrow \quad U^\top y = \tilde{\Sigma}z \quad \Leftrightarrow \quad \begin{bmatrix} U_r^\top y \\ U_{nr}^\top y \end{bmatrix} = \begin{bmatrix} \Sigma z_1 \\ 0 \end{bmatrix}.$$

The lower block of these relations ($U_{nr}^\top y = 0$) just says that $y \in \mathcal{R}(A)$, while the first block defines the shape of the image set inside $\mathcal{R}(A)$. Since $\|z\|_2 \le 1$ implies $\|z_1\|_2 \le 1$, we thus have that

$$\mathcal{E}_y = \{y \in \mathcal{R}(A) : U_r^\top y = \Sigma z_1, z_1 \in \mathcal{B}^r\}.$$

Since Σ is invertible, we can also write

$$z_1 = \Sigma^{-1} U_r^\top y,$$

hence

$$z_1^\top z_1 \le 1 \quad \Leftrightarrow \quad y^\top H y \le 1, \quad H = U_r \Sigma^{-2} U_r^\top,$$

and $H = U_r \Sigma^{-2} U_r^\top = A^{\dagger\top} A^\dagger$, thus finally

$$\mathcal{E}_y = \{y \in \mathcal{R}(A) : y^\top H y \le 1\}, \quad H = A^{\dagger\top} A^\dagger. \qquad (6.24)$$

The set in (6.24) is a bounded but possibly degenerate ellipsoid (which is flat on the subspace orthogonal to $\mathcal{R}(A)$), with the axis directions given by the right singular vectors u_i and with the semi-axis lengths given by σ_i, $i = 1, \dots, n$, see Figure 6.12, and Section 9.2.2 for a discussion on representations of ellipsoids.

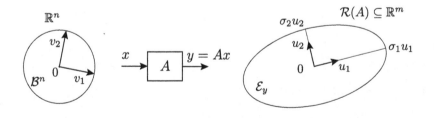

Figure 6.12 The image of a unit ball $\mathcal{B}^n$ under a linear map is an ellipsoid $\mathcal{E}_y$.

The ellipsoid $\mathcal{E}_y$ takes a very simple shape, if we represent it in a proper orthogonal basis for $\mathcal{R}(A)$. That is, expressing every $y \in \mathcal{R}(A)$ as $y = U_r x$, the image set $\mathcal{E}_y$ is simply given by

$$\mathcal{E}_y = \{y = U_r z : z \in \mathcal{E}_x\}, \quad \mathcal{E}_x = \{x \in \mathbb{R}^r : x^\top \Sigma^{-2} x \le 1\},$$

where $\mathcal{E}_x$ is a non-degenerate ellipsoid with axes aligned to the standard axes of $\mathbb{R}^r$, and semi-axis lengths given by σ_i, $i = 1, \dots, r$. When

A is full rank and "wide" (that is, $n \geq m$, $r = \text{rank } A = m$), then $A^\dagger = A^\top(AA^\top)^{-1}$, and $H = A^{\dagger\top}A^\dagger = (AA^\top)^{-1}$, thus $\mathcal{E}_y$ is the bounded and non-degenerate (full-dimensional) ellipsoid

$$\mathcal{E}_y = \{y \in \mathbb{R}^m : y^\top(AA^\top)^{-1}y \leq 1\}.$$

6.6.2   Inverse problem

We tackle the inverse problem in a similar way: suppose we know that the output y is in the unit ball $\mathcal{B}^m$, and we ask what is the set of input vectors x that would yield such a set as output. Formally, we seek the pre-image of the unit ball under the linear map (6.22), that is

$$\mathcal{E}_x = \{x \in \mathbb{R}^n : Ax \in \mathcal{B}^m\}.$$

Since $Ax \in \mathcal{B}^m$ if and only if $x^\top A^\top Ax \leq 1$, we immediately obtain that the pre-image set we seek is the full-dimensional (but possibly unbounded) ellipsoid (see Section 9.2.2)

$$\mathcal{E}_x = \{x \in \mathbb{R}^n : x^\top(A^\top A)x \leq 1\}.$$

This ellipsoid is unbounded along directions x in the nullspace of A (clearly, if $x \in \mathcal{N}(A)$, then $y = Ax = 0 \in \mathcal{B}^m$). If A is "skinny" and full rank (i.e., $n \leq m$ and $r = \text{rank}(A) = n$), then $A^\top A \succ 0$ and the ellipsoid $\mathcal{E}_x$ is bounded. The axes of $\mathcal{E}_x$ are along the directions of the left singular vectors v_i, and the semi-axis lengths are given by σ_i^{-1}, $i = 1, \ldots, n$, see Figure 6.13.

Figure 6.13 Pre-image of an output unit ball under the linear map $y = Ax$.

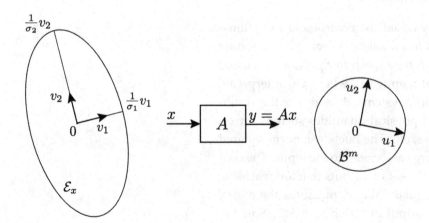

We summarize our findings in the next lemma.

Lemma 6.4 (Image and pre-image of a unit ball under a linear map)
Given $A \in \mathbb{R}^{m,n}$, let $A = U_r \Sigma V_r^\top$ be a compact SVD of A, with $r = \mathrm{rank}(A)$, and let $\mathcal{B}^q \doteq \{z \in \mathbb{R}^q : \|z\|_2 \leq 1\}$ be the unit Euclidean ball in $\mathbb{R}^q$. Let

$$
\begin{aligned}
\mathcal{E}_y &= \{y \in \mathbb{R}^m : y = Ax, \; x \in \mathcal{B}^n\}, \\
\mathcal{E}_x &= \{x \in \mathbb{R}^n : Ax \in \mathcal{B}^m\}.
\end{aligned}
$$

Then

1. *The image set $\mathcal{E}_y$ is a bounded ellipsoid having as axes the right singular vectors of A: u_i, $i = 1, \ldots, m$. The lengths of the semi-axes are given by $\sigma_i > 0$, for $i = 1, \ldots, r$, and are zero for $i = r+1, \ldots, m$. That is, the ellipsoid is degenerate (flat) if $r < m$. Moreover, if $r = m \leq n$, then $\mathcal{E}_y$ is bounded and full-dimensional, and it has the explicit representation*

$$
\mathcal{E}_y = \{y \in \mathbb{R}^m : y^\top (AA^\top)^{-1} y \leq 1\}, \quad AA^\top \succ 0.
$$

2. *The pre-image set $\mathcal{E}_x$ is a non-degenerate ellipsoid having as axes the left singular vectors of A: v_i, $i = 1, \ldots, n$. The lengths of the semi-axes are given by $\sigma_i^{-1} > 0$, for $i = 1, \ldots, r$, and are infinite for $i = r+1, \ldots, n$. That is, the ellipsoid is unbounded (i.e., cylindrical) along the directions $v_{r+1}, \ldots, v_n$, if $r < n$. Moreover, if $r = n \leq m$, then $\mathcal{E}_x$ is bounded, and it has the explicit representation*

$$
\mathcal{E}_x = \{x \in \mathbb{R}^n : x^\top (A^\top A) x \leq 1\}, \quad A^\top A \succ 0.
$$

6.6.3 *The control and estimation ellipsoids*

The ellipsoids $\mathcal{E}_y$, $\mathcal{E}_x$ are usually called the control and the estimation ellipsoid, respectively. This terminology reflects two important engineering interpretations that can be given to $\mathcal{E}_y$, $\mathcal{E}_x$, as discussed next. Consider the input/output map (6.22), where x is interpreted as the *actuator* input given to the "system" A, and y is the resulting output; x usually represents physical quantities such as forces, torques, voltages, pressures, flows, etc. The Euclidean norm squared $\|x\|_2^2$ has the interpretation of *energy* associated to the input. The control ellipsoid $\mathcal{E}_y$ thus represents the set of outputs that are reachable with unit input energy. The singular value σ_i measures the *control authority* of the input along the output direction $u_i \in \mathbb{R}^m$. Small σ_i means that it is hard (i.e., energy expensive) to reach outputs along the direction u_i. If $r < m$, then there is actually no control authority along the directions $u_{r+1}, \ldots, u_m$, i.e., the output subspace spanned by these vectors is unreachable.

Conversely, suppose one needs to determine the value of an unknown parameter θ via noisy linear measurements on θ of the form

$$a_i^\top \theta = \hat{y}_i + y_i, \quad i = 1, \ldots, m,$$

where $\hat{y}_i$ is the nominal reading of the measurement, and y_i represents the uncertainty (e.g., noise) on the i-th measurement. In vector format, we have

$$A\theta = \hat{y} + y,$$

where we assume that the uncertainty vector y is bounded in norm: $\|y\|_2 \leq 1$. Let $\hat{\theta}$ be the parameter value corresponding to the nominal readings, that is

$$A\hat{\theta} = \hat{y},$$

then the actual (unknown) value of θ is $\theta = \hat{\theta} + x$, where x represents the uncertainty on θ induced by the uncertainty in the measurements. It clearly holds that $A(\hat{\theta} + x) = \hat{y} + y$, whence

$$Ax = y, \quad \|y\|_2 \leq 1.$$

The estimation ellipsoid $\mathcal{E}_x$ thus provides the uncertainty region around the nominal parameter $\hat{\theta}$, caused by the uncertainty y on the nominal measurement readings. The axes of this ellipsoid are along the directions of the right singular vectors v_i of matrix A, with semi-axis lengths given by σ_i^{-1}, $i = 1, \ldots, r$: the smaller σ_i^{-1} is, the smaller is the confidence interval around the θ parameter along direction v_i. The shortest axis $\sigma_1^{-1} v_1$ denotes the direction in input space which is least sensitive to measurement errors; the largest axis $\sigma_r^{-1} v_r$ denotes the direction in input space which is most sensitive to measurement errors.

Example 6.7 (*Force generation*) Consider again Example 6.3 relative to a rigid body equipped with n thrusters that need to impress to the body a desired resultant force and torque. Here, the input vector x contains the individual forces generated by the thrusters, and the output y is the resultant force/torque. As a numerical example we consider $n = 6$ thrusters positioned at angles $\theta_1 = 0$, $\theta_2 = \pi/16$, $\theta_3 = (15/16)\pi$, $\theta_4 = \pi$, $\theta_5 = (17/16)\pi$, $\theta_6 = (31/16)\pi$, and take as output only the components of the force: $y = [f_x \ f_y]^\top$. Thus, we have $y = Ax$, with

$$A = \begin{bmatrix} \cos\theta_1 & \cdots & \cos\theta_6 \\ \sin\theta_1 & \cdots & \sin\theta_6 \end{bmatrix}$$

$$= \begin{bmatrix} 1 & 0.9808 & -0.9808 & -1 & -0.9808 & 0.9808 \\ 0 & 0.1951 & 0.1951 & 0 & -0.1951 & -0.1951 \end{bmatrix}.$$

The set of output resultant forces y obtainable with inputs x such that $\|x\|_2 \leq 1$ is given by the control ellipsoid

$$\mathcal{E}_y = \{y : y^\top P^{-1} y \leq 1\},$$

where

$$P = AA^\top = \begin{bmatrix} 5.8478 & 0.0 \\ 0.0 & 0.1522 \end{bmatrix}.$$

The control ellipsoid has in this case its axes aligned with the standard axes of $\mathbb{R}^2$, and semi-axis lengths given by $\sigma_1 = 2.4182$ (horizontal axis) and $\sigma_2 = 0.3902$ (vertical axis), see Figure 6.14. Generating an output force in the vertical direction thus requires about 6.2 times the control effort than generating an output force in the horizontal direction.

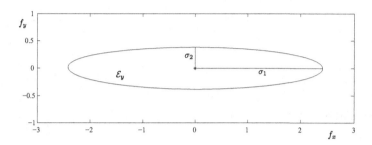

Figure 6.14 Control ellipsoid of resultant forces reachable with unit-norm input.

Example 6.8 (*Trilateration*) Consider Example 6.2, dealing with a problem of localization of an object's coordinates from measurements of the distance of the object from three known beacons. This is indeed an estimation problem, where the unknown position is to be determined from (possibly noisy) distance measurements. Calling a_1, a_2, a_3 the points representing the three beacons, the A matrix relating the unknown position with the vector of measured data is

$$A = \begin{bmatrix} 2(a_2 - a_1)^\top \\ 2(a_3 - a_1)^\top \end{bmatrix}.$$

The estimation ellipsoid $\mathcal{E}_x$ gives us a precise assessment of how errors in the measurement vector impact the uncertainty on the object position. If the measurement errors are bounded in the unit Euclidean ball, the uncertainty region around the nominal object position is given by the estimation ellipsoid

$$\mathcal{E}_x = \{x : x^\top H x \leq 1\}, \quad H = A^\top A.$$

For a numerical example, we consider two situations. In the first scenario the beacons are located at

$$a_1 = \begin{bmatrix} 0 \\ 0 \end{bmatrix}, \ a_2 = \begin{bmatrix} 4 \\ -1 \end{bmatrix}, \ a_3 = \begin{bmatrix} 5 \\ 0 \end{bmatrix},$$

while in the second scenario they are located at

$$a_1 = \begin{bmatrix} 0 \\ 0 \end{bmatrix}, \ a_2 = \begin{bmatrix} 0 \\ 3 \end{bmatrix}, \ a_3 = \begin{bmatrix} 5 \\ 0 \end{bmatrix}.$$

Figure 6.15 shows the estimation ellipsoids in the two scenarios. The estimation ellipsoids represent in this example the confidence regions in space where the object's actual position is located, given the measurements' uncertainty. We observe that in the first scenario we have large uncertainty along the v_2 axis (which is almost vertical), hence in this situation we would have good confidence on the object's horizontal location and poor confidence on the vertical one. The second scenario yields a more "spherical" ellipsoid, with more balanced uncertainty in the horizontal and vertical axes.

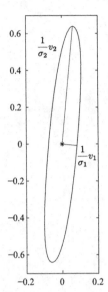
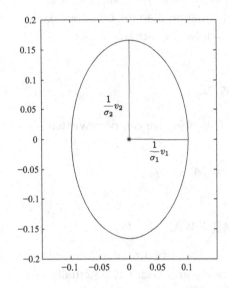

Figure 6.15 Confidence ellipsoid for the actual object position in the first scenario (left) and in the second scenario (right).

6.7   Variants of the least-squares problem

6.7.1   Linear equality-constrained LS

A generalization of the basic LS problem (6.5) allows for the addition of linear equality constraints on the x variable, resulting in the constrained problem

$$\min_{x} \|Ax - y\|_2^2, \quad \text{s.t.: } Cx = d,$$

where $C \in \mathbb{R}^{p,n}$ and $d \in \mathbb{R}^p$. This problem can be converted into a standard LS one, by "eliminating" the equality constraints, via a standard procedure described in Section 12.2.6.1. Alternatively, the multiplier-based method described in Section 9.6.1 can be used to solve the problem via an augmented system of normal equations.

6.7.2 Weighted LS

The standard LS objective is a sum of squared equation residuals

$$\|Ax - y\|_2^2 = \sum_{i=1}^{m} r_i^2, \quad r_i = a_i^\top x - y_i,$$

where $a_i^\top$, $i = 1, \ldots, m$, are the rows of A. In some cases, the equation residuals may not be given the same importance (for example, it is more important to satisfy, say, the first equation than the other ones), and this relative importance can be modeled by introducing *weights* into the LS objective, that is

$$f_0(x) = \sum_{i=1}^{m} w_i^2 r_i^2,$$

where $w_i \geq 0$ are the given weights. This objective can be rewritten in the form

$$f_0(x) = \|W(Ax - y)\|_2^2 = \|A_w x - y_w\|_2^2,$$

where

$$W = \mathrm{diag}\,(w_1, \ldots, w_m), \quad A_w \doteq WA, \; y_w = Wy.$$

Hence, the weighted LS problem still has the structure of a standard LS problem, with row-weighted matrix A_w and vector y_w. Actually, the weight matrix may also be generalized from diagonal to symmetric positive definite. Assuming $W \succ 0$ means giving different weights to different *directions* in the residual space. Indeed, let $r = Ax - y$ denote the vector of residuals, then the weighted LS objective is

$$f_0(x) = \|W(Ax - y)\|_2^2 = \|Wr\|_2^2 = r^\top (W^\top W)r,$$

and the unit-level set in residual space $\mathcal{E}_r = \{r : r^\top (W^\top W)r \leq 1\}$ is an ellipsoid with axes aligned to the eigenvectors u_i of W, and semi-axis lengths given by λ_i^{-1}, where λ_i, $i = 1, \ldots, m$, are the eigenvalues of W. The largest semi-axis, having length $\lambda_{\min}^{-1}$, is the direction along which the cost function f_0 is *least* sensitive to residuals: the residual vector can be large along this direction, and still remain in the unit-level cost set $\mathcal{E}_r$. Similarly, the smallest semi-axis, having length $\lambda_{\max}^{-1}$, is the direction along which the cost function f_0 is *most* sensitive to residuals. Therefore, $\lambda_{\max} = \lambda_1 \geq \cdots \geq \lambda_m = \lambda_{\min}$ act as residual weights, along the corresponding directions $u_1, \ldots, u_m$.

The solution of the weighted LS problem is thus analogous to the solution of a standard LS problem, and it amounts to solving the weighted normal equations $A_w^\top A_w x = A_w^\top y_w$, that is

$$A^\top PAx = A^\top Py, \quad P = W^\top W, \; W \succeq 0.$$

6.7.3 ℓ_2-regularized LS

Regularized LS refers to a class of problems of the form

$$\min_x \; \|Ax - y\|_2^2 + \phi(x),$$

where a "regularization," or *penalty*, term $\phi(x)$ is added to the usual LS objective. In the most usual cases, ϕ is proportional either to the ℓ_1 or to the ℓ_2 norm of x. The ℓ_1-regularized case gives rise to the LASSO problem, which is discussed in more detail in Section 9.6.2; the ℓ_2-regularized case is instead discussed next.

To motivate the idea of ℓ_2-regularization, we use a "control" interpretation. Consider the linear map $f(x) = Ax$, where x is the input, and let y be a given, desired, output, see Figure 6.16.

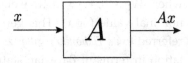

Figure 6.16 A linear control problem: find x such that Ax is close to target output y.

The standard LS problem can be interpreted as a *control* problem: we want to determine a suitable input x so that the output is as close as possible (in Euclidean norm sense) to the assigned desired output y. Here, all the focus is on matching the desired output as closely as possible. However, doing so requires an effort in the input, that is it requires a certain energy in the "actuators," and this energy can be measured by the squared norm of the input vector, $\|x\|_2^2$. A natural tradeoff thus arises, since on the one hand we want the output matching error to be small, and on the other hand we want to spend little input energy for achieving this goal. We may formalize such a tradeoff by defining a "mixed" objective

$$f_0(x) = \|Ax - y\|_2^2 + \gamma\|x\|_2^2, \quad \gamma \geq 0,$$

where $\gamma \geq 0$ defines the relative weights between the two competing criteria: small γ biases the problem towards solutions with good output matching but possibly large input energy; large γ biases the problem instead towards solutions with small input energy but possibly poor output matching. The resulting problem

$$\min_x \; \|Ax - y\|_2^2 + \gamma\|x\|_2^2. \tag{6.25}$$

is an ℓ_2-regularized LS problem. Recalling that the squared Euclidean norm of a block-partitioned vector is equal to the sum of the squared norms of the blocks, i.e.,

$$\left\| \begin{bmatrix} a \\ b \end{bmatrix} \right\|_2^2 = \|a\|_2^2 + \|b\|_2^2,$$

we see that the regularized LS problem can be rewritten in the format of a standard LS problem as follows:

$$\|Ax - y\|_2^2 + \gamma\|x\|_2^2 = \|\tilde{A}x - \tilde{y}\|_2^2,$$

where

$$\tilde{A} \doteq \begin{bmatrix} A \\ \sqrt{\gamma} I_n \end{bmatrix}, \quad \tilde{y} \doteq \begin{bmatrix} y \\ 0_n \end{bmatrix}. \tag{6.26}$$

A more general format for regularized LS allows for the introduction of weighting matrices on the output matching residuals and on the deviation of the input term from a nominal value, resulting in the following problem:

$$\min_x \; \|W_1(Ax - y)\|_2^2 + \|W_2(x - x_0)\|_2^2, \tag{6.27}$$

where $W_1 \succ 0$, $W_2 \succ 0$ are weighting matrices, and x_0 is some given nominal value for x. This generalized regularization is sometimes referred to as *Tikhonov regularization* (see Section 6.7.4 for an interpretation in terms of Bayesian statistics). This regularized LS problem can still be cast in the format of a standard LS problem as follows:

$$\min_x \; \|\tilde{A}x - \tilde{y}\|_2^2,$$

where

$$\tilde{A} \doteq \begin{bmatrix} W_1 A \\ W_2 \end{bmatrix}, \quad \tilde{y} \doteq \begin{bmatrix} W_1 y \\ W_2 x_0 \end{bmatrix}. \tag{6.28}$$

Clearly, (6.25) is a special case of (6.27), with $W_1 = I_m$, $W_2 = \sqrt{\gamma} I_n$, $x_0 = 0_n$.

6.7.3.1 Sensitivity of the regularized solution. We saw in Section 6.5.4 that the sensitivity of an LS solution to changes in the y parameter is dictated by the condition number $\kappa(A^\top A) = \kappa^2(A)$. We next show that indeed regularization improves (i.e., decreases) the condition number, and hence it reduces the sensitivity of the solution to perturbations in the problem data. Consider for simplicity the full-column rank case, where $r = \text{rank}(A) = n$. In such a case, the compact SVD of A can be written as $A = U_r \Sigma V^\top$, with $VV^\top = I_n$. The regularized problem (6.25) is a standard LS problem with the $\tilde{A}$ matrix given in (6.26), for which we have

$$\begin{aligned} \tilde{A}^\top \tilde{A} &= A^\top A + \gamma I_n = V\Sigma^2 V^\top + \gamma I_n \\ &= V(\Sigma^2 + \gamma I_n)V^\top. \end{aligned}$$

The singular values of $\tilde{A}^\top \tilde{A}$ are thus given by $\sigma_i^2 + \gamma$, where σ_i, $i = 1, \ldots, n$, are the singular values of A. The condition number $\kappa(\tilde{A}^\top \tilde{A})$ is then given by

$$\begin{aligned} \kappa(\tilde{A}^\top \tilde{A}) &= \frac{\sigma_{\max}(\tilde{A}^\top \tilde{A})}{\sigma_{\min}(\tilde{A}^\top \tilde{A})} = \frac{\sigma_{\max}^2(A) + \gamma}{\sigma_{\min}^2(A) + \gamma} = \frac{\kappa^2(A) + L}{1 + L} \\ &= \frac{\kappa^2(A)}{1 + L} + \frac{L}{1 + L}, \end{aligned}$$

where

$$L \doteq \frac{\gamma}{\sigma_{\min}^2(A)}.$$

Whenever $L \gg 1$, the second term in the expression of $\kappa(\tilde{A}^\top \tilde{A})$ is close to one, whereas the first term is much smaller than $\kappa^2(A)$, therefore

$$L \gg 1 \quad \Rightarrow \quad \kappa(\tilde{A}^\top \tilde{A}) \ll \kappa(A^\top A),$$

which means that regularization may greatly improve the condition number. Notice further that even in the case when $A^\top A$ is singular (recall that the LS solution is non-unique in this case), the regularized matrix $\tilde{A}^\top \tilde{A}$ is always nonsingular, hence the solution of the regularized problem is always unique (for $\gamma > 0$).

6.7.4 Bayesian interpretation of Tikhonov regularization

There is an important interpretation of the regularized problem (6.27) in the context of statistical estimation. Consider a generalization of the linear, noisy measurements model introduced in Section 6.3.4.5, Eq. (6.11):

$$y = Ax + z_m, \tag{6.29}$$

where z_m represents a vector of random measurement errors having zero mean and covariance matrix

$$\mathrm{var}\{z_m z_m^\top\} = \Sigma_m \succ 0.$$

When we discussed linear unbiased estimation (BLUE) in Section 6.3.4.5, our goal was to build a minimum variance estimator of x as a linear function of the measurements y. In the so-called Bayesian approach to statistical estimation, however, one always assumes, before making any actual measurement on the unknown parameter x, that one has some *prior belief* on the value of this parameter. Such a prior belief may come from past experience or other pre-existing information, and it is quantified by assuming a prior probability distribution on x. Here, we assume that our prior information is synthesized by the relation

$$x_0 = x + z_p, \tag{6.30}$$

where x_0 is a given vector quantifying our *a priori* belief on x (i.e., our prior knowledge makes us believe that a reasonable value for x may be x_0), and z_p is a zero-mean random vector quantifying the *uncertainty* we have on our *a priori* belief. We suppose that z_p is statistically independent of z_m, and we only know the covariance matrix of z_p:

$$\mathrm{var}\{z_p\} = \Sigma_p \succ 0.$$

Now, we build a linear estimator of x which takes into account *both* the information coming from the measurements (6.29) and the prior information (6.30), that is

$$\hat{x} = K_m y + K_p x_0, \tag{6.31}$$

where K_m is the measurement gain of the estimator, and K_p is the prior information gain. For this estimator to be BLUE we need to determine K_m, K_p so that $\mathbb{E}\{\hat{x}\} = x$ and $\mathrm{var}\{\hat{x}\}$ is minimal. Notice that

$$\hat{x} = K\bar{y}, \quad \bar{y} \doteq \begin{bmatrix} y \\ x_0 \end{bmatrix}, \ K \doteq \begin{bmatrix} K_m & K_p \end{bmatrix},$$

where we may consider $\bar{y}$ as an augmented measurement

$$\bar{y} = \bar{A}x + z, \quad \bar{A} \doteq \begin{bmatrix} A \\ I_n \end{bmatrix}, \ z \doteq \begin{bmatrix} z_m \\ z_p \end{bmatrix},$$

with

$$\Sigma \doteq \mathrm{var}\{z\} = \begin{bmatrix} \Sigma_m & 0 \\ 0 & \Sigma_p \end{bmatrix},$$

since z_m and z_p are assumed to be independent. Then, we have that

$$\mathbb{E}\{\hat{x}\} = \mathbb{E}\{K\bar{y}\} = \mathbb{E}\{K(\bar{A}x + z)\} = K\bar{A}x,$$

hence for the estimator to be unbiased we must have $\mathbb{E}\{\hat{x}\} = x$, thus

$$K\bar{A} = I.$$

If we let

$$\Sigma^{1/2} = \begin{bmatrix} \Sigma_m^{1/2} & 0 \\ 0 & \Sigma_p^{1/2} \end{bmatrix}$$

be the matrix square-root of $\Sigma \succ 0$, then the previous relation can also be written as

$$K\Sigma^{1/2}\Sigma^{-1/2}\bar{A} = I,$$

which means that $K\Sigma^{1/2}$ should be a left inverse of $\Sigma^{-1/2}\bar{A}$. Any such left inverse can be written as

$$K\Sigma^{1/2} = (\Sigma^{-1/2}\bar{A})^\dagger + Q, \tag{6.32}$$

where Q is any matrix such that

$$Q(\Sigma^{-1/2}\bar{A}) = 0. \tag{6.33}$$

Now,

$$\hat{x} - \mathbb{E}\{\hat{x}\} = \hat{x} - x = Kz,$$

hence

$$\text{var}\{\hat{x}\} \;=\; \mathbb{E}\{Kzz^\top K^\top\} = K\Sigma K^\top = K\Sigma^{1/2}\Sigma^{1/2}K^\top$$

$$[\text{using (6.32), (6.33)}] \;=\; (\Sigma^{-1/2}\bar{A})^\dagger(\Sigma^{-1/2}\bar{A})^{\dagger\top} + QQ^\top.$$

Since for any Q we have $\text{var}\{\hat{x}\} \succeq (\Sigma^{-1/2}\bar{A})^\dagger(\Sigma^{-1/2}\bar{A})^{\dagger\top}$, we have that the minimal variance is attained for $Q = 0$, thus the BLUE estimator gain is given by (6.32) with $Q = 0$, i.e.,

$$\begin{aligned}
\hat{x} &= K\bar{y} = (\Sigma^{-1/2}\bar{A})^\dagger\Sigma^{-1/2}\bar{y} = \begin{bmatrix} \Sigma_m^{-1/2}A \\ \Sigma_p^{-1/2} \end{bmatrix}^\dagger \begin{bmatrix} \Sigma_m^{-1/2}y \\ \Sigma_p^{-1/2}x_0 \end{bmatrix} \\
&= \left(A^\top\Sigma_m^{-1}A + \Sigma_p^{-1}\right)^{-1} \begin{bmatrix} A^\top\Sigma_m^{-1/2} & \Sigma_p^{-1/2} \end{bmatrix} \begin{bmatrix} \Sigma_m^{-1/2}y \\ \Sigma_p^{-1/2}x_0 \end{bmatrix} \\
&= \left(A^\top\Sigma_m^{-1}A + \Sigma_p^{-1}\right)^{-1} \left(A^\top\Sigma_m^{-1}y + \Sigma_p^{-1}x_0\right). \qquad (6.34)
\end{aligned}$$

Letting

$$\Sigma_+ \doteq \text{var}\{\hat{x}\} = \left(A^\top\Sigma_m^{-1}A + \Sigma_p^{-1}\right)^{-1},$$

we obtain the two matrix gains in (6.31) as

$$\begin{aligned}
K_m &= \Sigma_+ A^\top\Sigma_m^{-1}, \\
K_p &= \Sigma_+\Sigma_p^{-1}.
\end{aligned}$$

It is immediate to verify from (6.34) that the BLUE estimator coincides with the solution of the regularized LS problem in (6.27)–(6.28), by setting weights $W_1 = \Sigma_m^{-1/2}$, $W_2 = \Sigma_p^{-1/2}$. Using the matrix inversion lemma (see (3.9)), we can also express Σ_+ as follows:

$$\Sigma_+ = \left(A^\top\Sigma_m^{-1}A + \Sigma_p^{-1}\right)^{-1} = \Sigma_p - \Sigma_p A^\top(\Sigma_m + A\Sigma_p A^\top)^{-1}A\Sigma_p.$$

Letting

$$Q \doteq \Sigma_m + A\Sigma_p A^\top,$$

we can express the estimator in the alternative form

$$\begin{aligned}
\hat{x} &= \Sigma_+\left(A^\top\Sigma_m^{-1}y + \Sigma_p^{-1}x_0\right) \qquad (6.35) \\
&= (I - \Sigma_p A^\top Q^{-1}A)\Sigma_p\left(A^\top\Sigma_m^{-1}y + \Sigma_p^{-1}x_0\right) \\
&\quad [\text{add and subtract } \Sigma_p A^\top Q^{-1}y] \\
&= x_0 + \Sigma_p A^\top Q^{-1}(y - Ax_0) + Zy,
\end{aligned}$$

where

$$Z = \Sigma_p A^\top\Sigma_m^{-1} - \Sigma_p A^\top Q^{-1}(A\Sigma_p A^\top\Sigma_m^{-1} + I).$$

Collecting Σ_p on the left and Σ_m^{-1} on the right in the expression for Z, we have that

$$
\begin{aligned}
Z &= \Sigma_p \left(A^\top - A^\top Q^{-1} (A\Sigma_p A^\top + \Sigma_m) \right) \Sigma_m^{-1} \\
&= \Sigma_p \left(A^\top - A^\top Q^{-1} Q \right) \Sigma_m^{-1} \\
&= 0,
\end{aligned}
$$

therefore the expression for the estimator in (6.35) simplifies to

$$
\hat{x} = x_0 + \Sigma_p A^\top Q^{-1} (y - Ax_0). \tag{6.36}
$$

This latter format of the estimator is referred to as the *innovation form*, where the term $y - Ax_0$ has indeed the interpretation of innovation, or new information, brought by the measurement y with respect to the *a priori* best guess of the output Ax_0.

The derivations presented in this paragraph form the basis of the *recursive* solution approach to linear estimation (and of recursive solution of LS problems): we started from prior information x_0, collected measurements y, and built the BLUE estimator $\hat{x}$ (in the form (6.31) or (6.36)). If now a new measurement, say y_{new}, becomes available, we can iterate the same idea by constructing an *updated* estimator that has the previous estimate $\hat{x}$ as prior information and y_{new} as measurement, and then iterate the mechanism as further new measurements become available. The explicit formulation of the recursive BLUE estimator is left to the reader as an exercise.

6.7.5 Total least squares

One of the interpretations of the standard LS problem mentioned in Section 6.3.4 was in terms of minimal perturbation δy in the y term necessary for the linear equations $Ax = y + \delta y$ to become feasible. The total least-squares (TLS) approach extends this idea by allowing the perturbation to act both on the y term and on the A matrix. That is, we seek a perturbation matrix $[\delta A \ \delta y] \in \mathbb{R}^{m,n+1}$ with minimal Frobenius norm, such that the equations $(A + \delta A)x = y + \delta y$ are feasible. Formally, we seek to solve the following optimization problem:

$$
\begin{aligned}
&\min_{\delta A, \delta y} \quad \|[\delta A \ \delta y]\|_F^2 \tag{6.37} \\
&\text{s.t.:} \quad y + \delta y \in \mathcal{R}(A + \delta A).
\end{aligned}
$$

Let

$$
D^\top \doteq [A \, y], \quad \delta D^\top \doteq [\delta A \, \delta y], \tag{6.38}
$$

and let us assume for simplicity that $\mathrm{rank}(D) = n + 1$, and let

$$D^\top = \sum_{i=1}^{n+1} \sigma_i u_i v_i^\top \tag{6.39}$$

be an SVD for $D^\top$. We shall make the further technical assumption that

$$\sigma_{n+1} < \sigma_{\min}(A). \tag{6.40}$$

Then, the feasibility condition $y + \delta y \in \mathcal{R}(A + \delta A)$ is equivalent to the existence of a vector $c \in \mathbb{R}^{n+1}$ (normalized without loss of generality with $\|c\|_2 = 1$) such that

$$(D^\top + \delta D^\top)c = 0, \quad c_{n+1} \neq 0, \ \|c\|_2 = 1, \tag{6.41}$$

and, in turn, since $D^\top$ is assumed full rank, this condition is equivalent to requiring that $D^\top + \delta D^\top$ becomes rank deficient (and that a vector can be found in the nullspace of this matrix with a nonzero component in the $(n+1)$-th position; but we'll see that this latter requirement is insured by condition (6.40)). Consider then the following problem:

$$\begin{aligned} \min \quad & \|\delta D^\top\|_F^2 \\ \text{s.t.:} \quad & \mathrm{rank}(D^\top + \delta D^\top) = n. \end{aligned}$$

A solution for this problem can be readily obtained, as discussed in Remark 5.2, and it is given by

$$\delta D^{\top *} = -\sigma_{n+1} u_{n+1} v_{n+1}^\top,$$

for which we have

$$D^\top + \delta D^{\top *} = \sum_{i=1}^{n} \sigma_i u_i v_i^\top,$$

whence a vector c in the nullspace of $D^\top + \delta D^{\top *}$ is $c = v_{n+1}$. We now check that this vector is such that $c_{n+1} \neq 0$, hence all conditions in (6.41) are satisfied. Indeed, suppose for contradiction that $c = v_{n+1} = [\tilde{v}^\top \ 0]^\top$ then, since v_{n+1} is an eigenvector of $DD^\top$ associated with σ_{n+1}^2,

$$DD^\top v_{n+1} = \sigma_{n+1}^2 v_{n+1} \quad \Leftrightarrow \quad \begin{bmatrix} A^\top \\ y^\top \end{bmatrix} \begin{bmatrix} A & y \end{bmatrix} \begin{bmatrix} \tilde{v} \\ 0 \end{bmatrix} = \begin{bmatrix} \sigma_{n+1}^2 \tilde{v} \\ 0 \end{bmatrix},$$

which would imply that $(A^\top A)\tilde{v} = \sigma_{n+1}^2 \tilde{v}$, which is in contradiction with the starting assumption (6.40).

Once the optimal perturbation $\delta D^{\top *} = [\delta A^* \ \delta y^*]$ is found, a solution x such that

$$(A + \delta A^*)x = y + \delta y^*$$

is easily found by considering that $[x \ -1]^\top$ is proportional to v_{n+1}. Hence once v_{n+1} is found from SVD of D, we can easily determine α such that

$$\begin{bmatrix} x \\ -1 \end{bmatrix} = \alpha v_{n+1}.$$

Also, x can be found directly by considering that $[x \ -1]^\top$ must be an eigenvector of $DD^\top$ associated with eigenvalue σ_{n+1}^2, that is

$$\begin{bmatrix} A^\top A & A^\top y \\ y^\top A & y^\top y \end{bmatrix} \begin{bmatrix} x \\ -1 \end{bmatrix} = \sigma_{n+1}^2 \begin{bmatrix} x \\ -1 \end{bmatrix}.$$

The upper block in these equations then yields the optimal solution x_{TLS}

$$x_{\text{TLS}} = (A^\top A - \sigma_{n+1}^2 I)^{-1} A^\top y.$$

We summarize the previous derivations in the following theorem.

Theorem 6.2 (Total least squares) *Given $A \in \mathbb{R}^{m,n}$, $y \in \mathbb{R}^m$, with the notation set in (6.38), (6.39), assume that $\text{rank}(D) = n + 1$, and that condition (6.40) is satisfied. Then the TLS problem (6.37) has a unique solution given by*

$$[\delta A^* \ \delta y^*] = -\sigma_{n+1} u_{n+1} v_{n+1}^\top.$$

Moreover, the optimal solution vector x_{TLS} such that

$$(A + \delta A^*) x_{\text{TLS}} = y + \delta y^*$$

is found as

$$x_{\text{TLS}} = (A^\top A - \sigma_{n+1}^2 I)^{-1} A^\top y.$$

Remark 6.2 Interestingly, the TLS problem is also closely connected to the hyperplane fitting problem discussed in Exercise 5.6. To see this connection, consider a hyperplane through the origin $\mathcal{H} = \{z \in \mathbb{R}^{n+1} : z^\top c = 0\}$ and let us seek a normal direction c, $\|c\|_2 = 1$, such that the mean-square distances from $\mathcal{H}$ to the data d_i forming the columns of matrix D in (6.38) is minimized. The sum of these squared distances is equal to $\|D^\top c\|^2$, which is minimized by the direction $c = v_{n+1}$, where v_{n+1} is the left singular vector of $D^\top$ corresponding to the smallest singular value σ_{n+1}. This is the same direction arising in the TLS problem in (6.41), from which the optimal TLS solution x_{TLS} can be found by normalizing the last entry in c to -1.

6.8 Exercises

Exercise 6.1 (Least squares and total least squares) Find the least-squares line and the total least-squares[3] line for the data points (x_i, y_i), $i = 1, \ldots, 4$, with $x = (-1, 0, 1, 2)$, $y = (0, 0, 1, 1)$. Plot both lines on the same set of axes.

[3] See Section 6.7.5.

Exercise 6.2 (Geometry of least squares) Consider a least-squares problem

$$p^* = \min_x \|Ax - y\|_2,$$

where $A \in \mathbb{R}^{m,n}$, $y \in \mathbb{R}^m$. We assume that $y \notin \mathcal{R}(A)$, so that $p^* > 0$. Show that, at optimum, the residual vector $r = y - Ax$ is such that $r^\top y > 0$, $A^\top r = 0$. Interpret the result geometrically. *Hint:* use the SVD of A. You can assume that $m \geq n$, and that A is full column rank.

Exercise 6.3 (Lotka's law and least squares) Lotka's law describes the frequency of publication by authors in a given field. It states that $X^a Y = b$, where X is the number of publications, Y the relative frequency of authors with X publications, and a and b are constants (with $b > 0$) that depend on the specific field. Assume that we have data points (X_i, Y_i), $i = 1, \ldots, m$, and seek to estimate the constants a and b.

1. Show how to find the values of a, b according to a linear least-squares criterion. Make sure to define the least-squares problem involved precisely.

2. Is the solution always unique? Formulate a condition on the data points that guarantees unicity.

Exercise 6.4 (Regularization for noisy data) Consider a least-squares problem

$$\min_x \|Ax - y\|_2^2,$$

in which the data matrix $A \in \mathbb{R}^{m,n}$ is noisy. Our specific noise model assumes that each row $a_i^\top \in \mathbb{R}^n$ has the form $a_i = \hat{a}_i + u_i$, where the noise vector $u_i \in \mathbb{R}^n$ has zero mean and covariance matrix $\sigma^2 I_n$, with σ a measure of the size of the noise. Therefore, now the matrix A is a function of the uncertain vector $u = (u_1, \ldots, u_n)$, which we denote by $A(u)$. We will write $\hat{A}$ to denote the matrix with rows $\hat{a}_i^\top$, $i = 1, \ldots, m$. We replace the original problem with

$$\min_x \mathbb{E}_u \{\|A(u)x - y\|_2^2\},$$

where $\mathbb{E}_u$ denotes the expected value with respect to the random variable u. Show that this problem can be written as

$$\min_x \|\hat{A}x - y\|_2^2 + \lambda \|x\|_2^2,$$

where $\lambda \geq 0$ is some regularization parameter, which you will determine. That is, regularized least squares can be interpreted as a way to take into account uncertainties in the matrix A, in the expected value sense. *Hint:* compute the expected value of $((\hat{a}_i + u_i)^\top x - y_i)^2$, for a specific row index i.

Exercise 6.5 (Deleting a measurement in least squares) In this exercise, we revisit Section 6.3.5, and assume now that we would like to *delete* a measurement, and update the least-squares solution accordingly.[4]

[4] This is useful in the context of cross-validation methods, as evoked in Section 13.2.2.

We are given a full column rank matrix $A \in \mathbb{R}^{m,n}$, with rows $a_i^\top$, $i = 1, \ldots, m$, a vector $y \in \mathbb{R}^m$, and a solution to the least-squares problem

$$x^* = \arg\min_x \sum_{i=1}^m (a_i^\top x - y_i)^2 = \arg\min_x \|Ax - y\|_2.$$

Assume now we delete the last measurement, that is, replace (a_m, y_m) by $(0,0)$. We assume that the matrix obtained after deleting any one of the measurements is still full column rank.

1. Express the solution to the problem after deletion, in terms of the original solution, similar to the formula (6.15). Make sure to explain why any quantities you invert are positive.

2. In the so-called leave-one-out analysis, we would like to efficiently compute all the m solutions corresponding to deleting one of the m measurements. Explain how you would compute those solutions computationally efficiently. Detail the number of operations (flops) needed. You may use the fact that to invert a $n \times n$ matrix costs $O(n^3)$.

Exercise 6.6 The Michaelis–Menten model for enzyme kinetics relates the rate y of an enzymatic reaction to the concentration x of a substrate, as follows:

$$y = \frac{\beta_1 x}{\beta_2 + x},$$

where β_i, $i = 1, 2$, are positive parameters.

1. Show that the model can be expressed as a linear relation between the values $1/y$ and $1/x$.

2. Use this expression to find an estimate $\hat{\beta}$ of the parameter vector β using linear least squares, based on m measurements (x_i, y_i), $i = 1, \ldots, m$.

3. The above approach has been found to be quite sensitive to errors in input data. Can you experimentally confirm this opinion?

Exercise 6.7 (Least norm estimation on traffic flow networks)   You want to estimate the traffic (in San Francisco for example, but we'll start with a smaller example). You know the road network as well as the historical average of flows on each road segment.

1. We call q_i the flow of vehicles on each road segment $i \in I$. Write down the linear equation that corresponds to the conservation of vehicles at each intersection $j \in J$. *Hint:* think about how you might represent the road network in terms of matrices, vectors, etc.

2. The goal of the estimation is to estimate the traffic flow on each of the road segments. The flow estimates should satisfy the conservation of vehicles exactly at each intersection. Among the solutions that satisfy this constraint, we are searching for the estimate that is the closest to the historical average, $\bar{q}$, in the ℓ_2-norm sense. The vector $\bar{q}$ has size I and the i-th element represents the average for the road segment i. Pose the optimization problem.

3. Explain how to solve this problem mathematically. Detail your answer (do not only give a formula but explain where it comes from).

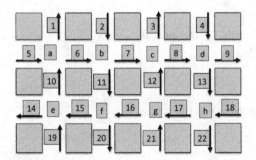

Figure 6.17 Example of the traffic estimation problem. The intersections are labeled a to h. The road segments are labeled 1 to 22. The arrows indicate the direction of traffic.

4. Formulate the problem for the small example of Figure 6.17 and solve it using the historical average given in Table 6.1. What is the flow that you estimate on road segments 1, 3, 6, 15, and 22?

5. Now, assume that besides the historical averages, you are also given some flow measurements on some of the road segments of

the network. You assume that these flow measurements are correct and want your estimate of the flow to match these measurements perfectly (besides matching the conservation of vehicles of course). The right column of Table 6.1 lists the road segments for which we have such flow measurements. Do you estimate a different flow on some of the links? Give the difference in flow you estimate for road segments 1, 3, 6, 15, and 22. Also check that your estimate gives you the measured flow on the road segments for which you have measured the flow.

segment	average	measured
1	2047.6	2028
2	2046.0	2008
3	2002.6	2035
4	2036.9	
5	2013.5	2019
6	2021.1	
7	2027.4	
8	2047.1	
9	2020.9	2044
10	2049.2	
11	2015.1	
12	2035.1	
13	2033.3	
14	2027.0	2043
15	2034.9	
16	2033.3	
17	2008.9	
18	2006.4	
19	2050.0	2030
20	2008.6	2025
21	2001.6	
22	2028.1	2045

Table 6.1 Table of flows: historical averages $\bar{q}$ (center column), and some measured flows (right column).

Exercise 6.8 (A matrix least-squares problem) We are given a set of points $p_1, \ldots, p_m \in \mathbb{R}^n$, which are collected in the $n \times m$ matrix $P = [p_1, \ldots, p_m]$. We consider the problem

$$\min_{X} F(X) \doteq \sum_{i=1}^{m} \|x_i - p_i\|_2^2 + \frac{\lambda}{2} \sum_{1 \leq i,j \leq m} \|x_i - x_j\|_2^2,$$

where $\lambda \geq 0$ is a parameter. In the above, the variable is an $n \times m$ matrix $X = [x_1, \ldots, x_m]$, with $x_i \in \mathbb{R}^n$ the i-th column of X, $i = 1, \ldots, m$. The above problem is an attempt at clustering the points p_i; the first term encourages the cluster center x_i to be close to the corresponding point p_i, while the second term encourages the x_is to be close to each other, with a higher grouping effect as λ increases.

1. Show that the problem belongs to the family of ordinary least-squares problems. You do not need to be explicit about the form of the problem.

2. Show that

$$\frac{1}{2} \sum_{1 \leq i,j \leq m} \|x_i - x_j\|_2^2 = \text{trace } XHX^\top,$$

where $H = mI_m - \mathbf{1}\mathbf{1}^\top$ is an $m \times m$ matrix, with I_m the $m \times m$ identity matrix, and $\mathbf{1}$ the vector of ones in $\mathbb{R}^m$.

3. Show that H is positive semidefinite.

4. Show that the gradient of the function F at a matrix X is the $n \times m$ matrix given by

$$\nabla F(X) = 2(X - P + \lambda XH).$$

Hint: for the second term, find the first-order expansion of the function $\Delta \to \text{trace}((X+\Delta)H(X+\Delta)^\top)$, where $\Delta \in \mathbb{R}^{n,m}$.

5. As mentioned in Remark 6.1, optimality conditions for a least-squares problem are obtained by setting the gradient of the objective to zero. Using the formula (3.10), show that optimal points

are of the form

$$x_i = \frac{1}{m\lambda + 1} p_i + \frac{m\lambda}{m\lambda + 1} \hat{p}, \quad i = 1, \ldots, m,$$

where $\hat{p} = (1/m)(p_1 + \ldots + p_m)$ is the center of the given points.

6. Interpret your results. Do you believe the model considered here is a good one to cluster points?

7
Matrix algorithms

IN THIS CHAPTER we present a compact selection of numerical algorithms for performing basic matrix computations. Specifically, we describe the power iteration method for computing eigenvalues and eigenvectors of square matrices (along with some of its variants, and a version suitable for computing SVD factors); we discuss iterative algorithms for solving square systems of linear equations, and we detail the construction of the QR factorization for rectangular matrices.

7.1 *Computing eigenvalues and eigenvectors*

7.1.1 *The power iteration method*

In this section we outline a technique for computing eigenvalues and eigenvectors of a diagonalizable matrix. The power iteration (PI) method is perhaps the simplest technique for computing one eigenvalue/eigenvector pair for a matrix. It has rather slow convergence and it is subject to some limitations. However, we present it here since it forms the building block of many other more refined algorithms for eigenvalue computation, such as the Hessenberg QR algorithm, and also because interest in the PI method has been recently revived by applications to very large-scale matrices, such as the ones arising in web-related problems (e.g., Google PageRank). Many other techniques exist for computing eigenvalues and eigenvectors, some of them tailored for matrices with special structure, such as sparse, banded, or symmetric. Such algorithms are described in standard texts on numerical linear algebra.

Let then $A \in \mathbb{R}^{n,n}$, assume A is diagonalizable, and denote by $\lambda_1, \ldots, \lambda_2$ the eigenvalues of A, ordered with decreasing modulus, that is $|\lambda_1| > |\lambda_2| \geq \cdots \geq |\lambda_n|$ (notice that we are assuming that $|\lambda_1|$

is strictly larger than $|\lambda_2|$, that is A has a "dominant" eigenvalue). Since A is diagonalizable, we can write it as $A = U\Lambda U^{-1}$, where we can assume without loss of generality that the eigenvectors $u_1, \ldots, u_n$ forming the columns of U are normalized so that $\|u_i\|_2 = 1$. Notice that from Lemma 3.3 we have that $A^k = U\Lambda^k U^{-1}$, that is

$$A^k U = U\Lambda^k.$$

Let now $x \in \mathbb{C}^n$ be a randomly chosen "trial" vector, with $\|x\|_2 = 1$, define $x = Uw$, and consider

$$A^k x = A^k Uw = U\Lambda^k w = \sum_{i=1}^n w_i \lambda_i^k u_i.$$

Observe that, if x is chosen at random (e.g., from a normal distribution and then normalized), then the first entry w_1 of w, is nonzero with probability one. Dividing and multiplying the previous expression by λ_1^k, we obtain

$$A^k x = \lambda_1^k \sum_{i=1}^n w_i \left(\frac{\lambda_i}{\lambda_1}\right)^k u_i = w_1 \lambda_1^k \left(u_1 + \sum_{i=2}^n \frac{w_i}{w_1}\left(\frac{\lambda_i}{\lambda_1}\right)^k u_i\right).$$

That is, $A^k x$ has a component $\alpha_k u_1$ along the span of u_1, and a component $\alpha_k z$ along the span of $u_2, \ldots, u_n$, i.e.,

$$A^k x = \alpha_k u_1 + \alpha_k z, \quad \alpha_k = w_1 \lambda_1^k \in \mathbb{C}, \ z = \sum_{i=2}^n \frac{w_i}{w_1}\left(\frac{\lambda_i}{\lambda_1}\right)^k u_i.$$

For the size of the z component, we have (letting $\beta_i \doteq w_i/w_1$)

$$
\begin{aligned}
\|z\|_2 &= \left\|\sum_{i=2}^n \beta_i \left(\frac{\lambda_i}{\lambda_1}\right)^k u_i\right\|_2 \leq \sum_{i=2}^n \left\|\beta_i \left(\frac{\lambda_i}{\lambda_1}\right)^k u_i\right\|_2 \\
&= \sum_{i=2}^n |\beta_i| \left|\frac{\lambda_i}{\lambda_1}\right|^k \|u_i\|_2 = \sum_{i=2}^n |\beta_i| \left|\frac{\lambda_i}{\lambda_1}\right|^k \\
&\leq \left|\frac{\lambda_2}{\lambda_1}\right|^k \sum_{i=2}^n |\beta_i|,
\end{aligned}
$$

where the last inequality follows from the ordering of the moduli of the eigenvalues. Since $|\lambda_2/\lambda_1| < 1$, we have that the size of the z component, relative to the component along u_1, goes to zero for $k \to \infty$, at a linear rate determined by the ratio $|\lambda_2|/|\lambda_1|$. Thus $A^k x \to \alpha_k u_1$, which means that $A^k x$ tends to be parallel to u_1, as $k \to \infty$. Therefore, by normalizing vector $A^k x$, we obtain that

$$\lim_{k\to\infty} \frac{A^k x}{\|A^k x\|_2} = u_1.$$

Define

$$x(k) = \frac{A^k x}{\|A^k x\|_2}, \qquad (7.1)$$

and notice also that $x(k) \to u_1$ implies that $Ax(k) \to Au_1 = \lambda_1 u_1$, hence $x^\star(k) Ax(k) \to \lambda_1 u_1^\star u_1$ (here, $\star$ denotes the transpose conjugate, since the u_i vectors can be complex valued). Therefore, recalling that $u_1^\star u_1 = \|u_1\|_2^2 = 1$, we have

$$\lim_{k \to \infty} x^\star(k) Ax(k) = \lambda_1,$$

that is the product $x^\star(k) Ax(k)$ converges to the largest-modulus eigenvalue of A. Evaluating

$$(A^k x)^\star A(A^k x) = \alpha_k^2 (\lambda_1 + u_1^\star Az + \lambda_1 z^\star u_1 + z^\star Az)$$

shows that the convergence of $x(k)^\star Ax(k)$ to λ_1 still happens at a linear rate dictated by the $|\lambda_2|/|\lambda_1|$ ratio.

The above reasoning suggests the following iterative Algorithm 1, where we notice that normalization in (7.1) only changes the norm of vector $x(k)$ and not its direction, therefore normalization can be performed at each step (as in step 4 in the algorithm) while still obtaining $x(k)$ in the form (7.1) at the k-th iteration.

Algorithm 1 Power iteration.

Require: $A \in \mathbb{R}^{n,n}$ diagonalizable, $|\lambda_1| > |\lambda_2|$, $x \in \mathbb{C}^n$, $\|x\|_2 = 1$

1: $k = 0$; $x(k) = x$

2: **repeat**

3:     $y(k+1) = Ax(k)$

4:     $x(k+1) = y(k+1)/\|y(k+1)\|_2$

5:     $\lambda(k+1) = x^\star(k+1) Ax(k+1)$

6:     $k = k+1$

7: **until** convergence

One big advantage of the power iteration is that the algorithm relies mostly on matrix–vector multiplications, for which any special structure of A, such as sparsity, can be exploited.

Two main drawbacks of the PI method are that (a) it determines only one eigenvalue (the one of maximum modulus) and the corresponding eigenvector, and (b) its convergence rate depends on the ratio $|\lambda_2|/|\lambda_1|$, hence performance can be poor when this ratio is close to one. One technique to overcome these issues is to apply the PI algorithm to a properly shifted version of matrix A, as discussed next.

7.1.2 Shift-invert power method

Given a complex scalar σ, and $A \in \mathbb{R}^{n,n}$ diagonalizable, consider the matrix

$$B_\sigma = (A - \sigma I)^{-1}.$$

By the spectral mapping theorem, see (3.15), B_σ has the same eigenvectors as A, and the eigenvalues of B_σ are $\mu_i = (\lambda_i - \sigma)^{-1}$, where λ_i, $i = 1, \ldots, n$ are the eigenvalues of A. The largest modulus eigenvalue of B_σ, μ_{max}, now corresponds to the eigenvalue λ_i which is closest to σ in the complex plane. Applying the PI method to B_σ we thus obtain the eigenvalue λ_i which is closest to the selected σ, as well as the corresponding eigenvector. The shift-invert power method is outlined in Algorithm 2.

Algorithm 2 Shift-invert power iteration.

Require: $A \in \mathbb{R}^{n,n}$ diagonalizable, $(A - \sigma I)^{-1}$ has a dominant eigenvalue, $x \in \mathbb{C}^n$, $\|x\|_2 = 1$, $\sigma \in \mathbb{C}$

1: $k = 0$; $x(k) = x$
2: **repeat**
3: $\quad y(k+1) = (A - \sigma I)^{-1} x(k)$
4: $\quad x(k+1) = y(k+1)/\|y(k+1)\|_2$
5: $\quad \lambda(k+1) = x^\star(k+1) A x(k+1)$
6: $\quad k = k + 1$
7: **until** convergence

The advantage of the shift-invert power method over the PI method is that we can now converge rapidly (but still at linear rate) to any desired eigenvalue, by choosing the "shift" σ sufficiently close to the target eigenvalue. However, the shift-invert method requires that one knows beforehand some good approximation of the target eigenvalue, to be used as a shift. If such a good approximation is not known in advance, a variation of the method would be to start the algorithm with some coarse approximation σ, and then at some point when a reasonable approximation of the eigenvector is obtained, to modify the shift dynamically, improving it iteratively as we improve the estimate of the eigenvector. This idea is discussed in the next paragraph.

7.1.3 Rayleigh quotient iterations

Suppose that at some step of the shift-invert power algorithm we have an approximate eigenvector $x(k) \neq 0$. Then, we look for some approximate eigenvalue σ_k, that is for a scalar that satisfies approximately the eigenvalue/eigenvector equation

$$x(k)\sigma_k \simeq Ax(k),$$

where by approximately we mean that we look for σ_k such that the squared norm of the equation residual is minimal, i.e., $\min \|x(k)\sigma_k - Ax(k)\|_2^2$. By imposing that the derivative of this function with respect to σ_k is zero, we obtain

$$\sigma_k = \frac{x^*(k)Ax(k)}{x^*(k)x(k)}, \tag{7.2}$$

a quantity known as a *Rayleigh quotient*, see also Section 4.3.1. If we adaptively choose shifts according to (7.2) in a shift-invert power algorithm, we obtain the so-called Rayleigh quotient iteration method, outlined in Algorithm 3. Unlike the PI method, the Rayleigh quotient iteration method can be proved to have locally quadratic convergence, that is, after a certain number of iterations, the convergence gap of the running solution at iteration $k + 1$ is proportional to the square of the gap of the solution at iteration k.

Algorithm 3 Rayleigh quotient iteration.

Require: $A \in \mathbb{R}^{n,n}$, $x \in \mathbb{C}^n$, $\|x\|_2 = 1$

Ensure: A is diagonalizable, x is an approximate eigenvector of A

1: $k = 0$; $x(k) = y$
2: **repeat**
3: $\quad \sigma_k = \frac{x^*(k)Ax(k)}{x^*(k)x(k)}$
4: $\quad y(k+1) = (A - \sigma_k I)^{-1}x(k)$
5: $\quad x(k+1) = y(k+1)/\|y(k+1)\|_2$
6: $\quad k = k+1$
7: **until** convergence

Example 7.1 As an example of application of the PI method and its variants, we consider the problem of computing the Google PageRank eigenvector discussed in Example 3.5. Here, we have

$$A = \begin{bmatrix} 0 & 0 & 1 & \frac{1}{2} \\ \frac{1}{3} & 0 & 0 & 0 \\ \frac{1}{3} & \frac{1}{2} & 0 & \frac{1}{2} \\ \frac{1}{3} & \frac{1}{2} & 0 & 0 \end{bmatrix}$$

and, due to the nature of the problem, we know in advance that the dominant eigenvalue is $\lambda_1 = 1$. We actually computed exactly (since this was a simple, toy-scale problem) the corresponding eigenvector in Example 3.5, $v = \frac{1}{31}[12\ 4\ 9\ 6]^\top$, which we can now use to quantify the distance of the approximate iterates produced by the algorithms from the actual exact eigenvector. Notice that the actual eigenvector normalization that is used in the algorithms is irrelevant: one may use the Euclidean norm, or any other normalization such as, for instance, requiring that the entries of the eigenvector sum up to one. Figure 7.1 shows the course of the error $e(k) = \|x(k)/\|x(k)\|_2 - v/\|v\|_2\|_2$ over 20 iterations of the basic PI algorithm (Algorithm 1), of the shift-invert algorithm (Algorithm 2, where we chose $\sigma = 0.9$), and of the Rayleigh quotient iteration algorithm (Algorithm 3, started with constant $\sigma = 0.9$, with adaptive Rayleigh quotient adjustments taking over after the first two plain iterations).

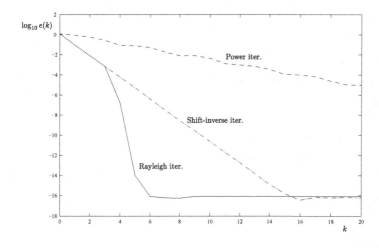

Figure 7.1 Approximation error for the dominant eigenvector of the PageRank matrix.

7.1.4 Computing the SVD using power iterations

The factors of the SVD of $A \in \mathbb{R}^{m,n}$ can be obtained by computing the spectral factorization of the two symmetric matrices $AA^\top$ and $A^\top A$. Indeed, we have seen in the proof of Theorem 5.1 that the V factor is the eigenvector matrix from the spectral factorization of

$$A^\top A = V\Lambda_n V^\top$$

and that the U factor has as columns the eigenvectors of

$$AA^\top = U\Lambda_m U^\top. \tag{7.3}$$

Λ_n and Λ_m are diagonal matrices whose first r diagonal entries are the squared singular values σ_i^2, $i = 1,\ldots,r$, and the remaining diagonal entries are zero.

In the following, we outline how a power iteration method can be used to determine the left and right singular vectors corresponding to the largest singular value of a matrix. The basic idea is to apply power iteration to the square (symmetric) matrix $A^\top A$, but in an implicit way, bypassing the explicit computation of that matrix, which would be in general dense.[1] Consider the following recursion for $k = 0, 1, 2, \ldots$:

[1] See Remark 3.1.

$$
\begin{aligned}
u(k+1) &= \frac{Av(k)}{\|Av(k)\|_2}, \\
v(k+1) &= \frac{A^\top u(k+1)}{\|A^\top u(k+1)\|_2}.
\end{aligned}
$$

Eliminating $u(k+1)$ leads to the fact that $v(k+1)$ is proportional to $A^\top Av(k)$. Since $v(k+1)$ has unit norm, we have

$$
v(k+1) = \frac{A^\top Av(k)}{\|A^\top Av(k)\|_2},
$$

hence we recognize the power iteration applied to the square (symmetric) matrix $A^\top A$. Similarly, the sequence of $u(k)$s correspond to the power iteration applied to $AA^\top$. Hence, the algorithm below computes the largest singular value of A, σ_1, and the associated left and right singular vectors u_1, v_1 (with $\sigma_1 = u_1^\top Av_1$), provided the largest singular value is separated from the second largest.

Algorithm 4 Power iteration for singular values.

Require: $A \in \mathbb{R}^{m,n}$, with $\sigma_1 > \sigma_2$, $v \in \mathbb{R}^n$, $\|v\|_2 = 1$

1: $k = 0$; $v(k) = v$
2: **repeat**
3:    $y(k+1) = Av(k)$
4:    $u(k+1) = y(k+1)/\|y(k+1)\|_2$
5:    $z(k+1) = A^\top u(k+1)$
6:    $v(k+1) = z(k+1)/\|z(k+1)\|_2$
7:    $k = k+1$
8: **until** convergence

This technique can then be applied recursively on a *deflated* version of the matrix A, in order to determine also the other singular values and their corresponding left and right singular vectors. More precisely, we define the matrix

$$
A_i = A_{i-1} - \sigma_i u_i v_i^\top, \quad i = 1, \ldots, r; \quad A_0 = A, \ \sigma_0 = 0,
$$

where $r = \operatorname{rank}(A)$, and apply Algorithm 4 to A_i, for $i = 1, \ldots, r$, in order to obtain all terms of the compact SVD of A (under the hypothesis that the singular values are well separated).

7.2   Solving square systems of linear equations

In this section we discuss numerical techniques for solving systems of linear equations of the form

$$Ax = y, \quad A \in \mathbb{R}^{n,n}, \quad A \text{ invertible.}$$

The general rectangular case can be dealt with via SVD, and it is discussed in Section 6.4.3.

7.2.1   Diagonal systems

We start by considering the simplest possible structure that a system of linear equations may have, that is, the diagonal structure. A square, diagonal, nonsingular system of linear equations has the form

$$
\begin{bmatrix}
a_{11} & 0 & \cdots & 0 \\
0 & a_{22} & 0 & \vdots \\
\vdots & \vdots & \ddots & \vdots \\
0 & \cdots & 0 & a_{nn}
\end{bmatrix}
x =
\begin{bmatrix}
y_1 \\
y_2 \\
\vdots \\
y_n
\end{bmatrix},
$$

with $a_{11}, a_{22}, \ldots, a_{nn} \neq 0$. It is rather obvious that the unique solution of such a system can be written immediately as

$$
x =
\begin{bmatrix}
y_1/a_{11} \\
y_2/a_{22} \\
\vdots \\
y_n/a_{nn}
\end{bmatrix}.
$$

7.2.2   Triangular systems

A second situation where the solution of a square nonsingular system is quite easy to obtain is when the A matrix has *triangular* structure, that is A is of the form

$$
A =
\begin{bmatrix}
a_{11} & a_{12} & \cdots & a_{1n} \\
0 & a_{22} & \cdots & a_{2n} \\
\vdots & \vdots & \ddots & \vdots \\
0 & \cdots & 0 & a_{nn}
\end{bmatrix}
\quad \text{(upper-triangular matrix),}
$$

or of the form

$$
A =
\begin{bmatrix}
a_{11} & 0 & \cdots & 0 \\
a_{21} & a_{22} & \cdots & 0 \\
\vdots & \vdots & \ddots & \vdots \\
a_{n1} & a_{n2} & \cdots & a_{nn}
\end{bmatrix}
\quad \text{(lower-triangular matrix),}
$$

with $a_{11}, a_{22}, \ldots, a_{nn} \neq 0$. Consider for instance the lower-triangular case:

$$\begin{bmatrix} a_{11} & 0 & \cdots & 0 \\ a_{21} & a_{22} & \cdots & 0 \\ \vdots & \vdots & \ddots & \vdots \\ a_{n1} & a_{n2} & \cdots & a_{nn} \end{bmatrix} x = \begin{bmatrix} y_1 \\ y_2 \\ \vdots \\ y_n \end{bmatrix}.$$

The solution can be obtained by a so-called *forward substitution* technique: start from the first equation and obtain $x_1 = y_1/a_{11}$ then, substituting this value in the second equation, we have

$$a_{21}x_1 + a_{22}x_2 = a_{21}y_1/a_{11} + a_{22}x_2 = y_2.$$

Hence, we obtain $x_2 = \frac{y_2 - a_{21}y_1/a_{11}}{a_{22}}$. We next substitute x_1, x_2 in the third equation to obtain x_3, and proceeding in the same way we eventually terminate by obtaining x_n. This scheme is summarized in Algorithm 5.

Algorithm 5 Forward substitution.

Require: $A \in \mathbb{R}^{n,n}$ nonsingular and lower triangular and $y \in \mathbb{R}^n$.

1: $x_1 = y_1/a_{11}$
2: **for** $i = 2$ to n **do**
3: $\quad s = y_i$
4: $\quad$ **for** $j = 1, \ldots, i-1$ **do**
5: $\quad\quad s = s - a_{ij}x_j$
6: $\quad$ **end for**
7: $\quad x_i = s/a_{ii}$
8: **end for**

An analogous algorithm can be readily devised for the solution of upper-triangular systems, as formally stated in Algorithm 6.

Algorithm 6 Backward substitution.

Require: $A \in \mathbb{R}^{n,n}$ nonsingular and upper triangular and $y \in \mathbb{R}^n$.

1: $x_n = y_n/a_{nn}$
2: **for** $i = n-1, \ldots, 1$ **do**
3: $\quad s = y_i$
4: $\quad$ **for** $j = i+1, \ldots, n$ **do**
5: $\quad\quad s = s - a_{ij}x_j$
6: $\quad$ **end for**
7: $\quad x_i = s/a_{ii}$
8: **end for**

Remark 7.1 *Operation count.* It is easy to determine the total number of algebraic operations (divisions, multiplications, and sums/subtractions) required to obtain the solution of a triangular system via backward substitution. At each stage $i = n, \ldots, 1$, the algorithm performs $n - i$ multiply-and-sum operations, plus one division. Therefore, the total count is

$$\sum_{i=n}^{1} 2(n - i) + 1 = n^2.$$

7.2.3 Gaussian elimination

As shown in Section 7.2.2, the solution of triangular nonsingular systems is very easy to obtain. But what about a generic nonsingular but possibly non-triangular matrix? The idea we illustrate in this section is to transform a generic system into an equivalent upper-triangular system by means of appropriate operations, and then to solve the resulting triangular system using backward substitution. Such an iterative triangularization technique is known as Gaussian elimination.

Example 7.2 (*Simple illustration of Gaussian elimination*) Consider the system

$$\begin{bmatrix} 1 & 2 & 3 \\ 2 & 8 & 7 \\ 4 & 4 & 4 \end{bmatrix} x = \begin{bmatrix} 1 \\ 3 \\ 2 \end{bmatrix}.$$

This system is nonsingular, but it is not in triangular form. However, if we multiply the first equation in the system by 2 then subtract it from the second equation and substitute the so-obtained equation in place of the second equation, we obtain an *equivalent* system

$$\begin{bmatrix} 1 & 2 & 3 \\ 0 & 4 & 1 \\ 4 & 4 & 4 \end{bmatrix} x = \begin{bmatrix} 1 \\ 1 \\ 2 \end{bmatrix}.$$

Further, if we multiply the first equation in the system by 4 then subtract it from the third equation and substitute the so-obtained equation in place of the third equation, we obtain

$$\begin{bmatrix} 1 & 2 & 3 \\ 0 & 4 & 1 \\ 0 & -4 & -8 \end{bmatrix} x = \begin{bmatrix} 1 \\ 1 \\ -2 \end{bmatrix},$$

where we note that the elements below the first entry in the first column have been zeroed out. Finally, if we now multiply the second equation by -1 then subtract it from the third equation and substitute the so-obtained equation in place of the third equation in the system, we obtain an equivalent system in upper triangular form:

$$\begin{bmatrix} 1 & 2 & 3 \\ 0 & 4 & 1 \\ 0 & 0 & -7 \end{bmatrix} x = \begin{bmatrix} 1 \\ 1 \\ -1 \end{bmatrix}.$$

This system can now be readily solved by backward substitution, obtaining $x_3 = 1/7$, $x_2 = 3/14$, $x_1 = 1/7$.

We now describe Gaussian elimination in more generality. Consider a square nonsingular system

$$\begin{bmatrix} a_{11} & a_{12} & \cdots & a_{1n} \\ a_{21} & a_{22} & \cdots & a_{2n} \\ \vdots & \vdots & \ddots & \vdots \\ a_{n1} & a_{n2} & \cdots & a_{nn} \end{bmatrix} x = \begin{bmatrix} y_1 \\ y_2 \\ \vdots \\ y_n \end{bmatrix}.$$

Substitute each equation from $j = 2$ onward with equation j minus equation 1 multiplied by a_{j1}/a_{11} (assuming $a_{11} \neq 0$), thus obtaining the equivalent system

$$\begin{bmatrix} a_{11} & a_{12} & a_{13} & \cdots & a_{1n} \\ 0 & a_{22}^{(1)} & a_{23}^{(1)} & \cdots & a_{2n}^{(1)} \\ 0 & a_{32}^{(1)} & a_{33}^{(1)} & \cdots & a_{3n}^{(1)} \\ \vdots & \vdots & \vdots & \ddots & \vdots \\ 0 & a_{n2}^{(1)} & a_{n3}^{(1)} & \cdots & a_{nn}^{(1)} \end{bmatrix} x = \begin{bmatrix} y_1 \\ y_2^{(1)} \\ y_3^{(1)} \\ \vdots \\ y_n^{(1)} \end{bmatrix},$$

where $a_{ij}^{(1)} = a_{ij} - a_{1j}a_{j1}/a_{11}$. Next, substitute each equation from $j = 3$ onward with equation j minus equation 2 multiplied by $a_{j2}^{(1)}/a_{22}^{(1)}$ (assuming $a_{22}^{(1)} \neq 0$), thus obtaining the equivalent system

$$\begin{bmatrix} a_{11} & a_{12} & a_{13} & \cdots & a_{1n} \\ 0 & a_{22}^{(1)} & a_{23}^{(1)} & \cdots & a_{2n}^{(1)} \\ 0 & 0 & a_{33}^{(2)} & \cdots & a_{3n}^{(2)} \\ \vdots & \vdots & \vdots & \ddots & \vdots \\ 0 & 0 & a_{n3}^{(2)} & \cdots & a_{nn}^{(2)} \end{bmatrix} x = \begin{bmatrix} y_1 \\ y_2^{(1)} \\ y_3^{(2)} \\ \vdots \\ y_n^{(2)} \end{bmatrix},$$

where $a_{ij}^{(2)} = a_{ij}^{(1)} - a_{2j}^{(1)}a_{j2}^{(1)}/a_{22}^{(1)}$. Clearly, proceeding in this same way for $n - 1$ times, we eventually determine a system which is equivalent to the original systems and which is in upper-triangular form:

$$\begin{bmatrix} a_{11} & a_{12} & a_{13} & \cdots & a_{1n} \\ 0 & a_{22}^{(1)} & a_{23}^{(1)} & \cdots & a_{2n}^{(1)} \\ 0 & 0 & a_{33}^{(2)} & \cdots & a_{3n}^{(2)} \\ \vdots & \vdots & \vdots & \ddots & \vdots \\ 0 & 0 & 0 & \cdots & a_{nn}^{(n-1)} \end{bmatrix} x = \begin{bmatrix} y_1 \\ y_2^{(1)} \\ y_3^{(2)} \\ \vdots \\ y_n^{(n-1)} \end{bmatrix}.$$

This latter system can then be solved by backward substitution.

Remark 7.2 *Elimination with pivoting.* Notice that the approach described above fails if, at any stage $k = 1, \ldots, n - 1$ of the procedure, we encounter a diagonal element $a_{kk}^{(k-1)} = 0$, since division by such an element is not possible. In practice, numerical problems shall also be expected if $|a_{kk}^{(k-1)}|$ is very small. To overcome this difficulty, the basic procedure can be modified so as to include partial or full *pivoting*. The idea of (full) pivoting is very simple: at stage k of the procedure we look for the largest-modulus element among $a_{ij}^{(k-1)}$, $i > k$, $j \geq k$. Such element is called the pivot, and the rows and columns of the current-stage matrix are exchanged so as to bring this element into position (k, k); then the elimination phase proceeds as previously described, and the process is repeated. Notice that when exchanging two rows of the matrix the elements in the vector y also need to be exchanged accordingly. Similarly, when exchanging two columns of the matrix the corresponding elements of x need to be exchanged. Partial pivoting works in a similar way, but only the elements in the column below element $a_{kk}^{(k-1)}$ are searched for a pivot, therefore only two rows need be exchanged in this case. Pivoting increases the numerical effort required for computing the solution, since a search over pivot element is involved at each stage, and memory management operations are required in order to exchange rows (and columns, in the case of full pivoting).

The next algorithm describes Gaussian elimination with partial pivoting.

Algorithm 7 Gaussian elimination with partial pivoting.

Require: $A \in \mathbb{R}^{n,n}$ nonsingular and $y \in \mathbb{R}^n$. Let $S = [A \, y]$.

1: **for** $i = 1, \ldots, n - 1$ **do**
2:    find i_p such that $|s_{i_p i}| \geq |s_{ki}|$ for all $k = i, \ldots, n$
3:    let $S \leftarrow$ exchange row i with row i_p of S
4:    **for** $k = i + 1, \ldots, n$ **do**
5:       **for** $j = i, \ldots, n + 1$ **do**
6:          $s_{kj} = s_{kj} - (s_{ki}/s_{ii})s_{ij}$
7:       **end for**
8:    **end for**
9: **end for**

Operations count. We next compute the number of elementary operations required for solving a square system via Gaussian elimination. Considering first the Gaussian elimination procedure, we see that at the first iteration of the process it takes $2n + 1$ operations to update the second row of matrix $S = [A \, y]$ (one division and n multiply-subtracts to find the new entries along the row). To zero out all the entries in the first column below the first entry and to update all the rows from the second onwards takes therefore $(n-1)(2n+1)$ operations. We next need $(n-2)(2n-1)$ operations to get the second column zeroed out and the matrix updated; for the third column we need $(n-3)(2n-3)$ operations, etc. The sum of all of these operations is:

$$
\begin{aligned}
\sum_{i=1}^{n-1} (n-i)(2(n-i+1)+1) &= \sum_{i=1}^{n-1} i(2i+3) = 2\sum_{i=1}^{n-1} i^2 + 3\sum_{i=1}^{n-1} i \\
&= 3\frac{n(n-1)(2n-1)}{6} + 2\frac{n(n-1)}{2} \\
&= \sim n^3
\end{aligned}
$$

(here, the notation $\sim$ denotes the leading term in the polynomial; this notation is more informative than the usual $O(\cdot)$ notation, since the coefficient of the leading term is indicated). We finally need to apply backward substitution to the transformed triangular system, which takes an additional n^2 operations. This leaves the leading complexity term unaltered, so the total number of operations for solving a generic nonsingular system is $\sim n^3$.

7.3   QR factorization

The QR factorization is a linear algebra operation that factors a matrix into an orthogonal component, which is a basis for the row space of the matrix, and a triangular component. In the QR factorization a matrix $A \in \mathbb{R}^{m,n}$, with $m \geq n$, rank$(A) = n$, is thus decomposed as:

$$
A = QR,
$$

where $Q \in \mathbb{R}^{m,n}$ has orthogonal columns (i.e., $Q^\top Q = I_n$) , and $R \in \mathbb{R}^{n,n}$ is upper triangular.

There are many ways of calculating the QR factorization, including the Householder transformation method, the modified Gram–Schmidt algorithm, and the fast Givens method. Here, we describe the method based on the modified Gram–Schmidt (MGS) procedure.

7.3.1   Modified Gram–Schmidt procedure

We recall from Section 2.3.3 that, given a set of linearly independent vectors $\{a^{(1)}, \ldots, a^{(n)}\}$, the Gram–Schmidt (GS) procedure constructs an orthonormal set of vectors $\{q^{(1)}, \ldots, q^{(n)}\}$ having the same span as the original set, as follows: for $k = 1, \ldots, n$,

$$\zeta^{(k)} = a^{(k)} - \sum_{i=1}^{k-1} \langle a^{(k)}, q^{(i)} \rangle q^{(i)}, \tag{7.4}$$

$$q^{(k)} = \frac{\zeta^{(k)}}{\|\zeta^{(k)}\|}.$$

Let $S_{k-1} = \text{span}\{a^{(1)}, \ldots, a^{(k-1)}\}$ and let $S_{k-1}^{\perp}$ denote the orthogonal complement of S_{k-1}. In Eq. (7.4) the GS procedure computes the projection of $a^{(k)}$ onto S_{k-1}, and then subtracts it from $a^{(k)}$, thus obtaining the projection of $a^{(k)}$ onto $S_{k-1}^{\perp}$. It is easy to see that the projection operation in (7.4) can be expressed in matrix form as follows:

$$\zeta^{(k)} = P_{S_{k-1}^{\perp}} a^{(k)}, \quad P_{S_{k-1}^{\perp}} = I - P_{S_{k-1}}, \quad P_{S_{k-1}} = \sum_{i=1}^{k-1} q^{(i)} q^{(i)\top}, \tag{7.5}$$

with $P_{S_0} = 0$, $P_{S_0^{\perp}} = I$. Further, the orthogonal projector matrix $P_{S_{k-1}^{\perp}} = I - P_{S_{k-1}}$ can be written as the product of elementary projections onto the subspaces orthogonal to each $q^{(1)}, \ldots, q^{(k-1)}$, that is

$$P_{S_{k-1}^{\perp}} = P_{q^{(k-1)\perp}} \cdots P_{q^{(1)\perp}}, \quad P_{q^{(i)\perp}} = I - q^{(i)} q^{(i)\top}, k > 1.$$

This fact can be easily verified directly: take for instance $k = 3$ (the general case follows from an identical argument), then

$$
\begin{aligned}
P_{q^{(2)\perp}} P_{q^{(1)\perp}} &= (I - q^{(2)} q^{(2)\top})(I - q^{(1)} q^{(1)\top}) \\
&= I - q^{(1)} q^{(1)\top} - q^{(2)} q^{(2)\top} + q^{(2)} q^{(2)\top} q^{(1)} q^{(1)\top} \\
\text{(since } q^{(2)\top} q^{(1)} = 0 \text{)} \quad &= I - q^{(1)} q^{(1)\top} - q^{(2)} q^{(2)\top} \\
&= I - P_{S_2} = P_{S_2^{\perp}}.
\end{aligned}
$$

In the MGS, each $\zeta^{(k)} = P_{q^{(1)\perp}} \cdots P_{q^{(k-1)\perp}} I a^{(k)}$ is thus computed recursively as follows:

$$
\begin{aligned}
\zeta^{(k)}(1) &= a^{(k)}, \\
\zeta^{(k)}(2) &= P_{q^{(1)\perp}} \zeta^{(k)}(1) = (I - q^{(1)} q^{(1)\top}) \zeta^{(k)}(1) \\
&= \zeta^{(k)}(1) - q^{(1)} q^{(1)\top} \zeta^{(k)}(1), \\
\zeta^{(k)}(3) &= P_{q^{(2)\perp}} \zeta^{(k)}(2) = \zeta^{(k)}(2) - q^{(2)} q^{(2)\top} \zeta^{(k)}(2), \\
&\vdots \quad \vdots \quad \vdots \\
\zeta^{(k)}(k) &= P_{q^{(k-1)\perp}} \zeta^{(k)}(k-1) \\
&= \zeta^{(k)}(k-1) - q^{(k-1)} q^{(k-1)\top} \zeta^{(k)}(k-1).
\end{aligned}
$$

Although the two formulations (GS and MGS) are mathematically equivalent, the latter can be proved to be more stable numerically. The MGS procedure is next formalized as an algorithm.

Algorithm 8 Modified Gram–Schmidt procedure.

Require: A set of l.i. vectors $\{a^{(1)}, \ldots, a^{(n)}\}$, $a^{(i)} \in \mathbb{R}^m$, $m \geq n$.

1: **for** $i = 1, \ldots, n$ **do**
2:    $\zeta^{(i)} = a^{(i)}$
3: **end for**
4: **for** $i = 1, \ldots, n$ **do**
5:    $r_{ii} = \|\zeta^{(i)}\|$
6:    $q^{(i)} = \zeta^{(i)} / r_{ii}$
7:    **for** $j = i+1, \ldots, n$ **do**
8:       $r_{ij} = q^{(i)\top} \zeta^{(j)}$, $\zeta^{(j)} = \zeta^{(j)} - r_{ij} q^{(i)}$
9:    **end for**
10: **end for**

Operations count. For large m, n, the computational work is dominated by the innermost loop of Algorithm 8: m multiply-add for computing $r_{ij} = q^{(i)\top} \zeta^{(j)}$, and m multiply-subtract for computing $\zeta^{(j)} = \zeta^{(j)} - r_{ij} q^{(i)}$, for a total of $4m$ operations per inner loop. The overall operation count for Algorithm 8 is thus approximately given by

$$\sum_{i=1}^{n} \sum_{j=1+1}^{n} 4m = \sum_{i=1}^{n} (n - i)4m = \left(n^2 - \frac{n(n+1)}{2}\right)4m \sim 2mn^2.$$

MGS as a QR decomposition. We next show that the MGS algorithm actually provides the Q and R factors of a QR factorization of A. Let $a^{(1)}, \ldots, a^{(n)}$ denote the columns of A. We see from (7.5) that $\zeta^{(1)} = a^{(1)}$ and, for $j > 1$,

$$\zeta^{(j)} = a^{(j)} - \sum_{i=1}^{j-1} q^{(i)} q^{(i)\top} a^{(j)}.$$

Let now $r_{jj} = \|\zeta^{(j)}\|$, $r_{ij} = q^{(i)\top} a^{(j)}$, and recall $q^{(j)} = \zeta^{(j)} / r_{jj}$. The previous equation becomes

$$r_{jj} q^{(j)} = a^{(j)} - \sum_{i=1}^{j-1} r_{ij} q^{(i)},$$

that is

$$a^{(j)} = r_{jj} q^{(j)} + \sum_{i=1}^{j-1} r_{ij} q^{(i)}.$$

This latter equation gives the desired factorization $A = QR$, with

$$[a^{(1)} \cdots a^{(n)}] = [q^{(1)} \cdots q^{(n)}] \begin{bmatrix} r_{11} & r_{12} & \cdots & r_{1n} \\ 0 & r_{22} & \cdots & r_{2n} \\ \vdots & \vdots & \ddots & \vdots \\ 0 & 0 & \cdots & r_{nn} \end{bmatrix}.$$

The above reasoning constitutes a constructive proof of the following fact.

Theorem 7.1 (QR factorization) *Any matrix $A \in \mathbb{R}^{m,n}$, with $m \geq n$, rank$(A) = n$, can be factored as $A = QR$, where $R \in \mathbb{R}^{n,n}$ is upper triangular with positive diagonal entries, and $Q \in \mathbb{R}^{m,n}$ has orthonormal columns (that is, it satisfies $Q^\top Q = I_n$).*

7.3.2   MGS and QR decomposition for rank-deficient matrices.

In the standard GS procedure we assumed that the vectors $\{a^{(1)}, \ldots, a^{(n)}\}$, $a^{(i)} \in \mathbb{R}^m$, are linearly independent, that is matrix $A = [a^{(1)} \cdots a^{(n)}] \in \mathbb{R}^{m,n}$ is full column rank. In this paragraph, we discuss how to generalize the GS procedure and the QR factorization to the case when A is not full rank, i.e., when $\{a^{(1)}, \ldots, a^{(n)}\}$ are not linearly independent. In this case, let $k \leq n$ be the smallest integer for which the vector $a^{(k)}$ is a linear combination of the previous vectors $\{a^{(1)}, \ldots, a^{(k-1)}\}$, that is

$$a^{(k)} = \sum_{i=1}^{k-1} \tilde{\alpha}_i a^{(i)},$$

for some scalars $\tilde{\alpha}_i$, $i = 1, \ldots, k - 1$. Since, by construction, $\{q^{(1)}, \ldots, q^{(k-1)}\}$ span the same subspace as $\{a^{(1)}, \ldots, a^{(k-1)}\}$, we also have that

$$a^{(k)} = \sum_{i=1}^{k-1} \alpha_i q^{(i)},$$

for some scalars α_i, $i = 1, \ldots, k - 1$. Therefore, since the vectors q_j, $j = 1, \ldots, k - 1$, are orthonormal,

$$\langle a^{(k)}, q^{(j)} \rangle = \sum_{i=1}^{k-1} \alpha_i \langle q^{(i)}, q^{(j)} \rangle = \alpha_j,$$

hence we see from (7.4) that $\zeta^{(k)} = 0$, thus the standard procedure cannot proceed further. The generalized procedure, however, proceeds by just discarding all vectors $a^{(k')}$, $k' \geq k$, for which $\zeta^{(k')} = 0$, until either the procedure is terminated, or a vector $a^{(k')}$ with $\zeta^{(k')} \neq 0$ is found, in which case the corresponding normalized vector $q^{(k')}$

is added to the orthonormal set, and the procedure is iterated. Upon termination, this modified procedure returns a set of $r = \text{rank } A$ orthonormal vectors $\{q^{(1)}, \dots, q^{(r)}\}$ which form an orthonormal basis for $\mathcal{R}(A)$. This procedure provides a generalized QR factorization, since each column of A is represented as a linear combination of the columns of $Q = [q^{(1)} \cdots q^{(r)}]$, with a non-decreasing number of nonzero coefficients. In particular, a first block of $n_1 \geq 1$ columns of A are written as a linear combination of $q^{(1)}$, a second block of $n_2 \geq 1$ columns of A are written as a linear combination of $q^{(1)}, q^{(2)}$, etc., till the r-th block of n_r columns of A, which is written as a linear combination of $q^{(1)}, \dots, q^{(r)}$, where $n_1 + n_2 + \cdots + n_r = n$. In formulas,

$$A = QR, \quad R = \begin{bmatrix} R_{11} & R_{12} & \cdots & R_{1r} \\ 0 & R_{22} & \cdots & R_{2r} \\ 0 & 0 & \ddots & \vdots \\ 0 & 0 & \cdots & R_{rr} \end{bmatrix}, \quad R_{ij} \in \mathbb{R}^{1,n_j},$$

the matrix R being in a block-upper-triangular form. The columns of R can then be reordered so that the column corresponding to the first element of R_{ii} is moved to column i, $i = 1, \dots, r$ (column pivoting) This corresponds to writing

$$A = QRE^\top, \quad R = [\tilde{R} \ M],$$

where E is a suitable column-permutation matrix (note that permutation matrices are orthogonal), $\tilde{R} \in \mathbb{R}^{r,r}$ is upper triangular and invertible, and $M \in \mathbb{R}^{r,n-r}$. Notice that an alternative, "full," form of the QR factorization uses all m columns in the Q matrix: $m - r$ orthonormal columns are added to $q^{(1)} \cdots q^{(r)}$ so as to complete an orthonormal basis of $\mathbb{R}^m$. Therefore, $m - r$ trailing rows with zeros are appended to the R matrix, to obtain

$$A = QRE^\top, \quad Q \in \mathbb{R}^{m,m}, Q^\top Q = I_m, \quad R = \begin{bmatrix} \tilde{R} & M \\ 0_{m-r,r} & 0_{m-r,n-r} \end{bmatrix}.$$

7.4 Exercises

Exercise 7.1 (Sparse matrix–vector product) Recall from Section 3.4.2 that a matrix is said to be sparse if most of its entries are zero. More formally, assume a $m \times n$ matrix A has sparsity coefficient $\gamma(A) \ll 1$, where $\gamma(A) \doteq d(A)/s(A)$, $d(A)$ is the number of nonzero elements in A, and $s(A)$ is the size of A (in this case, $s(A) = mn$).

1. Evaluate the number of operations (multiplications and additions) that are required to form the matrix–vector product Ax, for any given vector $x \in \mathbb{R}^n$ and generic, non-sparse A. Show that this number is reduced by a factor $\gamma(A)$, if A is sparse.

2. Now assume that A is not sparse, but is a rank-one modification of a sparse matrix. That is, A is of the form $\tilde{A} + uv^\top$, where $\tilde{A} \in \mathbb{R}^{m,n}$ is sparse, and $u \in \mathbb{R}^m$, $v \in \mathbb{R}^m$ are given. Devise a method to compute the matrix–vector product Ax that exploits sparsity.

Exercise 7.2 (A random inner product approximation) Computing the standard inner product between two vectors $a, b \in \mathbb{R}^n$ requires n multiplications and additions. When the dimension n is huge (say, e.g., of the order of 10^{12}, or larger), even computing a simple inner product can be computationally prohibitive.

Let us define a random vector $r \in \mathbb{R}^n$ constructed as follows: choose uniformly at random an index $i \in \{1, \ldots, n\}$, and set $r_i = 1$, and $r_j = 0$ for $j \neq i$. Consider the two scalar random numbers $\tilde{a}, \tilde{b}$ that represent the "random projections" of the original vectors a, b along r:

$$\tilde{a} \doteq r^\top a = a_i,$$
$$\tilde{b} \doteq r^\top b = b_i.$$

Prove that

$$n\mathbb{E}\{\tilde{a}\tilde{b}\} = a^\top b,$$

that is, $n\tilde{a}\tilde{b}$ is an unbiased estimator of the value of the inner product $a^\top b$. Observe that computing $n\tilde{a}\tilde{b}$ requires very little effort, since it is just equal to na_ib_i, where i is the randomly chosen index. Notice, however, that the variance of such an estimator can be large, as it is given by

$$\text{var}\{n\tilde{a}\tilde{b}\} = n \sum_{k=1}^{n} a_i^2 b_i^2 - \left(a^\top b\right)^2$$

(prove also this latter formula). *Hint:* let e_i denote the i-th standard basis vector of $\mathbb{R}^n$; the random vector r has discrete probability distribution $\text{Prob}\{r = e_i\} = 1/n$, $i = 1, \ldots, n$, hence $\mathbb{E}\{r\} = \frac{1}{n}\mathbf{1}$. Further, observe that the products $r_k r_j$ are equal to zero for $k \neq j$ and that the vector $r^2 \doteq [r_1^2, \ldots, r_n^2]^\top$ has the same distribution as r.

Generalizations of this idea to random projections onto k-dimensional subspaces are indeed applied for matrix-product approximation, SVD factorization, and PCA on huge-scale problems. The key theoretical tool underlying these results is known as the Johnson–Lindenstrauss lemma.

Exercise 7.3 (Power iteration for SVD with centered, sparse data)
In many applications such as principal component analysis (see Section 5.3.2), one needs to find the few largest singular values of a centered data matrix. Specifically, we are given a $n \times m$ matrix $X = [x_1, \ldots, x_m]$ of m data points in $\mathbb{R}^n$, $i = 1, \ldots, m$, and define the centered matrix $\tilde{X}$ to be

$$\tilde{X} = [\tilde{x}_1 \cdots \tilde{x}_m], \quad \tilde{x}_i \doteq x_i - \bar{x}, \ i = 1, \ldots, m,$$

with $\bar{x} = \frac{1}{m} \sum_{i=1}^{m} x_i$ the average of the data points. In general, $\tilde{X}$ is dense, even if X itself is sparse. This means that each step of the power iteration method involves two matrix–vector products, with a dense matrix. Explain how to modify the power iteration method in order to exploit sparsity, and avoid dense matrix–vector multiplications.

Exercise 7.4 (Exploiting structure in linear equations) Consider the linear equation in $x \in \mathbb{R}^n$

$$Ax = y,$$

where $A \in \mathbb{R}^{m,n}, y \in \mathbb{R}^m$. Answer the following questions to the best of your knowledge.

1. The time required to solve the general system depends on the sizes m, n and the entries of A. Provide a rough estimate of that time as a function of m, n only. You may assume that m, n are of the same order.

2. Assume now that $A = D + uv^\top$, where D is diagonal, invertible, and $u \in \mathbb{R}^m$, $v \in \mathbb{R}^n$. How would you exploit this structure to solve the above linear system, and what is a rough estimate of the complexity of your algorithm?

3. What if A is upper-triangular?

Exercise 7.5 (Jacobi method for linear equation) Let $A = (a_{ij}) \in \mathbb{R}^{n,n}$, $b \in \mathbb{R}^n$, with $a_{ii} \neq 0$ for every $i = 1, \ldots, n$. The *Jacobi method* for solving the square linear system

$$Ax = b$$

consists of decomposing A as a sum: $A = D + R$, where $D = \text{diag}(a_{11}, \ldots, a_{nn})$, and R contains the off-diagonal elements of A, and then applying the recursion

$$x^{(k+1)} = D^{-1}(b - Rx^{(k)}), \quad k = 0, 1, 2, \ldots,$$

with initial point $\hat{x}(0) = D^{-1}b$.

The method is part of a class of methods known as *matrix splitting*, where A is decomposed as a sum of a "simple" invertible matrix and another matrix; the Jacobi method uses a particular splitting of A.

1. Find conditions on D, R that guarantee convergence from an arbitrary initial point. *Hint:* assume that $M \doteq -D^{-1}R$ is diagonalizable.

2. The matrix A is said to be strictly row diagonally dominant if

$$\forall\, i = 1, \ldots, n\ :\ |a_{ii}| > \sum_{j \neq i} |a_{ij}|.$$

   Show that when A is strictly row diagonally dominant, the Jacobi method converges.

Exercise 7.6 (Convergence of linear iterations) Consider linear iterations of the form

$$x(k+1) = Fx(k) + c, \quad k = 0, 1, \ldots, \tag{7.6}$$

where $F \in \mathbb{R}^{n,n}$, $c \in \mathbb{R}^n$, and the iterations are initialized with $x(0) = x_0$. We assume that the iterations admit a stationary point, i.e., that there exists $\bar{x} \in \mathbb{R}^n$ such that

$$(I - F)\bar{x} = c. \tag{7.7}$$

In this exercise, we derive conditions under which $x(k)$ tends to a finite limit for $k \to \infty$. We shall use these results in Exercise 7.7, to set up a linear iterative algorithm for solving systems of linear equations.

1. Show that the following expressions hold for all $k = 0, 1, \ldots$:

$$\begin{aligned} x(k+1) - x(k) &= F^k(I - F)(\bar{x} - x_0), & (7.8) \\ x(k) - \bar{x} &= F^k(x_0 - \bar{x}). & (7.9) \end{aligned}$$

2. Prove that, for all x_0, $\lim_{k\to\infty} x(k)$ converges to a finite limit if and only if F^k is convergent (see Theorem 3.5). When $x(k)$ converges, its limit point $\bar{x}$ satisfies (7.7).

Exercise 7.7 (A linear iterative algorithm) In this exercise we introduce some "equivalent" formulations of a system of linear equations

$$Ax = b, \quad A \in \mathbb{R}^{m,n}, \tag{7.10}$$

and then study a linear recursive algorithm for solution of this system.

1. Consider the system of linear equations

$$Ax = AA^\dagger b, \qquad\qquad (7.11)$$

where $A^\dagger$ is any pseudoinverse of A (that is, a matrix such that $AA^\dagger A = A$). Prove that (7.11) always admits a solution. Show that every solution of equations (7.10) is also a solution for (7.11). Conversely, prove that if $b \in \mathcal{R}(A)$, then every solution to (7.11) is also a solution for (7.10).

2. Let $R \in \mathbb{R}^{n,m}$ be any matrix such that $\mathcal{N}(RA) = \mathcal{N}(A)$. Prove that

$$A^\dagger \doteq (RA)^\dagger R$$

is indeed a pseudoinverse of A.

3. Consider the system of linear equations

$$RAx = Rb, \qquad\qquad (7.12)$$

where $R \in \mathbb{R}^{n,m}$ is any matrix such that $\mathcal{N}(RA) = \mathcal{N}(A)$ and $Rb \in \mathcal{R}(RA)$. Prove that, under these hypotheses, the set of solutions of (7.12) coincides with the set of solutions of (7.11), for $A^\dagger = (RA)^\dagger R$.

4. Under the setup of the previous point, consider the following linear iterations: for $k = 0, 1, \ldots,$

$$x(k+1) = x(k) + \alpha R(b - Ax(k)), \qquad\qquad (7.13)$$

where $\alpha \neq 0$ is a given scalar. Show that if $\lim_{k\to\infty} x(k) = \bar{x}$, then $\bar{x}$ is a solution for the system of linear equations (7.12). State appropriate conditions under which $x(k)$ is guaranteed to converge.

5. Suppose A is positive definite (i.e., $A \in S^n$, $A \succ 0$). Discuss how to find a suitable scalar α and matrix $R \in \mathbb{R}^{n,n}$ satisfying the conditions of point 3, and such that the iterations (7.13) converge to a solution of (7.12). *Hint:* use Exercise 4.7.

6. Explain how to apply the recursive algorithm (7.13) for finding a solution to the linear system $\tilde{A}x = \tilde{b}$, where $\tilde{A} \in \mathbb{R}^{m,n}$ with $m \geq n$ and rank $\tilde{A} = n$. *Hint:* apply the algorithm to the normal equations.

II

Convex optimization models

8

Convexity

WE HAVE SEEN in Section 6.3.1 that the ordinary least-squares problem can be solved using standard linear algebra tools. This is thus a case where the solution of a minimization problem can be found efficiently and globally, i.e., we are guaranteed that no other point besides the LS-optimal solutions may possibly yield a better LS objective. Such desirable properties actually extend essentially to a wider class of optimization problems. The key feature that renders an optimization problem "nice" is a property called *convexity*, which is introduced in this chapter. In particular, we shall characterize convex *sets* and convex *functions*, and define the class of convex optimization *problems* as those where a convex objective function is minimized over a convex set. Engineering problems that can be modeled in this convexity framework are typically amenable to an efficient numerical solution. Further, for certain types of convex models having particular structure, such as linear, convex quadratic, or convex conic, specialized algorithms are available that are efficient to the point of providing the user with a reliable "technology" for modeling and solving practical problems.

8.1 Convex sets

8.1.1 Open and closed sets, interior, and boundary

We start by recalling informally some basic topological notions of subsets of $\mathbb{R}^n$. A set $\mathcal{X} \subseteq \mathbb{R}^n$ is said to be *open* if for any point $x \in \mathcal{X}$

there exist a ball centered in x which is contained in $\mathcal{X}$. Precisely, for any $x \in \mathbb{R}^n$ and $\epsilon > 0$ define the Euclidean ball of radius r centered at x

$$B_\epsilon(x) \doteq \{z : \|z - x\|_2 < \epsilon\}.$$

Then $\mathcal{X} \subseteq \mathbb{R}^n$ is open if

$$\forall x \in \mathcal{X}, \quad \exists \epsilon > 0 : B_\epsilon(x) \subset \mathcal{X}.$$

A set $\mathcal{X} \subseteq \mathbb{R}^n$ is said to be *closed* if its complement $\mathbb{R}^n \setminus \mathcal{X}$ is open. The whole space $\mathbb{R}^n$ and the empty set $\emptyset$ are declared open by definition. However, they are also closed (open and closed are not necessarily mutually exclusive attributes), according to the definition.

The *interior* of a set $\mathcal{X} \subseteq \mathbb{R}^n$ is defined as

$$\text{int}\,\mathcal{X} = \{z \in \mathcal{X} : B_\epsilon(z) \subseteq \mathcal{X}, \text{ for some } \epsilon > 0\}.$$

The *closure* of a set $\mathcal{X} \subseteq \mathbb{R}^n$ is defined as

$$\overline{\mathcal{X}} = \{z \in \mathbb{R}^n : z = \lim_{k \to \infty} x_k, \ x_k \in \mathcal{X}, \forall k\},$$

i.e., the closure of $\mathcal{X}$ is the set of limits of sequences in $\mathcal{X}$. The *boundary* of $\mathcal{X}$ is defined as

$$\partial\mathcal{X} = \overline{\mathcal{X}} \setminus \text{int}\,\mathcal{X}.$$

A set $\mathcal{X} \subseteq \mathbb{R}^n$ is open if and only if $\mathcal{X} = \text{int}\,\mathcal{X}$. An open set does not contain any of its boundary points; a closed set contains all of its boundary points. Unions and intersections of open (resp. closed) sets are open (resp. closed).

A set $\mathcal{X} \subseteq \mathbb{R}^n$ is said to be *bounded* if it is contained in a ball of finite radius, that is if there exist $x \in \mathbb{R}^n$ and $r > 0$ such that $\mathcal{X} \subseteq B_r(x)$. If $\mathcal{X} \subseteq \mathbb{R}^n$ is closed and bounded, then it is said to be *compact*.

Example 8.1 (*Intervals on the real line*) Let $a, b \in \mathbb{R}$, $a < b$. The interval $[a, b]$ is a closed set. Its boundary is the discrete set $\{a, b\}$ and its interior is the (open) set $\{x : a < x < b\}$. The interval $[a, b)$ is neither closed nor open. The interval (a, b) is open. A semi-infinite interval of the form $[a, +\infty)$ is closed, since its complement $(-\infty, a)$ is open.[1]

8.1.2 Combinations and hulls

Given a set of points (vectors) in $\mathbb{R}^n$, $\mathcal{P} = \{x^{(1)}, \dots, x^{(m)}\}$, the linear hull (subspace) generated by these points is the set of all possible *linear combinations* of the points, that is the set of points of the form

$$x = \lambda_1 x^{(1)} + \cdots + \lambda_m x^{(m)}, \tag{8.1}$$

[1] Note that, technically, $+\infty$ is *not* a boundary point of the interval $[a, +\infty)$ as a subset of $\mathbb{R}$, since the definition of closure of a set (and hence of boundary) requires the boundary point to belong to the underlying space $\mathbb{R}$, which does not contain $+\infty$. If one considers instead intervals as subsets of the extended real line $\mathbb{R} \cup \{-\infty, +\infty\}$, then $[a, +\infty]$ is closed, $(a, +\infty)$ is open, and $[a, +\infty)$ is neither open nor closed.

for $\lambda_i \in \mathbb{R}, i = 1, \dots, m$. The *affine hull*, aff $\mathcal{P}$, of $\mathcal{P}$ is the set generated by taking all possible linear combinations (8.1) of the points in $\mathcal{P}$, under the restriction that the coefficients λ_i sum up to one, that is $\sum_{i=1}^m \lambda_i = 1$. aff $\mathcal{P}$ is the smallest affine set containing $\mathcal{P}$.

A *convex combination* of the points is a special type of linear combination $x = \lambda_1 x^{(1)} + \cdots + \lambda_m x^{(m)}$ in which the coefficients λ_i are restricted to be non-negative and to sum up to one, that is $\lambda_i \geq 0$ for all i, and $\sum_{i=1}^m \lambda_i = 1$. Intuitively, a convex combination is a weighted average of the points, with weights given by the λ_i coefficients. The set of all possible convex combination is called the *convex hull* of the point set:

$$\mathrm{co}(x^{(1)}, \dots, x^{(m)}) = \left\{ x = \sum_{i=1}^m \lambda_i x^{(i)} : \lambda_i \geq 0, i = 1, \dots, m; \sum_{i=1}^m \lambda_i = 1 \right\}.$$

Similarly, we define a *conic* combination of a set of points as a linear combination where the coefficients are restricted to be non-negative. Correspondingly, the *conic hull* of a set of points is defined as

$$\mathrm{conic}(x^{(1)}, \dots, x^{(m)}) = \left\{ x = \sum_{i=1}^m \lambda_i x^{(i)} : \lambda_i \geq 0, i = 1, \dots, m \right\}.$$

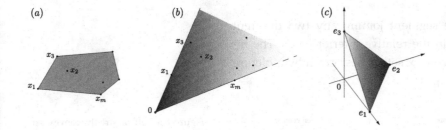

(a) (b) (c)

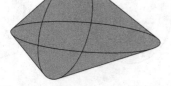

Figure 8.1 Convex hull (a) and conic hull (b) of a set of points in $\mathbb{R}^2$; (c) shows the convex hull of the standard unit basis vectors of $\mathbb{R}^3$, $\{e_1, e_2, e_3\}$. The conic hull of this same set is the entire positive orthant $\mathbb{R}^3_+$.

Figure 8.1 gives examples of the convex hull and of the conic hull of a set of points.

Convex and conic hulls may be defined not only for discrete collections of points, but also for generic sets. The convex (resp. affine, conic) hull of a set $C \in \mathbb{R}^n$ is the set of all possible convex (resp. affine, conic) combinations of its points, see Figure 8.2 for an example.

Figure 8.2 The convex hull of the union of two ellipses.

8.1.3 Convex sets

A subset $C \subseteq \mathbb{R}^n$ is said to be *convex* if it contains the line segment between any two points in it:

$$x_1, x_2 \in C, \lambda \in [0, 1] \quad \Rightarrow \quad \lambda x_1 + (1 - \lambda) x_2 \in C.$$

The *dimension d* of a convex set $C \subseteq \mathbb{R}^n$ is defined as the dimension of its affine hull. Notice that it can happen that $d < n$. For example, the set $C = \{x = [\alpha\ 0]^\top : \alpha \in [0, 1]\}$ is a convex subset of $\mathbb{R}^2$ (the "ambient space," of dimension $n = 2$), but its affine dimension is $d = 1$. The *relative interior*, relint C, of a convex set C is the interior of C relative to its affine hull. That is, x belongs to relint C if there exists an open ball of dimension d contained in aff C, centered at x, and with positive radius, which is contained in C, see Figure 8.3 for a pictorial explanation. The relative interior coincides with the usual interior when C is "full-dimensional," that is when its affine dimension coincides with the dimension of the ambient space.

Subspaces and affine sets, such as lines, planes, and higher-dimensional "flat" sets are obviously convex, as they contain the entire line passing through any two points, not just the line segment. Half-spaces are also convex, as geometrical intuition suggests. A set C is said to be a *cone* if it has the property that if $x \in C$, then $\alpha x \in C$, for every $\alpha \geq 0$. A set C is said to be a *convex cone* if it is convex and it is a cone. The conic hull of a set is a convex cone.

A set C is said to be *strictly convex* if it is convex, and

$$x_1 \neq x_2 \in C, \ \lambda \in (0, 1) \quad \Rightarrow \quad \lambda x_1 + (1 - \lambda)x_2 \in \text{relint } C,$$

that is, the interior of the line segment joining any two different points $x_1, x_2 \in C$ is contained in the relative interior of C. The intuitive idea of convex and non-convex sets is best given by a picture in two dimensions, see Figure 8.4.

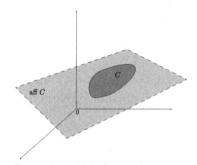

Figure 8.3 In this figure, the convex set $C \subseteq \mathbb{R}^3$ has an affine hull of dimension $d = 2$. Thus, C has no "regular" interior. However, its relative interior, relint C, is given by the darker shaded region in the picture.

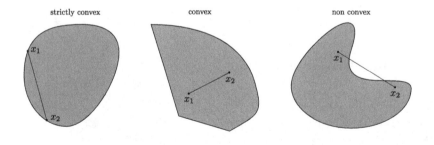

Figure 8.4 Left, a strictly convex set $\mathbb{R}^2$. Middle, convex set in $\mathbb{R}^2$: for any two pair of points x_1, x_2, the line segment joining the two points is entirely in the set. Right, a non-convex set: there exist a pair of points such that the segment joining them is not entirely included in the set.

8.1.4 Operations that preserve convexity

Certain operations on convex sets, such as intersection, projection, perspective transformation, etc., preserve convexity.

8.1.4.1 Intersection. If $C_1, \ldots, C_m$ are convex sets, then their intersection

$$C = \bigcap_{i=1,\ldots,m} C_i$$

is still a convex set. This fact can be easily proved by direct application of the definition of convexity. Indeed, consider any two points $x^{(1)}, x^{(2)} \in C$ (notice that this implies that $x^{(1)}, x^{(2)} \in C_i$, $i = 1, \ldots, m$) and take any $\lambda \in [0, 1]$. Then, by convexity of C_i, the point $x = \lambda x^{(1)} + (1 - \lambda)x^{(2)}$ belongs to C_i, and this is true for all $i = 1, \ldots, m$, therefore $x \in C$. The intersection rule actually holds for possibly infinite families of convex sets: if $C(\alpha)$, $\alpha \in \mathcal{A} \subseteq \mathbb{R}^q$, is a family of convex sets, parameterized by α, then the set

$$C = \bigcap_{\alpha \in \mathcal{A}} C_\alpha$$

is convex. This rule is often useful to prove convexity of a set, as in the following examples.

Example 8.2 (*Polyhedra*) A half-space

$$\mathcal{H} = \{x \in \mathbb{R}^n : c^\top x \le d\}, \quad c \neq 0$$

is a convex set. The intersection of m half-spaces $\mathcal{H}_i$, $i = 1, \ldots, m$, is a convex set called a *polyhedron*, see Figure 8.5. More on polyhedra and polytopes (which are bounded polyhedra) in Section 9.2.

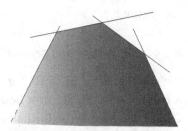

Figure 8.5 The intersection of half-spaces is a convex polyhedron. In the figure is an example in $\mathbb{R}^2$.

Example 8.3 (*The second-order cone*) The set in $\mathbb{R}^{n+1}$

$$\mathcal{K}_n = \{(x, t), x \in \mathbb{R}^n, t \in \mathbb{R} : \|x\|_2 \le t\}$$

is a convex cone, called the *second-order cone*. In $\mathbb{R}^3$ a second-order cone is described by the triples (x_1, x_2, t) that satisfy the equation

$$x_1^2 + x_2^2 \le t^2, \quad t \ge 0,$$

see Figure 8.6. Horizontal sections of this set at level $t \ge 0$ are disks of radius t.

$\mathcal{K}_n$ is a cone, since it is non-negative invariant (i.e., $z \in \mathcal{K}_n$ implies $\alpha z \in \mathcal{K}_n$, for all $\alpha \ge 0$). The fact that $\mathcal{K}_n$ is convex can be proven directly from the basic definition of a convex set. Alternatively, we may express $\mathcal{K}_n$ as a continuous intersection of half-spaces, as follows. From the Cauchy–Schwartz inequality, we have that

$$t \ge \|x\|_2 \iff \forall u, \|u\|_2 \le 1 : t \ge u^\top x,$$

whence

$$\mathcal{K}_n = \bigcap_{u: \|u\|_2 \le 1} \left\{ (x, t) \in \mathbb{R}^{n+1} : t \ge u^\top x \right\}.$$

Each one of the sets involved in the intersection, for fixed u, represents a half-space in (x, t), which is a convex set.

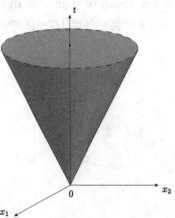

Figure 8.6 The second-order cone in $\mathbb{R}^3$.

8.1.4.2 Affine transformation. If a map $f : \mathbb{R}^n \to \mathbb{R}^m$ is affine, and $C \subset \mathbb{R}^n$ is convex, then the image set

$$f(C) = \{f(x) : x \in C\}$$

is convex. This fact is easily verified: any affine map has a matrix representation

$$f(x) = Ax + b.$$

Then, for any $y^{(1)}, y^{(2)} \in f(C)$ there exist $x^{(1)}, x^{(2)}$ in C such that $y^{(1)} = Ax^{(1)} + b, y^{(2)} = Ax^{(2)} + b$. Hence, for $\lambda \in [0,1]$, we have that

$$\lambda y^{(1)} + (1 - \lambda)y^{(2)} = A(\lambda x^{(1)} + (1 - \lambda)x^{(2)}) + b = f(x),$$

where $x = \lambda x^{(1)} + (1 - \lambda)x^{(2)} \in C$. In particular, the projection of a convex set C onto a subspace is representable by means of a linear map, see, e.g., Section 5.2, hence the projected set is convex, see Figure 8.7.

8.1.5 Supporting and separating hyperplanes

We say that $\mathcal{H} = \{x \in \mathbb{R}^n : a^\top x = b\}$ is a *supporting hyperplane* for a convex set $C \subseteq \mathbb{R}^n$ at a boundary point $z \in \partial C$, if $z \in \mathcal{H}$, and $C \subseteq \mathcal{H}_-$, where $\mathcal{H}_-$ is the half-space

$$\mathcal{H}_- = \{x \in \mathbb{R}^n : a^\top x \leq b\}.$$

A key result of convex analysis[2] states that a convex set always admits a supporting hyperplane, at any boundary point, see Figure 8.8.

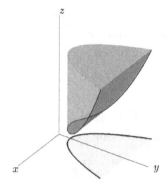

Figure 8.7 Convex set $\{(x,y,z) : y \geq x^2, z \geq y^2\}$ and its projection on the space of (x,y) variables. The projection turns out to be the set $\{(x,y) : y \geq x^2\}$.

[2] See, for instance, Corollary 11.6.2 in the classical book by T. Rockafellar, *Convex Analysis*, Princeton University Press, 1970.

Figure 8.8 (a) Supporting hyperplanes of a convex set at two different boundary points; (b) illustration of the separation theorem.

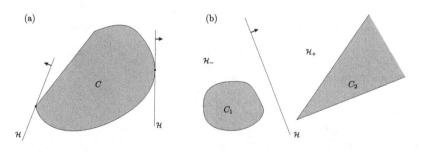

Theorem 8.1 (Supporting hyperplane theorem) *If $C \subseteq \mathbb{R}^n$ is convex and $z \in \partial C$, then there exists a supporting hyperplane for C at z.*

Given two subsets C_1, C_2 of $\mathbb{R}^n$, we say that $\mathcal{H}$ *separates* the two sets, if $C_1 \subseteq \mathcal{H}_-$ and $C_2 \subseteq \mathcal{H}_+$, where

$$\mathcal{H}_+ = \{x \in \mathbb{R}^n : a^\top x \geq b\}.$$

We say that $\mathcal{H}$ *strictly separates* C_1, C_2 if $C_1 \subseteq \mathcal{H}_{--}$ and $C_2 \subseteq \mathcal{H}_{++}$, where

$$\mathcal{H}_{--} = \{x \in \mathbb{R}^n : a^\top x < b\}, \quad \mathcal{H}_{++} = \{x \in \mathbb{R}^n : a^\top x > b\}.$$

Another fundamental result of convex analysis states that any two disjoint convex sets can be separated by a hyperplane,[3] see Figure 8.8.

[3] See, for instance, Corollary 11.4.2 in Rockafellar's book.

Theorem 8.2 (Separating hyperplane theorem) *Let C_1, C_2 be convex subsets of $\mathbb{R}^n$ having empty intersection ($C_1 \cap C_2 = \varnothing$). Then there exists a separating hyperplane $\mathcal{H}$ for C_1, C_2. Furthermore, if C_1 is closed and bounded and C_2 is closed, then C_1, C_2 can be strictly separated.*

Example 8.4 (*Farkas lemma*) An important application of the separating hyperplane theorem is in the proof of the so-called Farkas lemma, which is stated next. Let $A \in \mathbb{R}^{m,n}$ and $y \in \mathbb{R}^m$. Then, one and only one of the following two conditions is satisfied:

1. the system of linear equations $Ax = y$ admits a non-negative solution $x \geq 0$;

2. there exist $z \in \mathbb{R}^m$ such that $z^\top A \geq 0$, $z^\top y < 0$.

Notice first that these two conditions cannot be true at the same time. This is easily seen by contradiction, since if 1 holds then, for all z, $z^\top A x = z^\top y$, for some $x \geq 0$. But if also 2 holds, then $z^\top A \geq 0$, hence $z^\top A x \geq 0$ (since $x \geq 0$), and from 1 we would have that $z^\top y \geq 0$, which contradicts $z^\top y < 0$. Next, it suffices to prove that if 1 doesn't hold then 2 must hold. Suppose then that there exists no $x \geq 0$ such that $Ax = y$. Since Ax, $x \geq 0$, denotes the set of conic combinations of the columns of A (which we here denote by $\mathrm{conic}(A)$), we have that $y \notin \mathrm{conic}(A)$. Since the singleton $\{y\}$ is convex closed and bounded, and $\mathrm{conic}(A)$ is convex and closed, we can apply the separating hyperplane theorem and claim there must exist a hyperplane $\{x : z^\top x = q\}$ that strictly separates y from $\mathrm{conic}(A)$, that is

$$z^\top y < q, \quad z^\top A v > q, \ \forall v \geq 0.$$

Now, the second condition implies that $q < 0$ (take $v = 0$), hence the first condition implies that $z^\top y < 0$. Furthermore, the condition $z^\top A v > q$ for all $v \geq 0$ implies that $z^\top A \geq 0$, which would conclude the proof. This latter fact is verified by contradiction, since suppose that $z^\top A$ had a negative component, say the i-th component. Then, one may take v to be all zero except for the i-th component, which is taken positive and sufficiently large so as to make $z^\top A v = v_i [z^\top A]_i$ smaller than q, so obtaining a contradiction.

An equivalent formulation of the Farkas lemma is obtained by considering that statement 2 above implies the negation of statement 1, and vice versa. That is, the following two conditions are equivalent:

1. there exists $x \geq 0$ such that $Ax = y$;

2. $z^\top y \geq 0, \quad \forall z : z^\top A \geq 0.$

This formulation yields the following interpretation in terms of systems of linear inequalities: let $a_i \in \mathbb{R}^m$, $i = 1, \ldots, n$, be the columns of A, then

$$y^\top z \geq 0, \quad \forall z : a_i^\top z \geq 0, \, i = 1, \ldots, n,$$

if and only if there exist *multipliers* $x_i \geq 0$, $i = 1, \ldots, n$ such that y is a conic combination of the a_is, that is, if and only if

$$\exists x_i \geq 0, \, i = 1, \ldots, m \quad \text{such that } y = a_1 x_1 + \ldots + a_n x_n.$$

8.2 Convex functions

8.2.1 Definitions

Consider a function $f : \mathbb{R}^n \to \mathbb{R}$. The *effective domain* (or, simply, the domain) of f is the set over which the function is well defined:

$$\text{dom} f = \{x \in \mathbb{R}^n : -\infty < f(x) < \infty\}.$$

For example, the function $f(x) = \log(x)$ has effective domain $\text{dom} f = \mathbb{R}_{++}$ (the strictly positive reals), and the function

$$f(x) = \frac{a^\top x + b}{c^\top x + d} \tag{8.2}$$

has effective domain $\text{dom} f = \{x : c^\top x + d \neq 0\}$.

A function $f : \mathbb{R}^n \to \mathbb{R}$ is *convex* if $\text{dom} f$ is a convex set, and for all $x, y \in \text{dom} f$ and all $\lambda \in [0, 1]$ it holds that

$$f(\lambda x + (1 - \lambda)y) \leq \lambda f(x) + (1 - \lambda)f(y). \tag{8.3}$$

We say that a function f is *concave* if $-f$ is convex. A visualization of convexity for a function of scalar variable is given in Figure 8.9.

A function f is *strictly convex* if inequality (8.3) holds strictly (i.e., with $<$ instead of $\leq$), for all $x \neq y$ in the domain and all $\lambda \in (0, 1)$. A function $f : \mathbb{R}^n \to \mathbb{R}$ is *strongly convex* if there exists an $m > 0$ such that

$$\tilde{f}(x) = f(x) - \frac{m}{2}\|x\|_2^2$$

is convex, that is if

$$f(\lambda x + (1 - \lambda)y) \leq \lambda f(x) + (1 - \lambda)f(y) - \frac{m}{2}\lambda(1 - \lambda)\|x - y\|_2^2.$$

Clearly, strong convexity implies strict convexity.

A fundamental fact about convex functions is that they are continuous, as specified by the next lemma, reported here without proof.[4]

[4] See Section 10 in Rockafellar's book.

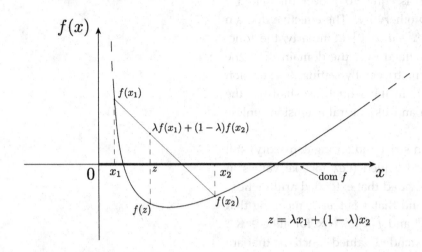

Lemma 8.1 (Continuity of convex functions) *If $f : \mathbb{R}^n \to \mathbb{R}$ is convex, then it is* continuous *over* int dom f. *Moreover, f is Lipschitz on every compact subset $\mathcal{X} \in$ int dom f, meaning that there exists a constant $M > 0$ such that*

$$|f(x) - f(y)| \leq M\|x - y\|_2, \quad \forall x, y \in \mathcal{X}.$$

Notice, however, that convex functions may have discontinuities at the boundary of the domain. For example, the function

$$f(x) = \begin{cases} x^2 & \text{if } x \in (-1, 1], \\ 2 & \text{if } x = -1, \\ +\infty & \text{otherwise} \end{cases}$$

has convex domain dom $f = [-1, 1]$, and it is continuous over the interior of its domain. However, it has discontinuities at the boundary points of the domain.

8.2.1.1 Extended-valued functions. Sometimes, it is useful to *extend* the range of values of f to include also $\pm\infty$. For instance, a natural extension $\bar{f}$ of $f(x) = \log(x)$ is

$$\bar{f}(x) = \begin{cases} \log(x) & \text{if } x > 0, \\ -\infty & \text{if } x \leq 0. \end{cases}$$

An arbitrary effective domain $\mathcal{X}$ can be imposed artificially on a function f, by defining an extended-valued function $\bar{f} : \mathbb{R}^n \to [-\infty, +\infty]$, which is equal to f for $x \in \mathcal{X}$ and equal to $+\infty$ otherwise. For example, for the linear-fractional function (8.2), we can define an extended-valued function $\bar{f}$ which is equal to f over the set $\{x : c^\top x + d > 0\}$, and is equal to $+\infty$ otherwise. The effective domain of such an $\bar{f}$ is $\mathrm{dom}\,\bar{f} = \{x : c^\top x + d > 0\}$. Similarly, the function $f(X) = \log \det(X)$ is often defined over the domain $\mathbb{S}^n_{++}$ (the set of symmetric, positive definite matrices), by setting its extension to be equal to $+\infty$ outside this set. In the sequel, we shall use the same symbol to denote a function and its natural extension, unless ambiguity arises.

The definitions of convexity (and strict and strong convexity) still hold if f is allowed to be extended-valued (i.e., to take values in the extended real line $[-\infty, \infty]$), provided that extended arithmetic is used to interpret the inequalities, and that f is *proper*, meaning that $\bar{f}(x) < +\infty$ for at least one $x \in \mathbb{R}^n$ and $\bar{f}(x) > -\infty$ for all $x \in \mathbb{R}^n$. We shall always consider convex extended-valued functions that are proper.

Note that for f to be convex, the convexity of the domain is required. For example, the function

$$f(x) = \begin{cases} x & \text{if } x \notin [-1, 1], \\ +\infty & \text{otherwise} \end{cases}$$

is not convex, although is it linear (hence, convex) on its (non-convex) domain $(-\infty, -1) \cup (1, +\infty)$. The function

$$f(x) = \begin{cases} \frac{1}{x} & \text{if } x > 0, \\ +\infty & \text{otherwise} \end{cases}$$

is convex over its (convex) domain $\mathrm{dom}\,f = \mathbb{R}_{++}$.

A useful extended-valued function is the *indicator function* of a convex set $\mathcal{X}$, defined as[5]

[5] See also Exercise 8.2.

$$I_\mathcal{X}(x) = \begin{cases} 0 & \text{if } x \in \mathcal{X}, \\ +\infty & \text{otherwise.} \end{cases}$$

8.2.1.2 Epigraph and sublevel sets. Given a function $f : \mathbb{R}^n \to (-\infty, +\infty]$, its epigraph (i.e., the set of points lying above the graph of the function) is the set

$$\mathrm{epi}\,f = \{(x, t), x \in \mathrm{dom}\,f, t \in \mathbb{R} : f(x) \le t\}.$$

It holds that f is a convex function if and only if epi f is a convex set. For $\alpha \in \mathbb{R}$, the α-sublevel set of f is defined as

$$S_\alpha = \{x \in \mathbb{R}^n : f(x) \le \alpha\}.$$

It can be easily verified that if f is a convex function, then S_α is a convex set, for any $\alpha \in \mathbb{R}$. Also, if f is strictly convex, then S_α is a strictly convex set. However, the converses of the latter two statements are not true in general. For instance, $f(x) = \ln(x)$ is not convex (it is actually concave), nevertheless its sublevel sets are the intervals $(0, e^{\alpha}]$, which are convex. A function f such that all its sublevel sets are convex is said to be *quasiconvex*.

8.2.1.3 *Closed functions.*

A function $f : \mathbb{R}^n \to (-\infty, \infty]$ is said to be *closed* if its epigraph is a closed set. In turn, this is equivalent to the fact that every sublevel set S_{α} of f, $\alpha \in \mathbb{R}$, is closed. Function f is said to be *lower semicontinuous*[6] (lsc) if for any $x_0 \in \text{dom } f$ and $\epsilon > 0$ there exists an open ball B centered in x_0 such that $f(x) \geq f(x_0) - \epsilon$ for all $x \in B$. All sublevel sets S_{α} are closed if and only if f is lsc. For a proper convex function, the concepts of closedness and lower semicontinuity are equivalent, i.e., a proper convex function is closed if and only if it is lsc.

[6] See, e.g., Section 7 in Rockafellar's book for further details.

If f is continuous and dom f is a closed set, then f is closed. As a particular case, since $\mathbb{R}^n$ is closed (as well as open), we have that if f is continuous and dom $f = \mathbb{R}^n$, then f is closed. Also, if f is continuous then it is closed and all sublevel sets are bounded if and only if f tends to $+\infty$ for any x that tends to the boundary of dom f, see Lemma 8.3.

8.2.1.4 *Interior and boundary points of sublevel sets.*

Let f be a proper, closed, and convex function with open domain, let $\alpha \in \mathbb{R}$, and consider the sublevel set $S_{\alpha} = \{x \in \mathbb{R}^n : f(x) \leq \alpha\}$. Then it holds that[7]

[7] See Corollary 7.6.1 in the cited book by Rockafellar.

$$\begin{aligned} S_{\alpha} &= \text{closure of } \{x \in \mathbb{R}^n : f(x) < \alpha\}, \\ \text{relint } S_{\alpha} &= \{x \in \mathbb{R}^n : f(x) < \alpha\}. \end{aligned}$$

For instance, when S_{α} is full-dimensional, this result implies that points such that $f(x) = \alpha$ are on the boundary of S_{α}, and points such that $f(x) < \alpha$ are in the interior of S_{α}. Therefore, in particular, if $f(x_0) < \alpha$, then there exists an open ball centered in x_0 which is contained in S_{α}; this fact is exploited for proving several properties of convex optimization problems.

8.2.1.5 *Sum of convex functions.*

If $f_i : \mathbb{R}^n \to \mathbb{R}$, $i = 1, \ldots, m$, are convex functions, then the function

$$f(x) = \sum_{i=1}^{m} \alpha_i f_i(x), \quad \alpha_i \geq 0, \ i = 1, \ldots, m$$

is also convex over $\cap_i \operatorname{dom} f_i$. This fact easily follows from the definition of convexity, since for any $x, y \in \operatorname{dom} f$ and $\lambda \in [0, 1]$,

$$
\begin{aligned}
f(\lambda x + (1 - \lambda)y) &= \sum_{i=1}^{m} \alpha_i f_i(\lambda x + (1 - \lambda)y) \\
&\leq \sum_{i=1}^{m} \alpha_i \left(\lambda f_i(x) + (1 - \lambda) f_i(y) \right) \\
&= \lambda f(x) + (1 - \lambda) f(y).
\end{aligned}
$$

For example, the negative entropy function

$$
f(x) = - \sum_{i=1}^{n} x_i \log x_i
$$

is convex over $\operatorname{dom} f = \mathbb{R}_{++}^n$ (the set of n-vectors with strictly positive entries), since it is the sum of functions that are convex over this domain (convexity of $-z \log z$ can be verified by checking that the second derivative of this function is positive over $\mathbb{R}_{++}$, see Section 8.2.2).

Similarly, we can easily verify that the sum of strictly convex functions is also strictly convex, that is

$$f, g \text{ strictly convex} \quad \Rightarrow \quad f + g \text{ strictly convex.}$$

Moreover, the sum of a convex function and a strongly convex function is strongly convex, that is

$$f \text{ convex}, g \text{ strongly convex} \quad \Rightarrow \quad f + g \text{ strongly convex.}$$

To prove this fact, observe from the definition that strong convexity of g means that there exist $m > 0$ such that $g(x) - (m/2)\|x\|_2^2$ is convex. Since the sum of two convex functions is convex, it holds that $f(x) + g(x) - (m/2)\|x\|_2^2$ is convex, which in turn implies that $f + g$ is strongly convex.

8.2.1.6 *Affine variable transformation.* Let $f : \mathbb{R}^n \to \mathbb{R}$ be convex, and define

$$g(x) = f(Ax + b), \quad A \in \mathbb{R}^{n,m}, \ b \in \mathbb{R}^n.$$

Then g is convex over $\operatorname{dom} g = \{x : Ax + b \in \operatorname{dom} f\}$. For example, $f(z) = -\log(z)$, is convex over $\operatorname{dom} f = \mathbb{R}_{++}$, hence $f(x) = -\log(ax + b)$ is also convex over $ax + b > 0$.

8.2.2   *Alternative characterizations of convexity*

Besides resorting to the definition, there are several other rules or conditions that can characterize convexity of a function. We here mention a few of them. Here and in the sequel, when mentioning convexity of a function it is implicitly assumed that $\operatorname{dom} f$ is convex.

8.2.2.1 First-order conditions. If f is differentiable (that is, dom f is open and the gradient exists everywhere on the domain), then f is convex if and only if

$$\forall x, y \in \text{dom}\, f, \quad f(y) \geq f(x) + \nabla f(x)^\top (y - x), \qquad (8.4)$$

and it is strictly convex if (8.4) holds with strict inequality for all $x, y \in \text{dom}\, f$, $x \neq y$.

To prove this fact, suppose f is convex. Then, the definition (8.3) implies that for any $\lambda \in (0, 1]$

$$\frac{f(x + \lambda(y - x)) - f(x)}{\lambda} \leq f(y) - f(x),$$

which, for $\lambda \to 0$, yields $\nabla f(x)^\top (y - x) \leq f(y) - f(x)$, proving one direction of the implication (strict convexity follows by simply replacing $\leq$ with $<$). Conversely, if (8.4) holds, then take any $x, y \in \text{dom}\, f$ and $\lambda \in [0, 1]$, and let $z = \lambda x + (1 - \lambda y)$. Then,

$$f(x) \geq f(z) + \nabla f(z)^\top (x - z),$$
$$f(y) \geq f(z) + \nabla f(z)^\top (y - z).$$

Taking a convex combination of these inequalities, we get

$$\lambda f(x) + (1 - \lambda) f(y) \geq f(z) + \nabla f(z)^\top 0 = f(z),$$

which concludes the proof.

The geometric interpretation of condition (8.4) is that the graph of f is bounded below everywhere by any one of its tangent hyperplanes or, equivalently, that any tangent hyperplane is a supporting hyperplane for epi$f$, see Figure 8.10.

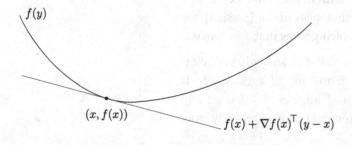

Figure 8.10 The epigraph of a differentiable convex function lies on the half-space defined by any of its tangent hyperplanes.

From Eq. (8.4) we also draw the following observation: the gradient of a convex function at a point $x \in \mathbb{R}^n$ (if it is nonzero) divides the whole space in two half-spaces:

$$\mathcal{H}_{++}(x) = \{y : \nabla f(x)^\top (y - x) > 0\},$$
$$\mathcal{H}_{-}(x) = \{y : \nabla f(x)^\top (y - x) \leq 0\},$$

and any point $y \in \mathcal{H}_{++}(x)$ is such that $f(y) > f(x)$.

8.2.2.2 Second-order conditions. If f is twice differentiable, then f is convex if and only if its Hessian matrix $\nabla^2 f$ is positive semidefinite everywhere on the (open) domain of f, that is if and only if $\nabla^2 f \succeq 0$ for all $x \in \text{dom} f$. This is perhaps the most commonly known characterization of convexity.

To see this fact, let $x_0 \in \text{dom} f$ and let $v \in \mathbb{R}^n$ be any direction. Since $\text{dom} f$ is open, the point $z = x_0 + \lambda v$ is still in $\text{dom} f$, for sufficiently small $\lambda > 0$, hence

$$f(z) = f(x_0) + \lambda \nabla f(x_0)^\top v + \frac{1}{2}\lambda^2 v^\top \nabla^2 f(x_0)v + O(\lambda^3),$$

which, by (8.4), implies

$$\frac{1}{2}\lambda^2 v^\top \nabla^2 f(x_0)v + O(\lambda^3) = f(z) - f(x_0) - \lambda \nabla f(x_0)^\top v \geq 0.$$

Dividing by $\lambda^2 > 0$, we have

$$\frac{1}{2}v^\top \nabla^2 f(x_0)v + \frac{O(\lambda^3)}{\lambda^2} \geq 0,$$

which, for $\lambda \to 0$, shows that $v^\top \nabla^2 f(x_0)v \geq 0$, proving the first part of the claim. Conversely, suppose $\nabla^2 f(x) \succeq 0$ for all $x \in \text{dom} f$. Then, for any $y \in \text{dom} f$, by a version of the second order-Taylor approximation theorem, it holds that

$$f(y) = f(x) + \nabla f(x)^\top (y - x) + \frac{1}{2}(y - x)^\top \nabla^2 f(z)(y - x),$$

where the Hessian is computed at some unknown point z lying in the segment between x and y. Since the last term in the above expression is non-negative (due to positive semidefiniteness of the Hessian), we obtain $f(y) \geq f(x) + \nabla f(x)^\top (y - x)$, which proves that f is convex.

By a similar argument one can prove that f is strongly convex if and only if $\nabla^2 f \succeq mI$, for some $m > 0$ and for all $x \in \text{dom} f$. It also holds that $\nabla^2 f \succ 0$ for all $x \in \text{dom} f$ implies that f is strictly convex (but the converse of this last statement is not necessarily true, in general; take for instance $f(x) = x^4$).

Example 8.5 *(Establishing convexity via Hessian)* We give three examples where convexity of a function is checked by checking the positive semi-definiteness of the Hessian matrix.

1. Consider two quadratic functions in two variables

$$\begin{aligned}
p(x) &= 4x_1^2 + 2x_2^2 + 3x_1 x_2 + 4x_1 + 5x_2 + 2 \times 10^5, \\
q(x) &= 4x_1^2 - 2x_2^2 + 3x_1 x_2 + 4x_1 + 5x_2 + 2 \times 10^5.
\end{aligned}$$

The Hessian of p is independent of x, and it is given by the constant matrix

$$\nabla^2 p = \begin{bmatrix} 8 & 3 \\ 3 & 4 \end{bmatrix}.$$

The eigenvalues of $\nabla^2 p$ are $\lambda_1 \simeq 2.39$, $\lambda_2 \simeq 9.6$: they are both positive, hence $\nabla^2 p$ is positive definite and therefore $p(x)$ is convex (strongly). Likewise, the Hessian of q is

$$\nabla^2 q = \begin{bmatrix} 8 & 3 \\ 3 & -4 \end{bmatrix},$$

whose eigenvalues are $\lambda_1 \simeq -4.71$, $\lambda_2 \simeq 8.71$, hence $\nabla^2 q$ is not positive semidefinite, thus $q(x)$ is not convex.

For a generic quadratic function of several variables, which can always be cast in the standard form

$$f(x) = \frac{1}{2} x^\top H x + c^\top x + d,$$

where H is a symmetric matrix, the Hessian is simply given by

$$\nabla^2 f = H,$$

hence f is convex if and only if H is positive semidefinite, and it is strongly convex if H is positive definite.

2. Consider the so-called *square-to-linear* function

$$f(x,y) = \begin{cases} \dfrac{x^\top x}{y} & \text{if } y > 0, \\ +\infty & \text{otherwise} \end{cases}$$

with domain

$$\operatorname{dom} f = \{(x,y) \in \mathbb{R}^n \times \mathbb{R} : y > 0\}.$$

This function is convex, since its domain is convex and, in the interior of the domain, the Hessian is given by

$$\nabla^2 f(x,y) = \frac{2}{y^3} \begin{bmatrix} y^2 I & -yx \\ -yx^\top & x^\top x \end{bmatrix}.$$

We check that the Hessian is indeed positive semidefinite: for any $w = (z,t) \in \mathbb{R}^n \times \mathbb{R}$, we have

$$\begin{aligned} \frac{y^3}{2} w^\top \nabla^2 f(x,y) w &= \begin{bmatrix} z \\ t \end{bmatrix}^\top \begin{bmatrix} y^2 I & -yx \\ -yx^\top & x^\top x \end{bmatrix} \begin{bmatrix} z \\ t \end{bmatrix} \\ &= \|yz - tx\|_2^2 \geq 0. \end{aligned}$$

3. The log-sum-exp (lse) function

$$f(x) = \ln \left(\sum_{i=1}^n e^{x_i} \right)$$

is monotonically increasing and convex over dom $f = \mathbb{R}^n$. Indeed, the Hessian of this function is (see Example 4.4)

$$\nabla^2 \mathrm{lse}(x) = \frac{1}{Z^2}\left(Z\mathrm{diag}\,(z) - zz^\top\right),$$

where

$$z = [e^{x_1} \cdots e^{x_n}], \quad Z = \sum_{i=1}^{n} z_i.$$

We now check that $\nabla^2 \mathrm{lse}(x)$ is positive semidefinite by verifying that $w^\top \nabla^2 \mathrm{lse}(x)w \geq 0$ for all $w \in \mathbb{R}^n$:

$$
\begin{aligned}
Z^2 w^\top \nabla^2 \mathrm{lse}(x)w &= Zw^\top \mathrm{diag}\,(z)\,w - w^\top zz^\top w \\
&= \left(\sum_{i=1}^{n} z_i\right)\left(\sum_{i=1}^{n} z_i w_i^2\right) - \left(\sum_{i=1}^{n} z_i w_i\right)^2 \geq 0,
\end{aligned}
$$

where the last passage follows from the Cauchy–Schwartz inequality:

$$\left\|\begin{bmatrix} w_1\sqrt{z_1} \\ \vdots \\ w_n\sqrt{z_n} \end{bmatrix}^\top \begin{bmatrix} \sqrt{z_1} \\ \vdots \\ \sqrt{z_n} \end{bmatrix}\right\|^2 \leq \left\|\begin{bmatrix} w_1\sqrt{z_1} \\ \vdots \\ w_n\sqrt{z_n} \end{bmatrix}\right\|_2^2 \left\|\begin{bmatrix} \sqrt{z_1} \\ \vdots \\ \sqrt{z_n} \end{bmatrix}\right\|_2^2.$$

8.2.2.3 Restriction to a line. A function f is convex if and only if its restriction to *any* line is convex. By restriction to a line we mean that for every $x_0 \in \mathbb{R}^n$ and $v \in \mathbb{R}^n$, the function of scalar variable t

$$g(t) = f(x_0 + tv)$$

is convex. This rule gives a very powerful criterion for proving convexity of certain functions.

Example 8.6 (*Log-determinant function*) The so-called *log-determinant* function $f(X) = -\log \det X$ is convex over the domain of symmetric, positive definite matrices, dom $f = S_{++}^n$. To verify this fact let $X_0 \in S_{++}^n$ be a positive definite matrix, $V \in S^n$, and consider the scalar-valued function

$$g(t) = -\log \det(X_0 + tV).$$

Since $X_0 \succ 0$, it can be factored (matrix square-root factorization) as $X_0 = X_0^{1/2}X_0^{1/2}$, hence

$$
\begin{aligned}
\det(X_0 + tV) &= \det(X_0^{1/2}X_0^{1/2} + tV) \\
&= \det\left(X_0^{1/2}(I + tX_0^{-1/2}VX_0^{-1/2})X_0^{1/2}\right) \\
&= \det X_0^{1/2}\det(I + tX_0^{-1/2}VX_0^{-1/2})\det X_0^{1/2} \\
&= \det X_0 \det(I + tX_0^{-1/2}VX_0^{-1/2}) \\
&= \det X_0 \prod_{i=1,\dots,n}(1 + t\lambda_i(Z)),
\end{aligned}
$$

where $\lambda_i(Z)$, $i = 1, \ldots, n$, are the eigenvalues of the matrix $Z = X_0^{-1/2} V X_0^{-1/2}$. Taking the logarithm, we thus obtain

$$g(t) = -\log \det X_0 + \sum_{i=1}^{n} -\log(1 + t\lambda_i(Z)).$$

The first term in the previous expression is a constant, and the second term is the sum of convex functions, hence $g(t)$ is convex for any $X_0 \in S_{++}^n$, $V \in S^n$, thus $-\log \det X$ is convex over the domain S_{++}^n.

8.2.2.4 *Pointwise supremum or maximum.* If $(f_\alpha)_{\alpha \in \mathcal{A}}$ is a family of convex functions indexed by the parameter α, and $\mathcal{A}$ is an arbitrary membership set for α, then the pointwise supremum function

$$f(x) = \sup_{\alpha \in \mathcal{A}} f_\alpha(x)$$

is convex over the domain $\{\cap_{\alpha \in \mathcal{A}} \operatorname{dom} f_\alpha\} \cap \{x : f(x) < \infty\}$. Note that the sup in the above definition can be substituted equivalently by a max whenever $\mathcal{A}$ is compact (i.e., closed and bounded). We next give a proof of this fact for a special case of the maximum of two convex functions. Let f_1, f_2 be convex, and let $f(x) = \max\{f_1(x), f_2(x)\}$, then for any $x, y \in \operatorname{dom} f_1 \cap \operatorname{dom} f_2$ and $\lambda \in [0, 1]$, we have

$$\begin{aligned}
f(\lambda x + (1-\lambda)y) &= \max\{f_1(\lambda x + (1-\lambda)y), f_2(\lambda x + (1-\lambda)y)\} \\
&\leq \max\{\lambda f_1(x) + (1-\lambda)f_1(y), \lambda f_2(x) + (1-\lambda)f_2(y)\} \\
&\leq \lambda \max\{f_1(x), f_2(x)\} + (1-\lambda)\max\{f_1(y), f_2(y)\} \\
&= \lambda f(x) + (1-\lambda)f(y).
\end{aligned}$$

There are many examples of application of this rule to prove convexity. For instance, given a norm $\|\cdot\|$, the *dual norm*[8] is defined as the function

$$f(x) = \|x\|^* = \max_{y: \|y\| \leq 1} y^\top x.$$

This function is convex over $\mathbb{R}^n$, since it is defined as the maximum of convex (in fact, linear) functions, indexed by the vector y. Similarly, the largest singular value of a matrix $X \in \mathbb{R}^{n,m}$

$$f(X) = \sigma_{\max}(X) = \max_{v: \|v\|_2 = 1} \|Xv\|_2$$

is convex over $\mathbb{R}^{n,m}$, since it is the pointwise maximum of convex functions which are the composition of the Euclidean norm with the affine function $X \to Xv$.

It actually turns out that not only is the supremum of convex functions convex, but also a converse statement holds (under the additional technical condition of closedness). That is, *every* closed convex

[8] For example, the dual of the Euclidean (ℓ_2) norm is the Euclidean norm itself (i.e., the Euclidean norm is self-dual)

$$\|x\|_2^* = \sup_{\|z\|_2 = 1} x^\top z = \frac{x^\top x}{\|x\|_2} = \|x\|_2.$$

The dual of the ℓ_∞-norm is the ℓ_1-norm:

$$\|x\|_\infty^* = \sup_{\|z\|_\infty = 1} x^\top z = \sum_{i=1}^{n} |x_i| = \|x\|_1.$$

The dual of the ℓ_1-norm is the ℓ_∞-norm (take the dual of the dual of the ℓ_∞-norm). More generally, the dual of the ℓ_p-norm is the ℓ_q-norm, where q satisfies

$$\frac{1}{p} + \frac{1}{q} = 1, \text{ i.e., } q = \frac{p}{p-1}.$$

function f can be expressed as the pointwise supremum of affine functions. In particular, f can be expressed[9] as the pointwise supremum over all affine functions that are global underestimators of f. Formally, let $f : \mathbb{R}^n \to \mathbb{R}$ be a closed convex function. Then $f = \bar{f}$, where

[9] See Section 12 in Rockafellar's book.

$$\bar{f}(x) = \sup\{a(x) : a \text{ is affine, and } a(z) \le f(z), \forall z\}. \quad (8.5)$$

If f is not closed, then the equality $f(x) = \bar{f}(x)$ still holds for any $x \in \operatorname{int} \operatorname{dom} f$.

8.2.2.5 *Partial minimization.*

If f is a convex function in (x, z) (i.e., it is jointly convex in the variables x and z), and Z is a nonempty and convex set, then the function

$$g(x) = \inf_{z \in Z} f(x, z)$$

is convex (provided that $g(x) > -\infty$ for all x).

Example 8.7 (*Schur complement lemma*) As an example of application of the partial minimization rule, we here give an alternative proof of the Schur complement rule for block-symmetric matrices, see Theorem 4.9. Let S be a symmetric matrix partitioned into blocks:

$$S = \begin{bmatrix} A & B \\ B^\top & C \end{bmatrix},$$

where both A and C are symmetric and square. Assume that $C \succ 0$ (i.e. C is positive definite). Then the following properties are equivalent:

- S is positive semidefinite;
- the *Schur complement* of C in S, defined as the matrix

$$A - BC^{-1}B^\top,$$

  is positive semidefinite.

To prove this fact we recall that $S \succeq 0$ if and only if $x^\top S x \ge 0$ for any vector x. Partitioning x as $x = (y, z)$, conformably to S, we have that $S \succeq 0$ if and only if

$$g(y, z) = \begin{bmatrix} y \\ z \end{bmatrix}^\top \begin{bmatrix} A & B \\ B^\top & C \end{bmatrix} \begin{bmatrix} y \\ z \end{bmatrix} \ge 0, \quad \forall(y, z).$$

This is equivalent to requiring that

$$0 \le f(y) = \min_z g(y, z)$$

holds for all y. To obtain a closed-form expression for f, we minimize g with respect to its second argument. Since this problem is unconstrained, we just set the gradient of g with respect to z to zero:

$$\nabla_z g(y, z) = 2(Cz + B^\top y) = 0,$$

which leads to the (unique) optimizer $z^*(y) = -C^{-1}B^\top y$ (notice that C^{-1} exists, since we are assuming $C \succ 0$). Plugging this value into g, we obtain

$$f(y) = g(y, z^*(y)) = y^\top (A - BC^{-1}B^\top)y.$$

Suppose now that $S \succeq 0$. Then the corresponding quadratic function g is jointly convex in its two arguments (y, z). Due to the partial minimization rule, we have that the pointwise minimum function $f(y)$ is convex as well, hence its Hessian must be positive semidefinite, and therefore $A - BC^{-1}B^\top \succeq 0$, as claimed. Conversely, if $A - BC^{-1}B^\top \succeq 0$, then $f(y) \geq 0$ for any y, which implies that $S \succeq 0$, thus concluding the proof.

8.2.2.6 *Composition with monotone convex/concave functions.*

The composition with another function does not always preserve convexity of a function. However, convexity is preserved under certain combinations of convexity/monotonicity properties, as discussed next.[10]

Consider first the case of functions of a scalar variable. If $f = h \circ g$, with h, g convex and h non-decreasing, then f is convex. Indeed, the condition $f(x) \leq z$ corresponds to $h(g(x)) \leq z$, which is equivalent to the existence of y such that

$$h(y) \leq z, \; g(x) \leq y.$$

This condition defines a convex set in the space of (x, y, z)-variables. The epigraph of f is thus the projection of this convex set onto the space of (x, z)-variables, hence it is convex, see Figure 8.11. In a similar way we can verify that if g is concave and h is convex and non-increasing, then f is convex. Analogous arguments hold for concavity of composite functions: f is concave if g is concave and h is concave and non-decreasing, or if g is convex and h is concave and non-increasing.

These rules have direct extensions to functions of a vector argument. For instance, if the component functions $g_i : \mathbb{R}^n \to \mathbb{R}$, $i = 1, \ldots, k$, are convex and $h : \mathbb{R}^k \to \mathbb{R}$ is convex and non-decreasing in each argument, with $\operatorname{dom} g_i = \operatorname{dom} h = \mathbb{R}$, then $x \to (h \circ g)(x) \doteq h(g_1(x), \ldots, g_k(x))$ is convex.

[10] For proofs of these results, see Chapter 3 in the book by S. Boyd and L. Vandenberghe, *Convex Optimization*, Cambridge University Press, 2004.

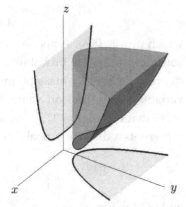

Figure 8.11 The set $\{(x, y, z) : h(y) \leq z, g(x) \leq y\}$ and its projection on the space of (x, z)-variables, which is the epigraph of f. The epigraph of g is the projection of the same set on the space of (x, y)-variables.

Example 8.8 Applying the composition rules, the reader may verify as an exercise the following facts:

- if g is convex then $\exp g(x)$ is convex;
- if g is concave and positive, then $\log g(x)$ is concave;
- if g is concave and positive, then $1/g(x)$ is convex;
- if g is convex and non-negative and $p \geq 1$, then $[g(x)]^p$ is convex;
- if g is convex then $-\log(-g(x))$ is convex on $\{x : g(x) < 0\}$;
- if $g_i, i = 1, \ldots, k$ are convex, then $\ln \sum_{i=1}^{k} e^{g_i(x)}$ is convex.

8.2.2.7 Jensen's inequality. Let $f : \mathbb{R}^n \to \mathbb{R}$ be a convex function, and let $z \in \mathbb{R}^n$ be a random variable such that $z \in \operatorname{int} \operatorname{dom} f$ with probability one. Then

$$f(\mathbb{E}\{z\}) \leq \mathbb{E}\{f(z)\} \quad (f \text{ convex}), \tag{8.6}$$

where $\mathbb{E}$ denotes the expected value of a random variable, and where it is assumed that the involved expectations exist.

To prove this key result, we use the fact that f can be represented as the pointwise supremum of affine functions, see (8.5). Indeed, for all $z \in \operatorname{int} \operatorname{dom} f$, we have

$$f(z) = \sup_{a \in \mathcal{A}} a(z) \geq a(z), \quad \forall a \in \mathcal{A},$$

where $\mathcal{A}$ is the set of affine functions that are global underestimators of f. Then, taking expectation of both sides of the previous equation, and recalling that the expectation operator is monotone, we have that

$$\mathbb{E}\{f(z)\} \geq \mathbb{E}\{a(z)\} = a\left(\mathbb{E}\{z\}\right), \quad \forall a \in \mathcal{A},$$

where the last equality follows from the fact that a is affine and $\mathbb{E}\{\cdot\}$ is a linear operator. This implies that

$$\mathbb{E}\{f(z)\} \geq \sup_{a \in \mathcal{A}} a\left(\mathbb{E}\{z\}\right) = f\left(\mathbb{E}\{z\}\right),$$

which proves (8.6). A special case of Jensen's inequality arises if one considers a discrete probability distribution for z, with support at m discrete probability-mass points $x^{(1)}, \ldots, x^{(m)} \in \mathbb{R}^n$. If the random variable z may take on value $x^{(i)}$ with probability θ_i, $i = 1, \ldots, m$ (note that $\theta_i \geq 0$, $i = 1, \ldots, m$, and $\sum_{i=1}^m \theta_i = 1$, since $\theta = [\theta_1 \cdots \theta_m]$ represents a discrete probability distribution), then

$$\mathbb{E}\{z\} = \sum_{i=1}^m \theta_i x^{(i)}, \quad \mathbb{E}\{f(z)\} = \sum_{i=1}^m \theta_i f(x^{(i)})$$

and (8.6) reads

$$f\left(\sum_{i=1}^m \theta_i x^{(i)}\right) \leq \sum_{i=1}^m \theta_i f(x^{(i)}), \tag{8.7}$$

which holds for any θ in the probability simplex, i.e., such that $\theta_i \geq 0$, $i = 1, \ldots, m$, and $\sum_{i=1}^m \theta_i = 1$. Observe further that $\mathbb{E}\{z\}$ is a point in the convex hull of $\{x^{(1)}, \ldots, x^{(m)}\}$, that is a convex combination of these points. Hence (8.7) says that a convex function f evaluated at any convex combination of points is no larger than the same combination of the values of f at the given points.

If f is a concave function, then Jensen's inequality clearly holds reversed:

$$\mathbb{E}\{f(z)\} \leq f(\mathbb{E}\{z\}) \quad (f \text{ concave}). \tag{8.8}$$

Example 8.9 *(Geometric–arithmetic mean inequality)* Given positive numbers $x_1, \ldots, x_n$, their *geometric mean* is defined as

$$f_g(x) = \left(\prod_{i=1}^{n} x_i \right)^{1/n},$$

whereas their standard, arithmetic, mean is

$$f_a(x) = \frac{1}{n} \sum_{i=1}^{n} x_i.$$

We next show that, for any $x > 0$, it holds that

$$f_g(x) \leq f_a(x).$$

To verify this fact, we take the logarithm of f_g, obtaining

$$
\begin{aligned}
\log f_g(x) &= \frac{1}{n} \sum_{i=1}^{n} \log(x_i) \\
&\leq \log \frac{\sum_{i=1}^{n} x_i}{n} = \log f_a(x),
\end{aligned}
$$

where the inequality follows from application of Jensen's inequality (8.8) to the concave function log. Finally, since log is monotone increasing, the inequality $\log f_g(x) \leq \log f_a(x)$ holds if and only if $f_g(x) \leq f_a(x)$, which is the statement we intended to prove.

Example 8.10 *(Young's inequality)* Given numbers $a, b \geq 0$ and $p, q > 0$ such that

$$\frac{1}{p} + \frac{1}{q} = 1,$$

it holds that

$$ab \leq \frac{1}{p}a^p + \frac{1}{q}b^q,$$

a relation known as Young's inequality. To prove this fact, we consider that

$$ab = e^{\ln ab} = e^{\ln a + \ln b} = e^{\frac{1}{p} \ln a^p + \frac{1}{q} \ln b^q}.$$

Then we observe that the exponential function is convex and, since $1/p + 1/q = 1$, apply Jensen's inequality to the last expression, obtaining that

$$ab \leq \frac{1}{p}e^{\ln a^p} + \frac{1}{q}e^{\ln b^q} = \frac{1}{p}a^p + \frac{1}{q}b^q,$$

which is the statement we wanted to prove.

8.2.3   *Subgradients and subdifferentials*

Consider again the characterization of a convex differentiable function given in (8.4). This relation states that at any point $x \in \text{dom } f$, the function $f(y)$ is lower bounded by an affine function of y, and that the bound is exact at x. That is,

$$f(y) \geq f(x) + g_x^\top (y - x), \quad \forall y \in \text{dom } f, \qquad (8.9)$$

where $g_x = \nabla f(x)$. Now, it turns out that even when f is non-differentiable (hence the gradient may not exist at some points) relation (8.9) may still hold for suitable vectors g_x. More precisely, if $x \in \text{dom } f$ and (8.9) holds for some vector $g_x \in \mathbb{R}^n$, then g_x is called a *subgradient* of f at x. The set of all subgradients of f at x is called the *subdifferential*, and it is denoted by $\partial f(x)$. A subgradient is a "surrogate" of the gradient: it coincides with the gradient, whenever a gradient exists, and it generalizes the notion of gradient at points where f is non-differentiable. A key result on subgradients is next stated without proof.[11]

Theorem 8.3 *Let* $f : \mathbb{R}^n \to \mathbb{R}$ *be convex and let* $x \in \text{relint dom } f$. *Then*

1. the subdifferential $\partial f(x)$ *is a closed, convex, nonempty and bounded set;*

2. if f *is differentiable at* x, *then* $\partial f(x)$ *contains only one element: the gradient of* f *at* x, *that is,* $\partial f(x) = \{\nabla f(x)\}$;

3. for any $v \in \mathbb{R}^n$ *it holds that*

$$f_v'(x) \doteq \lim_{t \to 0^+} \frac{f(x + tv) - f(x)}{t} = \max_{g \in \partial f(x)} v^\top g,$$

where $f_v'(x)$ *is the* directional derivative *of* f *at* x *along the direction* v.

In words, this theorem states that, for a convex f, a subgradient always exists at all points in the relative interior of the domain. Moreover, f is *directionally* differentiable at all such points. For a convex function f it thus holds that, for all $x \in \text{relint dom } f$,

$$f(y) \geq f(x) + g_x^\top (y - x), \quad \forall y \in \text{dom } f, \forall g_x \in \partial f(x).$$

Example 8.11 Consider the absolute value function (see Figure 8.12)

$$f(x) = |x|, \quad x \in \mathbb{R}.$$

For $x > 0$, f is differentiable, hence $\partial f(x) = \{\nabla f(x)\} = \{1\}$. For $x < 0$, f is also differentiable, and $\partial f(x) = \{\nabla f(x)\} = \{-1\}$. On the contrary, f is non-differentiable at $x = 0$. However, for all $y \in \mathbb{R}$ we can write

$$f(y) = |y| = \max_{|g| \leq 1} gy \geq gy, \quad \forall g : |g| \leq 1,$$

[11] For reference on this and other results on subgradients and subdifferentials, see Section 1.2 of the book by N. Z. Shor, *Minimization Methods for Non-differentiable Functions*, Springer, 1985; or Chapter D in J.-B. Hiriart-Urruty and C. Lemaréchal, *Fundamentals of Convex Analysis*, Springer, 2001.

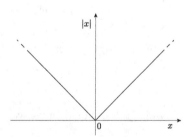

Figure 8.12 Absolute value function $f(x) = |x|, x \in \mathbb{R}$.

hence, for all $y \in \mathbb{R}$, we have

$$f(y) \geq f(0) + g(y - 0), \quad \forall g : |g| \leq 1,$$

which, compared to (8.9), shows that all $g \in [-1, 1]$ are subgradients of f at zero. Actually, these are all the possible subgradients, thus the interval $[-1, 1]$ is the subdifferential of f at zero. Thus, we have that

$$\partial|x| = \begin{cases} \text{sgn}(x) & \text{if } x \neq 0, \\ [-1, 1] & \text{if } x = 0. \end{cases}$$

Similarly, considering the ℓ_1 norm function

$$f(x) = \|x\|_1, \quad x \in \mathbb{R}^n,$$

we can write

$$f(y) = \|y\|_1 = \sum_{i=1}^{n} |y_i| = \sum_{i=1}^{n} \max_{|g_i| \leq 1} g_i y_i$$

$$\leq \sum_{i=1}^{n} g_i y_i, \quad \forall g : \|g\|_\infty \leq 1.$$

Hence, for all $y \in \mathbb{R}^n$ it holds that

$$f(y) \geq f(0) + g^\top(y - 0), \quad \forall g : \|g\|_\infty \leq 1.$$

All vectors $g \in \mathbb{R}^n$ such that $\|g\|_\infty \leq 1$ are thus subgradients of f at zero, and indeed it holds that $\partial f(0) = \{g : \|g\|_\infty \leq 1\}$.

8.2.3.1 Subgradient calculus. Besides resorting to the definition, as we did in the previous example, there exist several useful "rules" for computing subgradients and subdifferentials of functions obtained by composition, sum, pointwise maximum, etc., of individual operands. We here summarize some of these rules.[12]

Chain rules. Let $q : \mathbb{R}^n \to \mathbb{R}^m$ and $h : \mathbb{R}^m \to \mathbb{R}$ be such that the composite function $f = h \circ q : \mathbb{R}^n \to \mathbb{R}$, with values $f(x) = h(q(x))$, is convex. Then:

1. if q is differentiable at x and $q(x) \in \text{dom}\, h$, then

$$\partial f(x) = J_q(x)^\top \partial_q h(q(x)),$$

where $J_q(x)$ is the Jacobian of q at x;

2. if $m = 1$ and h is differentiable at $q(x) \in \text{dom}\, h$, then

$$\partial f(x) = \frac{\mathrm{d}h(q(x))}{\mathrm{d}q(x)} \partial q(x).$$

[12] For details, see the previously cited reference by Shor, as well as Section 2.3 of F. H. Clarke, *Optimization and Nonsmooth Analysis*, SIAM Classics in Applied Mathematics, 1990.

Affine variable transformation. As a particular case of the first of the previous chain rules, let $h : \mathbb{R}^m \to \mathbb{R}$ be convex and let $q(x) = Ax + b$, where $A \in \mathbb{R}^{m,n}$, $b \in \mathbb{R}^m$. Then the function from $\mathbb{R}^n$ to $\mathbb{R}$

$$f(x) = h(q(x)) = h(Ax + b)$$

has subdifferential

$$\partial f(x) = A^\top \partial_q h(q(x))$$

for all x such that $q(x) \in \mathrm{dom}\, h$.

Example 8.12 Consider the function

$$f(x) = |a^\top x - b|, \quad a \in \mathbb{R}^n, \ b \in \mathbb{R}.$$

This function is the composition of $h(x) = |x|$ with the affine function $q(x) = a^\top x - b$. Therefore, applying the affine variable transformation rule, we have that

$$\partial |a^\top x - b| = a \cdot \partial h(a^\top x - b) = \begin{cases} a \cdot \mathrm{sgn}(a^\top x - b) & \text{if } a^\top x - b \neq 0, \\ a \cdot [-1, \ 1] & \text{if } a^\top x - b = 0. \end{cases}$$

Sum or linear combination. Let $h : \mathbb{R}^n \to \mathbb{R}$, $q : \mathbb{R}^n \to \mathbb{R}$ be convex functions, let $\alpha, \beta \geq 0$, and let

$$f(x) = \alpha h(x) + \beta q(x).$$

Then, for any $x \in \mathrm{relint}\, \mathrm{dom}\, h \cap \mathrm{relint}\, \mathrm{dom}\, q$ it holds that

$$\partial f(x) = \alpha \partial h(x) + \beta \partial q(x).$$

Example 8.13 Consider the function

$$f(x) = \sum_{i=1}^{m} |a_i^\top x - b_i|, \quad a_i \in \mathbb{R}^n, \ b_i \in \mathbb{R}.$$

Applying the sum rule, we have

$$\partial f(x) = \sum_{i=1}^{m} \partial |a_i^\top x - b_i| = \sum_{i=1}^{m} \begin{cases} a_i \cdot \mathrm{sgn}(a_i^\top x - b_i) & \text{if } a_i^\top x - b_i \neq 0, \\ a_i \cdot [-1, \ 1] & \text{if } a_i^\top x - b_i = 0. \end{cases}$$

A special case is given by the ℓ_1 norm function $f(x) = \|x\|_1 = \sum_{i=1}^{n} |x_i|$, for which we obtain

$$\partial \|x\|_1 = \sum_{i=1}^{n} \partial |x_i| = \sum_{i=1}^{n} \begin{cases} e_i \cdot \mathrm{sgn}(x_i) & \text{if } x_i \neq 0, \\ e_i \cdot [-1, \ 1] & \text{if } x_i = 0, \end{cases}$$

where e_i denotes the i-th standard basis vector of $\mathbb{R}^n$.

Pointwise maximum.   Let $f_i : \mathbb{R}^n \to \mathbb{R}$, $i = 1, \ldots, m$, be convex functions, and let

$$f(x) = \max_{i=1,\ldots,m} f_i(x).$$

Then, for $x \in \text{dom } f$ it holds that

$$\partial f(x) = \text{co}\{\partial f_i(x) : i \in a(x)\},$$

where $a(x)$ is the set of indices of the functions f_i that are "active" at x, that is the ones that attain the maximum in the definition of f, hence $f(x) = f_i(x)$, for $i \in a(x)$.

This property also has an extension to pointwise maxima of arbitrary (possibly uncountable) families of convex functions, under some additional technical assumptions. More precisely, let

$$f(x) = \sup_{\alpha \in \mathcal{A}} f_\alpha(x),$$

where f_α are convex and closed functions. Then, for any $x \in \text{dom } f$ it holds that

$$\partial f(x) \supseteq \text{co}\{\partial f_\alpha(x) : \alpha \in a(x)\},$$

where $a(x) = \{\alpha \in \mathcal{A} : f(x) = f_\alpha(x)\}$. Moreover, if $\mathcal{A}$ is compact and the map $\alpha \to f_\alpha$ is closed, then equality holds in the previous inclusion, i.e.,

$$\partial f(x) = \text{co}\{\partial f_\alpha(x) : \alpha \in a(x)\}. \tag{8.10}$$

Example 8.14 (*Polyhedral functions*) Consider the polyhedral function (see Section 9.3.1)

$$f(x) = \max_{i=1,\ldots,m} a_i^\top x - b_i.$$

Here, the component functions $f_i(x) = a_i^\top x - b_i$ are differentiable, hence $\partial f_i(x) = \{\nabla f_i(x)\} = \{a_i\}$, thus we have

$$\partial f(x) = \text{co}\{a_i : i \in a(x)\}.$$

Similarly, let

$$f(x) = \|Ax - b\|_\infty = \max_{i=1,\ldots,m} |a_i^\top x - b_i|,$$

where $a_i^\top \in \mathbb{R}^n$ denote the rows of $A \in \mathbb{R}^{m,n}$. Applying the max rule for the subdifferential, for $f_i = |a_i^\top x - b_i|$, we have

$$\partial f(x) = \text{co}\{\partial f_i(x) : i \in a(x)\},$$

where

$$\partial f_i(x) = \begin{cases} a_i \cdot \text{sgn}(a_i^\top x - b_i) & \text{if } a_i^\top x - b_i \neq 0, \\ a_i \cdot [-1, 1] & \text{if } a_i^\top x - b_i = 0. \end{cases}$$

Example 8.15 (ℓ_1, ℓ_2, *and* ℓ_∞ *norms*) Subdifferentials of typical ℓ_p norms can be obtained by expressing the norm in the form of a supremum of linear functions over a suitable set, and then applying the sup rule for the subdifferentials. Specifically, we observe that

$$\|x\|_1 = \max_{\|v\|_\infty \leq 1} v^\top x,$$

$$\|x\|_2 = \max_{\|v\|_2 \leq 1} v^\top x,$$

$$\|x\|_\infty = \max_{\|v\|_1 \leq 1} v^\top x.$$

The ℓ_1 norm case has already been considered, so let us study the two other cases. For the ℓ_2 norm case, using (8.10) we have, for $f_v \doteq v^\top x$,

$$\partial \|x\|_2 = \mathrm{co}\{\partial f_v(x) : v \in a(x)\} = \mathrm{co}\{v : v \in a(x)\} = \mathrm{co}\{a(x)\},$$

where $a(x) = \{v : \|x\|_2 = v^\top x, \|v\|_2 \leq 1\}$. For $x \neq 0$, $\|x\|_2$ is attained for $v = x/\|x\|_2$, hence $a(x)$ is the singleton $\{x/\|x\|_2\}$, and $\partial \|x\|_2 = \{x/\|x\|_2\}$. For $x = 0$, we have instead $\|x\|_2 = 0$ which is attained for all feasible v, hence $a(x) = \{v : \|v\|_2 \leq 1\}$, and $\partial \|x\|_2 = \{v : \|v\|_2 \leq 1\}$. Summarizing,

$$\partial \|x\|_2 = \begin{cases} \dfrac{x}{\|x\|_2} & \text{if } x \neq 0, \\ \{g \in \mathbb{R}^n : \|g\|_2 \leq 1\} & \text{if } x = 0. \end{cases}$$

For the ℓ_∞ norm case, we have similarly

$$\partial \|x\|_\infty = \mathrm{co}\{a(x)\},$$

where $a(x) = \{v : \|x\|_\infty = v^\top x, \|v\|_1 \leq 1\}$. For $x \neq 0$, $\|x\|_\infty$ is attained at the vectors $v_j = e_j \mathrm{sgn}(x_j)$, where $j \in J(x)$, $J(x)$ being the set of indices of the largest entries in $|x|$ (there is only one element in $J(x)$, if x has a single entry of maximum modulus, or more than one element, if there are several entries with the same maximum value). Hence, for $x \neq 0$,

$$\partial \|x\|_\infty = \mathrm{co}\{e_j \mathrm{sgn}(x_j), j \in J(x)\}.$$

For $x = 0$, we have instead $\|x\|_\infty = 0$ which is attained for all feasible v, hence $a(x) = \{v : \|v\|_1 \leq 1\}$, and $\partial \|x\|_\infty = \{v : \|v\|_1 \leq 1\}$. Summarizing,

$$\partial \|x\|_\infty = \begin{cases} \mathrm{co}\{e_j \cdot \mathrm{sgn}(x_j), j \in J(x)\} & \text{if } x \neq 0, \\ \{g \in \mathbb{R}^n : \|g\|_1 \leq 1\} & \text{if } x = 0. \end{cases}$$

Example 8.16 (*Largest eigenvalue of a symmetric matrix*) Consider a symmetric matrix $A(x)$ whose entries are affine functions of a vector of variables $x \in \mathbb{R}^n$:

$$A(x) = A_0 + x_1 A_1 + \cdots + x_n A_n,$$

where $A_i \in \mathbb{S}^m$, $i = 0, \ldots, n$, and let

$$f(x) = \lambda_{\max}(A(x)).$$

To determine the subdifferential of f at x we exploit Rayleigh's variational characterization (see Theorem 4.3), which states that

$$f(x) = \lambda_{\max}(A(x)) = \max_{z:\,\|z\|_2=1} z^\top A(x)z$$

$$= \max_{z:\,\|z\|_2=1} z^\top A_0 z + \sum_{i=1}^{n} x_i z^\top A_i z.$$

$f(x)$ is thus expressed as the max over z (on the unit sphere) of functions $f_z(x) = z^\top A(x)z$ which are affine in x (hence, f is indeed convex). The active set

$$a(x) = \{z :\, \|z\|_2 = 1,\ f_z(x) = f(x)\}$$

is composed of the eigenvectors of $A(x)$ associated with the largest eigenvalue (and normalized with unit norm). We hence have that

$$\partial f(x) = \mathrm{co}\{\nabla f_z(x) :\, A(x)z = \lambda_{\max}(A(x))z,\ \|z\|_2 = 1\},$$

where

$$\nabla f_z(x) = [z^\top A_1 z \,\cdots\, z^\top A_n z]^\top.$$

In particular, f is differentiable at x whenever the eigenspace associated with $\lambda_{\max}(A(x))$ has dimension one, in which case $\partial f(x) = \{\nabla f_z(x)\}$, where z is the unique (up to a sign, which is, however, irrelevant) normalized eigenvector of $A(x)$ associated with $\lambda_{\max}(A(x))$.

8.3 Convex problems

An optimization problem of the form

$$p^* = \min_{x \in \mathbb{R}^n}\ f_0(x) \tag{8.11}$$
$$\text{subject to:}\quad f_i(x) \le 0,\quad i = 1,\dots,m, \tag{8.12}$$
$$h_i(x) = 0,\quad i = 1,\dots,q \tag{8.13}$$

is called a convex optimization problem, if

- the objective function f_0 is convex;

- the functions defining the inequality constraints, f_i, $i = 1,\dots,m$, are convex;

- the functions defining the equality constraints, h_i, $i = 1,\dots,q$, are affine.

The *feasible set* of this problem is the set of points x that satisfy the constraints:

$$\mathcal{X} = \{x \in \mathbb{R}^n :\, f_i(x) \le 0,\, i = 1,\cdots,m;\, h_i(x) = 0,\, i = 1,\cdots,q\}.$$

Note that the sets $\{x :\, f_i(x) \le 0\}$ are the 0-sublevel sets[13] of a convex function, hence they are convex. Also, the sets $\{x :\, h_i(x) = 0\}$, where

[13] We shall typically and tacitly assume in most of the derivations in the rest of this book that functions f_i are proper and closed, and that their sublevel sets are full-dimensional, i.e., their relative interior coincides with the standard interior.

h_i is an affine function, are flats, hence convex. Therefore, $\mathcal{X}$ is the intersection of convex sets and it is thus convex.

The linear equality constraints are usually expressed more compactly in matrix form as $Ax = b$, hence a generic format of a convex optimization problem is

$$p^* = \min_{x \in \mathbb{R}^n} \quad f_0(x)$$
$$\text{subject to:} \quad f_i(x) \leq 0, \quad i = 1, \ldots, m,$$
$$Ax = b,$$

where $A \in \mathbb{R}^{q,n}$ and $b \in \mathbb{R}^q$.

A convex optimization problem can equivalently be defined as a problem where one minimizes a convex objective function, subject to the restriction $x \in \mathcal{X}$, i.e. that the decision variable must belong to a convex set $\mathcal{X}$:

$$p^* = \min_{x \in \mathcal{X}} f_0(x). \tag{8.14}$$

The problem is said to be *unconstrained* when $\mathcal{X} = \mathbb{R}^n$.

Solving the optimization problem means finding the optimal minimal value p^* of the objective, and possibly also a *minimizer*, or optimal solution, that is a vector $x^* \in \mathcal{X}$ such that $f_0(x^*) = p^*$. If $\mathcal{X}$ is the empty set, we say that the problem is *infeasible*: no solution that satisfies the constraints exists. In such a case it is customary to set $p^* = +\infty$. When $\mathcal{X}$ is nonempty, we say that the problem is *feasible*. If the problem is feasible and $p^* = -\infty$ we say that the problem is *unbounded below*. Notice that it can also happen that the problem is feasible but still no optimal solution exists, in which case we say that the optimal value p^* is not *attained* at any finite point. The *optimal set* (or set of solutions) is defined as the set of feasible points for which the objective function attains the optimal value:

$$\mathcal{X}_{\text{opt}} = \{x \in \mathcal{X} : f_0(x) = p^*\}.$$

We shall also write, using the argmin notation,

$$\mathcal{X}_{\text{opt}} = \arg\min_{x \in \mathcal{X}} f_0(x).$$

If $x^* \in \mathcal{X}_{\text{opt}}$ is such that $f_i(x^*) < 0$, we say that the i-th inequality constraint is *inactive* (or *slack*) at the optimal solution x^*. Conversely, if $f_i(x^*) = 0$, we say that the i-th inequality constraint is *active* at x^*. Similarly, if x^* is in the relative interior of $\mathcal{X}$, we say that the whole feasible set $\mathcal{X}$ is inactive (see Figure 8.13), and if x^* is on the boundary of $\mathcal{X}$ we say that $\mathcal{X}$ is active at the optimum (see Figure 8.14).

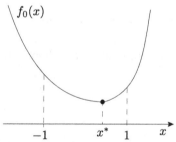

Figure 8.13 In this example, $\mathcal{X} = \{x : |x| \leq 1\}$ is inactive at the optimum for the problem $\min_{x \in \mathcal{X}} f_0(x)$.

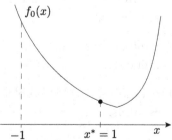

Figure 8.14 In this example, $\mathcal{X} = \{x : |x| \leq 1\}$ is active at the optimum for the problem $\min_{x \in \mathcal{X}} f_0(x)$.

Feasibility problems. In some cases, it may not be of interest to actually minimize an objective function, rather one is just interested in verifying if the problem is feasible or not and, in the positive case, to determine *any* point in the feasible set. This is called a *feasibility problem*:

$$\text{find } x \in \mathcal{X} \text{ or prove that } \mathcal{X} \text{ is empty.}$$

We next provide some simple examples illustrating an (incomplete) taxonomy of several cases that may arise in a convex optimization problem.

Example 8.17

- Consider the problem

$$\begin{aligned} p^* = \min_{x \in \mathbb{R}^2} \quad & x_1 + x_2 \\ \text{subject to:} \quad & x_1^2 \le 2, \\ & x_2^2 \le 1. \end{aligned}$$

The feasible set $\mathcal{X}$ for this problem is nonempty and it is given by the rectangle $[-\sqrt{2}, \sqrt{2}] \times [-1, 1]$. The optimal objective value is $p^* = -\sqrt{2} - 1$, which is attained at the unique optimal point

$$x^* = [-\sqrt{2} \ \ -1]^\top,$$

see Figure 8.15.

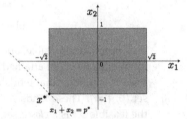

Figure 8.15 A feasible problem with unique optimal solution.

- Consider the problem

$$\begin{aligned} p^* = \min_{x \in \mathbb{R}^2} \quad & x_2 \qquad\qquad\qquad (8.15) \\ \text{subject to:} \quad & (x_1 - 2)^2 \le 1, \\ & x_2 \ge 0. \end{aligned}$$

The feasible set $\mathcal{X}$ for this problem is nonempty and it is depicted in Figure 8.16. The optimal objective value is $p^* = 0$, which is attained at infinitely many optimal points: the optimal set is

$$\mathcal{X}_{\text{opt}} = \{[x_1 \ 0]^\top : x_1 \in [1, 3]\}.$$

- The problem

$$\begin{aligned} p^* = \min_{x \in \mathbb{R}^2} \quad & e^{x_1} \\ \text{subject to:} \quad & x_2 \ge (x_1 - 1)^2 + 1, \\ & x_2 - x_1 + \frac{1}{2} \le 0 \end{aligned}$$

is unfeasible, thus, by convention, $p^* = +\infty$.

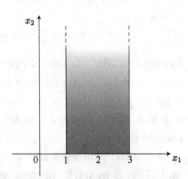

Figure 8.16 A feasible problem with multiple optimal solutions.

- The problem

$$\begin{aligned} p^* = \min_{x \in \mathbb{R}} \quad & e^{-x} \\ \text{subject to:} \quad & x \ge 0 \end{aligned}$$

is feasible, and the optimal objective value is $p^* = 0$. However, the optimal set is empty, since p^* is not attained at any finite point (it is only attained in the limit, as $x \to \infty$).

- Consider the problem

$$p^* = \min_{x \in \mathbb{R}^2} \quad x_1$$
$$\text{subject to:} \quad x_1 + x_2 \geq 0.$$

The feasible set is a half-space, and the problem is unbounded below ($p^* = -\infty$). No optimal solution exists, since the optimal value is attained asymptotically for $x_1 \to -\infty$, $x_1 \geq -x_2$, see Figure 8.17.

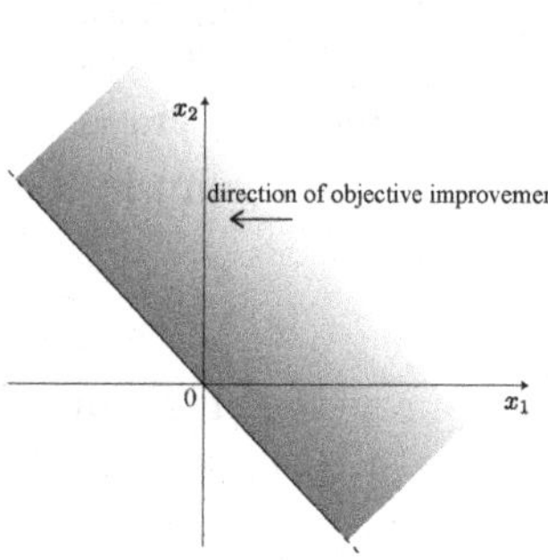

- Consider the problem

$$p^* = \min_{x \in \mathbb{R}} \quad (x+1)^2.$$

This is an unconstrained problem, for which $p^* = 0$ is attained at the (unique) optimal point $x^* = -1$. Consider next a constrained version of the problem, where

$$p^* = \min_{x \in \mathbb{R}} \quad (x+1)^2$$
$$\text{subject to:} \quad x > 0.$$

This problem is feasible, and has optimal value $p^* = 1$. However, this optimal value is not attained by any feasible point: it is attained in the limit by a point x that tends to 0, but 0 does not belong to the feasible set $(0, +\infty)$. These kinds of "subtleties" are avoided if we insure that the feasible set is a *closed* set, see the discussion in Section 8.3.2.

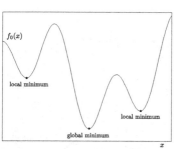

Figure 8.17 A feasible problem with unbounded-below objective.

8.3.1   Local and global optima

A point z is said to be a *local optimum* for the optimization problem $\min_{x \in \mathcal{X}} f_0(x)$, if there exists $r > 0$ such that z is optimal for the problem

$$\min_{x \in \mathcal{X}} f_0(x) \text{ subject to: } \|x - z\|_2 < r.$$

In other words, z is locally optimal if there exists a ball B_r of radius $r > 0$ centered in z such that for all points $x \in B_r \cap \mathcal{X}$ it holds that $f_0(x) \geq f_0(z)$. That is, z minimizes f_0 locally in a ball of radius r. If z is a global optimum point (i.e., a point in $\mathcal{X}_{\text{opt}}$), then it holds instead that $f_0(x) \geq f_0(z)$, for all $x \in \mathcal{X}$. A key fact is that in convex optimization problems any local optimal point is also globally optimal. This is in stark contrast with generic non-convex optimization problems, which are plagued by the possible existence of many local optima that are not globally optimal, see Figure 8.18. Numerical optimization algorithms may be trapped in local minima and hence often fail to converge to the global optimum, if the problem is not convex. The following key theorem holds.

Figure 8.18 A non-convex objective f_0 may have local minima that are not globally optimal.

Theorem 8.4 *Consider the optimization problem*

$$\min_{x \in \mathcal{X}} f_0(x).$$

If f_0 is a convex function and $\mathcal{X}$ is a convex set, then any locally optimal solution is also globally optimal. Moreover, the set $\mathcal{X}_{\text{opt}}$ of optimal points is convex.

Proof Let $x^* \in \mathcal{X}$ be a local minimizer of f_0, let $p^* = f_0(x^*)$, and consider any point $y \in \mathcal{X}$. We need to prove that $f_0(y) \geq f_0(x^*) = p^*$. By convexity of f_0 and $\mathcal{X}$ we have that, for $\theta \in [0, 1]$, $x_\theta = \theta y + (1 - \theta)x^* \in \mathcal{X}$, and

$$f_0(x_\theta) \leq \theta f_0(y) + (1 - \theta)f_0(x^*).$$

Subtracting $f_0(x^*)$ from both sides of this equation, we obtain

$$f_0(x_\theta) - f_0(x^*) \leq \theta(f_0(y) - f_0(x^*)).$$

Since x^* is a local minimizer, the left-hand side in this inequality is non-negative for all small enough values of $\theta > 0$. We thus conclude that the right-hand side is also non-negative, i.e., $f_0(y) \geq f_0(x^*)$, as claimed. Also, the optimal set is convex, since it can be expressed as the p^*-sublevel set of a convex function:

$$\mathcal{X}_{\text{opt}} = \{x \in \mathcal{X} : f_0(x) \leq p^*\},$$

which ends our proof. □

8.3.2 Existence of solutions

A sufficient condition for the existence of a solution for problem (8.11) is essentially established via the classical Weierstrass theorem.

Theorem 8.5 (Weierstrass extreme value theorem) *Every continuous function $f : \mathbb{R}^n \to \mathbb{R}$ on a nonempty compact (i.e., closed and bounded) set attains its extreme values on that set.*

Applying this theorem to our optimization problem (8.11) or (8.14), we obtain the following lemma.

Lemma 8.2 *If the set $\mathcal{X} \subseteq \text{dom} f_0$ is nonempty and compact, and f_0 is continuous on $\mathcal{X}$, then problem (8.14) attains an optimal solution x^*.*

Note that, since convex functions are continuous on open sets, we have that if f_0 is convex and $\mathcal{X} \subseteq \text{int dom} f_0$, then f_0 is continuous on $\mathcal{X}$ and the hypotheses of Lemma 8.2 hold.

While very useful, Lemma 8.2 still only provides a sufficient condition for existence of the solution, meaning, for instance, that a solution may well exist also when the feasible set $\mathcal{X}$ is not compact. The most typical case is for unconstrained minimization of f_0, where

$\mathcal{X} = \mathbb{R}^n$, which is obviously not compact. The next lemma provides another sufficient condition for existence of a solution, in cases when $\mathcal{X}$ is not compact (i.e., open and/or unbounded). To this end, we need to introduce the notion of a *coercive* function.

Definition 8.1 (Coercive function) *A function $f : \mathbb{R}^n \to \mathbb{R}$ is said to be coercive if for any sequence $\{x_k\} \subset$ int dom f tending to the boundary of dom f it holds that the corresponding value sequence $\{f(x_k)\}$ tends to $+\infty$.*

The next lemma states that the sublevel sets of a continuous coercive function are compact sets.

Lemma 8.3 *A continuous function $f : \mathbb{R}^n \to \mathbb{R}$ with open domain is coercive if and only if all its sublevel sets $S_\alpha = \{x : f(x) \leq \alpha\}$, $\alpha \in \mathbb{R}$, are compact.*

Proof Notice first that continuity of f immediately implies that S_α is closed. We next show that this set must also be bounded, if f is coercive. For the purpose of contradiction, suppose there exists an $\alpha \in \mathbb{R}$ such that S_α is unbounded. Since, by definition of effective domain, it holds that $S_\alpha \subseteq$ dom f, then S_α unbounded implies that also dom f is unbounded. Then there would exist a sequence $\{x_k\} \subset S_\alpha \subseteq$ dom f such that $\{\|x_k\|\} \to \infty$ for $k \to \infty$. But by coercitivity this would imply that also $\{f(x_k)\} \to +\infty$, which contradicts the hypothesis that $f(x_k) \leq \alpha$, for all $x_k \in S_\alpha$. We thus conclude that the sublevel set must be bounded.

Conversely, suppose all S_α are compact. Consider any sequence $\{x_k\} \subset$ dom f that tends to a boundary point of dom f and suppose, for the purpose of contradiction, that the corresponding value sequence $\{f(x_k)\}$ remains bounded, i.e., there exist a finite $\bar{\alpha} \in \mathbb{R}$ such that $f(x_k) \leq \bar{\alpha}$, for all k. Then, $\{x_k\} \subset S_{\bar{\alpha}}$ and, since $S_{\bar{\alpha}}$ is compact, $\{x_k\}$ has a limit $\bar{x} \in S_{\bar{\alpha}}$. But this would imply that the limit $\bar{x}$ belongs to the effective domain of f, since $f(\bar{x}) \leq \bar{\alpha}$, hence $f(\bar{x})$ is finite. We thus have that the sequence $\{x_k\} \subset$ dom f (which by assumption tends to a boundary point of dom f) has a limit $\bar{x} \in$ dom f, which contradicts the hypothesis that dom f is an open set. $\qquad\square$

We then have the following result on existence of solutions to the unconstrained version of problem (8.14).

Lemma 8.4 *If $\mathcal{X} = \mathbb{R}^n$ (unconstrained optimization), and f_0 is continuous and coercive, then problem (8.14) attains an optimal solution x^*.*

Proof To verify this result, take an $\alpha \in \mathbb{R}$ such that the sublevel set S_α is nonempty. By Lemma 8.3 the set S_α is compact, hence by the Weierstrass theorem f_0 attains a global minimum x^* on S_α. $\qquad\square$

A further sufficient condition for existence of a solution is obtained by combining the results in Lemma 8.2 and Lemma 8.4, as stated in the next result, whose proof is left as an exercise to the reader.

Lemma 8.5 *If $\mathcal{X} \subseteq \mathrm{dom}\, f_0$ is nonempty and closed, and f_0 is continuous on $\mathcal{X}$ and coercive, then problem (8.14) attains an optimal solution x^*.*

8.3.3   Uniqueness of the optimal solution

We warn the reader not to confuse the concept of global optimality with that of uniqueness of the optimal solution. For any convex optimization problem any locally optimal solution is also globally optimal, but this does not mean, in general, that the optimal solution is *unique*. A simple example of this fact is given, for instance, in problem (8.15), which is convex but admits infinitely many optimal solutions: all points with coordinates $(x_1, 0)$, with $x_1 \in [1, 3]$, are globally-optimal solutions. Intuitively, one may observe that such a lack of uniqueness is in this case due to the "flatness" of the objective function around the optimal points. Indeed, since flatness is ruled out by strict convexity, one can prove the following *sufficient* condition under which the optimal solution of a convex optimization problem is unique.

Theorem 8.6 *Consider the optimization problem (8.14). If f_0 is a strictly convex function, $\mathcal{X}$ is a convex set, and x^* is an optimal solution to the problem, then x^* is the unique optimal solution, that is $\mathcal{X}_{\mathrm{opt}} = \{x^*\}$.*

Proof Suppose, for the purpose of contradiction, that x^* is optimal for (8.14), and there exists another point $y^* \neq x^*$ which is also optimal for this problem. That is, both x^* and y^* are feasible and $f_0(x^*) = f_0(y^*) = p^*$. Let then $\lambda \in (0, 1)$ and consider the point $z = \lambda y^* + (1 - \lambda)x^*$. By convexity of $\mathcal{X}$, point z is feasible. Moreover, by strict convexity of f_0 it holds that $f_0(z) < \lambda f_0(y^*) + (1 - \lambda)f_0(x^*) = p^*$, which would imply that z has a better objective value than x^*, which is impossible since x^* is globally optimal. □

Another sufficient condition for uniqueness of the optimal solution can be stated for the class of convex programs with *linear* objective function (actually, we shall later show that *any* convex optimization problem can be converted into an equivalent problem having linear objective) and strictly convex feasible set. We first state a simple preliminary lemma, which establishes that any optimal solution of a convex optimization problem with a linear objective must be on the boundary of the feasible set.

Lemma 8.6 *Consider the optimization problem (8.14), let f_0 be a non-constant linear function (thus, $f_0 = c^\top x$, for some nonzero $c \in \mathbb{R}^n$), and let $\mathcal{X}$ be convex and closed. If x^* is an optimal solution to the problem, then x^* belongs to the boundary of $\mathcal{X}$.*

Proof Suppose, for the purpose of contradiction, that x^* is optimal and belongs to the interior of the feasible set $\mathcal{X}$, see Figure 8.19.

Let $p^* = c^\top x^*$ be the optimal objective level. By definition of an interior point, there exists an open ball of radius $r > 0$ centered at x^* and entirely contained in $\mathcal{X}$. That is, all points z such that $\|z - x^*\|_2 < r$ are feasible. Choose then $z = x^* - \alpha c$ with $\alpha = 0.5r/\|c\|_2$. It is immediate to check that this z is inside the above-mentioned ball, hence it is a feasible point. Moreover, $f_0(z) = c^\top z = c^\top x^* - \alpha c^\top c = p^* - \frac{r}{2\|c\|_2} < p^*$, hence the objective level of z is better (lower) than p^*, which contradicts the assumption that x^* is globally optimal. □

We can now establish the following sufficient condition for uniqueness of an optimal solution for convex programs with linear objective.

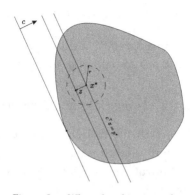

Figure 8.19 When the objective is linear, an optimal solution x^* cannot be in the interior of the feasible set, for otherwise it would be possible to move away from x^* while remaining feasible and improving the objective.

Theorem 8.7 *Consider the optimization problem (8.14), let f_0 be a non-constant linear function (thus, $f_0 = c^\top x$, for some nonzero $c \in \mathbb{R}^n$), and let $\mathcal{X}$ be closed, full-dimensional, and strictly convex. If the problem admits an optimal solution x^*, then this solution is unique.*

Proof Suppose, again for the purpose of contradiction, that x^* is optimal but non-unique. Then there exists $y^* \neq x^*$ which is feasible and such that $p^* = c^\top x^* = c^\top y^*$. Consider then a point z in the open segment joining x^* and y^*: $z = \lambda y^* + (1 - \lambda)x^*$, $\lambda \in (0, 1)$. By strict convexity of $\mathcal{X}$, the point z is in the relative interior of $\mathcal{X}$, which coincides with the interior of $\mathcal{X}$, since we assumed $\mathcal{X}$ to be full-dimensional. Hence, z belongs to the interior of $\mathcal{X}$, and

$$f_0(z) = c^\top z = \lambda c^\top y^* + (1 - \lambda)c^\top x^* = p^*,$$

thus, z is optimal. But according to Lemma 8.6 no optimal solution can be in the interior of $\mathcal{X}$ (it must be on the boundary), hence we found a contradiction. □

Remark 8.1 In some situations one can "regularize" a convex optimization problem by slightly modifying it in order to insure strict convexity of the objective or of the feasible set $\mathcal{X}$. For instance, if f_0 is convex (but not strictly), then one may consider a problem with a modified objective

$$\tilde{f}_0(x) = f_0(x) + \gamma\|x - c\|_2^2,$$

for some small $\gamma > 0$ and $c \in \mathbb{R}^n$. The addition of the strongly convex term $\gamma\|x - c\|_2^2$ makes $\tilde{f}_0(x)$ also strongly convex, and hence strictly convex. Thus, Theorem 8.6 would guarantee uniqueness of the optimal solution, for the modified problem. Similarly, a convex problem with linear objective and inequality constraints $f_i(x) \leq 0$, $i = 1, \ldots, m$, where f_i are convex (but not strictly convex), may be modified by adding a strongly convex term to the left-hand side of the constraints, thus making the feasible set strictly convex.

8.3.4 Problem transformations

An optimization problem can be transformed, or reformulated, into an *equivalent* one by means of several useful "tricks," such as: monotone transformation of the objective (e.g., scaling, logarithm, squaring) and constraints; change of variables; addition of slack variables; epigraphic reformulation; replacement of equality constraints with inequality ones; elimination of inactive constraints; etc.

By the term "equivalent" referred to two optimization problems, we here mean informally that the optimal objective value and optimal solutions (if they exist) of one problem can be easily obtained from the optimal objective value and optimal solutions of the other problem, and vice versa. We next analyze each of the mentioned transformation tricks.

8.3.4.1 Monotone objective transformation.
Consider an optimization problem of the form (8.11). Let $\varphi : \mathbb{R} \to \mathbb{R}$ be a continuous and strictly increasing function over $\mathcal{X}$, and consider the transformed problem

$$g^* = \min_{x \in \mathbb{R}^n} \quad \varphi(f_0(x)) \tag{8.16}$$
$$\text{subject to:} \quad f_i(x) \leq 0, \quad i = 1, \ldots, m,$$
$$h_i(x) = 0, \quad i = 1, \ldots, q.$$

Clearly, problems (8.11) and (8.16) have the same feasible set. We next show that they also have the same set of optimal solutions. Indeed, suppose x^* is optimal for problem (8.11), i.e., $f_0(x^*) = p^*$. Then, x^* is feasible for problem (8.16), thus it holds that

$$\varphi(f_0(x^*)) = \varphi(p^*) \geq g^*. \tag{8.17}$$

Assume next that $\tilde{x}^*$ is optimal for problem (8.16), i.e., $\varphi(f_0(\tilde{x}^*)) = g^*$. Then, $\tilde{x}^*$ is feasible for problem (8.11), thus it holds that

$$f_0(\tilde{x}^*) \geq p^*. \tag{8.18}$$

Now, since φ is continuous and strictly increasing over $\mathcal{X}$, it has a well-defined inverse φ^{-1}, thus we may write

$$\varphi(f_0(\tilde{x}^*)) = g^* \quad \Leftrightarrow \quad \varphi^{-1}(g^*) = f_0(\tilde{x}^*),$$

which, substituted in (8.18), yields

$$\varphi^{-1}(g^*) \geq p^*.$$

Since φ is strictly increasing and $\varphi(\varphi^{-1}(g^*)) = g^*$, the latter relation also implies that $g^* \geq \varphi(p^*)$, which, together with (8.17), implies that it must be $\varphi(p^*) = g^*$. This means that for any optimal solution x^* of problem (8.11) it holds that

$$\varphi(f_0(x^*)) = g^*,$$

which implies that x^* is also optimal for problem (8.16). Vice versa, for any optimal solution $\tilde{x}^*$ of problem (8.16) it holds that

$$f_0(\tilde{x}^*) = \varphi^{-1}(g^*) = p^*,$$

which implies that $\tilde{x}^*$ is also optimal for problem (8.11).

An example of a frequently encountered transformation is the logarithmic one: since $\varphi(\cdot) = \log(\cdot)$ is strictly increasing (for non-negative argument), then if f_0 is non-negative we can substitute the original objective with the transformed one, obtaining an equivalent problem.

Preserving/creating convexity. If the original problem (8.11) is convex, then the transformed problem (8.16) is also convex, provided that φ is convex (besides being strictly increasing).

A common convexity-preserving objective transformation consists of "squaring" a (non-negative) objective. Indeed, $\varphi(\cdot) = (\cdot)^2$ is convex and strictly increasing (for non-negative argument), hence if f_0 is *non-negative* and convex, then we can apply the previous equivalence result, preserving the convexity of the objective. Another elementary convexity-preserving objective transformation is simply given by multiplication of the objective function by a positive scalar, that is, a problem with objective f_0 is equivalent to a problem with objective αf_0, for $\alpha > 0$, and convexity is preserved by this transformation.

If some transformations preserve convexity, some other transformations can actually be used for "creating" convexity from an originally non-convex objective. This convexity-inducing technique typically works in conjunction with the change of variable and constraint transformation tricks described in the next sections.

8.3.4.2 Monotone constraint transformation. Strictly monotone functions can also be used to transform functional constraints into equivalent ones. If a constraint in a problem can be expressed as

$$\ell(x) \leq r(x),$$

and φ is a continuous and strictly increasing function over $\mathcal{X}$, then this constraint is equivalent to

$$\varphi(\ell(x)) \leq \varphi(r(x)),$$

where equivalent here means that the set of x satisfying the first constraint coincides with the set of x satisfying the second constraint. Similarly, if φ is continuous and strictly decreasing over $\mathcal{X}$, then the constraint is equivalent to $\varphi(\ell(x)) \geq \varphi(r(x))$.

8.3.4.3 Change of variables. Consider an optimization problem of the form (8.11), and let $F : X \to Y$ be an invertible mapping (i.e., for every $y \in Y$ there exist a unique $x \in X$ such that $F(x) = y$), describing a change of variables of the form

$$y = F(x) \quad \Leftrightarrow \quad x = F^{-1}(y),$$

where the set X includes the intersection of the domain of f_0 with the feasible set $\mathcal{X}$ of the problem. Then, problem (8.11) can be reformulated in the new variable y as

$$p^* = \min_{y \in \mathbb{R}^n} \quad g_0(y) \tag{8.19}$$

$$\text{subject to:} \quad g_i(y) \leq 0, \quad i = 1, \ldots, m,$$
$$s_i(y) = 0, \quad i = 1, \ldots, q,$$

where $g_i(y) = f_i(F^{-1}(y))$, $i = 0, 1, \ldots, m$, and $s_i(y) = h_i(F^{-1}(y))$, $i = 1, \ldots, q$. Clearly, if x^* is optimal for problem (8.11) then $y^* = F(x^*)$ is optimal for problem (8.19). Vice versa, if y^* is optimal for problem (8.19) then $x^* = F^{-1}(y^*)$ is optimal for problem (8.11).

If the original problem (8.11) is convex, then the problem in transformed variables (8.19) is convex whenever the variable transformation is linear or affine, that is if

$$y = F(x) = Bx + c,$$

where B is an invertible matrix. Sometimes, a well-chosen variable transformation may transform a non-convex problem into a convex one. A notable example is illustrated next.

Example 8.18 (*Optimization problems involving power laws*) Many problems dealing with area, volume, and size of basic geometric objects; birth and survival rates of (say) bacteria as functions of concentrations of chemicals; heat flows and losses in pipes, as functions of the pipe geometry; analog circuit properties as functions of circuit parameters; etc., involve quantities that are described by *power laws*, that is monomials of the form

$\alpha x_1^{a_1} x_2^{a_2} \cdots x_n^{a_n}$, where $\alpha > 0$, $x_i > 0$, $i = 1, \ldots, n$, and a_i are given real numbers. An optimization problem of the form

$$
\begin{aligned}
p^* = \min_x \quad & \alpha_0 x_1^{a_1^{(0)}} x_2^{a_2^{(0)}} \cdots x_n^{a_n^{(0)}} \\
\text{s.t.:} \quad & \alpha_j x_1^{a_1^{(j)}} x_2^{a_2^{(j)}} \cdots x_n^{a_n^{(j)}} \leq b_j, \quad j = 1, \ldots, m, \\
& x_i > 0, \quad\quad\quad\quad\quad\quad\quad i = 1, \ldots, n,
\end{aligned}
$$

is non-convex in the variables $x_1, \ldots, x_n$. However, applying a logarithmic transformation to the objective and constraint functions, we obtain an equivalent problem in the form

$$
\begin{aligned}
g^* = \min_x \quad & \log \alpha_0 + \sum_{i=1}^n a_i^{(0)} \log x_i \\
\text{s.t.:} \quad & \log \alpha_j + \sum_{i=1}^n a_i^{(j)} \log x_i \leq \log b_j, \quad j = 1, \ldots, m.
\end{aligned}
$$

Then, introducing the change of variables $y_i = \log x_i$, $i = 1, \ldots, n$, over the domain $x_i > 0$, the last problem is rewritten equivalently in the y variables as

$$
\begin{aligned}
g^* = \min_y \quad & \log \alpha_0 + \sum_{i=1}^n a_i^{(0)} y_i \\
\text{s.t.:} \quad & \log \alpha_j + \sum_{i=1}^n a_i^{(j)} y_i \leq \log b_j, \quad j = 1, \ldots, m,
\end{aligned}
$$

which is a convex (and in particular, linear) programming problem in y.

8.3.4.4 Addition of slack variables. Equivalent problem formulations are also obtained by introducing new "slack" variables into the problem. We here describe a typical case that arises when the objective involves the sum of terms, as in the following problem

$$
\begin{aligned}
p^* = \min_x \quad & \sum_{i=1}^r \varphi_i(x) \\
\text{s.t.:} \quad & x \in \mathcal{X}.
\end{aligned}
\tag{8.20}
$$

Introducing slack variables t_i, $i = 1, \ldots, p$, we reformulate this problem as

$$
\begin{aligned}
g^* = \min_{x,t} \quad & \sum_{i=1}^r t_i \\
\text{s.t.:} \quad & x \in \mathcal{X} \\
& \varphi_i(x) \leq t_i \quad i = 1, \ldots, r,
\end{aligned}
\tag{8.21}
$$

where this new problem has the original variable x, plus the vector of slack variables $t = (t_1, \ldots, t_r)$. These two problems are equivalent in the following sense:

1. if x is feasible for (8.20), then x, $t_i = \varphi_i(x)$, $i = 1,\ldots,r$, is feasible for (8.21);

2. if x, t is feasible for (8.21), then x is feasible for (8.20);

3. if x^* is optimal for (8.20), then x^*, $t_i^* = \varphi_i(x^*)$, $i = 1,\ldots,r$, is optimal for (8.21);

4. if x^*, t^* is optimal for (8.21), then x^* is optimal for (8.20);

5. $g^* = p^*$.

The first two points are immediate to prove. To check point 3 observe first that, by point 1, x^*, $t_i^* = \varphi_i(x^*)$, $i = 1,\ldots,r$, is feasible for (8.21). If it were not optimal for this problem, then there would exist another feasible pair $y^* \in \mathcal{X}$, $\tau^* = (\tau_1^*,\ldots,\tau_r^*)$ with a better objective, i.e., such that $\sum_{i=1}^r \tau_i^* < \sum_{i=1}^r t_i^* = \sum_{i=1}^r \varphi_i(x^*)$. But such a y^* is feasible for problem (8.20) and since $\varphi_i(y^*) \leq \tau_i^*$ for $i = 1,\ldots,r$, we would have that $\sum_{i=1}^r \varphi_i(y^*) \leq \sum_{i=1}^r \tau_i^* < \sum_{i=1}^r \varphi_i(x^*)$, which would contradict the fact that x^* is optimal for problem (8.20). Points 4 and 5 follow from an analogous reasoning.

A similar approach can also be followed when the problem has a constraint of the form

$$\sum_{i=1}^r \varphi_i(x) \leq 0,$$

which can be equivalently substituted by the constraints

$$\sum_{i=1}^r t_i \leq 0, \quad \varphi_i(x) \leq t_i,\ i = 1,\ldots,r,$$

involving x and the slack variable $t \in \mathbb{R}^r$.

Remark 8.2 *Generality of the linear objective.* A common use of the slack variable "trick" described above consists of transforming a convex optimization problem of the form (8.11), with generic convex objective f_0, into an *equivalent* convex problem having *linear* objective.

Introducing a new slack variable $t \in \mathbb{R}$, problem (8.11) is reformulated as

$$
\begin{aligned}
t^* = \min_{x \in \mathbb{R}^n, t \in \mathbb{R}} \quad & t \qquad\qquad\qquad\qquad (8.22)\\
\text{subject to:} \quad & f_i(x) \leq 0, \quad i = 1,\ldots,m,\\
& h_i(x) = 0, \quad i = 1,\ldots,q,\\
& f_0(x) \leq t.
\end{aligned}
$$

Problem (8.22) has a linear objective in the augmented variables (x,t), and it is usually referred to as the *epigraphic* reformulation of the original problem (8.11). Essentially, the "price to pay" for having a linear objective is the addition of one scalar variable t to the problem. Any convex optimization problem can thus be reformulated in the form of an equivalent convex problem with linear objective.

8.3.4.5   Substituting equality constraints with inequality constraints. In certain cases, we can substitute an equality constraint of the form $b(x) = u$ with an inequality constraint $b(x) \leq u$. This can be useful, in some cases, for gaining convexity. Indeed, if $b(x)$ is a convex function, then the set described by the *equality* constraint $\{x : b(x) = u\}$ is non-convex in general (unless b is affine); on the contrary, the set described by the *inequality* constraint $\{x : b(x) \leq u\}$ is the sublevel set of a convex function, hence it is convex. We give a sufficient condition under which such a substitution can be safely performed.

Consider a (not necessarily convex) problem of the form

$$p^* = \min_{x \in \mathcal{X}} \quad f_0(x) \tag{8.23}$$
$$\text{s.t.:} \quad b(x) = u,$$

where u is a given scalar, together with the related problem in which the equality constraint is substituted by an inequality one:

$$g^* = \min_{x \in \mathcal{X}} \quad f_0(x) \tag{8.24}$$
$$\text{s.t.:} \quad b(x) \leq u.$$

Clearly, since the feasible set of the first problem is included in the feasible set of the second problem, it always holds that $g^* \leq p^*$. We next prove that it actually holds that $g^* = p^*$, under the following conditions: (*i*) f_0 is non-increasing over $\mathcal{X}$, (*ii*) b is non-decreasing over $\mathcal{X}$, and (*iii*) the optimal value p^* is attained at some optimal point x^*, and the optimal value g^* is attained at some optimal point $\tilde{x}^*$.

The first condition means that, for $x, y \in \mathcal{X}$,

$$f_0(x) \leq f_0(y) \quad \Leftrightarrow \quad x \geq y,$$

where the vector inequality is to be intended element-wise. Similarly, the second condition means that, for $x, y \in \mathcal{X}$,

$$b(x) \geq b(y) \quad \Leftrightarrow \quad x \geq y.$$

To prove that under these assumptions it holds that $g^* = p^*$, we deny the statement and assume that $g^* < p^*$. Then it must be that $b(\tilde{x}^*) < u$ (for if $b(\tilde{x}^*) = u$, then $\tilde{x}^*$ would be feasible also for the equality constrained problem, and it would hold that $g^* = p^*$), thus

$$b(x^*) = u > b(\tilde{x}^*),$$

which, by monotonicity of b implies that $x^* \geq \tilde{x}^*$. In turn, by monotonicity of f_0, this implies that $f_0(x^*) \leq f_0(\tilde{x}^*)$, that is $p^* \leq g^*$, which contradicts the initial statement.

An interpretation of this setup and assumptions is obtained by considering the x variable as a vector of "resources" (e.g., money, labor, etc.), the objective f_0 as an index representing the performance achievable with the given resources, and the constraint $b(x) = c$ as a budget constraint, where b measures the resource consumption. The monotonicity hypothesis on f_0 is typically satisfied when the objective models a situation in which the more resources we put, the better performance we achieve. It is intuitively clear that, under such assumptions, the inequality constraint will always be saturated at optimum, since from the point of view of the objective, it is better to consume all the available resources, up until when their budget is saturated.

Analogously, if the problem is in maximization form

$$\max_{x \in \mathcal{X}} \ f_0(x) \quad \text{s.t.:} \quad b(x) = u,$$

then a sufficient condition for replacing the equality constraint with an inequality one is that both f_0 and b are non-decreasing over $\mathcal{X}$.

Remark 8.3 Observe that while we have $p^* = g^*$ (under the stated hypotheses) and every optimal solution of (8.23) is also optimal for (8.24), the converse is not necessarily true, that is problem (8.24) may have optimal solutions which are not feasible for the original problem (8.23). However, this converse implication holds if the objective is strictly monotone.

Example 8.19 (*Budget constraint in portfolio optimization*) A typical problem in portfolio optimization (see also Example 2.6 and Example 4.3) amounts to determining a portfolio mix $x \in \mathbb{R}^n$ such that the expected return is maximized, and the portfolio volatility remains under a fixed level. Formally,

$$\begin{aligned}
\max_{x} \quad & \hat{r}^\top x \\
\text{s.t.:} \quad & x^\top \Sigma x \leq \sigma^2, \\
& \mathbf{1}^\top x + \phi(x) = 1,
\end{aligned}$$

where $\hat{r} \in \mathbb{R}^n$ is the vector of expected returns of the component assets, $\Sigma \in \mathbb{S}_+^n$ is the returns covariance matrix, σ^2 is the given upper bound on portfolio volatility, and $\phi(x)$ is a function measuring the cost of transaction, which is a non-decreasing function of x. The last constraint expresses the fact that, assuming a unit initial capital in cash, the sum of the invested amounts, $\mathbf{1}^\top x$, must be equal to the initial capital, minus the expense for transaction costs.

Assuming that $\hat{r} > 0$, the objective function in the above problem is increasing in x, and the left-hand side of the equality constraint is non-decreasing in x. Hence, we can write the problem equivalently by substituting the equality constraint with the inequality one $\mathbf{1}^\top x + \phi(x) \leq 1$.

If ϕ is a convex function, the modified problem is convex, whereas the original formulation is not convex in general (unless ϕ is affine).

Remark 8.4 *Problems with a single constraint.* Consider a convex optimization problem with linear objective and a *single* inequality constraint

$$\min_{x \in \mathbb{R}^n} \ c^\top x \quad \text{s.t.:} \ b(x) \leq u, \tag{8.25}$$

and assume that $c \neq 0$, that the feasible set $\{x : b(x) \leq u\}$ is closed,[14] and that the problem attains an optimal solution x^*. Then, from Lemma 8.6, we have that this solution must belong to the boundary of the feasible set, which means that it must hold that $b(x^*) = u$. Therefore, whenever problem (8.25) attains an optimal solution, this solution is also optimal for the *equality* constrained problem

$$\min_{x \in \mathbb{R}^n} \ c^\top x \quad \text{s.t.:} \ b(x) = u.$$

[14] The feasible set is closed if b is continuous. Actually, the weaker condition of b being *lower semi-continuous* (or *closed*, as we usually assume tacitly), is sufficient to ensure closedness of the feasible set.

8.3.4.6 *Elimination of inactive constraints.*

Consider a generic convex optimization problem of the form

$$
\begin{aligned}
p^* = \min_{x \in \mathbb{R}^n} \ & f_0(x) \\
\text{s.t.:} \ & f_i(x) \leq 0, \quad i = 1, \ldots, m, \\
& Ax = b,
\end{aligned} \tag{8.26}
$$

and suppose the optimum is attained at some point x^*. The inequality constraints that are active at x^* are defined as those that hold with equality at that optimal point. We can thus define the set of indices that correspond to active constraints as

$$\mathcal{A}(x^*) = \{i : f_i(x^*) = 0, \ i = 1, \ldots, m\}.$$

Similarly, we define the complementary set of indices corresponding to constraints that are inactive at optimum:

$$\bar{\mathcal{A}}(x^*) = \{i : f_i(x^*) < 0, \ i = 1, \ldots, m\}.$$

It is a rather intuitive fact (although the proof is not entirely trivial) that all the inactive constraints can be removed from the original problem, without changing the optimal solution. More precisely, the following proposition holds.

Proposition 8.1 *Consider the convex problem (8.26) and assume it admits an optimal solution x^*. Then, x^* is also optimal for the problem*

$$
\begin{aligned}
\min_{x \in \mathbb{R}^n} \ & f_0(x) \\
\text{s.t.:} \ & f_i(x) \leq 0, \quad i \in \mathcal{A}(x^*), \\
& Ax = b.
\end{aligned} \tag{8.27}
$$

Proof Let

$$
\begin{aligned}
\mathcal{X} &\doteq \{x : f_i(x) \leq 0, i = 1, \ldots, m; \ Ax = b\}, \\
\mathcal{X}_\mathcal{A} &\doteq \{x : f_i(x) \leq 0, i \in \mathcal{A}(x^*); \ Ax = b\}
\end{aligned}
$$

denote the convex feasible sets of the original problem (8.26) and of the reduced problem (8.27), respectively, and let

$$
\mathcal{I} \doteq \{x : f_i(x) \leq 0, i \in \bar{\mathcal{A}}(x^*)\}.
$$

Since $x^* \in \mathcal{X} \subseteq \mathcal{X}_\mathcal{A}$, x^* is feasible for problem (8.27). We suppose, for the purpose of contradiction, that x^* is not optimal for this problem, which means that there exists $x \in \mathcal{X}_\mathcal{A}$ such that $f_0(x) < f_0(x^*)$.

Next observe that $x^* \in \mathcal{X} \subseteq \mathcal{I}$ and actually, since $f_i(x^*) < 0$ for $i \in \bar{\mathcal{A}}(x^*)$, x^* belongs to the interior of $\mathcal{I}$ (see Section 8.2.1.4). Therefore, there exists an open ball B centered in x^* that is also contained in $\mathcal{I}$. Consider now a point of the form $z(\lambda) \doteq (1 - \lambda)x^* + \lambda x$, for $\lambda \in [0,1]$ (such points are a convex combination of x^* and x). One can choose $\lambda \neq 0$ sufficiently small so that $z(\lambda)$ is close to x^* and belongs to B. Hence, for such λ, $z(\lambda) \in B \subseteq \mathcal{I}$. Notice that, since $x^* \in \mathcal{X} \subseteq \mathcal{X}_\mathcal{A}$ and $x \in \mathcal{X}_\mathcal{A}$, by convexity, all points along the segment $\{z(\lambda), \lambda \in [0, 1]\}$ are also in $\mathcal{X}_\mathcal{A}$. But for the chosen λ we also have $z(\lambda) \in \mathcal{I}$, hence

$$
z(\lambda) \in \mathcal{X}_\mathcal{A} \cap \mathcal{I} \equiv \mathcal{X},
$$

which means that $z(\lambda)$ is feasible for the original problem (8.26). By Jensen's inequality, we then have that

$$
\begin{aligned}
f_0(z(\lambda)) &\leq (1 - \lambda)f_0(x^*) + \lambda f_0(x) \\
&< (1 - \lambda)f_0(x^*) + \lambda f_0(x^*) = f_0(x^*) = p^*,
\end{aligned}
$$

where the second inequality in the above chain follows from the position $f_0(x) < f_0(x^*)$ that we are taking for the purpose of arriving at a contradiction. Indeed, we found a point $z(\lambda)$ which is feasible for problem (8.26) and which yields an objective value $f_0(z(\lambda)) < p^*$, which is impossible, since p^* is the optimal value of the problem. $\square$

Remark 8.5 A few remarks are in order regarding the discussed technique of elimination of inactive constraints. First, notice that, for using this technique in practice, one should *know in advance* which constraints are inactive at optimum. However, unfortunately, these are typically only known once the original problem is solved! Nevertheless, there are particular cases where an *a priori* analysis of the problem may help determine which constraints will necessarily be inactive at optimum, and these constraints can be effectively removed, thus reducing the "'size" of the problem one needs to solve numerically. An illustration of such a situation is described in Example 8.23.

A second warning concerns the fact that while all optimal solutions of the original problem are also optimal for the reduced problem, the converse is not true, in general. To convince oneself of this fact, one may consider minimizing the univariate convex function shown in Figure 8.20 with constraints $x \geq 0$ and $x \leq 1$.

Finally, the result only holds for convex problems, and may fail, in general, when the objective or constraints are not convex (consider, for instance, the minimization of the univariate function shown in Figure 8.21, with constraints $x \geq 0$ and $x \leq 1$).

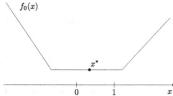

Figure 8.20 Constrained minimization of a convex function: all constrained-optimal points with inactive constraints are also optimal for the unconstrained problem, but not vice versa.

8.3.5 Special classes of convex models

The class of convex optimization problems encompassed by the formulation (8.11) is quite broad, allowing for a general type of convex objective and constraint functions. In this book, however, we concentrate on some more specialized optimization models, obtained by further qualifying the objective and constraint functions in (8.11). For such specialized models there exist well-established and efficient numerical solvers that provide the user with a reliable *technology* for effectively solving in practice most design problems that can be formulated. The fundamental convex models, which will be treated in detail in forthcoming chapters, are the following ones.

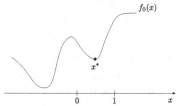

Figure 8.21 Constrained minimization of a non-convex function: if inactive constraints are removed the constrained-optimal solution may no longer be optimal for the unconstrained problem.

8.3.5.1 Linear and quadratic programs. Linear programs (LPs) are a specialized case of (8.11), where all the functions involved in the problem description are linear (or affine). They hence have the standard form

$$p^* = \min_{x \in \mathbb{R}^n} \quad c^\top x$$
$$\text{s.t.:} \quad a_i^\top x - b_i \leq 0, \quad i = 1, \ldots, m,$$
$$A_{\text{eq}} x = b_{\text{eq}}.$$

Quadratic programs (QPs, QCQPs) are also a specialized case of (8.11), in which the functions describing the objective and the inequality constraints are quadratic, i.e., they are polynomials in the x variable of degree at most two: $f_i(x) = x^\top H_i x + c_i^\top x + d_i$, where H_i are $n \times n$ symmetric matrices. Quadratic problems are convex if and only if $H_i \succeq 0$ for $i = 0, 1, \ldots, m$. Clearly, quadratic programs include linear programs as a special case, when $H_i = 0$, $i = 0, 1, \ldots, m$. Linear and quadratic programs are discussed in detail in Chapter 9.

8.3.5.2 Geometric programs. Geometric programming (GPs) is an optimization model in which the variables are non-negative, and the objective and constraints are sums of powers of those variables, with non-negative weights. Although GPs are not convex in their natural formulation, they can be transformed, via a logarithmic change of variables, into convex problems. GPs arise naturally in the context of geometric design, or with models of processes that are well approximated with power laws. They also arise (via their convex representation) when trying to fit discrete probabilities models in the context of classification. The objective and constraint functions involved in GPs are so-called *posynomials*, that is non-negative sums of monomials:

$$f_i(x) = \sum_{j=1}^{k} c_j^{(i)} x_1^{a_{j,1}^{(i)}} \cdots x_n^{a_{j,n}^{(i)}}, \quad i = 0, 1, \ldots, m,$$

with $c_j^{(i)} \geq 0$, and domain $x > 0$. Geometric programs are thus an extension of the optimization problems involving power laws (monomials) discussed in Example 8.18; they are further discussed in Section 9.7.

8.3.5.3 Second-order cone programs.
Second-order cone programs (SOCP) further extend convex quadratic programs, by dealing with constraints of the form

$$f_i(x) = \|A_i x + b_i\|_2 \leq c_i^\top x + d_i, \quad i = 1, \ldots, m,$$

where $A_i \in \mathbb{R}^{m_i, n}$ are given matrices, $b_i \in \mathbb{R}^{m_i}$, $c_i \in \mathbb{R}^n$ are given vectors, and $d_i \in \mathbb{R}$ are scalars. SOCPs arise, for example, in several geometrical optimization problems, as well as robust counterparts of linear programs when the data is affected by deterministic unknown-but-bounded or stochastic uncertainty, and in financial optimization problems. They are treated in detail in Chapter 10.

8.3.5.4 Semidefinite programs.
Semidefinite programs (SDPs) are convex optimization problems that involve minimization of a linear objective subject to a constraint that imposes positive semidefiniteness of a symmetric matrix that depends affinely on a vector of variables $x \in \mathbb{R}^n$. Specifically, given symmetric matrices $F_i \in \mathbb{S}^m, i = 0, 1, \ldots, n$, a semidefinite program usually takes the following form:

$$p^* = \min_{x \in \mathbb{R}^n} \quad c^\top x$$
$$\text{s.t.:} \quad F(x) \succeq 0,$$

where $F(x) \doteq F_0 + \sum_{i=1}^{n} x_i F_i$. Since

$$F(x) \succeq 0 \quad \text{if and only if} \quad \lambda_{\min}(F(x)) \geq 0,$$

the above SDP can be expressed in the usual form of (8.11), as

$$p^* = \min_{x \in \mathbb{R}^n} \quad c^\top x$$
$$\text{s.t.:} \quad f(x) \leq 0,$$

with $f(x) \doteq -\lambda_{\min}(F(x))$. SDPs encompass as special cases SOCPs, QPs and LPs. They are discussed in Chapter 11.

8.4   Optimality conditions

We here provide conditions characterizing optimality of a feasible point in a convex optimization problem. The following result holds.

Proposition 8.2 *Consider the optimization problem* $\min_{x \in \mathcal{X}} f_0(x)$, *where* f_0 *is convex and differentiable, and* $\mathcal{X}$ *is convex. Then,*

$$x \in \mathcal{X} \text{ is optimal} \quad \Leftrightarrow \quad \nabla f_0(x)^\top (y - x) \geq 0, \quad \forall y \in \mathcal{X}. \qquad (8.28)$$

Proof We know from (8.4) that for every $x, y \in \operatorname{dom} f_0$ it holds that

$$f_0(y) \geq f_0(x) + \nabla f_0(x)^\top (y - x). \qquad (8.29)$$

The implication from right to left in (8.28) is immediate, since $\nabla f_0(x)^\top (y - x) \geq 0$ for all $y \in \mathcal{X}$ implies, from (8.29), that $f_0(y) \geq f_0(x)$ for all $y \in \mathcal{X}$, i.e., that x is optimal. Conversely, suppose that x is optimal, we show that it must be that $\nabla f_0(x)^\top (y - x) \geq 0$ for all $y \in \mathcal{X}$. If $\nabla f_0(x) = 0$, then the claim holds trivially. Consider then the case when $\nabla f_0(x) \neq 0$, and suppose, by the purpose of contradiction, that x is optimal but there exists $y \in \mathcal{X}$ such that $\nabla f_0(x)^\top (y - x) < 0$. Then any point $x_\theta = \theta y + (1 - \theta) x$, $\theta \in [0, 1]$, along the segment from x to y is feasible and, for sufficiently small θ, x_θ is in a neighborhood of x where the sign of the first-order term of the Taylor expansion of f_0 prevails over all other terms, hence

$$
\begin{aligned}
f_0(x_\theta) &= f_0(x) + \nabla f_0(x)^\top (x_\theta - x) + o(\|x_\theta - x\|) \\
&= f_0(x) + \theta \nabla f_0(x)^\top (y - x) + o(\theta \|y - x\|) \\
&= f_0(x) + \text{negative term,}
\end{aligned}
$$

which implies that for such a θ we would have $f_0(x_\theta) < f_0(x)$, which contradicts the optimality of x. $\qquad \square$

If $\nabla f_0(x) \neq 0$, then Eq. (8.28) states that $\nabla f_0(x)$ is a normal direction defining a hyperplane $\{y : \nabla f_0(x)^\top (y - x) = 0\}$ such that (i) x is on the boundary of the feasible set $\mathcal{X}$, and (ii) the whole feasible set lies on one side of this hyperplane (see Figure 8.22), that is in the half-space defined by

$$\mathcal{H}_+(x) = \{y : \nabla f_0(x)^\top (y - x) \geq 0\}.$$

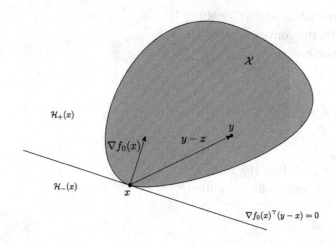

Figure 8.22 Geometry of the first-order condition for optimality: all points y in the feasible set are such that $y - x$ is a direction of increase for f_0 (for $\nabla f_0(x) \neq 0$).

Notice that the gradient vector $\nabla f_0(x)$ defines two sets of directions: for directions v_+ such that $\nabla f_0(x)^\top v_+ > 0$ (i.e., directions that have positive inner product with the gradient) we have that if we make a move away from x in the direction v_+ then the objective f_0 *increases*. Similarly, for directions v_- such that $\nabla f_0(x)^\top v_+ < 0$ (i.e., *descent* directions, that have negative inner product with the gradient) we have that if we make a sufficiently small move away from x in the direction v_- then the objective f_0 locally *decreases*. Condition (8.28) then says that x is an optimal point if and only if there is no feasible direction along which we may improve (decrease) the objective.

8.4.1 Optimality conditions for unconstrained problems

When the problem is unconstrained, i.e., $\mathcal{X} = \mathbb{R}^n$, then the optimality condition (8.28) requires that $\nabla f_0(x)^\top (y - x) \geq 0$ for all $y \in \mathbb{R}^n$. This implies that the condition must be satisfied for any y_1, and it should also be satisfied for any $y_2 = 2x - y_1$, which would imply that $\nabla f_0(x)^\top (y_1 - x) \geq 0$ and $-\nabla f_0(x)^\top (y_1 - x) \geq 0$, and this is only possible if $\nabla f_0(x) = 0$. We thus proved the following proposition.

Proposition 8.3 *In a convex unconstrained problem with differentiable objective, x is optimal if and only if*

$$\nabla f_0(x) = 0. \tag{8.30}$$

8.4.2 Optimality conditions for equality-constrained problems

Consider the linear-equality constrained optimization problem

$$\min_x f_0(x) \quad \text{s.t.: } Ax = b, \tag{8.31}$$

where $f_0 : \mathbb{R}^n \to \mathbb{R}$ is convex and differentiable, and $A \in \mathbb{R}^{m,n}$, $b \in \mathbb{R}^m$ define the equality constraints. Here, the convex feasible set is the affine set of solutions of the linear equations:

$$\mathcal{X} = \{x : Ax = b\}.$$

Using (8.28), we have that $x \in \mathcal{X}$ is optimal if and only if

$$\nabla f_0(x)^\top (y - x) \geq 0, \quad \forall y \in \mathcal{X}.$$

Now, all vectors $y \in \mathcal{X}$ (that is, all vectors y such that $Ay = b$) can be written as $y = x + z$, $z \in \mathcal{N}(A)$, hence the optimality condition becomes

$$\nabla f_0(x)^\top z \geq 0, \quad \forall z \in \mathcal{N}(A).$$

Since $z \in \mathcal{N}(A)$ if and only if $-z \in \mathcal{N}(A)$, we see that the condition actually implies

$$\nabla f_0(x)^\top z = 0, \quad \forall z \in \mathcal{N}(A).$$

In words, this means that $\nabla f_0(x)$ must be orthogonal to $\mathcal{N}(A)$, i.e.

$$\nabla f_0(x) \in \mathcal{N}(A)^\perp \equiv \mathcal{R}(A^\top),$$

where the last line follows from the fundamental theorem of linear algebra, see Section 3.2. The condition $\nabla f_0(x) \in \mathcal{R}(A^\top)$ is equivalent to the existence of a vector of coefficients $v \in \mathbb{R}^m$ such that $\nabla f_0(x) = A^\top v$. In conclusion, we have that x is optimal for problem (8.31) if and only if

$$Ax = b, \quad \text{and } \exists v \in \mathbb{R}^m : \nabla f_0(x) = A^\top v.$$

Since the same reasoning can be repeated using $-v$ instead of v, the above condition is also often expressed equivalently as in the following proposition.

Proposition 8.4 *A point x is optimal for problem (8.31) if and only if*

$$Ax = b, \quad \text{and } \exists v \in \mathbb{R}^m : \nabla f_0(x) + A^\top v = 0.$$

8.4.3 Optimality conditions for inequality-constrained problems

The following result gives sufficient conditions for optimality of a convex inequality constrained problem.

Proposition 8.5 *Consider the convex optimization problem $\min_{x \in \mathcal{X}} f_0(x)$, where f_0 is differentiable, and the feasible set $\mathcal{X}$ is defined via convex inequalities as*

$$\mathcal{X} = \{x \in \mathbb{R}^n : f_i(x) \leq 0, i = 1, \ldots, m\},$$

where f_i, $i = 1, \ldots, m$, are convex and continuosly differentiable. Let $x \in \mathcal{X}$ be a feasible point, and let $\mathcal{A}(x)$ denote the set of indices of the constraints that are active at x, that is

$$\mathcal{A}(x) = \{i : f_i(x) = 0, i = 1, \ldots, m\}.$$

If there exist $\lambda_i \geq 0$, $i \in \mathcal{A}(x)$, such that

$$\nabla f_0(x) + \sum_{i \in \mathcal{A}(x)} \lambda_i \nabla f_i(x) = 0, \tag{8.32}$$

then x is optimal.

Proof Consider first the case when $\mathcal{A}(x)$ is empty. Then x is feasible and condition (8.32) simply prescribes that $\nabla f_0(x) = 0$, in which case x coincides with an unconstrained minimum of f_0.

Assume next that $\mathcal{A}(x)$ is nonempty, and notice that, from the convexity of f_i, it follows that, for all y,

$$f_i(y) \geq f_i(x) + \nabla f_i(x)^\top (y - x) = \nabla f_i(x)^\top (y - x), \quad \text{for } i \in \mathcal{A}(x). \tag{8.33}$$

Therefore, for any $x \in \mathcal{X}$ such that $\mathcal{A}(x)$ is nonempty, the hyperplane $\{\nabla f_i(x)^\top (y - x) = 0\}$, $i \in \mathcal{A}(x)$, divides the whole space into two complementary half-spaces

$$\mathcal{H}_{++}^{(i)} \doteq \{y : \nabla f_i(x)^\top (y - x) > 0\}$$
$$\mathcal{H}_{-}^{(i)} \doteq \{y : \nabla f_i(x)^\top (y - x) \leq 0\},$$

and the feasible set $\mathcal{X}$ is not contained[15] in $\mathcal{H}_{++}^{(i)}$, thus it must be contained in $\mathcal{H}_{-}^{(i)}$, i.e., $\mathcal{X} \subseteq \mathcal{H}_{-}^{(i)}$. Since this is true for all $i \in \mathcal{A}(x)$, we obtain that

> [15] Since, for $y \in \mathcal{H}_{++}^{(i)}$, Eq. (8.33) implies that $f_i(y) > 0$, hence y is infeasible.

$$\mathcal{X} \subseteq \mathcal{P}(x) \doteq \bigcap_{i \in \mathcal{A}(x)} \mathcal{H}_{-}^{(i)}.$$

Now, the statement (8.32) is equivalent to

$$\nabla f_0(x)^\top (y - x) \geq - \sum_{i \in \mathcal{A}(x)} \lambda_i \nabla f_i(x)^\top (y - x), \quad \forall y.$$

In particular, since $\lambda_i \geq 0$, it holds that

$$\nabla f_0(x)^\top (y - x) \geq 0, \quad \forall y \in \mathcal{P}(x),$$

and since $\mathcal{X} \subseteq \mathcal{P}(x)$, this implies that

$$\nabla f_0(x)^\top (y - x) \geq 0, \quad \forall y \in \mathcal{X},$$

which, from Proposition 8.2, means that x is optimal. $\square$

8.4.4   Optimality conditions for non-differentiable problems

All the previous conditions were stated for the case when f_0 is differentiable. However, similar results do hold also for non-differentiable problems, provided that subgradients (and subdifferentials) are suitably used instead of gradients. Specifically, the equivalent of condition (8.28) for non-differentiable f_0 is[16]

$$x \in \mathcal{X} \text{ is optimal} \quad \Leftrightarrow \quad \exists g_x \in \partial f_0(x) : g_x^\top (y - x) \geq 0, \ \forall y \in \mathcal{X}.$$

The implication from right to left is immediate, applying the definition of a subgradient; the converse direction is slightly more involved and it is not proved here. For unconstrained problems, the optimality condition becomes

$$x \in \mathbb{R}^n \text{ is optimal} \quad \Leftrightarrow \quad 0 \in \partial f_0(x).$$

Example 8.20 Consider the unconstrained minimization problem

$$\min_{x \in \mathbb{R}^n} f_0(x), \quad f_0(x) = \max_{i=1,\dots,m} a_i^\top x + b_i.$$

Here, the objective is a polyhedral function, which is non-differentiable. The subdifferential of f_0 at x is given by

$$\partial f_0(x) = \mathrm{co}\{a_i : a_i^\top x + b_i = f_0(x)\},$$

and the optimality condition $0 \in \partial f_0(x)$ hence requires that

$$0 \in \mathrm{co}\{a_i : a_i^\top x + b_i = f_0(x)\}.$$

---

[16] See, e.g., Section 1.2 in N. Z. Shor, *Minimization Methods for Non-differentiable Functions*, Springer, 1985.

8.5   Duality

Duality is a central concept in optimization. Essentially, duality provides a technique for transforming an optimization problem (the *primal* problem) into another related optimization problem (the *dual* problem), which can provide useful information about the primal. In particular, the dual problem is always a convex optimization problem (even when the primal is not), and its optimal value provides a *lower bound* on the optimal objective value of the primal. When also the primal is convex and under some *constraint qualification* conditions, the primal and dual objective values actually coincide. Moreover, under some further assumptions, the value of the primal optimal variables can be recovered from the optimal dual variables. This feature is useful whenever the dual problem happens to be "easier" to solve than the primal. Further, duality plays an important role in certain algorithms for the solution of convex problems (see, e.g., Section 12.3.1), since it permits us to control, at each iteration of the algorithm, the level of suboptimality of the current solution candidate. Duality is also a key element in decomposition methods for distributed optimization (see Section 12.6.1).

Consider the optimization problem in standard form (8.11)–(8.13), which is recalled below for clarity and is here denoted as the *primal problem*:

$$p^* = \min_{x \in \mathbb{R}^n} \quad f_0(x) \tag{8.34}$$
$$\text{subject to:} \quad f_i(x) \le 0, \quad i = 1, \ldots, m, \tag{8.35}$$
$$h_i(x) = 0, \quad i = 1, \ldots, q, \tag{8.36}$$

and let $\mathcal{D}$ denote the domain of this problem, that is the intersection of the domains of the objective and the constraint functions, which is assumed to be nonempty. We shall build a new function, called the *Lagrangian*, defined as a weighted sum of the problem objective and the constraint functions, namely $\mathcal{L} : \mathcal{D} \times \mathbb{R}^m \times \mathbb{R}^q \to \mathbb{R}$, with

$$\mathcal{L}(x, \lambda, v) = f_0(x) + \sum_{i=1}^{m} \lambda_i f_i(x) + \sum_{i=1}^{q} v_i h_i(x),$$

where $\lambda = [\lambda_1 \cdots \lambda_m]$ is the vector of weights relative to the inequality constraints, and $v = [v_1 \cdots v_q]$ is the vector of weights relative to the equality constraints; λ and v are called the vectors of *Lagrange multipliers*, or dual variables, of the problem. Notice that we are not assuming convexity of $f_0, f_1, \ldots, f_m$ or of $h_1, \ldots, h_q$, for the time being.

8.5.1   The Lagrange dual function

Suppose that the value of the multipliers λ, v is fixed, with $\lambda \ge 0$ (element-wise, i.e., $\lambda_i \ge 0$, $i = 1, \ldots, m$). We may then consider the minimum (infimum) of the Lagrangian over the x variable:

$$
\begin{aligned}
g(\lambda, v) &= \inf_{x \in \mathcal{D}} \mathcal{L}(x, \lambda, v) \\
&= \inf_{x \in \mathcal{D}} \left(f_0(x) + \sum_{i=1}^{m} \lambda_i f_i(x) + \sum_{i=1}^{q} v_i h_i(x) \right). \tag{8.37}
\end{aligned}
$$

For given (λ, v), if $\mathcal{L}(x, \lambda, v)$ is unbounded below w.r.t. x, then the minimization over x yields $g(\lambda, v) = -\infty$; otherwise, $g(\lambda, v)$ is a finite value. The function $g(\lambda, v) : \mathbb{R}^m \times \mathbb{R}^p \to \mathbb{R}$ is called the *(Lagrange) dual function* of the problem (8.34)–(8.36). We next state two key properties of the dual function.

Proposition 8.6 (Lower bound property of the dual function)

The dual function $g(\lambda, v)$ is jointly concave in (λ, v). Moreover, it holds that

$$g(\lambda, v) \leq p^*, \quad \forall \lambda \geq 0, \ \forall v. \tag{8.38}$$

Proof To prove this proposition, first notice that, for each fixed x, the function $\mathcal{L}(x, \lambda, v)$ is affine in (λ, v), hence it is also concave in (λ, v) (recall that linear and affine functions are both convex and concave). But then, $g(\lambda, v)$ is defined as the pointwise infimum of concave functions (parameterized by x), hence it is concave (this immediately follows by applying the pointwise maximum rule for convex functions to the function $-\mathcal{L}$, which is convex in (λ, v) for any fixed x). Notice that we just proved that $g(\lambda, v)$ is concave, *regardless* of the fact that $f_i, i = 0, \ldots, m$ are convex or not; i.e., $g(\lambda, v)$ is *always* concave!

The second part of Proposition 8.6 can be proved by considering that any point x which is feasible for problem (8.34)–(8.36) must, by definition, satisfy the constraints of the problem, hence $f_i(x) \leq 0$, $i = 1, \ldots, m$, and $h_i(x) = 0, i = 1, \ldots, p$. Thus, since $\lambda_i \geq 0$, we have that $\lambda_i f_i(x) \leq 0$ and $v_i h_i(x) = 0$, therefore

$$
\begin{aligned}
\mathcal{L}(x, \lambda, v) &= f_0(x) + \sum_{i=1}^{m} \lambda_i f_i(x) + \sum_{i=1}^{q} v_i h_i(x) \\
&= f_0(x) + \sum_{i=1}^{m} \lambda_i f_i(x) \\
&\leq f_0(x), \quad \forall \text{ feasible } x, \ \forall v, \ \forall \lambda \geq 0. \tag{8.39}
\end{aligned}
$$

Now, since the infimum of a function is no larger than the value of that function evaluated at any given point x, we have that

$$g(\lambda, v) = \inf_{x \in \mathcal{D}} \mathcal{L}(x, \lambda, v) \leq \mathcal{L}(x, \lambda, v), \quad \forall x \in \mathcal{D}. \tag{8.40}$$

Therefore, by combining (8.39) and (8.40), we have that for any point $x \in \mathcal{D}$ which is feasible for problem (8.34)–(8.36), it holds that

$$g(\lambda, v) \leq \mathcal{L}(x, \lambda, v) \leq f_0(x), \quad \forall \text{ feasible } x \in \mathcal{D}, \ \forall v, \ \forall \lambda \geq 0.$$

Further, since this inequality holds for the value of $f_0(x)$ at all feasible x, it also holds for the optimal value of f_0, which is p^*, from which we conclude that $g(\lambda, v) \leq p^*, \forall v, \forall \lambda \geq 0$. $\square$

8.5.2 *The dual optimization problem*

From (8.38) we have that, for any v and $\lambda \geq 0$, the value of $g(\lambda, v)$ provides a lower bound on the primal optimal objective value p^*. It is then natural to try to find the *best possible* of such lower bounds.

This can be done by seeking for the maximum of $g(\lambda, \nu)$ over ν and $\lambda \geq 0$. Since $g(\lambda, \nu)$ is always a concave function, this is a concave maximization problem (which is equivalent to a convex minimization problem, where we minimize $-g$). Finding the best possible lower bound d^* on p^*, obtainable from the dual function, is called the *dual problem*:

$$d^* = \max_{\lambda, \nu} g(\lambda, \nu) \text{ s.t.: } \lambda \geq 0. \tag{8.41}$$

It is a remarkable fact that the dual problem is always a convex optimization problem, even when the primal problem is not convex. Moreover, it follows from Proposition 8.6 that

$$d^* \leq p^*.$$

This property is usually referred to as the *weak duality* property, and the quantity
$$\delta^* = p^* - d^*$$
is called the *duality gap*, which represents the "error" with which the dual optimal objective d^* approximates from below the primal optimal objective p^*.

8.5.3 Constraint qualification and strong duality

Under some additional hypotheses (such as convexity of the primal problem, plus "constraint qualification") a stronger relation actually holds between the optimal values of the primal and of the dual. We say that *strong duality* holds when $d^* = p^*$, that is, when the duality gap is zero. The following proposition establishes a sufficient condition for strong duality to hold.[17]

Proposition 8.7 (Slater's conditions for convex programs) *Let f_i, $i = 0, \ldots, m$, be convex functions, and let h_i, $i = 1, \ldots, q$, be affine functions. Suppose further that the first $k \leq m$ of the f_i functions, $i = 1, \ldots, k$ are affine (or let $k = 0$, if none of the $f_1, \ldots, f_m$ is affine). If there exists a point $x \in \text{relint } \mathcal{D}$ such that*

$$f_1(x) \leq 0, \ldots, f_k(x) \leq 0; \quad f_{k+1}(x) < 0, \ldots, f_m(x) < 0;$$
$$h_1(x) = 0, \ldots, h_q(x) = 0,$$

then strong duality holds between the primal problem (8.34) and the dual problem (8.41), that is $p^* = d^*$. Moreover, if $p^* > -\infty$, then the dual optimal value is attained, that is, there exist λ^*, μ^* such that $g(\lambda^*, \mu^*) = d^* = p^*$.

[17] See, e.g., Section 5.2.3 of S. Boyd and L. Vandenberghe, *Convex Optimization*, Cambridge University Press, 2004.

In words, Proposition 8.7 states that for convex programs we have strong duality whenever there exists a point that satisfies the affine inequality constraints and affine equality constraints, and that satisfies *strictly* (i.e., with strict inequality) the other (non-affine) inequality constraints.

Example 8.21 (*Dual of a linear program*) Consider the following optimization problem with linear objective and linear inequality constraints (a so-called linear program in standard inequality form, see Section 9.3)

$$p^* = \min_{x} \quad c^\top x \qquad (8.42)$$
$$\text{s.t.:} \quad Ax \leq b,$$

where $A \in \mathbb{R}^{m,n}$ is a matrix of coefficients, and the inequality $Ax \leq b$ is to be intended element-wise. The Lagrangian for this problem is

$$\mathcal{L}(x,\lambda) = c^\top x + \lambda^\top (Ax - b) = (c + A^\top \lambda)^\top x - \lambda^\top b.$$

In order to determine the dual function $g(\lambda)$ we next need to minimize $\mathcal{L}(x,\lambda)$ w.r.t. x. But $\mathcal{L}(x,\lambda)$ is affine in x, hence this function is unbounded below, unless the vector coefficient of x is zero (i.e., $c + A^\top \lambda = 0$), and it is equal to $-\lambda^\top b$ otherwise. That is,

$$g(\lambda) = \begin{cases} -\infty & \text{if } c + A^\top \lambda \neq 0, \\ -\lambda^\top b & \text{if } c + A^\top \lambda = 0. \end{cases}$$

The dual problem then amounts to maximizing $g(\lambda)$ over $\lambda \geq 0$. Clearly, if $g(\lambda) = -\infty$, then there is nothing to maximize, therefore in the dual problem we make explicit the condition that we maximize over those λ for which $g(\lambda)$ is not identically equal to $-\infty$. This results in the following explicit dual problem formulation:

$$d^* = \max_{\lambda} \quad -\lambda^\top b \qquad (8.43)$$
$$\text{s.t.:} \quad c + A^\top \lambda = 0,$$
$$\lambda \geq 0,$$

and, from weak duality, we have that $d^* \leq p^*$. Actually, Proposition 8.7 guarantees that strong duality holds, whenever problem (8.42) is feasible. We may also rewrite the dual problem into an equivalent minimization form, by changing the sign of the objective, which results in

$$-d^* = \min_{\lambda} \quad b^\top \lambda$$
$$\text{s.t.:} \quad A^\top \lambda + c = 0,$$
$$\lambda \geq 0,$$

and this is again an LP, in standard conic form (see Section 9.3). It is interesting to next derive the dual of this dual problem. The Lagrangian is in this case

$$\mathcal{L}_d(\lambda,\eta,\nu) = b^\top \lambda - \eta^\top \lambda + \nu^\top (A^\top \lambda + c),$$

where η is the vector of Lagrange multipliers relative to the inequality constraints $\lambda \geq 0$, and v is the vector of Lagrange multipliers relative to the equality constraints $A^\top \lambda + c = 0$. The "dual dual" function can then be obtained by minimizing $\mathcal{L}_d$ w.r.t. λ, that is

$$q(\eta, v) = \inf_\lambda \mathcal{L}_d(\lambda, \eta, v) = \begin{cases} -\infty & \text{if } b + Av - \eta \neq 0, \\ c^\top v & \text{if } b + Av - \eta = 0. \end{cases}$$

The "dual dual" problem thus amounts to maximizing $q(\eta, v)$ over v and over $\eta \geq 0$, that is

$$\begin{aligned} -dd^* = \max_{v, \eta} \quad & c^\top v \\ \text{s.t.:} \quad & b + Av = \eta, \\ & \eta \geq 0. \end{aligned}$$

Notice further that the combined constraints $b + Av = \eta$ and $\eta \geq 0$ are equivalent to the single constraint $b + Av \geq 0$, where the variable η is eliminated, therefore

$$\begin{aligned} -dd^* = \max_{v} \quad & c^\top v \\ \text{s.t.:} \quad & b + Av \geq 0, \end{aligned}$$

and weak duality prescribes that $-dd^* \leq -d^*$, i.e.,

$$dd^* \geq d^*.$$

Furthermore, Proposition 8.7 states that strong duality holds if problem (8.43) is feasible, in which case we have $dd^* = d^*$. Rewriting the "dual dual" problem in equivalent minimization form, and changing variable $(-v) \to x$, we also observe that, in this special case of LP, we recover exactly the primal problem:

$$\begin{aligned} dd^* = \min_{x} \quad & c^\top x \\ \text{s.t.:} \quad & Ax \leq b, \end{aligned}$$

whence $dd^* = p^*$.

To summarize, if the primal is feasible then strong duality holds, and we have $p^* = d^*$. Also, if the dual is feasible, then we have $dd^* = d^*$, but since the "dual dual" is equivalent to the primal we have $dd^* = p^*$, hence again $p^* = d^*$. This means that, in LP, strong duality holds between p^* and d^* when either the primal or the dual are feasible (equivalently, strong duality may fail only in the "pathological" situation when both the primal and the dual are unfeasible).

8.5.4 Recovering primal variables from dual variables

There are various reasons for being interested in considering the dual of an optimization problem. A first reason arises when the primal is some difficult non-convex optimization problem: if we are able to determine explicitly the dual of this problem, since the dual is always convex (more precisely, a concave maximization), we can compute efficiently a lower bound d^* on the optimal primal value p^*, and such a lower approximation of p^* may be of interest in many practical situations. A second reason is related to the fact that the dual problem has as many variables as there are constraints in the primal. For example, consider an LP as in (8.42) with $A \in \mathbb{R}^{m,n}$, where n is very large compared to m, that is the number n of variables in x is much larger than the number m of constraints. Whenever strong duality holds, the optimal value p^* can be computed by solving the dual problem in λ, with a much smaller number m of decision variables, which may be advantageous in some cases. A third reason is due to the fact that certain dual problems may have a particular structure that makes them either solvable explicitly (e.g., analytically), or amenable to specific solution algorithms that exploit the special structure of the dual (such as, for instance, the fact that the dual constraints may be a simple restriction of the dual variables to the positive orthant). Once the dual-optimal solution is found, under strong duality, it is in some cases possible to recover from it a primal-optimal solution, as explained next.

Assume that strong duality holds between problems (8.34) and (8.41), and that both the primal and the dual optimal values are attained at x^* and (λ^*, v^*), respectively. Then, since $p^* = f_0(x^*)$, $d^* = g(\lambda^*, v^*)$, and $p^* = d^*$, we have that

$$
\begin{aligned}
f_0(x^*) &= g(\lambda^*, v^*) = \inf_{x \in \mathcal{D}} \mathcal{L}(x, \lambda^*, v^*) \tag{8.44} \\
&\leq \mathcal{L}(x, \lambda^*, v^*), \quad \forall x \in \mathcal{D}.
\end{aligned}
$$

Since the last inequality holds for all $x \in \mathcal{D}$, it must hold also for x^*, hence we obtain that

$$
\begin{aligned}
f_0(x^*) &= \inf_{x \in \mathcal{D}} \mathcal{L}(x, \lambda^*, v^*) \\
&\leq \mathcal{L}(x^*, \lambda^*, v^*) = f_0(x^*) + \sum_{i=1}^{m} \lambda_i^* f_i(x^*) + \sum_{i=1}^{q} v_i^* h_i(x^*) \\
&\leq f_0(x^*),
\end{aligned}
$$

where the last inequality follows from the fact that x^* is optimal, hence feasible, for the primal problem, therefore $f_i(x^*) \leq 0, h_i(x^*) = 0$, and λ^* is optimal, hence feasible, for the dual, therefore $\lambda_i^* \geq 0$, whereby each term $\lambda_i^* f_i(x^*)$ is ≤ 0, while each term $v_i^* h_i(x^*)$ is zero. Observing the last chain of inequalities, since the first and the last terms are equal, we must conclude that all inequalities must actually hold with equality, that is

$$
f_0(x^*) = \inf_{x \in \mathcal{D}} \mathcal{L}(x, \lambda^*, v^*) = \mathcal{L}(x^*, \lambda^*, v^*),
$$

which has two consequences:

1. it must hold that $\lambda_i^* f_i(x^*) = 0$, for $i = 1, \ldots, m$; a property known as *complementary slackness*;

2. the primal-optimal point x^* is a minimizer of $\mathcal{L}(x, \lambda^*, \nu^*)$.

We say that an inequality constraint is *slack* when it is satisfied with strict inequality at the optimum. Conversely, we say that the constraint is *active* when it is satisfied with equality at the optimum. The complementary slackness property prescribes (for a problem where strong duality holds) that a primal and the corresponding dual inequality cannot be slack simultaneously, that is, if $f_i(x^*) < 0$, then it must be that $\lambda_i^* = 0$, and if $\lambda_i^* > 0$, then it must be that $f_i(x^*) = 0$. By looking at the dual variables λ_i^* we can hence determine which of the primal constraints are saturated (active) at the optimum.

The second consequence (i.e., the fact that x^* is a minimizer of $\mathcal{L}(x, \lambda^*, \nu^*)$) can, in some cases, be used to recover a primal-optimal variable from the dual-optimal variables. First observe that if the primal problem is convex, then $\mathcal{L}(x, \lambda^*, \nu^*)$ is also convex in x. Global minimizers of this function can then be determined by unconstrained minimization techniques. For instance, if $\mathcal{L}(x, \lambda^*, \nu^*)$ is differentiable, a necessary condition for x to be a global minimizer is determined by the zero-gradient condition

$$\nabla_x \mathcal{L}(x, \lambda^*, \nu^*) = 0,$$

that is,

$$\nabla_x f_0(x) + \sum_{i=1}^{m} \lambda_i^* \nabla_x f_i(x) + \sum_{i=1}^{q} \nu_i^* \nabla_x h_i(x) = 0. \qquad (8.45)$$

However, $\mathcal{L}(x, \lambda^*, \nu^*)$ may have multiple global minimizers, and it is *not* guaranteed that *any* global minimizer of $\mathcal{L}$ is a primal-optimal solution (what is guaranteed is that the primal-optimal solution x^* is among the global minimizers of $\mathcal{L}$). Therefore, in general, care should be exerted when recovering the primal-optimal solution from the dual. A particular case arises when $\mathcal{L}(x, \lambda^*, \nu^*)$ has an *unique* minimizer (which happens, for instance, when $\mathcal{L}(x, \lambda^*, \nu^*)$ is strictly convex). In this case the unique minimizer x^* of $\mathcal{L}$ is either primal feasible, and hence it is the primal-optimal solution, or it is not primal feasible, and then we can conclude that the no primal-optimal solution exists. We summarize this fact in the following proposition.

Proposition 8.8 (primal-optimal from dual-optimal solution)
Assume strong duality holds for problem (8.34), assume the dual attains a dual-optimal solution (λ^*, ν^*), and assume that $\mathcal{L}(x, \lambda^*, \nu^*)$ has a unique

minimizer x^*. If x^* is feasible for (8.34), then x^* is primal optimal, otherwise the primal problem does not admit an optimal solution.

Example 8.22 *(Dual of minimum-norm solution of linear equations)* Consider the problem of determining the minimum Euclidean norm solution of a system of underdetermined linear equations:

$$p^* = \min_x \quad \|x\|_2^2$$
$$\text{s.t.:} \quad Ax = b,$$

where $A \in \mathbb{R}^{m,n}$, $n \geq m$, is full rank. The Lagrangian for this problem is

$$\mathcal{L}(x,v) = x^\top x + v^\top (Ax - b).$$

This is a (strictly) convex quadratic function, which is minimized when

$$\nabla_x \mathcal{L}(x,v) = 2x + A^\top v = 0,$$

that is for the (unique) minimizer $x^*(v) = -(1/2)A^\top v$, hence

$$g(v) = \inf_x \mathcal{L}(x,v) = \mathcal{L}(x^*(v),v) = -\frac{1}{4}v^\top(AA^\top)v - v^\top b.$$

The dual problem is thus

$$d^* = \max_v \quad -\frac{1}{4}v^\top(AA^\top)v - v^\top b.$$

Since $AA^\top \succ 0$, this is a (strictly) concave quadratic maximization problem, and the maximum is attained at the unique point

$$v^* = -2(AA^\top)^{-1}b,$$

whence

$$d^* = b^\top(AA^\top)^{-1}b.$$

Since the primal problem has only linear equality constraints and it is feasible (due to the assumption that A is full rank), Slater's conditions guarantee that strong duality holds, hence $p^* = d^*$. Since further the Lagrangian is strictly convex in x, we can recover the primal-optimal solution from the dual-optimal solution, as the unique minimizer of $\mathcal{L}(x,v^*)$, which results in

$$x^* = x^*(v^*) = A^\top(AA^\top)^{-1}b.$$

8.5.5 Karush–Kuhn–Tucker optimality conditions

For an optimization problem with differentiable objective and constraint functions, for which strong duality holds, we can derive a set of necessary conditions for optimality, known as the Karush–Kuhn–Tucker (KKT) conditions.

Consider problem (8.34)–(8.36) and its dual (8.41), let strong duality hold, and let x^*, (λ^*, v^*) be a primal- and a dual-optimal solution, respectively. Then, it holds that

1. (primal feasibility) $f_i(x^*) \le 0$, $i = 1, \ldots, m$, and $h_i(x^*) = 0$, $i = 1, \ldots, q$;

2. (dual feasibility) $\lambda^* \ge 0$;

3. (complementary slackness) $\lambda_i^* f_i(x^*) = 0$, $i = 1, \ldots, m$;

4. (Lagrangian stationarity) $\nabla_x \mathcal{L}(x, \lambda^*, \nu^*)_{x=x^*} = 0$.

The first two items are obvious, since x^*, λ^* are primal-dual optimal, hence they must be primal-dual feasible. The last two items follow from the reasoning previously exposed from Eq. (8.44) to Eq. (8.45).

We next show that, for a convex primal problem, the above conditions are actually also sufficient for optimality. Indeed, the first two conditions imply that x^* is primal feasible, and λ^* is dual feasible. Further, since $\mathcal{L}(x, \lambda^*, \nu^*)$ is convex in x, the fourth condition states that x^* is a global minimizer of $\mathcal{L}(x, \lambda^*, \nu^*)$, hence

$$
\begin{aligned}
g(\lambda^*, \nu^*) &= \inf_{x \in \mathcal{D}} \mathcal{L}(x, \lambda^*, \nu^*) = \mathcal{L}(x^*, \lambda^*, \nu^*) \\
&= f_0(x^*) + \sum_{i=1}^m \lambda_i^* f_i(x^*) + \sum_{i=1}^q \nu_i^* h_i(x^*) \\
&= f_0(x^*),
\end{aligned}
$$

where the last equality follows from the fact that $h_i(x^*) = 0$ (from primal feasibility), and that $\lambda_i^* f_i(x^*) = 0$ (from complementary slackness). The above proves that the primal-dual feasible pair x^*, (λ^*, ν^*) is optimal, since the corresponding duality gap is zero.

8.5.6 Sensitivity of the optimal value

The optimal dual variables have an interesting interpretation as *sensitivities* of the optimal value p^* to perturbations in the constraints. To make things more precise, consider the following perturbed version of the primal problem (8.34),

$$
\begin{aligned}
p^*(u, v) = \min_{x \in \mathbb{R}^n} \quad & f_0(x) \\
\text{subject to:} \quad & f_i(x) \le u_i, \quad i = 1, \ldots, m, \\
& h_i(x) = v_i, \quad i = 1, \ldots, q,
\end{aligned}
$$

where the right-hand sides of the inequality and equality constraints have been modified from zero to u, v, respectively. For $u = 0$, $v = 0$ we recover the original, unperturbed, problem, i.e., $p^*(0, 0) = p^*$. A positive u_i has the interpretation of relaxing the i-th inequality constraint, while negative u_i means that we tightened the i-th inequality constraint. We are interested in determining how perturbations in the constraints impact the optimal value of the problem. We here state some results without formal proof. First, it can be proved that, if the primal problem (8.34) is convex, then so is the function $p^*(u,v)$; see Section 8.5.8. Further, under the hypothesis that strong duality holds and that the dual optimum is attained, it holds that

$$p^*(u,v) \geq p^*(0,0) - \begin{bmatrix} \lambda^* \\ v^* \end{bmatrix}^\top \begin{bmatrix} u \\ v \end{bmatrix},$$

a condition that can also be expressed by saying that $-(\lambda^*, \mu^*)$ provides a subgradient for $p^*(u,v)$ at $u = 0$, $v = 0$. When the function $p^*(u,v)$ is differentiable (which happens under some technical conditions, e.g., when the Lagrangian is smooth and strictly convex), the optimal dual variables actually provide the derivatives of $-p^*(u,v)$ at $u = 0$, $v = 0$, that is

$$\lambda_i^* = -\frac{\partial p^*(0,0)}{\partial u_i}, \quad v_i^* = -\frac{\partial p^*(0,0)}{\partial v_i}, \qquad (8.46)$$

hence λ^*, μ^* have the interpretation of sensitivities of the optimal value p^* to perturbation of the constraints.

To better understand in practice the importance of this interpretation, consider a problem with inequality constraints only, assuming the conditions for (8.46) to hold are satisfied. Inequality constraints often have the interpretation of resource constraints, thus a perturbation $u_i > 0$ means that we relax the i-th constraint, i.e., that we allow for more resources. On the contrary, a perturbation $u_i < 0$ means that we tighten the constraint, i.e., that we cut the resources. Now, if $\lambda_i^* = 0$, then it means that, up to first-order approximation, the problem objective is insensitive to the perturbation (indeed, at least in non-degenerate problems, $\lambda_i^* = 0$ implies that $f_i(x^*) < 0$, that is the constraint is slack, which means that it is not resource-critical). Conversely, if $\lambda_i^* > 0$, then complementary slackness implies that $f_i(x^*) = 0$, which means that the constraint is active, i.e., resource critical. In such a situation, changing the resource level (i.e., perturbing the right-hand side of the constraint to a small level u_i) will have an impact in the optimal value, and this perturbation is given (up to first order) by $-\lambda_i^* u_i$.

8.5.7 Max-min inequality and saddle points

From (8.37) and (8.41) it follows that the dual optimal objective can be expressed as the following max-min value:

$$d^* = \max_{\lambda \geq 0, v} \min_{x \in \mathcal{D}} \mathcal{L}(x, \lambda, v).$$

Also, notice that

$$\max_{\lambda \geq 0, \nu} \mathcal{L}(x, \lambda, \nu) = \begin{cases} f_0(x) & \text{if } f_i(x) \leq 0, \, h_i(x) = 0, \forall i, \\ +\infty & \text{otherwise}, \end{cases}$$

which means that the primal optimal value p^* can be expressed as the min-max value $p^* = \min_{x \in \mathcal{D}} \max_{\lambda \geq 0, \nu} \mathcal{L}(x, \lambda, \nu)$. The weak duality relation $d^* \leq p^*$ is then equivalent to the following min-max/max-min inequality

$$\max_{\lambda \geq 0, \nu} \min_{x \in \mathcal{D}} \mathcal{L}(x, \lambda, \nu) \leq \min_{x \in \mathcal{D}} \max_{\lambda \geq 0, \nu} \mathcal{L}(x, \lambda, \nu), \qquad (8.47)$$

while the strong duality condition $d^* = p^*$ is equivalent to the fact that the above min-max and max-min values are the same:

$$d^* = p^* \iff \max_{\lambda \geq 0, \nu} \min_{x \in \mathcal{D}} \mathcal{L}(x, \lambda, \nu) = \min_{x \in \mathcal{D}} \max_{\lambda \geq 0, \nu} \mathcal{L}(x, \lambda, \nu).$$

This means that, if strong duality holds, the max and min operations can be *exchanged* without changing the value of the problem. Also, if p^* and d^* are attained, we say that $\mathcal{L}(x, \lambda, \nu)$ has a *saddle point* at the primal and dual optimal values (x^*, λ^*, ν^*), see Figure 8.23.

Equality (8.47) is actually a special case of a more general inequality, known as the max-min inequality, which can be stated as follows: for any function $\varphi : \mathbb{R}^n \times \mathbb{R}^m \to \mathbb{R}$ and for any nonempty subsets $X \subseteq \mathbb{R}^n, Y \subseteq \mathbb{R}^m$ it holds that

$$\sup_{y \in Y} \inf_{x \in X} \varphi(x, y) \leq \inf_{x \in X} \sup_{y \in Y} \varphi(x, y). \qquad (8.48)$$

To prove this relation, notice that for any $\bar{y} \in Y$ it holds that

$$\inf_{x \in X} \varphi(x, \bar{y}) \leq \inf_{x \in X} \sup_{y \in Y} \varphi(x, y),$$

hence (8.48) follows by taking the sup over $\bar{y} \in Y$ of the left-hand side.

It is interesting to study under which conditions equality holds in equation (8.48), hence the max and min operations can be safely exchanged. The following theorem, adapted from an original result due to M. Sion, provides a sufficient condition for max-min/min-max equality to hold.

Theorem 8.8 (Minimax theorem) *Let $X \subseteq \mathbb{R}^n$ be convex and compact, and let $Y \subseteq \mathbb{R}^m$ be convex. Let $\varphi : X \times Y \to \mathbb{R}$ be a function such that for every $y \in Y$, $\varphi(\cdot, y)$ is convex and continuous over X, and for every $x \in X$, $\varphi(x, \cdot)$ is concave and continuous over Y. Then equality holds in (8.48), i.e.,*

$$\sup_{y \in Y} \min_{x \in X} \varphi(x, y) = \min_{x \in X} \sup_{y \in Y} \varphi(x, y).$$

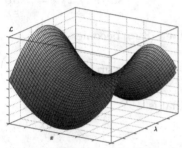

Figure 8.23 A function $\mathcal{L}$ with a saddle point.

We observe that the theorem can be easily reformulated for the case when it is the set Y, instead of X, that is compact. To see this fact it suffices to consider the function $\tilde{\varphi}(y, x) = -\varphi(x, y)$ and apply Theorem 8.8. Indeed, if φ satisfies the hypotheses of the theorem, then $\tilde{\varphi}(y, \cdot)$ is concave over X for every $y \in Y$, and $\tilde{\varphi}(\cdot, x)$ is convex over Y for every $x \in X$. Therefore, if X, Y are convex and Y is compact, applying Theorem 8.8 we obtain

$$\sup_{x \in X} \min_{y \in Y} \tilde{\varphi}(y, x) = \min_{y \in Y} \sup_{x \in X} \tilde{\varphi}(y, x). \tag{8.49}$$

Since

$$\min_{y \in Y} \tilde{\varphi}(y, x) = \min_{y \in Y} -\varphi(x, y) = -\max_{y \in Y} \varphi(x, y),$$

$$\sup_{x \in X} \tilde{\varphi}(y, x) = \sup_{x \in X} -\varphi(x, y) = -\inf_{x \in X} \varphi(x, y),$$

then (8.49) becomes

$$\sup_{x \in X} \left(-\max_{y \in Y} \varphi(x, y) \right) = \min_{y \in Y} \left(-\inf_{x \in X} \varphi(x, y) \right),$$

that is

$$\inf_{x \in X} \max_{y \in Y} \varphi(x, y) = \max_{y \in Y} \inf_{x \in X} \varphi(x, y),$$

which constitutes an alternative statement of Theorem 8.8, to be used when Y, instead of X, is compact.

Example 8.23 (*Square-root LASSO*) As an illustrative example, we consider a version of least squares that is sometimes referred to as the "square-root LASSO"

$$p^* = \min_x \|Ax - b\|_2 + \lambda \|x\|_1.$$

Here, $A \in \mathbb{R}^{m,n}$, $b \in \mathbb{R}^m$ and the parameter $\lambda > 0$ are given. The above problem is a useful variant of least squares, in which the ℓ_1-norm penalty encourages sparsity (number of nonzeros) in the solution (we cover this topic in more detail in Section 9.6.2). We can express the problem as

$$p^* = \min_x \max_{(u,v) \in Y} u^\top (Ax - b) + v^\top x,$$

where $Y = \{(u, v) : \|u\|_2 \le 1, \|v\|_\infty \le \lambda\}$ is compact. Theorem 8.8 applies here, and exchanging the min and max leads to

$$p^* = \max_{(u,v) \in Y} \min_x u^\top (Ax - b) + v^\top x.$$

Note that the infimum over x of the term $(A^\top u + v)^\top x$ would be $-\infty$, unless the coefficient $A^\top u + v$ is zero, hence

$$p^* = \max_{u,v} -b^\top u \ : \ v + A^\top u = 0, \ \|u\|_2 \le 1, \ \|v\|_\infty \le \lambda$$

$$= \max_u -b^\top u \ : \ \|u\|_2 \le 1, \ \|A^\top u\|_\infty \le \lambda.$$

The above problem is a kind of dual to the original problem, which is useful for its analysis, and algorithm design. In particular, the above dual can be used to eliminate variables from the original problem, based on a simple evaluation of the norms of the columns $a_1, \ldots, a_n$ of A. Let us write the problem as

$$\max_u \; -b^\top u \; : \; \|u\|_2 \leq 1, \; |a_i^\top u| \leq \lambda, \; i = 1, \ldots, n.$$

Now, if $\|a_i\|_2 < \lambda$ for some i, with a_i the i-th column of A, then

$$|a_i^\top u| < \lambda, \quad \forall u : \|u\|_2 \leq 1.$$

This means that the constraint $|a_i^\top u| \leq \lambda$ is not active, and that we can safely eliminate it from the dual.[18] In other words, the optimal value p^* remains the same if we remove the i-th column of A, or equivalently, set $x_i = 0$. The simple test $\|a_i\|_2 < \lambda$ thus allows us to predict that $x_i = 0$ at optimum, without solving for p^*.

[18] See the section on elimination of constraints 8.3.4.6.

8.5.8 Subgradients of the primal value function

Consider a primal convex problem of the form

$$p^*(u) = \min_{x \in \mathcal{D}} \; f_0(x) \tag{8.50}$$
$$\text{subject to:} \quad f_i(x) \leq u_i, \quad i = 1, \ldots, m,$$

where u is a vector of parameters. We consider the case of inequality constraints only, leaving the extension to the case when also equality constraints are present as an easy exercise to the reader. We next show that $p^*(u)$ is a convex function of u, and that subgradients for this function can be obtained from the Lagrange multipliers associated with (8.50). The Lagrangian of (8.50) is

$$\mathcal{L}(x, \lambda, u) = f_0(x) + \sum_{i=1}^m \lambda_i f_i(x) - \lambda^\top u,$$

with dual function $g(\lambda, u) = \min_{x \in \mathcal{D}} \mathcal{L}(x, \lambda, u)$, and dual problem

$$d^*(u) = \max_{\lambda \geq 0} g(\lambda, u) = \max_{\lambda \geq 0} \min_{x \in \mathcal{D}} \mathcal{L}(x, \lambda, u). \tag{8.51}$$

As already observed in the previous section, we have that

$$\max_{\lambda \geq 0} \mathcal{L}(x, \lambda, u) \;=\; \max_{\lambda \geq 0} f_0(x) + \sum_{i=1}^m \lambda_i(f_i(x) - u_i)$$

$$= \begin{cases} f_0(x) & \text{if } f_i(x) \leq u_i, \; i = 1, \ldots, m, \\ +\infty & \text{otherwise,} \end{cases}$$

therefore, the primal problem can be written equivalently as

$$p^*(u) = \min_{x \in \mathcal{D}} \max_{\lambda \geq 0} \mathcal{L}(x, \lambda, u).$$

Notice that $\mathcal{L}(x, \lambda, u)$ is convex in (x, u), for each given λ ($\mathcal{L}$ is actually convex in x and linear in u). Therefore, by the pointwise max rule (see Section 8.2.2.4), the function $\max_{\lambda \geq 0} \mathcal{L}(x, \lambda, u)$ is still convex in (x, u). Then, applying the property of partial minimization (Section 8.2.2.5) allows us to conclude that the optimal primal value function $p^*(u)$ is convex in u. Considering now the identity

$$\sum_{i=1}^{m} \lambda_i f_i(x) = \min_{v \in \mathbb{R}^m} \sum_{i=1}^{m} \lambda_i v_i, \text{ s.t.: } f_i(x) \leq v_i, \ i = 1, \ldots, m,$$

we have that

$$
\begin{aligned}
g(\lambda, u) &= \min_{x \in \mathcal{D}} \ f_0(x) + \sum_{i=1}^{m} \lambda_i f_i(x) - \lambda^\top u \\
&= \min_{v \in \mathbb{R}^m} \min_{x \in \mathcal{D}} \ f_0(x) + \lambda^\top (v - u), \text{ s.t.: } f_i(x) \leq v_i, \ i = 1, \ldots, m \\
&= \min_{v \in \mathbb{R}^m} p^*(v) + \lambda^\top (v - u).
\end{aligned}
$$

Now, suppose that $p^*(u)$ is finite and strong duality holds, i.e. $d^*(u) = p^*(u)$, and let λ_u be the optimal Lagrange multiplier achieving the optimum in (8.51). Then we have

$$
\begin{aligned}
p^*(u) &= d^*(u) = g(\lambda_u, u) = \min_{v \in \mathbb{R}^m} p^*(v) + \lambda_u^\top (v - u) \\
&\leq p^*(v) + \lambda_u^\top (v - u), \quad \forall v.
\end{aligned}
$$

This latter inequality proves that $-\lambda_u$ is a subgradient of p^* at u, that is

$$p^*(v) \geq p^*(u) - \lambda_u(v - u), \quad \forall v.$$

This property is exploited, for instance, in the primal decomposition methods described in Section 12.6.1.2.

8.5.9 Subgradients of the dual function

Consider a primal problem of the form

$$
\begin{aligned}
p^* = \min_{x \in \mathcal{D}} \ & f_0(x) \\
\text{subject to: } \ & f_i(x) \leq 0, \quad i = 1, \ldots, m,
\end{aligned}
$$

with is dual $d^* = \max_{\lambda \geq 0} g(\lambda)$, where $g(\lambda)$ is the dual function

$$g(\lambda) = \min_{x \in \mathcal{D}} \ \mathcal{L}(x, \lambda),$$

with $\mathcal{L}(x, \lambda) = f_0(x) + \sum_{i=1}^{m} \lambda_i f_i(x)$. We again consider the case of inequality constraints only, the extension to the case of equality constraints being straightforward.

We already know that g is a concave function (irrespective of convexity of $f_0, f_i, i = 1, \ldots, m$); we next show that we can readily find a subgradient for this function (or, to be more precise, for the convex function $-g$). For a given λ, let

$$x_\lambda = \arg\min_{x \in \mathcal{D}} \mathcal{L}(x, \lambda),$$

so that $g(\lambda) = \mathcal{L}(x_\lambda, \lambda)$, if such minimizer exists. For all $z \in \mathbb{R}^m$, it holds that

$$
\begin{aligned}
g(z) &= \min_{x \in \mathcal{D}} \mathcal{L}(x, z) \\
&\leq \mathcal{L}(x_\lambda, z) = f_0(x_\lambda) + \sum_{i=1}^m z_i f_i(x_\lambda) \\
&= f_0(x_\lambda) + \sum_{i=1}^m \lambda_i f_i(x_\lambda) + \sum_{i=1}^m (z_i - \lambda_i) f_i(x_\lambda) \\
&= g(\lambda) + \sum_{i=1}^m (z_i - \lambda_i) f_i(x_\lambda).
\end{aligned}
$$

Letting

$$F(x) = [f_1(x)\, f_2(x)\, \cdots\, f_m(x)]^\top,$$

the previous inequality becomes

$$g(z) \leq g(\lambda) + F(x_\lambda)^\top (z - \lambda), \quad \forall z,$$

which means that $F(x_\lambda)$ is a subgradient[19] for g at λ. Such a subgradient is thus obtained essentially at no additional cost, once we evaluate $g(\lambda)$ by minimizing $\mathcal{L}(x, \lambda)$ with respect to x. Summarizing, we have that

$$[f_1(x_\lambda)\, f_2(x_\lambda)\, \cdots\, f_m(x_\lambda)]^\top \in \partial g(\lambda),$$

for any x_λ that minimizes $\mathcal{L}(x, \lambda)$ over $x \in \mathcal{D}$. This property is exploited, for instance, in the dual decomposition methods described in Section 12.6.1.1.

Further, if $\mathcal{D}$ is assumed to be compact and nonempty, $f_0, f_i, i = 1, \ldots, m$ are continuous, and the minimizer x_λ is unique for every λ, then one can prove that $g(\lambda)$ is continuously differentiable, hence the above subgradient is unique and actually provides the gradient of g at λ:

$$\nabla g(\lambda) = [f_1(x_\lambda)\, f_2(x_\lambda)\, \cdots\, f_m(x_\lambda)]^\top, \quad \forall \lambda.$$

[19] There is a slight abuse of terminology here, since subgradients are defined for convex functions, while for concave functions we should more properly talk about supergradients, intending that a vector h is a supergradient for a concave function f if $-h$ is a subgradient of $-f$.

8.6 Exercises

Exercise 8.1 (Quadratic inequalities) Consider the set defined by the following inequalities:

$$(x_1 \geq x_2 - 1 \text{ and } x_2 \geq 0) \text{ or } (x_1 \leq x_2 - 1 \text{ and } x_2 \leq 0).$$

1. Draw the set. Is it convex?

2. Show that it can be described as a single quadratic inequality of the form $q(x) = x^\top A x + 2b^\top x + c \leq 0$, for a matrix $A = A^\top \in \mathbb{R}^{2,2}$, $b \in \mathbb{R}^2$ and $c \in \mathbb{R}$ which you will determine.

3. What is the convex hull of this set?

Exercise 8.2 (Closed functions and sets) Show that the indicator function $I_{\mathcal{X}}$ of a convex set $\mathcal{X}$ is convex. Show that this function is closed whenever $\mathcal{X}$ is a closed set.

Exercise 8.3 (Convexity of functions)

1. For x, y both positive scalars, show that

$$ye^{x/y} = \max_{\alpha > 0} \alpha(x + y) - y\alpha \cdot \ln \alpha.$$

Use the above result to prove that the function f defined as

$$f(x, y) = \begin{cases} ye^{x/y} & \text{if } x > 0, \ y > 0, \\ +\infty & \text{otherwise,} \end{cases}$$

is convex.

2. Show that for $r \geq 1$, the function $f_r : \mathbb{R}^m_+ \to \mathbb{R}$, with values

$$f_r(v) = \left(\sum_{j=1}^m v_j^{1/r} \right)^r$$

is concave. *Hint:* show that the Hessian of $-f$ takes the form $\kappa \text{diag}(y) - zz^\top$ for appropriate vectors $y \geq 0$, $z \geq 0$, and scalar $\kappa \geq 0$, and use Schur complements[20] to prove that the Hessian is positive semidefinite.

[20] See Section 4.4.7.

Exercise 8.4 (Some simple optimization problems) Solve the following optimization problems. Make sure to determine an optimal primal solution.

1. Show that, for given scalars α, β,

$$f(\alpha, \beta) \doteq \min_{d > 0} \alpha d + \frac{\beta^2}{d} = \begin{cases} -\infty & \text{if } \alpha \leq 0, \\ 2|\beta|\sqrt{\alpha} & \text{otherwise.} \end{cases}$$

2. Show that for an arbitrary vector $z \in \mathbb{R}^m$,

$$\|z\|_1 = \min_{d > 0} \frac{1}{2} \sum_{i=1}^m \left(d_i + \frac{z_i^2}{d_i} \right). \tag{8.52}$$

3. Show that for an arbitrary vector $z \in \mathbb{R}^m$, we have

$$\|z\|_1^2 = \min_d \sum_{i=1}^m \frac{z_i^2}{d_i} \; : \; d > 0, \; \sum_{i=1}^m d_i = 1.$$

Exercise 8.5 (Minimizing a sum of logarithms) Consider the following problem:

$$p^* = \max_{x \in \mathbb{R}^n} \sum_{i=1}^n \alpha_i \ln x_i$$
$$\text{s.t.:} \quad x \geq 0, \quad \mathbf{1}^\top x = c,$$

where $c > 0$ and $\alpha_i > 0$, $i = 1, \ldots, n$. Problems of this form arise, for instance, in maximum-likelihood estimation of the transition probabilities of a discrete-time Markov chain. Determine in closed-form a minimizer, and show that the optimal objective value of this problem is

$$p^* = \alpha \ln(c/\alpha) + \sum_{i=1}^n \alpha_i \ln \alpha_i,$$

where $\alpha \doteq \sum_{i=1}^n \alpha_i$.

Exercise 8.6 (Monotonicity and locality) Consider the optimization problems (no assumption of convexity here)

$$p_1^* \doteq \min_{x \in \mathcal{X}_1} f_0(x),$$
$$p_2^* \doteq \min_{x \in \mathcal{X}_2} f_0(x),$$
$$p_{13}^* \doteq \min_{x \in \mathcal{X}_1 \cap \mathcal{X}_3} f_0(x),$$
$$p_{23}^* \doteq \min_{x \in \mathcal{X}_2 \cap \mathcal{X}_3} f_0(x),$$

where $\mathcal{X}_1 \subseteq \mathcal{X}_2$.

1. Prove that $p_1^* \geq p_2^*$ (i.e., enlarging the feasible set cannot worsen the optimal objective).

2. Prove that, if $p_1^* = p_2^*$, then it holds that

$$p_{13}^* = p_1^* \quad \Rightarrow \quad p_{23}^* = p_2^*.$$

3. Assume that all problems above attain unique optimal solutions. Prove that, under such a hypothesis, if $p_1^* = p_2^*$, then it holds that

$$p_{23}^* = p_2^* \quad \Rightarrow \quad p_{13}^* = p_1^*.$$

Exercise 8.7 (Some matrix norms) Let $X = [x_1, \ldots, x_m] \in \mathbb{R}^{n,m}$, and $p \in [1, +\infty]$. We consider the problem

$$\phi_p(X) \doteq \max_u \ \|X^\top u\|_p \ : \ u^\top u = 1.$$

If the data is centered, that is, $X\mathbf{1} = 0$, the above amounts to finding a direction of largest "deviation" from the origin, where deviation is measured using the l_p-norm.

1. Is ϕ_p a (matrix) norm?

2. Solve the problem for $p = 2$. Find an optimal u.

3. Solve the problem for $p = \infty$. Find an optimal u.

4. Show that
$$\phi_p(X) = \max_v \ \|Xv\|_2 \ : \ \|v\|_q \le 1,$$

   where $1/p + 1/q = 1$ (hence, $\phi_p(X)$ depends only on $X^\top X$). *Hint:* you can use the fact that the norm dual to the l_p-norm is the l_q-norm and vice versa, in the sense that, for any scalars $p \ge 1, q \ge 1$ with $1/p + 1/q = 1$, we have

$$\max_{v: \|v\|_q \le 1} \ u^\top v = \|u\|_p.$$

Exercise 8.8 (Norms of matrices with non-negative entries) Let $X \in \mathbb{R}_+^{n,m}$ be a matrix with non-negative entries, and $p, r \in [1, +\infty]$, with $p \ge r$. We consider the problem

$$\phi_{p,r}(X) = \max_v \ \|Xv\|_r \ : \ \|v\|_p \le 1.$$

1. Show that the function $f_X : \mathbb{R}_+^m \to \mathbb{R}$, with values

$$f_X(u) = \sum_{i=1}^n \left(\sum_{j=1}^m X_{ij} u_j^{1/p} \right)^r$$

   is concave when $p \ge r$.

2. Use the previous result to formulate an efficiently solvable convex problem that has $\phi_{p,r}(X)^r$ as optimal value.

Exercise 8.9 (Magnitude least squares) For given n-vectors $a_1, \ldots, a_m$, we consider the problem

$$p^* = \min_x \ \sum_{i=1}^m \left(|a_i^\top x| - 1 \right)^2.$$

1. Is the problem convex? If so, can you formulate it as an ordinary least-squares problem? An LP? A QP? A QCQP? An SOCP? None of the above? Justify your answers precisely.

2. Show that the optimal value p^* depends only on the matrix $K = A^\top A$, where $A = [a_1, \ldots, a_m]$ is the $n \times m$ matrix of data points (that is, if two different matrices A_1, A_2 satisfy $A_1^\top A_1 = A_2^\top A_2$, then the corresponding optimal values are the same).

Exercise 8.10 (Eigenvalues and optimization) Given an $n \times n$ symmetric matrix Q, define

$$w_1 = \arg \min_{\|x\|_2=1} x^\top Q x, \quad \text{and } \mu_1 = \min_{\|x\|_2=1} x^\top Q x,$$

and for $k = 1, 2, \ldots, n-1$:

$$w_{k+1} = \arg \min_{\|x\|_2=1} x^\top Q x \quad \text{such that } w_i^\top x = 0, i = 1, \ldots, k,$$

$$\mu_{k+1} = \min_{\|x\|_2=1} x^\top Q x \quad \text{such that } w_i^\top x = 0, i = 1, \ldots, k.$$

Using optimization principles and theory:

1. show that $\mu_1 \leq \mu_2 \leq \cdots \leq \mu_n$;

2. show that the vectors $w_1, \ldots, w_n$ are linearly independent, and form an orthonormal basis of $\mathbb{R}^n$;

3. show how μ_1 can be interpreted as a Lagrange multiplier, and that μ_1 is the smallest eigenvalue of Q;

4. show how $\mu_2, \ldots, \mu_n$ can also be interpreted as Lagrange multipliers. *Hint:* show that μ_{k+1} is the smallest eigenvalue of $W_k^\top Q W_k$, where $W_k = [w_{k+1}, \ldots, w_n]$.

Exercise 8.11 (Block norm penalty) In this exercise we partition vectors $x \in \mathbb{R}^n$ into p blocks $x = (x_1, \ldots, x_p)$, with $x_i \in \mathbb{R}^{n_i}$, $n_1 + \cdots + n_p = n$. Define the function $\rho : \mathbb{R}^n \to \mathbb{R}$ with values

$$\rho(x) = \sum_{i=1}^{p} \|x_i\|_2.$$

1. Prove that ρ is a norm.

2. Find a simple expression for the "dual norm," $\rho_*(x) \doteq \sup_{z: \rho(z)=1} z^\top x$.

3. What is the dual of the dual norm?

4. For a scalar $\lambda \geq 0$, matrix $A \in \mathbb{R}^{m,n}$ and vector $y \in \mathbb{R}^m$, we consider the optimization problem

$$p^*(\lambda) \doteq \min_x \|Ax - y\|_2 + \lambda \rho(x).$$

Explain the practical effect of a high value of λ on the solution.

5. For the problem above, show that $\lambda > \sigma_{\max}(A_i)$ implies that we can set $x_i = 0$ at optimum. Here, $A_i \in \mathbb{R}^{m,n_i}$ corresponds to the i-th block of columns in A, and $\sigma_{\max}$ refers to the largest singular value.

9
Linear, quadratic, and geometric models

IN THIS CHAPTER WE STUDY THREE CLASSES of optimization models. The first two classes (linear and quadratic programs) are characterized by the fact that the functions involved in the problem definition are either linear or quadratic. The third class of problems (geometric programs) can be viewed as an extension of linear programming problems, obtained under a suitable logarithmic transformation.

A quadratic function in a vector of variables $x = [x_1\, x_2\, \cdots\, x_n]$ is a polynomial in x where the maximum degree of the monomials is equal to two. Such a degree-two polynomial can be written generically as follows

$$f_0(x) = \frac{1}{2}x^\top H x + c^\top x + d \quad \text{(a quadratic function)}, \qquad (9.1)$$

where $d \in \mathbb{R}$ is a scalar constant term, $c \in \mathbb{R}^n$ is a vector containing the coefficients of the terms of degree one, and $H \in \mathbb{R}^{n,n}$ is a symmetric matrix that contains the coefficients of the monomials of degree two. A linear function is of course a special case of a quadratic function, obtained considering $H = 0$:

$$f_0(x) = c^\top x + d \quad \text{(a linear function)}.$$

Linear and quadratic models treated in this chapter take the form

$$\begin{aligned} \text{minimize} \quad & f_0(x) & (9.2)\\ \text{subject to:} \quad & A_{\text{eq}}x = b_{\text{eq}}, \\ & f_i(x) \le 0, \quad i = 1, \dots, m, \end{aligned}$$

where $f_0, \dots, f_m$ are either quadratic or linear functions. More precisely, we shall be mainly concerned with the case when these functions are *convex*, which happens if and only if the Hessian matrices of f_i, $i = 0, 1, \ldots, m$, are positive semidefinite, see Example 8.5.

9.1   Unconstrained minimization of quadratic functions

Let us start our discussion by examining the unconstrained case, that is problem (9.2) when no constraints are present, thus x is unrestricted: $x \in \mathbb{R}^n$. Consider first the linear case, $f_0(x) = c^\top x + d$:

$$p^* = \min_{x \in \mathbb{R}^n} \ c^\top x + d.$$

It is an intuitive fact that $p^* = -\infty$ (i.e., the objective is unbounded below) whenever $c \neq 0$, and $p^* = d$, otherwise. Indeed, for $c \neq 0$ one may take $x = -\alpha c$, for any $\alpha > 0$ large at will, and drive f_0 to $-\infty$. On the contrary, for $c = 0$ the function is actually constant and equal to d. We have therefore, for a linear function,

$$p^* = \begin{cases} d & \text{if } c = 0, \\ -\infty & \text{otherwise.} \end{cases}$$

Consider next the general quadratic case

$$p^* = \min_{x \in \mathbb{R}^n} \ \frac{1}{2} x^\top H x + c^\top x + d.$$

Several situations are possible, depending on the sign of the eigenvalues of H (which we recall are all real, since H is symmetric).

(a) H has a negative eigenvalue $\lambda < 0$. Then let u be the corresponding eigenvector and take $x = \alpha u$, with $\alpha \neq 0$. Since $Hu = \lambda u$, we have

$$\begin{aligned} f_0(x) &= \frac{1}{2} x^\top H x + c^\top x + d = \frac{1}{2} \alpha^2 u^\top H u + \alpha c^\top u + d \\ &= \frac{1}{2} \lambda \alpha^2 \|u\|^2 + \alpha (c^\top u) + d, \end{aligned}$$

which tends to $-\infty$, for $\alpha \to \infty$. Hence, in this case f_0 is unbounded below, i.e., $p^* = -\infty$.

(b) All eigenvalues of H are non-negative: $\lambda_i \geq 0$, $i = 1, \ldots, n$. In this case f_0 is convex. We know from Eq. (8.30) that the minima are characterized by the condition that the gradient of the function is zero, that is

$$\nabla f_0(x) = Hx + c = 0. \tag{9.3}$$

The minimizer should thus satisfy the system of linear equations $Hx = -c$, and the following sub-cases are possible.

(b.1) If $c \notin \mathcal{R}(H)$, then there is no minimizer. Indeed, this case implies that H is singular, hence it has an eigenvalue $\lambda = 0$ and a corresponding eigenvector u. Thus, taking $x = \alpha u$ we have that, along direction u, $f_0(x) = \alpha(c^\top u) + d$. But $c^\top u \neq 0$ since $u \in \mathcal{N}(H)$ and $c \notin \mathcal{R}(H)$ must have a nonzero component along $\mathcal{N}(H)$, therefore $f_0(x)$ is unbounded below.

(b.2) If $c \in \mathcal{R}(H)$, then f_0 has a finite global minimum value $p^* = 0.5\,c^\top x^* + d$ (the fact that the minimum is global is a consequence of convexity), which is attained at any minimizer x^* such that $Hx^* = -c$. All such minimizers are of the form

$$x^* = -H^\dagger c + \zeta, \quad \zeta \in \mathcal{N}(H).$$

Since $\mathcal{N}(H) \perp \mathcal{R}(H^\top) \equiv \mathcal{R}(H)$, and since $c \in \mathcal{R}(H)$, it holds that $c^\top \zeta = 0$, for all $\zeta \in \mathcal{N}(H)$, hence

$$
\begin{aligned}
p^* &= -1/2\,c^\top H^\dagger c + d + 1/2\,c^\top \zeta \\
&= -\frac{1}{2} c^\top H^\dagger c + d.
\end{aligned}
$$

(c) All eigenvalues of H are positive: $\lambda_i > 0$, $i = 1, \ldots, n$. Then H is invertible, and there is a *unique* minimizer at

$$x^* = -H^{-1}c, \tag{9.4}$$

with corresponding optimal objective value

$$p^* = -\frac{1}{2} c^\top H^{-1} c + d.$$

Summarizing, the minimum value p^* of the quadratic function (9.1) is characterized as follows

$$p^* = \begin{cases} -\frac{1}{2} c^\top H^\dagger c + d & \text{if } H \succeq 0 \text{ and } c \in \mathcal{R}(H), \\ -\infty & \text{otherwise.} \end{cases}$$

Example 9.1 (*Least squares*) We have actually already encountered a special case of a quadratic minimization problem, in the context of the least-squares approximate solution of linear equations, see Section 6.3.1. Indeed, the LS problem amounts to minimizing $f_0(x) = \|Ax - y\|_2^2$, hence

$$f_0(x) = (Ax - y)^\top (Ax - y) = x^\top A^\top A x - 2y^\top A x + y^\top y,$$

which is a quadratic function in the standard form (9.1), with

$$H = 2(A^\top A), \quad c = -2A^\top y, \quad d = y^\top y.$$

Note that f_0 is always convex, since $A^\top A \succeq 0$. The solution is given by the first-order optimality condition in (9.3). Since $c \in \mathcal{R}(H)$, an LS solution satisfying these conditions always exists. Further, if A is full rank, then $A^\top A \succ 0$, then the solution is unique and it is given by the well-known formula

$$x^* = -H^{-1}c = (A^\top A)^{-1} A^\top y.$$

Example 9.2 (*Quadratic minimization under linear equality constraints*) The linear equality-constrained problem

$$\text{minimize} \quad f_0(x)$$
$$\text{subject to:} \quad Ax = b,$$

with $f_0(x) = \frac{1}{2} x^\top H x + c^\top x + d$, can be readily converted into unconstrained form by *eliminating* the equality constraints. To this end, we parameterize all x such that $Ax = b$ as

$$x = \bar{x} + Nz,$$

where $\bar{x}$ is one specific solution of $Ax = b$, N is a matrix containing by columns a basis for the nullspace of A, and z is a vector of free variables. Then we substitute x in f_0 and obtain a problem which is unconstrained in the variable z:

$$\min_z \varphi_0(z) = \frac{1}{2} z^\top \tilde{H} z + \tilde{c}^\top z + \tilde{d},$$

where

$$\tilde{H} = N^\top H N, \quad \tilde{c} = N^\top (c + H\bar{x}), \quad \tilde{d} = d + c^\top \bar{x} + \frac{1}{2} \bar{x}^\top H \bar{x}.$$

Another approach is also possible, when f_0 is convex, by exploiting the necessary and sufficient conditions in (8.31). That is, we seek $x \in \mathbb{R}^n$ and $\lambda \in \mathbb{R}^m$ such that

$$Ax = b, \quad \nabla f_0(x) = A^\top \lambda.$$

Since

$$\nabla f_0(x) = Hx + c,$$

we see that the optimal solution of the equality-constrained convex quadratic problem can be obtained by solving the following system of linear equations in the variables x, λ:

$$\begin{bmatrix} A & 0 \\ H & -A^\top \end{bmatrix} \begin{bmatrix} x \\ \lambda \end{bmatrix} = \begin{bmatrix} b \\ -c \end{bmatrix}. \tag{9.5}$$

9.2   Geometry of linear and convex quadratic inequalities

9.2.1   Linear inequalities and polyhedra

The set of points $x \in \mathbb{R}^n$ satisfying a linear inequality $a_i^\top x \leq b_i$ is a closed half-space; the vector a_i is normal to the boundary of the half-space and points outwards, see Section 2.4.4.3. A collection of m linear inequalities

$$a_i^\top x \le b_i, \quad i = 1, \ldots, m, \tag{9.6}$$

thus defines a region in $\mathbb{R}^m$ which is the intersection of m half-spaces and it is named a *polyhedron*. Notice that depending on the actual inequalities, this region can be unbounded, or bounded; in the latter case it is called a *polytope*, see Figure 9.1. It is often convenient to group several linear inequalities using matrix notation: we define

$$A = \begin{bmatrix} a_1^\top \\ a_2^\top \\ \vdots \\ a_m^\top \end{bmatrix}, \quad b = \begin{bmatrix} b_1 \\ b_2 \\ \vdots \\ b_m \end{bmatrix},$$

and then write inequalities (9.6) in the equivalent matrix form

$$Ax \le b,$$

where inequality is intended component-wise.

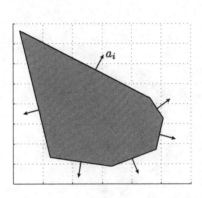 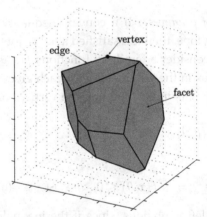

The polytope on the left of Figure 9.1 is described by six linear inequalities, each defining a half-space:

Figure 9.1 Example of polytopes in $\mathbb{R}^2$ (left) and in $\mathbb{R}^3$ (right).

$$\begin{bmatrix} 0.4873 & -0.8732 \\ 0.6072 & 0.7946 \\ 0.9880 & -0.1546 \\ -0.2142 & -0.9768 \\ -0.9871 & -0.1601 \\ 0.9124 & 0.4093 \end{bmatrix} x \le \begin{bmatrix} 1 \\ 1 \\ 1 \\ 1 \\ 1 \\ 1 \end{bmatrix}.$$

A polytope represented as the intersection of half-spaces is called an $\mathcal{H}$-polytope. Any $\mathcal{H}$-polytope also admits an equivalent representation as the convex hull of its *vertices*, in which case it is called a $\mathcal{V}$-polytope. For example, the two-dimensional polytope in Figure 9.1

can be represented as the convex hull of its vertices, given as columns in the following matrix:

$$V = \begin{bmatrix} 0.1562 & 0.9127 & 0.8086 & 1.0338 & -1.3895 & -0.8782 \\ -1.0580 & -0.6358 & 0.6406 & 0.1386 & 2.3203 & -0.8311 \end{bmatrix}.$$

The intersection of a polytope P with a supporting hyperplane H is called a *face* of P, which is again a convex polytope. Vertices are indeed the faces of dimension 0; the faces of dimension 1 are the *edges* of P, and the faces of dimension $\dim P - 1$ are called the *facets*, see Figure 9.1.

The intersection of a polytope with an affine set (such as the set of points satisfying linear equalities $A_{eq}x = b_{eq}$) is still a polytope. Indeed, the set

$$P = \{x : Ax \leq b, A_{eq}x = b_{eq}\}$$

can be expressed equivalently in "inequality-only" form as follows:

$$P = \{x : Ax \leq b, A_{eq}x \leq b_{eq}, -A_{eq}x \leq -b_{eq}\}.$$

Convex inequalities at vertices of polytopes. If a convex inequality is satisfied at the vertices of a polytope, then it is satisfied at all points of the polytope. More precisely, consider a family of functions $f(x, \theta)$: $\mathbb{R}^n \times \mathbb{R}^m \to \mathbb{R}$ which are convex in the parameter θ for each given x, and let $\theta^{(1)}, \ldots, \theta^{(p)}$ be given points in $\mathbb{R}^m$ defining the vertices of a $\mathcal{V}$-polytope

$$\Theta = \text{co}\{\theta^{(1)}, \ldots, \theta^{(p)}\}.$$

Then it holds that

$$f(x, \theta^{(i)}) \leq 0, \ i = 1, \ldots, p \quad \Leftrightarrow \quad f(x, \theta) \leq 0, \ \forall \theta \in \Theta. \tag{9.7}$$

The implication from right to left is obvious, since if the inequality is satisfied at all points of the polytope Θ, it is satisfied also at its vertices. The converse implication also follows easily from convexity: any $\theta \in \Theta$ is written as a convex combination of the vertices

$$\theta = \sum_{i=1}^{p} \alpha_i \theta^{(i)}, \quad \sum_i \alpha_i = 1, \ \alpha_i \geq 0, \ i = 1, \ldots, p,$$

then

$$f(x, \theta) \ = \ f\left(x, \sum_{i=1}^{p} \alpha_i \theta^{(i)}\right) \leq \sum_{i=1}^{p} \alpha_i f(x, \theta^{(i)}) \leq 0,$$

where the first inequality is the Jensen's inequality for convex functions. This simple property has many useful applications in engineering design problems, where inequalities typically have the meaning of specifications that need to be satisfied by the design. In such a context, if one needs to guarantee that some specification (expressed as a convex inequality) holds over a whole polytopic region for a parameter θ, then it suffices to insure that the specification is met at a finite number of points, namely at the vertices of the polytope.

Example 9.3 (*The probability simplex*) The probability simplex is the polytope defined as

$$P = \{x \in \mathbb{R}^n : x \geq 0, \sum_{i=1}^n x_i = 1\}.$$

The name suggests the fact that any x in the probability simplex has a natural interpretation of a discrete probability distribution, i.e. the x_is are non-negative and they sum up to one. The probability simplex in $\mathbb{R}^n$ has n vertices, which correspond to the standard orthonormal basis vectors for $\mathbb{R}^n$, that is

$$P = \text{co}\{e^{(1)}, \ldots, e^{(n)}\},$$

where $e^{(i)}$ is an n-vector whose entries are all zeros, except for a one in position i. The probability simplex in $\mathbb{R}^3$ is depicted in Figure 9.2.

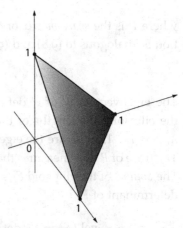

Figure 9.2 The probability simplex $\{x \in \mathbb{R}^3 : x \geq 0, 1^\top x = 1\}$.

Example 9.4 (*The ℓ_1-norm ball*) The ℓ_1-norm ball is the set $\{x \in \mathbb{R}^n : \|x\|_1 \leq 1\}$, that is the set where $\sum_{i=1}^n |x_i| \leq 1$. This set is indeed a polytope, since the previous inequality can be verified to be equivalent to a collection of 2^n linear inequalities. To see this fact, consider sign variables $s_i \in \{-1, 1\}$, $i = 1, \ldots, n$. Then

$$\sum_{i=1}^n |x_i| = \max_{s_i \in \{-1,1\}} \sum_{i=1}^n s_i x_i.$$

Therefore $\|x\|_1 \leq 1$ if and only if $\max_{s_i \in \{-1,1\}} \sum_{i=1}^n s_i x_i \leq 1$, which is in turn equivalent to requiring that

$$\sum_{i=1}^n s_i x_i \leq 1, \quad \text{for all } s_i \in \{-1, 1\}, \ i = 1, \ldots, m.$$

This is a collection of 2^n linear inequalities, corresponding to all the possible combinations of n sign variables. For example, for $n = 3$ the ℓ_1-norm ball is the octahedron depicted in Figure 9.3.

9.2.2 *Quadratic inequalities and ellipsoids*

Consider the zero-level set of a quadratic inequality, i.e., the set of $x \in \mathbb{R}^n$ such that

$$f_0(x) = \frac{1}{2} x^\top H x + c^\top x + d \leq 0. \tag{9.8}$$

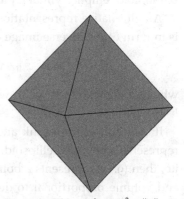

Figure 9.3 The set $\{x \in \mathbb{R}^3 : \|x\|_1 \leq 1\}$.

This set is convex if $H \succeq 0$, in which case it is a (possibly unbounded) ellipsoid. When $H \succ 0$ and $d \leq (1/4)c^\top H^{-1}c$, then this zero-level set is a bounded and full-dimensional ellipsoid with center in $\hat{x} = -\frac{1}{2}H^{-1}c$, whence (9.8) is rewritten as

$$f_0(x) = \frac{1}{2}(x - \hat{x})^\top H(x - \hat{x}) - \frac{1}{4}c^\top H^{-1}c + d \leq 0. \qquad (9.9)$$

A bounded, full-dimensional ellipsoid is also usually represented in the form

$$\mathcal{E} = \{x : (x - \hat{x})^\top P^{-1}(x - \hat{x}) \leq 1\}, \quad P \succ 0, \qquad (9.10)$$

where P is the *shape matrix* of the ellipsoid. Clearly, this representation is analogous to (9.8) and (9.9), with

$$H = 2P^{-1}, \quad c^\top H^{-1}c/4 - d = 1.$$

The eigenvectors u_i of P define the directions of the semi-axes of the ellipsoid; the lengths of the semi-axes are given by $\sqrt{\lambda_i}$, where $\lambda_i > 0$, $i = 1, \ldots, n$, are the eigenvalues of P, see Figure 9.4.

The trace of P thus measures the sum of the squared semi-axis lengths. The *volume* of the ellipsoid $\mathcal{E}$ is proportional to the square-root of the determinant of P:

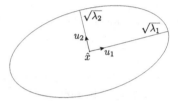

Figure 9.4 A two-dimensional ellipsoid.

$$\mathrm{vol}\,(\mathcal{E}) = \alpha_n (\det P)^{1/2}, \quad \alpha_n = \frac{2\pi^{n/2}}{n\Gamma(n/2)},$$

where Γ is the Gamma function (α_n denotes the volume of the unit Euclidean ball in $\mathbb{R}^n$). When $H \succeq 0$, if H has a zero eigenvalue, then the ellipsoid is unbounded along the directions of the eigenvectors associated with the zero eigenvalues. The zero-level set can in this case assume a variety of geometrical shapes, such as elliptic paraboloid, elliptic cylinder, parabolic cylinder, etc., see Figure 9.5.

An alternative representation of a (possibly unbounded) ellipsoid is in term of the inverse image of a unit ball, i.e.

$$\mathcal{E} = \{x \in \mathbb{R}^n : \|Ax - b\|_2^2 \leq 1\}, \quad A \in \mathbb{R}^{m,n}, \qquad (9.11)$$

which is equivalent to (9.8), with $H = 2A^\top A$, $c^\top = -2b^\top A$, $d = b^\top b - 1$.

If A is full column rank and $n \leq m$, then $A^\top A \succ 0$, and (9.11) represents a bounded ellipsoid. If A is symmetric and positive definite, then (9.11) represents a bounded ellipsoid with center $\hat{x} = A^{-1}b$ and volume proportional to $\det A^{-1}$. A further representation of a bounded (and possibly flat) ellipsoid is in terms of the image of the unit ball under an affine transformation, that is

$$\mathcal{E} = \{x \in \mathbb{R}^n : x = Bz + \hat{x} : \|z\|_2 \leq 1\}, \quad B \in \mathbb{R}^{n,m}. \qquad (9.12)$$

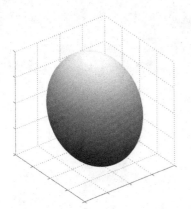
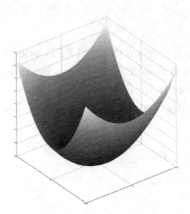
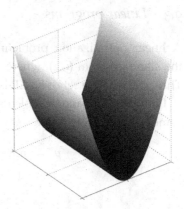

Figure 9.5 Examples of zero-level sets of convex quadratic functions in $\mathbb{R}^3$. From left to right: a compact ellipsoid, an elliptic paraboloid, and a parabolic cylinder.

Such an ellipsoid is *flat* (degenerate, or not fully dimensional) whenever $\mathcal{R}(B) \neq \mathbb{R}^n$. If B is full row rank and $n \leq m$, then (9.12) is equivalent to (9.10) and it represents a bounded full-dimensional ellipsoid with shape matrix $P = BB^\top \succ 0$.

Example 9.5 (*Zero-level sets of convex quadratic functions*) We give three simple examples of the geometrical shape of the zero-level sets of a convex quadratic function. Consider equation (9.8), with

$$H = \begin{bmatrix} 2/9 & 0 & 0 \\ 0 & 1/2 & 0 \\ 0 & 0 & 2 \end{bmatrix}, \quad c^\top = [0\,0\,0], \quad d = -1,$$

then the zero-level set is a bounded ellipsoid, see Figure 9.5. If we take instead

$$H = \begin{bmatrix} 1 & 0 & 0 \\ 0 & 2 & 0 \\ 0 & 0 & 0 \end{bmatrix}, \quad c^\top = [0\,0\,-1], \quad d = 0,$$

then the zero-level set is the epigraph of an elliptic paraboloid, see Figure 9.5. Further, for

$$H = \begin{bmatrix} 1 & 0 & 0 \\ 0 & 0 & 0 \\ 0 & 0 & 0 \end{bmatrix}, \quad c^\top = [0\,0\,-1], \quad d = 0,$$

the zero-level set is the epigraph of a parabolic cylinder, see again Figure 9.5. Notice finally that if $H = 0$, then the zero-level set is a half-space.

The previous discussion suggests that the family of sets of $x \in \mathbb{R}^n$ satisfying a collection of m convex quadratic inequalities

$$\frac{1}{2}x^\top H_i x + c_i^\top x + d_i \leq 0, \; H_i \succeq 0, \quad i = 1, \ldots, m$$

includes the family of polyhedra and polytopes, but it is much richer; Figure 9.6 shows an example in $\mathbb{R}^2$.

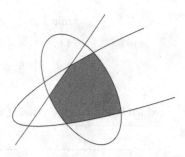

Figure 9.6 Intersection of the feasible sets of three quadratic inequalities in $\mathbb{R}^2$.

9.3   Linear programs

A linear optimization problem (or, linear program, LP) is one of the standard form (9.2), where every function $f_0, f_1, \ldots, f_m$ is affine. Thus, the feasible set of an LP is a polyhedron.

Linear optimization problems admits several standard forms. One comes directly from (9.2):

$$
\begin{aligned}
p^* = \min_{x} \quad & c^\top x + d \\
\text{s.t.:} \quad & A_{\text{eq}} x = b_{\text{eq}}, \\
& A x \le b,
\end{aligned}
$$

where the inequalities are understood component-wise; we shall denote this form as the *inequality form* of the LP. The constant term d in the objective function is, of course, immaterial: it offsets the value of the objective but it has no influence on the minimizer.

Remark 9.1 *Conic form of LP.* Another standard form, frequently used in several off-the-shelf algorithms for LP, is the so-called *conic* form:

$$
\begin{aligned}
p^* = \min_{x} \quad & c^\top x + d \\
\text{s.t.:} \quad & A_{\text{eq}} x = b_{\text{eq}}, \\
& x \ge 0.
\end{aligned}
$$

Clearly, the conic form is a special case of the inequality form, obtained by taking $A = -I$, and $b = 0$. However, we can also go the other way, and reformulate any standard inequality-form LP into conic form. To this end, consider an inequality-form LP, let $x = x_+ - x_-$, where $x_+ = \max(x, 0)$, $x_- = \max(-x, 0)$ are respectively the positive and the negative part of x, and let $\xi = b - Ax$. Then the inequality constraints can be written as $\xi \ge 0$, and it must be that $x_+ \ge 0$, $x_- \ge 0$. Thus introducing the augmented variable $z = [x_+ \ x_- \ \xi]$, we write the problem as follows:

$$
\begin{aligned}
\min_{z} \quad & [c^\top \ -c^\top \ 0]^\top z + d \\
\text{s.t.:} \quad & \begin{bmatrix} A_{\text{eq}} & -A_{\text{eq}} & 0 \end{bmatrix} z = b_{\text{eq}}, \\
& \begin{bmatrix} A & -A & I \end{bmatrix} z = b, \\
& z \ge 0,
\end{aligned}
$$

which is an LP in the variable z, in conic standard form.

Remark 9.2 *Geometric interpretation of LP.* The set of points that satisfy the constraints of an LP (i.e., the feasible set) is a polyhedron (or a polytope, when it is bounded):

$$
\mathcal{X} = \{ x \in \mathbb{R}^n : A_{\text{eq}} x = b_{\text{eq}}, \ A x \le b \}.
$$

Let $x_f \in \mathcal{X}$ be a feasible point. With such point is associated the objective level $c^\top x_f$ (from now on, we assume without loss of generality that $d = 0$). A point $x_f \in \mathcal{X}$ is an optimal point, hence a solution of our LP, if and only if there is no other point $x \in \mathcal{X}$ with lower objective, that is:

$$x_f \in \mathcal{X} \text{ is optimal for LP} \iff c^\top x \geq c^\top x_f, \ \forall y \in \mathcal{X},$$

see also the discussion in Section 8.4. Vice versa, the objective can be improved if one can find $x \in \mathcal{X}$ such that $c^\top(x - x_f) < 0$. Geometrically, this condition means that there exists a point x in the intersection of the feasible set $\mathcal{X}$ and the open half-space $\{x : c^\top(x - x_f) < 0\}$, i.e., that we can move away from x_f in a direction that forms a negative inner product with direction c (descent direction), while maintaining feasibility. At an optimal point x^* there is no feasible descent direction, see Figure 9.7.

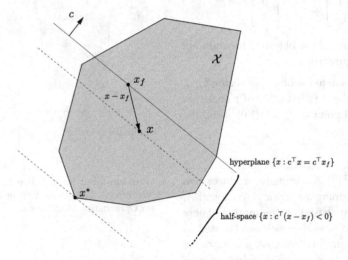

Figure 9.7 LP: move as far as possible in the direction $-c$, while maintaining feasibility. At the optimum x^* there are no feasible moves that improve the objective.

The geometric interpretation suggests that the following situations may arise in an LP.

- If the feasible set is empty (i.e., the linear equalities and inequalities have an empty intersection), then there is no feasible and hence no optimal solution; we assume in this case by convention that the optimal objective is $p^* = +\infty$.

- If the feasible set is nonempty and bounded, then the LP attains an optimal solution, and the optimal objective value p^* is finite. In this case, any optimal solution x^* is on a vertex, edge or facet of the feasible polytope. In particular, the optimal solution is unique if the optimal cost hyperplane $\{x : c^\top x = p^*\}$ intersects the feasible polytope only at a vertex.

- If the feasible set is nonempty but unbounded, then the LP may or may not attain an optimal solution, depending on the cost direction c, and there exist directions c such that the LP is unbounded below, i.e. $p^* = -\infty$ and the solution x^* "drifts" to infinity, see Figure 9.8.

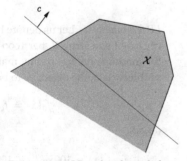

Figure 9.8 An LP with unbounded optimal objective.

Example 9.6 (*A linear program in two dimensions*) Consider the optimization problem

$$\min_{x \in \mathbb{R}^2} 3x_1 + 1.5x_2 \text{ subject to: } -1 \leq x_1 \leq 2, \, 0 \leq x_2 \leq 3.$$

The problem is an LP, and it can be put in standard inequality form:

$$\min : \min_{x \atop x \in \mathbb{R}^2} 3x_1 + 1.5x_2 \text{ subject to: } -x_1 \leq 1, \, x_1 \leq 2, \, -x_2 \leq 0, \, x_2 \leq 3$$

or, using matrix notation, $\min_x c^\top x$ subject to $Ax \leq b$, with

$$c^\top = [3 \; 1.5], \quad A = \begin{bmatrix} -1 & 0 \\ 1 & 0 \\ 0 & -1 \\ 0 & 1 \end{bmatrix}, \quad b = \begin{bmatrix} 1 \\ 2 \\ 0 \\ 3 \end{bmatrix}.$$

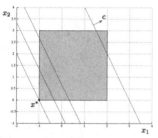

The level curves (curves of constant value) of the objective function are shown, along with the feasible set, in Figure 9.9.

The level curves are straight lines orthogonal to the objective vector, $c^\top = [3 \; 1.5]$. The problem amounts to finding the smallest value of p such that $p = c^\top x$ for some feasible x. The optimal point is $x^* = [-1 \; 0]^\top$, and the optimal objective value is $p^* = -3$.

Figure 9.9 A toy LP.

Example 9.7 (*A drug production problem.*[1]) A company produces two kinds of drugs, Drug I and Drug II, containing a specific active agent A, which is extracted from raw materials purchased on the market. There are two kinds of raw material, Raw I and Raw II, which can be used as sources of the active agent. The related production, cost, and resource data are given in Tables 9.1–9.3. The goal is to find the production plan which maximizes the profit for the company.

[1] Problem taken from A. Ben-Tal, A. Nemirovski, *Lectures on Modern Convex Optimization*, SIAM, 2001.

LP formulation. Let us denote by $x_{\text{DrugI}}, x_{\text{DrugII}}$ the amounts respectively of Drug I and Drug II, per 1000 packs produced. Let $x_{\text{RawI}}, x_{\text{RawII}}$ denote the amounts (in Kg) of raw materials to be purchased. According to the problem data, the objective to be minimized in this problem has the form

$$f_0(x) = f_{\text{costs}}(x) - f_{\text{income}}(x),$$

where

$$f_{\text{costs}}(x) = 100x_{\text{RawI}} + 199.90x_{\text{RawII}} + 700x_{\text{DrugI}} + 800x_{\text{DrugII}}$$

represents the purchasing and operational costs, and

$$f_{\text{income}}(x) = 6500x_{\text{DrugI}} + 7100x_{\text{DrugII}}$$

represents the income from selling the drugs.

Further, we have a total of five inequality constraints, and additional sign constraints on the variables.

Parameter	Drug I	Drug II
Selling price (USD) per 1000 packs	6,500	7,100
Content of agent A (grams) per 1000 packs	0.500	0.600
Manpower required (hrs) per 1000 packs	90.0	100.0
Equipment required (hrs) per 1000 packs	40.0	50.0
Operational costs (USD) per 1000 packs	700	800

Table 9.1 Drug production data.

Material	Purch. price (USD/kg)	Agent content (g/kg)
Raw I	100.00	0.01
Raw II	199.90	0.02

Table 9.2 Contents of raw materials.

Budget (USD)	Manpw. (hrs)	Equip. (hrs)	storage cap. (kg)
100,000	2,000	800	1,000

Table 9.3 Resources.

- Balance of active agent:

$$0.01x_{\text{RawI}} + 0.02x_{\text{RawII}} - 0.5x_{\text{DrugI}} - 0.6x_{\text{DrugII}} \geq 0.$$

This constraint says that the amount of raw material must be enough to produce the drugs.

- Storage constraint:

$$x_{\text{RawI}} + x_{\text{RawII}} \leq 1000.$$

This constraint says that the capacity of storage for the raw materials is limited.

- Manpower constraint:

$$90.0x_{\text{DrugI}} + 100.0x_{\text{DrugII}} \leq 2000,$$

which expresses the fact that the resources in manpower are limited: we cannot allocate more than 2,000 hours to the project.

- Equipment constraint:

$$40.0x_{\text{DrugI}} + 50.0x_{\text{DrugII}} \leq 800.$$

This says that the resources in equipment are limited.

- Budget constraint:

$$100.0x_{\text{RawI}} + 199.90x_{\text{RawII}} + 700x_{\text{DrugI}} + 800x_{\text{DrugII}} \leq 100,000.$$

This limits the total budget.

- Sign constraints:

$$x_{\text{RawI}} \geq 0, \ x_{\text{RawII}} \geq 0, \ x_{\text{DrugI}} \geq 0, \ x_{\text{DrugII}} \geq 0.$$

Solving this problem (e.g., via the Matlab `linprog` command, or via CVX) we obtain the following optimal value and a corresponding optimal solution:

$$p^* = -14085.13, \ x^*_{\text{RawI}} = 0, \ x^*_{\text{RawII}} = 438.789, \ x^*_{\text{DrugI}} = 17.552, \ x^*_{\text{DrugII}} = 0.$$

Note that both the budget and the balance constraints are active (i.e., they hold with equality at the optimum), which means that the production process utilizes the entire budget and the full amount of active agent contained in the raw materials. The solution promises the company a quite respectable profit of about 14%.

9.3.1 *LPs and polyhedral functions*

We say that a function $f : \mathbb{R}^n \to \mathbb{R}$ is polyhedral if its epigraph is a polyhedron, that is if

$$\operatorname{epi} f = \left\{ (x,t) \in \mathbb{R}^{n+1} : f(x) \le t \right\}$$

can be represented as

$$\operatorname{epi} f = \left\{ (x,t) \in \mathbb{R}^{n+1} : C \begin{bmatrix} x \\ t \end{bmatrix} \le d \right\}, \qquad (9.13)$$

for some matrix $C \in \mathbb{R}^{m,n+1}$ and vector $d \in \mathbb{R}^m$.

Polyhedral functions include in particular functions that can be expressed as a maximum of a finite number of affine functions:

$$f(x) = \max_{i=1,\dots,m} a_i^\top x + b_i,$$

where $a_i \in \mathbb{R}^n$, $b_i \in \mathbb{R}$, $i = 1,\dots,m$. Observing that for any family of functions $f_\alpha(x)$ parameterized by $\alpha \in \mathcal{A}$ it holds that

$$\max_{\alpha \in \mathcal{A}} f_\alpha(x) \le t \quad \Leftrightarrow \quad f_\alpha(x) \le t, \ \forall \alpha \in \mathcal{A},$$

we see that the epigraph of f

$$\operatorname{epi} f = \left\{ (x,t) \in \mathbb{R}^{n+1} : \max_{i=1,\dots,m} a_i^\top x + b_i \le t \right\}$$

can be expressed as the polyhedron

$$\operatorname{epi} f = \left\{ (x,t) \in \mathbb{R}^{n+1} : a_i^\top x + b_i \le t, \ i = 1,\dots,m \right\}.$$

Example 9.8 (*The ℓ_∞-norm function*) The ℓ_∞-norm function $f(x) = \|x\|_\infty$, $x \in \mathbb{R}^n$, is polyhedral since it can be written as the maximum of $2n$ affine functions:

$$f(x) = \max_{i=1,\dots,n} \max(x_i, -x_i).$$

Polyhedral functions also include functions that can be expressed as a sum of functions which are themselves maxima of affine functions, i.e.

$$f(x) = \sum_{j=1}^q \max_{i=1,\dots,m} a_{ij}^\top x + b_{ij},$$

for given vectors $a_{ij} \in \mathbb{R}^n$ and scalars b_{ij}. Indeed, the condition $(x, t) \in \text{epi } f$ is equivalent to the existence of a vector $u \in \mathbb{R}^q$ such that

$$\sum_{j=1}^{q} u_j \le t, \quad a_{ij}^\top x + b_{ij} \le u_j, \ i = 1, \ldots, m; \ j = 1, \ldots, q, \quad (9.14)$$

hence, epi f is the projection (on the space of (x, t)-variables) of a polyhedron, which is itself a polyhedron.

Example 9.9 (*The ℓ_1-norm function*) The ℓ_1-norm function $f(x) = \|x\|_1$, $x \in \mathbb{R}^n$, is polyhedral since it can be written as the sum of maxima of affine functions:

$$f(x) = \sum_{i=1,\ldots,n} \max(x_i, -x_i).$$

Example 9.10 (*Sum of largest components in a vector*) For $x \in \mathbb{R}^n$, the sum of the k-largest components in x is written as

$$s_k(x) \doteq \sum_{i=1}^{k} x_{[i]},$$

where $x_{[i]}$ is the i-th largest component of x. The function $s_k(x)$ is convex, since it is the pointwise maximum of linear (hence, convex) functions:

$$s_k(x) = \max_{(i_1,\ldots,i_k) \in \{1,\ldots,n\}^k} x_{i_1} + \ldots + x_{i_k}. \quad (9.15)$$

The functions inside the maximum are linear, hence the function s_k is polyhedral. Notice that there are $C_{n,k} = \binom{n}{k}$ linear functions[2] involved in the maximum, each obtained by choosing a particular subset of k indices in $\{1, \ldots, n\}$. For example, with $k = 2$ and $n = 4$,

$$s_2(x) = \max(x_1 + x_2, x_2 + x_3, x_3 + x_1, x_1 + x_4, x_2 + x_4, x_3 + x_4).$$

Hence, to represent the constraint $s_k(x) \le \alpha$ based on the above representation, we need to consider six constraints

$$x_1 + x_2 \le \alpha, \quad x_2 + x_3 \le \alpha, \quad x_3 + x_1 \le \alpha,$$
$$x_1 + x_4 \le \alpha, \quad x_2 + x_4 \le \alpha, \quad x_3 + x_4 \le \alpha.$$

The number of constraints grows very quickly with n, k. For instance, for $n = 100, k = 10$, we would need more than 10^{13} linear constraints!

A more efficient representation is possible based on the following expression (which we next prove)

$$s_k(x) = \min_t kt + \sum_{i=1}^{n} \max(0, x_i - t). \quad (9.16)$$

Using this form, the constraint $s_k(x) \le \alpha$ can be expressed as follows: there exist a scalar t and an n-vector u such that

$$kt + \sum_{i=1}^{n} u_i \le \alpha, \quad u \ge 0, \quad u_i \ge x_i - t, \ i = 1, \ldots, n.$$

[2] The *binomial coefficient* $\binom{n}{k}$ denotes the number of distinct k-element subsets that can be formed from a set containing n elements. It holds that

$$\binom{n}{k} = \frac{n!}{k!(n-k)!},$$

where ! denotes the factorial of an integer.

The above expression shows that s_k is convex, since the set of points (x, α) such that the above holds for some u, t is a polyhedron. The representation (9.16) is much more efficient than (9.15), as it involves a polyhedron with $2n + 1$ constraints. The price to pay is a moderate increase in the number of variables, which is now $2n + 1$ instead of n. The lesson here is that a polyhedron in an n-dimensional space, with an exponential number of facets, can be represented as a polyhedron in a higher dimensional space with a moderate number of facets. By adding just a few dimensions to the problem we are able to deal (implicitly) with a very high number of constraints.

We next provide a proof for the expression (9.16). We can assume without loss of generality that the elements of x are in decreasing order: $x_1, \ldots, x_n$. Then, $s_k(x) = x_1 + \cdots + x_k$. Now choose t such that $x_k \geq t \geq x_{k+1}$. We have

$$kt + \sum_{i=1}^{n} \max(0, x_i - t) = kt + \sum_{i=1}^{k}(x_i - t) = \sum_{i=1}^{k} x_i = s_k(x).$$

Since $s_k(x)$ is attained for a particular choice of t, we obtain that $s_k(x)$ is bounded below by the minimum over t:

$$s_k(x) \geq \min_t \left(kt + \sum_{i=1}^{n} \max(0, x_i - t) \right).$$

On the other hand, for every t, we have

$$
\begin{aligned}
s_k(x) &= \sum_{i=1}^{k}(x_i - t + t) = kt + \sum_{i=1}^{k}(x_i - t) \\
&\leq kt + \sum_{i=1}^{k} \max(0, x_i - t) \leq kt + \sum_{i=1}^{n} \max(0, x_i - t).
\end{aligned}
$$

Since the upper bound above is valid for every t, it remains valid when minimizing over t, and we have

$$s_k(x) \leq \min_t \; kt + \sum_{i=1}^{n} \max(0, x_i - t),$$

which concludes the proof.

9.3.1.1 *Minimization of polyhedral functions.* The problem of minimizing a polyhedral function, under linear equality or inequality constraints, can be cast as an LP. Indeed, consider the problem

$$\min_x f(x) \quad \text{s.t.:} \quad Ax \leq b,$$

with f polyhedral. We formally cast this problem as

$$\min_{x,t} t \quad \text{s.t.:} \, Ax \leq b, \; (x, t) \in \text{epi } f.$$

Since epi f is a polyhedron, it can be expressed as in (9.13), hence the problem above is an LP. Notice, however, that explicit representation of the LP in a standard form may require the introduction of additional slack variables, which are needed for representation of the epigraph, as was done for instance in (9.14).

Example 9.11 (ℓ_1 and ℓ_∞-norm regression problems) The concept of an approximate solution of an inconsistent system of linear equations $Ax = b$ has been introduced in Section 6.3.1 in the context of least squares, where we seek a vector x that minimizes the ℓ_2-norm of the residual vector $r = Ax - b$. Depending on the context, however, it may be sensible to seek approximate solutions that minimize other norms of the residual. We next show that two frequently encountered cases, namely those where the ℓ_∞ or the ℓ_1 norm is employed as a measure of residual error, can be cast as linear programs and hence solved efficiently using LP numerical codes. The choice of the norm reflects the sensitivity of the solution to *outliers* in the data and on the distribution of the residuals. Consider for example a regression problem with random data $A \in \mathbb{R}^{1000,100}$, $b \in \mathbb{R}^{1000}$: the ℓ_1 norm tends to encourage sparsity (number of nonzeros) of the residual vector, whereas the ℓ_∞ norm tends to equalize the magnitude of the residuals, see Figure 9.10.

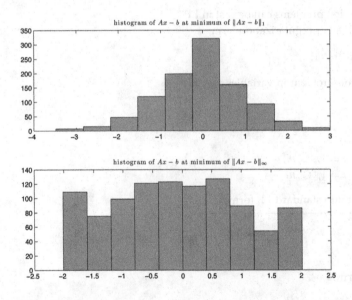

Figure 9.10 Histograms of ℓ_1 (top) and ℓ_∞ (bottom) residuals on a randomly generated problem with $A \in \mathbb{R}^{1000,100}$.

Consider first the minimization of the ℓ_∞ residual:

$$\min_x \|Ax - b\|_\infty, \quad A \in \mathbb{R}^{m,n}, b \in \mathbb{R}^m. \tag{9.17}$$

The problem may be first rewritten in epigraphic form, adding a slack scalar variable t:

$$\min_{x,t} t \quad \text{s.t.: } \|Ax - b\|_\infty \leq t,$$

and then we observe that

$$\|Ax - b\|_\infty \leq t \;\Leftrightarrow\; \max_{i=1,\ldots,m} |a_i^\top x - b_i| \leq t \;\Leftrightarrow\; |a_i^\top x - b_i| \leq t,\; i = 1,\ldots,m.$$

Hence, problem (9.17) is equivalent to the following LP in variables $x \in \mathbb{R}^n$ and $t \in \mathbb{R}$:

$$\begin{aligned}
\min_{x,t} \quad & t \\
\text{s.t.:} \quad & a_i^\top x - b_i \leq t, \quad i = 1,\ldots,m, \\
& a_i^\top x - b_i \geq -t, \quad i = 1,\ldots,m.
\end{aligned}$$

Similarly, for the minimization of the ℓ_1 residual, we have that

$$\min_x \|Ax - b\|_1, \quad A \in \mathbb{R}^{m,n}, b \in \mathbb{R}^m.$$

is equivalent to a problem with a vector u of additional slack variables $u \in \mathbb{R}^m$:

$$\min_{x,u} \sum_{i=1}^m u_i \quad \text{s.t.:} |a_i^\top x - b_i| \leq u_i,\; i = 1,\ldots,m,$$

which is in turn easily cast as a standard LP as follows:

$$\begin{aligned}
\min_{x,u} \quad & \mathbf{1}^\top u \\
\text{s.t.:} \quad & a_i^\top x - b_i \leq u_i, \quad i = 1,\ldots,m, \\
& a_i^\top x - b_i \geq -u_i, \quad i = 1,\ldots,m.
\end{aligned}$$

Finally, note that also *mixed* ℓ_∞/ℓ_1 regression problems can be cast in LP form. For instance, an ℓ_∞ regression with an ℓ_1 regularization term

$$\min_x \|Ax - b\|_\infty + \gamma\|x\|_1$$

is equivalent to the following optimization problem in variables $x \in \mathbb{R}^n$ and slacks $u \in \mathbb{R}^n$, $t \in \mathbb{R}$:

$$\begin{aligned}
\min_{x,t,u} \quad & t + \gamma \sum_{i=1}^n u_i \\
\text{s.t.:} \quad & |a_i^\top x - b_i| \leq t, \quad i = 1,\ldots,m, \\
& |x_i| \leq u_i, \quad i = 1,\ldots,n.
\end{aligned}$$

This latter problem is in turn readily cast into standard LP format.

9.3.2 LP duality

Consider a "primal" LP in inequality form:

$$\begin{aligned}
p^* = \min_x \quad & c^\top x \\
\text{s.t.:} \quad & A_{\text{eq}} x = b_{\text{eq}}, \\
& Ax \leq b.
\end{aligned}$$

The Lagrangian for this problem is

$$\begin{aligned}
\mathcal{L}(x,\lambda,\mu) &= c^\top x + \lambda^\top(Ax - b) + \mu^\top(A_{\text{eq}} x - b_{\text{eq}}) \\
&= (c + A^\top\lambda + A_{\text{eq}}^\top\mu)^\top x - \lambda^\top b - \mu^\top b_{\text{eq}}.
\end{aligned}$$

The dual function $g(\lambda, \mu)$ is obtained by minimizing $\mathcal{L}(x, \lambda, \mu)$ w.r.t. x. But $\mathcal{L}(x, \lambda)$ is affine in x, hence $g(\lambda, \mu)$ is unbounded below, unless the vector coefficient of x is zero (i.e., $c + A^\top \lambda + A_{eq}^\top \mu = 0$), and it is equal to $-\lambda^\top b - \mu^\top b_{eq}$ otherwise. That is,

$$g(\lambda, \mu) = \begin{cases} -\lambda^\top b - \mu^\top b_{eq} & \text{if } c + A^\top \lambda + A_{eq}^\top \mu = 0, \\ -\infty & \text{otherwise.} \end{cases}$$

The dual problem then amounts to maximizing $g(\lambda, \mu)$ over $\lambda \geq 0$ and μ. Clearly, if $g(\lambda, \mu) = -\infty$, then there is nothing to maximize, therefore in the dual problem we make explicit the condition that we maximize over those λ, μ for which $g(\lambda, \mu)$ is not identically equal to $-\infty$. This results in the following explicit dual problem formulation:

$$\begin{aligned} d^* = \max_{\lambda, \mu} \quad & -\lambda^\top b - \mu^\top b_{eq} \\ \text{s.t.:} \quad & c + A^\top \lambda + A_{eq}^\top \mu = 0, \\ & \lambda \geq 0, \end{aligned}$$

By changing the sign of the objective, we rewrite the dual in minimization form as follows

$$\begin{aligned} -d^* = \min_{\lambda, \mu} \quad & b^\top \lambda + b_{eq}^\top \mu \\ \text{s.t.:} \quad & A^\top \lambda + A_{eq}^\top \mu + c = 0, \\ & \lambda \geq 0, \end{aligned}$$

which is again an LP in the variables (λ, μ). From Proposition 8.7, and from a discussion analogous to that presented in Example 8.21, we have that strong duality holds between the primal and dual LP (i.e., $p^* = d^*$), provided that at least one of the two problems is feasible.

9.4 Quadratic programs

A quadratic optimization problem (or quadratic program, QP) is one of the standard form (9.2), where f_0 is a quadratic function (9.1) and $f_1, \ldots, f_m$ are affine functions. Thus, the feasible set of QP is a polyhedron (as in LP), but the objective is quadratic, rather than linear. If the H matrix in (9.1) is positive semidefinite, then we have a convex QP. The standard form of a QP is thus

$$\begin{aligned} p^* = \min_{x} \quad & \frac{1}{2} x^\top H x + c^\top x \\ \text{s.t.:} \quad & A_{eq} x = b_{eq}, \\ & Ax \leq b. \end{aligned}$$

Example 9.12 (*a QP in two variables*) Consider the problem

$$\min_{x} \frac{1}{2}\left(x_1^2 - x_1 x_2 + 2x_2^2\right) - 3x_1 - 1.5x_2, \text{ s.t.: } -1 \le x_1 \le 2, \; 0 \le x_2 \le 3.$$

$$(9.18)$$

This is a QP that can be cast in standard form with

$$H = \begin{bmatrix} 1 & -1/2 \\ -1/2 & 2 \end{bmatrix}, \; c^\top = [-3 \;\; -1.5], \; A = \begin{bmatrix} 1 & 0 \\ -1 & 0 \\ 0 & 1 \\ 0 & -1 \end{bmatrix}, \; b = \begin{bmatrix} 2 \\ 1 \\ 3 \\ 0 \end{bmatrix}.$$

We can inspect that the eigenvalues of H are non-negative

$$H = U\Lambda U^\top, \quad U = \begin{bmatrix} -0.3827 & 0.9239 \\ 0.9239 & 0.3827 \end{bmatrix}, \quad \Lambda = \begin{bmatrix} 2.2071 & 0 \\ 0 & 0.7929 \end{bmatrix},$$

hence the considered QP is a convex one. The optimal solution of the QP is (in general, we need to use a QP solver to find this) $x^* = [2 \;\; 1.25]^\top$, and the optimal objective value is $p^* = -5.5625$, see Figure 9.11.

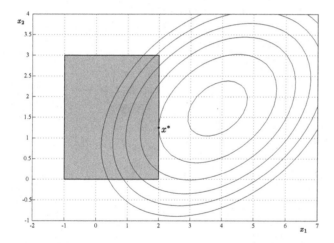

Figure 9.11 Level curves of the objective function and optimal solution of the QP (9.18).

9.4.1   *Constrained least squares*

Quadratic programs arise naturally from least-squares problems when linear equality or inequality constraints need be enforced on the decision variables. This is indeed the case in many situations where the variables have physical significance (they represent for instance lengths, volumes, concentrations, inertias, relative proportions, etc.) and constraints such as lower and upper limits on the variable values are naturally introduced. A linearly-constrained LS problem takes the form

$$p^* = \min_x \quad \|Rx - y\|_2^2$$
$$\text{s.t.:} \quad A_{eq}x = b_{eq},$$
$$Ax \leq b.$$

This is clearly a convex QP, having objective (neglecting a constant term $d = \|y\|^2$)

$$f_0(x) = \frac{1}{2}x^\top H x + c^\top x,$$

with $H = 2R^\top R \succeq 0$, $c^\top = -2y^\top R$.

Example 9.13 (*Tracking a financial index*) As an applicative example, consider a financial portfolio design problem, where the entries of $x \in \mathbb{R}^n$ represent the fractions of an investor's total wealth invested in each of n different assets, and where $r(k) \in \mathbb{R}^n$ represents the vector of simple returns of the component assets during the k-th period of time $[(k-1)\Delta, k\Delta]$, where Δ is a fixed duration, e.g., one month; see also Example 2.6. Suppose that the component y_k of the vector $y \in \mathbb{R}^T$ represents the return of some target financial index over the k-th period, for $k = 1, \ldots, T$: the so-called *index tracking* problem is to construct a portfolio x so as to track as close as possible the "benchmark" index returns y. Since the vector of portfolio returns over the considered time horizon is

$$z = Rx, \quad R \doteq \begin{bmatrix} r^\top(1) \\ \vdots \\ r^\top(T) \end{bmatrix} \in \mathbb{R}^{T,n},$$

we may seek for the portfolio x with minimum LS tracking error, by minimizing $\|Rx - y\|_2^2$. However, we need to take into account the fact that the elements of x represent relative *weights*, that is they are nonnegative and they sum up to one. The index tracking problem is therefore a constrained LS problem, thus a convex QP:

$$p^* = \min_x \quad \|Rx - y\|_2^2 \qquad (9.19)$$
$$\text{s.t.:} \quad \mathbf{1}^\top x = 1$$
$$x \geq 0.$$

As a numerical example, we consider again the financial data previously used in Example 5.3, consisting of 169 monthly return data of six indices: the MSCI US index, the MSCI EUR index, the MSCI JAP index, the MSCI PACIFIC index, the MSCI BOT liquidity index, and the MSCI WORLD index, as shown in Figure 5.6. The problem is to track the target index MSCI WORLD, using a portfolio composed of the other five indices.

Solving the convex QP in (9.19) with this data, we obtain the optimal portfolio composition

$$x^* = [0.5138 \ 0.3077 \ 0.0985 \ 0.0374 \ 0.0426]^\top,$$

and hence the optimal-tracking portfolio return sequence $z^* = Rx^*$, with tracking error $\|Rx^* - y\|_2^2 = 2.6102 \times 10^{-4}$. Figure 9.12 shows the result of investing one euro into each of the component indices and benchmark index, and into the tracking-optimal portfolio. As expected, the value sequence generated by the optimal portfolio is the closest one to the target index.

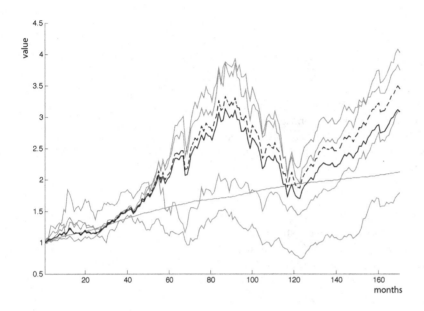

Figure 9.12 Light gray lines show the time value of component indices; the solid black line is the value of the target index, and the dashed black line is the value of the tracking-optimal portfolio.

9.4.2 Quadratic constrained quadratic programs

A generalization of the QP model is obtained by allowing quadratic (rather than merely linear) equality and inequality constraints. A quadratic constrained quadratic program (QCQP) thus takes the form

$$p^* = \min_x \quad x^\top H_0 x + 2c_0^\top x + d_0 \tag{9.20}$$
$$\text{s.t.:} \quad x^\top H_i x + 2c_i^\top x + d_i \leq 0, \quad i \in \mathcal{I},$$
$$x^\top H_j x + 2c_j^\top x + d_j = 0, \quad j \in \mathcal{E},$$

where $\mathcal{I}, \mathcal{E}$ denote the index sets relative to inequality constraints and equality constraints, respectively. A QCQP is convex if and only if $H_0 \succeq 0$, $H_i \succeq 0$ for $i \in \mathcal{I}$, and $H_j = 0$ for $j \in \mathcal{E}$. In other words, a QCQP is convex whenever the functions describing the objective and the inequality constraints are convex quadratic, and all the equality constraints are actually affine.

Example 9.14 (*Minimizing a linear function under an ellipsoidal constraint*)
Consider the following special case of a QCQP, where a linear objective
is minimized under an ellipsoidal constraint:

$$\min_{x \in \mathbb{R}^n} \quad c^\top x$$
$$\text{s.t.:} \quad (x - \hat{x})^\top P^{-1}(x - \hat{x}) \leq 1,$$

where $P \succ 0$. Here, the feasible set $\mathcal{X}$ is the ellipsoid $\{x : (x - \hat{x})^\top P^{-1}(x - \hat{x}) \leq 1\}$, and the geometrical interpretation of the problem is to move as
far as possible in the direction $-c$ while remaining in the ellipsoid, see
Figure 9.13(a).

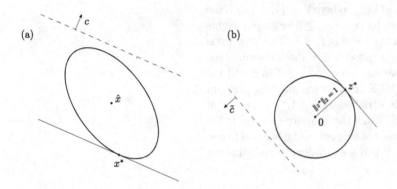

Figure 9.13 Minimization of a linear objective under an ellipsoidal constraint.

This optimization problem admits a "closed form" solution. To obtain
this solution we first perform the following change of variables:

$$z = E^{-1}(x - \hat{x}), \quad \text{i.e., } x = Ez + \hat{x},$$

where $P = E^2$ is the symmetric square-root factorization of P. Then the
original problem transforms in the new variable z to

$$\min_{z \in \mathbb{R}^n} \quad \tilde{c}^\top z + d$$
$$\text{s.t.:} \quad z^\top z \leq 1,$$

where $\tilde{c} = Ec$, and $d = c^\top \hat{x}$ is a constant term. In the new variable z
the problem has a simpler interpretation: move as far as possible in the
direction $-\tilde{c}$, maintaining a distance from the origin at most one, see
Figure 9.13(b). Clearly, the optimal solution is simply a vector lying in
direction $-\tilde{c}$ having unit norm, that is

$$z^* = -\frac{\tilde{c}}{\|\tilde{c}\|_2} = -\frac{Ec}{\|Ec\|_2}.$$

Going back to the original x variables we then obtain the optimal solution

$$x^* = Ez^* + \hat{x} = -\frac{Pc}{\sqrt{c^\top Pc}} + \hat{x}.$$

While convex QCQP are relatively "easy" to solve, non-convex QCQPs are a class of very hard optimization problems. This fact should not be surprising, since non-convex QCQPs represent a bridge between continuous and discrete optimization. For instance, a quadratic equality constraint such as $x_i^2 = 1$ implies that $x_i \in \{-1, 1\}$; similarly, the constraint $x_i^2 = x_i$ implies that $x_i \in \{0, 1\}$, i.e., that x_i is a binary variable. Consequently, for instance, hard combinatorial graph partitioning problems can be cast in non-convex QCQP format, see the next example.

Example 9.15 (*The max-cut problem*) The so-called max-cut problem is defined as follows: given a graph $G = (V, E)$, where $V = \{1, \ldots, n\}$ is the set of vertices and E is the set of edges, let $w_{ij} = w_{ji} \geq 0$ be given weights defined on the edges $(i, j) \in E$, with $w_{ij} = 0$ if $(i, j) \notin E$. Then, the max-cut problem amounts to determining a subset $S \subset V$ that maximizes the sum of weights over those edges that have one end point in S and the other in the complementary set $\bar{S} = V \setminus S$. In order to model the problem in the QCQP setting, we define node variables x_i, $i = 1, \ldots, n$ such that $x_i = 1$ if $i \in S$ and $x_i = -1$ if $i \in \bar{S}$. Then the quantity $(1 - x_i x_j)/2$ is equal to zero if i, j are in the same subset of vertices, and it is equal to one otherwise. Therefore, the max-cut problem is equivalent to the following non-convex QCQP

$$p^* = \min_x \quad \frac{1}{2} \sum_{i<j} w_{ij}(1 - x_i x_j)$$
$$\text{s.t.:} \quad x_i^2 = 1, \qquad\qquad i = 1, \ldots, n.$$

Convex approximations, or *relaxations*, of non-convex quadratic programs are discussed in Section 11.3.3.

9.4.3   Quadratic programming duality

We next derive the explicit dual for some special classes of quadratic programming models.

9.4.3.1   *Dual of convex QP.* We consider first the case of a primal convex QP with linear inequality constraints:

$$p^* = \min_x \quad x^\top H_0 x + 2c_0^\top x + d_0$$
$$\text{s.t.:} \quad\quad Ax \leq b,$$

with $H_0 \succeq 0$. The Lagrangian for this problem is

$$\begin{aligned} \mathcal{L}(x, \lambda) &= x^\top H_0 x + 2c_0^\top x + d_0 + \lambda^\top (Ax - b) \\ &= x^\top H_0 x + (2c_0 + A^\top \lambda)^\top x + d_0 - b^\top \lambda. \end{aligned}$$

According to point (b) in Section 9.1 there are now two possibilities. If $2c_0 + A^\top\lambda$ is in the range of H_0, that is, if there exists z such that

$$H_0 z = 2c_0 + A^\top\lambda,$$

then $\mathcal{L}(x, \lambda)$ has a finite minimum value (w.r.t. x), given by

$$g(\lambda) = -\frac{1}{4}(2c_0 + A^\top\lambda)^\top H_0^\dagger (2c_0 + A^\top\lambda) + d_0 - b^\top\lambda.$$

If instead $2c_0 + A^\top\lambda$ is not in the range of H_0, then $g(\lambda) = -\infty$. The dual problem can thus be written as follows:

$$
\begin{aligned}
d^* = \max_{\lambda, z} \quad & -\frac{1}{4}(2c_0 + A^\top\lambda)^\top H_0^\dagger (2c_0 + A^\top\lambda) + d_0 - b^\top\lambda \\
\text{s.t.:} \quad & H_0 z = 2c_0 + A^\top\lambda, \\
& \lambda \geq 0.
\end{aligned}
$$

Substituting the equality constraint in the objective, and observing that $H_0 H_0^\dagger H_0 = H_0$, we may simplify this problem to

$$
\begin{aligned}
d^* = \max_{\lambda, z} \quad & -\frac{1}{4}z^\top H_0 z + d_0 - b^\top\lambda \\
\text{s.t.:} \quad & H_0 z = 2c_0 + A^\top\lambda, \\
& \lambda \geq 0,
\end{aligned}
$$

which is again a convex QP. According to Proposition 8.7, the strong duality condition $p^* = d^*$ holds whenever the primal problem is feasible. Notice that if $H_0 \succ 0$, then the dual problem simply becomes

$$
\begin{aligned}
d^* = \max_{\lambda} \quad & -\frac{1}{4}(2c_0 + A^\top\lambda)^\top H_0^{-1}(2c_0 + A^\top\lambda) + d_0 - b^\top\lambda \\
\text{s.t.:} \quad & \lambda \geq 0.
\end{aligned}
$$

9.4.3.2 *Dual of convex QCQP.* Consider a primal convex QCQP of the form (9.20) where, for simplicity, we assume only inequality constraints are present, specifically

$$
\begin{aligned}
p^* = \min_{x} \quad & x^\top H_0 x + 2c_0^\top x + d_0 \\
\text{s.t.:} \quad & x^\top H_i x + 2c_i^\top x + d_i \leq 0, \quad i = 1, \ldots, m.
\end{aligned}
$$

Further, we assume that the objective is strictly convex, that is $H_0 \succ 0$, while $H_i \succeq 0, i = 1, \ldots, m$. The Lagrangian for this problem is

$$
\begin{aligned}
\mathcal{L}(x, \lambda) &= x^\top H_0 x + 2c_0^\top x + d_0 + \sum_{i=1}^{m} \lambda_i (x^\top H_i x + 2c_i^\top x + d_i) \\
&= x^\top H(\lambda) x + 2c(\lambda)^\top x + d(\lambda),
\end{aligned}
$$

where we defined

$$H(\lambda) = H_0 + \sum_{i=1}^{m} \lambda_i H_i, \quad c(\lambda) = c_0 + \sum_{i=1}^{m} \lambda_i c_i, \quad d(\lambda) = d_0 + \sum_{i=1}^{m} \lambda_i d_i.$$

Since we assumed $H_0 \succ 0$, we have that $H(\lambda) \succ 0$ for any $\lambda \geq 0$, hence the unique unconstrained minimum over x of $\mathcal{L}(x, \lambda)$ can be expressed explicitly using Eq. (9.4) as

$$x^*(\lambda) = -H(\lambda)^{-1} c(\lambda),$$

whence

$$g(\lambda) = \mathcal{L}(x^*(\lambda), \lambda) = -c(\lambda)^\top H(\lambda)^{-1} c(\lambda) + d(\lambda).$$

The dual thus assumes the form

$$d^* = \max_{\lambda} \quad -c(\lambda)^\top H(\lambda)^{-1} c(\lambda) + d(\lambda)$$
$$\text{s.t.:} \quad \lambda \geq 0,$$

or, in equivalent minimization form,

$$-d^* = \min_{\lambda} \quad c(\lambda)^\top H(\lambda)^{-1} c(\lambda) - d(\lambda)$$
$$\text{s.t.:} \quad \lambda \geq 0.$$

Further, using an epigraphic reformulation, we have

$$-d^* = \min_{\lambda, t} \quad t$$
$$\text{s.t.:} \quad c(\lambda)^\top H(\lambda)^{-1} c(\lambda) - d(\lambda) \leq t,$$
$$\lambda \geq 0.$$

The first constraint can be expressed equivalently, using the Schur complement rule, in the form of a positive-semidefinite constraint

$$\begin{bmatrix} t + d(\lambda) & c(\lambda)^\top \\ c(\lambda) & H(\lambda) \end{bmatrix} \succeq 0$$

The dual problem is hence finally expressed in the form

$$-d^* = \min_{\lambda, t} \quad t$$
$$\text{s.t.:} \quad \begin{bmatrix} t + d(\lambda) & c(\lambda)^\top \\ c(\lambda) & H(\lambda) \end{bmatrix} \succeq 0,$$
$$\lambda \geq 0.$$

This problem belongs to the class of so-called semidefinite programming models, which are studied in detail in Chapter 11. According to Proposition 8.7, strong duality is guaranteed if the primal problem is strictly feasible, that is if there exists an x satisfying the inequality constraints with strict inequality.

9.4.3.3 *Dual of non-convex QCQP with a single constraint.* Finally, we consider the case of a possibly non-convex QCQP, with a *single* inequality constraint, that is

$$p^* = \min_x \quad x^\top H_0 x + 2c_0^\top x + d_0$$
$$\text{s.t.:} \quad x^\top H_1 x + 2c_1^\top x + d_1 \leq 0.$$

The Lagrangian for this problem is given by

$$\mathcal{L}(x, \lambda) = x^\top (H_0 + \lambda H_1) x + 2(c_0 + \lambda c_1)^\top x + (d_0 + \lambda d_1).$$

According to points (a) and (b) in Section 9.1, the minimum of $\mathcal{L}(x, \lambda)$ w.r.t. x is $-\infty$, unless the following conditions are satisfied:

$$H_0 + \lambda H_1 \succeq 0,$$
$$c_0 + \lambda c_1 \in \mathcal{R}(H_0 + \lambda H_1).$$

Under these conditions, the minimum gives the value of the dual function at λ:

$$g(\lambda) = -(c_0 + \lambda c_1)^\top (H_0 + \lambda H_1)^\dagger (c_0 + \lambda c_1) + d_0 + \lambda d_1.$$

The dual problem is then

$$d^* = \max_\lambda \quad -(c_0 + \lambda c_1)^\top (H_0 + \lambda H_1)^\dagger (c_0 + \lambda c_1) + d_0 + \lambda d_1$$
$$\text{s.t.:} \quad H_0 + \lambda H_1 \succeq 0,$$
$$c_0 + \lambda c_1 \in \mathcal{R}(H_0 + \lambda H_1),$$
$$\lambda \geq 0.$$

Next, we reformulate the problem in epigraphic minimization form, as follows:

$$-d^* = \min_{\lambda, t} \quad t$$
$$\text{s.t.:} \quad (c_0 + \lambda c_1)^\top (H_0 + \lambda H_1)^\dagger (c_0 + \lambda c_1) - (d_0 + \lambda d_1) \leq t,$$
$$H_0 + \lambda H_1 \succeq 0,$$
$$c_0 + \lambda c_1 \in \mathcal{R}(H_0 + \lambda H_1),$$
$$\lambda \geq 0.$$

Then, applying a general version of the Schur complement rule (see Section 11.2.3.2), we may equivalently rewrite the first three constraints of this problem in the form of a positive semidefiniteness condition on a suitable matrix, obtaining

$$-d^* = \min_{\lambda, t} \quad t$$
$$\text{s.t.:} \quad \begin{bmatrix} t + d_0 + \lambda d_1 & (c_0 + \lambda c_1)^\top \\ (c_0 + \lambda c_1) & H_0 + \lambda H_1 \end{bmatrix} \succeq 0,$$

$$\lambda \geq 0.$$

The dual of the considered non-convex QP is thus a convex semidefinite program (see Chapter 11).

An important result can be proved on strong duality for the problem under consideration.[3] Specifically, if the primal problem is strictly feasible, then it holds that $p^* = d^*$. Notice that this statement cannot be claimed by simply appealing to Proposition 8.7, since the primal problem is not assumed to be convex here. This result is also connected with the so-called $\mathcal{S}$-procedure for quadratic functions, discussed in Section 11.3.3.1.

[3] The proof is not elementary; see, e.g., page 657 in S. Boyd and L. Vandenberghe, *Convex Optimization*, Cambridge University Press, 2004.

9.5   Modeling with LP and QP

9.5.1   Problems involving cardinality and their ℓ_1 relaxations

Many engineering applications require the determination of solutions that are *sparse*, that is possess only few nonzero entries (low-cardinality solutions). The quest for low-cardinality solutions often has a natural justification in terms of the general principle of *parsimony* of the ensuing design. However, finding minimum cardinality solutions (i.e., solutions with small ℓ_0 norm) is hard in general, from a computational point of view. For this reason, several *heuristics* are often used in order to devise tractable numerical schemes that provide low (albeit possibly not minimal) cardinality solutions. One of these schemes involves replacing the ℓ_0 norm with the ℓ_1 norm. This use is justified by extensive numerical evidence showing that, indeed, the ℓ_1 heuristic is effective for obtaining low-cardinality solutions; an application of this idea is developed in a linear binary classification application in Section 13.3. In the present section we further discuss this issue, trying to provide some analytical support for this heuristic.

The *convex envelope* env f of a function $f : C \to \mathbb{R}$ is the largest convex function that is an underestimator of f on C, i.e., env $f(x) \leq f(x)$ for all $x \in C$ and no other convex function is uniformly larger than env f on C, that is

$$\text{env } f = \sup\{\phi : C \to \mathbb{R} : \phi \text{ is convex and } \phi \leq f\}.$$

Intuitively, the epigraph of the convex envelope of f corresponds to convex hull of the epigraph of f, see Figure 9.14.

Finding the convex envelope of a function is a hard problem in general. However, some special cases are well known. For instance, if $C = [0, 1]^n$ (the unit hypercube) and f is a monomial $f = x_1 x_2 \cdots x_n$,

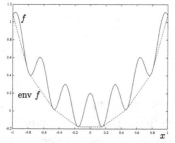

Figure 9.14 A non-convex function and its convex envelope (dashed) on $[-1, 1]$.

then

$$\text{env}\, f \;=\; \max\left(0,\, 1 - n + \sum_{i=1}^{n} x_i\right),$$

$$\text{env}\,(-f) \;=\; \max_{i=i,\dots,n} -x_i,$$

i.e., the convex envelopes for f and $-f$ on the unit hypercube are polyhedral functions. Also, for $x \in \mathbb{R}$ the convex envelope of card (x) on $[-1, 1]$ is env $f = |x|$ and, for the vector case $x \in \mathbb{R}^n$,

$$\text{env card}\,(x) = \frac{1}{R}\|x\|_1, \quad \text{on } C = \{x : \|x\|_\infty \le R\}.$$

This fact justifies the use of the ℓ_1 heuristic, at least in the case where the domain is bounded in the ℓ_∞ ball, since the ℓ_1 norm yields the best convex approximation of card (x) from below.

Another interesting relation between the ℓ_1 norm of $x \in \mathbb{R}^n$ and its cardinality is obtained via the Cauchy–Schwartz inequality applied to the inner product of $|x|$ and nz (x), where $|x|$ is the vector whose entries are the absolute values of x, and nz (x) is the vector whose i-th entry is one whenever $x_i \ne 0$, and is zero otherwise. Indeed, we have that, for all $x \in \mathbb{R}^n$,

$$\|x\|_1 = \text{nz}\,(x)^\top |x| \le \|\text{nz}\,(x)\|_2 \cdot \|x\|_2 = \|x\|_2 \sqrt{\text{card}\,(x)},$$

hence

$$\text{card}\,(x) \le k \quad \Rightarrow \quad \|x\|_1^2 \le k\|x\|_2^2.$$

This latter relation can be used to obtain convex relaxations on certain cardinality-constrained problems. For instance, consider the problem

$$p^* = \min_{x \in \mathbb{R}^n} c^\top x + \|x\|_2^2 \quad \text{s.t.:}\; Ax \le b,\; \text{card}\,(x) \le k.$$

Then, the objective of this problem is lower bounded (under the cardinality constraint) as follows: $c^\top x + \|x\|_2^2 \ge c^\top x + \|x\|_1^2 / k$. Therefore, we can obtain a lower bound for p^* by solving the problem

$$\tilde{p}^* = \min_{x \in \mathbb{R}^n} c^\top x + \|x\|_1^2 / k \quad \text{s.t.:}\; Ax \le b,$$

which can be expressed as a (convex) QP, by introducing a vector of slack variables $u \in \mathbb{R}^n$ and a scalar t:

$$\tilde{p}^* = \min_{x,u,t} \; c^\top x + t^2/k$$
$$\text{s.t.:}\; Ax \le b,$$
$$\sum_{i=1}^{n} u_i \le t,$$
$$-u_i \le x_i \le u_i, \quad i = 1,\dots,n.$$

Another classical problem where a cardinality penalty term is replaced by an ℓ_1-norm term is the LASSO problem, discussed in Section 9.6.2.

Example 9.16 (*Piece-wise constant fitting*) Suppose one observes a noisy time-series which is almost piece-wise constant. The goal in piece-wise constant fitting is to find what the constant levels are. In biological or medical applications, such levels might have interpretations of "states" of the system under observation. Formally, let $x \in \mathbb{R}^n$ denote the signal vector (which is unknown) and let $y \in \mathbb{R}^n$ denote the vector of noisy signal observations (i.e., y is true signal x, plus noise). Given y, we seek an estimate $\hat{x}$ of the original signal x, such that $\hat{x}$ has as few changes in consecutive time steps as possible. We model the latter requirement by minimizing the cardinality of the difference vector $D\hat{x}$, where $D \in \mathbb{R}^{n-1,n}$ is the difference matrix

$$D = \begin{bmatrix} -1 & 1 & 0 & \cdots & 0 \\ 0 & -1 & 1 & \cdots & 0 \\ \vdots & & & \ddots & \\ 0 & \cdots & 0 & -1 & 1 \end{bmatrix},$$

so that $D\hat{x} = [\hat{x}_2 - \hat{x}_1, \hat{x}_3 - \hat{x}_2, \ldots, \hat{x}_n - \hat{x}_{n-1}]^\top$. We are thus led to the problem

$$\min_{\hat{x}} \|y - \hat{x}\|_2^2 \quad \text{s.t.: } \operatorname{card}(D\hat{x}) \leq k,$$

where k is an estimate on the number of jumps in the signal. Here, the objective function in the problem is a measure of the error between the noisy measurement and its estimate $\hat{x}$. We can relax this hard problem via the ℓ_1-norm heuristic, by replacing the cardinality constraint with an ℓ_1 constraint, thus obtaining the QP

$$\min_{\hat{x}} \|y - \hat{x}\|_2^2 \quad \text{s.t.: } \|D\hat{x}\|_1 \leq q, \tag{9.21}$$

for some suitably chosen q (note that choosing $q = k$ need not imply that the solution resulting from the relaxed problem is such that $\operatorname{card}(D\hat{x}) \leq k$). Alternatively, one may cast a problem with a weighted objective:

$$\min_{\hat{x}} \|y - \hat{x}\|_2^2 + \gamma \|D\hat{x}\|_1,$$

for some suitable trade-off parameter $\gamma \geq 0$.

Figure 9.15 shows an example of signal reconstruction via piece-wise fitting. The top panel in Figure 9.15 shows the unknown signal x (dashed) and its available noisy measurement y; the center panel shows the unknown signal x (dashed) and its reconstruction $\hat{x}$ obtained via the ℓ_1 heuristic in (9.21); the bottom panel shows the unknown signal x (dashed) and its reconstruction $\hat{x}$ obtained by solving a variation on (9.21) where the ℓ_2 norm is used instead of the ℓ_1 norm in the constraint. We notice that the ℓ_1 heuristic is successful in eliminating the noise from the signal, while preserving sharp transitions in the phase (level) changes in the signal. On the contrary, with an ℓ_2 heuristic, noise elimination only comes at the price of sluggish phase transitions.

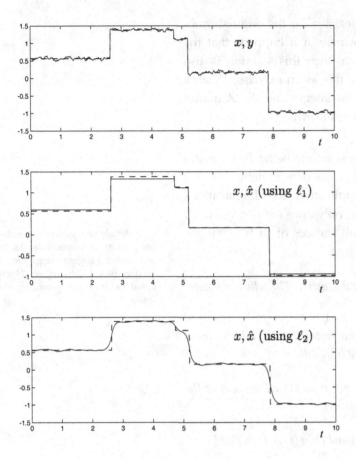

Figure 9.15 Example of reconstructing a piece-wise constant signal (top) from noisy measurements, using ℓ_1 (center) or ℓ_2 (bottom) heuristics.

9.5.2   LP relaxations of Boolean problems

A Boolean optimization problem is one where the variables are constrained to be Boolean, i.e. to take on values in $\{0, 1\}$. For example, a Boolean LP takes the form

$$p^* = \min_x \ c^\top x \quad \text{s.t.:} \ Ax \le b, \ x \in \{0,1\}^n.$$

Such problems are usually very hard to solve exactly, since they potentially require combinatorial enumeration of all the the 2^n possible point arrangements in $\{0,1\}^n$. A tractable *relaxation* of a Boolean problem is typically obtained by replacing the discrete set $\{0,1\}^n$ with the hypercube $[0,1]^n$, which is a convex set. For instance, the relaxation of the previous Boolean LP yields the standard LP

$$\tilde{p}^* = \min_x \ c^\top x \quad \text{s.t.:} \ Ax \le b, \ x \in [0,1]^n.$$

Since the feasible set of the relaxed problem is larger than (i.e., it includes) the feasible set of the original problem, the relaxation provides a lower bound on the original problem: $\tilde{p}^* \le p^*$. The solution of the LP relaxation is not necessarily feasible for the original problem (i.e., it may not be Boolean). However, if it happens that the solution of the LP relaxation is Boolean, then this solution is also optimal for the original problem (prove this as an exercise). Such a situation arises for instance when b is an integer and the A matrix has a particular property called *total unimodularity*.

9.5.2.1 Total unimodularity. A matrix A is said to be *totally unimodular* (TUM) if every square submatrix of A has determinant -1, 1, or 0. This matrix concept has interesting applications for LP relaxations of Boolean problems, due to the fact that polytopes defined via TUM matrices have *integer* vertices,[4] that is all vertices of such polytopes have integer entries.

Theorem 9.1 *Let $A \in \mathbb{R}^{m,n}$ be an integral matrix. The following statements hold:*

(1) A is TUM if and only if for any integral vector $b \in \mathbb{R}^n$ all vertices of the polyhedron $\{x : Ax \leq b, x \geq 0\}$ are integral;

(2) if A is TUM then for any integral vector $b \in \mathbb{R}^n$ all vertices of the polyhedron $\{x : Ax = b, x \geq 0\}$ are integral;

(3) A is TUM if and only if $A^\top$ is TUM if and only if $[A\ I]$ is TUM.

Also, the following corollary provides a useful sufficient condition for TUM.

Corollary 9.1 *A matrix $A \in \mathbb{R}^{m,n}$ is TUM if all the following conditions are satisfied:*

(a) each entry of A is -1, 1, or 0;

(b) each column of A contains at most two nonzero entries;

(c) the rows of A can be partitioned into two subsets $R_1 \cup R_2 = \{1, \ldots, m\}$ such that in each column j with two nonzero entries it holds that $\sum_{i \in R_1} a_{ij} = \sum_{i \in R_2} a_{ij}$.

An immediate consequence of the previous corollary is that a $(0, 1, -1)$ matrix is TUM if it contains no more than one 1 and no more than one -1 in each column. Such a particular situation actually arises in several optimization problems on graphs, where A is the incidence matrix of a directed graph or the incidence matrix of a bipartite graph, as illustrated in the following examples.

[4] For details and proofs of results related to total unimodular matrices and linear programming, see, e.g., Section 19 of A. Schrijver, *Theory of Linear and Integer Programming*, Wiley, 1998.

Example 9.17 (*Weighted bipartite matching*) A weighted bipartite matching problem arises when n agents need to be assigned to n tasks, in a one-to-one fashion, and the cost of matching agent i to task j is w_{ij}, see Figure 9.16.

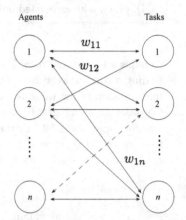

Figure 9.16 Bipartite matching problem.

Defining variables x_{ij} such that $x_{ij} = 1$ if agent i is assigned to task j and $x_{ij} = 0$ otherwise, the problem is written in the form of a Boolean LP:

$$
\begin{aligned}
p^* = \min_{x} \quad & \sum_{i,j=1}^{n} w_{ij} x_{ij} \\
\text{s.t.:} \quad & x_{ij} \in \{0,1\} \quad \forall i,j = 1,\dots,n, \\
& \sum_{i=1}^{n} x_{ij} = 1 \quad \forall j = 1,\dots,n \ \text{(one agent for each task)}, \\
& \sum_{j=1}^{n} x_{ij} = 1 \quad \forall i = 1,\dots,n \ \text{(one task for each agent)}.
\end{aligned}
$$

An LP relaxation is obtained by dropping the integrality constraint on the x_{ij} variables, thus obtaining

$$
\begin{aligned}
\tilde{p}^* = \min_{x} \quad & \sum_{i,j=1}^{n} w_{ij} x_{ij} \\
\text{s.t.:} \quad & x_{ij} \geq 0 \quad \forall i,j = 1,\dots,n, \\
& \sum_{i=1}^{n} x_{ij} = 1 \quad \forall j = 1,\dots,n, \\
& \sum_{j=1}^{n} x_{ij} = 1 \quad \forall i = 1,\dots,n.
\end{aligned} \tag{9.22}
$$

Although, in general, the optimal solution of the relaxed problem is not guaranteed to be Boolean, in the present special case it is possible to prove that any vertex solution of the relaxed problem is Boolean. Indeed, the constraints in problem (9.22) can be written more compactly as

$$
x \geq 0, \quad Ax = \mathbf{1}, \tag{9.23}
$$

where $A \in \mathbb{R}^{2n,n^2}$ is the (undirected) incidence matrix of the undirected bipartite graph in Figure 9.16, and $x \in \mathbb{R}^{n^2}$ is a vector containing a column vectorization of the matrix of variables $[x_{ij}], i,j = 1,\dots,n$. The rows of A correspond to nodes in the graph, say the first n nodes are the agents and the last n are the tasks. The columns of A represents the edges in the graph, and $A_{ie} = 1$ if edge e is incident on node i, and $A_{ie} = 0$ otherwise. The polytope represented by (9.23), known as the *bipartite perfect matching polytope*, thus has integer vertices, due to total unimodularity of A, hence for the weighted bipartite matching problem the LP relaxation actually provides an optimal solution to the original Boolean LP.

As a numerical example, consider matching $n = 4$ agents to respective tasks, with costs described by the matrix W:

$$
W = \begin{bmatrix} 5 & 1 & 2 & 2 \\ 1 & 0 & 5 & 3 \\ 2 & 1 & 2 & 1 \\ 1 & 1 & 2 & 3 \end{bmatrix}.
$$

The LP relaxation provides the optimal agent/task matching $(1,3)$, $(2,2)$, $(3,4)$, $(4,1)$, with associated optimal cost $p^* = 4$.

Example 9.18 (*Shortest path*) The shortest path problem is the problem of finding a path between two vertices (or nodes) in a directed graph such that the sum of the weights along the edges in the path is minimized. Consider a directed graph with nodes $V = \{1,\dots,n\}$ and edge set E, let t be the target node, let s be the source node, and let w_e denote the cost for traveling along edge $e \in E$. Then, the shortest (minimum cost) path can be found by solving the following Boolean LP:

$$p^* = \min_x \; \sum_{e \in E} w_e x_e$$
$$\text{s.t.:} \quad x_e \in \{0,1\} \quad \forall e \in E,$$
$$Ax = b,$$

where $A \in \mathbb{R}^{n,|E|}$ is the (directed) incidence matrix of the graph (i.e. $A_{ie} = 1$ if edge $e \in E$ starts at node i, $A_{ie} = -1$ if edge $e \in E$ ends at node i, and $A_{ie} = 0$ otherwise), and $b \in \mathbb{R}^n$ is a vector such that $b_s = 1$, $b_t = -1$, and $b_i = 0$, $\forall i \neq s,t$. Again, matrix A is TUM, hence the standard LP relaxation actually yields an optimal Boolean solution for this problem.

9.5.3 Network flows

Consider a network described by a directed graph with m nodes connected by n directed edges, as in Figure 9.17.

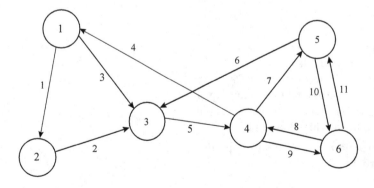

Figure 9.17 Example of a directed graph with $m = 6$ nodes and $n = 11$ edges.

The (directed) arc–node incidence matrix $A \in \mathbb{R}^{m,n}$ of the network is defined as in (3.2) and, for the example in Figure 9.17, we have

$$
A = \begin{bmatrix}
1 & 0 & 1 & -1 & 0 & 0 & 0 & 0 & 0 & 0 & 0 \\
-1 & 1 & 0 & 0 & 0 & 0 & 0 & 0 & 0 & 0 & 0 \\
0 & -1 & -1 & 0 & 1 & -1 & 0 & 0 & 0 & 0 & 0 \\
0 & 0 & 0 & 1 & -1 & 0 & 1 & -1 & 1 & 0 & 0 \\
0 & 0 & 0 & 0 & 0 & 1 & -1 & 0 & 0 & 1 & -1 \\
0 & 0 & 0 & 0 & 0 & 0 & 0 & 1 & -1 & -1 & 1
\end{bmatrix}.
$$

A *flow* (of traffic, information, charge) is represented by a vector $x \in \mathbb{R}^n$ of signed edge variables ($x_i \geq 0$ if the flow is in the direction of the arc, and $x_i < 0$ if it is in the opposite direction). The *net flow out* of node i is

$$
\text{out-flow}_i = (Ax)_i = \sum_{j=1}^{n} A_{ij} x_j.
$$

Suppose that with each edge flow x_i is associated a convex cost function $\phi_i(x_i)$: a *minimum-cost flow problem* amounts to determining minimal-cost flows that satisfy given supply/demand requirements, as well as capacity constraints on the links, that is:

$$
\min_{x} \sum_{i=1}^{n} \phi_i(x_i) \quad \text{s.t.:} \quad Ax = b, \ l \leq x \leq u,
$$

where b is the vector of *external supplies* at nodes ($b_j > 0$ if node j is a source, $b_j < 0$ if node j is a sink, and $b_j = 0$ otherwise) which satisfies a flow conservation equality $\mathbf{1}^\top b = 0$ so that the total supply equals the total demand, and l, u are lower and upper bounds on the flows, respectively (for instance, $l = 0$ if we want to impose that flows must follow the direction of the arcs). The constraint $Ax = b$ represents the balance equations of the network. In the special case where ϕ_i are linear functions, the above problem is an LP with objective $\phi^\top x$, where ϕ_i now represents the unit cost of a flow through link i.

9.5.3.1 *Maximum flow problems.*

A related problem arises when there is one single source node s and one single sink node t, and one seeks to maximize the flow between s and t. Letting the (unknown) external supply vector be $b = \gamma e$, where $e_s = 1$, $e_t = -1$, and $e_i = 0$ for $i \neq s, t$, we then seek to maximize γ while satisfying the flow balance and capacity constraints, that is

$$
\max_{x, \gamma} \gamma \quad \text{s.t.:} \quad Ax = \gamma e, \ l \leq x \leq u,
$$

which is an LP.

9.5.4   Nash equilibria in zero-sum games

Game theory models the behavior of conflicting rational players aiming at maximizing their payoff (or at minimizing their cost) via actions that take into account counter-actions from the other players. A central role in the theory of games is played by the so-called two-person zero-sum games, that is games involving two players A, B with perfectly conflicting objectives, i.e., such that if the payoff for player A is p then the payoff for player B is $-p$. In the discrete case, the possible "moves" or choices for each player are discrete and finite, hence they can be listed by rows and columns (say, actions for player A in the rows and actions for player B in the columns). These types of games are also known as *matrix games*. If A has m available actions and B has n available actions, then the game can be represented via a *payoff matrix* $P \in \mathbb{R}^{m,n}$, where the entry P_{ij} represents the payoff for player A when she plays the i-th action and B plays the j-th action (choices of actions by the two players are assumed to be taken simultaneously). Since the game is zero sum, it is enough to specify in P the payoffs for player A, since the ones for player B are simply $-P$.

As a first example, consider two hot-dog vendors on the same street, competing for the same market. Each vendor has a fixed cost of \$200, and must choose a high price (\$2 per sandwich) or a low price (\$1 per sandwich). At a price of \$2, vendors can sell 200 sandwiches, for a total revenue of \$400. At a price of \$1, vendors can sell 400 sandwiches, for a total revenue of \$400. If both vendors fix the same price, they split the sales evenly between them, otherwise the vendor with the lower price sells the whole amount and the vendor with the higher price sells nothing. The payoff table (payoffs are profits: revenue minus fixed cost) is given in Table 9.4.

$A \backslash B$	price \$1	price \$2	min row
price \$1	$\boxed{0}$	200	0
price \$2	-200	0	-200
max col	0	200	

Table 9.4 Payoff matrix P for the hot-dog vendors game.

For such a game it is rational for each player to choose the strategy that maximizes her minimum payoff:

$$\text{best strategy for A:} \quad \max_i \min_j P_{ij},$$

$$\text{best strategy for B:} \quad \min_j \max_i P_{ij},$$

that is, player A looks at the minimum value over each row, finding the vector of minimal payoffs for her actions (shown as the rightmost column in the table), then she chooses the action corresponding to the largest entry in this column. The safe strategy for A is thus to set the price to $1. Similarly, player B looks at the maximum value over each column (since her payoffs are the negative of the entries in the table), finding the vector of minimal payoffs for her actions (shown as the lowermost row in the table), then she chooses the action corresponding to the smallest entry in this row. The safe strategy for B is also to set the price to $1. Notice that in this case both the A player and the B player strategies lead to an equal payoff. When this happens, we have found an *equilibrium solution* for the game in pure strategies. Such a solution is characterized by the fact that the common payoff represents a *saddle point* of the payoff matrix, that is an entry such that

$$\max_i \min_j P_{ij} = \min_j \max_i P_{ij},$$

the numerical value of such a common payoff is called the *value* of the game. A game may have multiple saddle points, but these will all have an equivalent value. A saddle-point equilibrium represents a decision by two players upon which neither can improve by unilaterally departing from it.

However, not all games posses a saddle point. As a second example consider the game of Odd-or-Even in which two players simultaneously call out one of the numbers: zero, one, or two. If the sum of the outcomes is odd, then the 'Odd' player wins from the other player an amount in dollars equal to the numerical value of the sum, and vice versa. The payoff matrix of the game is

Odd\Even	'0'	'1'	'2'	min row
'0'	0	1	−2	−2
'1'	1	−2	3	−2
'2'	−2	3	−4	−4
max col	1	3	3	

Table 9.5 Payoff matrix P for the Odd-or-Even game.

It is immediate to check that this game has no saddle point, hence no pure strategy yields an equilibrium solution:

$$-2 = \max_i \min_j P_{ij} \leq \min_j \max_i P_{ij} = 1.$$

In such cases, each player can resort to *mixed strategies*, that is she can choose a decision at random, according to an assigned probability distribution over the decisions. Players now reason precisely as in the previous case, but focusing on the *expected* payoffs. The problem then amounts to finding suitable distributions so that a saddle point for the expected payoff matrix exists (equalizing strategy). Suppose then that player A plays decision i with probability $q_i^{(A)}$, and player

B plays decision j with probability $q_j^{(B)}$, and let $q^{(A)}$, $q^{(B)}$ be vectors representing the probability distributions on the strategies of players A and B, respectively. Then the vector of expected payoffs for player A corresponding to her possible strategies is $Pq^{(B)}$ (the i-th entry in this column vector is the average (according to the probabilities on the strategies of player B) of the payoffs of player A if she chooses the i-th action). The overall expected payoff for player A, considering that she also randomizes upon her strategies, is therefore

$$q^{(A)\top}Pq^{(B)}.$$

Now, since player A wants to maximize her worst-case expected payoff against all possible choices of $q^{(B)}$ (each player knows the payoff matrix P, but does not know the opponent's randomization strategy), player A's problem amounts to solving

$$V_A = \max_{q^{(A)}\in S^m} \min_{q^{(B)}\in S^n} q^{(A)\top}Pq^{(B)},$$

where S^m, S^n denote respectively the probability simplex in m and n dimensions. Player B reasons in a dual way, hence the problem for player B amounts to solving

$$V_B = \min_{q^{(B)}\in S^n} \max_{q^{(A)}\in S^m} q^{(A)\top}Pq^{(B)}. \tag{9.24}$$

A fundamental result of Von Neumann actually guarantees that $V_A = V_B = V$, the *value* of the game, i.e., that any matrix game has a min-max equilibrium solution, either in pure or mixed strategy. Under such an equilibrium strategy, the expected payoff for player A is at least V, no matter what player B does, and the expected loss for player B is at most V, no matter what player A does. If V is zero we say the game is fair. If V is positive, we say the game favors player A, while if V is negative, we say the game favors player B. In the next paragraph we outline a connection between games and optimization problems, and show how the optimal mixed strategy for a matrix game can be computed via linear programming.

9.5.4.1 LP solution of matrix games. We consider the problem of computing the optimal mixed strategy for player B, the case of player A being completely analogous. Start by rewriting problem (9.24) in epigraphic form

$$V_B = \min_{\gamma\in\mathbb{R},q^{(B)}\in S^n} \gamma$$
$$\text{s.t.:} \quad \max_{q^{(A)}\in S^m} q^{(A)\top}Pq^{(B)} \leq \gamma.$$

Then observe that $\max_{y \in \mathcal{Y}} f(y) \leq \gamma$ if and only if $f(y) \leq \gamma$ for all $y \in \mathcal{Y}$, therefore the problem is rewritten as

$$V_B = \min_{\gamma \in \mathbb{R}, q^{(B)} \in S^n} \gamma$$
$$\text{s.t.:} \quad q^{(A)\top} P q^{(B)} \leq \gamma, \quad \forall q^{(A)} \in S^m.$$

Now, simplex S^m is a polytope having as vertices the standard basis $e_1, \ldots, e_m$ of $\mathbb{R}^m$. Hence, applying the vertex result in (9.7), we have equivalently

$$V_B = \min_{\gamma \in \mathbb{R}, q^{(B)} \in S^n} \gamma$$
$$\text{s.t.:} \quad e_i^\top P q^{(B)} \leq \gamma, \quad i = 1, \ldots, m.$$

Considering that $e_i^\top P$ is nothing but the i-th row of P, and writing explicitly the condition $q^{(B)} \in S^n$ as $q^{(B)} \geq 0, \mathbf{1}_n^\top q^{(B)} = 1$, we finally obtain a problem in standard LP format:

$$V_B = \min_{\gamma \in \mathbb{R}, q^{(B)} \in \mathbb{R}^n} \gamma$$
$$\text{s.t.:} \quad P q^{(B)} \leq \gamma \mathbf{1}_m, \ q^{(B)} \geq 0, \ \mathbf{1}_n^\top q^{(B)} = 1.$$

An analogous reasoning shows that the equilibrium strategy for player A may be found by solving the following LP:

$$V_A = \max_{\gamma \in \mathbb{R}, q^{(A)} \in \mathbb{R}^m} \gamma$$
$$\text{s.t.:} \quad P^\top q^{(A)} \geq \gamma \mathbf{1}_n, \ q^{(A)} \geq 0, \ \mathbf{1}_m^\top q^{(A)} = 1.$$

Using the LP approach, the reader may verify, for instance, that the Odd-or-Even game in Table 9.5 is a fair game ($V = 0$) and has optimal mixed strategies

$$q^{(A)} = \begin{bmatrix} 1/4 \\ 1/2 \\ 1/4 \end{bmatrix}, \quad q^{(B)} = \begin{bmatrix} 1/4 \\ 1/2 \\ 1/4 \end{bmatrix}.$$

9.6 LS-related quadratic programs

A major source of quadratic problems comes from LS problems and their variants. We have already seen in Example 9.1 that the standard LS objective

$$f_0(x) = \|Ax - y\|_2^2$$

is a convex quadratic function, which can be written in the standard form

$$f_0(x) = \frac{1}{2} x^\top H x + c^\top x + d,$$

with

$$H = 2(A^\top A), \quad c = -2A^\top y, \quad d = y^\top y.$$

Finding the unconstrained minimum of f_0 is just a linear algebra problem, which amounts to finding a solution for the system of linear equations (the *normal equations*) resulting from the optimality condition $\nabla f_0(x) = 0$ (see Section 8.4.1):

$$A^\top A x = A^\top y.$$

We next illustrate some variants of the basic LS problem, some of which are also amenable to a simple linear-algebra based solution.

9.6.1 Equality constrained LS

Some variants on the basic LS problem have been discussed in Section 6.7. Here, we briefly discuss the case of an LS problem with additional linear equality constraints on the variables. It has been shown in Example 9.2 that minimizing a convex quadratic function under linear equality constraints is equivalent to solving an augmented system of linear equations. Therefore, solving the linear equality constrained LS problem

$$\begin{aligned} \min_x \quad & \|Ax - y\|_2^2 \\ \text{s.t.:} \quad & Cx = d \end{aligned}$$

is equivalent to solving the following linear equations in x, λ (see Eq. (9.5)):

$$\begin{bmatrix} C & 0 \\ A^\top A & C^\top \end{bmatrix} \begin{bmatrix} x \\ \lambda \end{bmatrix} = \begin{bmatrix} d \\ A^\top y \end{bmatrix}.$$

9.6.2 ℓ_1 regularization and the LASSO problem

Regularized LS problems, with an ℓ_2 regularization term, have been discussed in Section 6.7.3. An important variation arises when the regularization term in (6.25) involves the ℓ_1 norm of x, instead of the ℓ_2 norm. This results in the following problem, known as the *basis pursuit denoising problem* (BPDN):

$$\min_{x \in \mathbb{R}^n} \|Ax - y\|_2^2 + \lambda \|x\|_1, \quad \lambda \geq 0, \tag{9.25}$$

where $\|x\|_1 = |x_1| + \cdots + |x_n|$. Problem (9.25) has received enormous attention in recent years from the scientific community, due to its relevance in the field of *compressed sensing* (CS). The basic idea behind (9.25) is that the ℓ_1 norm of x is used as a proxy for the cardinality of x (the number of nonzero entries in x), see Section 9.5.1

for a justification of this fact. The interpretation of (9.25) is then that it formalizes a tradeoff between the accuracy with which Ax approximates y and the *complexity* of the solution, intended as the number of nonzero entries in x. The larger λ is, the more problem (9.25) is biased towards finding low-complexity solutions, i.e., solutions with many zeros. Such kinds of solution are of paramount importance, for instance, in signal and image compression. Suppose for example that $y \in \mathbb{R}^m$ is a vector representing the gray-scale levels of pixels in a multi-megapixel image (thus m can be extremely large, e.g., in the order of a few million), and let A contain by columns n fixed feature vectors (i.e., what is usually referred to as a *dictionary*): we seek to approximate the original image as a linear combination Ax of the columns of A with few nonzero coefficients. Then, instead of transmitting the whole bulky image vector y, one could transmit only the few nonzero coefficients in x and still the receiver (who knows the feature basis A) can approximately reconstruct the image as Ax.

We already encountered a problem similar to (9.25) in the context of piecewise constant fitting, see Example 9.16. Problem (9.25) can be cast in the form of a standard QP by introducing slack variables $u \in \mathbb{R}^n$:

$$\min_{x,u \in \mathbb{R}^n} \quad \|Ax - y\|_2^2 + \lambda \sum_{i=1}^{n} u_i$$
$$\text{s.t.:} \quad |x_i| \leq u_i, \ i = 1,\ldots,n.$$

An essentially analogous version of problem (9.25) is obtained by imposing a constraint on the ℓ_1 norm of x (instead of inserting this term in the objective as a penalty), resulting in the so-called *least absolute shrinkage and selection operator* (LASSO) problem:[5]

$$\min_{x \in \mathbb{R}^n} \quad \|Ax - y\|_2^2$$
$$\text{s.t.:} \quad \|x\|_1 \leq \alpha.$$

[5] Often, in the literature, the term LASSO is also used to refer to problem (9.25).

The LASSO problem can readily be cast in the standard QCQP format by introducing slack variables. Yet another version in which the problem can be formulated is in the form of minimization of $\|x\|_1$ subject to a constraint on the residual norm, that is

$$\min_{x \in \mathbb{R}^n} \quad \|x\|_1$$
$$\text{s.t.:} \quad \|Ax - y\|_2 \leq \epsilon,$$

which can also be cast as a QCQP. All these variations on the LASSO problem yield convex optimization models that can be solved by standard efficient algorithms for QCQP, at least in principle. Notice, however, that the typical applications where LASSO-type problems arise may involve a very large number of variables, hence several specialized algorithms have been developed to solve ℓ_1-regularized problems with maximal efficiency; see Section 12.3.3.8 and Section 12.3.4 for a discussion on some of these algorithms.

Example 9.19 (*Image compression in a wavelet basis*) A gray-scale image, represented by a vector $y \in \mathbb{R}^m$, typically admits an essentially sparse representation, in a suitable basis. This means that, for an appropriate dictionary matrix $A \in \mathbb{R}^{m,n}$, the image y can be well approximated by a linear combination Ax of the feature vectors, where the coefficients x of the combination are sparse. Usual dictionary matrices employed in image analysis include discrete fourier transform (DFT) bases, and wavelet (WT) bases. Wavelet bases, in particular, have been recognized to be quite effective in providing sparse representations of standard images (they are used, for instance in the Jpeg2000 compression protocol).

Figure 9.18 "Boat" original 256×256 gray-scale image.

Consider, for example, the 256×256 gray-scale image shown in Figure 9.18. Each pixel in this image is represented by an integer value y_i in the range $[0, 255]$, where the 0 level is for black, and 255 is for white. The histogram of y for the original image is shown in Figure 9.19; clearly, in this representation, the image is not sparse.

However, if we consider the image representation in the wavelet transform domain (which implicitly amounts to considering a suitable dictionary matrix A containing by columns the wavelet bases), we obtain a vector representation $\tilde{y}$ whose absolute value has the histogram shown in Figure 9.20. For this example, we are using a Daubechies orthogonal wavelet transform, hence A is a $65,536 \times 65,536$ orthogonal matrix.

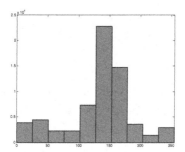

Figure 9.19 Histogram of y, for the boat image.

Figure 9.20 shows that the wavelet representation $\tilde{y}$ of the image contains very few large coefficients, while most of the coefficients are relatively small (however, $\tilde{y}$ is not yet sparse, since its elements are not exactly zero). If all these small coefficients are retained, then $\tilde{y}$ carries the same information as y, that is, it is a *lossless* encoding of the original image, in the wavelet domain: $y = A\tilde{y}$. However, if we allow for this equality to be relaxed to approximate equality $y \simeq Ax$, we may tradeoff some accuracy for a representation x in the wavelet domain which has many zero coefficients, i.e., a sparse representation. Such a sparse tradeoff can typically be obtained by solving the LASSO problem (9.25) for suitable λ, that is $\min_x \frac{1}{2}\|Ax - y\|_2^2 + \lambda\|x\|_1$. In our specific situation, since A is orthogonal, we have that the above problem is equivalent to

$$\min_x \frac{1}{2}\|x - \tilde{y}\|_2^2 + \lambda\|x\|_1,$$

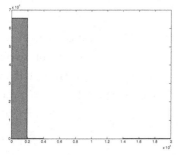

Figure 9.20 Histogram of the wavelet transform $\tilde{y}$, for the boat image.

where $\tilde{y} \doteq A^\top y$ is the image representation in the wavelet domain. Interestingly, this problem is *separable*, i.e., it can be reduced to a series of univariate minimization problems, since

$$\frac{1}{2}\|x - \tilde{y}\|_2^2 + \lambda\|x\|_1 = \sum_{i=1}^{m} \frac{1}{2}(x_i - \tilde{y}_i)^2 + \lambda|x_i|.$$

Moreover, each of the single-variable problems

$$\min_{x_i} \frac{1}{2}(x_i - \tilde{y}_i)^2 + \lambda|x_i|$$

admits a simple closed-form solution as (see Section 12.3.3.5)

$$x_i^* = \begin{cases} 0 & \text{if } |\tilde{y}_i| \leq \lambda, \\ \tilde{y}_i - \lambda\text{sgn}(\tilde{y}_i) & \text{otherwise.} \end{cases}$$

In words, this means that all coefficients $\tilde{y}_i$ in the wavelet basis are thresholded to zero if their modulus is smaller than λ, and are offset by λ, otherwise (soft thresholding). Once we have computed x^*, we can reconstruct an actual image in the standard domain, by computing the inverse wavelet transform (i.e., ideally, we construct the product Ax^*).

(a)   (b)   (c)

In our current example, solving the LASSO problem with $\lambda = 30$ we obtained a representation x^* in the wavelet domain that has only $4,540$ nonzero coefficients (against the $65,536$ nonzero coefficients present in $\tilde{y}$ or in y). We have therefore a compression factor of about 7%, meaning that the size of the compressed image is only 7% of the size of the original image. Reducing the regularization parameter to $\lambda = 10$, we obtained instead a representation x^* in the wavelet domain with $11,431$ nonzero coefficients, and thus a compression factor of about 17%. Figure 9.21 shows the original image along with the reconstructed compressed images obtained, respectively, by choosing $\lambda = 10$ and $\lambda = 30$ in the LASSO problem.

Figure 9.21 Comparison of original boat image (a), wavelet compression with $\lambda = 10$ (b), and wavelet compression with $\lambda = 30$ (c).

9.7 Geometric programs

Geometric programming (GP) deals with optimization models where the problem variables are positive (they typically represent physical quantities such as pressure, areas, prices, concentrations, energy, etc.) and appear in the objective and constraint functions in the form of non-negative linear combinations of positive monomials, see Example 8.18 and Section 8.3.5.2.

9.7.1 Monomials and posynomials

For two vectors $x \in \mathbb{R}^n$, $a \in \mathbb{R}^n$, and scalar $c > 0$, we use the following notation to represent a positive monomial:

$$cx^a \doteq cx_1^{a_1} x_2^{a_2} \cdots x_n^{a_n}, \quad x > 0.$$

A *posynomial* is then defined as a function $f : \mathbb{R}_{++}^n \to \mathbb{R}$ which is a non-negative linear combination of positive monomials:

$$f(x) = \sum_{i=1}^{K} c_i x^{a_{(i)}}, \quad x > 0, \tag{9.26}$$

where $c_i > 0$ and $a_{(i)} \in \mathbb{R}^n$, $i = 1, \ldots, K$. Further, a *generalized posynomial* is any function obtained from posynomials via addition, multiplication, pointwise maximum, and raising to a constant power. For example, the function $f : \mathbb{R}_{++}^3 \to \mathbb{R}$ with values

$$f(x) = \max \left(2x_1^{2.3} x_2^7, x_1 x_2 x_3^{3.14}, \sqrt{x_1 + x_2^3} \right)$$

is a generalized posynomial.

Example 9.20 (*Construction and operating costs of a storage tank*) Consider a cylindrical liquid storage tank with height h and diameter d, as shown in Figure 9.22.

The tank includes a base, which is made of a different material from the tank itself. In our model, the base's height does not depend on the tank's height. This is a reasonable approximation for heights not exceeding a certain value.

The costs to manufacture, and then operate during a given period of time (say, a year) the tank, include the following.

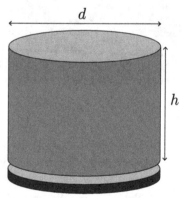

Figure 9.22 A liquid storage tank.

- *Filling costs* represent the costs associated with supplying a volume of liquid (say, water) in the given time period. These costs depend on the ratio $V_{\text{supp}}/V_{\text{tank}}$, where V_{supp} is the volume to be supplied, and V_{tank} is the volume of the tank (the smaller the volume of the tank is with respect to the volume to be supplied, the more often we have to refill the tank, and the larger the cost). Thus, the filling cost is inversely proportional to the tank's volume:

$$C_{\text{fill}}(d, h) = \alpha_1 \frac{V_{\text{supp}}}{V_{\text{tank}}} = c_1 h^{-1} d^{-2},$$

where α_1 is some positive constant, expressed in (say) dollars, and $c_1 = 4\alpha_1 V_{\text{supp}}/\pi$.

• *Construction costs* include the costs associated with building a base for the tank, and costs associated with building the tank itself. In our model, the first type of cost depends only on the base area $\pi d^2/4$, while the second type of cost depends on the surface of the tank, πdh (this assumes that we can use the same base height for a variety of tank heights). Thus, the total construction cost can be written as

$$C_{\text{constr}}(d,h) = C_{\text{base}}(d,h) + C_{\text{tank}}(d,h) = c_2 d^2 + c_3 dh,$$

where $c_2 = \alpha_2 \pi/4$, $c_3 = \alpha_3 \pi$, with $\alpha_2 > 0$, $\alpha_3 > 0$ constants expressed in dollars per square meter.

The total manufacturing and operating cost function is thus the posynomial function

$$C_{\text{total}}(d,h) = C_{\text{fill}}(d,h) + C_{\text{constr}}(d,h) = c_1 h^{-1} d^{-2} + c_2 d^2 + c_3 dh. \quad (9.27)$$

Assuming the numerical values $V_{\text{supp}} = 8 \times 10^5$ litres, $\alpha_1 = 10$ \$, $\alpha_2 = 6$ \$/m^2, $\alpha_3 = 2$ \$/m^2, we obtain the plot shown in Figure 9.23 for the level curves of the total cost $C_{\text{total}}(d,h)$. The sublevel sets are non-convex, hence $C_{\text{total}}(d,h)$ is non-convex.

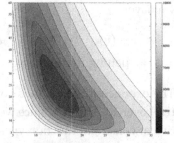

Figure 9.23 Some level curves of the tank cost $C_{\text{total}}(d,h)$.

Example 9.21 (*Signal-to-noise ratio in wireless communications*) Consider a cellular wireless network with n transmitter/receiver pairs. Transmit powers are denoted by $p_1, \ldots, p_n$. Transmitter i is supposed to transmit to receiver i but, due to interference, some signal from the other transmitters is also present. In addition, there is (self-) noise power in each receiver. To measure this, we form the signal to interference plus noise ratio (SINR) at each receiver, which takes the form

$$\gamma_i = \frac{S_i}{I_i + \sigma_i}, \quad i = 1, \ldots, n,$$

where S_i is a measure of the (desired) signal power received from transmitter i, I_i is the total signal power received from all the other receivers, and σ_i is a measure of the receiver noise. The SINR is a (in general, complicated) function of the power used at the transmitters. We can express the SINRs at the receivers in terms of the powers $p_1, \ldots, p_n$ more explicitly, by assuming that the received powers S_i, are linear functions of the transmitted powers $p_1, \ldots, p_n$. This model, also known as the Rayleigh fading model, states that

$$S_i = G_{ii} p_i, \quad i = 1, \ldots, n,$$

and $I_i = \sum_{j \neq i} G_{ij} p_j$, where the coefficients G_{ij}, $1 \leq i, j \leq n$, are known as the *path gains* from transmitter j to receiver i. The SINR functions

$$\gamma_i(p) = \frac{S_i}{I_i + \sigma_i} = \frac{G_{ii} p_i}{\sigma_i + \sum_{j \neq i} G_{ij} p_j}, \quad i = 1, \ldots, n,$$

are not posynomials, but their inverses are indeeed posynomial functions in the powers $p_1, \ldots, p_n$.

$$\gamma_i^{-1}(p) = \frac{\sigma_i}{G_{ii}} p_i^{-1} + \sum_{j \neq i} \frac{G_{ij}}{G_{ii}} p_j p_i^{-1}, \quad i = 1, \ldots, n.$$

9.7.2   *Convex representation of posynomials*

Monomials and (generalized) posynomials are not convex. However, we can obtain a convex representation, via a simple change of variables, plus a logarithmic transformation. Consider first a simple positive monomial function

$$f(x) = cx^a = cx_1^{a_1} x_2^{a_2} \cdots x_n^{a_n}, \quad x \in \mathbb{R}^n_{++}, c > 0.$$

Taking a logarithmic change of variables

$$y_i = \ln x_i, \quad i = 1, \ldots, n, \tag{9.28}$$

we have

$$\begin{aligned}
\tilde{g}(y) = f(x(y)) &= ce^{a_1 y_1} \cdots e^{a_n y_n} = ce^{a_1 y_1 + \cdots + a_n y_n} \\
[\text{letting } b \doteq \ln c] &= e^{a^\top y + b}.
\end{aligned}$$

The exponential function is convex, hence $\tilde{g}(y)$ is convex (and positive) over $\mathbb{R}^n$. Further, since we have seen that transforming an objective or constraint function via an increasing function yields an equivalent problem (see, e.g., Section 8.3.4.1), we can consider instead of $\tilde{g}(y)$ the function

$$g(y) \doteq \ln \tilde{g}(y) = a^\top y + b,$$

which turns out, in this case, to be also convex in y. This further logarithmic transformation has the additional advantage of actually resulting in a *linear* function of the variable y (notice also that dealing directly with the exponential function $\tilde{g}(y)$ may raise numerical problems, since this function, although convex, may take very large values). Optimization models in which only positive monomials appear in the objective and constraints can thus be transformed into equivalent linear programs, as shown also in Example 8.18.

Consider next f to be a posynomial, with the notation defined in (9.26). Using again the change of variables (9.28), and letting $b_i \doteq \ln c_i$, we have

$$\tilde{g}(y) = f(x(y)) = \sum_{i=1}^{K} e^{a_{(i)}^\top y + b_i},$$

which is a sum of convex functions, hence it is convex in the variable y. To avoid dealing with the large range of numerical values that exponential functions typically attain, we further take the logarithm of $\tilde{g}$, obtaining the function

$$g(y) \doteq \ln \tilde{g}(y) = \ln \left(\sum_{i=1}^{K} e^{a_{(i)}^\top y + b_i} \right) = \mathrm{lse}(Ay + b),$$

where we defined $A \in \mathbb{R}^{K,n}$ as the matrix having rows $a_{(i)}^{\top}$, $i = 1,\ldots,K$, $b = [b_1 \cdots, b_K]^{\top}$, and where lse is the log-sum-exp function defined in Example 2.14 (see also Example 4.4 and Example 8.5), which is a convex function. Thus, we can view a posynomial as the log-sum-exp function of an affine combination of the logarithm of the original variables. Since the lse function is convex, this transformation will allow us to use convex optimization to solve models based on posynomials.

Remark 9.3 *Convex representation of generalized posynomials.* Adding variables, and with the logarithmic change of variables seen above, we can also transform generalized posynomial inequalities into convex ones. Consider, for example, the posynomial f with values

$$f(x) = \max(f_1(x), f_2(x)),$$

where f_1, f_2 are posynomials. For $t > 0$, the constraint $f(x) \leq t$ can be expressed as two posynomial constraints in (x,t), namely $f_1(x) \leq t$, $f_2(x) \leq t$.

Likewise, for $t > 0$, $\alpha > 0$, consider the power constraint

$$(f(x))^{\alpha} \leq t,$$

where f is an ordinary posynomial. Since $\alpha > 0$, the above is equivalent to

$$f(x) \leq t^{1/\alpha},$$

which is in turn equivalent to the posynomial constraint in (x,t)

$$g(x,t) \doteq t^{-1/\alpha} f(x) \leq 1.$$

Hence, by adding as many variables as necessary, we can express a generalized posynomial constraint as a set of ordinary posynomial ones.

9.7.3 Standard forms of GP

A geometric program is an optimization problem involving generalized posynomial objective and inequality constraints, and (possibly) monomial equality constraints. In standard form, a GP can be written as

$$\begin{aligned}
\min_{x} \quad & f_0(x) \\
\text{s.t.:} \quad & f_i(x) \leq 1, \quad i = 1,\ldots,m, \\
& h_i(x) = 1, \quad i = 1,\ldots,p,
\end{aligned}$$

where $f_0,\ldots,f_m$ are generalized posynomials, and h_i, $i = 1,\ldots,p$, are positive monomials.

Assuming for simplicity that $f_0,\ldots,f_m$ are standard posynomials, we can express a GP explicitly, in the so-called standard form:

$$\min_{x} \quad \sum_{k=1}^{K_0} c_{k0} x^{a_{(k0)}}$$

$$\text{s.t.:} \quad \sum_{k=1}^{K_i} c_{ki} x^{a_{(ki)}} \leq 1, \quad i = 1, \dots, m,$$

$$g_i x^{r_{(i)}} = 1, \quad i = 1, \dots, p,$$

where $a_{(k0)}, \dots, a_{(km)}$, $r_{(1)}, \dots, r_{(p)}$ are vectors in $\mathbb{R}^n$, and c_{ki}, g_i are positive scalars.[6]

Using the logarithmic transformations described in Section 9.7.2, we may rewrite the above GP (which is non-convex) into an equivalent convex formulation, as follows:

> [6] Converting the problem into the standard form when the original formulation involves generalized posynomials entails adding new variables and constraints; see Remark 9.3.

$$\min_{y} \quad \text{lse}(A_0 y + b_0)$$

$$\text{s.t.:} \quad \text{lse}(A_i y + b_i) \leq 0, \quad i = 1, \dots, m,$$

$$R y + h = 0, \tag{9.29}$$

where A_i is a matrix with rows $a_{(1i)}^\top, \dots, a_{(K_i i)}^\top$, $i = 0, 1, \dots, m$; b_i is a vector with elements $c_{1i}, \dots, c_{K_i i}$, $i = 0, 1, \dots, m$; R is a matrix with rows $r_{(1)}^\top, \dots, r_{(p)}^\top$, and h is a vector with elements $\ln g_1, \dots, \ln g_p$.

Example 9.22 (*Optimization of a liquid storage tank*) We consider again the liquid storage tank model introduced in Example 9.20. The problem is to find the diameter d and height h of the tank, so as to minimize the cost $C_{\text{total}}(d, h)$ in (9.27), which is a posynomial function, subject to constraints. The constraints involve upper and lower bounds on the variables

$$0 < d \leq d_{\max}, \quad 0 < h \leq h_{\max}.$$

We might also include an upper bound $\kappa_{\max}$ on the *aspect ratio* of the tank (the aspect ratio constraint is useful, for instance, to take into account structural resistance to wind)

$$h \leq \kappa_{\max} d.$$

The problem takes the standard GP form

$$\min_{d,h} \quad c_1 h^{-1} d^{-2} + c_2 d^2 + c_3 dh$$

$$\text{s.t.:} \quad 0 < d_{\max}^{-1} d \leq 1,$$

$$0 < h_{\max}^{-1} h \leq 1,$$

$$\kappa_{\max}^{-1} h d^{-1} \leq 1.$$

Using the numerical values $V_{\text{supp}} = 8 \times 10^5$ litres, $\alpha_1 = 10$ \$, $\alpha_2 = 6$ \$/m^2, $\alpha_3 = 2$ \$/m^2, and bounds $d_{\max} = 20$ m, $h_{\max} = 30$ m, $\kappa_{\max} = 3$, we can solve this problem numerically, obtaining the optimal solution

$$d^* = 14.84, \quad h^* = 22.26,$$

with corresponding optimal objective value $C_{\text{total}}^* = 5191.18$.

9.8   Exercises

Exercise 9.1 (Formulating problems as LPs or QPs) Formulate the problem

$$p_j^* \doteq \min_x f_j(x),$$

for different functions f_j, $j = 1, \ldots, 5$, with values given in Table 9.6, as QPs or LPs, or, if you cannot, explain why. In our formulations, we always use $x \in \mathbb{R}^n$ as the variable, and assume that $A \in \mathbb{R}^{m,n}$, $y \in \mathbb{R}^m$, and $k \in \{1, \ldots, m\}$ are given. If you obtain an LP or QP formulation, make sure to put the problem in standard form, stating precisely what the variables, objective, and constraints are. *Hint:* for the last one, see Example 9.10.

$f_1(x)$	$=$	$\|Ax - y\|_\infty + \|x\|_1$
$f_2(x)$	$=$	$\|Ax - y\|_2^2 + \|x\|_1$
$f_3(x)$	$=$	$\|Ax - y\|_2^2 - \|x\|_1$
$f_4(x)$	$=$	$\|Ax - y\|_2^2 + \|x\|_2^2$
$f_5(x)$	$=$	$\sum_{i=1}^k |Ax - y|_{[i]} + \|x\|_2^2$

Table 9.6 Table of the values of different functions f. $|z|_{[i]}$ denotes the element in a vector z that has the i-th largest magnitude.

Exercise 9.2 (A slalom problem) A two-dimensional skier must slalom down a slope, by going through n parallel gates of known position (x_i, y_i), and of width c_i, $i = 1, \ldots, n$. The initial position (x_0, y_0) is given, as well as the final one, (x_{n+1}, y_{n+1}). Here, the x-axis represents the direction down the slope, from left to right, see Figure 9.24.

1. Find the path that minimizes the total length of the path. Your answer should come in the form of an optimization problem.

2. Try solving the problem numerically, with the data given in Table 9.7.

Exercise 9.3 (Minimum distance to a line segment) The line segment linking two points $p, q \in \mathbb{R}^n$ (with $p \neq q$) is the set $\mathcal{L} = \{\lambda p + (1 - \lambda)q : 0 \le \lambda \le 1\}$.

1. Show that the minimum distance D_* from a point $a \in \mathbb{R}^n$ to the line segment $\mathcal{L}$ can be written as a QP in one variable:

$$\min_\lambda \|\lambda c + d\|_2^2 : 0 \le \lambda \le 1,$$

for appropriate vectors c, d, which you will determine. Explain why we can always assume $a = 0$.

2. Prove that the minimum distance is given by[7]

$$D_*^2 = \begin{cases} q^\top q - \frac{(q^\top(p-q))^2}{\|p-q\|_2^2} & \text{if } p^\top q \le \min(q^\top q, p^\top p), \\ q^\top q & \text{if } p^\top q > q^\top q, \\ p^\top p & \text{if } p^\top q > p^\top p. \end{cases}$$

3. Interpret the result geometrically.

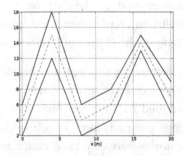

Figure 9.24 Slalom problem with $n = 5$ obstacles. "Uphill" (resp. "downhill") is on the left (resp. right) side. The middle path is dashed, initial and final positions are not shown.

i	x_i	y_i	c_i
0	0	4	N/A
1	4	5	3
2	8	4	2
3	12	6	2
4	16	5	1
5	20	7	2
6	24	4	N/A

Table 9.7 Problem data for Exercise 9.2.

[7] Notice that the conditions expressing D_*^2 are mutually exclusive, since $|p^\top q| \le \|p\|_2 \|q\|_2$.

Exercise 9.4 (Univariate LASSO) Consider the problem

$$\min_{x \in \mathbb{R}} f(x) \doteq \frac{1}{2}\|ax - y\|_2^2 + \lambda|x|,$$

where $\lambda \geq 0$, $a \in \mathbb{R}^m$, $y \in \mathbb{R}^m$ are given, and $x \in \mathbb{R}$ is a scalar variable. This is a univariate version of the LASSO problem discussed in Section 9.6.2. Assume that $y \neq 0$ and $a \neq 0$, (since otherwise the optimal solution of this problem is simply $x = 0$). Prove that the optimal solution of this problem is

$$x^* = \begin{cases} 0 & \text{if } |a^\top y| \leq \lambda, \\ x_{ls} - \text{sgn}(x_{ls})\frac{\lambda}{\|a\|_2^2} & \text{if } |a^\top y| > \lambda, \end{cases}$$

where

$$x_{ls} \doteq \frac{a^\top y}{\|a\|_2^2}$$

corresponds to the solution of the problem for $\lambda = 0$. Verify that this solution can be expressed more compactly as $x^* = \text{sthr}_{\lambda/\|a\|_2^2}(x_{ls})$, where sthr is the *soft threshold* function defined in (12.66).

Exercise 9.5 (An optimal breakfast) We are given a set of $n = 3$ types of food, each of which has the nutritional characteristics described in Table 9.8. Find the optimal composition (amount of servings per each food) of a breakfast having minimum cost, number of calories between 2,000 and 2,250, amount of vitamin between 5,000 and 10,000, and sugar level no larger than 1,000, assuming that the maximum number of servings is 10.

Food	Cost	Vitamin	Sugar	Calories
Corn	0.15	107	45	70
Milk	0.25	500	40	121
Bread	0.05	0	60	65

Table 9.8 Food costs and nutritional values per serving.

Exercise 9.6 (An LP with wide matrix) Consider the LP

$$p^* = \min_x c^\top x : l \leq Ax \leq u,$$

where $A \in \mathbb{R}^{m,n}$, $c \in \mathbb{R}^n$, and $l, u \in \mathbb{R}^m$, with $l \leq u$. We assume that A is wide, and full rank, that is: $m \leq n$, $m = \text{rank}(A)$. We are going to develop a closed-form solution to the LP.

1. Explain why the problem is always feasible.

2. Assume that $c \notin \mathcal{R}(A^\top)$. Using the result of Exercise 6.2, show that $p^* = -\infty$. *Hint:* set $x = x_0 + tr$, where x_0 is feasible, r is such that $Ar = 0$, $c^\top r > 0$, and let $t \to -\infty$.

3. Now assume that there exists $d \in \mathbb{R}^m$ such that $c = A^\top d$. Using the fundamental theorem of linear algebra (see Section 3.2.4), any vector x can be written as $x = A^\top y + z$ for some pair (y, z) with $Az = 0$. Use this fact, and the result of the previous part, to express the problem in terms of the variable y only.

4. Reduce further the problem to one of the form

$$\min_v \, d^\top v \, : \, l \leq v \leq u.$$

Make sure to justify any change of variable you may need. Write the solution to the above in closed form. Make sure to express the solution steps of the method clearly.

Exercise 9.7 (Median versus average) For a given vector $v \in \mathbb{R}^n$, the average can be found as the solution to the optimization problem

$$\min_{x \in \mathbb{R}} \|v - x\mathbf{1}\|_2^2, \qquad (9.30)$$

where $\mathbf{1}$ is the vector of ones in $\mathbb{R}^n$. Similarly, it turns out that the median (any value x such that there is an equal number of values in v above or below x) can be found via

$$\min_{x \in \mathbb{R}} \|v - x\mathbf{1}\|_1. \qquad (9.31)$$

We consider a robust version of the average problem (9.30):

$$\min_x \, \max_{u \,:\, \|u\|_\infty \leq \lambda} \|v + u - x\mathbf{1}\|_2^2, \qquad (9.32)$$

in which we assume that the components of v can be independently perturbed by a vector u whose magnitude is bounded by a given number $\lambda \geq 0$.

1. Is the robust problem (9.32) convex? Justify your answer precisely, based on expression (9.32), and without further manipulation.

2. Show that problem (9.32) can be expressed as

$$\min_{x \in \mathbb{R}} \sum_{i=1}^{n} (|v_i - x| + \lambda)^2.$$

3. Express the problem as a QP. State precisely the variables, and constraints if any.

4. Show that when λ is large, the solution set approaches that of the median problem (9.31).

5. It is often said that the median is a more robust notion of "middle" value than the average, when noise is present in v. Based on the previous part, justify this statement.

Exercise 9.8 (Convexity and concavity of optimal value of an LP)
Consider the linear programming problem

$$p^* \doteq \min_x c^\top x \ : \ Ax \le b,$$

where $c \in \mathbb{R}^n$, $A \in \mathbb{R}^{m,n}$, $b \in \mathbb{R}^m$. Prove the following statements, or provide a counter-example.

1. The objective function p^* is a concave function of c.

2. The objective function p^* is a convex function of b (you may assume that the problem is feasible).

3. The objective function p^* is a concave function of A.

Exercise 9.9 (Variational formula for the dominant eigenvalue)
Recall from Exercise 3.11 that a positive matrix $A > 0$ has a dominant eigenvalue $\lambda = \rho(A) > 0$, and corresponding left eigenvector $w > 0$ and right eigenvector $v > 0$ (i.e., $w^\top A = \lambda w^\top$, $Av = \lambda v$) which belong to the probability simplex $S = \{x \in \mathbb{R}^n \ : \ x \ge 0, \mathbf{1}^\top x = 1\}$. In this exercise, we shall prove that the dominant eigenvalue has an optimization-based characterization, similar in spirit to the "variational" characterization of the eigenvalues of symmetric matrices. Define the function $f : S \to \mathbb{R}_{++}$ with values

$$f(x) \doteq \min_{i=1,\dots,n} \frac{a_i^\top x}{x_i}, \quad \text{for } x \in S,$$

where $a_i^\top$ is the i-th row of A, and we let $\frac{a_i^\top x}{x_i} \doteq +\infty$ if $x_i = 0$.

1. Prove that, for all $x \in S$ and $A > 0$, it holds that

$$Ax \ge f(x)x \ge 0.$$

2. Prove that

$$f(x) \le \lambda, \quad \forall x \in S.$$

3. Show that $f(v) = \lambda$, and hence conclude that

$$\lambda = \max_{x \in S} f(x),$$

which is known as the Collatz–Wielandt formula for the dominant eigenvalue of a positive matrix. This formula actually holds more generally for non-negative matrices,[8] but you are not asked to prove this fact.

[8] For a non-negative matrix $A \ge 0$ an extension of the results stated in Exercise 3.11 for positive matrices holds. More precisely, if $A \ge 0$, then $\lambda = \rho(A) \ge 0$ is still an eigenvalue of A, with a corresponding eigenvector $v \ge 0$ (the difference here being that λ could be zero, and not simple, and that v may not be strictly positive). The stronger results of $\lambda > 0$ and simple, and $v > 0$ are recovered under the additional assumption that $A \ge 0$ is *primitive*, that is there exists an integer k such that $A^k > 0$ (Perron–Frobenius theorem).

Exercise 9.10 (LS with uncertain A matrix) Consider a linear least-squares problem where the matrix involved is random. Precisely, the residual vector is of the form $A(\delta)x - b$, where the $m \times n$ A matrix is affected by stochastic uncertainty. In particular, assume that

$$A(\delta) = A_0 + \sum_{i=1}^{p} A_i \delta_i,$$

where δ_i, $i = 1, \ldots, p$ are i.i.d. random variables with zero mean and variance σ_i^2. The standard least-squares objective function $\|A(\delta)x - b\|_2^2$ is now random, since it depends on δ. We seek to determine x such that the expected value (with respect to the random variable δ) of $\|A(\delta)x - b\|_2^2$ is minimized. Is such a problem convex? If yes, to which class does it belong (LP, LS, QP, etc.)?

10

Second-order cone and robust models

SECOND-ORDER CONE PROGRAMMING (SOCP) is a generalization of linear and quadratic programming that allows for affine combinations of variables to be constrained inside a special convex set, called a *second-order cone*. The SOCP model includes as special cases LPs, as well as problems with convex quadratic objective and constraints. SOCP models are particularly useful in geometry problems, approximation problems, as well as in probabilistic (chance-constrained) approaches to linear optimization problems in which the data is affected by random uncertainty. Data uncertainty also motivates the introduction in this chapter of *robust optimization* models, which enable the user to obtain solutions that are resilient (robust) against the uncertainty that is in practice often present in the description of an optimization problem.

10.1   Second-order cone programs

The second-order cone (SOC) in $\mathbb{R}^3$ is the set of vectors (x_1, x_2, t) such that $\sqrt{x_1^2 + x_2^2} \leq t$. Horizontal sections of this set at level $\alpha \geq 0$ are disks of radius α. A visualization of the SOC in $\mathbb{R}^3$ is given in Figure 10.1.

This definition can actually be extended to arbitrary dimension: an $(n+1)$-dimensional SOC is the following set:

$$\mathcal{K}_n = \{(x,t),\, x \in \mathbb{R}^n,\, t \in \mathbb{R} : \|x\|_2 \leq t\}. \tag{10.1}$$

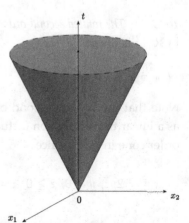

Figure 10.1 The second-order cone in $\mathbb{R}^3$.

Example 10.1 (*Magnitude constraints on affine complex vectors*) Many design problems involve complex variables and magnitude constraints. Such constraints can often be represented via SOCs. The basic idea is that the magnitude of a complex number $z = z_R + jz_I$, with z_R, z_I the real and imaginary parts, can be expressed as the Euclidean norm of (z_R, z_I):

$$|z| = \sqrt{z_R^2 + z_I^2} = \left\| \begin{bmatrix} z_R \\ z_I \end{bmatrix} \right\|_2.$$

For example, consider a problem involving a magnitude constraint on a complex number $f(x)$, where $x \in \mathbb{R}^n$ is a design variable, and the complex-valued function $f : \mathbb{R}^n \to \mathbb{C}$ is affine. The values of such a function can be written as

$$f(x) = (a_R^\top x + b_R) + j(a_I^\top x + b_I),$$

where $a_R, a_I \in \mathbb{R}^n$, $b_R, b_I \in \mathbb{R}$. For $t \in \mathbb{R}$, the magnitude constraint

$$|f(x)| \leq t$$

can be written as

$$\left\| \begin{bmatrix} a_R^\top x + b_R \\ a_I^\top x + b_I \end{bmatrix} \right\|_2 \leq t,$$

which is a second-order cone constraint on (x, t).

10.1.1 Geometry

An SOC is a convex cone. First, the set $\mathcal{K}_n$ in (10.1) is convex, since it can be expressed as the intersection of (an infinite number of) half-spaces:

$$\mathcal{K}_n = \bigcap_{u: \|u\|_2 \leq 1} \left\{ (x, t), x \in \mathbb{R}^n, t \in \mathbb{R} : x^\top u \leq t \right\}.$$

Second, it is a cone, since for any $z \in \mathcal{K}_n$ it holds that $\alpha z \in \mathcal{K}_n$, for any $\alpha \geq 0$.

10.1.1.1 The rotated second-order cone.

The rotated second-order cone in $\mathbb{R}^{n+2}$ is the set

$$\mathcal{K}_n^r = \left\{ (x, y, z), x \in \mathbb{R}^n, y \in \mathbb{R}, z \in \mathbb{R} : x^\top x \leq 2yz, \ y \geq 0, \ z \geq 0 \right\}.$$

Note that the rotated second-order cone in $\mathbb{R}^{n+2}$ can be expressed as a linear transformation (actually, a rotation) of the (plain) second-order cone in $\mathbb{R}^{n+2}$, since

$$\|x\|_2^2 \leq 2yz, \ y \geq 0, \ z \geq 0 \iff \left\| \begin{bmatrix} x \\ \frac{1}{\sqrt{2}}(y - z) \end{bmatrix} \right\|_2 \leq \frac{1}{\sqrt{2}}(y + z). \tag{10.2}$$

That is, $(x, y, z) \in \mathcal{K}_n^r$ if and only if $(w, t) \in \mathcal{K}_n$, where

$$w = (x, (y - z)/\sqrt{2}), \quad t = (y + z)/\sqrt{2}.$$

These two sets of variables are related by a rotation matrix R:

$$\begin{bmatrix} x \\ \frac{1}{\sqrt{2}}(y-z) \\ \frac{1}{\sqrt{2}}(y+z) \end{bmatrix} = R \begin{bmatrix} x \\ y \\ z \end{bmatrix}, \quad R = \begin{bmatrix} I_n & 0 & 0 \\ 0 & \frac{1}{\sqrt{2}} & -\frac{1}{\sqrt{2}} \\ 0 & \frac{1}{\sqrt{2}} & \frac{1}{\sqrt{2}} \end{bmatrix},$$

which proves that rotated second-order cones are also convex. Constraints of the form $\|x\|_2^2 \leq 2yz$, as appearing in (10.2), are usually referred to as *hyperbolic* constraints.

10.1.1.2 Quadratic constraints. Second-order cone constraints can be used to express convex quadratic inequalities. Precisely, if $Q = Q^\top \succeq 0$, the constraint

$$x^\top Q x + c^\top x \leq t \tag{10.3}$$

is equivalent to the existence of w, y, z such that

$$w^\top w \leq 2yz, \; z = 1/2, \; w = Q^{1/2}x, \; y = t - c^\top x, \tag{10.4}$$

where $Q^{1/2}$ is the matrix square-root of the positive semidefinite matrix Q. In the space of (x, w, y) variables, the above constraints represent the intersection of a rotated second-order cone constraint $(w, y, z) \in \mathcal{K}_n^r$, with the affine sets

$$z = 1/2, \; \{(x, w) : w = Q^{1/2}x\}, \; \{(x, y) : y = t - c^\top x\}.$$

10.1.1.3 Second-order cone inequalities. The standard format of a second-order cone constraint on a variable $x \in \mathbb{R}^n$ expresses the condition that $(y, t) \in \mathcal{K}_m$, with $y \in \mathbb{R}^m$, $t \in \mathbb{R}$, where y, t are some affine functions of x. Formally, these affine functions can be expressed as $y = Ax + b$, $t = c^\top x + d$, hence the condition $(y, t) \in \mathcal{K}_m$ becomes

$$\|Ax + b\|_2 \leq c^\top x + d, \tag{10.5}$$

where $A \in \mathbb{R}^{m,n}$, $b \in \mathbb{R}^m$, $c \in \mathbb{R}^n$, and $d \in \mathbb{R}$.

For example, the convex quadratic constraint (10.3) can be expressed in standard SOC form by first writing it in the rotated conic form (10.4), and then applying (10.2), which results in the SOC

$$\left\| \begin{bmatrix} \sqrt{2}Q^{1/2}x \\ t - c^\top x - 1/2 \end{bmatrix} \right\|_2 \leq t - c^\top x + 1/2.$$

10.1.2 SOCP in standard form

A second-order cone program is a convex optimization problem having linear objective and SOC constraints. When the SOC constraints have the standard form (10.5), we have a SOCP in *standard inequality form*:

$$
\begin{aligned}
\min_{x \in \mathbb{R}^n} \quad & c^\top x \\
\text{s.t.:} \quad & \|A_i x + b_i\|_2 \le c_i^\top x + d_i, \quad i = 1, \ldots, m,
\end{aligned}
\tag{10.6}
$$

where $A_i \in \mathbb{R}^{m_i, n}$ are given matrices, $b_i \in \mathbb{R}^{m_i}$, $c_i \in \mathbb{R}^n$ are vectors, and d_i are given scalars. An equivalent formulation, called *conic standard form*, makes the conic inclusions explicit in the constraints:

$$
\begin{aligned}
\min_{x \in \mathbb{R}^n} \quad & c^\top x \\
\text{s.t.:} \quad & (A_i x + b_i, c_i^\top x + d_i) \in \mathcal{K}_{m_i}, \quad i = 1, \ldots, m.
\end{aligned}
$$

SOCPs are representative of a quite large class of convex optimization problems. Indeed, LPs, convex QPs, and convex QCQPs can all be represented as SOCPs, as illustrated next.

Example 10.2 (*Square-root LASSO as an SOCP*) Return to the square-root LASSO problem mentioned in Example 8.23:

$$
p^* = \min_x \|Ax - b\|_2 + \lambda \|x\|_1,
$$

where $A \in \mathbb{R}^{m,n}$, $b \in \mathbb{R}^m$, and the parameter $\lambda > 0$ are given. The problem can be expressed as an SOCP, namely

$$
p^* = \min_{x, t, u} t + \lambda \sum_{i=1}^n u_i \ : \ t \ge \|Ax - b\|_2, \ u_i \ge |x_i|, \ i = 1, \ldots, n.
$$

Linear programs as SOCPs. The linear program (LP) in standard inequality form

$$
\begin{aligned}
\min_x \quad & c^\top x \\
\text{s.t.:} \quad & a_i^\top x \le b_i, \quad i = 1, \ldots, m,
\end{aligned}
$$

can be readily cast in SOCP form as

$$
\begin{aligned}
\min_x \quad & c^\top x \\
\text{s.t.:} \quad & \|C_i x + d_i\|_2 \le b_i - a_i^\top x, \quad i = 1, \ldots, m,
\end{aligned}
$$

where $C_i = 0$, $d_i = 0$, $i = 1, \ldots, m$.

Quadratic programs as SOCPs.   The quadratic program (QP)

$$\min_{x} \quad x^\top Q x + c^\top x$$
$$\text{s.t.:} \quad a_i^\top x \le b_i, \qquad i = 1, \ldots, m,$$

where $Q = Q^\top \succeq 0$, can be cast as an SOCP as follows. First, we set $w = Q^{1/2}x$ and introduce a slack variable y, thus rewriting the problem as

$$\min \quad c^\top x + y$$
$$\text{s.t.:} \quad w^\top w \le y,$$
$$w = Q^{1/2}x,$$
$$a_i^\top x \le b_i, \qquad i = 1, \ldots, m.$$

Then we observe that $w^\top w \le y$ can be expressed in rotated conic form by introducing a further slack variable z: $w^\top w \le yz$, with z linearly constrained so that $z = 1$. Therefore, we have

$$\min \quad c^\top x + y$$
$$\text{s.t.:} \quad w^\top w \le yz,$$
$$z = 1$$
$$w = Q^{1/2}x,$$
$$a_i^\top x \le b_i, \qquad i = 1, \ldots, m,$$

which, using (10.2), can be further rewritten as

$$\min_{x,y} \quad c^\top x + y$$
$$\text{s.t.:} \quad \left\| \begin{bmatrix} 2Q^{1/2}x \\ y - 1 \end{bmatrix} \right\|_2 \le y + 1,$$
$$a_i^\top x \le b_i, \quad i = 1, \ldots, m.$$

Quadratic-constrained quadratic programs as SOCPs.   The convex quadratic-constrained quadratic program (QCQP)

$$\min_{x} \quad x^\top Q_0 x + a_0^\top x$$
$$\text{s.t.:} \quad x^\top Q_i x + a_i^\top x \le b_i, \quad i = 1, \ldots, m,$$

with $Q_i = Q_i^\top \succeq 0$, $i = 0, 1, \ldots, m$, can be cast as an SOCP as follows. First, we introduce a slack variable t and rewrite the problem in epigraph form as

$$\min \quad a_0^\top x + t$$
$$\text{s.t.:} \quad x^\top Q_0 x \le t,$$
$$x^\top Q_i x + a_i^\top x \le b_i, \quad i = 1, \ldots, m.$$

Now, each constraint of the form $x^\top Q x + a^\top x \le b$ is equivalent to the existence of w, y, z such that

$$w^\top w \le yz, \ z = 1, \ w = Q^{1/2}x, \ y = b - a^\top x$$

Therefore, the QCQP is rewritten as

$$
\begin{aligned}
\min \quad & a_0^\top x + t \\
\text{s.t.:} \quad & w_0^\top w_0 \leq t, \ w_0 = Q_0^{1/2} x, \\
& w_i^\top Q_i w_i \leq b_i - a_i^\top x, \ w_i = Q_i^{1/2} x, \quad i = 1, \ldots, m,
\end{aligned}
$$

which, using (10.2), can be further rewritten in explicit SOCP format as

$$
\begin{aligned}
\min_{x,t} \quad & a_0^\top x + t \\
\text{s.t.:} \quad & \left\| \begin{bmatrix} 2Q_0^{1/2} x \\ t - 1 \end{bmatrix} \right\|_2 \leq t + 1, \\
& \left\| \begin{bmatrix} 2Q_i^{1/2} x \\ b_i - a_i^\top x - 1 \end{bmatrix} \right\|_2 \leq b_i - a_i^\top x + 1, \quad i = 1, \ldots, m.
\end{aligned}
$$

Remark 10.1 There is an alternative, and simpler, way to convert a quadratic constraint of the form

$$
x^\top Q x + a^\top x \leq b \tag{10.7}
$$

into an SOCP constraint, when the matrix Q is positive definite (hence Q and $Q^{1/2}$ are invertible). Since

$$
x^\top Q x + a^\top x = \left(Q^{1/2} x + \frac{1}{2} Q^{-1/2} a \right)^\top \left(Q^{1/2} x + \frac{1}{2} Q^{-1/2} a \right) - \frac{1}{4} a^\top Q^{-1} a,
$$

we have that (10.7) is equivalent to

$$
\left\| Q^{1/2} x + \frac{1}{2} Q^{-1/2} a \right\|_2^2 \leq b + \frac{1}{4} a^\top Q^{-1} a,
$$

which is in turn equivalent to the SOCP constraint

$$
\left\| Q^{1/2} x + \frac{1}{2} Q^{-1/2} a \right\|_2 \leq \sqrt{ b + \frac{1}{4} a^\top Q^{-1} a }.
$$

10.1.3 SOCP duality

Consider the SOCP in standard form (10.6):

$$
\begin{aligned}
p^* = \min_x \quad & c^\top x \\
\text{s.t.:} \quad & \| A_i x + b_i \|_2 \leq c_i^\top x + d_i, \quad i = 1, \ldots, m.
\end{aligned}
$$

We have

$$
\begin{aligned}
p^* &= \min_x \max_{\lambda \geq 0} \ c^\top x + \sum_{i=1}^m \lambda_i (\| A_i x + b_i \|_2 - c_i^\top x - d_i) \\
&= \min_x \max_{\|u_i\|_2 \leq \lambda_i, \, i=1,\ldots,m} \ c^\top x \\
& \qquad\qquad\qquad + \sum_{i=1}^m \left(u_i^\top (A_i x + b_i) - \lambda_i (c_i^\top x + d_i) \right),
\end{aligned}
$$

where we have used the dual representation of the Euclidean norm in the second line. Applying the max-min inequality (8.48), we obtain $p^* \geq d^*$, where

$$d^* = \max_{\|u_i\|_2 \leq \lambda_i,\, i=1,\ldots,m} \min_x c^\top x + \sum_{i=1}^m \left(u_i^\top (A_i x + b_i) - \lambda_i (c_i^\top x + d_i) \right).$$

Solving for x, we obtain the dual problem:

$$d^* = \max_{u_i, \lambda_i,\, i=1,\ldots,m} \left(\sum_{i=1}^m u_i^\top b_i - \lambda_i d_i \right)$$
$$\text{s.t.:} \quad \sum_{i=1}^m \left(A_i^\top u_i - \lambda_i d_i \right) = c,$$
$$\|u_i\|_2 \leq \lambda_i, \qquad\qquad i = 1,\ldots,m.$$

Note that the dual problem is also an SOCP. From Slater's conditions for strong duality, it turns out that if the primal problem is strictly feasible, then $p^* = d^*$, and there is no duality gap.

Example 10.3 (*Dual of square-root LASSO*) Return to the square-root LASSO problem of Example 10.2. Since it can be written as an SOCP, we can apply the result above; strong duality results from the primal problem being strictly feasible (in a trivial way, as it has no constraints). Alternatively, we can directly represent the primal as a minimax problem:

$$p^* = \min_x \max_{u,v} u^\top (b - Ax) + v^\top x \,:\, \|u\|_2 \leq 1,\ \|v\|_\infty \leq \lambda.$$

Applying the minimax Theorem 8.8, we obtain

$$p^* = \max_u u^\top b \,:\, \|u\|_2 \leq 1,\ \|A^\top u\|_\infty \leq \lambda.$$

From this, we can observe that if $\|A^\top u\|_\infty < \lambda$ for every u with $\|u\|_2 \leq 1$, then the second constraint is inactive at optimum. This implies that $p^* = \|b\|_2$, hence $x = 0$ is optimal for the primal problem.

10.2 SOCP-representable problems and examples

10.2.1 Sums and maxima of norms

SOCPs are useful to tackle problems involving minimization of sums or maxima of Euclidean norms. The problem

$$\min_x \sum_{i=1}^p \|A_i x - b_i\|_2,$$

where $A_i \in \mathbb{R}^{m,n}$, $b_i \in \mathbb{R}^m$ are given data, can be readily cast as an SOCP by introducing auxiliary scalar variables $y_1,\ldots,y_p$ and rewriting the problem as

$$\min_{x,y} \quad \sum_{i=1}^{p} y_i$$
$$\text{s.t.:} \quad \|A_i x - b_i\|_2 \leq y_i, \quad i = 1, \ldots, p.$$

Similarly, the problem

$$\min_{x} \quad \max_{i=1,\ldots,p} \|A_i x - b_i\|_2$$

can be cast in SOCP format, by introducing a scalar slack variable y, as follows:

$$\min_{x,y} \quad y$$
$$\text{s.t.:} \quad \|A_i x - b_i\|_2 \leq y, \quad i = 1, \ldots, p.$$

10.2.2 Block sparsity

We have seen in Section 9.5.1 and Section 9.6.2 that the addition of a regularization term of the form $\|x\|_1$ in a problem objective has the effect of encouraging *sparsity* in the solution, that is it promotes solutions with low cardinality. In many problems, however, one may be interested in seeking solutions that are *block sparse*. By a block-sparse solution we here mean that the solution vector $x \in \mathbb{R}^n$ is partitioned into blocks

$$x = \begin{bmatrix} x_1 \\ \vdots \\ x_p \end{bmatrix}, \quad x_i \in \mathbb{R}^{n_i}, \ i = 1, \ldots, p; \ n_1 + \cdots + n_p = n,$$

and we wish many of these blocks to be zero, while the nonzero blocks need not be sparse. Clearly,

$$\sum_{i=1}^{p} \|x_i\|_2 = \left\| \begin{bmatrix} \|x_1\|_2 \\ \vdots \\ \|x_p\|_2 \end{bmatrix} \right\|_1,$$

that is, the sum of ℓ_2-norms of the blocks is precisely the ℓ_1-norm of the vector whose components are $\|x_i\|_2$. Therefore, including in the objective a term proportional to $\sum_{i=1}^{p} \|x_i\|_2$ will promote solutions where many blocks x_i are zero and few blocks are nonzero and possibly full. As an example, consider an LS problem where one seeks to minimize $\|Ax - y\|_2$ and A is structured in blocks as $A = [A_1 \cdots A_p]$, where $A_i \in \mathbb{R}^{m,n_i}$. Then, a block-sparse regularization of the problem can be expressed as

$$\min_{x} \|Ax - y\|_2 + \gamma \sum_{i=1}^{p} \|x_i\|_2,$$

where $\gamma \geq 0$ is a penalty weight (the higher γ the more incentive is put on finding block-sparse solutions). This problem is put in SOCP form by introducing slack scalar variables z and $t_1, \ldots, t_p$:

$$
\begin{aligned}
\min_{x,z,t} \quad & z + \gamma \sum_{i=1}^{p} t_i \\
\text{s.t.:} \quad & \|Ax - y\|_2 \leq z, \\
& \|x_i\|_2 \leq t_i, \qquad i = 1, \ldots, p.
\end{aligned}
$$

10.2.3   Quadratic-over-linear problems

Consider the optimization problem

$$
\begin{aligned}
\min_{x} \quad & \sum_{i=1}^{p} \frac{\|A_i x - b_i\|_2^2}{c_i^\top x + d_i} \\
\text{s.t.:} \quad & c_i^\top x + d_i > 0, \qquad i = 1, \ldots, p.
\end{aligned}
$$

By introducing auxiliary scalar variables $t_1, \ldots, t_p$, we first rewrite it as

$$
\begin{aligned}
\min_{x,t} \quad & \sum_{i=1}^{p} t_i \\
\text{s.t.:} \quad & \|A_i x - b_i\|_2^2 \leq (c_i^\top x + d_i) t_i, \quad i = 1, \ldots, p, \\
& c_i^\top x + d_i > 0, \qquad\qquad\quad i = 1, \ldots, p.
\end{aligned}
$$

Then, we apply (10.2) to obtain the SOCP formulation

$$
\begin{aligned}
\min_{x,t} \quad & \sum_{i=1}^{p} t_i \\
\text{s.t.:} \quad & \left\| \begin{bmatrix} 2(A_i x - b_i) \\ c_i^\top x + d_i - t_i \end{bmatrix} \right\|_2 \leq c_i^\top x + d_i + t_i, \quad i = 1, \ldots, p.
\end{aligned}
$$

10.2.4   Log-Chebyshev approximation

Consider the problem of finding an approximate solution to an over-determined system of linear equations

$$
a_i^\top x \simeq y_i, \quad y_i > 0, \; i = 1, \ldots, m.
$$

In certain applications it makes sense to use an error criterion based on the maximum logarithm of the ratios $(a_i^\top x)/y_i$, instead of the usual LS criterion. The log-Chebyshev approximation problem thus amounts to solving

$$
\begin{aligned}
\min_{x} \quad & \max_{i=1,\ldots,m} \left| \log(a_i^\top x) - \log y_i \right| \\
\text{s.t.:} \quad & a_i^\top x > 0, \qquad\qquad i = 1, \ldots, m.
\end{aligned}
$$

Now, for $a_i^\top x > 0$, it holds that

$$\left| \log(a_i^\top x) - \log y_i \right| = \left| \log \frac{a_i^\top x}{y_i} \right| = \log \max \left\{ \frac{a_i^\top x}{y_i}, \frac{y_i}{a_i^\top x} \right\},$$

and, since log is monotone increasing,

$$\max_{i=1,\dots,m} \left| \log(a_i^\top x) - \log y_i \right| = \log \max_{i=1,\dots,m} \max \left\{ \frac{a_i^\top x}{y_i}, \frac{y_i}{a_i^\top x} \right\}.$$

Again, since log is monotone increasing, we have that minimizing the log of a function over some set is equivalent to taking the log of the minimum of the function over the same set, hence the log-Chebyshev approximation problem is equivalent to

$$\min_{x} \quad \max_{i=1,\dots,m} \max \left\{ \frac{a_i^\top x}{y_i}, \frac{y_i}{a_i^\top x} \right\}$$
$$\text{s.t.:} \quad a_i^\top x > 0, \qquad\qquad i = 1,\dots,m.$$

Then, introducing an auxiliary scalar variable t and expressing the problem in epigraphic form, we have

$$\min_{x,t} \quad t$$
$$\text{s.t.:} \quad (a_i^\top x)/y_i \le t, \qquad i = 1,\dots,m,$$
$$(a_i^\top x)/y_i \ge 1/t, \quad i = 1,\dots,m,$$
$$a_i^\top x > 0, \qquad\quad i = 1,\dots,m.$$

The latter set of constraints are of the hyperbolic form $1 \le t \cdot (a_i^\top x)/y_i$, hence applying (10.2) we finally express the problem in SOCP form

$$\min_{x,t} \quad t$$
$$\text{s.t.:} \quad (a_i^\top x)/y_i \le t, \qquad\qquad\qquad i = 1,\dots,m,$$
$$\left\| \begin{bmatrix} 2 \\ t - (a_i^\top x)/y_i \end{bmatrix} \right\|_2 \le t + (a_i^\top x)/y_i, \quad i = 1,\dots,m.$$

10.2.5 Chance-constrained LP

Chance-constrained linear programs arise naturally from standard LPs, when some of the data describing the linear inequalities is uncertain and random. More precisely, consider an LP in standard inequality form:

$$\min_{x} \quad c^\top x$$
$$\text{s.t.:} \quad a_i^\top x \le b_i, \quad i = 1,\dots,m.$$

Suppose now that the problem data vectors a_i, $i = 1,\dots,m$, are not known precisely. Rather, all is known is that a_i are random vectors, with normal (Gaussian) distribution with mean value $\mathbb{E}\{a_i\} = \bar{a}_i$ and covariance matrix $\text{var}\{a_i\} = \Sigma_i \succ 0$. In such a case, also the scalar $a_i^\top x$ is a random variable; precisely, it is a normal random variable with

$$\mathbb{E}\{a_i^\top x\} = \bar{a}_i^\top x, \quad \text{var}\{a_i^\top x\} = x^\top \Sigma_i x.$$

It makes therefore no sense to impose a constraint of the form $a_i^\top x \leq b_i$, since the left-hand side of this expression is a normal random variable, which can assume any value, so such a constraint would always be violated by some outcomes of the random data a_i. It then appears natural to ask that the constraint $a_i^\top x \leq b_i$ be satisfied up to a given level of probability $p_i \in (0, 1)$. This level is chosen *a priori* by the user, and represents the probabilistic *reliability* level at which the constraint will remain satisfied in spite of random fluctuations in the data. The probability-constrained (or chance-constrained) counterpart of the nominal LP is therefore

$$\min_x \quad c^\top x \tag{10.8}$$
$$\text{s.t.:} \quad \text{Prob}\{a_i^\top x \leq b_i\} \geq p_i, \quad i = 1, \ldots, m, \tag{10.9}$$

where p_i are the assigned reliability levels. The reader should be warned that this chance-constrained problem may be very hard to solve, and it is not guaranteed to be convex, in the general case. However, in the specific case of concern (namely, when a_i, $i = 1, \ldots, m$, are independent normal random vectors, and $p_i > 0.5$, $i = 1, \ldots, m$), one may prove that the chance-constrained problem is indeed convex, and it can actually be recast in the form of an SOCP.

Proposition 10.1 *Consider problem (10.8)–(10.9), under the assumptions that $p_i > 0.5$, $i = 1, \ldots, m$, and that a_i, $i = 1, \ldots, m$, are independent normal random vectors with expected values $\bar{a}_i$ and covariance matrices $\Sigma_i \succ 0$. Then, (10.8)–(10.9) is equivalent to the SOCP*

$$\min_x \quad c^\top x$$
$$\text{s.t.:} \quad \bar{a}_i^\top x \leq b_i - \Phi^{-1}(p_i)\|\Sigma_i^{1/2}x\|_2, \quad i = 1, \ldots, m, \tag{10.10}$$

where $\Phi^{-1}(p)$ is the inverse cumulative probability distribution of a standard normal variable.

Proof We start by observing that

$$a_i^\top x \leq b_i \quad \Leftrightarrow \quad \frac{a_i^\top x - \bar{a}_i^\top x}{\sqrt{x^\top \Sigma_i x}} \leq \frac{b_i - \bar{a}_i^\top x}{\sqrt{x^\top \Sigma_i x}}, \tag{10.11}$$

which follows from subtracting on both sides of the equation $a_i^\top x \leq b_i$ the expected value $\mathbb{E}\{a_i^\top x\} = \bar{a}_i^\top x$, and then dividing both sides by the positive quantity $\sigma_i(x) \doteq \sqrt{x^\top \Sigma_i x}$. This latter quantity is nothing but the standard deviation of $a_i^\top x$ and, using the matrix square-root factorization $\Sigma_i = \Sigma_i^{1/2} \Sigma_i^{1/2}$, it holds that

$$\sigma_i(x) = \sqrt{x^\top \Sigma_i x} = \|\Sigma_i^{1/2} x\|_2. \qquad (10.12)$$

Defining

$$z_i(x) \doteq \frac{a_i^\top x - \bar{a}_i^\top x}{\sigma_i(x)}, \qquad (10.13)$$

$$\tau_i(x) \doteq \frac{b_i - \bar{a}_i^\top x}{\sigma_i(x)}, \qquad (10.14)$$

we have from (10.11) that

$$\text{Prob}\{a_i^\top x \leq b_i\} = \text{Prob}\{z_i(x) \leq \tau_i(x)\}. \qquad (10.15)$$

Now, observe that $z_i(x)$ is a standardized normal random variable (that is, a normal variable with zero mean and unit variance), and let $\Phi(\zeta)$ denote the standard normal cumulative probability distribution function, i.e.,

$$\Phi(\zeta) \doteq \text{Prob}\{z_i(x) \leq \zeta\}. \qquad (10.16)$$

Function $\Phi(\zeta)$ is well known and tabulated (also, it is related to the so-called *error function*, $\text{erf}(\zeta)$, for which it holds that $\Phi(\zeta) = 0.5(1 + \text{erf}(\zeta/\sqrt{2}))$); a plot of this function is shown in Figure 10.2.

It then follows from (10.15) and (10.16) that

$$\text{Prob}\{a_i^\top x \leq b_i\} = \Phi(\tau_i(x)), $$

hence each constraint in (10.9) is equivalent to

$$\Phi(\tau_i(x)) \geq p_i. \qquad (10.17)$$

Since Φ is monotone increasing, denoting by Φ^{-1} the inverse cumulative distribution function, we have that (10.17) holds if and only if

$$\tau_i(x) \geq \Phi^{-1}(p_i), $$

where, for $p_i > 0.5$, $\Phi^{-1}(p_i)$ is a positive number. Hence, recalling (10.14) and (10.12), we have that each constraint in (10.9) is equivalent to the SOC constraint

$$b_i - \bar{a}_i^\top x \geq \Phi^{-1}(p_i)\|\Sigma_i^{1/2} x\|_2, $$

whereby the claim follows immediately. $\qquad \qquad \square$

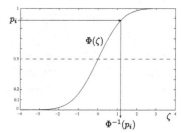

Figure 10.2 Plot of the standard normal cumulative probability distribution function $\Phi(\zeta)$.

Example 10.4 *(Value at risk (VaR) in portfolio optimization)*   Let us recall the financial portfolio model introduced in Section 4.3. The vector $r \in \mathbb{R}^n$ contains the random rate of returns of n assets, and $x \in \mathbb{R}^n$ is the vector describing the fractions of investor's wealth allocated in each of the assets. A classical assumption in portfolio management (albeit a debatable one) is that r is a normal random vector, with expected value $\hat{r} = \mathbb{E}\{r\}$ and covariance $\Sigma = \text{var}\{r\} \succ 0$, which are assumed to be known.

A popularly embraced technique for measuring downside risk in a portfolio is the so-called *value at risk* (VaR). VaR is defined as the α percentile of the portfolio return, typically for low α values (e.g., 1, 5 or 10 percent). Thus, VaR measures the potential loss in value of a risky portfolio over a defined period, for a given confidence level. For example, if the VaR on a portfolio return is 80% at $\alpha = 0.05$, it means that there is at most 5% chance that the portfolio value will drop more than 80% over the given period. Formally, let us define by

$$\varrho(x) = r^\top x$$

the random return of the portfolio over the fixed period of time over which r is defined. Then, the *loss* of the portfolio is simply defined as $\ell(x) = -\varrho(x)$, and the value at risk α is defined, in general, as

$$
\begin{aligned}
\text{VaR}_\alpha(x) &= \inf_{\gamma} : \text{Prob}\{\ell(x) \geq \gamma\} \leq \alpha \\
&= -\sup_{\zeta} : \text{Prob}\{\varrho(x) \leq \zeta\} \leq \alpha.
\end{aligned}
$$

In the case when the cumulative distribution function of $\ell(x)$ is continuous and strictly increasing (as it is assumed in the case considered here, where $\ell(x)$ is normal), then

$$\sup_{\zeta} : \text{Prob}\{\varrho(x) \leq \zeta\} \leq \alpha \equiv \inf_{\zeta} : \text{Prob}\{\varrho(x) \leq \zeta\} \geq \alpha,$$

and the expression on the right is the definition of the inverse cumulative distribution function of $\varrho(x)$ (also known as the *quantile* function)

$$F_{\varrho(x)}^{-1}(\alpha) = \inf_{\zeta} : \text{Prob}\{\varrho(x) \leq \zeta\} \geq \alpha,$$

therefore,

$$\text{VaR}_\alpha(x) = -F_{\varrho(x)}^{-1}(\alpha),$$

and for $v > 0$ we have that

$$\text{VaR}_\alpha(x) \leq v \Leftrightarrow F_{\varrho(x)}^{-1}(\alpha) \geq -v \Leftrightarrow F_{\varrho(x)}(-v) \leq \alpha,$$

where the last condition reads $\text{Prob}\{r^\top x \leq -v\} \leq \alpha$, or, equivalently, taking the complementary event,

$$\text{Prob}\{r^\top x + v \geq 0\} \geq p, \quad p \doteq 1 - \alpha$$

(note that we used $\geq$ instead of $>$, since this makes no difference here, due to continuity of the probability distribution). This constraint is of the form (10.9), hence, for $\alpha < 0.5$, it can be expressed equivalently in SOC form as in (10.10), and therefore

$$\text{VaR}_\alpha(x) \leq v \quad \Leftrightarrow \quad \hat{r}^\top x + v \geq \Phi^{-1}(1-\alpha)\|\Sigma^{1/2}x\|_2,$$

where Φ is the standard normal cumulative distribution function.

A typical portfolio allocation problem may thus be posed as follows: given a target desired return μ on the investment and a risk level $\alpha < 0.5$ (e.g., $\alpha = 0.02$), find a portfolio composition (assume that no short selling is allowed, thus $x \geq 0$) with expected return at least μ and minimal VaR_α. Such a problem is easily posed as an SOCP as follows:

$$\begin{aligned}
\min_{x,v} \quad & v \\
\text{s.t.:} \quad & x \geq 0, \\
& \sum_{i=1}^n x_i = 1, \\
& \hat{r}^\top x \geq \mu, \\
& \hat{r}^\top x + v \geq \Phi^{-1}(1-\alpha)\|\Sigma^{1/2}x\|_2.
\end{aligned}$$

10.2.6 Facility location problems

Consider the problem of locating a warehouse to serve a number of service locations. The design variable is the location of the warehouse, $x \in \mathbb{R}^2$, while the service locations are given by the vector $y_i \in \mathbb{R}^2$, $i = 1, \ldots, m$. One possible location criterion is to determine x so as to minimize the maximum distance from the warehouse to any location. This amounts to considering a minimization problem of the form

$$\min_x \max_{i=1,\ldots,m} \|x - y_i\|_2,$$

which is readily cast in SOCP form as follows:

$$\begin{aligned}
\min_{x,t} \quad & t \\
\text{s.t.:} \quad & \|x - y_i\|_2 \leq t, \quad i = 1, \ldots, m.
\end{aligned}$$

An alternative location criterion, which is a good proxy for the average transportation cost, is the *average* distance from the warehouse to the facilities. This leads to the problem

$$\min_x \frac{1}{m} \sum_{i=1}^m \|x - y_i\|_2,$$

which can be cast as the SOCP

$$\begin{aligned}
\min_{x,t} \quad & \frac{1}{m} \sum_{i=1}^m t_i \\
\text{s.t.:} \quad & \|x - y_i\|_2 \leq t_i, \quad i = 1, \ldots, m.
\end{aligned}$$

Figure 10.3 shows the results of an example with $m = 10$ randomly chosen service locations, for both the min-max criterion and the min-average criterion. The optimal objective was 0.5310 for the max-distance case, and 0.3712 for the average-distance case.

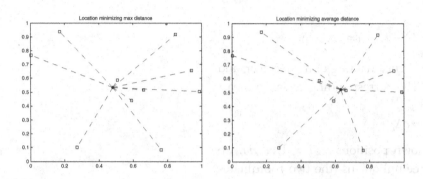

Figure 10.3 Warehouse position for min-max criterion (left) and min-average criterion (right).

10.2.7 GPS localization with time synchronization

We consider here a more realistic version of the planar trilateration problem discussed in Example 6.2. In that schematic problem, we wished to determine the 2D position coordinates of a point $x \in \mathbb{R}^2$, using range (distance) measurements from three anchor points (radio beacons, satellites, etc.) whose geographic coordinates are exactly known. In navigation systems such as GPS (Global Positioning System), however, these distances are computed indirectly from "time-of-flight" measurements, as explained next. In simplified terms, each beacon (or satellite) transmits packets, each containing a time stamp with the precise time instant at which the packet left the transmitter; let t_i^T denote the time at which the packet left transmitter i, as measured by the clock onboard satellite i. All satellites have extremely precise atomic clocks on board, hence all satellite clocks are perfectly synchronized on the Coordinated Universal Time (UTC). The user at point x has a GPS receiver that receives packets, and the receiver has a local clock on board. If a packet from satellite i is received at time t_i^R (this time is as measured in the local receiver clock), then the receiver may compute the time of flight of the packet as the difference between t_i^R and t_i^T. Then, assuming the packet travels at the speed of light c, the receiver may convert the time-of-flight information into distance information. However, the GPS receiver typically has a cheap clock on board, which is not synchronized with the satellites' clocks. Therefore, for correct time-of-flight evaluation, one must convert the local clock time to UTC, i.e.,

$$[t_i^R]_{UTC} = t_i^R + \delta,$$

where δ is the offset between the UTC and local time (notice that δ could be of the order of several seconds). The time of flight of the packet is therefore given by

$$f_i = [t_i^R]_{UTC} - t_i^T = t_i^R - t_i^T + \delta = \Delta_i + \delta,$$

where $\Delta_i \doteq t_i^R - t_i^T$ is the time difference reading of the receiver, and δ is the (unknown) clock offset. The corresponding distance measurement is then

$$d_i = c f_i = c\Delta_i + c\delta.$$

If m satellites are available, at known positions a_i, $i = 1, \ldots, m$, we may write m equations in the three unknowns (the two coordinates x_1, x_2 of x, and the sync parameter δ):

$$\|x - a_i\|_2 = c\Delta_i + c\delta, \quad i = 1, \ldots, m. \tag{10.18}$$

If $m = 4$ satellites are available, by squaring each equation, and then taking the difference between the first three equations and the fourth, we obtain a system of three linear equations in the three unknowns:

$$
\begin{aligned}
2(a_4 - a_1)^\top x &= d_1^2 - d_4^2 + \|a_4\|_2^2 - \|a_1\|_2^2, \\
2(a_4 - a_2)^\top x &= d_2^2 - d_4^2 + \|a_4\|_2^2 - \|a_2\|_2^2, \\
2(a_4 - a_3)^\top x &= d_3^2 - d_4^2 + \|a_4\|_2^2 - \|a_3\|_2^2.
\end{aligned}
$$

That is

$$
\begin{aligned}
2(a_4 - a_1)^\top x + 2c^2(\Delta_4 - \Delta_1)\delta &= c^2(\Delta_1^2 - \Delta_4^2) + \|a_4\|_2^2 - \|a_1\|_2^2, \\
2(a_4 - a_2)^\top x + 2c^2(\Delta_4 - \Delta_2)\delta &= c^2(\Delta_2^2 - \Delta_4^2) + \|a_4\|_2^2 - \|a_2\|_2^2, \\
2(a_4 - a_3)^\top x + 2c^2(\Delta_4 - \Delta_3)\delta &= c^2(\Delta_3^2 - \Delta_4^2) + \|a_4\|_2^2 - \|a_3\|_2^2.
\end{aligned}
$$

The solutions to the original system of nonlinear equations (10.18) are contained in the solution set of the above system of linear equations. A solution of this system of linear equations, which is *a posteriori* checked to also satisfy $\|x - a_4\|_2 = c\Delta_4 + c\delta$, yields the position estimate x, as well as the clock synchronization parameter δ. However, this approach needs $m = 4$ satellites.

If only $m = 3$ satellites are available, one may still seek for a solution of the three nonlinear equations (10.18) (three equations in three unknowns), although the solution of this system of equations is not guaranteed to be unique. We can actually find a solution to (10.18) with $m = 3$ using convex optimization, as follows. Starting from

(10.18), we write a system of three equivalent equations, where the first two equations are the difference between the first two equations in (10.18) squared and the third equation squared. That is, system (10.18) is equivalent to

$$2(a_3 - a_1)^\top x + 2c^2(\Delta_3 - \Delta_1)\delta = c^2(\Delta_1^2 - \Delta_3^2) + \|a_3\|_2^2 - \|a_1\|_2^2,$$
$$2(a_3 - a_2)^\top x + 2c^2(\Delta_3 - \Delta_2)\delta = c^2(\Delta_2^2 - \Delta_3^2) + \|a_3\|_2^2 - \|a_2\|_2^2,$$
$$\|x - a_3\|_2 = c\Delta_3 + c\delta.$$

A consistent solution can then be found by finding the minimum δ such that the previous equations are satisfied. Such a minimization problem would not be convex, due to the last nonlinear equality constraint. However, we can relax this latter constraint to an inequality one, since we are guaranteed that, at the optimum, equality will actually hold (due to the fact that the relaxed problem has a single inequality constraint, see Remark 8.4). Therefore, a solution to system (10.18), with $m = 3$, can be found by solving the following SOCP

$$\min_{x,\delta} \quad \delta$$
$$\text{s.t.:} \quad 2(a_3 - a_1)^\top x + 2c^2(\Delta_3 - \Delta_1)\delta = c^2(\Delta_1^2 - \Delta_3^2) + \|a_3\|_2^2 - \|a_1\|_2^2,$$
$$2(a_3 - a_2)^\top x + 2c^2(\Delta_3 - \Delta_2)\delta = c^2(\Delta_2^2 - \Delta_3^2) + \|a_3\|_2^2 - \|a_2\|_2^2,$$
$$\|x - a_3\|_2 \leq c\Delta_3 + c\delta$$

(notice that another solution to the same system of equations can be found by maximizing δ, instead of minimizing it).

As numerical examples, we considered instances of $m = 3$ randomly positioned beacons and unknown point on the plane, obtaining the results shown in Figure 10.4.

Sometimes, however, since the solution of the system is not unique, the solution obtained by minimizing δ does not correspond to the correct sync and point position (see Figure 10.4(b)). In such cases, the correct solution could actually be found by maximizing δ. The intuitive reason for this fact is that if we parameterize all solutions x satisfying the first two linear equations, they will be a linear function of δ. Then substituting these x into the equation $\|x - a_3\|_2^2 = (c\Delta_3 + c\delta)^2$ we would obtain a quadratic equation in δ, which has at most two roots, indeed corresponding to the maximum and minimum values of δ subject to the constraints in the above SOCP. Localization and synchronization can then be performed by solving for both min δ and max δ, under the above constraints, and then deciding which solution is the correct one using previous knowledge (such as approximate prior knowledge of position, additional bearing measurements, WiFi signals, etc.).

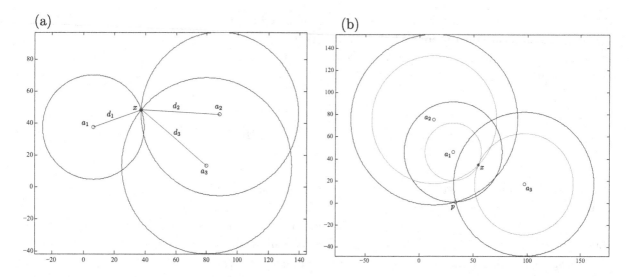

Figure 10.4 (a) Correct sync and localization; (b) wrong sync (light gray circles): p is the true point position, x is the estimated one; in this case, the correct sync (black circles) can be found by maximizing δ, instead of minimizing it.

10.2.8   Separation of ellipsoids

We next consider the problem of finding a hyperplane that separates two ellipsoids in $\mathbb{R}^n$, or determining that no such hyperplane exists. We start first by seeking a condition under which a sphere $\mathcal{B}$ of unit radius centered at zero, $\mathcal{B} = \{x : \|x\|_2 \leq 1\}$, is entirely contained in the half-space

$$\mathcal{H} = \left\{x : a^\top x \leq b\right\},$$

where $a \in \mathbb{R}^n$ and $b \in \mathbb{R}$ are given. The containment condition requires that the inequality defining the half-space remains satisfied for all points in the sphere, that is

$$\mathcal{B} \subset \mathcal{H} \quad \Leftrightarrow \quad a^\top x \leq b, \ \forall x : \|x\|_2 \leq 1.$$

The latter condition is in turn equivalent to

$$b \geq \max_{x: \|x\|_2 \leq 1} a^\top x.$$

The maximum of the inner product on the right-hand side of this inequality is attained for a unit-norm x aligned to a (see Section 2.2.2.4), that is for $x = a/\|a\|_2$, for which the condition simplifies to

$$\mathcal{B} \subset \mathcal{H} \quad \Leftrightarrow \quad b \geq \|a\|_2.$$

Now consider the case with an ellipsoid instead of a sphere. An ellipsoid $\mathcal{E}$ can be described as an affine transformation of the unit sphere:

$$\mathcal{E} = \{x = \hat{x} + Ru : \|u\|_2 \leq 1\},$$

where $\hat{x} \in \mathbb{R}^n$ is the center, and R is a matrix that determines the shape of the ellipsoid. The containment condition $\mathcal{E} \subset \mathcal{H}$ is thus equivalent to $b \geq \max_{u:\,\|u\|_2 \leq 1} a^\top(\hat{x} + Ru)$. Using the same argument as before, we easily obtain that

$$\mathcal{E} \subset \mathcal{H} \quad \Leftrightarrow \quad b \geq a^\top \hat{x} + \|R^\top a\|_2.$$

Note that the condition that $\mathcal{E}$ should be contained in the complementary half-space $\bar{\mathcal{H}} = \{x : a^\top x \geq b\}$ is readily obtained by changing a, b into $-a, -b$, that is:

$$\mathcal{E} \subset \bar{\mathcal{H}} \quad \Leftrightarrow \quad b \leq a^\top \hat{x} - \|R^\top a\|_2.$$

Next, consider two ellipsoids

$$\mathcal{E}_i = \{\hat{x}_i + R_i u_i : \|u_i\|_2 \leq 1\},\ i = 1, 2,$$

where $\hat{x}_i \in \mathbb{R}^n$ are the centers, and $R_i,\ i = 1, 2$, are the shape matrices of the ellipsoids. The hyperplane $\{x : a^\top x = b\}$ separates (possibly not strictly) the two ellipsoids if and only if $\mathcal{E}_1 \in \mathcal{H}$ and $\mathcal{E}_2 \in \bar{\mathcal{H}}$, or vice versa. Thus,

$$b_1 \doteq a^\top \hat{x}_1 + \|R_1^\top a\|_2 \leq b \leq a^\top \hat{x}_2 - \|R_2^\top a\|_2 \doteq b_2.$$

Thus, the existence of a separating hyperplane is equivalent to the existence of $a \in \mathbb{R}^n$ such that

$$a^\top(\hat{x}_2 - \hat{x}_1) \geq \|R_1^\top a\|_2 + \|R_2^\top a\|_2.$$

Exploiting the homogeneity with respect to a (i.e., the fact that a can be multiplied by any nonzero scalar, without modifying the condition), we can always normalize this condition so that $a^\top(\hat{x}_2 - \hat{x}_1) = 1$. This results in an SOCP in variables $a \in \mathbb{R}^n$, $t \in \mathbb{R}$:

$$
\begin{aligned}
p^* = \quad &\min_{a, t} \ t \\
\text{s.t.:} \quad &\|R_1^\top a\|_2 + \|R_2^\top a\|_2 \leq t, \\
&a^\top(\hat{x}_2 - \hat{x}_1) = 1.
\end{aligned}
$$

The ellipsoids are separable if and only if $p^* \leq 1$. In this case, an appropriate value of b is $b = \frac{b_1 + b_2}{2}$.

10.2.9 Minimum surface area problems

Consider a surface in $\mathbb{R}^3$ that is described by a function from the square $C = [0, 1] \times [0, 1]$ into $\mathbb{R}$. The corresponding surface area is defined via the integral

$$A(f) = \int_C \sqrt{1 + \|\nabla f(x, y)\|_2^2}\, dx dy.$$

The minimum surface area problem is to find the function f that minimizes the area $A(f)$, subject to boundary values on the contour of C. To be specific, we will assume that we are given values of f on the lower and upper sides of the square, that is

$$f(x,0) = l(x), \quad f(x,1) = u(x), \quad x \in [0, 1],$$

where $l : \mathbb{R} \to \mathbb{R}$ and $u : \mathbb{R} \to \mathbb{R}$ are two given functions.

The above is an infinite dimensional problem, in the sense that the unknown is a function, not a finite-dimensional vector. To find an approximate solution, we here resort to *discretization*. That is, we discretize the square C with a grid of points $((i-1)h, (j-1)h)$, $1 \leq i, j \leq K+1$, where K is an integer, and where $h = 1/K$ is the (uniform) spacing of the grid. We represent the variable of our problem, f, as a matrix F in $\mathbb{R}^{K+1,K+1}$ with elements $F_{i,j} = f((i-1)h, (j-1)h)$. Similarly, we represent the boundary conditions as vectors L and U.

To approximate the gradient, we start from the first-order expansion of a function of two variables, valid for some small increment h:

$$\frac{\partial f(x,y)}{\partial x} \simeq \frac{1}{h}(f(x+h,y) - f(x,y)).$$

We thus obtain that the gradient of f at a grid point can be approximated as

$$G_{i,j} = \nabla f((i-1)h, (j-1)h) \simeq \begin{bmatrix} K(F_{i+1,j} - F_{i,j}) \\ K(F_{i,j+1} - F_{i,j}) \end{bmatrix}, \quad 1 \leq i, j \leq K. \tag{10.19}$$

Now, approximating the integral with a summation over all grid points, we obtain a discretized version of the problem in SOCP format, as follows:

$$\min_{F \in \mathbb{R}^{K+1,K+1}} \quad \frac{1}{K^2} \sum_{1 \leq i,j \leq K} \left\| \begin{bmatrix} K(F_{i+1,j} - F_{i,j}) \\ K(F_{i,j+1} - F_{i,j}) \\ 1 \end{bmatrix} \right\|_2$$

$$\text{s.t.:} \quad F_{i,1} = L_i = l((i-1)h), \qquad 1 \leq i \leq K+1,$$
$$F_{i,K+1} = U_i = u((i-1)h), \qquad 1 \leq i \leq K+1.$$

10.2.10 *Total variation image restoration*

A discretization technique similar to the one introduced in the previous section can also be used in the context of digital image restoration. Digital images always contain noise, and the objective of image restoration is to filter out the noise. Early methods involved a least-squares approach, but the solutions exhibited the "ringing" phenomenon, with spurious oscillations near edges in the restored image. To address this phenomenon, one may add to the objective of the least-squares problem a term which penalizes the variations in the image.

We may represent a given (noisy) image as a function from the square $C = [0, 1] \times [0, 1]$ into $\mathbb{R}$. We define the image restoration problem as minimizing, over functions $\hat{f} : C \to \mathbb{R}$, the objective

$$\int_C \|\nabla \hat{f}(x)\|_2 dx + \lambda \int_C (\hat{f}(x) - f(x))^2 dx,$$

where the function $\hat{f}$ is our restored image. The first term penalizes functions which exhibit large variations, while the second term accounts for the distance from the estimate to the noisy image f. This is an infinite dimensional problem, in the sense that the variable is a function, not a finite-dimensional vector. Therefore, we tackle it approximately via discretization. We can discretize the square C with a square grid, as in Section 10.2.9:

$$x_{ij} = \left[\begin{array}{c} \frac{i-1}{K} \\ \frac{j-1}{K} \end{array} \right], \ 1 \le i, j \le K + 1.$$

We represent the data of our problem, f, as a matrix $F \in \mathbb{R}^{K+1,K+1}$, with elements $F_{ij} = f(x_{ij})$. Similarly, we represent the variable $\hat{f}$ of our problem with a $(K + 1) \times (K + 1)$ matrix $\hat{F}$ containing the values of $\hat{f}$ at the grid points x_{ij}. We then approximate the gradient $\nabla \hat{f}(x)$ as in (10.19):

$$\hat{G}_{i,j} = \nabla \hat{f}(x_{ij}) \simeq \left[\begin{array}{c} K(\hat{F}_{i+1,j} - \hat{F}_{i,j}) \\ K(\hat{F}_{i,j+1} - \hat{F}_{i,j}) \end{array} \right], \quad 1 \le i, j \le K.$$

The discretized version of our problem is thus written in SOCP format as

$$\min_{\hat{f}} \frac{1}{K^2} \sum_{1 \le i,j \le K} \left(\left\| \left[\begin{array}{c} K(\hat{F}_{i+1,j} - \hat{F}_{i,j}) \\ K(\hat{F}_{i,j+1} - \hat{F}_{i,j}) \end{array} \right] \right\|_2 + \lambda (\hat{F}_{ij} - F_{ij})^2 \right): \quad (10.20)$$

As an example, Figure 10.5(a) shows an original gray-scale image of 256×256 pixels. Each pixel has a value from 0 (black) to 255 (white), corresponding to its luminance level. Figure 10.5(b) is obtained by adding Gaussian noise with standard deviation $\sigma = 12$ to each pixel of the original image. This noisy image constitutes the F matrix. Given F, we aim to restore (i.e., de-noise) the image by solving problem (10.20) for the $\hat{F}$ variable. Notice that this problem can be quite "large scale:" in our small example with a 256×256 image we have $65,536$ variables in the optimization problem. De-noising a

$1,024 \times 1,024$ image would already involve a number of variable of the order of a million. For this reason, specialized and fast convex optimization algorithms should be used for this type of problem. In this example, we used the TFOCS package (see `cvxr.com/tfocs/`) to solve the problem numerically. In particular, for the choice of the parameter $\lambda = 8$, we obtained the restored image shown in Figure 10.5(c).

(a) Original

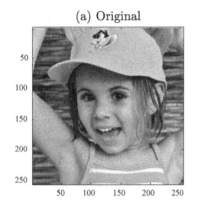

(b) Noise ($\sigma = 12$) added

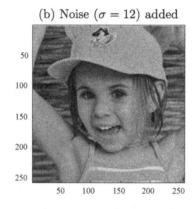

(c) TV recovered

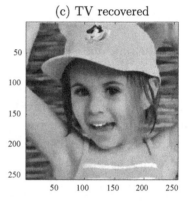

Figure 10.5 (a) Original image; (b) image with noise added; (c) de-noised image obtained via solution of (10.20), with $\lambda = 8$.

10.3   Robust optimization models

In this section we introduce the reader to models and techniques that permit us to take into account the presence of *uncertainty* in the data describing an optimization problem, and to obtain solutions that are *robust* against this uncertainty. Section 10.3.1 introduces the main ideas, while Section 10.3.2 illustrates how to deal with uncertain data in linear programs. Section 10.3.3 discusses a robust least-squares model, and Section 10.3.4 presents an approximate approach for obtaining robust solutions to general uncertain optimization problems.

10.3.1   Robust LP

Consider a linear program in standard inequality form:

$$
\begin{aligned}
\min_{x} \quad & c^\top x \\
\text{s.t.:} \quad & a_i^\top x \le b_i, \quad i = 1, \dots, m.
\end{aligned}
\tag{10.21}
$$

In many practical cases, the data of the linear program (contained in the vectors c, b, and a_i, $i = 1, \dots, m$) are not known precisely. For example, the coefficient matrix $A \in \mathbb{R}^{m,n}$ (whose rows are $a_i^\top$, $i = 1, \dots, m$) may be given by a known nominal matrix $\hat{A}$, plus a perturbation Δ, which is only known to be bounded in norm as

$\|\Delta\|_F \leq \rho$. A robust LP in this case seeks a solution that minimizes the objective, while guaranteeing constraint satisfaction *for all possible* values of the uncertainty term, that is

$$\min_x \quad c^\top x$$
$$\text{s.t.:} \quad (\hat{A} + \Delta)x \leq b, \quad \forall \Delta : \|\Delta\|_F \leq \rho.$$

We anticipate that this robust LP is equivalent to the following SOCP (this follows as a particular case of the ellipsoidal uncertainty model discussed later in Section 10.3.2.3)

$$\min_x \quad c^\top x$$
$$\text{s.t.:} \quad \hat{a}_i^\top x + \rho \|x\|_2 \leq b_i, \quad i = 1, \ldots, m.$$

Solving the LP without taking into account the uncertainty in the problem's data might make the supposedly "optimal" solution become suboptimal, or even infeasible. This idea is discussed and developed in the next example and in the following sections.

Example 10.5 (*Uncertainty in the drug production problem*) Let us reconsider the drug production problem discussed in Example 9.7, assuming now that a small variation may occur in some data of the problem, due to uncertainty. Specifically, we assume that the content of the active agent in the raw materials is subject to variation, with a margin of relative error of 0.5% for raw material I, and of 2% for raw material II. The possible values of the coefficients are shown as intervals in Table 10.1.

Material	agent content (g/kg)
Raw I	$[0.00995, 0.01005]$
Raw II	$[0.0196, 0.0204]$

Table 10.1 Uncertainty on agent contents in raw materials.

Let us now check the impact of this uncertainty on the optimal solution previously computed in Example 9.7, when uncertainty was ignored:

$$p^* = -8819.658, \ x^*_{RawI} = 0, \ x^*_{RawII} = 438.789, \ x^*_{DrugI} = 17.552, \ x^*_{DrugII} = 0.$$

The uncertainty affects the constraint on the balance of the active agent. In the nominal problem, this constraint was

$$0.01x_{RawI} + 0.02x_{RawII} - 0.05x_{DrugI} - 0.600x_{DrugII} \geq 0.$$

At optimum, this constraint is active. Therefore, even a tiny error in the first and second coefficients makes the constraint become invalid, i.e., the optimal solution computed on the nominal problem (which ignored uncertainty) may result in being infeasible on the actual data.

An adjustment policy. To remedy the problem, there is a simple solution: adjust the levels of drug production so as to satisfy the balance constraint. Let us adjust the production of Drug I, since that of Drug II is zero according to the original plan. Clearly, if the actual content of active ingredient increases, the balance constraint will remain valid. In such a case, there is nothing to adjust, and the original production plan is still valid (feasible) on the actual uncertain problem and nominally optimal. The balance constraint does become invalid only if "nature is against us," that is when the level of active agent is less than originally thought. Since the original optimal production plan recommends us to purchase only the raw material II, a change in the corresponding coefficient (nominally set at 0.02) to the lesser value 0.0196 results, if we are to adopt the above simple "adjustment policy," in a variation in the amount of production of Drug I from 17,552 packs (the nominal value) to the (2% less) value of 17,201 packs. Accordingly, the cost function will decrease from the nominal value of 8,820 to the 21% (!) smaller value of 6,929. This shows that, for this problem, even a tiny variation in a single coefficient can result in a substantial decrease in the profit predicted by the model.

If we are to believe that the uncertain coefficients are actually random, and take their extreme values, say, with 1/2 probability each, then the expected value of the cost (still with the above adjustment policy) will be also random, with expected value (8,820+6,929)/2 = 7,874. Thus, the expected loss due to random uncertainty is still high, at 11%.

Uncertainty can originate also from implementation errors. Often, the optimal variable x^* corresponds to some action or implementation process, which may be fraught with errors. For example, in a manufacturing process, the planned production amounts are never exactly implemented due to, say, production plant failures or fixed sizing of the production lots. Implementation errors may result in catastrophic behavior, in the sense that when the optimal variable x^* is replaced by its error-affected version, the constraints may become violated, or the cost function may become much worse (higher) than thought.

10.3.1.1 Robust optimization: main idea. In robust optimization, we overcome the mentioned issues by taking into account the fact that the data in the LP may be imprecisely known right since the modeling phase. This will in turn provide us with solutions that are *guaranteed* to "work" (i.e., to remain feasible), no matter what the uncertainty does. To this end, we postulate that a *model of the uncertainty* is known. In its simplest version, this model assumes that the individual rows a_is are known to belong to given sets $U_i \subseteq \mathbb{R}^n$, $i = 1, \ldots, m$. We can think of those sets as sets of confidence for the coefficients of the linear program. The main idea in robust LP is to impose that each constraint be satisfied in a worst-case sense, that is each constraint should hold *for all* possible values of $a_i \in U_i$.

The robust counterpart to the original LP is thus defined as follows:

$$\min_{x} \quad c^\top x$$
$$\text{s.t.:} \quad a_i^\top x \le b_i, \ \forall a_i \in U_i, \quad i = 1, \dots, m.$$

The interpretation of the above problem is that it attempts to find a solution that is feasible for any choice of the coefficient vectors a_i within their respective sets of confidence U_i.

The robust counterpart LP is always convex, independent of the shape of the sets of confidence U_i. Indeed, for each i, the set

$$\mathcal{X}_i = \{x : \ a_i^\top x \le b_i, \ \forall a_i \in U_i\}$$

is representable as the intersection of (possibly infinitely many) convex sets (namely, half-spaces)

$$\mathcal{X}_i = \bigcap_{a_i \in U_i} \{x : \ a_i^\top x \le b_i\},$$

hence $\mathcal{X}_i$ is convex. Therefore, robust LPs are still convex optimization problems. However, depending on the type and structure of the uncertainty sets U_i, and due to the fact that these sets may contain a dense infinity of elements, it may be difficult to express the robust program in some usable explicit form. There are, however, notable exceptions for which this is possible, and for these classes of uncertainty sets U_i we shall say that the robust LP counterpart is "computationally tractable." We shall next detail three such tractable cases, which are important in applications. To this end, we notice first that the *robust half-space constraint*

$$a^\top x \le b, \ \forall a \in U$$

(which is, as we previously remarked, convex irrespective of the set U) can be expressed in terms of an *inner* optimization problem, as follows:

$$\max_{a \in U} a^\top x \le b. \tag{10.22}$$

What makes the robust LP "tractable" is indeed the possibility of solving the inner problem (10.22) explicitly, as detailed in the next section.

10.3.2   Tractable robust LP counterparts

We here discuss three cases for which the robust counterpart of an LP with uncertain data leads to a tractable convex optimization problem, namely the cases of (i) discrete uncertainty, (ii) box (or interval) uncertainty sets, and (iii) ellipsoidal uncertainty sets.

10.3.2.1 *Discrete uncertainty.* In the discrete uncertainty model, the uncertainty on each coefficient vector a_i is described by a finite set of points:

$$U_i = \left\{ a_i^{(1)}, \ldots, a_i^{(K_i)} \right\},$$

where each vector $a_i^{(k)} \in \mathbb{R}^n, k = 1, \ldots, K_i$, corresponds to a particular "scenario," or possible outcome of the uncertainty. The robust half-space constraint

$$a_i^\top x \le b, \quad \forall a_i \in U_i$$

can simply be expressed as a set of K_i affine inequalities:

$$a_i^{(k)\top} x \le b, \; k = 1, \ldots, K_i. \tag{10.23}$$

Note that the discrete uncertainty model actually enforces more than feasibility at points $a_i^{(k)}$. In fact, the constraints (10.23) imply that

$$\left(\sum_{k=1}^{K_i} \lambda_k a_i^{(k)\top} \right) x \le b$$

also holds for any set of non-negative weights $\lambda_1, \ldots, \lambda_{K_i}$ summing to one. Therefore, satisfaction of the discrete inequalities (10.23) implies satisfaction of the inequality $a_i^\top x \le b$ for all a_i belonging to the convex hull of the set U_i.

With discrete uncertainty, the robust counterpart to the original LP (10.21) becomes

$$\min_x \quad c^\top x$$
$$\text{s.t.:} \quad a_i^{(k)\top} x \le b, \; k = 1, \ldots, K_i; \; i = 1, \ldots, m.$$

Thus, the discrete-robust counterpart of an LP is still an LP, with a total of $m(K_1 + \cdots + K_m)$ constraints, where K_i is the number of elements in the discrete set U_i. The discrete uncertainty model is attractive for its simplicity, since the robust counterpart retains the same structure (LP) of the original problem. However, such a model may become impractical to solve in case of a very large number of discrete points.

10.3.2.2 *Box uncertainty.* The box uncertainty model assumes that every coefficient vector a_i is only known to lie in a "box," or, more generally, a hyper-rectangle in $\mathbb{R}^n$. In its simplest case, this uncertainty model has the following form:

$$U_i = \{a_i : \|a_i - \hat{a}_i\|_\infty \le \rho_i\}, \tag{10.24}$$

where $\rho_i \ge 0$ is a measure of the size of the uncertainty, and $\hat{a}_i$ represents the nominal value of the coefficient vector. The set U_i is a "box" (hypercube) of half-side length ρ_i around the center $\hat{a}_i$. The condition $a_i \in U_i$ can be equivalently expressed as

$$a_i = \hat{a}_i + \rho_i \delta_i, \quad \|\delta_i\|_\infty \leq 1, \tag{10.25}$$

where δ_i represents the uncertainty around the nominal value $\hat{a}_i$. Note that the robust counterpart to the robust half-space constraint

$$a_i^\top x \leq b_i, \quad \forall a_i \in U_i$$

can be handled as a discrete model, by considering as discrete uncertainty points the vectors $a_i^{(k)} = \hat{a}_i + \rho_i v^k$, where v^k, $k = 1, \ldots, 2^n$, represent the vertices of the unit box (that is, vectors having ± 1 as components). Indeed, enforcing the constraint $a_i^{(k)\top} x \leq b_i$ at the vertices of the hypercube U_i implies that $a_i^\top x \leq b$ holds for every a_i in the convex hull of the vertices, that is in all U_i. This approach, however, may not be practical, since the number of vertices (hence the number of constraints in the scenario counterpart of the LP) grows geometrically with the dimension n. There is actually a much more effective reformulation of the robust counterpart in the case of LP with box uncertainty, which can be obtained by examining the inner optimization problem in (10.22). We have[1]

[1] See Section 2.2.2.4.

$$
\begin{aligned}
b \;\geq\; \max_{a_i \in U_i} a_i^\top x &= \max_{\|\delta_i\|_\infty \leq 1} \hat{a}_i^\top x + \rho_i \delta_i^\top x \\
&= \hat{a}_i^\top x + \rho_i \max_{\|\delta_i\|_\infty \leq 1} \delta_i^\top x \\
&= \hat{a}_i^\top x + \rho_i \|x\|_1.
\end{aligned}
$$

Therefore, the robust counterpart of the original LP (10.21) under box uncertainty (10.24) can be written as an optimization problem with polyhedral constraints:

$$
\begin{aligned}
\min_x \quad & c^\top x \\
\text{s.t.:} \quad & \hat{a}_i^\top x + \rho_i \|x\|_1 \leq b_i \quad i = 1, \ldots, m.
\end{aligned}
$$

This problem can in turn be expressed in standard LP form by introducing a slack vector $u \in \mathbb{R}^n$:

$$
\begin{aligned}
\min_{x,u} \quad & c^\top x \\
\text{s.t.:} \quad & \hat{a}_i^\top x + \rho_i \sum_{j=1}^n u_j \leq b_i, \quad i = 1, \ldots, m, \\
& -u_j \leq x_j \leq u_i, \quad j = 1, \ldots, n.
\end{aligned}
$$

Notice that Eq. (10.25) implies that each entry of a_i is bounded in an interval centered in the corresponding entry of $\hat{a}_i$ and of half-width equal to ρ_i. Thus, according to this model, all entries in the vector a_i have the same uncertainty radius ρ_i. The model can be easily generalized to include the case where each entry of a_i is bounded in an interval of possibly different length, by assuming that

$$a_i = \hat{a}_i + \rho_i \odot \delta_i, \quad \|\delta_i\|_\infty \le 1,$$

where now $\rho_i \in \mathbb{R}^n$ is a vector containing the half-lengths of the uncertainty intervals in each entry of a_i, and $\odot$ denotes the Hadamard (entry-by-entry) product of two vectors. The reader may verify as an exercise that the robust counterpart of the original LP, in this setting, is given by the following LP:

$$\begin{aligned}
\min_{x,u} \quad & c^\top x \\
\text{s.t.:} \quad & \hat{a}_i^\top x + \rho_i^\top u \le b_i, \quad i = 1, \dots, m \\
& -u_j \le x_j \le u_i, \quad j = 1, \dots, n.
\end{aligned}$$

An example of application of robust linear programs with interval uncertainty, in the context of inventory control, is presented in Section 16.5.

10.3.2.3 Ellipsoidal uncertainty. In the ellipsoidal uncertainty model each vector a_i is contained in an ellipsoid U_i of the form

$$U_i = \{a_i = \hat{a}_i + R_i \delta_i : \|\delta_i\|_2 \le 1\}, \tag{10.26}$$

where $R_i \in \mathbb{R}^{n,p}$ is a matrix which describes the "shape" of the ellipsoid around its center $\hat{a}_i$. If $R_i = \rho_i I$ for some $\rho_i \ge 0$, then this set is simply a hypersphere of radius ρ_i centered at $\hat{a}_i$ and we refer to this special case as the spherical uncertainty model. Ellipsoidal uncertainty models are useful to "couple" uncertainties across different components of the coefficient vector a_i. This is in contrast with the previous "box" model, which allows uncertainties to take their largest values independently of each other (the box model is also referred to as the "independent intervals" model). With an ellipsoidal model, the robust half-space constraint becomes[2]

[2] See Section 2.2.2.4.

$$\begin{aligned}
b \ge \; & \max_{a_i \in U_i} a_i^\top x = \max_{\|\delta_i\|_2 \le 1} \hat{a}_i^\top x + \delta_i^\top R_i^\top x \\
= \; & \hat{a}_i^\top x + \max_{\|\delta_i\|_2 \le 1} \delta_i^\top (R_i^\top x) \\
= \; & \hat{a}_i^\top x + \|R_i^\top x\|_2.
\end{aligned}$$

The robust counterpart of the original LP (10.21) under ellipsoidal uncertainty (10.26) is thus the following SOCP:

$$\begin{aligned}
\min_{x} \quad & c^\top x \\
\text{s.t.:} \quad & \hat{a}_i^\top x + \|R_i^\top x\|_2 \le b_i \quad i = 1, \dots, m.
\end{aligned}$$

10.3.3 Robust least squares

Let us start from a standard LS problem:

$$\min_x \ \|Ax - y\|_2,$$

where $A \in \mathbb{R}^{m,n}$, $y \in \mathbb{R}^m$. Now assume that the matrix A is imperfectly known. A simple way to model the uncertainty in the data matrix A is to assume that A is only known to be within a certain "distance" (in matrix space) to a given "nominal" matrix $\hat{A}$. Precisely, let us assume that

$$\|A - \hat{A}\| \leq \rho,$$

where $\| \cdot \|$ denotes the largest singular value norm, and $\rho \geq 0$ measures the size of the uncertainty. Equivalently, we may say that

$$A = \hat{A} + \Delta,$$

where Δ is the uncertainty, which satisfies $\|\Delta\| \leq \rho$. We now address the *robust least-squares* problem:

$$\min_x \ \max_{\|\Delta\| \leq \rho} \ \|(\hat{A} + \Delta)x - y\|_2.$$

The interpretation of this problem is that we aim at minimizing (with respect to x) the worst-case value (with respect to the uncertainty Δ) of the residual norm. For fixed x, and using the fact that the Euclidean norm is convex, we have that

$$\|(\hat{A} + \Delta)x - y\|_2 \leq \|\hat{A}x - y\|_2 + \|\Delta x\|_2.$$

By definition of the largest singular value norm, and given our bound on the size of the uncertainty, we have

$$\|\Delta x\|_2 \leq \|\Delta\| \cdot \|x\|_2 \leq \rho \|x\|_2.$$

Thus, we have a bound on the objective value of the robust LS problem:

$$\max_{\|\Delta\| \leq \rho} \ \|(\hat{A} + \Delta)x - y\|_2 \leq \|\hat{A}x - y\|_2 + \rho \|x\|_2.$$

It turns out that the upper bound is actually attained by some choice of the matrix Δ, specifically for the following dyadic matrix:

$$\Delta = \frac{\rho}{\|\hat{A}x - y\|_2 \cdot \|x\|_2} (\hat{A}x - y)x^\top.$$

Hence, the robust LS problem is equivalent to

$$\min_x \ \|\hat{A}x - y\|_2 + \rho \|x\|_2.$$

This is a regularized LS problem, which can be cast in SOCP format as follows:

$$\min_{x,u,v} \ u + \rho v, \quad \text{s.t.:} \ u \geq \|\hat{A}x - y\|_2, \ v \geq \|x\|_2.$$

Further linear equality or inequality constraints can also be easily added to the problem, while retaining its SOCP structure.

10.3.4 The scenario approach to robustness

Generic convex optimization problems, subject to generic uncertainty, do not admit exact tractable robust reformulations. When an uncertain problem does not fall into one of the categories discussed in the previous sections (or into some other tractable class, not discussed in this book), one can resort to a general approach to robustness that has the advantage of being completely general, at the price of being approximate, in a sense to be specified next. Consider a generic convex optimization problem, subject to uncertainty:

$$\min_{x \in \mathbb{R}^n} \ f_0(x)$$
$$\text{s.t.:} \ f_i(x, \delta) \leq 0, \quad i = 1, \ldots, m,$$

where f_i, $i = 0, \ldots, m$, are convex in x for each value of $\delta \in \mathcal{U}$, while they can be arbitrary functions of δ. Also, $\mathcal{U}$ is a generic uncertainty set. The robust counterpart of this problem,

$$\min_{x \in \mathbb{R}^n} \ f_0(x) \tag{10.27}$$
$$\text{s.t.:} \ f_i(x, \delta) \leq 0, \quad i = 1, \ldots, m, \quad \forall \delta \in \mathcal{U},$$

does not admit a tractable reformulation, in general.

Let us then assume that the uncertainty δ is random, with some probability distribution. The actual distribution of the uncertainty need not be known, all we need is a number N of independent and identically distributed (iid) samples $\delta^{(1)}, \ldots, \delta^{(N)}$ of the uncertainty, generated according to this distribution. Using these samples of the uncertainty, called *scenarios*, we act "as if" the uncertainty was discrete, and consider a problem (the so-called *scenario problem*) that is robust only with respect to the discrete sampled scenarios:

$$\min_{x \in \mathbb{R}^n} \ f_0(x) \tag{10.28}$$
$$\text{s.t.:} \ f_i(x, \delta^{(j)}) \leq 0, \quad i = 1, \ldots, m, \quad j = 1, \ldots, N.$$

Clearly, the solution obtained from the scenario problem is not guaranteed to be robust in the sense of problem (10.27), since it cannot be guaranteed, in general, to be feasible *for all* values of the uncertainty $\delta \in \mathcal{U}$. However, a remarkable fact is that, if N is chosen to be suitably large, then the scenario solution can be proved to be *probabilistically robust*, up to a pre-specified level of probability $\alpha \in (0, 1)$. This means that we can fix a desired level of probabilistic robustness α, and then find a value of N such that the ensuing scenario problem will provide an optimal solution x^* with the property[3] that

$$\text{Prob}\{\delta : f_i(x^*, \delta) \leq 0, \ i = 1, \dots, m\} \geq \alpha. \qquad (10.29)$$

That is, the scenario solution is indeed robustly feasible, up to a probability level α. The tradeoff between the desired level α of robustness and the required number N of required scenarios is captured by the following simple formula[4]

$$N \geq \frac{2}{1 - \alpha}(n + 10). \qquad (10.30)$$

10.4 Exercises

Exercise 10.1 (Squaring SOCP constraints) When considering a second-order cone constraint, a temptation might be to square it in order to obtain a classical convex quadratic constraint. This might not always work. Consider the constraint

$$x_1 + 2x_2 \geq \|x\|_2,$$

and its squared counterpart:

$$(x_1 + 2x_2)^2 \geq \|x\|_2^2.$$

Is the set defined by the second inequality convex? Discuss.

Exercise 10.2 (A complicated function) We would like to minimize the function $f : \mathbb{R}^3 \to \mathbb{R}$, with values:

$$f(x) = \max\left(x_1 + x_2 - \min\left(\min(x_1 + 2, x_2 + 2x_1 - 5), x_3 - 6\right), \frac{(x_1 - x_3)^2 + 2x_2^2}{1 - x_1}\right),$$

with the constraint $\|x\|_\infty < 1$. Explain precisely how to formulate the problem as an SOCP in standard form.

Exercise 10.3 (A minimum time path problem) Consider Figure 10.6, in which a point in 0 must move to reach point $p = [4 \ \ 2.5]^\top$, crossing three layers of fluids having different densities.

In the first layer, the point can travel at a maximum speed v_1, while in the second layer and third layers it may travel at lower maximum

[3] To be precise, since x^* is itself random, *a priori*, the probabilistic robustness property (10.29) itself holds only up to a certain level of *confidence*. However, this confidence level can be made so high that (10.29) can be considered a certain event, for any practical purpose.

[4] This formula is a simplified one, obtained under the assumptions that problem (10.28) is feasible and admits a unique optimal solution. The constant 10 appearing in the formula is related to the (hidden) level of confidence of statement (10.29), and refers to a confidence level of about $1 - 1.7 \times 10^{-5}$. If higher confidence is required, then this constant grows just a little. For instance, for confidence level $1 - 7.6 \times 10^{-10}$, the constant becomes 20. Details and proofs of results related to the scenario approach can be found in G. Calafiore, Random convex programs, *SIAM J. Optimization*, 2010.

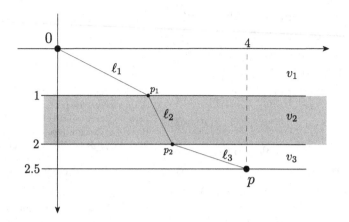

Figure 10.6 A minimum-time path problem.

speeds, respectively $v_2 = v_1/\eta_2$, and $v_3 = v_1/\eta_3$, with $\eta_2, \eta_3 > 1$. Assume $v_1 = 1$, $\eta_2 = 1.5$, $\eta_3 = 1.2$. You have to determine what is the fastest (i.e., minimum time) path from 0 to p. *Hint:* you may use path leg lengths ℓ_1, ℓ_2, ℓ_3 as variables, and observe that, in this problem, equality constraints of the type $\ell_i =$ "something" can be equivalently substituted by inequality constraints $\ell_i \geq$ "something" (explain why).

Exercise 10.4 (k-ellipses) Consider k points $x_1, \ldots, x_k$ in $\mathbb{R}^2$. For a given positive number d, we define the k-ellipse with radius d as the set of points $x \in \mathbb{R}^2$ such that the sum of the distances from x to the points x_i is equal to d.

1. How do k-ellipses look when $k = 1$ or $k = 2$? *Hint:* for $k = 2$, show that you can assume $x_1 = -x_2 = p$, $\|p\|_2 = 1$, and describe the set in an orthonormal basis of $\mathbb{R}^n$ such that p is the first unit vector.

2. Express the problem of computing the *geometric median*, which is the point that minimizes the sum of the distances to the points x_i, $i = 1, \ldots, k$, as an SOCP in standard form.

3. Write a code with input $X = (x_1, \ldots, x_k) \in \mathbb{R}^{2,k}$ and $d > 0$ that plots the corresponding k-ellipse.

Exercise 10.5 (A portfolio design problem) The returns on $n = 4$ assets are described by a Gaussian (normal) random vector $r \in \mathbb{R}^n$, having the following expected value $\hat{r}$ and covariance matrix Σ:

$$\hat{r} = \begin{bmatrix} 0.12 \\ 0.10 \\ 0.07 \\ 0.03 \end{bmatrix}, \quad \Sigma = \begin{bmatrix} 0.0064 & 0.0008 & -0.0011 & 0 \\ 0.0008 & 0.0025 & 0 & 0 \\ -0.0011 & 0 & 0.0004 & 0 \\ 0 & 0 & 0 & 0 \end{bmatrix}.$$

The last (fourth) asset corresponds to a risk-free investment. An investor wants to design a portfolio mix with weights $x \in \mathbb{R}^n$ (each weight x_i is non-negative, and the sum of the weights is one) so as to obtain the best possible expected return $\hat{r}^\top x$, while guaranteeing that: (i) no single asset weights more than 40%; (ii) the risk-free assets should not weight more than 20%; (iii) no asset should weight less than 5%; (iv) the probability of experiencing a return lower than $q = -3\%$ should be no larger than $\epsilon = 10^{-4}$. What is the maximal achievable expected return, under the above constraints?

Exercise 10.6 (A trust-region problem) A version of the so-called (convex) *trust-region* problem amounts to finding the minimum of a convex quadratic function over a Euclidean ball, that is

$$\min_x \quad \frac{1}{2} x^\top H x + c^\top x + d$$
$$\text{s.t.:} \quad x^\top x \le r^2,$$

where $H \succ 0$, and $r > 0$ is the given radius of the ball. Prove that the optimal solution to this problem is unique and is given by

$$x(\lambda^*) = -(H + \lambda^* I)^{-1} c,$$

where $\lambda^* = 0$ if $\|H^{-1}c\|_2 \le r$, or otherwise λ^* is the unique value such that $\|(H + \lambda^* I)^{-1} c\|_2 = r$.

Exercise 10.7 (Univariate square-root LASSO) Consider the problem

$$\min_{x \in \mathbb{R}} f(x) \doteq \|ax - y\|_2 + \lambda |x|,$$

where $\lambda \ge 0$, $a \in \mathbb{R}^m$, $y \in \mathbb{R}^m$ are given, and $x \in \mathbb{R}$ is a scalar variable. This is a univariate version of the square-root LASSO problem introduced in Example 8.23. Assume that $y \neq 0$ and $a \neq 0$, (since otherwise the optimal solution of this problem is simply $x = 0$). Prove that the optimal solution of this problem is

$$x^* = \begin{cases} 0 & \text{if } |a^\top y| \le \lambda \|y\|_2, \\ x_{\text{ls}} - \text{sgn}(x_{\text{ls}}) \frac{\lambda}{\|a\|_2^2} \sqrt{\frac{\|a\|_2^2 \|y\|_2^2 - (a^\top y)^2}{\|a\|_2^2 - \lambda^2}} & \text{if } |a^\top y| > \lambda \|y\|_2, \end{cases}$$

where

$$x_{\text{ls}} \doteq \frac{a^\top y}{\|a\|_2^2}.$$

Exercise 10.8 (Proving convexity via duality) Consider the function $f : \mathbb{R}^n_{++} \to \mathbb{R}$, with values

$$f(x) = 2 \max_t t - \sum_{i=1}^{n} \sqrt{x_i + t^2}.$$

1. Explain why the problem that defines f is a convex optimization problem (in the variable t). Formulate it as an SOCP.

2. Is f convex?

3. Show that the function $g : \mathbb{R}^n_{++} \to \mathbb{R}$, with values

$$g(y) = \sum_{i=1}^{n} \frac{1}{y_i} - \frac{1}{\displaystyle\sum_{i=1}^{n} y_i}$$

is convex. *Hint:* for a given $y \in \mathbb{R}^n_{++}$, show that

$$g(y) = \max_{x>0} \ -x^T y - f(x).$$

Make sure to justify any use of strong duality.

Exercise 10.9 (Robust sphere enclosure) Let B_i, $i = 1, \ldots, m$, be m given Euclidean balls in $\mathbb{R}^n$, with centers x_i and radii $\rho_i \geq 0$. We wish to find a ball B of minimum radius that contains all the B_i, $i = 1, \ldots, m$. Explain how to cast this problem into a known convex optimization format.

11

Semidefinite models

SEMIDEFINITE PROGRAMMING (SDP) is an optimization model with vector or matrix variables, where the objective to be minimized is linear, and the constraints involve affine combinations of symmetric matrices that are required to be positive (or negative) semidefinite. SDPs include as special cases LPs, QCQPs, and SOCPs; they are perhaps the most powerful class of convex optimization models with specific structure, for which efficient and well-developed numerical solution algorithms are currently available.

SDPs arise in a wide range of applications. For example, they can be used as sophisticated relaxations (approximations) of nonconvex problems, such as Boolean problems with quadratic objective, or rank-constrained problems. They are useful in the context of stability analysis or, more generally, in control design for linear dynamical systems. They are also used, to mention just a few, in geometric problems, in system identification, in algebraic geometry, and in matrix completion problems under sparsity constraints.

11.1 *From linear to conic models*

In the late 1980s, researchers were trying to generalize linear programming. At that time, LP was known to be solvable efficiently, in time roughly cubic in the number of variables or constraints. The new interior-point methods for LP had just become available, and their excellent practical performance matched the theoretical complexity bounds. It seemed, however, that, beyond linear problems, one encountered a wall. Except for a few special problem classes, such as QP, it appeared that as soon as a problem contained nonlinear terms, one could no longer hope to recover the nice practical and theoretical efficiency found in LP. In previous decades it had been noted that convex problems could be efficiently solved in theory (under some mild conditions), but the known numerical methods were extremely slow in practice. It seemed, however, that to harness the power of interior-point methods and apply them to problems other than LP, one had to look closely at convex optimization. A breakthrough occurred by rethinking the role of the set of non-negative vectors, which is the basic object in LP. In the standard conic form, a generic LP can be written as

$$\min_{x} : c^\top x, \text{ s.t.: } Ax = b, \quad x \in \mathbb{R}^n_+,$$

where $\mathbb{R}^n_+$ is the set of non-negative vectors in $\mathbb{R}^n$, i.e., the positive orthant. Researchers asked: what are the basic characteristics of $\mathbb{R}^n_+$ that make interior-point methods work so well? In other words, are there other sets that could be used in place of $\mathbb{R}^n_+$, and still allow efficient methods? It turns out that the key characteristic of interest in $\mathbb{R}^n_+$ is that it is a *convex cone* (i.e., a convex set that is invariant under positive scaling of its elements), and that many of the desirable features of LP can be extended to problems involving as constraint sets some specific convex cones, other than $\mathbb{R}^n_+$. This idea yields a broad class of convex optimization models of the form

$$\min_{x} : c^\top x, \text{ s.t.: } Ax = b, \quad x \in \mathcal{K},$$

where $\mathcal{K}$ is a convex cone. For example, when $\mathcal{K}$ is the second-order cone (or any combination of second-order cones, arising when, say, some variables are in a cone, others in another, and all are coupled by affine equalities) then the above model specializes to an SOCP model. The SDP model class is instead obtained when x is a matrix variable, $\mathcal{K}$ is the cone of positive semidefinite matrices, and we minimize a linear function of x under affine equality constraints on the entries of x. Efficient interior-point solution methods can indeed be extended from LP to SOCP and SDP, although the numerical complexity of SOCP, and especially of SDP models, remains higher than that of the LP model. The practical consequence of this is that the *scale* of SOCP and SDP problems that can be solved numerically on a standard workstation remains smaller than that of LP models. Current technology permits us to solve generic LPs with a number of variables and constraints on the order of millions, and generic SOCP/SDP models two or three orders of magnitude smaller.

11.2 Linear matrix inequalities

11.2.1 The cone of positive semidefinite matrices

We recall from Section 4.4 that an $n \times n$ symmetric matrix F is positive semidefinite (PSD, denoted by $F \succeq 0$) if and only if all of its eigenvalues are non-negative. An alternative and equivalent condition for F to be PSD is that the associated quadratic form is non-negative:

$$z^\top F z \geq 0, \quad \forall z \in \mathbb{R}^n.$$

The set S^n_+ of PSD matrices is a convex cone. Indeed, S^n_+ is a cone, since $F \in S^n_+$ implies that $\alpha F \in S^n_+$, for any $\alpha \geq 0$. Moreover, S^n_+ is convex, since for any $F_1, F_2 \in S^n_+$ and $\gamma \in [0, 1]$, we have that

$$z^\top(\gamma F_1 + (1 - \gamma)F_2)z = \gamma z^\top F_1 z + (1 - \gamma)z^\top F_2 z \geq 0, \quad \forall z \in \mathbb{R}^n.$$

Example 11.1 (*PSD matrices*)

- The matrix

$$F = \begin{bmatrix} 29 & 19 & -4 \\ 19 & 28 & 7 \\ -4 & 7 & 15 \end{bmatrix}$$

  is symmetric, and its eigenvalues are

$$\lambda_1 = 4.8506, \quad \lambda_2 = 2.1168, \quad \lambda_3 = 0.3477.$$

  All the eigenvalues are non-negative, hence F is PSD (actually, all eigenvalues are strictly positive in this case, hence $F \succ 0$).

- For any vector $v \in \mathbb{R}^n$, the dyad $F = vv^\top$ is PSD, since the associated quadratic form is a perfect square: $q(x) = x^\top(vv^\top)x = (v^\top x)^2 \geq 0$.

- More generally, for any, possibly rectangular, matrix A, the matrices $A^\top A$ and $AA^\top$ are both PSD. The converse is also true: any PSD matrix F can be factored as $F = A^\top A$, for some appropriate matrix A.

11.2.2 Linear matrix inequalities

11.2.2.1 Definition. A *linear matrix inequality* (LMI) in standard form is a constraint on a vector of variables $x \in \mathbb{R}^m$ of the form

$$F(x) = F_0 + \sum_{i=1}^m x_i F_i \succeq 0, \tag{11.1}$$

where the $n \times n$ *coefficient matrices* $F_0, \ldots, F_m$ are symmetric. Sometimes, these matrices are not explicitly defined. That is, if $F : \mathbb{R}^m \to S^n$ is an affine map that takes its values in the set of symmetric matrices of order n, then $F(x) \succeq 0$ is an LMI.

Example 11.2 (*Representation of an LMI in standard form*) Quite often linear matrix inequality constraints are imposed on *matrix* variables, rather than on vector variables. The following one is a typical example. For a given square matrix $A \in \mathbb{R}^{n,n}$ and positive definite matrix $P \in S^n_{++}$, the so-called Lyapunov inequality

$$-I - A^\top P - PA \succeq 0 \qquad (11.2)$$

is an LMI in the matrix variable P. To express this LMI in the standard format (11.1) involving a vector of variables x, one may define a suitable one-to-one mapping between the symmetric matrix variable P and a vector $x \in \mathbb{R}^m$ containing the $m = n(n+1)/2$ free entries of P. Usually, we take x as a *vectorization* of matrix P; for instance, x contains first the diagonal elements of P, then the elements in the first upper diagonal, etc. For example, if $n = 3$, then $m = 6$, and

$$P = \begin{bmatrix} p_{11} & p_{12} & p_{13} \\ p_{12} & p_{22} & p_{23} \\ p_{13} & p_{23} & p_{33} \end{bmatrix} \quad \Rightarrow \quad x(P) = \begin{bmatrix} p_{11} \\ p_{22} \\ p_{33} \\ p_{12} \\ p_{23} \\ p_{13} \end{bmatrix};$$

$$x = \begin{bmatrix} x_1 \\ x_2 \\ x_3 \\ x_4 \\ x_5 \\ x_6 \end{bmatrix} \quad \Rightarrow \quad P(x) = \begin{bmatrix} x_1 & x_4 & x_6 \\ x_4 & x_2 & x_5 \\ x_6 & x_5 & x_3 \end{bmatrix}.$$

The coefficient matrix F_0 can then be easily obtained by setting $x = 0$ and plugging $P(x)$ into (11.2), obtaining $F_0 = -I$. The coefficient matrices F_i, $i = 1, \ldots, m$, can be obtained by setting $x_i = 1$, $x_j = 0$, $j \neq i$, plugging $P(x)$ into (11.2), and then subtracting F_0 from the result:

$$F_1 = -\begin{bmatrix} 2a_{11} & a_{12} & a_{13} \\ a_{12} & 0 & 0 \\ a_{13} & 0 & 0 \end{bmatrix}, \quad F_2 = -\begin{bmatrix} 0 & a_{21} & 0 \\ a_{21} & 2a_{22} & a_{23} \\ 0 & a_{23} & 0 \end{bmatrix},$$

$$F_3 = -\begin{bmatrix} 0 & 0 & a_{31} \\ 0 & 0 & a_{32} \\ a_{31} & a_{32} & 2a_{33} \end{bmatrix}, \quad F_4 = -\begin{bmatrix} 2a_{21} & a_{11}+a_{22} & a_{23} \\ a_{11}+a_{22} & 2a_{12} & a_{13} \\ a_{23} & a_{13} & 0 \end{bmatrix},$$

$$F_5 = -\begin{bmatrix} 0 & a_{31} & a_{21} \\ a_{31} & 2a_{32} & a_{33}+a_{22} \\ a_{21} & a_{22}+a_{33} & 2a_{23} \end{bmatrix}, \quad F_6 = -\begin{bmatrix} 2a_{31} & a_{32} & a_{11}+a_{33} \\ a_{32} & 0 & a_{12} \\ a_{11}+a_{33} & a_{12} & 2a_{13} \end{bmatrix}.$$

The operation of rewriting a generic LMI into the standard format (11.1) is elementary conceptually, although it can be very tedious in practice; usually, it is done automatically and internally by parsers/solvers for convex optimization, such as CVX or Yalmip, so that it remains completely transparent to the user, who can express the problem in the most natural format.

11.2.2.2 *Geometry and convexity of LMI sets.* Let us denote by $\mathcal{X}$ the set of points $x \in \mathbb{R}^m$ that satisfy an LMI:

$$\mathcal{X} \doteq \left\{ x \in \mathbb{R}^m : F_0 + \sum_{i=1}^m x_i F_i \succeq 0 \right\}. \qquad (11.3)$$

The set $\mathcal{X}$ is convex. To verify this fact, recall that $F(x) \succeq 0$ if and only if $z^\top F(x)z \geq 0$, for all $z \in \mathbb{R}^n$. Since

$$z^\top F(x)z = f_0(z) + \sum_{i=1}^m x_i f_i(z),$$

where $f_i(z) \doteq z^\top F_i z$, $i = 0, \dots, m$, we obtain that the points x such that $F(x) \succeq 0$ belong to the intersection of an infinite number of half-spaces

$$\mathcal{H}_z = \{ x \in \mathbb{R}^m : a_z^\top x + b_z \geq 0 \}, \qquad (11.4)$$
$$a_z \doteq [f_1(z) \cdots f_m(z)], \ b_z \doteq f_0(z),$$

each parameterized by $z \in \mathbb{R}^n$, which proves that $\mathcal{X}$ is a convex set; see Figure 11.1.

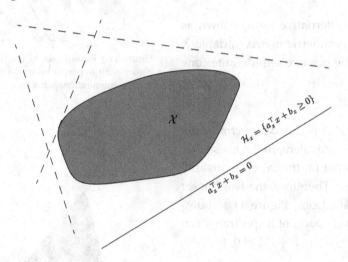

Figure 11.1 For each $z \in \mathbb{R}^n$, the LMI feasible set $\mathcal{X}$ belongs to a half-space $\mathcal{H}_z$ defined in (11.4).

For some specific vectors $z \in \mathbb{R}^n$, it may happen that the boundary of the half-space $\mathcal{H}_z$ is a supporting hyperplane for the feasible set $\mathcal{X}$, see Section 8.1.5. Indeed, a *necessary* condition (but not sufficient, in general) for the hyperplane $\{ x \in \mathbb{R}^n : a_z^\top x + b_z = 0 \}$ to have a point x_0 in common with the set $\mathcal{X}$ is that $z^\top F(x_0)z = 0$, that is $F(x_0)$ is singular, and z is an eigenvector associated with the null eigenvalue of $F(x_0)$. Therefore, a necessary condition for (a_z, b_z) to define a supporting hyperplane for $\mathcal{X}$ is that z belongs to the nullspace of $F(x)$ for some $x \in \mathbb{R}^m$.

Example 11.3 Figure 11.2 shows the set $\mathcal{X}$ of points $x \in \mathbb{R}^2$ that satisfy the LMI

$$F(x) = x_1 F_1 + x_2 F_2 - I \preceq 0, \tag{11.5}$$

with

$$F_1 = \begin{bmatrix} -1.3 & -4.2 & -0.1 & 2.1 & -1 \\ -4.2 & -0.1 & -1.7 & -4.5 & 0.9 \\ -0.1 & -1.7 & 2.3 & -4.4 & -0.4 \\ 2.1 & -4.5 & -4.4 & 3.3 & -1.7 \\ -1 & 0.9 & -0.4 & -1.7 & 4.7 \end{bmatrix},$$

$$F_2 = \begin{bmatrix} 1.6 & 3.9 & 1.6 & -5.3 & -4 \\ 3.9 & -1.8 & -4.7 & 1 & 2.9 \\ 1.6 & -4.7 & -1.3 & 1.6 & -2.6 \\ -5.3 & 1 & 1.6 & 2.7 & 2.6 \\ -4 & 2.9 & -2.6 & 2.6 & -3.4 \end{bmatrix}.$$

Figure 11.3 shows some of the hyperplanes $a_z^\top x + b_z = 0$, obtained for a few randomly chosen directions $z \in \mathbb{R}^5$.

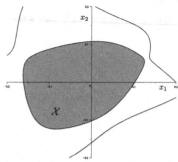

Figure 11.2 The feasible set (gray region) of the LMI (11.5). Contour lines (in black) show the locus of points $x = (x_1, x_2)$ for which $\det F(x) = 0$.

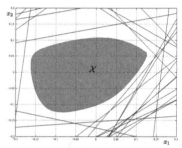

Figure 11.3 Feasible set of (11.5), together with hyperplanes $a_z^\top x + b_z = 0$, for some randomly chosen z.

11.2.2.3   The conic standard form. In an alternative form, known as the *conic* standard form, an LMI in the symmetric matrix variable X can be expressed as the intersection of the positive semidefinite cone with an affine set:

$$X \in \mathcal{A}, \; X \succeq 0, \tag{11.6}$$

where $\mathcal{A}$ is an affine set. The set $S = \mathbb{S}_+^n \cap \mathcal{A}$ is called a *spectrahedron*. The standard forms (11.1) and (11.6) are equivalent, in the sense that one can always transform one into the other (at the expense possibly of adding new variables and constraints). Therefore, the feasible set $\mathcal{X}$ of a standard-form LMI is a spectrahedron. Figure 11.4 shows an example of a 3D plot, in the (x_1, x_2, x_3) space, of a spectrahedron corresponding to the combined LMIs $P \succeq 0$, $A^\top P + PA \preceq 0$, for

$$P = \begin{bmatrix} x_1 & x_3 \\ x_3 & x_2 \end{bmatrix}, \quad A = \begin{bmatrix} -1 & 0.5 \\ 0 & -0.8 \end{bmatrix}.$$

Geometrically, the boundary of the LMI set $\mathcal{X}$ in (11.3) is defined by a multivariate polynomial surface in the x variable. Indeed, it is a known fact that a symmetric matrix $F(x)$ is PSD if and only if the sums $g_k(x)$ of the *principal minors* of $F(x)$ of order k, $k = 1, \dots, n$, are non-negative (a principal minor of order k is the determinant of a submatrix of $F(x)$ obtained by considering a subset $J \subseteq \{1, \dots, n\}$ of its rows and columns of cardinality k).

Figure 11.4 A 3D plot of a spectrahedron corresponding to the Lyapunov inequalities $P \succeq 0$, $A^\top P + PA \preceq 0$, $A \in \mathbb{R}^{2,2}$.

Since $F(x)$ is affine in x, the functions $g_k(x)$, $k = 1, \ldots, n$, are polynomials of degree at most k in the variable x, and the set $\mathcal{X}$ is described by the system of polynomial inequalities

$$\mathcal{X} = \{x \in \mathbb{R}^m : g_k(x) \geq 0, k = 1, \ldots, n\},$$

which is a closed *semialgebraic* set. Notice that $g_1(x) = \text{trace}\, F(x)$, and $g_n(x) = \det F(x)$. In particular, the boundary of the LMI feasible region $\mathcal{X}$ is described by the determinant $\{x : g_n(x) = \det F(x) \geq 0\}$, while the other polynomials $g_k(x)$ only isolate the convex connected component of this region.

Example 11.4 Consider the LMI in $x \in \mathbb{R}^2$

$$F(x) = \begin{bmatrix} 1 + x_1 & x_1 - x_2 & x_1 \\ x_1 - x_2 & 1 - x_2 & 0 \\ x_1 & 0 & 1 + x_2 \end{bmatrix} \succeq 0. \qquad (11.7)$$

The feasible set $\mathcal{X}$ is described by the points in the intersection of the following polynomial inequalities, and it is depicted in Figure 11.5.

$$\begin{aligned} g_1(x) &= \text{trace}\, F(x) = 3 + x_1 \geq 0, \\ g_2(x) &= 3 + 2x_1 + 2x_1x_2 - 2x_1^2 - 2x_2^2 \geq 0, \\ g_3(x) &= \det F(x) = 1 + x_1 - 2x_1^2 - 2x_2^2 + 2x_1x_2 + x_1x_2^2 - x_2^3 \geq 0. \end{aligned}$$

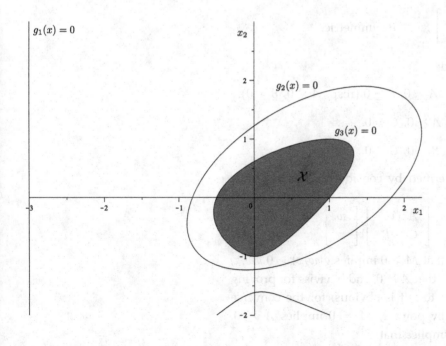

Figure 11.5 Intersection of the regions where $g_k(x) \geq 0$, $k = 1, 2, 3$, for the LMI (11.7).

11.2.3   *Useful "tricks" for LMI manipulation*

Several manipulations are often useful in order to represent constraints in a suitable LMI format. We here discuss some of these "LMI tricks."

11.2.3.1   Multiple LMIs. Multiple LMI constraints can be combined into a single LMI constraint. Consider two affine maps from $\mathbb{R}^m$ to the space of symmetric matrices of order n_1 and n_2, respectively: $F_1 : \mathbb{R}^m \to \mathbb{S}^{n_1}$, $F_2 : \mathbb{R}^m \to \mathbb{S}^{n_2}$. Then the two LMIs

$$F_1(x) \succeq 0, \quad F_2(x) \succeq 0$$

are equivalent to one LMI, involving a larger matrix of size $(n_1 + n_2) \times (n_1 + n_2)$, having F_1, F_2 as diagonal blocks:

$$F(x) = \begin{bmatrix} F_1(x) & 0 \\ 0 & F_2(x) \end{bmatrix} \succeq 0.$$

This rule is an immediate consequence of the fact that the eigenvalues of a block-diagonal matrix are the union of the eigenvalues of each of the diagonal blocks.

11.2.3.2   Block matrices and the Schur complement rule. Consider a symmetric block matrix

$$M \doteq \begin{bmatrix} A & C^\top \\ C & B \end{bmatrix}, \quad A, B \text{ symmetric.}$$

The following implications hold:

1. $M \succeq 0$ (resp. $M \succ 0$) $\quad \Rightarrow \quad A \succeq 0, B \succeq 0$ (resp. $A \succ 0, B \succ 0$).

2. If $B = 0$, then $M \succeq 0 \quad \Leftrightarrow \quad A \succeq 0, C = 0$.

3. If $A = 0$, then $M \succeq 0 \quad \Leftrightarrow \quad B \succeq 0, C = 0$.

These three rules can all be verified by considering the quadratic form

$$g(w, z) = \begin{bmatrix} w \\ z \end{bmatrix}^\top \begin{bmatrix} A & C^\top \\ C & B \end{bmatrix} \begin{bmatrix} w \\ z \end{bmatrix}.$$

Point 1 is proved by observing that $M \geq 0$ implies $g(w, z) \geq 0 \; \forall w, z$, hence $g(w, 0) \geq 0$, which implies that $A \succeq 0$, and likewise for proving that $B \succeq 0$. Point 2 from right to left is obvious; for the converse implication first observe that, by point 1, $M \succeq 0$ implies $A \succeq 0$. Moreover, $M \succeq 0$ (with $B = 0$) implies that

$$g(w, z) \doteq w^\top A w + w^\top C z \geq 0, \quad \forall w, z.$$

The first term is always non-negative, while for any $w \neq 0$ the second term is unbounded below in z, unless $C = 0$; hence $g(w,z) \geq 0 \ \forall w, z$ implies $C = 0$. Point 3 follows from an analogous reasoning.

The Schur complement rule (see Theorem 4.9) provides necessary and sufficient conditions for positive definiteness of the block matrix M in terms of suitable conditions on its blocks; it is also very useful for converting certain nonlinear matrix inequalities (e.g., quadratic) into LMI form. The standard Schur complement rule states that, if $B \succ 0$, then

$$M \succeq 0 \quad \Leftrightarrow \quad A - C^\top B^{-1} C \succeq 0.$$

Or, equivalently, if $A \succ 0$, then

$$M \succeq 0 \quad \Leftrightarrow \quad B - C A^{-1} C^\top \succeq 0.$$

For (strict) positive definiteness, the Schur rule states that

$$M \succ 0 \ \Leftrightarrow \ B \succ 0, \ A - C^\top B^{-1} C \succ 0,$$
$$M \succ 0 \ \Leftrightarrow \ A \succ 0, \ B - C A^{-1} C^\top \succ 0.$$

There also exist a (more involved) version of the Schur rule, that can be applied when neither A nor B is strictly positive definite, namely:

$$M \succeq 0 \ \Leftrightarrow \ B \succeq 0, \ A - C^\top B^\dagger C \succeq 0, \ (I - BB^\dagger)C = 0,$$
$$M \succeq 0 \ \Leftrightarrow \ A \succeq 0, \ B - C A^\dagger C^\top \succeq 0, \ (I - AA^\dagger)C^\top = 0.$$

Observe, for instance, the first of these rules: since $I - BB^\dagger$ is a projector onto $\mathcal{R}(B)^\perp$, the condition $(I - BB^\dagger)C = 0$ is equivalent to requiring that $\mathcal{R}(C) \subseteq \mathcal{R}(B)$. Similarly, the condition in the second rule $(I - AA^\dagger)C^\top = 0$ is equivalent to requiring that $\mathcal{R}(C^\top) \subseteq \mathcal{R}(A)$.

11.2.3.3 *Congruence transformations.* Given a symmetric matrix M, a congruence transformation on M is a matrix obtained by pre- and post-multiplying M by a matrix factor R, that is $G = R^\top M R$. We have from Theorem 4.7 that

$$M \succeq 0 \ \Rightarrow \ R^\top M R \succeq 0,$$
$$M \succ 0 \ \Rightarrow \ R^\top M R \succ 0 \quad \text{(if } R \text{ is full column rank).}$$

Further, if R is square and nonsingular, then the implication holds in both ways:

$$M \succeq 0 \ \Leftrightarrow \ R^\top M R \succeq 0, \quad \text{(if } R \text{ is nonsingular).}$$

11.2.3.4 Finsler's lemma and variable elimination. The following set of matrix inequality equivalences is generally known under the name of Finsler's lemma.

Lemma 11.1 (Finsler) *Let $A \in S^n$ and $B \in \mathbb{R}^{m,n}$. The following statements are equivalent:*

1. $z^\top A z > 0$ for all $z \in \mathcal{N}(B)$, $z \neq 0$;

2. $B_\perp^\top A B_\perp \succ 0$, where $B_\perp$ is a matrix containing by columns a basis for $\mathcal{N}(B)$, that is a matrix of maximum rank such that $BB_\perp = 0$;

3. *there exists a scalar $\lambda \in \mathbb{R}$ such that $A + \lambda B^\top B \succ 0$;*

4. *there exists a matrix $Y \in \mathbb{R}^{n,m}$ such that $A + YB + B^\top Y^\top \succ 0$.*

Proof The implication 1 $\leftrightarrow$ 2 follows by considering that any $z \in \mathcal{N}(B)$ can be written as $z = B_\perp v$, for some vector v, hence the statement in 1 means that $v^\top B_\perp^\top A B_\perp v > 0$ for all $v \neq 0$, which is equivalent to 2. The implications 3 $\rightarrow$ 2 and 4 $\rightarrow$ 2 both follow from the congruence transformation rule, by multiplying the respective matrix inequalities by $B_\perp$ on the right and by $B_\perp^\top$ on the left.

The proof of 1 $\rightarrow$ 3 is slightly more involved, and can be skipped if the reader is not interested in these technical details. Let 1 hold and write any vector $y \in \mathbb{R}^n$ as the sum of two orthogonal components, one along $\mathcal{N}(B)$ and one along $\mathcal{N}^\perp(B) = \mathcal{R}(B^\top)$:

$$y = z + w, \quad z \in \mathcal{N}(B), \; w \in \mathcal{R}(B^\top).$$

If we let $B_\perp$, $B_\top$ denote matrices containing by columns a basis for $\mathcal{N}(B)$ and a basis for $\mathcal{R}(B^\top)$, respectively, we may write $z = B_\perp v$, $w = B_\top \xi$, for some free vectors v, ξ, hence

$$y = z + w = B_\perp v + B_\top \xi. \tag{11.8}$$

Now observe that, by point 2, $B_\perp^\top A B_\perp \succ 0$, hence

$$z^\top A z = v^\top (B_\perp^\top A B_\perp) v \geq \eta_a \|v\|_2^2,$$

where $\eta_a > 0$ denotes the smallest eigenvalue of $B_\perp^\top A B_\perp$ (which is strictly positive, since this matrix is positive definite). Similarly, notice that $Bw \neq 0$ for $w \neq 0$, since by definition w cannot be in the nullspace of B, therefore $w^\top B^\top B w > 0$ for all $w = B_\top \xi$, which means that $B_\top^\top B^\top B B_\top \succ 0$, whence

$$w^\top (B^\top B) w = \xi^\top (B_\top^\top B^\top B B_\top) \xi \geq \eta_b \|\xi\|_2^2, \tag{11.9}$$

where $\eta_b > 0$ denotes the smallest eigenvalue of $B_\top^\top B^\top B B_\top$. Now consider the following quadratic form:

$$
\begin{aligned}
y^\top (A + \lambda B^\top B) y &= (z+w)^\top (A + \lambda B^\top B)(z+w) \\
\text{[since } Bz = 0] &= z^\top A z + 2 z^\top A w + w^\top (A + \lambda B^\top B) w \\
\text{[(11.8)--(11.9)]} &\geq \eta_a v^\top v + 2 v^\top B_\perp^\top A B_\top \xi + \xi^\top (B_\top^\top A B_\top + \lambda \eta_b I) \xi \\
\text{[} R \doteq B_\perp^\top A B_\top] &= \begin{bmatrix} v \\ \xi \end{bmatrix}^\top \begin{bmatrix} \eta_a I & R \\ R^\top & B_\top^\top A B_\top + \lambda \eta_b I \end{bmatrix} \begin{bmatrix} v \\ \xi \end{bmatrix}.
\end{aligned}
$$

We next show that one can always find a value for λ such that the matrix in the last expression is positive definite, which would imply that $y^\top (A + \lambda B^\top B) y > 0$ for all $y \neq 0$, which would in turn prove the desired statement, i.e., that there exists a λ such that $A + \lambda B^\top B \succ 0$. To this end, notice that, by the Schur complement rule, we have that

$$
\begin{bmatrix} \eta_a I & R \\ R^\top & B_\top^\top A B_\top + \lambda \eta_b I \end{bmatrix} \succ 0 \quad \Leftrightarrow \quad B_\top^\top A B_\top + \lambda \eta_b I - \frac{1}{\eta_a} R^\top R \succ 0,
$$

and, by the eigenvalue shift rule, the latter condition is equivalent to

$$
\lambda > \frac{1}{\eta_b} \lambda_{\max} \left(\frac{1}{\eta_a} R^\top R - B_\top^\top A B_\top \right),
$$

which concludes this part of the proof. Finally, the implication $3 \to 4$ follows immediately by choosing $Y = \frac{\lambda}{2} B$, whereby all implications in the lemma are proved. □

A generalized form of the equivalence between points 2 and 4 in Finsler's lemma is usually known in the LMI lingo as the "variable elimination lemma." Let $A(x)$ be an affine function of a vector of variables x, and let Y be an additional matrix variable, not depending on x. Then, the elimination lemma[1] states that

$$
\exists x, Y : \; A(x) + C Y B + B^\top Y^\top C^\top \succ 0 \quad \Leftrightarrow \quad \begin{cases} C_\perp A(x) C_\perp^\top \succ 0, \\ B_\perp^\top A(x) B_\perp \succ 0. \end{cases}
$$

This rule is useful for converting the condition on the left, containing both variables x and Y, into the conditions on the right, where the variable Y has been eliminated.

11.2.3.5 LMI robustness lemma. Another useful rule deals with LMIs that depend affinely on a matrix Y of uncertain parameters. In this case, the condition that the LMI holds robustly, i.e., for all values of Y in a norm-bounded set, can be converted into a standard LMI condition. More precisely, let $A(x) \in \mathbb{S}^n$, be a matrix that depends affinely on a vector x of variables, and let $B \in \mathbb{R}^{m,n}$, $C \in \mathbb{R}^{n,p}$. Then, we have that[2] the LMI in variable x:

[1] See, e.g., Section 2.6 in S. Boyd, L. El Ghaoui, E. Feron, V. Balakrishnan, *Linear Matrix Inequalities in System and Control Theory*, SIAM, 1994; or Chapter 2 in R. E. Skelton, T. Iwasaki and K. Grigoriadis, *A Unified Algebraic Approach to Linear Control Design*, CRC Press, 1998.

[2] See again the two books mentioned above.

$$A(x) + CYB + B^\top Y^\top C^\top \succeq 0$$

holds robustly for all $Y \in \mathbb{R}^{p,m} : \|Y\|_2 \leq 1$, if and only if the following LMI in x and $\lambda \in \mathbb{R}$ holds:

$$\begin{bmatrix} A(x) - \lambda CC^\top & B^\top \\ B & \lambda I_m \end{bmatrix} \succeq 0,$$

or, equivalently, if and only if the following LMI in x and $\lambda \in \mathbb{R}$ holds:

$$\begin{bmatrix} A(x) - \lambda B^\top B & C \\ C^\top & \lambda I_p \end{bmatrix} \succeq 0.$$

11.2.4 Linear, quadratic, and conic inequalities in LMI form

Many special cases of convex inequalities, such as affine, quadratic, and second-order cone inequalities can be represented in LMI format.

Affine inequalities. Consider a single affine inequality in $x \in \mathbb{R}^n$: $a^\top x \leq b$, where $a \in \mathbb{R}^n$, $b \in \mathbb{R}$. This is a trivial special case of an LMI, where the coefficient matrices are scalar: $F_0 = b$, $F_i = -a_i$, $i = 1, \ldots, n$. Using the previous rule on multiple LMIs, we obtain that a set of ordinary affine inequalities

$$a_i^\top x \leq b_i, \quad i = 1, \ldots, m$$

can be cast as a single LMI $F(x) \succeq 0$, where

$$\begin{aligned} F(x) &= \operatorname{diag}\left(b_1 - a_1^\top x, \cdots, b_m - a_m^\top x\right) \\ &= \begin{bmatrix} b_1 - a_1^\top x & & \\ & \ddots & \\ & & b_m - a_m^\top x \end{bmatrix}. \end{aligned}$$

Quadratic inequalities. Consider a convex quadratic inequality

$$f(x) \doteq x^\top Q x + c^\top x + d \leq 0, \quad Q \succeq 0.$$

If $f(x)$ is strictly convex (i.e., $Q \succ 0$), then the inequality $f(x) \leq 0$ can be expressed in the form of an LMI, using the Schur complement rule, as

$$\begin{bmatrix} -c^\top x - d & x^\top \\ x & Q^{-1} \end{bmatrix} \succeq 0.$$

If instead $f(x)$ is convex, but not strictly, then $Q \succeq 0$, and we can factor it as $Q = E^\top E$, hence, using the Schur complement rule again, we obtain that $f(x) \leq 0$ is equivalent to the LMI

$$\begin{bmatrix} -c^\top x - d & (Ex)^\top \\ (Ex) & I \end{bmatrix} \succeq 0.$$

Second-order cone inequalities. Also second-order cone (SOC) inequalities can be represented as LMIs. To verify this fact, let us start with the elementary SOC inequality $\|y\|_2 \leq t$, with $y \in \mathbb{R}^n$ and $t \in \mathbb{R}$. This SOC inequality is equivalent to the LMI

$$\begin{bmatrix} t & y^\top \\ y & tI_n \end{bmatrix} \succeq 0.$$

Indeed, the equivalence is immediate for $t = 0$. If instead $t > 0$, then for every $z \in \mathbb{R}^n$ and every $\alpha \in \mathbb{R}$, we have that

$$t \begin{bmatrix} \alpha \\ z \end{bmatrix}^\top \begin{bmatrix} t & y^\top \\ y & tI_n \end{bmatrix} \begin{bmatrix} \alpha \\ z \end{bmatrix} = \|tz + \alpha y\|_2^2 + \alpha^2(t^2 - \|y\|_2^2).$$

Therefore, if $\|y\|_2 \leq t$ then the previous expression is ≥ 0 for all (z, α), and, conversely, if this expression is non-negative for all (z, α) then it is non-negative for $z = y$, $\alpha = -t$, which implies that $\|y\|_2 \leq t$. More generally, a second-order cone inequality of the form

$$\|Ax + b\|_2 \leq c^\top x + d, \tag{11.10}$$

with $A \in \mathbb{R}^{m,n}$, $b \in \mathbb{R}^m$, $c \in \mathbb{R}^n$, $d \in \mathbb{R}$, can be expressed as the LMI

$$\begin{bmatrix} c^\top x + d & (Ax + b)^\top \\ (Ax + b) & (c^\top x + d)I_n \end{bmatrix} \succeq 0. \tag{11.11}$$

To verify this fact observe first that (11.11) implies that both the diagonal blocks are PSD, that is $c^\top x + d \geq 0$. Now, suppose first that $c^\top x + d > 0$; then, the Schur complement rule insures that (11.11) is equivalent to (11.10). If instead $c^\top x + d = 0$, then (11.11) implies that it must be that $Ax + b = 0$, hence (11.11) and (11.10) are still equivalent in this case.

11.3 Semidefinite programs

11.3.1 Standard forms

A semidefinite program (SDP) is a convex optimization problem, where one minimizes a linear objective function under an LMI constraint. In standard inequality form an SDP is expressed as

$$\min_{x \in \mathbb{R}^m} \quad c^\top x \tag{11.12}$$
$$\text{s.t.:} \quad F(x) \succeq 0,$$

where

$$F(x) = F_0 + \sum_{i=1}^m x_i F_i,$$

F_i, $i = 0, \ldots, m$, are given $n \times n$ symmetric matrices, $c \in \mathbb{R}^m$ is the given objective direction, and $x \in \mathbb{R}^m$ is the optimization variable. The geometric interpretation is, as usual, that we move as far as possible in direction $-c$, while remaining feasible; an optimal point for problem (11.12) is thus a farthest point in the feasible set $\mathcal{X} = \{x : F(x) \succeq 0\}$ along the direction opposite to c, see Figure 11.6.

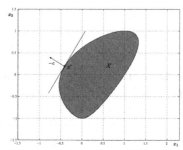

Figure 11.6 Example: SDP (11.12), with $c^\top = [1 \; -0.6]$, and $F(x)$ given by (11.7). The optimal solution is $x^* = [-0.4127 \; 0.1877]^\top$.

Standard conic form. The standard conic form of an SDP derives from the corresponding conic representation of its LMI constraint. Denoting by $X \in \mathbb{S}^n$ the matrix variable, the conic LMI formulation (11.6) imposes that $X \succeq 0$ and that X should belong to an affine set $\mathcal{A}$. This latter affine set is specified by imposing a number of affine constraints on the entries of X, using the standard inner product for a matrix space, that is $\langle A, B \rangle = \text{trace}\, A^\top B$, hence $\mathcal{A} = \{X \in \mathbb{S}^n : \text{trace}\, A_i X = b_i, \; i = 1, \ldots, m\}$, where A_i are given symmetric $n \times n$ matrices, and b_i are scalars, $i = 1, \ldots, m$. Similarly, the linear objective function is expressed via the inner product $\text{trace}\, CX$, where $C \in \mathbb{S}^n$. A generic conic-form SDP is thus expressed as

$$\min_{X \in \mathbb{S}^n} \quad \text{trace}\, CX \tag{11.13}$$
$$\text{s.t.:} \quad X \succeq 0,$$
$$\text{trace}\, A_i X = b_i, \quad i = 1, \ldots, m.$$

11.3.2 SDP duality

To obtain the dual of an SDP, we first establish the following SDP-based characterization of the largest eigenvalue of a symmetric matrix X:

$$\lambda_{\max}(X) = v(X) \doteq \max_Z \text{trace}\, ZX \; : \; Z \succeq 0, \; \text{trace}\, Z = 1. \tag{11.14}$$

Indeed, let $X = U\Lambda U^\top$ be a spectral decomposition of X, with $\Lambda = \text{diag}(\lambda_1, \ldots, \lambda_n)$ containing the eigenvalues of X arranged in decreasing order, so that $\lambda_1 = \lambda_{\max}(X)$. Using the change of variable

$Z \to U^\top Z U$, and exploiting the fact that $U U^\top$ is the identity, as well as properties of the trace operator,[3] we obtain $v(X) = v(\Lambda)$. Hence,

[3] See Section 3.1.4.

$$
\begin{aligned}
v(X) \;&=\; \max_Z \; \operatorname{trace} Z\Lambda \;:\; Z \succeq 0, \;\; \operatorname{trace} Z = 1 \\[4pt]
&=\; \max_Z \; \sum_{i=1}^{n} \lambda_i Z_{ii} \;:\; Z \succeq 0, \;\; \operatorname{trace} Z = 1 \quad (11.15) \\[4pt]
&=\; \max_z \; \lambda^\top z \;:\; z \geq 0, \;\; \sum_{i=1}^{n} z_i = 1 \quad\quad (11.16) \\[4pt]
&=\; \max_{1 \leq i \leq n} \; \lambda_i = \lambda_{\max}(X).
\end{aligned}
$$

The third line stems from the fact that z is feasible for problem (11.16) if and only if $\operatorname{diag}(z)$ is feasible for problem (11.15).[4]

Now consider the SDP in standard conic form (11.13). We express it as

[4] Note that, at optimum, $Z = U^\top e_1 e_1^\top U = u_1 u_1^\top$, with e_1 (resp. u_1) the first unit vector (resp. column of U). This shows that Z is rank-one at optimum, that is, it is of the form $zz^\top$ for some z with $z^\top z = 1$. Thus:
$$\lambda_{\max}(X) = \max_z \; z^\top X z \;:\; z^\top z = 1,$$
which is the Rayleigh quotient representation, as given in Theorem 4.3.

$$
\begin{aligned}
\min_{X \in \mathbb{S}^n} \quad & \operatorname{trace} CX \\
\text{s.t.:} \quad & \lambda_{\max}(-X) \leq 0, \\
& \operatorname{trace} A_i X = b_i, \quad i = 1, \dots, m.
\end{aligned}
$$

Using the following variational representation of eigenvalues above, we obtain

$$
\begin{aligned}
p^* \;=\;& \min_{X \in \mathbb{S}^n} \max_{\lambda, \nu, Z} \; \operatorname{trace} CX + \sum_{i=1}^{m} \nu_i (b_i - \operatorname{trace} A_i X) - \lambda \operatorname{trace} ZX \\[4pt]
& \text{s.t.:} \quad \lambda \geq 0, \; \nu, \; Z \succeq 0, \; \operatorname{trace} Z = 1 \\[4pt]
\geq\;& \max_{\lambda, \nu, Z} \; \nu^\top b \;:\; C - \sum_{i=1}^{m} \nu_i A_i + \lambda Z = 0, \;\; Z \succeq 0, \;\; \lambda \geq 0 \\[4pt]
=\;& \max_{\lambda, \nu, Z} \; \nu^\top b \;:\; C - \sum_{i=1}^{m} \nu_i A_i + Z = 0, \;\; Z \succeq 0 \\[4pt]
=\;& \max_{\lambda, \nu, Z} \; \nu^\top b \;:\; C - \sum_{i=1}^{m} \nu_i A_i \preceq 0.
\end{aligned}
$$

In the second line, we have used the max-min inequality (8.48); in the third line, we have absorbed the variable λ into Z; and eliminated the latter in the final step.

As with LP and SOCPs, the dual problem is also an SDP. A similar derivation shows that the dual to the dual SDP above is nothing other than the primal we started with. From Slater's conditions for strong duality,[5] it turns out that if the primal problem is strictly feasible, then $p^* = d^*$, and there is no duality gap.

[5] See Proposition 8.7.

Example 11.5 (*Variational characterization of the maximal eigenvalue*) The dual of the variational characterization of the largest eigenvalue (11.14) turns out to be

$$\min_{\nu} \nu \ : \ \nu I \succeq X.$$

The value of the above problem can be directly shown to be indeed the largest eigenvalue, after spectral decomposition of X. In this case, there is no duality gap, as guaranteed by the strict feasibility of the primal problem (11.14).

11.3.3 SDP relaxation of non-convex quadratic problems

Consider an optimization problem in which both the objective and constraint functions are (not necessarily convex) quadratic, as introduced in Section 9.4.2:

$$
\begin{aligned}
p^* = \min_{x} \quad & x^\top H_0 x + 2c_0^\top x + d_0 \\
\text{s.t.:} \quad & x^\top H_i x + 2c_i^\top x + d_i \leq 0, \quad i \in \mathcal{I}, \\
& x^\top H_j x + 2c_j^\top x + d_j = 0, \quad j \in \mathcal{E}.
\end{aligned}
$$

Here, H_0 and H_i, $i \in \mathcal{I}, \mathcal{E}$ are symmetric matrices. In general, the above problem, which we referred to as a quadratically constrained quadratic problem (QCQP), is non-convex, and hard to solve. Not surprisingly, there are many applications for this rich class, some of which are described in the exercises.

Semidefinite optimization may be used to obtain bounds on such hard QCQP problems, via a technique called rank relaxation. The basic idea is to first express the problem in terms of the variable x and an additional symmetric matrix variable $X \doteq xx^\top$. We can rewrite the above in an equivalent way:

$$
\begin{aligned}
p^* = \min_{x} \quad & \text{trace}\, H_0 X + 2c_0^\top x + d_0 \\
\text{s.t.:} \quad & \text{trace}\, H_i X + 2c_i^\top x + d_i \leq 0, \quad i \in \mathcal{I}, \\
& \text{trace}\, H_j X + 2c_j^\top x + d_j = 0, \quad j \in \mathcal{E}, \\
& X = xx^\top.
\end{aligned}
$$

Here we have exploited the fact that $\text{trace}\, AB = \text{trace}\, BA$ for any matrices A, B of compatible size. We can now relax the last equality constraint $X = xx^\top$ into a (convex) inequality one $X \succeq xx^\top$, which in turn can be written as an LMI in X, x:

$$
\begin{bmatrix} X & x \\ x^\top & 1 \end{bmatrix} \succeq 0.
$$

Since we have relaxed a constraint into a more general, convex one in the context of a minimization problem, we obtain a *lower* bound $p^* \geq q^*$, where q^* is the optimal value of the convex problem:

$$q^* \doteq \min_{x,X} \quad \text{trace}\, H_0 X + 2c_0^\top x + d_0$$
$$\text{s.t.:} \quad \text{trace}\, H_i X + 2c_i^\top x + d_i \leq 0, \quad i \in \mathcal{I},$$
$$\text{trace}\, H_j X + 2c_j^\top x + d_j = 0, \quad j \in \mathcal{E},$$
$$\begin{bmatrix} X & x \\ x^\top & 1 \end{bmatrix} \succeq 0.$$

We further observe that the objective function is linear; the constraints except the last are all linear equalities and inequalities; and the last one is an LMI. Hence, the above is an SDP.

The approach can be pushed further to provide not only a bound on the original hard problem, but also quality guesses as to an optimal solution. However, there are no guarantees in general that such guesses are even feasible. In particular, an optimal solution x^* for the SDP above is not even guaranteed to be feasible for the original problem. One case when such guarantees exist is the so-called $\mathcal{S}$-procedure, which is discussed in Section 11.3.3.1.

Another case when the approach works well, and can be further analyzed with success, relates to quadratic Boolean optimization. Precisely, consider the special case of a non-convex QCQP:

$$p^* = \max_{x} \quad x^\top H x$$
$$\text{s.t.:} \quad x_i^2 = 1, \quad i = 1, \dots, n.$$

Here, H is a given $n \times n$ symmetric matrix. Applying the relaxation approach above leads to an upper bound, as we begin with a maximization problem: $p^* \leq q^*$, where

$$q^* \doteq \min_{x,X} \quad \text{trace}\, H X$$
$$\text{s.t.:} \quad X_{ii} = 1, \quad i = 1, \dots, n,$$
$$\begin{bmatrix} X & x \\ x^\top & 1 \end{bmatrix} \succeq 0.$$

We note that the variable x only appears in the last (LMI) constraint. Since the latter holds for some $x \in \mathbb{R}^n$ if and only if $X \succeq 0$, we can further reduce our relaxed problem to

$$q^* \doteq \min_{X} \quad \text{trace}\, H X$$
$$\text{s.t.:} \quad X_{ii} = 1, \quad i = 1, \dots, n,$$
$$X \succeq 0.$$

Several interesting results are known for the above bound. First, the quality of the bound is bounded, independently of the problem size n. Precisely, we have[6]

[6] This result, originally due to Yu. Nesterov, is given as Theorem 3.4.2 in Ben-Tal and Nemirovski, *Lectures on Modern Convex Optimization*, SIAM, 2001.

$$\frac{2}{\pi}q^* \leq p^* \leq q^*.$$

In addition, there exists a method to generate points x that are feasible for the original problem (that is, Boolean vectors), and such that the corresponding objective $x^\top H x$ achieves the lower bound $\frac{2}{\pi}q^*$.

11.3.3.1   The $\mathcal{S}$-procedure. The so-called $\mathcal{S}$-procedure establishes an equivalence between a certain LMI condition and an implication between two quadratic functions.[7] More precisely, let $f_0(x)$, $f_1(x)$ be two quadratic functions:

[7] See also the discussion in Section 9.4.3.3.

$$f_0(x) = x^\top F_0 x + 2g_0^\top x + h_0 = \begin{bmatrix} x \\ 1 \end{bmatrix}^\top \begin{bmatrix} F_0 & g_0 \\ g_0^\top & h_0 \end{bmatrix} \begin{bmatrix} x \\ 1 \end{bmatrix},$$

$$f_1(x) = x^\top F_1 x + 2g_1^\top x + h_1 = \begin{bmatrix} x \\ 1 \end{bmatrix}^\top \begin{bmatrix} F_1 & g_1 \\ g_1^\top & h_1 \end{bmatrix} \begin{bmatrix} x \\ 1 \end{bmatrix},$$

where $F_0, F_1 \in \mathbb{S}^n$, $g_0, g_1 \in \mathbb{R}^n$, and $h_0, h_1 \in \mathbb{R}$. We do not assume convexity, that is F_0, F_1 are not required to be positive semidefinite. Assume that the constraint $f_1(x) \leq 0$ is strictly feasible, i.e., that there exists a point $\tilde{x} \in \mathbb{R}^n$ such that $f_1(\tilde{x}) < 0$. Then, the following two statements (a) and (b) are equivalent:

(a) $f_1(x) \leq 0 \quad \Rightarrow \quad f_0(x) \leq 0;$

(b) there exists a scalar $\tau \geq 0$ such that

$$\begin{bmatrix} F_0 & g_0 \\ g_0^\top & h_0 \end{bmatrix} \preceq \tau \begin{bmatrix} F_1 & g_1 \\ g_1^\top & h_1 \end{bmatrix}.$$

Notice that statement (a) can be interpreted in terms of inclusion of the zero-sublevel set of f_1 in that of f_0, i.e., $\mathcal{X}_1 \subseteq \mathcal{X}_0$, where $\mathcal{X}_1 \doteq \{x \in \mathbb{R}^n : f_1(x) \leq 0\}$, $\mathcal{X}_0 \doteq \{x \in \mathbb{R}^n : f_0(x) \leq 0\}$. Also, statement (b) can be formulated equivalently as

$$\exists \tau \geq 0 : \quad f_0(x) - \tau f_1(x) \leq 0, \quad \forall x.$$

The implication from (b) to (a) is immediate to prove. Indeed, if (b) holds, then by multiplying the LMI in (b) on the left by $[x^\top \ 1]$ and on the right by its transpose, we obtain that $f_0(x) \leq \tau f_1(x)$ for some $\tau \geq 0$. Therefore, if $f_1(x) \leq 0$, then also $f_0(x) \leq 0$, which is the statement in (a). The converse implication, from (a) to (b), is more difficult to prove, and it needs the assumption of strict feasibility on f_1. This latter part of the proof is not reported here.

The implication from (b) to (a) can also be easily extended to an arbitrary number of quadratic functions. Indeed, defining

$$f_i(x) \doteq \begin{bmatrix} x \\ 1 \end{bmatrix}^\top \begin{bmatrix} F_i & g_i \\ g_i^\top & h_i \end{bmatrix} \begin{bmatrix} x \\ 1 \end{bmatrix}, \quad i = 0, 1, \ldots, m,$$

it is easy to check that the statement:

$$\exists \tau_1, \ldots, \tau_m \geq 0 : \quad \begin{bmatrix} F_0 & g_0 \\ g_0^\top & h_0 \end{bmatrix} \preceq \sum_{i=1}^m \tau_i \begin{bmatrix} F_i & g_i \\ g_i^\top & h_i \end{bmatrix} \quad (11.17)$$

implies the statement

$$\begin{cases} f_1(x) \leq 0 \\ \quad \vdots \\ f_m(x) \leq 0 \end{cases} \Rightarrow \quad f_0(x) \leq 0.$$

In terms of zero-sublevel sets $\mathcal{X}_i \doteq \{x \in \mathbb{R}^n : f_i(x) \leq 0\}$, $i = 0, 1, \ldots, m$, the above implication states equivalently that (11.17) implies that

$$\bigcap_{i=1,\ldots,m} \mathcal{X}_i \subseteq \mathcal{X}_0,$$

that is, $\mathcal{X}_0$ contains the intersection of the sets $\mathcal{X}_i$, $i = 1, \ldots, m$. Also, condition (11.17) can be expressed equivalently as

$$\exists \tau_1, \ldots, \tau_m \geq 0 : \quad f_0(x) - \sum_{i=1}^m \tau_i f_i(x) \leq 0, \quad \forall x.$$

11.4 Examples of SDP models

SDP models arise in a wide variety of application contexts. Here, we expose a necessarily small selection of them; further examples are discussed in some of the application chapters; see, in particular, Chapter 15.

11.4.1 Some matrix problems

Semidefinite programs arise often as extensions of matrix problems from linear algebra, involving matrices that depend affinely on a vector of variables x. We next describe some of these problems.

11.4.1.1 Minimization of the spectral norm. Let $A(x) \in \mathbb{R}^{p,n}$ be a matrix whose entries are affine functions of a vector of variables $x \in \mathbb{R}^m$. This means that $A(x)$ can be written as

$$A(x) = A_0 + x_1 A_1 + \cdots + x_m A_m.$$

The problem of minimizing the spectral norm of $A(x)$,

$$\min_x \; \|A(x)\|_2, \tag{11.18}$$

can be cast as an SDP problem as follows. First recall that $\|A(x)\|_2 = \sigma_1(A(x))$, where $\sigma_1(A(x))$ is the largest singular value of $A(x)$, which coincides with the square-root of the largest eigenvalue of $A^\top(x)A(x)$, see Corollary 5.1. Then we have that

$$\|A(x)\|_2 \leq t \;\Leftrightarrow\; \|A(x)\|_2^2 \leq t^2 \;\Leftrightarrow\; \lambda_1(A^\top(x)A(x)) \leq t^2,$$

and the latter condition holds if and only if

$$\lambda_i(A^\top(x)A(x)) \leq t^2, \; i = 1,\dots,n.$$

Since, by the eigenvalue shift rule (3.13), it holds that

$$\lambda_i(A^\top(x)A(x) - t^2 I_n) = \lambda_i(A^\top(x)A(x)) - t^2, \quad i = 1,\dots,n,$$

we have that

$$\lambda_i(A^\top(x)A(x)) \leq t^2, \; \forall i \quad\Leftrightarrow\quad \lambda_i(A^\top(x)A(x) - t^2 I_n) \leq 0, \; \forall i,$$

and the latter condition is equivalent to requiring that

$$A^\top(x)A(x) - t^2 I_n \preceq 0.$$

Using the Schur complement rule, this matrix inequality is further rewritten in LMI format (in variables t^2 and x) as

$$\begin{bmatrix} t^2 I_n & A^\top(x) \\ A(x) & I_p \end{bmatrix} \succeq 0.$$

Since $t = 0$ if and only if $A(x) = 0$, this LMI is also equivalent to

$$\begin{bmatrix} t I_n & A^\top(x) \\ A(x) & t I_p \end{bmatrix} \succeq 0,$$

which is obtained via congruence, pre- and post-multiplying by the diagonal matrix $\mathrm{diag}\left(1/\sqrt{t}, \sqrt{(t)}\right)$, assuming $t > 0$. Problem (11.18) is thus equivalent to the following SDP in the variables x, t:

$$\min_{x \in \mathbb{R}^m, t \in \mathbb{R}} \quad t$$
$$\text{s.t.:} \quad \begin{bmatrix} t I_n & A^\top(x) \\ A(x) & t I_p \end{bmatrix} \succeq 0.$$

11.4.1.2 *Minimization of the Frobenius norm.* Let again $A(x) \in \mathbb{R}^{p,n}$ be a matrix whose entries are affine functions of a vector of variables $x \in \mathbb{R}^m$. The problem of minimizing the Frobenius norm (squared) of $A(x)$,

$$\min_x \|A(x)\|_F^2, \tag{11.19}$$

can be formulated in SDP format as follows

$$\min_{x \in \mathbb{R}^m, Y \in \mathbb{S}^p} \quad \text{trace}\, Y \tag{11.20}$$

$$\text{s.t.:} \quad \begin{bmatrix} Y & A(x) \\ A^\top(x) & I_n \end{bmatrix} \succeq 0. \tag{11.21}$$

To verify this equivalence, we first observe that

$$\|A(x)\|_F^2 = \text{trace}\, A(x) A^\top(x),$$

and that, by the Schur complement rule,

$$\begin{bmatrix} Y & A(x) \\ A^\top(x) & I_n \end{bmatrix} \succeq 0 \quad \Leftrightarrow \quad A(x) A^\top(x) \preceq Y,$$

and, since $X \preceq Y$ implies $\text{trace}\, X \leq \text{trace}\, Y$, the constraint (11.21) implies that $\|A(x)\|_F^2 \leq \text{trace}\, Y$. Now, if x^* is optimal for problem (11.19), then x^* and $Y^* = A(x^*) A^\top(x^*)$ are feasible for problem (11.20), and are indeed optimal for this problem, since the objective value of (11.20) is $\text{trace}\, Y^* = \text{trace}\, A(x^*) A^\top(x^*)$ and it cannot be improved, for otherwise x^* would not be optimal for problem (11.19). Conversely, if x^* and $Y^* = A(x^*) A^\top(x^*)$ are optimal for problem (11.20), then x^* is also optimal for problem (11.19), for otherwise there would exist a point $\tilde{x} \neq x^*$ such that $\|A(\tilde{x})\|_F^2 < \|A(x^*)\|_F^2$, and this would imply that $\tilde{x}$, $\tilde{Y} = A(\tilde{x}) A^\top(\tilde{x})$ improve the objective of (11.20) with respect to x^*, which would contradict the optimality of x^*.

11.4.1.3 *Minimization of the condition number of a PD matrix.* Let $F(x)$ be a symmetric $n \times n$ matrix whose entries are affine functions of a vector of variables $x \in \mathbb{R}^m$. We address the problem of determining x such that $F(x) \succ 0$ and the condition number of $F(x)$ is minimized. The condition number is defined as

$$\kappa(F(x)) = \frac{\sigma_1(F(x))}{\sigma_n(F(x))},$$

where σ_1 and σ_n are the largest and the smallest singular values of $F(x)$, respectively. Under the condition that $F(x) \succ 0$, however, the singular values of $F(x)$ coincide with the eigenvalues of $F(x)$, hence

$$\kappa(F(x)) \leq \gamma \quad \Leftrightarrow \quad \frac{\lambda_{\max}(F(x))}{\lambda_{\min}(F(x))} \leq \gamma.$$

We thus want to solve the problem

$$\gamma^* = \min_{x,\gamma} \quad \gamma \qquad\qquad (11.22)$$
$$\text{s.t.:} \quad F(x) \succ 0,$$
$$\kappa(F(x)) \leq \gamma.$$

Observe that $F(x) \succ 0$ if and only if there exists a scalar $\mu > 0$ such that $F(x) \succeq \mu I$. Moreover, for $\mu > 0$, $\gamma \geq 1$, it holds that

$$F(x) \succeq \mu I \quad \Leftrightarrow \quad \lambda_{\min}(F(x)) \geq \mu \quad \Leftrightarrow \quad \frac{1}{\lambda_{\min}(F(x))} \leq \frac{1}{\mu},$$
$$F(x) \preceq \gamma\mu I \quad \Leftrightarrow \quad \lambda_{\max}(F(x)) \leq \gamma\mu,$$

therefore, the constraints in problem (11.22) are equivalent to

$$\mu > 0, \ \mu I \preceq F(x) \preceq \gamma\mu I. \qquad\qquad (11.23)$$

Notice, however, that these constraints are not in LMI form (w.r.t. all variables x, γ, μ), due to the presence of the product term $\gamma\mu$. This issue cannot be eliminated (in general) and, indeed, problem (11.22) cannot be converted into a single SDP, unless $F(x)$ is *linear* in x. Let us first consider this special case: if $F(x)$ is linear in x (i.e., $F(0) = 0$), then condition (11.23) is homogeneous in (x, μ), meaning that it holds for some (x, μ) if and only if it holds for $(\alpha x, \alpha\mu)$, for any scalar $\alpha > 0$. We can then divide all terms in (11.23) by μ, and obtain the equivalent condition

$$I \preceq F(x) \preceq \gamma I. \qquad\qquad (11.24)$$

Therefore, if $F(x)$ is linear in x, then problem (11.22) is equivalent to the SDP

$$[\text{for } F(0) = 0] \quad \gamma^* = \min_{x,\gamma} \quad \gamma$$
$$\text{s.t.:} \quad I \preceq F(x) \preceq \gamma I.$$

When $F(x)$ is not linear in x (but, of course, still affine in x), then problem (11.22) cannot be converted into a single SDP problem. However, it remains computationally tractable, since it can be easily solved via a *sequence* of SDP problems. More precisely, select a *fixed* value of $\gamma \geq 1$, and consider the following SDP problem:

$$\tilde{\mu} = \min_{x \in \mathbb{R}^m, \mu > 0} \quad \mu \qquad\qquad (11.25)$$
$$\text{s.t.:} \quad \mu I \preceq F(x) \preceq \gamma\mu I,$$

and let $\tilde{\mu}, \tilde{x}$ denote its optimal variables. If (11.25) is feasible, then it means that we have found an $\tilde{x}$ such that $\kappa(F(\tilde{x})) \leq \gamma$, thus the value of γ may not be optimal, and it may be decreased. On the other hand, if (11.25) is infeasible (by convention in this case we set $\tilde{\mu} = \infty$), then it means that the selected γ was too small, and it should be increased, since for sure $\gamma^* > \gamma$. We can therefore find the optimal γ^* by proceeding iteratively according, for instance, to a *bisection* technique:[8]

[8] See also Exercise 12.3.

1. initialization: find any $\tilde{x}$ such that $F(\tilde{x}) \succ 0$, and set $\gamma_{\text{low}} = 1$, $\gamma_{\text{up}} = \kappa(F(\tilde{x}))$;

2. if $\gamma_{\text{up}} - \gamma_{\text{low}} \leq \epsilon$, then return $x^* = \tilde{x}$, $\gamma^* = \gamma_{\text{up}}$, and exit.

3. set $\gamma = \frac{1}{2}(\gamma_{\text{low}} + \gamma_{\text{up}})$;

4. solve SDP problem in (11.25), and find its optimal variables $\tilde{x}$, $\tilde{\mu}$;

5. if $\tilde{\mu} < \infty$ (problem was feasible), then set $\gamma_{\text{up}} = \gamma$;

6. If $\tilde{\mu} = \infty$ (problem was infeasible), then set $\gamma_{\text{low}} = \gamma$;

7. go to 2.

Clearly, at iteration $k = 0$ we know that the optimal γ is located in an interval of length $\ell = \kappa(F(0)) - 1$; at the first iteration it is located in an interval of length $\ell/2$, at the second iteration in an interval of length $\ell/2^2$, etc. Thus, if the procedure exits after k iterations (number of passages through point 3 in the procedure), then we know that the optimal γ is located in an interval of length $\ell/2^k$. The above iterative procedure thus exits with an ϵ-suboptimal solution as soon as $\ell/2^k \leq \epsilon$, that is, taking base-2 logarithms, when k is the smallest integer such that $k \geq \log_2(\ell/\epsilon)$, i.e., for

$$k = \left\lceil \log_2 \frac{\ell}{\epsilon} \right\rceil.$$

Example 11.6 Consider a variation on the localization problem based on trilateration as discussed in Example 6.2 and in Example 6.8. Supposing there are $m + 1$ beacons (with $m \geq 2$), the localization problem can be written in the form $Ap = y$, where $p^\top = [p_1 \ p_2]$ is the vector of planar coordinates of the object that we want to localize, $y \in \mathbb{R}^m$ is a known term vector that depends on the distance measurements from the object to the beacons, and

$$A^\top = \begin{bmatrix} \delta_1 & \cdots & \delta_m \end{bmatrix}, \quad \delta_i = a_{i+1} - a_1, \ i = 1, \ldots, m,$$

where $a_i \in \mathbb{R}^n$ are the vectors containing the coordinates of the beacons, and δ_i are the beacon positions relative to a reference beacon a_1. It has been discussed in Example 6.8 that if the measurement vector y

is affected by spherical uncertainty (errors), then this uncertainty is reflected into uncertainty in the localization of p, and the uncertainty region around the nominal position is described by the estimation ellipsoid

$$E_p = \{p : p^\top (A^\top A)p \leq 1\}.$$

The lengths of the semi-axes of this ellipsoid are given by $\sigma_1^{-1}, \sigma_2^{-1}$, where σ_1, σ_2 are the singular values of A. Here we consider an "experiment design" type of problem: we assume that the position of the anchors is not known completely, and we aim at finding good anchor positions, so that the error region around the nominal estimated position is "as spherical as possible." The rationale behind this criterion is that we want to avoid having certain directions with large uncertainty and other directions with small uncertainty; in other words, uncertainty in the localization should be distributed as evenly as possible along all directions. This criterion is quantified by the *eccentricity* of the ellipsoid, which is simply the ratio between the largest and the smallest semi-axes, thus it coincides with the condition number of A.

To consider a tractable problem formulation, we assume that the reference beacon a_1 is fixed (e.g., $a_1 = 0$), and that also the *directions* of the relative beacon positions δ_i are given. That is, we assume that

$$\delta_i = \rho_i v_i, \quad \|v_i\|_2 = 1, \rho_i \geq 0, i = 1, \ldots, m,$$

where the directions v_i are given and fixed, while the distances ρ_i from the reference beacon are the variables to be determined. The problem then becomes to find $x_i \doteq \rho_i^2 \geq 0, i = 1, \ldots, m$, so as to minimize the (squared) condition number of $A(x)$, which coincides with the condition number of the symmetric matrix

$$F(x) = A^\top A = \begin{bmatrix} \delta_1 & \cdots & \delta_m \end{bmatrix} \begin{bmatrix} \delta_1 & \cdots & \delta_m \end{bmatrix}^\top = \sum_{i=1}^m x_i v_i v_i^\top.$$

Since $F(x)$ is linear in x, we have that there exists an x such that $\kappa(F(x)) \leq \gamma$ if and only if there exists an x such that condition (11.24) is satisfied. Therefore, our beacon placement problem can be expressed in the form of the following SDP

$$\begin{aligned}
\min_{x \in \mathbb{R}^m, \gamma \in \mathbb{R}} \quad & \gamma \\
\text{s.t.:} \quad & \sum_{i=1}^m x_i v_i v_i^\top \preceq \gamma I, \\
& \sum_{i=1}^m x_i v_i v_i^\top \succeq I, \\
& x_i \geq 0, \qquad i = 1, \ldots, m.
\end{aligned}$$

For example, choosing $m = 5$ random directions v_i (columns in the matrix below)

$$\begin{bmatrix} -0.5794 & 0.7701 & 0.2323 & -0.1925 & -0.9880 \\ -0.8151 & -0.6379 & -0.9727 & 0.9813 & 0.1543 \end{bmatrix},$$

we obtain optimal beacon placements for $x^* = [0.6711\ 0.2642\ 0.2250\ 0.2277\ 0.6120]^\top$ (or positive scalar multiples of this vector, due to homogeneity), with $\gamma^* = 1$, see Figure 11.7. The matrix $F(x^*)$ at the optimum results to be, in this case, the identity matrix.

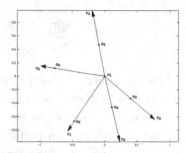

Figure 11.7 Optimal beacon placement.

11.4.1.4 *Matrix completion problems.*

Matrix completion problems refer to a class of problems in which one aims at recovering a matrix X from incomplete information about its entries (plus some *a priori* knowledge or assumptions). For example, suppose that only a few entries of a matrix $X \in \mathbb{R}^{m,n}$ are revealed (i.e., we know that $X_{ij} = d_{ij}$, for $(i,j) \in J$, where J is a set of indices of cardinality $q < mn$, where d_{ij} are given numbers), and we wish to guess what the whole matrix X is. Indeed, as such, this may look like an "impossible" problem, as it really is. Recovering the entire matrix from incomplete information is not possible, unless we have some additional information, or we make some assumptions on X. To make a simple example, consider the incomplete matrix X shown in Figure 11.8. Is it possible to recover the whole X from this incomplete information? The answer is no, if we have no additional information. The answer is instead yes, if we know that this is a Sudoku matrix...

In our treatment we shall not consider Sudoku-type completion problems, since they involve integer variables, and are thus typically non-convex and computationally hard. However, we consider other types of completion problems, where the prior assumption is that the unknown matrix should have minimal *rank*. These problems are therefore named *minimum-rank matrix completion* problems. A famous example in this class is the so-called "Netflix problem." This is a problem arising in a recommender system, where rows in matrix X represent users and columns represent movies. Users are given the opportunity of giving a rate mark on the movies. However, each user rates only a few movies (if any), hence only a few entries of the X matrix are known. Yet, it would be interesting to complete this matrix by "guessing" the missing entries, so that the vendor might recommend movies that a user may like and be willing to order. Here, one may observe that users' preferences on the movies are a function of a few factors (such as genre, country of production, filmmaker, etc.), hence each row in X may be written as the product of a row vector with few terms (the factors) times a large "factor loading" matrix, which implies that X will have rank no larger than the number of factors. This suggests the idea that X can be completed by finding the matrix of minimum rank which is compatible with the available entries. Formally, this gives rise to the following optimization problem:[9]

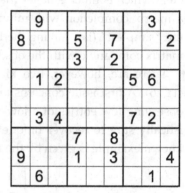

Figure 11.8 An incomplete 9×9 matrix.

[9] For a full treatment of this problem and pointers to related literature, see the paper by E. Candés, B. Recht, Exact matrix completion via convex optimization, *Foundations of Computational Mathematics*, 2009.

$$\min_{X \in \mathbb{R}^{m,n}} \quad \text{rank } X \tag{11.26}$$
$$\text{s.t.:} \quad X_{ij} = d_{ij}, \quad \text{for } (i,j) \in J,$$

where the index set J has cardinality $q < mn$. Observe indeed that if the prior information on X is that $\text{rank } X = r < \min(m,n)$, then X is defined by a number of free terms (or degrees of freedom) which is smaller than mn. In particular, considering the compact SVD of $X = \sum_{i=1}^{r} \sigma_i u_i v_i^\top$, we have that X has $r(m+n-r)$ degrees of freedom, corresponding to $rm - r(r+1)/2$ parameters for the u_i vectors (the $r(r+1)/2$ term corresponds to the degrees of freedom to be subtracted due to the orthonormality conditions among the u_i vectors), plus $rn - r(r+1)/2$ parameters for the v_i vectors, plus r parameters for the singular values σ_i. It is thus apparent that when r is small compared to m, n then the matrix X is defined by a number of parameters which is much smaller than the number of its entries. Finding a matrix completion with minimum rank thus amounts to finding the "simplest" (i.e., having the least number of degrees of freedom) matrix compatible with the observed entries.

There are, however, two kinds of difficulty related to problem (11.26). A first issue concerns *uniqueness* of the recovered matrix: by observing few entries of a low-rank matrix we cannot be deterministically sure to recover uniquely the hidden matrix itself. A well-known example is given by the rank-one matrix

$$X = \begin{bmatrix} 0 & 0 & \cdots & 1 \\ 0 & 0 & \cdots & 0 \\ \vdots & \vdots & \ddots & \vdots \\ 0 & 0 & \cdots & 0 \end{bmatrix},$$

whose entries are all zero, except for the one in the upper-right corner. Clearly, such a matrix could not be recovered from observing a generic subset of its entries (unless our observation set contains the "1" in the upper-right corner). We shall not dwell too much on this uniqueness issue, which has been the subject of extensive research in the "compressive sensing" literature. We only mention that to go round this problem we should take a probabilistic point of view, by considering matrices that are generated by certain random matrix *ensembles*, as well as random policies for selecting the entries to be observed (e.g., uniformly). Under such hypotheses there exist results that, roughly speaking, guarantee *with high probability* that the actual hidden matrix can be recovered from observation of a subset J of its entries, provided that the cardinality of J is large enough.

The second difficulty with problem (11.26) is instead of a computational nature: rank X is not a convex function of X, and problem (11.26) is indeed a very hard (technically, NP-hard) optimization problem: known algorithms for finding an exact solution to this problem take a time which is a double exponential function of the matrix size, which makes them basically useless, as soon as the dimension grows. There exist, however, *relaxations* (i.e., approximations) of the above problem which are amenable to efficient solution. We next discuss one such approximation. Consider the SVD of a generic matrix $X \in \mathbb{R}^{m,n}$: $X = \sum_{i=1}^{n} \sigma_i u_i v_i^\top$. The rank of X coincides with the number of its nonzero singular values, that is with the *cardinality* of the vector $s(X) \doteq [\sigma_1 \dots, \sigma_n]^\top$. We can then say that

$$\operatorname{rank} X = \|s(X)\|_0,$$

where $\|s(X)\|_0$ denotes indeed the ℓ_0 (pseudo) "norm" of the vector $s(X)$, which counts the number of nonzero entries in $s(X)$. Now, since $\|s(X)\|_0$ is non-convex, we may substitute it with the ℓ_1-norm, as justified in Section 9.5.1, and hence minimize the ℓ_1 norm of $s(X)$, $\|s(X)\|_1 = \sum_{i=1}^{n} \sigma_i(X)$, instead of rank X. The ℓ_1 norm of the vector of singular values is actually a matrix norm; hence, it is convex in its argument. It is known as the *nuclear norm*:[10]

$$\|X\|_* \doteq \|s(X)\|_1 = \sum_{i=1}^{n} \sigma_i(X).$$

The nuclear-norm heuristic thus amounts to solving, instead of (11.26), the convex optimization problem

$$\min_{X \in \mathbb{R}^{m,n}} \quad \|X\|_*$$
$$\text{s.t.:} \quad X_{ij} = d_{ij}, \quad \text{for } (i,j) \in J.$$

The interesting fact, which is not proven here,[11] is that the nuclear-norm minimization problem can be expressed in the form of the following SDP:

$$\min_{X \in \mathbb{R}^{m,n}, Y \in \mathbb{S}^m, Z \in \mathbb{S}^n} \quad \operatorname{trace} Y + \operatorname{trace} Z$$
$$\text{s.t.:} \quad X_{ij} = d_{ij}, \qquad \text{for } (i,j) \in J,$$
$$\begin{bmatrix} Y & X \\ X^\top & Z \end{bmatrix} \succeq 0,$$

(when X is symmetric, one may take $Y = Z$ in the above problem, thus eliminating one matrix variable, without loss of generality). Moreover, it can be proved that the solution from this heuristic "often" (in the probabilistic sense mentioned above) coincides with

[10] When X is symmetric, and positive semidefinite, the nuclear norm reduces to the trace.

[11] For a proof see, e.g., M. Fazel, *Matrix Rank Minimization with Applications*, Ph.D. thesis, Stanford University, 2002.

the solution of the original problem (11.26). Under appropriate hypotheses (essentially, that X comes from a random ensemble of matrices, that entries are sampled at random – for example uniformly – over rows and columns, and that q is sufficiently large), there thus exists a regime in which, with high probability, the solution to the computationally hard rank minimization problem is unique, and it coincides with the solution to the nuclear-norm minimization problem. One estimate for how large q needs to be for the recovery regime to hold prescribes that q should be of the order of $d^{5/4}r \log d$, where $d = \max(m, n)$ and r is the rank of X.

The minimum-rank completion problem in (11.26) is actually a special case of a more general class of problems, called *affine rank minimization* problems, where one minimizes the rank of a matrix X subject to affine constraints on the matrix entries, that is

$$\min_{X \in \mathbb{R}^{m,n}} \quad \text{rank } X \tag{11.27}$$
$$\text{s.t.:} \quad \mathcal{A}(X) = b,$$

where $\mathcal{A} : \mathbb{R}^{m,n} \to \mathbb{R}^q$ is a given linear map, and $b \in \mathbb{R}^q$ is a given vector. In this context, each entry b_i of vector b can be interpreted as a *linear measurement* on the entries of X, and the problem amounts to reconstructing the unknown X from the given q linear measurements. A convex relaxation of problem (11.27) is obtained by replacing the rank function with the nuclear norm function, which in turn yields an SDP formulation of the relaxed problem.

Example 11.7 (*Completion of Euclidean distance matrices*) Consider a set of points $p_1, \ldots, p_m \in \mathbb{R}^r$, $m \geq r$, and define the *Euclidean distance matrix* (EDM) for these points as the symmetric matrix $D \in \mathbb{S}^m$ whose (i, j)-th entry is the Euclidean distance between p_i and p_j, that is

$$D_{ij} = \|p_i - p_j\|_2^2, \quad i, j = 1, \ldots, m.$$

A typical problem, arising for instance in autonomous agents localization, cartography, computer graphics, and molecular geometry endeavors, is to determine (up to an orthogonal transformation and absolute offset) the configuration of points p_i, $i = 1, \ldots, m$, from possibly incomplete information on the distance matrix[12] D.

Let $P = [p_1 \cdots p_m] \in \mathbb{R}^{r,m}$, and let $\tilde{P}$ denote the matrix of centered data points:

$$\tilde{P} = PE, \quad E \doteq \left(I_m - \frac{1}{m} \mathbf{1}\mathbf{1}^\top \right),$$

where each column $\tilde{p}_i$ of $\tilde{P}$ is equal to $p_i - \bar{p}$, where $\bar{p}$ is the centroid of the data points. Observe that we can express each entry of D in terms of the Gram matrix $G \doteq P^\top P$, since

$$D_{ij} = \|p_i\|_2^2 + \|p_j\|_2^2 - 2p_i^\top p_j = G_{ii} + G_{jj} - 2G_{ij}, \quad i, j = 1, \ldots, m.$$

[12] The paper by A. Y. Alfakih, On the uniqueness of Euclidean distance matrix completions, *Linear Algebra and its Applications*, 2003, gives a full treatment of Euclidean matrix completions, and also contains references to most of the related literature.

Hence

$$D = \operatorname{diag}(G)\mathbf{1}^\top + \mathbf{1}\operatorname{diag}(G)^\top - 2G, \qquad (11.28)$$

where $\operatorname{diag}(G)$ here denotes the column vector containing the diagonal entries of G. Now notice that, since $E\mathbf{1} = 0$, it holds that

$$EDE = -2EGE = -2\tilde{P}^\top \tilde{P}. \qquad (11.29)$$

We may draw two conclusions from the last two equations. First, equation (11.29) indicates that if we know the distance matrix D, then we can recover the configuration of centered points (up to an orthogonal transformation) by computing a matrix square-root factorization of $-\frac{1}{2}EDE$. Second, equation (11.28) implies (by using Lemma 3.1) that $\operatorname{rank} D \le \operatorname{rank} G + 2$, hence (since $G = P^\top P$, $P \in \mathbb{R}^{r,m}$ implies $\operatorname{rank} G \le r$)

$$\operatorname{rank} D \le r + 2.$$

Since r is the embedding dimension of the data points, the previous bound implies that $\operatorname{rank} D$ is typically small compared to m, especially in geo-localization problems, where $r = 2$ (planar localization) or $r = 3$ (3D localization). This fact suggests that, at least with high probability, when $r < m$ we may recover the full Euclidean distance matrix D (and hence the centered configurations), from a sufficiently large number q of randomly selected observations of the entries of D.

For a numerical test, we considered several randomly generated configuration matrices P with $r = 2$, $m = 30$, where p_i, $i = 1,\ldots,m$, are extracted from a standard normal distribution. For each instance of P, we constructed the corresponding Euclidean distance matrix D, and we solved the following (symmetric) nuclear norm minimization problem

$$\min_{X \in \mathbb{S}^m, Y \in \mathbb{S}^m} \quad \operatorname{trace} Y \qquad (11.30)$$
$$\text{s.t.:} \quad X_{ij} = D_{ij}, \qquad \text{for } (i,j) \in J,$$
$$\begin{bmatrix} Y & X \\ X^\top & Y \end{bmatrix} \succeq 0,$$

where J is an index set of cardinality q indicating the q randomly selected entries of D that are revealed to the solver. The number q is to be compared with the number of free entries in the symmetric matrix X, which is $n_v \doteq m(m+1)/2$. Clearly, if the ratio $\eta = q/n_v$ is high (close to one), then we are revealing most of the entries, hence we expect correct recovery (i.e., that $X = D$) with high probability, while we expect lower recovery rates for lower η. The numerical experiment was conducted for values of η in $\{0.6, 0.7, 0.8, 0.9, 0.95\}$. For each η, we solved problem (11.30) $N = 50$ times, each time with a randomly extracted P matrix. We declared D recovered when the optimal X was such that $\|X - D\|_F / \|D\|_F \le 10^{-3}$, and we kept record of the average successful recoveries. Results are shown in Figure 11.9.

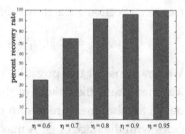

Figure 11.9 Rates of recovery of the full EDM, for $r = 2$, $m = 30$, as a function of the fraction η of the revealed entries.

11.4.2   Geometrical problems

Several geometric problems involving spheres, ellipsoids, and polytopes can be posed in terms of convex programs and in particular

SDPs,[13] as exemplified next.

[13] Many of these problems are also discussed in Chapter 8 of Boyd and Vandenberghe's book.

11.4.2.1 Largest ellipsoid contained in a polytope. Let us describe a bounded ellipsoid as the image of a unit ball under an affine map, that is

$$\mathcal{E} = \{x \in \mathbb{R}^n : x = Qz + c, \ \|z\|_2 \le 1\}, \qquad (11.31)$$

where $c \in \mathbb{R}^n$ is the center of the ellipsoid, and $Q \in S_+^n$ is the square-root of the shape matrix P of the ellipsoid, see Section 9.2.2. The lengths of the semi-axes of $\mathcal{E}$ are given by the singular values of Q, $\sigma_i(Q)$, $i = 1,\ldots,n$ (see Lemma 6.4), which, since $Q \succeq 0$, coincide with the eigenvalues of Q, $\lambda_i(Q)$, $i = 1,\ldots,n$.

Let then $\mathcal{P}$ denote a polytope, described as the intersection of m given half-spaces:

$$\mathcal{P} = \{x \in \mathbb{R}^n : a_i^\top x \le b_i, \ i = 1,\ldots,m\}.$$

The containment condition $\mathcal{E} \subseteq \mathcal{P}$ means that the inequalities $a_i^\top x \le b_i$, $i = 1,\ldots,m$, must be satisfied for all $x \in \mathcal{E}$, that is, for $i = 1,\ldots,m$, it must hold that

$$a_i^\top x \le b_i, \ x = Qz + c, \ \forall z : \|z\|_2 \le 1,$$

see Figure 11.10.

Substituting $x = Qz + c$ into the inequalities $a_i^\top x \le b_i$, we have, for $i = 1,\ldots,m$,

$$a_i^\top Qz + a_i^\top c \le b_i, \ \forall z : \|z\|_2 \le 1 \quad \Leftrightarrow \quad \max_{\|z\|_2 \le 1} a_i^\top Qz + a_i^\top c \le b_i.$$

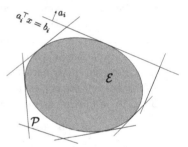

Figure 11.10 Ellipsoid $\mathcal{E}$ inscribed in a polytope $\mathcal{P}$.

Since the max in the equation above is attained for $z = Qa_i / \|Qa_i\|_2$ (see, e.g., Section 2.2.2.4), we obtain that

$$\mathcal{E} \subseteq \mathcal{P} \quad \Leftrightarrow \quad \|Qa_i\|_2 + a_i^\top c \le b_i, \ i = 1,\ldots,m. \qquad (11.32)$$

We now consider two possible "measures" of the size of $\mathcal{E}$: a first typical measure is the *volume* of the ellipsoid, which is proportional to $\det Q$; another measure is sum of the semi-axis lengths, which coincides with the trace of Q. A maximum-volume ellipsoid contained in $\mathcal{P}$ is obtained by maximizing $\det Q$, $Q \succ 0$, under the constraints in (11.32). However, since log is a monotone increasing function, we can equivalently maximize $\log \det Q$, which has the advantage of being a *concave* function of Q, over the domain $Q \succ 0$ (see Example 8.6). A maximum-volume ellipsoid contained in $\mathcal{P}$ can thus be obtained by solving the following convex optimization problem (it involves the maximization of a concave objective $f_0 = \log \det Q$, which is equivalent to minimizing the convex objective $-f_0$):

$$\max_{Q \in S^n, c \in \mathbb{R}^n} \quad \log \det Q$$
$$\text{s.t.:} \quad Q \succ 0,$$
$$\|Qa_i\|_2 + a_i^\top c \le b_i. \quad i = 1, \dots, m.$$

This problem is convex, but it is not in SDP format, due to the $\log \det Q$ objective. However, it can be reformulated into an equivalent SDP format, although this reformulation is not detailed here.[14] Standard software for convex optimization, such as CVX, automatically recognize a $\log \det$-type objective, and transform it internally into an equivalent SDP approximation. The maximum-volume ellipsoid contained in $\mathcal{P}$ is called the *Löwner–John ellipsoid* of $\mathcal{P}$; every full-dimensional convex set has a *unique* Löwner–John ellipsoid. Moreover, if the Löwner–John ellipsoid is scaled by a factor n around its center, then one obtains an ellipsoid that contains $\mathcal{P}$. That is, if $\mathcal{E}^*$ is a Löwner–John ellipsoid for $\mathcal{P}$, then it is the unique maximum volume ellipsoid such that $\mathcal{E}^* \subseteq \mathcal{P}$, and it holds that

$$\mathcal{P} \subseteq n\mathcal{E}^*.$$

Further, if the $\mathcal{P}$ set is symmetric around its center, then one may improve the scaling factor from n to $\sqrt{n}$.

A maximum-sum-of-semiaxis lengths ellipsoid contained in $\mathcal{P}$ is instead obtained by maximizing $\text{trace } Q$, $Q \succeq 0$, under the constraints in (11.32). Since $\text{trace } Q$ is linear (hence concave) in Q, we directly obtain the following SDP problem:

$$\max_{Q \in S^n, c \in \mathbb{R}^n} \quad \text{trace } Q \tag{11.33}$$
$$\text{s.t.:} \quad Q \succeq 0,$$
$$\begin{bmatrix} b_i - a_i^\top c & a_i^\top Q \\ Qa_i & (b_i - a_i^\top c)I \end{bmatrix} \succeq 0, \quad i = 1, \dots, m,$$

where the last LMI constraints were obtained by applying (11.11) to the SOC constraints $\|Qa_i\|_2 + a_i^\top c \le b_i$.

A relevant special case arises when we constrain *a priori* $\mathcal{E}$ to be a sphere, that is we set $Q = rI_n$, where $r \ge 0$ represents the radius of the sphere. In this case, problem (11.33) specializes to

$$\max_{r \ge 0, c \in \mathbb{R}^n} \quad r$$
$$\text{s.t.:} \quad r\|a_i\|_2 + a_i^\top c \le b_i, \quad i = 1, \dots, m,$$

which is simply an LP. The center c of the largest sphere inscribed in $\mathcal{P}$ is usually known as the *Chebyshev center* of the polytope.

[14] See Section 4.2 in A. Ben-Tal and A. Nemirovski, *Lectures on Modern Convex Optimization*, SIAM, 2001.

11.4.2.2   Smallest ellipsoid containing a polytope.   We next consider the problem of finding a minimum-size ellipsoid $\mathcal{E}$ that contains a polytope $\mathcal{P}$ defined by means of its vertices (see Figure 11.11):

$$\mathcal{P} = \mathrm{co}\{x^{(1)}, \ldots, x^{(p)}\},$$

We describe the ellipsoid $\mathcal{E}$ by means of the following representation:

$$\mathcal{E} = \left\{ x \in \mathbb{R}^n : \begin{bmatrix} 1 & (x-c)^\top \\ (x-c) & P \end{bmatrix} \succeq 0 \right\}, \qquad (11.34)$$

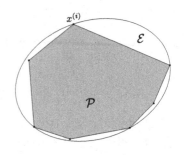

Figure 11.11 Ellipsoid $\mathcal{E}$ circumscribing a polytope $\mathcal{P}$.

where c is the center of the ellipsoid, and P is the shape matrix. Notice that, by the (non-strict) Schur complement rule, the LMI condition in the above representation is equivalent to the conditions

$$P \succeq 0, \quad (x-c) \in \mathcal{R}(P), \quad (x-c)^\top P^\dagger (x-c) \leq 1.$$

If we let $P = QQ^\top$ be a full-rank factorization of P, where $Q \in \mathbb{R}^{n,m}$, $\mathrm{rank}\, P = m \leq n$, then $P^\dagger = Q^{\top\dagger} Q^\dagger$, $Q^\dagger Q = I_m$, and $\mathcal{R}(P) = \mathcal{R}(Q)$. Therefore, the condition $(x-c) \in \mathcal{R}(P)$ means that there exists $z \in \mathbb{R}^m$ such that $x - c = Qz$, and

$$(x-c)^\top P^\dagger (x-c) = z^\top Q^\top Q^{\top\dagger} Q^\dagger Q z = z^\top z,$$

whence

$$(x-c)^\top P^\dagger (x-c) \leq 1 \quad \Leftrightarrow \quad \|z\|_2 \leq 1.$$

Therefore, the representation in (11.34) is equivalent to the representation in (11.31)

$$\mathcal{E} = \{x \in \mathbb{R}^n : x = Qz + c, \ \|z\|_2 \leq 1\},$$

with $Q \in \mathbb{R}^{n,m}$ full column rank but possibly rectangular. This representation allows for the description of bounded ellipsoids that may be "flat" along some directions, i.e., ellipsoids that are contained in an affine space of dimension lower than the embedding dimension n. Whenever an ellipsoid is flat, its volume is identically zero, hence minimizing the volume measure may be inappropriate for possibly flat ellipsoids. A frequently used measure of size, that can be used also for flat ellipsoids, is instead the sum of the *squared* semi-axis lengths, which is given by the trace of P. A minimum-trace ellipsoid containing $\mathcal{P}$ can thus be computed as follows. Observe that $\mathcal{P} \subseteq \mathcal{E}$ if and only if $x^{(i)} \in \mathcal{E}$ for $i = 1, \ldots, p$, hence

$$\mathcal{P} \subseteq \mathcal{E} \quad \Leftrightarrow \quad \begin{bmatrix} 1 & (x^{(i)} - c)^\top \\ (x^{(i)} - c) & P \end{bmatrix} \succeq 0, \ i = 1, \ldots, p,$$

from which we obtain the SDP

$$\min_{P \in S^n, c \in \mathbb{R}^n} \quad \text{trace } P \tag{11.35}$$

$$\text{s.t.:} \quad \begin{bmatrix} 1 & (x^{(i)} - c)^\top \\ (x^{(i)} - c) & P \end{bmatrix} \succeq 0, \quad i = 1, \dots, p.$$

If the LMI condition in (11.34) is instead strengthened to a strict inequality, then the condition $P \succ 0$ implies that the ellipsoid $\mathcal{E}$ is full-dimensional, and we can factor $P = Q^2$, with $Q \succ 0$. Then the representation in (11.34), under strict inequality, is equivalent to the representation

$$\mathcal{E} = \{x \in \mathbb{R}^n : (x - c)^\top Q^{-1} Q^{-1} (x - c) \leq 1\}, \quad Q \succ 0.$$

Posing $A = Q^{-1}$ and $b = Ac$, this is also equivalent to

$$\mathcal{E} = \{x \in \mathbb{R}^n : \|Ax - b\|_2^2 \leq 1\}, \quad A \succ 0.$$

The volume of $\mathcal{E}$ is in this case proportional to $\det(Q) = \det(A^{-1})$. A minimum-volume ellipsoid containing $\mathcal{P}$ can then be obtained by minimizing $\det(A^{-1})$, $A \succ 0$, under the containment condition $\mathcal{P} \subseteq \mathcal{E}$. However, the objective function $\det(A^{-1})$ is not convex over $A \succ 0$. To overcome this issue, we simply consider the logarithm of the volume as the minimization objective: this gives the same minimizer as the original problem, since the log function is monotonic increasing, and has the advantage that the modified objective

$$f_0 = \log \det(A^{-1}) = -\log \det(A)$$

is convex over the domain $A \succ 0$, see Example 8.6. A minimum-volume ellipsoid containing $\mathcal{P}$ can thus be obtained by solving the following convex optimization problem:

$$\min_{A \in S^n, b \in \mathbb{R}^n} \quad -\log \det(A)$$

$$\text{s.t.:} \quad A \succ 0; \ \|Ax^{(i)} - b\|_2^2 \leq 1, \quad i = 1, \dots, p.$$

In the special case where we seek for the minimum-size *sphere* that contains $\mathcal{P}$, then both the minimum-volume and the minimum-trace problems are equivalent to the following SOCP:

$$\min_{r \in \mathbb{R}, c \in \mathbb{R}^n} \quad r \quad \text{s.t.:} \ \|x^{(i)} - c\|_2 \leq r, \ i = 1, \dots, p.$$

Example 11.8 (*Inner and outer ellipsoidal approximations of a polytope*) Consider a polytope $\mathcal{P}$ with vertices described by the columns of the following matrix (see Figure 11.12):

$$V = \begin{bmatrix} -2.7022 & -0.1863 & 0.2824 & 0.8283 & 1.5883 \\ -1.4118 & -0.9640 & 1.3385 & 0.9775 & 0.3340 \end{bmatrix}.$$

The same polytope also has the inequality representation as the set of $x \in \mathbb{R}^2$ such that

$$
\begin{bmatrix}
0.1752 & -0.9845 \\
0.5904 & -0.8071 \\
0.6462 & 0.7632 \\
0.5516 & 0.8341 \\
-0.6777 & 0.7354
\end{bmatrix} x \leq
\begin{bmatrix}
0.9164 \\
0.6681 \\
1.2812 \\
1.2722 \\
0.7930
\end{bmatrix}.
$$

Figure 11.12 shows the polytope $\mathcal{P}$, together with the minimum volume (solid line) and the minimum trace (dashed line) circumscribed ellipsoids, and the maximum volume (solid line) and maximum sum-of-axis lengths (dashed line) inscribed ellipsoids.

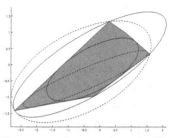

Figure 11.12 Inner and outer ellipsoidal bounding of a polytope.

11.4.2.3 Smallest volume ellipsoid containing the union of ellipsoids.

We here consider the problem of finding a minimal ellipsoid covering a set of m given ellipsoids, see Figure 11.13.

Let the given ellipsoids be specified in the form of convex quadratic inequalities

$$
\mathcal{E}_i = \{x \in \mathbb{R}^n : f_i(x) \leq 0\}, \quad i = 1, \ldots, m, \tag{11.36}
$$

where, for $i = 1, \ldots, m$,

$$
f_i(x) \doteq x^\top F_i x + 2g_i^\top x + h_i =
\begin{bmatrix} x \\ 1 \end{bmatrix}^\top
\begin{bmatrix} F_i & g_i \\ g_i^\top & h_i \end{bmatrix}
\begin{bmatrix} x \\ 1 \end{bmatrix},
$$

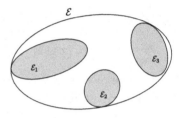

Figure 11.13 An ellipsoid $\mathcal{E}$ circumscribing $m = 3$ ellipsoids $\mathcal{E}_i$, $i = 1, 2, 3$.

with $F_i \succ 0$, and $h_i < g_i^\top F_i^{-1} g_i$. These ellipsoids have centers at $x^{(i)} = -F_i^{-1} g_i$, and the inequalities $f_i(x) \leq 0$ are strictly feasible, since $f_i(x^{(i)}) < 0$. Let the outer-bounding ellipsoid (to be determined) be parameterized as

$$
\mathcal{E} = \{x \in \mathbb{R}^n : f_0(x) \leq 0\}, \tag{11.37}
$$

where

$$
f_0(x) = \|Ax + b\|_2^2 - 1 =
\begin{bmatrix} x \\ 1 \end{bmatrix}^\top
\begin{bmatrix} F_0 & g_0 \\ g_0^\top & h_0 \end{bmatrix}
\begin{bmatrix} x \\ 1 \end{bmatrix},
$$

with $A \succ 0$, $F_0 = A^2$, $g_0 = Ab$, $h_0 = b^\top b - 1$ (notice that we used this description for $\mathcal{E}$ in order to avoid a homogeneous parameterization: here h_0 is not a free variable). Then, from the $\mathcal{S}$-procedure we have that

$$
\mathcal{E}_i \subseteq \mathcal{E} \quad \Leftrightarrow \quad \exists \tau_i \geq 0 :
\begin{bmatrix} F_0 & g_0 \\ g_0^\top & h_0 \end{bmatrix} \preceq \tau_i
\begin{bmatrix} F_i & g_i \\ g_i^\top & h_i \end{bmatrix}.
$$

Elaborating on this latter condition, with the position $\tilde{b} \doteq Ab$, we obtain

$$
\begin{bmatrix} A^2 - \tau_i F_i & Ab - g_i \\ b^\top A - g_i^\top & b^\top b - 1 - \tau_i h_i \end{bmatrix} = \begin{bmatrix} A^2 - \tau_i F_i & \tilde{b} - g_i \\ \tilde{b}^\top - g_i^\top & \tilde{b}^\top A^{-2}\tilde{b} - 1 - \tau_i h_i \end{bmatrix}
$$

$$
= \begin{bmatrix} A^2 - \tau_i F_i & \tilde{b} - g_i \\ \tilde{b}^\top - g_i^\top & -1 - \tau_i h_i \end{bmatrix} - \begin{bmatrix} 0 \\ \tilde{b}^\top \end{bmatrix}(-A^{-2})\begin{bmatrix} 0 \\ \tilde{b}^\top \end{bmatrix}^\top \preceq 0.
$$

Using the Schur complement rule, the latter matrix inequality is equivalent to

$$
\begin{bmatrix} F_0 - \tau_i F_i & \tilde{b} - g_i & 0 \\ \tilde{b}^\top - g_i^\top & -1 - \tau_i h_i & \tilde{b}^\top \\ 0 & \tilde{b} & -F_0 \end{bmatrix} \preceq 0, \qquad (11.38)
$$

which is an LMI condition in the variables $F_0, \tilde{b}$.

Since the volume of $\mathcal{E}$ is proportional to $\det^{1/2} F_0^{-1}$, we may find a minimum-volume outer-bounding ellipsoid by minimizing the logarithm of $\det^{1/2} F_0^{-1}$ (which is a convex function of $F_0 \succ 0$) under the LMI constraints in (11.38), obtaining the convex optimization problem

$$
\min_{F_0 \in \mathbb{S}^n, \tilde{b} \in \mathbb{R}^n, \tau_1,\ldots,\tau_m \in \mathbb{R}} \quad -\frac{1}{2}\log\det(F_0)
$$
$$
\text{s.t.:} \quad \tau_i \geq 0, \qquad i = 1,\ldots,m
$$
$$
(11.38), \qquad i = 1,\ldots,m.
$$

Remark 11.1 *Smallest sum-of-semiaxis lengths ellipsoid containing the union of ellipsoids.* An ellipsoid with minimal sum-of-semiaxis lengths covering the union of ellipsoids can be found by solving an SDP problem. Such a formulation can be obtained by considering the representation $\mathcal{E}_i = \{x : x = x^{(i)} + E_i z_i, \|z_i\|_2 \leq 1\}$ for the given ellipsoids $\mathcal{E}_i$, $i = 1,\ldots,m$, and the representation in (11.34) for the covering ellipsoid $\mathcal{E}$. The precise formulation can then be obtained by applying the LMI robustness lemma to (11.35), which is left to the reader as an exercise.

11.4.2.4 *Maximum volume ellipsoid contained in the intersection of ellipsoids.* We here consider the problem of finding a maximal-volume ellipsoid contained in the intersection of a set of m given ellipsoids, see Figure 11.14.

Let the given ellipsoids be represented as

$$
\mathcal{E}_i = \left\{ x \in \mathbb{R}^n : \begin{bmatrix} 1 & (x - x^{(i)})^\top \\ (x - x^{(i)}) & P_i \end{bmatrix} \succeq 0 \right\}, \quad i = 1,\ldots,m,
$$

where $x^{(i)}$ are the centers and $P_i \succeq 0$ are the shape matrices of the given ellipsoids, for $i = 1,\ldots,m$. Let instead the to-be-determined

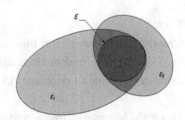

Figure 11.14 An ellipsoid $\mathcal{E}$ inscribed in the intersection of two ellipsoids.

inscribed ellipsoid be represented as

$$\mathcal{E} = \{x \in \mathbb{R}^n : x = Qz + c, \ \|z\|_2 \leq 1\},$$

with $c \in \mathbb{R}^n$, $Q \in \mathbb{S}^n_{++}$. Now, we observe that $\mathcal{E} \subseteq \mathcal{E}_i$ if and only if

$$\begin{bmatrix} 1 & (x - x^{(i)})^\top \\ (x - x^{(i)}) & P_i \end{bmatrix} \succeq 0, \quad \forall x \in \mathcal{E},$$

that is, if and only if

$$\begin{bmatrix} 1 & (c - x^{(i)})^\top + z^\top Q \\ (c - x^{(i)}) + Qz & P_i \end{bmatrix} \succeq 0, \quad \forall z : \|z\|_2 \leq 1.$$

We then rewrite this last condition as

$$\begin{bmatrix} 1 & (c - x^{(i)})^\top \\ (c - x^{(i)}) & P_i \end{bmatrix}$$
$$+ \begin{bmatrix} 1 \\ 0 \end{bmatrix} z^\top \begin{bmatrix} 0 & Q \end{bmatrix} + \left(\begin{bmatrix} 1 \\ 0 \end{bmatrix} z^\top \begin{bmatrix} 0 & Q \end{bmatrix} \right)^\top \succeq 0,$$
$$\forall z : \|z\|_2 \leq 1,$$

and apply the LMI robustness lemma to obtain the following LMI condition on Q, c, λ_i:

$$\begin{bmatrix} 1 - \lambda_i & (c - x^{(i)})^\top & 0 \\ (c - x^{(i)}) & P_i & Q \\ 0 & Q & \lambda_i I_n \end{bmatrix} \succeq 0. \tag{11.39}$$

Since the volume of $\mathcal{E}$ is proportional to $\det Q$, we obtain a maximum-volume ellipsoid inscribed in the intersection of the $\mathcal{E}_i$s, by solving the following convex optimization problem:

$$\max_{Q \in \mathbb{S}^n, c \in \mathbb{R}^n, \lambda \in \mathbb{R}^m} \log \det(Q) \tag{11.40}$$
$$\text{s.t.:} \quad Q \succ 0,$$
$$(11.39), \quad i = 1, \ldots, m.$$

Clearly, a maximal sum-of-semiaxis-lengths ellipsoid contained in $\cap_i \mathcal{E}_i$ can be obtained analogously, by maximizing $\operatorname{trace} Q$, under the same constraints of problem (11.40).

11.4.2.5 Minimal ellipsoid containing the intersection of ellipsoids. On the contrary to all the previous minimal and maximal ellipsoidal covering problems, the problem of finding a minimal ellipsoid $\mathcal{E}$ containing the intersection of m given ellipsoids $\mathcal{E}_i$ (see Figure 11.15) is computationally hard. This problem does not admit an exact solution computable via convex optimization. However, we can find a *suboptimal* solution via convex optimization, based on a sufficient condition for the containment $\cap_i \mathcal{E}_i \subseteq \mathcal{E}$.

Let the given ellipsoids $\mathcal{E}_i$ be represented as in (11.36) and let the to-be-determined covering ellipsoid $\mathcal{E}$ be represented as in (11.37). Then we have from the $\mathcal{S}$-procedure that a *sufficient* condition for the containment $\cap_i \mathcal{E}_i \subseteq \mathcal{E}$ is that there exist scalars $\tau_1, \ldots, \tau_m \geq 0$ such that (11.17) holds, that is (using the fact that $F_0 = A^2$, $g_0 = Ab$, $h_0 = b^\top b - 1$, $\tilde{b} = Ab$):

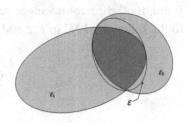

Figure 11.15 An ellipsoid $\mathcal{E}$ circumscribing the intersection of two ellipsoids.

$$\begin{bmatrix} A^2 & Ab \\ b^\top A & b^\top b - 1 \end{bmatrix} - \sum_{i=1}^{m} \tau_i \begin{bmatrix} F_i & g_i \\ g_i^\top & h_i \end{bmatrix} \preceq 0.$$

Rewriting the previous inequality in the form

$$\begin{bmatrix} F_0 & \tilde{b} \\ \tilde{b}^\top & -1 \end{bmatrix} - \sum_{i=1}^{m} \tau_i \begin{bmatrix} F_i & g_i \\ g_i^\top & h_i \end{bmatrix} + \begin{bmatrix} 0 \\ \tilde{b}^\top \end{bmatrix} F_0^{-1} \begin{bmatrix} 0 \\ \tilde{b}^\top \end{bmatrix}^\top \preceq 0$$

and applying the Schur complement rule, we obtain the LMI condition

$$\begin{bmatrix} F_0 - \sum_{i=1}^{m} \tau_i F_i & \tilde{b} - \sum_{i=1}^{m} \tau_i g_i & 0 \\ \tilde{b}^\top - \sum_{i=1}^{m} \tau_i g_i^\top & -1 - \sum_{i=1}^{m} \tau_i h_i & \tilde{b}^\top \\ 0 & \tilde{b} & -F_0 \end{bmatrix} \preceq 0. \qquad (11.41)$$

Then, a suboptimal minimum volume ellipsoid can be computed by solving the following convex optimization problem:

$$\min_{F_0 \in \mathbb{S}^n, \tilde{b} \in \mathbb{R}^n, \tau_1, \ldots, \tau_m \in \mathbb{R}} \quad -\frac{1}{2} \log \det(F_0)$$
$$\text{s.t.:} \quad \tau_i \geq 0, \qquad i = 1, \ldots, m,$$
$$(11.41).$$

A suboptimal ellipsoid can also be easily computed for the case of the trace criterion (we minimize the sum of the squared semi-axis lengths). To this end, we describe $\mathcal{E}$ as

$$\mathcal{E} = \{x \in \mathbb{R}^n : (x - c)^\top P^{-1}(x - c) \leq 1\},$$

which corresponds to the previous representation for $F_0 = P^{-1}$, $g_0 = P^{-1}c$, $h_0 = c^\top P^{-1}c - 1$. Hence, the sufficient condition (11.17) becomes

$$\begin{bmatrix} P^{-1} & P^{-1}c \\ c^\top P^{-1} & c^\top P^{-1}c - 1 \end{bmatrix} - \sum_{i=1}^{m} \tau_i \begin{bmatrix} F_i & g_i \\ g_i^\top & h_i \end{bmatrix} \preceq 0,$$

which we rewrite as

$$\begin{bmatrix} -\sum_{i=1}^{m} \tau_i F_i & -\sum_{i=1}^{m} \tau_i g_i \\ -\sum_{i=1}^{m} \tau_i g_i^\top & -1 - \sum_{i=1}^{m} \tau_i h_i \end{bmatrix} + \begin{bmatrix} I \\ c^\top \end{bmatrix} P^{-1} \begin{bmatrix} I & c \end{bmatrix} \preceq 0.$$

Using the Schur complement rule, this latter condition is equivalent to the following LMI in P, c and $\tau_1, \ldots, \tau_m$:

$$
\begin{bmatrix}
-\sum_{i=1}^{m} \tau_i F_i & -\sum_{i=1}^{m} \tau_i g_i & I \\
-\sum_{i=1}^{m} \tau_i g_i^\top & -1 - \sum_{i=1}^{m} \tau_i h_i & c^\top \\
I & c & -P
\end{bmatrix} \preceq 0. \tag{11.42}
$$

Then a suboptimal minimum trace ellipsoid can be computed by solving the following SDP:

$$
\min_{P \in \mathbb{S}^n, c \in \mathbb{R}^n, \tau_1, \ldots, \tau_m \in \mathbb{R}} \quad \mathrm{trace}(P)
$$
$$
\text{s.t.:} \quad \tau_i \geq 0, \quad i = 1, \ldots, m,
$$
$$
(11.42).
$$

11.5   Exercises

Exercise 11.1 (Minimum distance to a line segment revisited)   In this exercise, we revisit Exercise 9.3, and approach it using the $\mathcal{S}$-procedure of Section 11.3.3.1.

1. Show that the minimum distance from the line segment $\mathcal{L}$ to the origin is above a given number $R \geq 0$ if and only if

$$
\|\lambda(p - q) + q\|_2^2 \geq R^2 \text{ whenever } \lambda(1 - \lambda) \geq 0.
$$

2. Apply the $\mathcal{S}$-procedure, and prove that the above is in turn equivalent to the LMI in $\tau \geq 0$:

$$
\begin{bmatrix}
\|p - q\|_2^2 + \tau & q^\top(p - q) - \tau/2 \\
q^\top(p - q) - \tau/2 & q^\top q - R^2
\end{bmatrix} \succeq 0.
$$

3. Using the Schur complement rule,[15] show that the above is consistent with the result given in Exercise 9.3.

[15] See Theorem 4.9.

Exercise 11.2 (A variation on principal component analysis)   Let $X = [x_1, \ldots, x_m] \in \mathbb{R}^{n,m}$. For $p = 1, 2$, we consider the problem

$$
\phi_p(X) \doteq \max_u \sum_{i=1}^{m} |x_i^\top u|^p \; : \; u^\top u = 1. \tag{11.43}
$$

If the data is centered, the case $p = 1$ amounts to finding a direction of largest "deviation" from the origin, where deviation is measured using the ℓ_1-norm; arguably, this is less sensitive to outliers than the case $p = 2$, which corresponds to principal component analysis.

1. Find an expression for ϕ_2, in terms of the singular values of X.

2. Show that the problem, for $p = 1$, can be approximated via an SDP, as $\phi_1(X) \leq \psi_1(X)$, where

$$\psi_1(X) \doteq \max_U \sum_{i=1}^m \sqrt{x_i^\top U x_i} \; : \; U \succeq 0, \; \operatorname{trace} U = 1.$$

Is ψ_1 a norm?

3. Formulate a dual to the above expression. Does strong duality hold? *Hint:* introduce new variables $z_i = x_i^\top U x_i, i = 1, \ldots, m$, and dualize the corresponding constraints.

4. Use the identity (8.52) to approximate, via weak duality, the problem (11.43). How does your bound compare with ψ_1?

5. Show that

$$\psi_1(X)^2 = \min_D \operatorname{trace} D \; : \; D \text{ diagonal}, D \succ 0, \; D \succeq X^\top X.$$

Hint: scale the variables in the dual problem and optimize over the scaling. That is, set $D = \alpha \tilde{D}$, with $\lambda_{\max}(X\tilde{D}^{-1}X^\top) = 1$ and $\alpha > 0$, and optimize over α. Then argue that we can replace the equality constraint on $\tilde{D}$ by a convex inequality, and use Schur complements to handle that corresponding inequality.

6. Show that

$$\phi_1(X) = \max_{v: \|v\|_\infty \leq 1} \|Xv\|_2.$$

Is the maximum always attained with a vector v such that $|v_i| = 1$ for every i? *Hint:* use the fact that

$$\|z\|_1 = \max_{v: \|v\|_\infty \leq 1} z^\top v.$$

7. A result by Yu. Nesterov[16] shows that for any symmetric matrix $Q \in \mathbb{R}^{m,m}$, the problem

$$p^* = \max_{v: \|v\|_\infty \leq 1} v^\top Q v$$

can be approximated within $\pi/2$ relative value via SDP. Precisely, $(2/\pi)d^* \leq p^* \leq d^*$, where

$$d^* = \min_D \operatorname{trace} D \; : \; D \text{ diagonal}, D \succeq Q. \qquad (11.44)$$

Use this result to show that

$$\sqrt{\frac{2}{\pi}}\psi_1(X) \leq \phi_1(X) \leq \psi_1(X).$$

That is, the SDP approximation is within $\approx 80\%$ of the true value, irrespective of the problem data.

[16] Yu. Nesterov, Quality of semidefinite relaxation for nonconvex quadratic optimization, discussion paper, CORE, 1997.

8. Discuss the respective complexity of the problems of computing ϕ_2 and ψ_1 (you can use the fact that, for a given $m \times m$ symmetric matrix Q, the SDP (11.44) can be solved in $O(m^3)$).

Exercise 11.3 (Robust principal component analysis) The following problem is known as robust principal component analysis:[17]

[17] See Section 13.5.4.

$$p^* \doteq \min_X \|A - X\|_* + \lambda \|X\|_1,$$

where $\|\cdot\|_*$ stands for the nuclear norm,[18] and $\|\cdot\|_1$ here denotes the sum of the absolute values of the elements of a matrix. The interpretation is the following: A is a given data matrix and we would like to decompose it as a sum of a low rank matrix and a sparse matrix. The nuclear norm and ℓ_1 norm penalties are respective convex heuristics for these two properties. At optimum, X^* will be the sparse component and $A - X^*$ will be the low-rank component such that their sum gives A.

[18] The nuclear norm is the sum of the singular values of the matrix; see Section 11.4.1.4 and Section 5.2.2.

1. Find a dual for this problem. *Hint:* we have, for any matrix W:

$$\|W\|_* = \max_Y \text{ trace } W^\top Y \; : \; \|Y\|_2 \leq 1,$$

where $\|\cdot\|_2$ is the largest singular value norm.

2. Transform the primal or dual problem into a known programming class (i.e. LP, SOCP, SDP, etc.). Determine the number of variables and constraints. *Hint:* we have

$$\|Y\|_2 \leq 1 \iff I - YY^\top \succeq 0,$$

where I is the identity matrix.

3. Using the dual, show that when $\lambda > 1$, the optimal solution is the zero matrix. *Hint:* if Y^* is the optimal dual variable, the complementary slackness condition states that $|Y_{ij}^*| < \lambda$ implies $X_{ij}^* = 0$ at optimum.

Exercise 11.4 (Boolean least squares) Consider the following problem, known as *Boolean least squares*:

$$\phi = \min_x \|Ax - b\|_2^2 \; : \; x_i \in \{-1, 1\}, \; i = 1, \ldots, n.$$

Here, the variable is $x \in \mathbb{R}^n$, where $A \in \mathbb{R}^{m,n}$ and $b \in \mathbb{R}^m$ are given. This is a basic problem arising, for instance, in digital communications. A brute force solution is to check all 2^n possible values of x, which is usually impractical.

1. Show that the problem is equivalent to

$$\phi = \min_{X,x} \quad \text{trace}(A^\top AX) - 2b^\top Ax + b^\top b$$
$$\text{s.t.:} \quad X = xx^\top,$$
$$X_{ii} = 1, \quad i = 1, \ldots, n,$$

in the variables $X = X^\top \in \mathbb{R}^{n,n}$ and $x \in \mathbb{R}^n$.

2. The constraint $X = xx^\top$, i.e., the set of rank-1 matrices, is not convex, therefore the previous problem formulation is still hard. Show that the following "'relaxed" problem, which is an SDP,

$$\phi_{\text{sdp}} = \min_X \quad \text{trace}(A^\top AX) - 2b^\top Ax + b^\top b$$
$$\text{s.t.:} \quad \begin{bmatrix} X & x \\ x^\top & 1 \end{bmatrix} \succeq 0,$$
$$X_{ii} = 1, \quad i = 1, \ldots, n,$$

produces a lower-bound to the original problem, i.e., $\phi \geq \phi_{\text{sdp}}$. Once this problem is solved, an approximate solution to the original problem can be obtained by rounding the solution: $x_{\text{sdp}} = \text{sgn}(x^*)$, where x^* is the optimal solution of the semidefinite relaxation.

3. Another approximation method is to relax the non-convex constraints $x_i \in \{-1, 1\}$ to convex interval constraints $-1 \leq x_i \leq 1$ for all i, which can be written as $\|x\|_\infty \leq 1$. Therefore, a different lower bound is given by:

$$\phi \geq \phi_{\text{int}} \doteq \min \|Ax - b\|_2^2 \; : \; \|x\|_\infty \leq 1.$$

Once this problem is solved, we can round the solution by $x_{\text{int}} = \text{sgn}(x^*)$ and evaluate the original objective value $\|Ax_{\text{int}} - b\|_2^2$. Which one of ϕ_{sdp} and ϕ_{int} produces the closest approximation to ϕ? Justify your answer carefully.

4. Use now 100 independent realizations with normally distributed data, $A \in \mathbb{R}^{10,10}$ (independent entries with mean zero) and $b \in \mathbb{R}^{10}$ (independent entries with mean one). Plot and compare the histograms of $\|Ax_{\text{sdp}} - b\|_2^2$ of part 2, $\|Ax_{\text{int}} - b\|_2^2$ of part 3, and the objective corresponding to a naïve method $\|Ax_{\text{ls}} - b\|_2^2$, where $x_{\text{ls}} = \text{sgn}\left((A^\top A)^{-1} A^\top b\right)$ is the rounded ordinary least-squares solution. Briefly discuss the accuracy and computation time (in seconds) of the three methods.

5. Assume that, for some problem instance, the optimal solution (x, X) found via the SDP approximation is such that x belongs

to the original non-convex constraint set $\{x : x_i \in \{-1,1\}, \ i = 1,\ldots,n\}$. What can you say about the SDP approximation in that case?

Exercise 11.5 (Auto-regressive process model) We consider a process described by the difference equation

$$y(t+2) = \alpha_1(t)y(t+1) + \alpha_2(t)y(t) + \alpha_3(t)u(t), \ \ t = 0,1,2,\ldots,$$

where the $u(t) \in \mathbb{R}$ is the input, $y(t) \in \mathbb{R}$ the output, and the coefficient vector $\alpha(t) \in \mathbb{R}^3$ is time-varying. We seek to compute bounds on the vector $\alpha(t)$ that are (a) independent of t, (b) consistent with some given historical data.

The specific problem we consider is: given the values of $u(t)$ and $y(t)$ over a time period $1 \le t \le T$, find the smallest ellipsoid $\mathcal{E}$ in $\mathbb{R}^3$ such that, for every t, $1 \le t \le T$, the equation above is satisfied for some $\alpha(t) \in \mathcal{E}$.

1. What is a geometrical interpretation of the problem, in the space of αs?

2. Formulate the problem as a semidefinite program. You are free to choose the parameterization, as well as the measure of the size of $\mathcal{E}$ that you find most convenient.

3. Assume we restrict our search to spheres instead of ellipsoids. Show that the problem can be reduced to a linear program.

4. In the previous setting, $\alpha(t)$ is allowed to vary with time arbitrarily fast, which may be unrealistic. Assume that a bound is imposed on the variation of $\alpha(t)$, such as $\|\alpha(t+1) - \alpha(t)\|_2 \le \beta$, where $\beta > 0$ is given. How would you solve the problem with this added restriction?

Exercise 11.6 (Non-negativity of polynomials) A second-degree polynomial with values $p(x) = y_0 + y_1 x + y_2 x^2$ is non-negative everywhere if and only if

$$\forall x : \begin{bmatrix} x \\ 1 \end{bmatrix}^\top \begin{bmatrix} y_0 & y_1/2 \\ y_1/2 & y_2 \end{bmatrix} \begin{bmatrix} x \\ 1 \end{bmatrix} \ge 0,$$

which in turn can be written as an LMI in $y = (y_0, y_1, y_2)$:

$$\begin{bmatrix} y_0 & y_1/2 \\ y_1/2 & y_2 \end{bmatrix} \succeq 0.$$

In this exercise, you show a more general result, which applies to any polynomial of even degree $2k$ (polynomials of odd degree can't be non-negative everywhere). To simplify, we only examine the case $k = 2$, that is, fourth-degree polynomials; the method employed here can be generalized to $k > 2$.

1. Show that a fourth-degree polynomial p is non-negative everywhere if and only if it is a sum of squares, that is, it can be written as

$$p(x) = \sum_{i=1}^{4} q_i(x)^2,$$

where q_is are polynomials of degree at most two. *Hint:* show that p is non-negative everywhere if and only if it is of the form

$$p(x) = p_0 \left((x - a_1)^2 + b_1^2 \right) \left((x - a_2)^2 + b_2^2 \right),$$

for some appropriate real numbers $a_i, b_i, i = 1, 2$, and some $p_0 \geq 0$.

2. Using the previous part, show that if a fourth-degree polynomial is a sum of squares, then it can be written as

$$p(x) = \begin{bmatrix} 1 & x & x^2 \end{bmatrix} Q \begin{bmatrix} 1 \\ x \\ x^2 \end{bmatrix} \qquad (11.45)$$

for some positive semidefinite matrix Q.

3. Show the converse: if a positive semidefinite matrix Q satisfies condition (11.45) for every x, then p is a sum of squares. *Hint:* use a factorization of Q of the form $Q = AA^\top$, for some appropriate matrix A.

4. Show that a fourth-degree polynomial $p(x) = y_0 + y_1 x + y_2 x^2 + y_3 x^3 + y_4 x^4$ is non-negative everywhere if and only if there exists a 3×3 matrix Q such that

$$Q \succeq 0, \quad y_{l-1} = \sum_{i+j=l+1} Q_{ij}, \quad l = 1, \ldots, 5.$$

Hint: equate the coefficients of the powers of x in the left and right sides of equation (11.45).

Exercise 11.7 (Sum of top eigenvalues) For $X \in S^n$, and $i \in \{1, \ldots, n\}$, we denote by $\lambda_i(X)$ the i-th largest eigenvalue of X. For $k \in \{1, \ldots, n\}$, we define the function $f_k : S^n \to \mathbb{R}$ with values

$$f_k(X) = \sum_{i=1}^{k} \lambda_i(X).$$

This function is an intermediate between the largest eigenvalue (obtained with $k = 1$) and the trace (obtained with $k = n$).

1. Show that for every $t \in \mathbb{R}$, we have $f_k(X) \leq t$ if and only if there exist $Z \in S^n$ and $s \in \mathbb{R}$ such that

$$t - ks - \text{trace}(Z) \geq 0, \quad Z \succeq 0, \quad Z - X + sI \succeq 0.$$

Hint: for the sufficiency part, think about the interlacing property[19] of the eigenvalues.

[19] See Eq. (4.6)

2. Show that f_k is convex. Is it a norm?

3. How would you generalize these results to the function that assigns the sum of the top k singular values to a general rectangular $m \times n$ matrix, with $k \leq \min(m, n)$? *Hint:* for $X \in \mathbb{R}^{m,n}$, consider the symmetric matrix

$$\tilde{X} \doteq \begin{bmatrix} 0 & X \\ X^\top & 0 \end{bmatrix}.$$

12

Introduction to algorithms

IN THIS CHAPTER WE ILLUSTRATE some iterative techniques (algorithms) for *solving numerically*, up to a given accuracy, different classes of optimization problems. These methods share a common general structure: some initial information (such as an initial candidate point $x_0 \in \mathbb{R}^n$) is given at iteration $k = 0$, together with a desired numerical *accuracy* $\epsilon > 0$. At each iteration $k = 0, 1, \ldots$, some information about the problem is collected at the current point x_k, and this information is used to *update* the candidate point according to algorithm-specific rules, thus obtaining a new point x_{k+1}. Then, a *stopping criterion* is checked (usually by verifying if the current solution meets the desired accuracy level ϵ). If yes, then the current point is returned as a numerical solution (to accuracy ϵ) of the problem; otherwise we set $k \leftarrow k + 1$, and iterate the process.

Algorithms differ from one another with respect to the type of information that is collected at point x_k, and to the way this information is used to update the current solution. A typical update rule takes the form of a simple recursion

$$x_{k+1} = x_k + s_k v_k, \tag{12.1}$$

where the scalar $s_k > 0$ is called the *stepsize*, and $v_k \in \mathbb{R}^n$ is the update (or search) *direction*. The meaning of Eq. (12.1) is that from the current point x_k we move away along direction v_k, and the length of the move is dictated by the stepsize s_k.

Some algorithms, such as the descent methods described in Section 12.2.1, can be applied to general (i.e., possibly non-convex) problems. However, no kind of "guarantee" of convergence can usually be given under such generality. On the contrary, if the problem is convex, then (possibly under some further technical assumptions) the algorithms presented in this chapter are typically guaranteed to converge to a global optimal solution (if such a solution exists). Also, under convexity assumptions, we can estimate the *rate* at which convergence happens, and in some cases predict *a priori* the number of iterations required to reach the desired numerical accuracy.

A basic classification of algorithms distinguishes between *first-order* and *second-order* methods. This terminology derives from classical unconstrained optimization of differentiable functions, and refers to whether only first derivatives (gradient) of the objective function are used to determine the search direction at each step (first-order methods), or also the second derivatives (Hessian) are used for the same purpose (second-order methods). In this chapter, we outline some standard first- and second-order methods for unconstrained minimization, and then discuss various extensions that enable us to account for the presence of constraints, for non-differentiability of the objective and the constraint functions, and for decentralized structures of the optimization process itself.

This chapter is organized as follows: Section 12.1 presents some technical preliminaries that are later needed in the analysis of the optimization algorithms. Section 12.2 discusses techniques for unconstrained minimization of differentiable non-convex or convex objective functions, presenting the gradient method, general descent methods, and the Newton and quasi-Newton algorithms for convex minimization, together with versions adapted to deal with the presence of linear equality constraints on the variables. Section 12.3 discusses techniques for dealing with differentiable convex optimization problems with inequality constraints, and presents in particular a second-order method (the barrier method, Section 12.3.1) and a first-order method based on the concept of proximal gradients (Section 12.3.2). This section also discusses some specialized techniques for the LASSO and related problems. Section 12.4 presents methods for constrained optimization of convex but possibly non-differentiable functions. It describes in particular the projected subgradient method (Section 12.4.1), the alternate subgradient method (Section 12.4.2), and the ellipsoid algorithm (Section 12.4.3). In Section 12.5 we discuss coordinate descent methods. Finally, in Section 12.6 we briefly outline decentralized optimization techniques, such as the primal and the dual decomposition methods.

Remark 12.1 Some parts of this chapter are rather technical. These technicalities are needed for the analysis of the various types of convergence of the algorithms. The reader who is mainly interested in the key ideas and in the general description of the algorithms may, however, safely skip all convergence proofs, as well as most of Section 12.1.

12.1 Technical preliminaries

Most of the proofs of convergence of the optimization algorithms described in the next sections hinge upon one or both of the following "regularity" properties of the functions involved in the problem description, namely Lipschitz continuity of the gradients and strong convexity. These properties and their consequences are summarized in the following subsections.

Assumption 1 (Working hypotheses) *In the rest of this chapter, we shall make the standing assumption that $f_0 : \mathbb{R}^n \to \mathbb{R}$ is a closed function, that is all the sublevel sets $S_\alpha = \{x : f_0(x) \leq \alpha\}$, $\alpha \in \mathbb{R}$, are closed sets. Further, we assume that f_0 is bounded below, and that it attains its (global) minimum value f_0^* at some point $x^* \in \mathrm{dom}\, f_0$. Given a point $x_0 \in \mathrm{dom}\, f_0$, we define S_0 as the sublevel set*

$$S_0 \doteq \{x : f_0(x) \leq f_0(x_0)\}.$$

12.1.1 Gradient Lipschitz continuity

A function $f_0 : \mathbb{R}^n \to \mathbb{R}$ is said to be Lipschitz continuous on a domain $S \subseteq \mathbb{R}^n$, if there exists a constant $R > 0$ (possibly depending on S) such that

$$|f_0(x) - f_0(y)| \leq R\|x - y\|_2, \quad \forall x, y \in S.$$

A differentiable function $f_0 : \mathbb{R}^n \to \mathbb{R}$ is said to have Lipschitz continuous gradient on S if there exists a constant $L > 0$ (possibly depending on S) such that

$$\|\nabla f_0(x) - \nabla f_0(y)\|_2 \leq L\|x - y\|_2, \quad \forall x, y \in S. \tag{12.2}$$

Intuitively, f_0 has Lipschitz continuous gradient if its gradient "does not vary too fast." Indeed, if f_0 is twice differentiable, the above condition is equivalent to a bound on the Hessian of f_0. The following proposition summarizes some useful implications of gradient Lipschitz continuity.[1]

Lemma 12.1

1. *If $f_0 : \mathbb{R}^n \to \mathbb{R}$ is twice continuously differentiable, then (12.2) holds if and only if f_0 has bounded Hessian on S, that is*

$$\|\nabla^2 f_0(x)\|_F \leq L, \quad \forall x \in S.$$

2. *If f_0 is continuously differentiable, then (12.2) implies that*

$$|f_0(x) - f_0(y) - \nabla f_0(y)^\top (x - y)| \leq \frac{L}{2}\|x - y\|_2^2, \quad \forall x, y \in S. \tag{12.3}$$

[1] Proofs of the following facts can be found, for instance, Yu. Nesterov, *Introductory Lectures on Convex Optimization: a Basic Course*, Springer, 2004.

3. *If f_0 is continuously differentiable and* convex, *then (12.2) implies that*

$$0 \leq f_0(x) - f_0(y) - \nabla f_0(y)^\top (x - y) \leq \frac{L}{2} \|x - y\|_2^2, \quad \forall x, y \in S,$$
(12.4)

and that the following inequality holds $\forall x, y \in S$:

$$\frac{1}{L} \|\nabla f_0(x) - \nabla f_0(y)\|_2^2 \leq (\nabla f_0(x) - \nabla f_0(y))^\top (x - y) \leq L \|x - y\|_2^2.$$

12.1.1.1 Quadratic upper bound. Inequality (12.3) implies that, for any given $y \in S$, $f_0(x)$ is upper bounded by a (strongly) convex quadratic function:

$$f_0(x) \leq f_0(y) + \nabla f_0(y)^\top (x - y) + \frac{L}{2} \|x - y\|_2^2, \quad \forall x, y \in S, \quad (12.5)$$

where the quadratic upper bound function is defined as

$$f_{\mathrm{up}}(x) \doteq f_0(y) + \nabla f_0(y)^\top (x - y) + \frac{L}{2} \|x - y\|_2^2. \quad (12.6)$$

12.1.1.2 Implications on the unconstrained minimum. Let $x^* \in \mathrm{dom}\, f_0$ be a global unconstrained minimizer of f_0. Then, clearly, it must hold that $x^* \in S_0$ (recall S_0 is the sublevel set defined in Assumption 1), hence we may write that

$$x^* \in \arg\min_{x \in S_0} f_0(x),$$

and, if f_0 is differentiable, the unconstrained optimality condition requires that $\nabla f_0(x^*) = 0$. If, further, f_0 has Lipschitz continuous gradient on S_0, then it holds that

$$\frac{1}{2L} \|\nabla f_0(x)\|_2^2 \leq f_0(x) - f_0^* \leq \frac{L}{2} \|x - x^*\|_2^2, \quad \forall x \in S_0. \quad (12.7)$$

The bound on the right in (12.7) is readily obtained by evaluating (12.5) at $y = x^*$, and recalling that $\nabla f_0(x^*) = 0$. The bound on the left in (12.7) is instead obtained by first evaluating (12.5) at $x = y - \frac{1}{L} \nabla f_0(y)$, which yields

$$f_0(x) \leq f_0(y) - \frac{1}{2L} \|\nabla f_0(y)\|_2^2, \quad \forall y \in S_0.$$

Then, since $f_0^* \leq f_0(x)$, $\forall x \in \mathrm{dom}\, f_0$, this last inequality also implies $f_0^* \leq f_0(y) - \frac{1}{2L} \|\nabla f_0(y)\|_2^2$, $\forall y \in S_0$, which is the desired bound.

12.1.1.3 Lipschitz constant for functions with compact sublevel sets. If f_0 is twice continuously differentiable, and the sublevel set $S_0 = \{x :$

$f_0(x) \leq f_0(x_0)\}$ is compact, then f_0 has Lipschitz continuous gradient on S_0. This is due to the fact that the Hessian is continuous, hence $\|\nabla^2 f_0(x)\|_F$ is continuous and, from the Weierstrass theorem, it attains a maximum over the compact set S_0. Therefore, applying point 1 of Lemma 12.1, we have that f_0 has Lipschitz continuous gradient on S_0, and a suitable Lipschitz constant is

$$L = \max_{x \in S_0} \|\nabla^2 f_0(x)\|_F.$$

Compactness of the sublevel set S_0 is guaranteed, for instance, if f_0 is coercive (see Lemma 8.3), or when f_0 is strongly convex (see the next section).

12.1.2 Strong convexity and its implications

We recall from the definition in Section 8.2.1 that a function $f_0 : \mathbb{R}^n \to \mathbb{R}$ is said to be strongly convex if there exists $m > 0$ such that

$$f_0(x) - \frac{m}{2}\|x\|_2^2$$

is convex. From this definition it also follows that if f_0 is strongly convex and twice differentiable, then

$$\nabla f_0(x) \succeq mI, \quad \forall x \in \text{dom } f_0.$$

Strong convexity has several other interesting implications, as discussed next.

12.1.2.1 Quadratic lower bound.
We know from the characterization in (8.4) that a differentiable function f is convex if and only if

$$\forall x, y \in \text{dom } f, \ f(y) \geq f(x) + \nabla f(x)^\top (y - x), \qquad (12.8)$$

which means that the linear function $f(x) + \nabla f(x)^\top (y - x)$ is a global lower bound on $f(y)$. Then, applying (12.8) to $f(x) = f_0(x) - \frac{m}{2}\|x\|_2^2$, we have that a differentiable f_0 is strongly convex if and only if

$$\forall x, y \in \text{dom } f_0, \ f_0(y) \geq f_0(x) + \nabla f_0(x)^\top (y - x) + \frac{m}{2}\|y - x\|_2^2,$$
$$(12.9)$$

which means geometrically that at any $x \in \text{dom } f_0$, there is a convex quadratic function

$$f_{\text{low}}(y) \doteq f_0(x) + \nabla f_0(x)^\top (y - x) + \frac{m}{2}\|y - x\|_2^2$$

that bounds from below the graph of f_0, that is such that $f_0(y) \geq f_{\text{low}}(y)$, for all $y \in \text{dom } f_0$, see Figure 12.1.

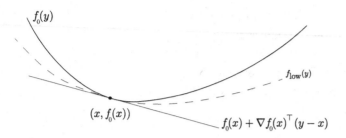

Figure 12.1 A differentiable strongly convex function f_0 admits a global quadratic lower bound function f_{low}, at any $x \in \text{dom } f_0$.

This quadratic lower bound property also holds for non-differentiable functions, using subgradients instead of gradients. Indeed, if f is convex, but possibly non-differentiable, then it holds that, for all $x \in \text{relint dom } f$,

$$f(y) \geq f(x) + h_x^\top (y - x), \quad \forall y \in \text{dom } f, \forall h_x \in \partial f(x),$$

where h_x is a subgradient of f at x. Thus, if f_0 is strongly convex, but possibly non-differentiable, applying the previous inequality to the convex function $f_0(x) - (m/2)\|x\|_2^2$, it holds that, $\forall x \in \text{relint dom } f_0$ and for $g_x \in \partial f_0(x)$,

$$f_0(y) - \frac{m}{2}\|y\|_2^2 \geq f_0(x) - \frac{m}{2}\|x\|_2^2 + (g_x - mx)^\top (y - x),$$

thus

$$
\begin{aligned}
f_0(y) &\geq f_0(x) + g_x^\top (y - x) - mx^\top (y - x) + \frac{m}{2}(\|y\|_2^2 - \|x\|_2^2) \\
&= f_0(x) + g_x^\top (y - x) + \frac{m}{2}\|y - x\|_2^2
\end{aligned}
$$

holds for all $y \in \text{dom } f$ and all $g_x \in \partial f_0(x)$. Thus, also in the non-differentiable case, a strongly convex function f_0 admits a quadratic lower bound, at any $x \in \text{relint dom } f_0$.

12.1.2.2 Quadratic upper bound.

We next show that if f_0 is strongly convex and twice continuously differentiable, then f_0 has Lipschitz continuous gradient over S_0, hence it can be upper bounded by a quadratic function.

We start by observing that, for any initial point $x_0 \in \text{dom } f_0$, if f_0 is strongly convex then the level set $S_0 = \{y : f_0(y) \leq f_0(x_0)\}$ is contained in a regular ellipsoid and hence it is bounded. To see this fact, it suffices to consider the strong convexity inequality (12.9), from which we obtain that

$$y \in S_0 \Rightarrow 0 \geq f_0(y) - f_0(x_0) \geq \nabla f_0(x_0)^\top (y - x_0) + \frac{m}{2}\|y - x_0\|_2^2,$$

where the region of y satisfying the inequality $\nabla f_0(x_0)^\top (y - x_0) + \frac{m}{2}\|y - x_0\|_2^2 \leq 0$ is a bounded ellipsoid. Since the Hessian of f_0 is assumed to be continuous, it remains bounded over bounded regions, which implies that there exists a finite constant $M > 0$ (possibly depending on x_0) such that

$$\nabla^2 f_0(x) \preceq MI, \quad \forall x \in S_0.$$

By Lemma 12.1, this implies in turn that f_0 has Lipschitz continuous gradient over S_0 (with Lipschitz constant M), hence it admits a quadratic upper bound, i.e.,

$$f_0(x) \leq f_0(x) + \nabla f_0(x)^\top (y - x) + \frac{M}{2}\|y - x\|_2^2, \quad \forall x, y \in S_0.$$

12.1.2.3 Bounds on the optimality gap. Summarizing the findings of the previous two sections, we have that for a strongly convex and twice differentiable function f_0 it holds that

$$mI \preceq \nabla^2 f_0(x) \preceq MI, \quad \forall x \in S_0,$$

and that f_0 is upper and lower bounded by two convex quadratic functions, as follows:

$$f_{\text{low}}(y) \leq f_0(y), \quad \forall x, y \in \operatorname{dom} f_0,$$
$$f_0(y) \leq f_{\text{up}}(y), \quad \forall x, y \in S_0,$$

where

$$f_{\text{low}}(y) \;=\; f_0(x) + \nabla f_0(x)^\top (y - x) + \frac{m}{2}\|y - x\|_2^2, \quad (12.10)$$

$$f_{\text{up}}(y) \;=\; f_0(x) + \nabla f_0(x)^\top (y - x) + \frac{M}{2}\|y - x\|_2^2. \quad (12.11)$$

From these two inequalities we can derive key bounds on the gap between the value of f_0 at any point $x \in \operatorname{dom} f_0$ and the global unconstrained minimum value f_0^*.

Let the minimum f_0^* over $\operatorname{dom} f_0$ be attained at some $x^* \in \operatorname{dom} f_0$ (such a minimizer is *unique*, since f_0 is strongly convex). As we already discussed, it must be that $x^* \in S_0$. Then, writing the inequality $f_{\text{low}}(y) \leq f_0(y)$ for $x = x^*$ we obtain (since $\nabla f_0(x^*) = 0$)

$$f_0(y) \geq f_0^* + \frac{m}{2}\|y - x^*\|_2^2, \quad \forall y \in S_0. \quad (12.12)$$

Further, the inequality $f_0(y) \geq f_{\text{low}}(y)$ $\forall y \in S_0$, implies that it also holds that $f_0(y) \geq \min_z f_{\text{low}}(z)$. Setting the gradient of $f_{\text{low}}(z)$ to zero yields the minimizer $z^* = x - \frac{1}{m}\nabla f_0(x)$, hence

$$f_0(y) \;\geq\; \min_z f_{\text{low}}(z) = f_{\text{low}}(z^*)$$

$$=\; f_0(x) - \frac{1}{2m}\|\nabla f_0(x)\|_2^2, \quad \forall x, y \in S_0.$$

Since this last inequality holds for all y, it also holds for $y = x^*$, thus

$$f_0^* \geq f_0(x) - \frac{1}{2m}\|\nabla f_0(x)\|_2^2, \quad \forall x \in S_0. \qquad (12.13)$$

Putting together (12.12) and (12.13), we obtain

$$\frac{m}{2}\|x - x^*\|_2^2 \leq f_0(x) - f_0^* \leq \frac{1}{2m}\|\nabla f_0(x)\|_2^2, \quad \forall x \in S_0, \qquad (12.14)$$

which shows that the gap $f_0(x) - f_0^*$ (as well as the distance from x to the minimizer x^*) is upper bounded by the norm of the gradient of f_0 at x. Also, following a reasoning analogous to that in Section 12.1.1.2, we obtain the "swapped" inequality

$$\frac{1}{2M}\|\nabla f_0(x)\|_2^2 \leq f_0(x) - f_0^* \leq \frac{M}{2}\|x - x^*\|_2^2, \quad \forall x \in S_0. \qquad (12.15)$$

12.2   Algorithms for smooth unconstrained minimization

The focus of this section is on iterative algorithms for solving numerically the unconstrained optimization problem

$$f_0^* = \min_{x \in \mathbb{R}^n} f_0(x),$$

where $f_0 : \mathbb{R}^n \to \mathbb{R}$ is the objective function. We shall here assume that Assumption 1 holds, that an initial point $x_0 \in \operatorname{dom} f_0$ is given, and that f_0 is continuosly differentiable.[2] Notice that no convexity assumption is yet made, at this point.

[2] We shall informally define as *smooth* optimization problems the ones with objective and constraint functions that are once or twice differentiable, and as *non-smooth* the other cases.

12.2.1   First-order descent methods

We start by discussing a simple class of first-order methods where the iterates are of the form (12.1), and the search direction v_k is computed on the basis of the gradient of f_0 at x_k.

12.2.1.1   Descent directions. Consider a point $x_k \in \operatorname{dom} f_0$ and a direction $v_k \in \mathbb{R}^n$. Using the first-order Taylor series expansion[3] for f_0, we have that

[3] See Section 2.4.5.2.

$$f_0(x_k + sv_k) \simeq f_0(x_k) + s\nabla f_0(x_k)^\top v_k, \quad \text{for } s \to 0.$$

The local rate of variation of f_0, in the neighborhood of x_k and along the direction v_k, is thus given by

$$\delta_k \doteq \lim_{s \to 0} \frac{f_0(x_k + sv_k) - f_0(x_k)}{s} = \nabla f_0(x_k)^\top v_k.$$

The local directional rate of variation δ_k is thus positive whenever $\nabla f_0(x_k)^\top v_k > 0$, that is for directions v_k that form a positive inner product with the gradient $\nabla f_0(x_k)$. On the contrary, directions v_k for which $\nabla f_0(x_k)^\top v_k < 0$ are called decrease (or *descent*) directions. The reason for this is that if the new point x_{k+1} is chosen according to (12.1) as $x_{k+1} = x_k + s v_k$, then

$$f_0(x_{k+1}) < f_0(x_k), \quad \text{for sufficiently small } s > 0. \qquad (12.16)$$

It is then natural to ask what is the direction of *maximum* local decrease: from the Cauchy–Schwartz inequality we have that

$$-\|\nabla f_0(x_k)\|_2 \|v_k\|_2 \le \nabla f_0(x_k)^\top v_k \le \|\nabla f_0(x_k)\|_2 \|v_k\|_2,$$

hence δ_k is minimal over all v_k with $\|v_k\|_2 = 1$, for

$$v_k = -\frac{\nabla f_0(x_k)}{\|\nabla f_0(x_k)\|_2}, \qquad (12.17)$$

i.e., when v_k points in the direction of the negative gradient. The direction v_k in (12.17) is thus called the *steepest descent* direction, with respect to the standard Euclidean norm.

12.2.1.2   A descent scheme. From the discussion in Section 12.2.1.1 follows quite naturally the idea of updating recursively the search points according to (12.1), choosing at each iteration the search direction as a descent direction, such as, for instance, the anti-gradient $v_k = -\nabla f_0(x_k)$. A generic descent method is outlined in Algorithm 9.

The behavior of this algorithm depends on the actual choice of the descent directions, and on the strategy used for determining the stepsizes s_k. It should be clear that the fact that v_k is a direction of descent does not imply that the function value at x_{k+1} will decrease *for any $s_k > 0$*, see Figure 12.2.

Indeed, the decrease guaranteed by (12.16) is only local (i.e., for infinitesimal s_k), hence a key problem is to find a finite stepsize that guarantees a sufficient decrease. A discussion on typical techniques for stepsize selection is given in the next section.

12.2.1.3   Stepsize selection. Consider the restriction of f_0 along the direction v_k:

$$\phi(s) \doteq f_0(x_k + s v_k), \quad s \ge 0.$$

Clearly, ϕ is a function of the scalar variable s, and $\phi(0) = f_0(x_k)$. Choosing a suitable stepsize amounts to finding $s > 0$ such that $\phi(s) < \phi(0)$. A natural approach would then be to compute s that *minimizes ϕ*, that is

$$s^* = \arg\min_{s \ge 0} \phi(s).$$

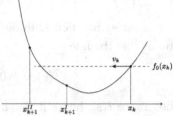

Figure 12.2 Even if f_0 is convex, the stepsize should be chosen with care, in order to insure a sufficient function decrease. Here, a local direction of decrease at x_k is given by a move to the "left." If $s_k > 0$ is small enough we may end up at point x_{k+1}^I, for which $f_0(x_{k+1}^I) < f_0(x_k)$. However, if s_k is too large, we may end up at point x_{k+1}^{II}, where the decrease condition is not satisfied.

Algorithm 9 Descent algorithm.

Require: $f_0 : \mathbb{R}^n \to \mathbb{R}$ differentiable, $x_0 \in \text{dom } f_0$, $\epsilon > 0$
1: Set $k = 0$
2: Determine a descent direction v_k
3: Determine step length $s_k > 0$
4: Update: $x_{k+1} = x_k + s_k v_k$
5: If accuracy ϵ is attained, then exit and return x_k, else let $k \leftarrow k+1$ and go to 2.

This method is called *exact line search*, and provides a stepsize s^* with the best possible decrease. However, finding s^* requires solving a univariate (and generically non-convex) optimization problem, which may be computationally demanding. For this reason, exact line search is rarely used in practical algorithms. A more practical alternative consists of searching for an s value guaranteeing a sufficient *rate* of decrease in ϕ. Consider the tangent line to ϕ at 0:

$$\phi(s) \simeq \ell(s) \doteq \phi(0) + s\delta_k, \quad \delta_k \doteq \nabla f_0(x_k)^\top v_k < 0, \ s \geq 0.$$

ℓ is a linear function with negative slope δ_k. Now, for $\alpha \in (0,1)$, it holds that the line

$$\bar{\ell}(s) \doteq \phi(0) + s(\alpha\delta_k), \quad s \geq 0$$

lies above $\ell(s)$, hence it also lies above $\phi(s)$, at least for small $s > 0$, see Figure 12.3.

Figure 12.3 Illustration of the Armijo condition and backtracking line search. Bold segments denote the regions where the condition is met.

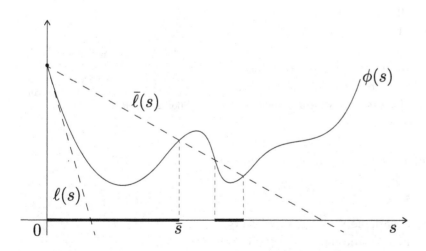

Since ϕ is bounded below while $\bar{\ell}$ is unbounded below, there must exist a point where $\phi(s)$ and $\bar{\ell}(s)$ cross; let $\bar{s}$ be the smallest of such points. All values of s for which $\phi(s) \leq \bar{\ell}(s)$ provide a sufficient rate of decrease, given by the slope $\alpha\delta_k$ of $\bar{\ell}$. This rate condition is known as the *Armijo condition*, stating that the valid stepsizes must satisfy

$$\phi(s) \leq \phi(0) + s(\alpha\delta_k),$$

or, more explicitly,

$$f_0(x_k + sv_k) \leq f_0(x_k) + s\alpha(\nabla f_0(x_k)^\top v_k),$$

for the chosen $\alpha \in (0, 1)$.

The Armijo condition is clearly satisfied by all $s \in (0, \bar{s})$, hence this condition alone is still not sufficient to insure that the stepsize is not chosen too small (which is necessary for convergence of the method). We shall see in Section 12.2.2.1 that, under the Lipschitz continuous gradient assumption, there exists a *constant* stepsize that satisfies the Armijo condition. However, such a constant stepsize is generally unknown in advance (or it may be too small for practical efficiency of the method), hence a usual practice amounts to employing a so-called *backtracking* approach, whereby an initial value of s is fixed to some constant value s_{init} (typically, $s_{\text{init}} = 1$), and then the value of s is iteratively decreased at a fixed rate $\beta \in (0, 1)$, until the Armijo condition is met, see Algorithm 10.

Algorithm 10 Backtracking line search.

Require: f_0 differentiable, $\alpha \in (0, 1)$, $\beta \in (0, 1)$, $x_k \in \text{dom} f_0$, v_k a descent direction, s_{init} a positive constant (typically, $s_{\text{init}} = 1$)
1: Set $s = s_{\text{init}}$, $\delta_k = \nabla f_0(x_k)^\top v_k$
2: If $f_0(x_k + sv_k) \leq f(x_k) + s\alpha\delta_k$, then return $s_k = s$
3: Else let $s \leftarrow \beta s$ and go to 2

12.2.2   *The gradient method*

In this section, we analyze more closely the convergence properties of Algorithm 9, for the most common case where the descent direction is simply the negative gradient (that is, the direction of steepest local descent). We take henceforth

$$v_k = -\nabla f_0(x_k).$$

Very little can actually be said about the properties of the gradient descent algorithm, unless we make some additional assumptions on the regularity of the objective function. More precisely, we shall assume that f_0 has Lipschitz continuous gradient on S_0, that is, there exists a positive constant L such that

$$\|\nabla f_0(x) - \nabla f_0(y)\|_2 \leq L \|x - y\|_2, \quad \forall x, y \in S_0.$$

12.2.2.1 Lower bound on the stepsize. Let now x_k be the current point in a gradient descent algorithm, and let

$$x = x_k - s\nabla f_0(x_k).$$

Evaluating f_0 and f_{up} in (12.6) at x we obtain the restrictions of these functions along the direction $v_k = -\nabla f_0(x_k)$:

$$\phi(s) = f_0(x_k - s\nabla f_0(x_k)),$$
$$\phi_{up}(s) = f_0(x_k) - s\|\nabla f_0(x_k)\|_2^2 + s^2 \frac{L}{2}\|\nabla f_0(x_k)\|_2^2,$$

where (12.5) clearly implies that

$$\phi(s) \leq \phi_{up}(s), \quad s \geq 0.$$

It can be readily observed that $\phi(s)$ and $\phi_{up}(s)$ have the same tangent line at $s = 0$, which is given by

$$\ell(s) \doteq f_0(x_k) - s\|\nabla f_0(x_k)\|_2^2.$$

Consider then the line previously defined for the Armijo rule:

$$\bar{\ell}(s) \doteq f_0(x_k) - s\alpha\|\nabla f_0(x_k)\|_2^2, \quad \alpha \in (0, 1).$$

This line intercepts the upper bound function $\phi_{up}(s)$ at the point

$$\bar{s}_{up} = \frac{2}{L}(1 - \alpha). \tag{12.18}$$

It is then clear that a *constant* stepsize $s_k = \bar{s}_{up}$ would satisfy the Armijo condition at each iteration of the algorithm, see Figure 12.4. Note, however, that one would need to know the numerical value of the Lipschitz constant L (or an upper bound on it), in order to implement such a stepsize in practice. This can be avoided by using backtracking. Indeed, suppose we initialize the backtracking procedure with $s = s_{init}$. Then, either this initial value satisfies the Armijo condition, or it is iteratively reduced until it does. The iterative reduction certainly stops at a value $s \geq \beta\bar{s}_{up}$, hence backtracking guarantees that

$$s_k \geq \min(s_{init}, \beta\bar{s}_{up}) \doteq s_{lb}. \tag{12.19}$$

To summarize, we have that, for both constant stepsizes and stepsizes computed according to a backtracking line search, there exists a constant $s_{lb} > 0$ such that

$$s_k \geq s_{lb}, \quad \forall k = 0, 1, \ldots \tag{12.20}$$

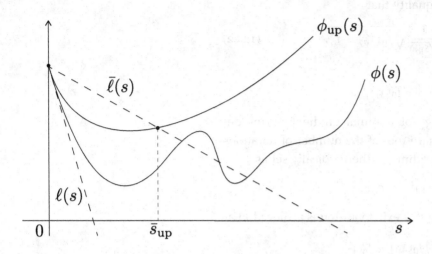

Figure 12.4 For a function with Lipschitz continuous gradient there is a constant stepsize $s = \bar{s}_{\text{up}}$ that satisfies the Armijio condition.

12.2.2.2 *Convergence to a stationary point.* Consider a gradient descent algorithm

$$x_{k+1} = x_k - s_k \nabla f_0(x_k),$$

with stepsizes s_k computed via a backtracking line search (or constant stepsizes), satisfying the Armijo condition

$$f_0(x_{k+1}) \leq f_0(x_k) - s_k \alpha \|\nabla f_0(x_k)\|_2^2. \tag{12.21}$$

Then

$$
\begin{aligned}
f_0(x_k) - f_0(x_{k+1}) &\geq s_k \alpha \|\nabla f_0(x_k)\|_2^2 \\
\text{[using (12.20)]} &\geq s_{\text{lb}} \alpha \|\nabla f_0(x_k)\|_2^2, \quad \forall k = 0, 1, \ldots
\end{aligned}
$$

Summing these inequalities from $0, 1, \ldots, k$, we obtain

$$s_{\text{lb}} \alpha \sum_{i=0}^{k} \|\nabla f_0(x_i)\|_2^2 \leq f_0(x_0) - f_0(x_{k+1}) \leq f_0(x_0) - f_0^*.$$

Since the summation on the left is bounded by a constant as $k \to \infty$, we conclude that it must be that

$$\lim_{k \to \infty} \|\nabla f_0(x_k)\|_2 = 0.$$

This means that the algorithm converges to a *stationary point* of f_0, that is to a point where the gradient of f_0 is zero. Notice that such a point is not necessarily a local minimum of f_0 (for instance, it could be an inflection point, see Figure 12.5).

Further, by noticing that

$$\sum_{i=0}^{k} \|\nabla f_0(x_i)\|_2^2 \geq (k+1) \min_{i=0,\ldots,k} \|\nabla f_0(x_i)\|_2^2,$$

Figure 12.5 Stationary points are points where the gradient is zero. These include extrema (max, min), as well as inflection, or saddle, points.

we obtain from the previous inequality that

$$g_k^* \leq \frac{1}{\sqrt{k+1}} \frac{1}{\sqrt{s_{\text{lb}}\alpha}} \sqrt{f_0(x_0) - f_0^*}, \tag{12.22}$$

where we defined

$$g_k^* \doteq \min_{i=0,\dots,k} \|\nabla f_0(x_i)\|_2.$$

This means that the sequence g_k^* of minimal gradient norms decreases at a *rate* given by the square-root of the number of iterations k. The stopping criterion in Algorithm 9 is then typically set as

$$\|\nabla f_0(x_k)\|_2 \leq \epsilon,$$

and, using (12.22), we obtain that this exit condition is achieved in at most

$$k_{\max} = \left\lceil \frac{1}{\epsilon^2} \frac{f_0(x_0) - f_0^*}{s_{\text{lb}}\alpha} \right\rceil$$

iterations, where the notation $\lceil \cdot \rceil$ denotes the smallest integer no smaller than the argument (i.e., the *ceiling* operator).

12.2.2.3 Analysis of the gradient method for convex functions. In the previous section we analyzed the convergence behavior of the gradient descent algorithm for generic, possibly non-convex, functions. We verified that, even under a Lipschitz continuity assumption on the gradient, only convergence to a stationary point can be guaranteed globally. In this section, we show that if the function f_0 is convex, then convergence to a global minimum can be guaranteed, and an explicit bound can be derived on the rate at which $f_0(x_k)$ converges to f_0^*. In the rest of this section we shall thus assume additionally that f_0 is *convex*.

First, we observe that, for convex f_0, x^* is a (global) minimizer if and only if $\nabla f_0(x^*) = 0$, see Section 8.4.1. Therefore, the gradient algorithm converges to a global minimum point. We next analyze at which *rate* this convergence is reached.

Consider the decrease in objective function obtained at one step of the gradient algorithm (we consider, for simplicity in the proofs, the backtracking parameter to be fixed at $\alpha = 1/2$): from (12.21) we have

$$f_0(x_{k+1}) \leq f_0(x_k) - s_k \alpha \|\nabla f_0(x_k)\|_2^2. \tag{12.23}$$

Since f_0 is convex, it holds that

$$f_0(y) \geq f_0(x) + \nabla f_0(x)^\top (y - x), \quad \forall x, y \in \text{dom } f_0,$$

which, for $y = x^*$, gives

$$f_0(x) \leq f_0^* + \nabla f_0(x)^\top (x - x^*), \quad \forall x \in \text{dom } f_0.$$

Substituting this into (12.23), we obtain

$$
\begin{aligned}
f_0(x_{k+1}) &\leq f_0(x_k) - s_k \alpha \|\nabla f_0(x_k)\|_2^2 \\
[\text{letting } \alpha = 1/2] &\leq f_0^* + \nabla f_0(x_k)^\top (x_k - x^*) - \frac{s_k}{2} \|\nabla f_0(x_k)\|_2^2 \\
&= f_0^* + \frac{1}{2s_k} \left(\|x_k - x^*\|_2^2 - \|x_k - x^* - s_k \nabla f_0(x_k)\|_2^2 \right) \\
&= f_0^* + \frac{1}{2s_k} \left(\|x_k - x^*\|_2^2 - \|x_{k+1} - x^*\|_2^2 \right) \\
[\text{since } s_k \geq s_{\mathrm{lb}}] &\leq f_0^* + \frac{1}{2s_{\mathrm{lb}}} \left(\|x_k - x^*\|_2^2 - \|x_{k+1} - x^*\|_2^2 \right).
\end{aligned}
$$

Considering this inequality for $k = 0, 1, \ldots$, we have

$$
\begin{aligned}
f_0(x_1) - f_0^* &\leq \frac{1}{2s_{\mathrm{lb}}} \left(\|x_0 - x^*\|_2^2 - \|x_1 - x^*\|_2^2 \right), \\
f_0(x_2) - f_0^* &\leq \frac{1}{2s_{\mathrm{lb}}} \left(\|x_1 - x^*\|_2^2 - \|x_2 - x^*\|_2^2 \right), \\
f_0(x_3) - f_0^* &\leq \frac{1}{2s_{\mathrm{lb}}} \left(\|x_2 - x^*\|_2^2 - \|x_3 - x^*\|_2^2 \right), \\
&\vdots \qquad \vdots
\end{aligned}
$$

Hence, summing the first k of these inequalities, we have that

$$
\begin{aligned}
\sum_{i=1}^{k} (f_0(x_i) - f_0^*) &\leq \frac{1}{2s_{\mathrm{lb}}} \left(\|x_0 - x^*\|_2^2 - \|x_k - x^*\|_2^2 \right) \\
&\leq \frac{1}{2s_{\mathrm{lb}}} \|x_0 - x^*\|_2^2.
\end{aligned}
$$

Now, since the sequence $f_0(x_k) - f_0^*$ is non-increasing with respect to k, its value is no larger than the average of the previous values of the sequence, that is

$$
f_0(x_k) - f_0^* \leq \frac{1}{k} \sum_{i=1}^{k} (f_0(x_i) - f_0^*) \leq \frac{1}{2s_{\mathrm{lb}}k} \|x_0 - x^*\|_2^2, \qquad (12.24)
$$

which proves that $f_0(x_k) \to f_0^*$ at least at a rate which is inversely proportional to k. We then achieve an accuracy ϵ' on the objective function, i.e.,

$$
f_0(x_k) - f_0^* \leq \epsilon'
$$

in at most

$$
k_{\max} = \left\lceil \frac{\|x_0 - x^*\|_2^2}{2\epsilon' s_{\mathrm{lb}}} \right\rceil
$$

iterations. We shall see next that this bound on the rate of convergence can be improved, at least for the class of strongly convex functions.

12.2.2.4 Analysis of the gradient method for strongly convex functions.
We obtain improved convergence results for the gradient method,
under an assumption of *strong convexity* on the objective function f_0,
i.e., there exists $m > 0$ such that

$$\tilde{f}_0(x) = f_0(x) - \frac{m}{2}\|x\|_2^2$$

is convex. We hence next assume, in addition to the hypotheses in
Assumption 1, that f_0 is twice continuously differentiable and it is
strongly convex (strong convexity also implies that f_0 has Lipschitz
continuous gradient on S_0, for some Lipschitz constant $M \geq m$).

Global convergence rate. We next derive a result on the convergence
rate of the gradient algorithm on a strongly convex objective, which
improves upon the generic estimate given in (12.24). To this end,
consider again the objective decrease in one iterate guaranteed by
(12.23), where for simplicity we set $\alpha = 1/2$:

$$
\begin{aligned}
f_0(x_{k+1}) &\leq & f_0(x_k) - s_k\alpha\|\nabla f_0(x_k)\|_2^2 \\
[\text{for } \alpha = 1/2] &= & f_0(x_k) - \frac{s_k}{2}\|\nabla f_0(x_k)\|_2^2 \\
[\text{since } s_k \geq s_{\text{lb}}] &\leq & f_0(x_k) - \frac{s_{\text{lb}}}{2}\|\nabla f_0(x_k)\|_2^2.
\end{aligned}
$$

Subtracting f_0^* on both sides of this inequality, and using the bound
previously derived in (12.13), we have that

$$
\begin{aligned}
f_0(x_{k+1}) - f_0^* &\leq & (f_0(x_k) - f_0^*) - \frac{s_{\text{lb}}}{2}\|\nabla f_0(x_k)\|_2^2 \\
&\leq & (f_0(x_k) - f_0^*) - 2m\frac{s_{\text{lb}}}{2}(f_0(x_k) - f_0^*) \\
&= & (1 - ms_{\text{lb}})(f_0(x_k) - f_0^*).
\end{aligned}
$$

Now, we recall from (12.18), (12.19) that, for $\alpha = 1/2$,

$$ms_{\text{lb}} = m\min(s_{\text{init}}, \beta\bar{s}_{\text{up}}) = \min(ms_{\text{init}}, \beta(m/M)).$$

Since $\beta < 1$ and $m/M \leq 1$, we have that $ms_{\text{lb}} < 1$, hence

$$(f_0(x_{k+1}) - f_0^*) \leq c(f_0(x_k) - f_0^*),$$

where $c = 1 - ms_{\text{lb}} \in (0, 1)$. Applying the last inequality recursively
from 0 to k, we obtain

$$f_0(x_k) - f_0^* \leq c^k(f_0(x_0) - f_0^*), \tag{12.25}$$

which proves that convergence happens at a geometric rate. This
type of convergence is usually named *linear*, and a reason for this is

that the logarithm of the optimality gap $f_0(x_k) - f_0^*$ decreases linearly with iterations, that is

$$\log(f_0(x_k) - f_0^*) \leq k \log c + d_0, \quad d_0 \doteq \log(f_0(x_0) - f_0^*).$$

We then see that an accuracy ϵ' is achieved on the objective function, i.e.,

$$f_0(x_k) - f_0^* \leq \epsilon'$$

in at most

$$k_{\max} = \left\lceil \frac{\log(1/\epsilon') + d_0}{\log(1/c)} \right\rceil$$

iterations. Further, from (12.25), together with Eq. (12.14) and (12.15) we have that

$$\frac{m}{2}\|x_k - x^*\|_2^2 \leq f_0(x_k) - f_0^* \leq c^k(f_0(x_0) - f_0^*) \leq c^k \frac{M}{2}\|x_0 - x^*\|_2^2,$$

hence

$$\|x_k - x^*\|_2 \leq c^{k/2}\sqrt{\frac{M}{m}}\|x_0 - x^*\|_2,$$

which provides an upper bound on the rate of convergence of x_k to x^*. Similarly, from the left inequality in (12.15) and the right inequality in (12.14) we obtain that

$$\frac{1}{2M}\|\nabla f_0(x_k)\|_2^2 \leq f_0(x_k) - f_0^* \leq c^k(f_0(x_0) - f_0^*) \leq c^k \frac{1}{2m}\|\nabla f_0(x_0)\|_2^2,$$

hence

$$\|\nabla f_0(x_k)\|_2 \leq c^{k/2}\sqrt{\frac{M}{m}}\|\nabla f_0(x_0)\|_2,$$

which provides an upper bound on the rate of convergence of the gradient to zero.

Also, from (12.14) we may derive useful stopping criteria in terms of the accuracy on the objective function value and on the minimizer. In fact, if at some iteration k we verify in a gradient algorithm that the condition $\|\nabla f_0(x_k)\| \leq \epsilon$ is met, then we can conclude that

$$f_0(x_k) - f_0^* \leq \frac{1}{2m}\|\nabla f_0(x_k)\|_2^2 \leq \frac{\epsilon^2}{2m},$$

i.e., the current objective value $f_0(x_k)$ is $\epsilon' = \epsilon^2/(2m)$ close to the minimum. Similarly, from (12.14) we have that

$$\|x_k - x^*\|_2 \leq \frac{1}{m}\|\nabla f_0(x_k)\|_2 \leq \frac{\epsilon}{m},$$

i.e., the current point x_k is $\epsilon'' = \epsilon/m$ close to the global minimizer x^*.

12.2.2.5   Summary of the convergence properties of the gradient method.
We here summarize the convergence properties for the gradient descent algorithm, under our standard Assumption 1 and that f_0 is continuously differentiable. Different types and rates of global convergence can be guaranteed, depending on additional properties of f_0.

1. If f_0 has Lipschitz continuous gradient on S_0, then the gradient algorithm (with exact line search, backtracking line search, or constant stepsizes) converges globally to a stationary point of f_0 (that is, a point where the gradient of f_0 is zero). Moreover, $\min_{i=0,\dots,k} \|\nabla f_0(x_i)\|_2$ goes to zero at a rate proportional to $1/\sqrt{k}$, where k is the iteration count of the algorithm.

2. If f_0 has Lipschitz continuous gradient and it is convex, then the gradient algorithm (with exact line search, backtracking line search, or constant stepsizes) converges to a global minimizer x^*. Further, it produces a sequence $f_0(x_k)$ that converges to the global minimum value f_0^* at a rate proportional to $1/k$.

3. If f_0 is *strongly* convex, then the gradient algorithm (with exact line search, backtracking line search, or constant stepsizes) converges to the (unique) global minimizer x^*. Further, the sequences $\|f_0(x_k) - f_0^*\|_2$, $\|x_k - x^*\|_2$, and $\|\nabla f_0(x_k)\|_2$ all converge to zero at a linear rate, that is the logarithm of these sequences tends linearly to $-\infty$.

It is important to remark that the important part in the analysis developed in the previous sections is to highlight the worst-case functional dependence of the accuracy of the iterates with respect to k. The precise numerical value of the accuracy bounds, however, is hardly useful in practice, since it requires the knowledge of several constants and quantities (e.g., L, M, m, $\|x_0 - x^*\|_2$, etc.) that are rarely known exactly.

12.2.3   Variable-metric descent methods

A variation on the gradient descent scheme is derived by considering descent directions that are obtained as a suitable linear transformation of the gradient. These methods employ a standard recursion of the type
$$x_{k+1} = x_k + s_k v_k,$$
where
$$v_k = -H_k \nabla f_0(x_k), \quad H_k \succ 0.$$

That is, the gradient is pre-multiplied by a symmetric positive definite matrix H_k. Clearly, v_k is a descent direction, since (due to positive definiteness of H_k)

$$\delta_k = \nabla f_0(x_k)^\top v_k = -\nabla f_0(x_k)^\top H_k \nabla f_0(x_k) < 0, \quad \forall k.$$

The name "variable metric" derives from the fact that $H_k \succ 0$ defines an inner product $\langle x, y \rangle_k = x^\top H_k y$ on $\mathbb{R}^n$, and hence induces a norm (or metric) $\|x\|_k = \sqrt{x^\top H_k x}$ at each iteration of the algorithm, and v_k is the steepest descent direction, with respect to this norm. If matrices H_k are chosen so that

$$H_k \succeq \omega I_n, \quad \text{for some } \omega > 0 \text{ and } \forall k, \tag{12.26}$$

then we have that

$$|\nabla f_0(x_k)^\top v_k| = \nabla f_0(x_k)^\top H_k \nabla f_0(x_k)^\top \geq \omega \|\nabla f_0(x_k)\|_2^2, \quad \forall k,$$

and

$$\|v_k\|_2 = \|H_k \nabla f_0(x_k)\|_2 \geq \omega \|\nabla f_0(x_k)\|_2, \quad \forall k.$$

It is then not difficult to see that all steps, onwards from Eq. (12.21), essentially follow in an analogous manner, by substituting stepsizes $s \leftarrow \omega s'$, where s' is the stepsize for the variable-metric method. Therefore, all the previous convergence results, as summarized in Section 12.2.2.5, apply to the variable metric methods as well (under the assumption (12.26)). However, although the results on the global convergence rate are the same as for the standard gradient method, using suitable matrices H_k may drastically change the *local* convergence rate of the algorithm. By local we here mean the rate at which the algorithm converges to a local minimum x^*, when the initial point x_0 is sufficiently close to x^*.

Under suitable hypotheses (such as the Hessian of f_0 at x^* is positive definite), it can be proved (we did not do it) that the standard gradient algorithm converges to x^* locally at a linear rate, that is

$$\|x_k - x^*\|_2 \leq K \cdot a^k, \quad \text{if } \|x_0 - x^*\| \text{ is small enough,}$$

where $a < 1$ and K is some constant (that depends on x_0). Notice that this result holds without the need of convexity or strong convexity assumptions. Incidentally, we did prove that, for strongly convex functions, the gradient algorithm converges *globally*, and not only locally, at a linear rate. This local linear convergence rate *can* indeed be improved, by using a suitable variable-metric method. For example, we shall see in Section 12.2.4 that the (damped) Newton method is nothing but a variable-metric algorithm, in which one chooses $H_k^{-1} = \nabla^2 f_0(x_k)$, and *superlinear* local convergence can be proved for this method. Other specific variable-metric algorithms are discussed in Section 12.2.5.

12.2.4   *The Newton algorithm*

The Newton method is a well-known iterative technique, originally used for finding a root of a nonlinear function of one variable, say $g : \mathbb{R} \to \mathbb{R}$. In order to determine a point for which $g(x) = 0$, one proceeds as follows: starting from a current candidate point x_k, we approximate g locally with its tangent line $\tilde{g}(x) = g(x_k) + g'(x_k)(x - x_k)$, and then say that the updated candidate x_{k+1} for the root is the point where $\tilde{g}(x) = 0$, that is

$$x_{k+1} = x_k - \frac{g(x_k)}{g'(x_k)}.$$

In our optimization context, we essentially adapt this idea to the multivariate setting, observing that the (unconstrained) minima we are seeking are nothing but the "roots" of the system of equations $\nabla f_0(x) = 0$. Informally, our function g is now the gradient of f_0, and the derivative of g is the Hessian matrix of f_0. Therefore, the Newton update formula becomes

$$x_{k+1} = x_k - [\nabla^2 f_0(x_k)]^{-1} \nabla f_0(x_k), \quad k = 0, 1, \ldots \qquad (12.27)$$

(this formula clearly implies that the Hessian should be non-singular, in order to be meaningful).

There is also an alternative and more formal interpretation of the recursion (12.27): consider the second-order Taylor approximation of f_0 around the current candidate point x_k:

$$
\begin{aligned}
f_0(x) &\simeq f_q^{(k)}(x) &&(12.28)\\
&\doteq f_0(x_k) + \nabla f_0(x_k)^\top (x - x_k) + \frac{1}{2}(x - x_k)^\top \nabla^2 f_0(x_k)(x - x_k),
\end{aligned}
$$

and assume that $\nabla^2 f_0(x_k) \succ 0$. The minimum of the quadratic approximation $f_q^{(k)}(x)$ is characterized by

$$\nabla f_q^{(k)}(x) = \nabla f_0(x_k) + \nabla^2 f_0(x_k)(x - x_k) = 0,$$

which is attained at $x = x_k - [\nabla^2 f_0(x_k)]^{-1} \nabla f_0(x_k)$, and the minimum value of the quadratic approximation is

$$\min_x f_q^{(k)}(x) = f_0(x_k) - \frac{1}{2}\lambda_k^2,$$

where

$$\lambda_k^2 \doteq \nabla f_0(x_k)^\top [\nabla^2 f_0(x_k)]^{-1} \nabla f_0(x_k),$$

and $\lambda_k \geq 0$ is called the Newton *decrement*, since it quantifies the gap between the current value of $f_0(x_k)$ and the minimum of the quadratic approximation:

$$f_0(x_k) - \min_x f_q^{(k)}(x) = \frac{1}{2}\lambda_k^2.$$

The interpretation of (12.27) is thus that the updated point x_{k+1} is the one which minimizes the local quadratic approximation of f_0 at x_k, see Figure 12.6.

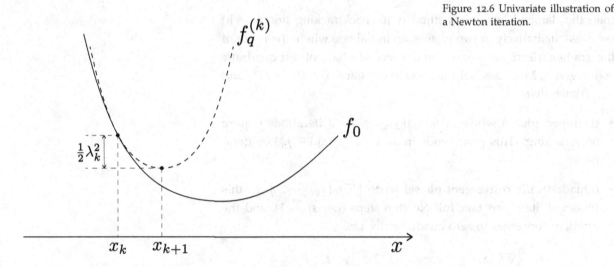

Figure 12.6 Univariate illustration of a Newton iteration.

The basic Newton iterations in (12.27) are, in general, not guaranteed to converge globally. To correct this issue, a stepsize $s_k > 0$ is introduced in the update, as follows:

$$x_{k+1} = x_k - s_k[\nabla^2 f_0(x_k)]^{-1}\nabla f_0(x_k), \quad k = 0,1,\ldots, \tag{12.29}$$

which defines the so-called *damped* Newton method. The Newton and damped Newton methods are *second-order* methods, since they need at each step second-order local information (the Hessian) about the objective function. The recursion in (12.29) can also be interpreted as a variable-metric descent algorithm, where the descent direction is given by $v_k = -H_k \nabla f_0(x_k)$, with $H_k = [\nabla^2 f_0(x_k)]^{-1}$, and the stepsizes can be chosen according to the usual rules (e.g., by backtracking). The Newton method (and its damped version) is particularly useful for minimizing strongly convex functions, since this class of functions guarantees that $\nabla^2 f_0(x) \succeq mI, \forall x$, for some $m > 0$ (hence the H_k matrices are well defined), and linear global convergence can be guaranteed. The Newton direction v_k is actually a steepest descent direction, in the metric induced by the local Hessian. This means that

$$v_k = -[\nabla^2 f_0(x_k)]^{-1} \nabla f_0(x_k) \qquad (12.30)$$

is the direction that minimizes $\delta_k = \nabla f_0(x_k)^\top v$ over all v such that $v^\top \nabla^2 f_0(x_k) v = \lambda_k^2$.

It can be proved[4] that under the assumptions of strong convexity and that the Hessian of f_0 is Lipschitz continuous over S_0, i.e., there exists a constant $L' > 0$ such that

$$\|\nabla^2 f_0(x) - \nabla^2 f_0(y)\|_2 \le L' \|x - y\|_2, \quad \forall x, y \in S_0,$$

then the damped Newton method (with backtracking line search) behaves qualitatively in two phases, an initial one where the norm of the gradient decreases linearly, and a second phase of fast quadratic convergence. More precisely, there exist constants $\eta \in (0, m^2/L')$ and $\gamma > 0$ such that

- **(Damped phase)** while $\|\nabla f_0(x_k)\|_2 \ge \eta$, most iterations require backtracking. This phase ends in at most $(f_0(x_0) - f_0^*)/\gamma$ iterations;

- **(Quadratically convergent phase)** when $\|\nabla f_0(x_k)\|_2 < \eta$. In this phase, all iterations take full Newton steps (i.e., $s_k = 1$), and the gradient converges to zero quadratically, i.e.,

$$\|\nabla f_0(x_t)\|_2 \le \text{const.} \cdot (1/2)^{2^{t-k}}, \quad t \ge k.$$

Overall, we may conclude that the damped Newton method reaches accuracy $f_0(x_k) - f_0^* \le \epsilon$ in at most

$$k_{\max} = \left\lceil \frac{f_0(x_0) - f_0^*}{\gamma} + \log_2 \log_2 \frac{\epsilon_0}{\epsilon} \right\rceil$$

iterations, where γ, ϵ_0 are constants that depend on m, L' and the initial point x_0. A more advanced analysis of the Newton method, based on the hypotheses of f_0 being strictly convex and *self-concordant*,[5] actually provides the following more useful bound, which does not depend on unknown quantities:

$$k_{\max} = \left\lceil \frac{f_0(x_0) - f_0^*}{\gamma} + \log_2 \log_2 \frac{1}{\epsilon} \right\rceil, \quad \gamma = \frac{\alpha\beta(1 - 2\alpha)^2}{20 - 8\alpha}, \quad \alpha < 1/2,$$

$$(12.31)$$

where α, β are the backtracking parameters. Also, it can be proved that, under the self-concordance hypothesis, $f_0(x_k) - f_0^* \le \lambda_k^2$ holds

[4] We will not do it here. See, for instance, Section 9.5 in Boyd and Vandenberghe's book.

[5] A convex function $f : \mathbb{R} \to \mathbb{R}$ is *self-concordant* if $|f'''(x)| \le 2[f''(x)]^{3/2}$, for all $x \in \text{dom} f$. A function $f : \mathbb{R}^n \to \mathbb{R}$ is self-concordant if it is self-concordant along any line in its domain, that is if $f(x + tv)$ is a self-concordant function of $t \in \mathbb{R}$, for all $x \in \text{dom} f$ and all $v \in \mathbb{R}^n$. Many useful convex functions have the self-concordance property, such as, for example, the so-called logarithmic barrier function for linear inequalities, $f(x) = -\sum_i \log(b_i - a_i^\top x)$, which is self-concordant over the domain where the inequalities $a_i^\top x < b_i$ are satisfied.

in the quadratically convergent phase of the Newton method (for $\lambda_k \leq 0.68$, a number that derives from an advanced analysis that we are not presenting here), hence the decrement λ_k gives a precise upper bound on the optimality gap, which can be used in the Newton algorithm as a stopping criterion, see Algorithm 11.

Algorithm 11 Damped Newton algorithm.

Require: f_0 twice continuously differentiable and either (*i*) strongly convex with Lipschitz continuous Hessian or (*ii*) strictly convex and self-concordant; $x_0 \in \text{dom } f_0$, $\epsilon > 0$

1: Set $k = 0$
2: Determine the Newton direction $v_k = -[\nabla^2 f_0(x_k)]^{-1} \nabla f_0(x_k)$, and (squared) decrement $\lambda_k^2 = -\nabla f_0(x_k)^\top v_k$
3: If $\lambda_k^2 \leq \epsilon$, then return x_k and quit
4: Determine step length $s_k > 0$ by backtracking
5: Update: $x_{k+1} = x_k + s_k v_k$, $k \leftarrow k + 1$, and go to 2

12.2.4.1 Cost of Newton iterations. Equation (12.31) shows that the number of Newton iterations necessary for achieving an ϵ accuracy on the objective value grows extremely slowly with respect to ϵ^{-1}. This dependence is "doubly logarithmic," and one can consider it constant to most practical purposes, since

$$\log_2 \log_2 \frac{1}{\epsilon} \leq 6, \quad \forall \epsilon \geq 10^{-19}.$$

However, the main limit in application of the Newton algorithm consists of the numerical cost for computing, at each iteration, the Newton direction v_k in (12.30). To compute this direction, one has to solve for the vector $v \in \mathbb{R}^n$ the system of linear equations (the Newton system)

$$[\nabla^2 f_0(x_k)]v = -\nabla f_0(x_k).$$

This system can be solved via Cholesky factorization[6] in $O(n^3/3)$ operations, for a generic unstructured Hessian. Moreover, at each step, one has to compute and store the whole Hessian matrix, and this can be a limiting factor when the problem dimension n is very large. To avoid recomputing the Hessian matrix at each step, several *approximate* approaches have been proposed, that avoid computation of second derivatives, as well as the solution of the Newton system. Some of these techniques, also known as *quasi-Newton* methods, are briefly discussed next.

[6] See Section 6.4.4.2.

12.2.5 *Quasi-Newton methods*

Quasi-Newton methods are variable-metric methods in which the H_k matrix is updated at each iteration according to some suitable rule, and it is used as a proxy for the Hessian matrix. The advantage over the "exact" Newton method is that calculation of second derivatives is avoided, together with the problem of inverting the Hessian. Observe that for a convex quadratic function

$$f(x) = \frac{1}{2}x^\top Ax + b^\top x + c, \quad A \succ 0,$$

it holds that $\nabla f(x) = Ax + b$, and $\nabla^2 f(x) = A$, thus the Hessian A satisfies the relation

$$\nabla f(x) - \nabla f(y) = A(x - y).$$

Multiplying both sides on the left by $H = A^{-1}$, we obtain the so-called *secant* condition for the inverse Hessian H:

$$H(\nabla f(x) - \nabla f(y)) = x - y.$$

The intuitive idea is thus that if a matrix H satisfies the secant condition, and if function f_0 can be approximated by a quadratic function, then its (inverse) Hessian matrix would be approximated by H. Quasi-Newton methods initialize $H_0 = I_n$, and then update this matrix according to various rules satisfying the secant condition

$$H_{k+1}(\nabla f(x_{k+1}) - \nabla f(x_k)) = x_{k+1} - x_k.$$

Typical update rules satisfying the secant condition are the following ones. Let $\Delta_k x \doteq x_{k+1} - x_k$, $\Delta_k g \doteq \nabla f_0(x_{k+1}) - \nabla f_0(x_k)$, $z_k \doteq H_k \Delta_k g$.

- Rank-one correction:

$$H_{k+1} = H_k + \frac{(\Delta_k x - z_k)(\Delta_k x - z_k)^\top}{(\Delta_k x - z_k)^\top \Delta_k g};$$

- Davidon–Fletcher–Powell (DFP) correction:

$$H_{k+1} = H_k + \frac{\Delta_k x \Delta_k x^\top}{\Delta_k g^\top \Delta_k x} - \frac{z_k z_k^\top}{\Delta_k g^\top z_k};$$

- Broyden–Fletcher–Goldfarb–Shanno (BFGS) correction:

$$H_{k+1} = H_k + \frac{(z_k \Delta_k x^\top) + (z_k \Delta_k x^\top)^\top}{\Delta_k g^\top z_k} - \nu_k \frac{z_k z_k^\top}{\Delta_k g^\top z_k},$$

where

$$\nu_k = 1 + \frac{\Delta_k g^\top \Delta_k x}{\Delta_k g^\top z_k}.$$

It can be proved that quasi-Newton methods have a global convergence rate analogous to that of the gradient method, while the local convergence happens at a superlinear rate.[7]

[7] See, for instance, Chapter 6 in J. Nocedal, S.J. Wright, *Numerical Optimization*, Springer, 2006.

12.2.6 Dealing with linear equality constraints

We here consider the problem of minimizing a convex objective f_0, under linear equality constraints, that is

$$p^* = \min_{x \in \mathbb{R}^n} f_0(x) \qquad (12.32)$$
$$\text{s.t.:} \quad Ax = b,$$

where $A \in \mathbb{R}^{m,n}$, with A full row rank: $\operatorname{rank} A = m$. There are essentially three ways of dealing with this problem. The first approach consists of *eliminating* the equality constraints, so transforming the problem into an unconstrained one, to which the methods described in the previous sections could be applied. A second technique consists of applying a descent technique to the problem (e.g., the gradient method, or the Newton method), computing the descent direction so that each iterate remains feasible. We next describe these two approaches.

12.2.6.1 Elimination of linear equality constraints.

Since $\operatorname{rank} A = m$, we can find a matrix $N \in \mathbb{R}^{n,n-m}$ containing by columns a basis for the nullspace of A. For example, N can be chosen to contain the last $n - m$ orthonormal columns of the V factor of the SVD of A, see Section 5.2.4. Then we know from Proposition 6.1 that all vectors x satisfying $Ax = b$ can be written in the form

$$x = \bar{x} + Nz,$$

where $\bar{x} \in \mathbb{R}^n$ is some fixed solution of $Ax = b$, and $z \in \mathbb{R}^{n-m}$ is a new free variable. Now, substituting x in f_0, we obtain a new objective function

$$\tilde{f}_0(z) = f_0(\bar{x} + Nz), \qquad (12.33)$$

in the new variable z. Problem (12.32) is then equivalent to the unconstrained problem

$$p^* = \min_{z \in \mathbb{R}^{n-m}} \tilde{f}_0(z).$$

Once this problem is solved (e.g., via one of the methods for unconstrained minimization discussed previously), and an optimal variable z^* is obtained, we can recover an optimal variable for the original problem as

$$x^* = \bar{x} + Nz^*.$$

One possible advantage of this approach is that the transformed unconstrained problem has $n - m$ variables, which can be much less than the original number of variables n. One drawback, however, appears when the matrix A is sparse; in general, a corresponding matrix N is dense. In that case, it may be more beneficial to work directly with the equality constraints, as is done in Section 12.2.6.3.

12.2.6.2 *Feasible update gradient algorithm.* A second approach for dealing with problem (12.32) consists of applying a descent method with a suitably chosen descent direction v_k, guaranteeing feasibility at each iteration. Let us first observe that the optimality conditions for problem (12.32) state that (see Example 8.4.2) x is optimal if and only if

$$Ax = b, \text{ and } \nabla f_0(x) = A^\top \lambda, \quad \text{for some } \lambda \in \mathbb{R}^m,$$

where the second condition states that the gradient at the optimum is orthogonal to the nullspace of A, i.e., the above characterization is equivalent to

$$x \text{ optimal} \iff Ax = b, \text{ and } \nabla f_0(x) \in \mathcal{N}^\perp(A). \tag{12.34}$$

We now adapt the gradient descent algorithm to problem (12.32). The idea is simply to take as update direction the gradient of f_0, projected onto the nullspace of A. That is, we take

$$v_k = -P\nabla f_0(x_k), \quad P = NN^\top,$$

where $N \in \mathbb{R}^{n,n-m}$ contains by columns an orthonormal basis for $\mathcal{N}(A)$. The P matrix is an orthogonal projector onto $\mathcal{N}(A)$ (see Section 5.2.4), which implies that for any vector $\xi \in \mathbb{R}^n$, the vector $P\xi$ lies in $\mathcal{N}(A)$. Then, if the current point x_k satisfies the constraints (i.e., $Ax_k = b$), then also the updated point

$$x_{k+1} = x_k + s_k v_k = x_k - s_k P\nabla f_0(x_k)$$

satisfies the constraints, since

$$Ax_{k+1} = Ax_k - s_k A(P\nabla f_0(x_k)) = b.$$

This guarantees that, if we start the algorithm with a feasible point x_0, all subsequent iterates will remain feasible. We next verify that v_k is indeed a direction of descent. Note that $P \succeq 0$, and moreover

$$z^\top P z = 0 \text{ iff } z \perp \mathcal{N}(A).$$

Decrease directions are characterized by the condition that $\nabla f_0(x_k)^\top v_k < 0$, and we have that

$$\nabla f_0(x_k)^\top v_k = -\nabla f_0(x_k) P \nabla f_0(x_k) \begin{cases} < 0 & \text{if } \nabla f_0(x_k) \notin \mathcal{N}^\perp(A), \\ = 0 & \text{if } \nabla f_0(x_k) \in \mathcal{N}^\perp(A), \end{cases}$$

which means that, at each iteration, either v_k is a decrease direction, or $\nabla f_0(x_k) \in \mathcal{N}^\perp(A)$, which, in view of (12.34), implies that x_k is optimal. This gradient projection algorithm converges with properties analogous to those of the standard gradient algorithm.

12.2.6.3   Feasible update Newton algorithm.   The Newton algorithm can also be easily adapted to deal with the linear equality constrained problem (12.32), by designing the Newton iterates to be feasible at each step. We have seen that the standard Newton update point is computed as the minimizer of the quadratic approximation $f_q^{(k)}$ of f_0 at x_k, see (12.28). The idea of the modified method is then to determine the update as the minimizer of the same quadratic approximation, under the equality constraints. That is, given the current feasible point x_k, the updated point should solve

$$\min_x \quad f_0(x_k) + \nabla f_0(x_k)^\top (x - x_k) + \tfrac{1}{2}(x - x_k)^\top \nabla^2 f_0(x_k)(x - x_k)$$
$$\text{s.t.:} \quad Ax = b.$$

The optimality conditions for this problem can be characterized explicitly (see Example 9.2) as

$$Ax = b, \text{ and } \nabla f_q^{(k)}(x) = A^\top \lambda, \quad \text{for some } \lambda \in \mathbb{R}^m.$$

Setting $\Delta_x = x - x_k$ (the full Newton step), and observing that $A\Delta_x = Ax - Ax_k = Ax - b$, and that $\nabla f_q^{(k)}(x) = \nabla f_0(x_k) + \nabla^2 f_0(x_k)(x - x_k)$, the previous conditions are rewritten in terms of Δ_x as

$$A\Delta_x = 0, \text{ and } \nabla f_0(x_k) + \nabla^2 f_0(x_k)\Delta_x = -A^\top \lambda, \quad \text{for some } \lambda \in \mathbb{R}^m$$

(here, we just renamed the vector $-\lambda$ to λ), which, in compact matrix notation, is written as

$$\begin{bmatrix} \nabla^2 f_0(x_k) & A^\top \\ A & 0 \end{bmatrix} \begin{bmatrix} \Delta_x \\ \lambda \end{bmatrix} = \begin{bmatrix} -\nabla f_0(x_k) \\ 0 \end{bmatrix}. \tag{12.35}$$

Solving the above system of linear equations (known as the KKT system for the linear equality constrained Newton method) yields the desired step Δ_x. The modified Newton method then updates the current point according to

$$x_{k+1} = x_k + s_k \Delta_x.$$

The Newton decrement is now defined as

$$\lambda_k^2 = \Delta_x^\top \nabla^2 f_0(x_k)\Delta_x,$$

and it holds that

$$f(x) - \min_{Ay=b} f_q^{(k)}(y) = \frac{1}{2}\lambda_k^2.$$

In the quadratically convergent phase it also typically holds that $f_0(x_k) - p^* \leq \lambda_k^2$, hence λ_k^2 can be used as a stopping criterion for the algorithm. The damped Newton algorithm with linear equality constraints is schematically described in Algorithm 12.

Algorithm 12 Damped Newton with linear equality constraints.

Require: f_0 twice continuously differentiable and either (*i*) strongly convex with Lipschitz continuous Hessian or (*ii*) strictly convex and self-concordant; $x_0 \in \operatorname{dom} f_0$, $Ax_0 = b$, $\epsilon > 0$

1: Set $k = 0$

2: Determine Newton step Δ_x by solving (12.35), and the (squared) decrement $\lambda_k^2 = \Delta_x^\top \nabla^2 f_0(x_k) \Delta_x$

3: If $\lambda_k^2 \leq \epsilon$, then return x_k and quit

4: Determine step length $s_k > 0$ by backtracking

5: Update: $x_{k+1} = x_k + s_k \Delta_x$, $k \leftarrow k + 1$, and go to 2

12.3 Algorithms for smooth convex constrained minimization

In this section, we discuss two techniques for dealing with differentiable convex constrained optimization. The problem under study is the form

$$p^* = \min_{x \in \mathbb{R}^n} f_0(x) \tag{12.36}$$
$$\text{s.t.:}\quad x \in \mathcal{X},$$

where f_0 is convex, and $\mathcal{X}$ is either some some "simple" convex constraint set (such as a norm ball, the positive orthant, etc.) or, more generally, it is of the form

$$\mathcal{X} = \{x \in \mathbb{R}^n : f_i(x) \leq 0, i = 1, \ldots, m\}, \tag{12.37}$$

where f_i are convex functions. We assume, without loss of generality, that no linear equality constraints are present (if there are such constraints, they can be eliminated via the procedure described in Section 12.2.6.1).

In Section 12.3.1, we describe a rather general technique for solving (12.36), based on the idea of barrier functions for the constraint set which, as we shall see, allows the solution of the constrained problem via a sequence of unconstrained minimization ones (this technique requires all functions f_i, $i = 0, 1, \ldots, m$, to be twice differentiable). In Section 12.3.2 we discuss instead an alternative method based on the concept of *proximal mappings*, which is suitable for cases where $\mathcal{X}$ is of "simple" form (we shall be more precise as to what simple means) and boils down to a scheme of gradient step followed by a projection onto the feasible set.

12.3.1 Barrier algorithm for convex constrained minimization

We next consider problem (12.36), where $\mathcal{X}$ is a closed set described as in (12.37), where f_i are convex, closed, and twice continuously differentiable. That is, we consider the convex optimization problem

$$\begin{aligned} p^* = \min_{x \in \mathbb{R}^n} & \; f_0(x) && (12.38) \\ \text{s.t.:} & \; f_i(x) \le 0, \quad i = 1, \dots, m. \end{aligned}$$

We further assume that p^* is finite and it is attained at some optimal point x^*, and that the problem is strictly feasible, that is there exists $\bar{x} \in \text{dom} f_0$ such that $f_i(\bar{x}) < 0$, $i = 1, \dots, m$. This latter assumption guarantees that Slater conditions are verified, hence strong duality holds and the optimum of the dual of (12.38) is attained (duality plays an important role in barrier methods, as it will become clear soon).

A continuous function $\phi : \mathbb{R}^n \to \mathbb{R}$ is said to be a convex *barrier function* for the set $\mathcal{X}$, if it is convex, and $\phi \to \infty$ as x approaches the boundary of $\mathcal{X}$. Typical examples of convex barrier functions for $\mathcal{X}$ are the following ones:

1. power barrier: $\sum_{i=1}^{m} (-f_i(x))^{-p}$, for $p \ge 1$;

2. logarithmic barrier: $-\sum_{i=1}^{m} \ln(-f_i(x))$;

3. exponential barrier: $\sum_{i=1}^{m} \exp(-1/f_i(x))$.

Here, we consider in particular the logarithmic barrier function

$$\phi(x) = -\sum_{i=1}^{m} \ln(-f_i(x)),$$

for which we have the explicit derivatives

$$\begin{aligned} \nabla \phi(x) &= \sum_{i=1}^{m} \frac{1}{-f_i(x)} \nabla f_i(x), && (12.39) \\ \nabla^2 \phi(x) &= \sum_{i=1}^{m} \frac{1}{[f_i(x)]^2} \nabla f_i(x) \nabla f_i(x)^\top + \sum_{i=1}^{m} \frac{1}{-f_i(x)} \nabla^2 f_i(x). \end{aligned}$$

The idea is to consider an unconstrained problem obtained by adding to f_0 a penalty term given by the logarithmic barrier, that is we consider the problem

$$\min_{x} \; f_0(x) + \frac{1}{t} \phi(x),$$

where $t > 0$ is a parameter weighting the importance of the original objective f_0 and the barrier in the new objective. We assume that an optimal solution $x^*(t)$ for this problem exists and it is unique, and that an initial strictly feasible point $x_0 \in \mathcal{X}$ is known (see Section 12.3.1.2 for a technique for determining a suitable initial feasible point). Multiplying the previous objective by t does not change the minimizer, so we can equivalently consider the problem

$$\min_x \ \psi_t(x) \doteq tf_0(x) + \phi(x), \quad t > 0, \qquad (12.40)$$

for which $x^*(t)$ is still the unique minimizer. Clearly, the role of $\phi(x)$ is to prevent the solution of this problem from drifting out of the feasible domain $\mathcal{X}$, i.e., ϕ acts indeed as a barrier for the feasible set $\mathcal{X}$: since $\phi(x)$ is equal to $+\infty$ outside the domain $\mathcal{X}$, and it tends to $+\infty$ as x approaches the boundary of the domain, the minimizer $x^*(t)$ is guaranteed to be strictly feasible, i.e., $f_i(x^*(t)) < 0$, $i = 1, \ldots, m$. The first-order optimality conditions for $\psi(t)$ then state that $\nabla \psi_t(x^*(t)) = 0$, that is, in view of (12.39),

$$t\nabla f_0(x^*(t)) + \sum_{i=1}^m \frac{1}{-f_i(x^*(t))} \nabla f_i(x^*(t)) = 0. \qquad (12.41)$$

Defining

$$\lambda^*(t) = \frac{1}{-tf_i(x^*(t))} > 0,$$

we see that the optimality condition in (12.41) equivalently states that the Lagrangian $\mathcal{L}(x, \lambda)$ of problem (12.38), evaluated for $\lambda = \lambda^*(t)$,

$$\mathcal{L}(x, \lambda^*(t)) = f_0(x) + \sum_{i=1}^m \lambda^*(t) f_i(x)$$

is minimized at $x^*(t)$, since it holds that $\nabla \mathcal{L}(x^*(t), \lambda^*(t)) = 0$. Hence, recalling that the dual function $g(\lambda) = \min_x \mathcal{L}(x, \lambda)$, for any $\lambda \geq 0$, is a lower bound for p^*, and evaluating g at $\lambda = \lambda^*(t)$, we obtain that

$$\begin{aligned} p^* &\geq g(\lambda^*(t)) = \mathcal{L}(x^*(t), \lambda^*(t)) = f_0(x^*(t)) + \sum_{i=1}^m \lambda^*(t) f_i(x^*(t)) \\ &= f_0(x^*(t)) + \sum_{i=1}^m \frac{1}{-tf_i(x^*(t))} f_i(x^*(t)) \\ &= f_0(x^*(t)) - \frac{m}{t}. \end{aligned}$$

This is the key inequality justifying the use of the barrier method for solving (12.38), since it states that the solution $x^*(t)$ of the unconstrained problem (12.40) is an ϵ-suboptimal solution to the original problem, that is, for given $\epsilon > 0$, it holds that

$$f_0(x^*(t)) - p^* \leq \epsilon, \quad \text{if } \frac{m}{t} \leq \epsilon,$$

which clearly implies that $f_0(x^*(t)) \to p^*$ as $t \to \infty$.

Ideally, one may fix a value $t \geq m/\epsilon$, and solve the unconstrained problem (12.40), using for instance the Newton method, to obtain an ϵ-suboptimal solution for (12.38). While this idea may work in principle, it may be problematic in practice, due to the fact that the initial point x_0 may be far from an optimal point x^* and, more critically, that the function $\psi_t(x)$ to be minimized tends to be ill conditioned for large t (its Hessian varies rapidly near the boundary of the feasible set). This implies that the Newton method may require many iterations to converge to $x^*(t)$. The usual approach is then to solve a *sequence* of unconstrained minimization problems, for increasing values of t, starting from an initial moderate value t_{init}, until the exit condition $m/t \leq \epsilon$ is met. The outline of such a sequential barrier method is given in Algorithm 13.

Algorithm 13 Sequential barrier method.

Require: x_0 strictly feasible, $t_{\text{init}} > 0$, $\mu > 1$, $\epsilon > 0$

1: Set $k = 0$, $t = t_{\text{init}}$, $x = x_0$
2: Solve $\min_z t f_0(z) + \phi(z)$, using the (damped) Newton method, with initial point x, and let x_k^* be the corresponding optimal solution
3: Update $x \leftarrow x_k^*$
4: If $m/t \leq \epsilon$, then return x and quit
5: Update $t \leftarrow \mu t$, $k \leftarrow k + 1$, and go to 2

Each iteration k of Algorithm 13 is called a *centering step* (or, an *outer iteration*), and x_k^* is the k-th *central point*. The curve traced by the minimizers of ψ_t, $\{x^*(t), t \geq 0\}$, is called the *central path*, and it is a curve lying in the interior of the feasible set $\mathcal{X}$. For this reason, the barrier algorithm belongs to the family of so-called *interior-point* methods. Each centering step requires a number of *inner iterations*, which are the iterations needed by the Newton method in order to compute x_k^* to a given accuracy. The numerical efficiency of the barrier method depends therefore on a tradeoff between the number of outer iterations (centering steps), and the effort required for each of these iterations, that is the number of inner iterations. As we discussed previously, setting $t_{\text{init}} \geq m/\epsilon$ would make Algorithm 13 terminate in just one outer iteration, but this may require a large number of inner iterations. Instead, increasing t progressively as $t_{k+1} = \mu t_k$, where t_k denotes the value of t used in the k-th centering step, allows us to reduce the number of inner iterations per outer iteration. This is mainly due to the fact that the Newton algorithm for the k-th centering step is started at an initial point x_{k-1}^* which is the minimizer of the previous objective $\psi_{t_{k-1}}$. Since t_k is not much larger than t_{k-1}, intuitively ψ_{t_k} should not change too much with respect to $\psi_{t_{k-1}}$, hence the new minimizer x_k^* should be "not too far" from the previous minimizer x_{k-1}^*. Overall, the number of centering steps required by Algorithm 13 for solving (12.38) to ϵ accuracy on the objective value is given by

$$\left\lceil \frac{\log(m\epsilon^{-1}/t_{\text{init}})}{\log \mu} + 1 \right\rceil,$$

and each centering step requires a number of inner iterations (i.e., iterations of the Newton method) that can be bounded above, for instance, by (12.31), if the corresponding assumptions on ψ_t are satisfied (namely, if ψ_t self-concordant).

12.3.1.1 *Self-concordant barriers for LP, QCQP, and SOCP.* We next illustrate specific barriers and their derivatives for the standard models of LP, QCQP, and SOCP. Consider a linear program in standard inequality form:

$$\begin{aligned} p^* = \ &\min_{x \in \mathbb{R}^n} \ c^\top x \\ \text{s.t.:} \ \ &a_i^\top x \le b_i, \quad i = 1, \dots, m. \end{aligned}$$

The logarithmic barrier for this problem is

$$\phi(x) = -\sum_{i=1}^m \ln(b_i - a_i^\top x),$$

which can be proved to be self-concordant. From (12.39), we have

$$\begin{aligned} \nabla \phi(x) &= \sum_{i=1}^m \frac{a_i}{b_i - a_i^\top x}, \\ \nabla^2 \phi(x) &= \sum_{i=1}^m \frac{a_i a_i^\top}{(b_i - a_i^\top x)^2}. \end{aligned}$$

A (convex) quadratically constrained quadratic program (9.20), with no equality constraints, has the standard form (12.38), with

$$\begin{aligned} f_0(x) &= \frac{1}{2} x^\top H_0 x + c_0^\top x + d_0, \\ f_i(x) &= \frac{1}{2} x^\top H_i x + c_i^\top x + d_i, \quad i = 1, \dots, m. \end{aligned}$$

The logarithmic barrier for this problem

$$\phi(x) = -\sum_{i=1}^{m} \ln\left(-\left(\frac{1}{2}x^{\top}H_i x + c_i^{\top}x + d_i\right)\right)$$

can be proved to be self-concordant, and we have that

$$\nabla\phi(x) = \sum_{i=1}^{m} \frac{H_i x + c_i}{-f_i(x)},$$

$$\nabla^2\phi(x) = \sum_{i=1}^{m} \frac{(H_i x + c_i)(H_i x + c_i)^{\top}}{[f_i(x)]^2} + \sum_{i=1}^{m} \frac{H_i}{-f_i(x)}.$$

A second-order cone program (10.6) in the standard form

$$p^* = \min_{x\in\mathbb{R}^n} \quad c^{\top}x$$
$$\text{s.t.:} \quad \|A_i x + b_i\|_2 \le c_i^{\top}x + d_i, \quad i = 1,\ldots,m,$$

can be formulated equivalently by "squaring" the constraints as

$$p^* = \min_{x\in\mathbb{R}^n} \quad c^{\top}x$$
$$\text{s.t.:} \quad f_i(x) \le 0, \quad i = 1,\ldots,m,$$
$$c_i^{\top}x + d_i \ge 0, \quad i = 1,\ldots,m,$$

with

$$f_i(x) = \|A_i x + b_i\|_2^2 - (c_i^{\top}x + d_i)^2, \quad i = 1,\ldots,m.$$

The logarithmic barrier for this equivalent problem can be proved to be self-concordant, and we have that

$$\nabla\phi(x) = \sum_{i=1}^{m} \frac{\nabla f_i(x)}{-f_i(x)} - \frac{c_i}{c_i^{\top}x + d_i},$$

$$\nabla^2\phi(x) = \sum_{i=1}^{m} \frac{\nabla f_i(x)\nabla f_i(x)^{\top}}{[f_i(x)]^2} + 2\frac{A_i^{\top}A_i - c_i c_i^{\top}}{-f_i(x)} + \frac{c_i^{\top}c_i}{(c_i^{\top}x + d_i)^2},$$

where $\nabla f_i(x) = 2(A_i^{\top}A_i - c_i c_i^{\top})x + 2(A_i^{\top}b + d_i c_i)$.

12.3.1.2 *Computing an initial feasible point.*

The barrier method needs to be initialized with a (strictly) feasible point x_0. Such an initial point can be determined by solving a preliminary optimization problem, usually called the *phase I* problem. The rationale behind the phase I method is to introduce a slack variable accounting for the violation of the original constraints of problem (12.38). That is, we substitute the original constraints $f_i(x) \le 0$ with relaxed constraints of the form $f_i(x) \le s$, where $s \in \mathbb{R}$ is a new variable, and consider the phase I optimization problem

$$s^* = \min_{x,s} \quad s$$
$$\text{s.t.:} \quad f_i(x) \leq s, \quad i = 1, \dots, m.$$

First, we observe that we can *always* easily find an initial strictly feasible point $\tilde{x}_0, s_0$ for this problem. To this end, it suffices to choose *any* point $\tilde{x}_0 \in \text{dom } f_0$, and then select any scalar s_0 such that

$$s_0 > \max_{i=1,\dots,m} f_i(\tilde{x}_0).$$

Starting with this initial feasible point, we can then solve the phase I problem using a barrier method, obtaining an optimal point $\tilde{x}^*$, and the optimal (minimal) violation s^*. Then we conclude that:

- if $s^* < 0$, then it means that $f_i(\tilde{x}^*) \leq s^* < 0$, hence $\tilde{x}^*$ is strictly feasible for the original problem (12.38);

- if $s^* = 0$, then the original problem (12.38) is feasible, but it does not admit a strictly feasible point (note, however, that from a numerical point of view one can never really say that a variable is exactly zero);

- if $s^* > 0$, then the original problem (12.38) is infeasible.

In practice, one solves the phase I problem, and if $s^* \leq -\epsilon$, for some reasonably small $\epsilon > 0$, then $\tilde{x}^*$ is strictly feasible for the original problem (12.38), and it can be used as an initial point for solving this problem via Algorithm 13.

12.3.2 *Proximal gradient algorithm*

In this section we discuss a first-order technique for solving constrained convex optimization problems of the form (12.36), when f_0 is convex and differentiable, and $\mathcal{X}$ is a convex set of simple structure (we shall soon define what we mean by "simple"). This method follows as a special case of a more general family of techniques used for solving a class of optimization problems with mixed differentiable plus non-differentiable objective, based on the concept of *proximal mapping*. This concept is reviewed next.

12.3.2.1 Proximal mapping and projections. Given a closed convex function $h : \mathbb{R}^n \to \mathbb{R}$ (not necessarily differentiable), we define the proximal mapping of h as follows:

$$\text{prox}_h(x) = \arg\min_z \left(h(z) + \frac{1}{2}\|z - x\|_2^2 \right).$$

Since $h(z)$ is convex and the additional term $\|z - x\|_2^2$ is strongly convex, then for each x the function $h(z) + 0.5\|z - x\|_2^2$ is also strongly convex (see Section 8.2.1.5). Moreover, convexity of h implies that $h(z) \geq h(x) + \eta_x^\top (z - x)$, for all x in the interior of $\mathrm{dom}\, h$, where η_x is a subgradient of h at x. Hence

$$h(z) + \frac{1}{2}\|z - x\|_2^2 \geq h(x) + \eta_x^\top (z - x) + \frac{1}{2}\|z - x\|_2^2,$$

which implies that the function on the left of this inequality is bounded below. This property, together with strong convexity, guarantees that the global minimizer $\mathrm{prox}_h(x)$ is well defined, since it exists and it is unique. An interesting special case arises when $h(z)$ is the indicator function of a closed convex set $\mathcal{X}$, i.e.,

$$h(z) = I_{\mathcal{X}}(z) \doteq \begin{cases} 0 & \text{if } z \in \mathcal{X}, \\ +\infty & \text{otherwise.} \end{cases}$$

In this case, we have

$$\mathrm{prox}_{I_{\mathcal{X}}}(x) = \arg\min_{z \in \mathcal{X}} \frac{1}{2}\|z - x\|_2^2, \tag{12.42}$$

hence $\mathrm{prox}_{I_{\mathcal{X}}}(x) = [x]_{\mathcal{X}}$ is the Euclidean projection of x onto $\mathcal{X}$. We next refer to as "simple" those functions h for which it is easy to compute the proximal mapping. Accordingly, we denote as simple a convex set $\mathcal{X}$ onto which it is computationally easy to determine a projection; examples of such sets are illustrated in Section 12.3.3.

We observe that the constrained minimization problem (12.36) can be rewritten into an unconstrained form

$$\min_x \; f_0(x) + I_{\mathcal{X}}(x), \tag{12.43}$$

where the indicator function $I_{\mathcal{X}}(x)$ acts as a non-differentiable barrier for the feasible set $\mathcal{X}$. In the next section we discuss an algorithm for solving a more general class of problems of the form

$$\min_x \; f_0(x) + h(x), \tag{12.44}$$

where f_0 is convex and differentiable, and $h(x)$ is convex and "simple." Problem (12.44) clearly includes (12.43), for $h(x) = I_{\mathcal{X}}(x)$.

12.3.2.2 Proximal gradient algorithm.
We address the solution of the problem

$$\min_x \; f(x), \tag{12.45}$$

where

$$f(x) \doteq f_0(x) + h(x), \tag{12.46}$$

via a modification of the gradient algorithm, adapted to the current situation where $h(x)$ may not be differentiable (hence its gradient may not exist).

Given a current point x_k, the approach is to first perform a standard gradient step (using the gradient of f_0 only), and then compute the new point x_{k+1} via the proximal map of h. In formulas, we take

$$x_{k+1} = \text{prox}_{s_k h}(x_k - s_k \nabla f_0(x_k)),$$

where $s_k > 0$ is a stepsize. We next show how this update can be interpreted in terms of a "modified" gradient step. By the definition of proximal map, we have

$$\begin{aligned}
x_{k+1} &= \text{prox}_{s_k h}(x_k - s_k \nabla f_0(x_k)) \\
&= \arg\min_z \left(s_k h(z) + \frac{1}{2}\|z - x_k + s_k \nabla f_0(x_k)\|_2^2 \right)
\end{aligned}$$

[dividing by s_k does not change the minimizer]

$$\begin{aligned}
&= \arg\min_z \left(h(z) + \frac{1}{2s_k}\|(z - x_k) + s_k \nabla f_0(x_k)\|_2^2 \right) \\
&= \arg\min_z \left(h(z) + \frac{1}{2s_k}\|z - x_k\|_2^2 + \nabla f_0(x_k)^\top (z - x_k) + \right. \\
&\qquad\qquad\qquad\qquad\qquad\qquad \left. + \frac{s_k}{2}\|\nabla f_0(x_k)\|_2^2 \right)
\end{aligned}$$

[adding the constant term $f_0(x_k) - \frac{s_k}{2}\|\nabla f_0(x_k)\|_2^2$ does not change the minimizer]

$$= \arg\min_z \left(h(z) + f_0(x_k) + \nabla f_0(x_k)^\top (z - x_k) + \frac{1}{2s_k}\|z - x_k\|_2^2 \right).$$

The interpretation of this last formulation is that the updated point x_{k+1} is the minimizer of $h(z)$ plus a local quadratic approximation of $f_0(z)$ at x_k, that is $x_{k+1} = \arg\min_z \psi_k(z)$, where

$$\begin{aligned}
\psi_k(z) &\doteq h(z) + q_k(z), \\
q_k(z) &\doteq f_0(x_k) + \nabla f_0(x_k)^\top (z - x_k) + \frac{1}{2s_k}\|z - x_k\|_2^2. \quad (12.47)
\end{aligned}$$

Let us define a vector $g_s(x)$ as follows:

$$g_s(x) \doteq \frac{1}{s}\left(x - \text{prox}_{sh}(x - s \nabla f_0(x))\right),$$

and also set, for notational simplicity,

$$g_k \doteq g_{s_k}(x_k) = \frac{1}{s_k}(x_k - x_{k+1}). \quad (12.48)$$

With the above notation, we can formally write our algorithm as

$$x_{k+1} = x_k - s_k g_k, \quad (12.49)$$

where g_k has the role of a "pseudo" gradient, and it is called the *gradient map* of f_0 on h at x_k. Indeed, g_k inherits some of the key properties of a standard gradient. For instance, we shall prove that the optimality condition for (12.45) is $g_s(x) = 0$. Also, if $h = 0$, then $g_k = \nabla f_0(x_k)$, hence g_k is simply the gradient of f_0 at x_k. If instead $h = I_\mathcal{X}$, then the geometrical meaning of the gradient map is illustrated in Figure 12.7.

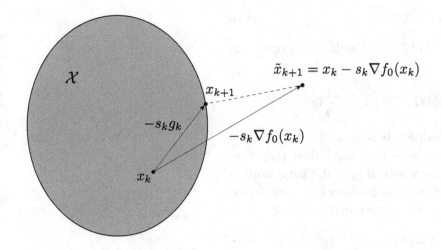

Figure 12.7 Illustration of a proximal gradient step, for the case when h is the indicator function of a closed convex set $\mathcal{X}$. In this case x_{k+1} is the Euclidean projection of $\tilde{x}_{k+1}$ onto $\mathcal{X}$.

A version of the proximal gradient algorithm, with constant stepsizes, is formally stated in Algorithm 14.

Algorithm 14 Proximal gradient algorithm (with constant stepsizes).

Require: f_0 convex, differentiable, bounded below, with Lipschitz continuous gradient (Lipschitz constant L); h convex and closed, $x_0 \in \text{dom } f_0$, $\epsilon > 0$

1: Set $k = 0$, $s = 1/L$
2: Update: $x_{k+1} = \text{prox}_{sh}(x_k - s\nabla f_0(x_k))$
3: If accuracy ϵ is attained (see, e.g., (12.60)), then exit and return x_k, else let $k \leftarrow k + 1$ and go to 2.

12.3.2.3 Convergence of the proximal gradient algorithm. This section is rather technical. The reader not interested in the details may just consider the result given at the end of this section, and move to the next section.

We shall next prove the convergence of Algorithm 14 under the hypothesis that f_0 is strongly convex, with Lipschitz continuous gradient on dom f_0. We start by observing that

$$\nabla q_k(z) = \nabla f_0(x_k) + \frac{1}{s_k}(z - x_k),$$

with q_k defined in (12.47), hence

$$\nabla q_k(x_{k+1}) = \nabla f_0(x_k) - g_k.$$

Also, we observe that a point x_k is optimal for problem (12.45), i.e., it minimizes $f = f_0 + h$, if and only if $g_k = 0$. To verify this fact, we recall that the optimality condition for problem (12.45) requires (noticing that h may not be differentiable)[8] that

$$0 \in \partial h(x_k) + \nabla f_0(x_k), \qquad (12.50)$$

where $\partial h(x)$ is the subdifferential of h at x. Similarly, the optimality condition for ψ_k requires

$$0 \in \partial h(z) + \nabla q_k(z) = \partial h(z) + \nabla f_0(x_k) + \frac{1}{s_k}(z - x_k),$$

and, in view of (12.50), this condition is satisfied for $z = x_k$, if x_k is optimal for f. Thus, if x_k is optimal for (12.45), then $x_{k+1} = x_k$, hence $g_k = 0$, due to (12.48). Conversely, if $g_k = 0$, then it must be that $x_{k+1} = x_k$ (again, due to (12.48)), and then since x_{k+1} minimizes $\psi_k(z)$, from the optimality conditions we have that

$$0 \in \partial h(x_{k+1}) + \nabla q_k(x_{k+1}) = \partial h(x_{k+1}) + \nabla f_0(x_k) - g_k,$$

that is

$$g_k \in \partial h(x_{k+1}) + \nabla f_0(x_k), \qquad (12.51)$$

which, for $x_{k+1} = x_k$, gives $0 = g_k \in \partial h(x_k) + \nabla f_0(x_k)$, which implies that x_k is optimal for f. Note that (12.51) means that there exists a subgradient $\eta_{k+1} \in \partial h(x_{k+1})$, such that

$$\nabla f_0(x_k) = g_k - \eta_{k+1}, \qquad (12.52)$$

where, by definition of a subgradient, it holds that

$$h(z) \geq h(x_{k+1}) + \eta_{k+1}^\top (z - x_{k+1}), \quad \forall z \in \operatorname{dom} h. \qquad (12.53)$$

The last two relations will be useful soon. Now, the assumptions of strong convexity and Lipschitz continuous gradient on f_0 imply that there exists $m, L > 0$, $L \geq m$, such that (see Eqs. (12.4), (12.10))

$$f_0(z) \geq f_0(x_k) + \nabla f_0(x_k)^\top (z - x_k) + \frac{m}{2}\|z - x_k\|_2^2, \quad \forall z \in \operatorname{dom} f_0,$$

$$f_0(z) \leq f_0(x_k) + \nabla f_0(x_k)^\top (z - x_k) + \frac{L}{2}\|z - x_k\|_2^2, \quad \forall z \in \operatorname{dom} f_0.$$

The second of these inequalities, evaluated at $z = x_{k+1}$, and for step-sizes such that $1/s_k \geq L$, yields

$$f_0(x_{k+1}) \leq q_k(x_{k+1}). \qquad (12.54)$$

From the first inequality (adding $h(z)$ on both sides) we have instead that, for all $z \in \text{dom } f_0$,

$$
\begin{aligned}
f(z) - \frac{m}{2}\|z - x_k\|_2^2 \;\geq\; & h(z) + f_0(x_k) + \nabla f_0(x_k)^\top (z - x_k) \\
= \; & h(z) + f_0(x_k) + \nabla f_0(x_k)^\top (x_{k+1} - x_k) \\
& + \nabla f_0(x_k)^\top (z - x_{k+1}) \\
\text{[using (12.52)]} \quad = \; & h(z) + f_0(x_k) + \nabla f_0(x_k)^\top (x_{k+1} - x_k) \\
& + g_k^\top (z - x_{k+1}) - \eta_{k+1}^\top (z - x_{k+1}) \\
\text{[from (12.53)]} \quad \geq \; & h(x_{k+1}) + f_0(x_k) + \nabla f_0(x_k)^\top (x_{k+1} - x_k) \\
& + g_k^\top (z - x_{k+1}) \\
\text{[from (12.47)]} \quad = \; & h(x_{k+1}) + q_k(x_{k+1}) - \frac{1}{2s_k}\|x_{k+1} - x_k\|_2^2 \\
& + g_k^\top (z - x_{k+1}) \\
\text{[from (12.48)]} \quad = \; & h(x_{k+1}) + q_k(x_{k+1}) - \frac{s_k}{2}\|g_k\|_2^2 + g_k^\top (z - x_{k+1}) \\
= \; & h(x_{k+1}) + q_k(x_{k+1}) - \frac{s_k}{2}\|g_k\|_2^2 \\
& + g_k^\top (z - x_k) + g_k^\top (x_k - x_{k+1}) \\
\text{[use (12.48)]} \quad = \; & h(x_{k+1}) + q_k(x_{k+1}) + \frac{s_k}{2}\|g_k\|_2^2 + g_k^\top (z - x_k) \\
\text{[$1/s_k \geq L$, (12.54)]} \quad \geq \; & h(x_{k+1}) + f_0(x_{k+1}) + \frac{s_k}{2}\|g_k\|_2^2 + g_k^\top (z - x_k) \\
\text{[from (12.46)]} \quad = \; & f(x_{k+1}) + \frac{s_k}{2}\|g_k\|_2^2 + g_k^\top (z - x_k). \qquad (12.55)
\end{aligned}
$$

Using this inequality for $z = x_k$, we obtain

$$
f(x_{k+1}) \leq f(x_k) - \frac{s_k}{2}\|g_k\|_2^2,
$$

which shows that the proximal gradient algorithm is a descent algorithm, and using again inequality (12.55) for $z = x^*$, where x^* is the minimizer of f we are seeking (hence $f(x_{k+1}) \geq f(x^*)$), we get

$$
g_k^\top (x_k - x^*) \geq \frac{s_k}{2}\|g_k\|_2^2 + \frac{m}{2}\|x_k - x^*\|_2^2. \qquad (12.56)
$$

Further, rewriting (12.55) as

$$
f(z) \geq f(x_{k+1}) + \frac{s_k}{2}\|g_k\|_2^2 + g_k^\top (z - x_k) + \frac{m}{2}\|z - x_k\|_2^2, \quad \forall z \in \text{dom } f_0, \qquad (12.57)
$$

and minimizing both sides over z (note that the minimum of the expression on the right is attained at $z = x_k - (1/m)g_k$), we obtain

$$
f(x_{k+1}) - f(x^*) \leq \frac{1}{2}\|g_k\|_2^2(1/m - s_k), \qquad (12.58)
$$

where $1/m - s_k \geq 0$, since this is implied by $L \geq m$ and $s_k \leq 1/L$.

Also, evaluating (12.57) at $z = x^*$, we obtain

$$
\begin{aligned}
f(x_{k+1}) - f(x^*) &\leq g_k^\top (x_k - x^*) - \frac{s_k}{2}\|g_k\|_2^2 - \frac{m}{2}\|x_k - x^*\|_2^2 \\
&\leq g_k^\top (x_k - x^*) - \frac{s_k}{2}\|g_k\|_2^2 \\
&= \frac{1}{2s_k}\left(\|x_k - x^*\|_2^2 - \|x_k - x^* - s_k g_k\|_2^2\right) \\
&= \frac{1}{2s_k}\left(\|x_k - x^*\|_2^2 - \|x_{k+1} - x^*\|_2^2\right). \quad (12.59)
\end{aligned}
$$

We next wrap up all these preliminaries. To derive our final result, let us consider, for simplicity, the case of constant stepsizes $s_k = s = 1/L$ (the proof can be adapted also for the case of stepsizes obtained via backtracking line search). Recalling (12.49), we have

$$
\begin{aligned}
\|x_{k+1} - x^*\|_2^2 &= \|(x_k - x^*) - s_k g_k\|_2^2 \\
&= \|x_k - x^*\|_2^2 + s_k^2\|g_k\|_2^2 - 2s_k g_k^\top (x_k - x^*) \\
\text{[using (12.56)]} \quad &\leq (1 - ms_k)\|x_k - x^*\|_2^2 \\
\text{[for } s_k = 1/L] \quad &= \left(1 - \frac{m}{L}\right)\|x_k - x^*\|_2^2,
\end{aligned}
$$

whence

$$
\|x_k - x^*\|_2^2 \leq \left(1 - \frac{m}{L}\right)^k \|x_0 - x^*\|_2^2,
$$

which shows that the proximal gradient algorithm converges to x^* at a linear rate. Also, if m, L are known, then (12.58) provides a stopping criterion for Algorithm 14, based on checking the norm of g_k, since, for $\epsilon \geq 0$,

$$
\|g_k\|_2^2 \leq 2\epsilon \frac{mL}{L - m} \quad \Rightarrow \quad f(x_{k+1}) - f(x^*) \leq \epsilon. \quad (12.60)
$$

Further, adding inequalities (12.59), we get

$$
\begin{aligned}
\sum_{i=1}^{k} f(x_i) - f(x^*) &\leq \frac{1}{2s}\sum_{i=1}^{k}\left(\|x_{i-1} - x^*\|_2^2 - \|x_i - x^*\|_2^2\right) \\
&= \frac{L}{2}\left(\|x_0 - x^*\|_2^2 - \|x_k - x^*\|_2^2\right) \\
&\leq \frac{L}{2}\|x_0 - x^*\|_2^2.
\end{aligned}
$$

Since $f(x_i)$ is non-increasing with i, the last value $f(x_k)$ is no larger than the average of the previous values, that is

$$
f(x_k) - f(x^*) \leq \frac{1}{k}\sum_{i=1}^{k}(f(x_i) - f(x^*)) \leq \frac{L}{2k}\|x_0 - x^*\|_2^2,
$$

which shows that $f(x_k) \to f(x^*)$ at rate $1/k$. Our findings are summarized in the next theorem.

Theorem 12.1 *For Algorithm 14 it holds that*

$$f(x_k) - f(x^*) \leq \frac{L}{2k} \|x_0 - x^*\|_2^2.$$

Moreover, under the additional hypothesis that f_0 is strongly convex (with strong convexity constant m), it also holds that

$$\|x_k - x^*\|_2^2 \;\leq\; \left(1 - \frac{m}{L}\right)^k \|x_0 - x^*\|_2^2,$$

$$f(x_{k+1}) - f(x^*) \;\leq\; \frac{1}{2}\|g_k\|_2^2(1/m - 1/L).$$

12.3.3   Computing proximal maps and projections

We here discuss several relevant cases of functions h for which the proximal maps are "easy" to compute. We recall that if h is the indicator function of a closed convex set $\mathcal{X}$, then the proximal map is just the Euclidean projection onto $\mathcal{X}$ (see (12.42)) hence, in this case, the proximal gradient algorithm solves the constrained optimization problem (12.36):

$$p^* = \min_{x \in \mathbb{R}^n} f_0(x) \tag{12.61}$$
$$\text{s.t.:}\quad x \in \mathcal{X}.$$

12.3.3.1   Projection onto a half-space.

Let $\mathcal{X}$ be a half-space

$$\mathcal{X} = \{x : a^\top x \leq b\}, \quad a \neq 0.$$

Then, for given $x \in \mathbb{R}^n$, we have

$$\text{prox}_{I_\mathcal{X}}(x) = \arg\min_{z \in \mathcal{X}} \|z - x\|_2^2 = [x]_\mathcal{X},$$

and the projection $[x]_\mathcal{X}$ is x, if $x \in \mathcal{X}$, or it is equal to the projection of x onto the hyperplane $\{x : a^\top x = b\}$, if $x \notin \mathcal{X}$. This latter projection is computed as discussed in Section 2.3.2.2, hence

$$[x]_\mathcal{X} = \begin{cases} x & \text{if } a^\top x \leq b, \\ x + \frac{b - a^\top x}{\|a\|_2^2} a & \text{if } a^\top x > b. \end{cases}$$

12.3.3.2   Projection onto the positive orthant.

Let

$$\mathcal{X} = \mathbb{R}_+^n = \{x \in \mathbb{R}^n : x \geq 0\}.$$

Then, for given $x \in \mathbb{R}^n$, we have

$$[x]_\mathcal{X} = \arg\min_{z \geq 0} \|z - x\|_2^2 = \arg\min_{z_i \geq 0} \sum_{i=1}^n (z_i - x_i)^2,$$

where we see that the optimal z should have components $z_i = x_i$, if $x_i \geq 0$, or $z_i = 0$ otherwise, hence

$$[x]_{\mathcal{X}} = [x]_+ = \max(0, x),$$

where the max is here intended element-wise.

12.3.3.3 Projection onto the standard simplex. Let $\mathcal{X}$ be the standard (probability) simplex

$$\mathcal{X} = \{x \in \mathbb{R}^n : x \geq 0, \mathbf{1}^\top x = 1\}.$$

Computing the projection $[x]_{\mathcal{X}}$ amounts to solving

$$\begin{array}{ll}
\min_{z} & \dfrac{1}{2}\|z - x\|_2^2 \\
\text{s.t.:} & z \geq 0 \\
& \mathbf{1}^\top z = 1.
\end{array}$$

Considering the (partial) Lagrangian for this problem, we have

$$\mathcal{L}(z, v) = \frac{1}{2}\|z - x\|_2^2 + v(\mathbf{1}^\top z - 1)$$

and the dual function

$$\begin{aligned}
g(v) &= \min_{z \geq 0} \mathcal{L}(z, v) = \min_{z \geq 0} \frac{1}{2}\|z - x\|_2^2 + v(\mathbf{1}^\top z - 1) \\
&= \min_{z \geq 0} \sum_{i=1}^{n} \left(\frac{1}{2}(z_i - x_i)_2^2 + v z_i \right) - v \\
&= \sum_{i=1}^{n} \min_{z_i \geq 0} \left(\frac{1}{2}(z_i - x_i)_2^2 + v z_i \right) - v.
\end{aligned}$$

The problem we need to solve for determining the dual function is *separable*, meaning that the optimal solution z is obtained by finding the optimal values of the individual components, which are obtained by solving a simple one-dimensional minimization:

$$z_i^*(v) = \arg\min_{z_i \geq 0} \frac{1}{2}(z_i - x_i)_2^2 + v z_i.$$

The function to be minimized here is a convex parabola, having its vertex at $v_i = x_i - v$. The minimizer $z_i^*(v)$ is thus the vertex, if $v_i \geq 0$, or it is zero otherwise. That is

$$z^*(v) = \max(x - v\mathbf{1}, 0).$$

The optimal value v^* of the dual variable v should be then obtained by maximizing $g(v)$ with respect to v. However, there exists only one value of v which makes $z^*(v)$ belong to the simplex (i.e., primal feasible: $\sum_i z_i^*(v) = 1$), hence this must be the optimal value of the dual variable. In summary, the projection is computed as

$$[x]_{\mathcal{X}} = z^*(v^*) = \max(x - v^* \mathbf{1}, 0),$$

where v^* is the solution of the scalar equation[9]

$$\sum_{i=1}^{n} \max(x_i - v, 0) = 1.$$

12.3.3.4 Projection onto the Euclidean ball.

Let $\mathcal{X}$ be the unit Euclidean ball

$$\mathcal{X} = \{x \in \mathbb{R}^n : \|x\|_2 \le 1\}.$$

Then, it is straightforward to verify that the projection of x onto $\mathcal{X}$ is

$$[x]_{\mathcal{X}} = \begin{cases} x & \text{if } \|x\|_2 \le 1, \\ \frac{x}{\|x\|_2} & \text{if } \|x\|_2 > 1. \end{cases}$$

12.3.3.5 Projection onto the ℓ_1-norm ball.

Let $\mathcal{X}$ be the unit ℓ_1 ball

$$\mathcal{X} = \{x \in \mathbb{R}^n : \|x\|_1 \le 1\}.$$

For a given $x \in \mathbb{R}^n$, computing the projection $[x]_{\mathcal{X}}$ amounts to solving

$$\min_{\|z\|_1 \le 1} \frac{1}{2} \|z - x\|_2^2. \tag{12.62}$$

The Lagrangian of this problem is

$$\mathcal{L}(z, \lambda) = \frac{1}{2} \|z - x\|_2^2 + \lambda(\|z\|_1 - 1),$$

hence the dual function is

$$\begin{aligned} q(\lambda) &= \min_z \mathcal{L}(z, \lambda) = \min_z \frac{1}{2} \|z - x\|_2^2 + \lambda(\|z\|_1 - 1) \\ &= \sum_{i=1}^{n} \min_{z_i} \left(\frac{1}{2}(z_i - x_i)^2 + \lambda |z_i| \right) - \lambda. \end{aligned} \tag{12.63}$$

We then see that the values $z_i^*(\lambda)$ that minimize the above function are found by solving the following univariate minimizations:

$$z_i^*(\lambda) = \arg\min_{z_i} \varphi(z_i, \lambda), \quad \varphi(z_i, \lambda) \doteq \frac{1}{2}(z_i - x_i)^2 + \lambda |z_i|, \quad i = 1, \ldots, n.$$

To solve this problem, we use the identity $|z_i| = \max_{|\varrho_i| \leq 1} \varrho_i z_i$, and write

$$\min_{z_i} \frac{1}{2}(z_i - x_i)^2 + \lambda|z_i| = \min_{z_i} \left(\frac{1}{2}(z_i - x_i)^2 + \lambda \max_{|\varrho_i| \leq 1} \varrho_i z_i \right)$$

$$= \min_{z_i} \max_{|\varrho_i| \leq 1} \frac{1}{2}(z_i - x_i)^2 + \lambda \varrho_i z_i$$

[using Theorem 8.8] $= \max_{|\varrho_i| \leq 1} \min_{z_i} \frac{1}{2}(z_i - x_i)^2 + \lambda \varrho_i z_i.$

The inner minimization (w.r.t. z_i) is readily solved by setting the derivative to zero, obtaining

$$z_i^*(\lambda) = x_i - \lambda\varrho_i, \qquad (12.64)$$

which, substituted back, yields

$$\min_{z_i} \frac{1}{2}(z_i - x_i)^2 + \lambda\varrho_i z_i = \lambda\left(\varrho_i x_i - \frac{1}{2}\lambda\varrho_i^2 \right).$$

Continuing the previous chain of equalities, we thus have that

$$\min_{z_i} \frac{1}{2}(z_i - x_i)^2 + \lambda|z_i| = \lambda \max_{|\varrho_i| \leq 1} \left(\varrho_i x_i - \frac{1}{2}\lambda\varrho_i^2 \right).$$

The function to be maximized here (w.r.t. ϱ_i) is a concave parabola, having its vertex at $v_i = x_i/\lambda$ (we let here $\lambda > 0$, since for $\lambda = 0$ the dual function is trivially zero). Hence, if $|v_i| \leq 1$ the maximum is attained at $\varrho_i^* = v_i = x_i/\lambda$. Otherwise, the maximum is attained at one of the extremes of the feasible interval $\varrho_i \in [-1, 1]$ and, in particular, at $\varrho_i^* = 1$ if $x_i \geq 0$, and at $\varrho_i^* = -1$ if $x_i < 0$. Therefore,

$$\varrho_i^* = \begin{cases} x_i/\lambda & \text{if } |x_i| \leq \lambda, \\ \text{sgn}(x_i) & \text{otherwise.} \end{cases}$$

Correspondingly, the minimizer $z_i^*(\lambda)$ of $\varphi(z_i, \lambda)$ is given by (12.64) as

$$z_i^*(\lambda) = x_i - \lambda\varrho_i^* = \begin{cases} 0 & \text{if } |x_i| \leq \lambda, \\ x_i - \lambda\text{sgn}(x_i) & \text{otherwise.} \end{cases} \qquad (12.65)$$

This can be more compactly written as

$$z_i^*(\lambda) = \text{sgn}(x_i)[|x_i| - \lambda]_+ \doteq \text{sthr}_\lambda(x_i), \quad i = 1, \dots, n \qquad (12.66)$$

where $[\cdot]_+$ denotes the projection onto the positive orthant (positive part of the argument). The function sthr_λ is known as the *soft threshold* function, or *shrinkage* operator, see Figure 12.8. When x is a vector,

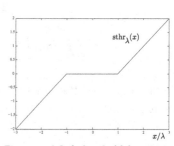

Figure 12.8 Soft threshold function.

we indicate with the notation $\text{sthr}_\lambda(x)$ a vector whose i-th entry is $\text{sthr}_\lambda(x_i)$, $i = 1, \ldots, n$.

Now, since strong duality holds for problem (12.62), and the solution to this problem is unique, the optimal primal variable $[x]_\mathcal{X}$ coincides with $z_i^*(\lambda^*)$, where $\lambda^* \geq 0$ is the value of the dual variable that maximizes $q(\lambda)$. We can, however, find the optimal λ^* via a simplified reasoning in the present case. Specifically, consider first the case when $\|x\|_1 \leq 1$, then in this case the projection is simply x itself, i.e., $[x]_\mathcal{X} = x$. Consider then the case $\|x\|_1 > 1$, then the projection $[x]_\mathcal{X}$ will be on the boundary of $\mathcal{X}$, which means that $\|[x]_\mathcal{X}\|_1 = 1$. We then use this condition to determine the optimal λ: λ^* is the solution of the scalar equation

$$\sum_{i=1}^n |z_i^*(\lambda)| = 1. \tag{12.67}$$

Notice that, from simple manipulation of (12.65), we have that

$$|z_i^*(\lambda)| = \begin{cases} 0 & \text{if } |x_i| \leq \lambda \\ |x_i| - \lambda & \text{if } |x_i| > \lambda \end{cases} = \max(|x_i| - \lambda, 0).$$

Hence, equation (12.67) becomes

$$\sum_{i=1}^n \max(|x_i| - \lambda, 0) = 1,$$

and λ^* is the value of λ that solves this equation. Once λ^* is found, the projection we seek is given by

$$[x]_\mathcal{X} = \text{sgn}(x_i)[|x_i| - \lambda^*]_+.$$

12.3.3.6 Projection onto the positive semidefinite cone. Consider the cone of positive semidefinite matrices

$$\mathcal{X} = \{X \in \mathbb{S}^n : X \succeq 0\} = \mathbb{S}_+^n.$$

Given a matrix $X \in \mathbb{S}^n$ we want to compute its projection onto $\mathcal{X}$. Since we are working in a matrix space, we shall define projections according to the Frobenius norm, that is

$$[X]_\mathcal{X} = \arg\min_{Z \in \mathcal{X}} \|Z - X\|_F^2.$$

Let now $X = U \Lambda U^\top$ be a spectral factorization for X, where U is an orthogonal matrix, and Λ is diagonal, containing the eigenvalues of X on the diagonal. Since the Frobenius norm is unitarily invariant, we have that

$$\begin{aligned} \|Z - X\|_F^2 &= \|Z - U\Lambda U^\top\|_F^2 = \|U(U^\top Z U - \Lambda)U^\top\|_F^2 \\ &= \|U^\top Z U - \Lambda\|_F^2 = \|\tilde{Z} - \Lambda\|_F^2, \end{aligned}$$

where we defined $\tilde{Z} \doteq U^\top Z U$. Since Λ is diagonal it is easy to see that the $\tilde{Z}$ that minimizes $\|\tilde{Z} - \Lambda\|_F^2$ is also diagonal, and

$$\tilde{Z}^* = \text{diag}\left([\lambda_1]_+, \ldots, [\lambda_n]_+\right) = [\Lambda]_+,$$

whence $Z^* = U\tilde{Z}^* U^\top$. In summary, the projection of $X = U\Lambda U^\top$ onto the positive semidefinite cone is given by

$$[X]_{S_+^n} = U[\Lambda]_+ U^\top.$$

12.3.3.7 *Proximal map of ℓ_1 regularization.* In many problems of practical relevance the function h in (12.46) is a scalar multiple of the ℓ_1 norm of x. For instance, in the ℓ_1-regularized least-squares problem (also known as the LASSO), we consider

$$\min_x \frac{1}{\gamma}\|Ax - b\|_2^2 + \|x\|_1, \tag{12.68}$$

which is of the form (12.45), with $f_0(x) = (1/\gamma)\|Ax - b\|_2^2$ strongly convex (assuming A is full rank), and $h(x) = \|x\|_1$ convex but non-differentiable. This class of problems is thus solvable by means of the proximal gradient algorithm. To this end, we need to be able to efficiently compute the proximal map of sh, where $s \geq 0$ is a scalar (the stepsize), namely

$$\text{prox}_{sh}(x) = \arg\min_z s\|x\|_1 + \frac{1}{2}\|z - x\|_2^2.$$

This is precisely the problem already considered in (12.63), for which we showed that the solution is given by the soft threshold function in (12.66), i.e.,

$$\text{prox}_{sh}(x) = \text{sthr}_s(x),$$

where the i-th component of the vector $\text{sthr}_s(x)$ is $\text{sgn}(x_i)[|x_i| - s]_+$.

12.3.3.8 *Proximal gradient algorithm for the LASSO.* Using the result in the previous section, we can specify the proximal gradient algorithm in the case of the LASSO problem in (12.68). Notice that we have in this case

$$\nabla f_0(x) = \frac{2}{\gamma}\left(A^\top A x - A^\top b\right),$$

$$\nabla^2 f_0(x) = \frac{2}{\gamma}(A^\top A),$$

from which it follows that the strong convexity constant for f_0 is

$$m = \frac{2}{\gamma}\sigma_{\min}(A^\top A). \tag{12.69}$$

Further, we have that

$$\|\nabla f_0(x) - \nabla f_0(y)\|_2 = \frac{2}{\gamma}\|A^\top A(x-y)\|_2 \le \frac{2}{\gamma}\sigma_{\max}(A^\top A)\|x-y\|_2,$$

from which we obtain a global Lipschitz constant for the gradient:

$$L = \frac{2}{\gamma}\sigma_{\max}(A^\top A). \tag{12.70}$$

If L and m can be computed (or at least estimated) as described above, the following proximal gradient algorithm (Algorithm 15), solves the LASSO problem (12.68) by returning a solution x guaranteeing that $f(x) - f(x^*) \le \epsilon$, where $\epsilon > 0$ is the required accuracy. This algorithm is also known as the ISTA (iterative shrinkage–thresholding algorithm).

Algorithm 15 Proximal gradient for LASSO (constant stepsizes).

Require: $\epsilon > 0$, x_0, A full rank.

1: Compute m, L according to (12.69), (12.70)
2: Set $k = 0$, $s = 1/L$
3: Compute gradient $\nabla f_0(x_k) = (2/\gamma)(A^\top A x_k - A^\top b)$
4: Update: $x_{k+1} = \mathrm{sthr}_s(x_k - s\nabla f_0(x_k))$
5: Compute $\|g_k\|_2 = \|x_k - x_{k+1}\|_2/s$
6: If $\|g_k\|_2^2 \le 2\epsilon m L/(L-m)$, then return $x = x_{k+1}$ and exit, else let $k \leftarrow k+1$ and go to 3.

We remark that in recent years there has been a tremendous activity in theory, algorithms and applications of ℓ_1 regularization and LASSO-related problems, especially in the context of the "compressive sensing" field. More sophisticated techniques thus exist for solving the LASSO and related problems. The key essential advance provided by these techniques consists of "accelerating" the basic proximal gradient scheme, so as to reach convergence rates of the order of $1/k^2$ (recall that the basic proximal gradient has convergence rate of the order of $1/k$ on objective value). Although a full coverage of these methodologies is out of the scope of this book, we shall discuss one of these "fast" methods in the next section.

12.3.4 Fast proximal gradient algorithm

The basic proximal gradient algorithm discussed in the previous section can be suitably modified so as to reach an accelerated convergence rate (in the objective function values) of order $1/k^2$. One of these modifications is the so-called FISTA type (fast iterative shrinkage–thresholding algorithm).[10] A version of this algorithm, with constant stepsizes, is reported in Algorithm 16.

[10] See A. Beck, M. Teboulle, A fast iterative shrinkage–thresholding algorithm for linear inverse problems, *SIAM Journal on Imaging Sciences*, 2009.

Algorithm 16 Fast proximal gradient (constant stepsizes).

Require: x_0, a Lipschitz constant L for ∇f_0

1: Set $k = 1, s = 1/L, y_1 = x_0, t_1 = 1$

2: Update: $x_k = \text{prox}_{sh}(y_k - s\nabla f_0(y_k))$

3: Update: $t_{k+1} = \frac{1+\sqrt{1+4t_k^2}}{2}$

4: Update: $y_{k+1} = x_k + \frac{t_k - 1}{t_{k+1}}(x_k - x_{k-1})$

5: If $\|x_k - x_{k-1}\|_2 \le \epsilon$, then return $x = x_k$ and exit, else let $k \leftarrow k+1$ and go to 2.

When applied to the specific LASSO problem in (12.68), step 2 in this algorithm simply reduces to soft thresholding:

$$\text{prox}_{sh}(y_k - s\nabla f_0(y_k)) = \text{sthr}_s(y_k - s\nabla f_0(y_k)).$$

The exit condition in step 5 here simply checks a minimal improvement on the optimal variable, and does not imply ϵ-suboptimality either on the optimal value or on the minimizer. However, the following result holds for the convergence of Algorithm 16.

Theorem 12.2 *For the sequence x_k, $k = 1, \ldots$, generated by Algorithm 16 it holds that*

$$f(x_k) - f(x^*) \le \frac{2L}{(k+1)^2}\|x_0 - x^*\|_2^2,$$

where x^* is any optimal solution to problem (12.45).

This result[11] shows that Algorithm 16 guarantees a convergence rate in objective function values of the order of $1/k^2$.

[11] See the cited paper by Beck and Teboulle for a proof.

When the Lipschitz constant L for ∇f_0 is not known *a priori*, the algorithm can be modified so as to include a backtracking step for incrementally adjusting the value of L, as specified in Algorithm 17.

12.4 *Algorithms for non-smooth convex optimization*

The proximal gradient algorithm discussed in Section 12.3.2 is applicable for solving constrained convex optimization problems of the form (12.36), when the objective f_0 is differentiable, and the feasible set $\mathcal{X}$ has "simple" structure. In this section we discuss algorithms that permit the treating of more general situations, in which either the objective function and/or the functions describing the inequality constraints of the problem are non-differentiable.

Specifically, Section 12.4.1 presents the projected subgradient algorithm, which can be used when f_0 is non-differentiable and $\mathcal{X}$ is "simple." This algorithm essentially has the structure of the (projected) gradient algorithm, but uses subgradients of f_0 instead of its

---

Algorithm 17 Fast proximal gradient (with backtracking).

Require: $x_0, L_0 > 0, \eta > 1$

1: Set $k = 1, y_1 = x_0, t_1 = 1$

2: Let $p_i = \text{prox}_{\frac{1}{M_i} h}(y_k - \nabla f_0(y_k)/M_i)$, and find the smallest non-negative integer i such that

$$f(p_i) \leq f_0(y_k) + \nabla^\top f_0(y_k)(x_k - y_k) + \frac{M_i}{2}\|x_k - y_k\|_2^2 + h(x_k)$$

holds for $M_i = \eta^i L_{k-1}$

3: Set $L_k = \eta^i L_{k-1}$

4: Update: $x_k = \text{prox}_{\frac{1}{L_k} h}(y_k - \nabla f_0(y_k)/L_k)$

5: Update: $t_{k+1} = \frac{1+\sqrt{1+4t_k^2}}{2}$

6: Update: $y_{k+1} = x_k + \frac{t_k - 1}{t_{k+1}}(x_k - x_{k-1})$

7: If $\|x_k - x_{k-1}\|_2 \leq \epsilon$, then return $x = x_k$ and exit, else let $k \leftarrow k+1$ and go to 2.

---

gradients. Despite this similarity, however, subgradient-type algorithms are not (in general) descent methods. Moreover, in order to guarantee convergence, the stepsizes must be chosen according to rules that are quite different from the ones used in the gradient algorithm. Also, the "price to pay" for the increased generality of subgradient methods is that convergence happens at a slower rate, which can be proved to be of the order of $1/\sqrt{k}$, where k is the iteration number. In Section 12.4.2 we then discuss the alternate subgradient algorithm, which is a modified version of the subgradient method that makes it possible to treat the case where f_0 is non-differentiable and $\mathcal{X}$ is not necessarily "simple" and it is described by a set of convex inequalities. Finally, Section 12.4.3 presents the ellipsoid algorithm, which is a classical method for non-differentiable constrained optimization.

12.4.1   The projected subgradient algorithm

We consider a constrained minimization problem of the form

$$p^* = \min_{x \in \mathcal{X}} f_0(x), \tag{12.71}$$

where $\mathcal{X} \subseteq \mathbb{R}^n$ is a convex and closed set, and f_0 is a convex function, with dom $f_0 = \mathbb{R}^n$. We assume that this problem admits an optimal solution x^*. For any point $x \in \text{int } \mathcal{X}$, by the definition of subgradient, we have that

$$f_0(z) \geq f_0(x) + g_x^\top(z - x), \quad \forall z \in \mathcal{X} \tag{12.72}$$

holds for any $g_x \in \partial f_0(x)$. Evaluating this inequality at $z = x^*$, we have that

$$g_x^\top (x - x^*) \geq f_0(x) - f_0(x^*) \geq 0, \qquad (12.73)$$

which is a key inequality for proving the convergence of the subgradient method. Also, any subgradient g_x of f_0 at x divides the whole space $\mathbb{R}^n$ into two half-spaces

$$\mathcal{H}_{++} = \{z : g_x^\top (z - x) > 0\}, \quad \mathcal{H}_- = \{z : g_x^\top (z - x) \leq 0\},$$

and we see from (12.72) that for all $z \in \mathcal{H}_{++}$ we have $f_0(z) > f_0(x)$, hence the optimal points cannot lie in $\mathcal{H}_{++}$, thus $x^* \in \mathcal{H}_-$, see Figure 12.9.

We next describe a subgradient algorithm for solving (12.71) when $\mathcal{X}$ is a "simple" closed convex set, where by simple we mean that it is easy to compute the Euclidean projection of a point onto $\mathcal{X}$. Consider problem (12.71), and the recursion

$$x_{k+1} = [x_k - s_k g_k]_\mathcal{X}, \quad k = 0, 1, \ldots, \qquad (12.74)$$

where $x_0 \in \mathcal{X}$ is a feasible initial point, g_k is any subgradient of f_0 at x_k, s_k are suitable stepsizes (to be specified later), and $[\cdot]_\mathcal{X}$ denotes the Euclidean projection onto $\mathcal{X}$. The following proposition holds.

Proposition 12.1 (Convergence of projected subgradient) *Assume that: (a) problem (12.71) attains an optimal solution x^*; (b) there exists a finite constant G such that $\|g\|_2 \leq G$ for all $g \in \partial f_0(x)$ and all x (equivalently, f_0 is Lipschitz with constant G); (c) a constant R is known such that $\|x_0 - x^*\| \leq R$. Let $f_{0,k}^*$ denote the best value achieved by $f_0(x_i)$ for i from 0 to k. Then*

$$f_{0,k}^* - p^* \leq \frac{R^2 + G^2 \sum_{i=0}^k s_i^2}{2 \sum_{i=0}^k s_i}.$$

In particular, if the s_k sequence is square-summable but non-summable (i.e., $\sum_{k=0}^\infty s_k^2 < \infty$, and $\sum_{k=0}^\infty s_k = \infty$; for example, $s_k = \gamma/(k+1)$ for some $\gamma > 0$), then

$$\lim_{k \to \infty} f_{0,k}^* - p^* = 0.$$

Proof Let

$$z_{k+1} = x_k - s_k g_k$$

denote an update in the direction of the negative subgradient, before the projection onto $\mathcal{X}$ is made. We have that

$$
\begin{aligned}
\|z_{k+1} - x^*\|_2^2 &= \|x_k - s_k g_k - x^*\|_2^2 \qquad (12.75) \\
&= \|x_k - x^*\|_2^2 + s_k^2 \|g_k\|_2^2 + 2 s_k g_k^\top (x^* - x_k) \\
\text{[using (12.73)]} \quad &\leq \|x_k - x^*\|_2^2 + s_k^2 \|g_k\|_2^2 + 2 s_k (p^* - f_0(x_k)),
\end{aligned}
$$

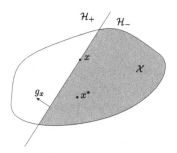

Figure 12.9 Any subgradient g_x of f_0 at $x \in \mathcal{X}$ defines a half-space $\mathcal{H}_- = \{z : g_x^\top (z - x) \leq 0\}$ containing the optimal points of problem (12.71).

where $p^* = f_0(x^*)$. Now, since x_{k+1} is the projection of z_{k+1} onto $\mathcal{X}$, we have

$$\|x_{k+1} - x\|_2 = \|[z_{k+1}]_{\mathcal{X}} - x\|_2 \leq \|z_{k+1} - x\|_2, \quad \forall x \in \mathcal{X},$$

hence, since $x^* \in \mathcal{X}$,

$$\|x_{k+1} - x^*\|_2 \leq \|z_{k+1} - x^*\|_2.$$

Therefore, combining this with (12.75), we obtain

$$\|x_{k+1} - x^*\|_2^2 \leq \|x_k - x^*\|_2^2 + s_k^2\|g_k\|_2^2 - 2s_k(f_0(x_k) - p^*). \quad (12.76)$$

Since $f_0(x_k) - p^* > 0$, the last inequality shows that each iteration of the projected subgradient method may locally decrease the distance from an optimal solution (if $s_k \geq 0$ is sufficiently small). Applying inequality (12.76) recursively from 0 to $k+1$ gives

$$\|x_{k+1} - x^*\|_2^2 \leq \|x_0 - x^*\|_2^2 + \sum_{i=0}^{k} s_i^2\|g_i\|_2^2 - 2\sum_{i=0}^{k} s_i(f(x_i) - p^*),$$

hence

$$2\sum_{i=0}^{k} s_i(f(x_i) - p^*) \leq \|x_0 - x^*\|_2^2 - \|x_{k+1} - x^*\|_2^2 + \sum_{i=0}^{k} s_i^2\|g_i\|_2^2$$

$$\leq R^2 + \sum_{i=0}^{k} s_i^2\|g_i\|_2^2. \quad (12.77)$$

Now, since $s_i \geq 0$, it holds that

$$\sum_{i=0}^{k} s_i(f(x_i) - p^*) \geq \left(\sum_{i=0}^{k} s_i\right) \min_{i=0,\dots,k}(f(x_i) - f^*) = \left(\sum_{i=0}^{k} s_i\right)(f_{0,k}^* - p^*),$$

where $f_{0,k}^*$ is the best (smallest) value achieved by $f_0(x_i)$ for i from 0 to k. Combining this inequality with (12.77), and recalling that $\|g_k\|_2 \leq G$, we obtain

$$f_{0,k}^* - p^* \leq \frac{R^2 + G^2 \sum_{i=0}^{k} s_i^2}{2\sum_{i=0}^{k} s_i}. \quad (12.78)$$

It is then easy to see that, for instance, the choice $s_i = \gamma/(i+1)$, with $\gamma > 0$, is such that, for $k \to \infty$, $\sum_{i=0}^{k} s_i^2 \to 0$ and $\sum_{i=0}^{k} s_i \to \infty$, thus the upper bound in (12.78) converges to zero. $\qquad \square$

Remark 12.2 *Optimal stepsizes for fixed number of iterations.* For fixed k, consider the right-hand side of (12.78) as a function of the stepsizes s_i, $i = 0, \dots, k$: we can seek for values of the stepsizes such that this bound is minimized. Taking the derivative of the bound with respect to s_j and setting it to zero gives the following relation for the optimal stepsizes:

$$s_j = \frac{(R/G)^2 + \sum_{i=0}^k s_i^2}{2\sum_{i=0}^k s_i}, \quad j = 0,\ldots,k.$$

This implies that the optimal stepsizes are constant, i.e., $s_j = c$, $j = 0,\ldots,k$, and the constant c is found from the above identity:

$$c = \frac{(R/G)^2 + (k+1)c^2}{2(k+1)c},$$

from which we obtain $c = (R/G)/\sqrt{k+1}$, thus

$$s_j = \frac{R}{G\sqrt{k+1}}, \quad j = 0,\ldots,k.$$

With this choice for the stepsizes, the bound in (12.78) becomes

$$f_{0,k}^* - p^* \le \frac{RG}{\sqrt{k+1}}.$$

This means that the number of iterations needed to reach accuracy $f_{0,k}^* - p^* \le \epsilon$ grows as $O(\epsilon^2)$.

Remark 12.3 *Subgradients and descent directions.* It is important to remark that the subgradient step in (12.74) does not in general decrease the objective value. This is due to the fact that a negative subgradient $-g_k$ need not be a descent direction for f_0 at x_k (contrary to what happens in the differentiable case, where the negative gradient $-\nabla f_0(x_k)$ always is the steepest descent direction). If desired, however, the method can be modified so that the subgradient step in (12.74), prior to projection, is indeed a descent step. In order to do this, we need to select an appropriate subgradient in the subdifferential, as explained next.

Recall the definition of directional derivative of a convex function $f : \mathbb{R}^n \to \mathbb{R}$ at x in direction $v \in \mathbb{R}^n$:

$$f_v'(x) = \lim_{h \to 0^+} \frac{f(x+hv) - f(x)}{h}.$$

If f is differentiable at x, then $f_v'(x) = v^\top \nabla f(x)$. If f is non-differentiable at x, then

$$f_v'(x) = \max_{g \in \partial f(x)} v^\top g.$$

Now, the direction v is a descent direction for f at x, if $f_v'(x) < 0$. Therefore, if we have the entire subdifferential of f, we can search this set for a specific subgradient which is also a descent direction. To this end, it suffices to minimize $f_v'(x)$ over all directions, and check if the minimum is negative:

$$
\begin{aligned}
\min_{\|v\|_2=1} f_v'(x) &= \min_{\|v\|_2=1} \max_{g \in \partial f(x)} v^\top g \\
\text{[using max-min theorem]} &= \max_{g \in \partial f(x)} \min_{\|v\|_2=1} v^\top g \\
\text{[min is achieved for } v^* = -g/\|g\|_2] &= \max_{g \in \partial f(x)} -\|g\|_2 \\
&= -\min_{g \in \partial f(x)} \|g\|_2.
\end{aligned}
$$

Therefore, we solve the convex optimization problem

$$g^* = \arg \min_{g \in \partial f(x)} \|g\|_2.$$

If $\|g^*\|_2 = 0$, then $0 \in \partial f(x)$, and this means that x is a critical point, so we have found an optimal point. If instead $\|g^*\|_2 > 0$, then we have found a descent direction:

$$v^* = -\frac{g^*}{\|g^*\|_2}.$$

12.4.2   The alternate subgradient algorithm

In this section we discuss a variation on the subgradient algorithm which is useful for dealing with problems where the feasible set $\mathcal{X}$ is described via a set of inequalities. More precisely, we consider a problem in the form (12.38)

$$p^* = \min_{x \in \mathbb{R}^n} f_0(x) \tag{12.79}$$
$$\text{s.t.:} \quad f_i(x) \leq 0, \quad i = 1, \ldots, m,$$

where $f_0, f_i, i = 1, \ldots, m$, are convex but possibly non-differentiable. Defining

$$h(x) = \max_{i=1,\ldots,m} f_i(x),$$

this problem can be equivalently written in a form with a single convex inequality constraint, as follows:

$$p^* = \min_{x \in \mathbb{R}^n} f_0(x) \quad \text{s.t.:} \, h(x) \leq 0. \tag{12.80}$$

For this problem, we consider a subgradient-type algorithm of the form

$$x_{k+1} = x_k - s_k g_k,$$

where $s_k \geq 0$ is a stepsize, and

$$g_k \in \begin{cases} \partial f_0(x_k) & \text{if } h(x_k) \leq 0, \\ \partial h(x_k) & \text{if } h(x_k) > 0. \end{cases} \tag{12.81}$$

Notice that a subgradient of h can be obtained from subgradients of the component functions f_i, by applying the max rule of subgradient calculus, see Section 8.2.3.1.

The idea behind this algorithm is that, at each iteration k, we perform a step in the direction of a negative subgradient of the objective function f_0, if x_k is feasible, or a step in the direction of a negative subgradient of the constraint function h, if x_k is infeasible. Let us denote again by $f_{0,k}^*$ the objective value of the best feasible point up to iteration k (note that this need not be the value of f_0 at iteration k)

$$f_{0,k}^* = \min\{f_0(x_i) : x_i \text{ is feasible, } i = 0, \dots, k\}.$$

We next show that, under suitable assumptions, $f_{0,k}^* \to p^*$ as $k \to \infty$.

Proposition 12.2 (Convergence of alternate subgradient)   *Assume that problem (12.80) admits a strictly feasible point x_{sf} and that it attains an optimal solution x^*. Assume further that there exist constants R, G such that $\|x_0 - x^*\|_2 \leq R$, $\|x_0 - x_{sf}\|_2 \leq R$, $\|g_k\|_2 \leq G$ for all k. Then, it holds that $f_{0,k}^* \to p^*$ as $k \to \infty$.*

Proof We proceed by contradiction. Assume that $f_{0,k}^*$ does not tend to p^*. Then there exists $\epsilon > 0$ such that $f_{0,k}^* \geq p^* + \epsilon$ for all k, hence

$$f_0(x_k) \geq p^* + \epsilon$$

for all k for which x_k is feasible. We next show that this leads to a contradiction. Let first

$$\tilde{x} = (1 - \theta)x^* + \theta x_{sf}, \quad \theta \in (0,1).$$

By convexity of f_0 we have

$$f_0(\tilde{x}) \leq (1 - \theta)p^* + \theta f_0(x_{sf}).$$

By choosing $\theta = \min\{1, (\epsilon/2)(f_0(x_{sf}) - p^*)^{-1}\}$, this implies that

$$f_0(\tilde{x}) \leq p^* + \epsilon/2, \tag{12.82}$$

that is, $\tilde{x}$ is $(\epsilon/2)$-suboptimal. Moreover,

$$h(\tilde{x}) \leq (1 - \theta)h(x^*) + \theta h(x_{sf}) \leq \theta h(x_{sf}),$$

therefore

$$h(\tilde{x}) \leq -\mu, \quad \mu \doteq -\theta h(x_{sf}) > 0. \tag{12.83}$$

Consider now an $i \in \{0, 1, \dots, k\}$ for which x_i is feasible. Then $g_i \in \partial f_0(x_i)$ and $f_0(x_i) \geq p^* + \epsilon$ which, from (12.82), gives

$$f_0(x_i) - f_0(\tilde{x}) \geq \epsilon/2,$$

and therefore

$$
\begin{aligned}
\|x_{i+1} - \tilde{x}\|_2^2 &= \|x_i - \tilde{x} - s_i g_i\|_2^2 \\
&= \|x_i - \tilde{x}\|_2^2 - 2s_i g_i^\top (x_i - \tilde{x}) + s_i^2 \|g_i\|_2^2 \\
&\quad [\text{since } f_0(\tilde{x}) \geq f_0(x_i) + g_i^\top (\tilde{x} - x_i)] \\
&\leq \|x_i - \tilde{x}\|_2^2 - 2s_i(f_0(x_i) - f_0(\tilde{x})) + s_i^2 \|g_i\|_2^2 \\
&\leq \|x_i - \tilde{x}\|_2^2 - s_i \epsilon + s_i^2 \|g_i\|_2^2.
\end{aligned}
$$

Suppose instead that $i \in \{0, 1, \ldots, k\}$ is such that x_i is infeasible. Then $g_i \in \partial h(x_i)$ with $h(x_i) > 0$, and from (12.83) we have that

$$h(x_i) - h(\tilde{x}) \geq \mu.$$

Therefore, a reasoning identical to the previous one yields

$$\|x_{i+1} - \tilde{x}\|_2^2 \quad \leq \quad \|x_i - \tilde{x}\|_2^2 - 2s_i\mu + s_i^2\|g_i\|_2^2.$$

Hence, for any iteration (feasible or infeasible), we have

$$\|x_{i+1} - \tilde{x}\|_2^2 \leq \|x_i - \tilde{x}\|_2^2 - s_i\beta + s_i^2\|g_i\|_2^2,$$

where we defined $\beta \doteq \min(\epsilon, 2\mu) > 0$. Applying the above inequality recursively for $i = 0, \ldots, k$, we obtain

$$\|x_{k+1} - \tilde{x}\|_2^2 \leq \|x_0 - \tilde{x}\|_2^2 - \beta \sum_{i=0}^{k} s_i + \sum_{i=0}^{k} s_i^2 \|g_i\|_2^2.$$

From this it follows that

$$\beta \sum_{i=0}^{k} s_i \leq R^2 + G^2 \sum_{i=0}^{k} s_i^2,$$

hence

$$\beta \leq \frac{R^2 + G^2 \sum_{i=0}^{k} s_i^2}{\sum_{i=0}^{k} s_i}.$$

Now, if the stepsize sequence is diminishing and non-summable, then the right-hand side of the previous expression goes to zero as $k \to \infty$, thus leading to a contradiction. $\qquad\square$

12.4.3   The ellipsoid algorithm

Consider the setup and notation in (12.79)–(12.80) and, for $x \in \mathbb{R}^n$, define, similarly to what we did for the alternate subgradient algorithm,

$$g_x \in \begin{cases} \partial f_0(x) & \text{if } h(x) \leq 0, \\ \partial h(x) & \text{if } h(x) > 0. \end{cases}$$

Now, if x is a feasible point, i.e., $x \in \mathcal{X}$, where $\mathcal{X} \doteq \{x : h(x) \leq 0\}$, then g_x is a subgradient of f_0 at x, and we know that it defines a hyperplane that divides the whole space into two half-spaces, with the optimal points x^* lying in the half-space $\mathcal{H}_- = \{z : g_x^\top(z - x) \leq 0\}$, see Figure 12.9. This means that we can restrict our search for optimal points to this half-space, while the complementary half-space $\mathcal{H}_{++}$ can be cut out from the search. Similarly, if x is infeasible, i.e.,

$h(x) > 0$, then g_x is a subgradient of h at x, and by definition of a subgradient we have that

$$h(z) \geq h(x) + g_x^\top (z - x).$$

This implies that for all z in the half-space $\mathcal{H}_{++} = \{z : g_x^\top (z - x) > 0\}$ it holds that $h(z) > h(x) > 0$, hence all $z \in \mathcal{H}_{++}$ are also infeasible. Again, we can cut $\mathcal{H}_{++}$ out of our search, and restrict our interest to the complementary half-space $\mathcal{H}_-$. In summary, we have that at any point x, feasible or infeasible, g_x gives us the information that the "interesting" points are localized in the half-space $\mathcal{H}_-$. The ellipsoid algorithm (EA) belongs to the class of so-called *localization* methods, which exploit this information by progressively reducing a localization set that is guaranteed to contain the optimal points, by intersecting it with the *cuts* provided by the subgradients of the objective or of the constraint function.

The EA uses ellipsoidal localization sets. It is initialized with an ellipsoid $\mathcal{E}_0$ that is guaranteed to contain an optimal point x^*, and then iteratively updates this set with ellipsoids whose volume decreases and that are always guaranteed to contain the optimal point. More precisely, let us define the ellipsoid

$$\mathcal{E}_k = \{z \in \mathbb{R}^n : (z - x_k)^\top A_k^{-1}(z - x_k) \leq 1\},$$

where $x_k \in \mathbb{R}^n$ is the center of the ellipsoid and $A_k \in \mathbb{S}^n_{++}$ is the shape matrix, and assume that $x^* \in \mathcal{E}_k$. Let g_k be defined as in (12.81), and define the half-space $\mathcal{H}^k_- = \{z : g_k^\top (z - x_k) \leq 0\}$. By the previous reasoning, we are guaranteed that the optimal point belongs to the intersection of $\mathcal{E}_k$ and $\mathcal{H}^k_-$, i.e.,

$$x^* \in \mathcal{E}_k \cap \mathcal{H}^k_-.$$

Then we update the ellipsoid by computing the minimum-volume ellipsoid $\mathcal{E}_{k+1}$ (with center x_{k+1} and shape matrix A_{k+1}) that contains $\mathcal{E}_k \cap \mathcal{H}^k_-$, and iterate; see Figure 12.10.

Interestingly, there exists an explicit closed-form formula for computing the updated ellipsoid (that is, the minimum volume ellipsoid $\mathcal{E}_{k+1}$ containing $\mathcal{E}_k \cap \mathcal{H}^k_-$). Specifically, the center x_{k+1} and shape matrix A_{k+1} of $\mathcal{E}_{k+1}$ are obtained as follows (for dimension $n > 1$):

$$x_{k+1} = x_k - \frac{1}{n+1} \frac{A_k g_k}{\sqrt{g_k^\top A_k g_k}}, \tag{12.84}$$

$$A_{k+1} = \frac{n^2}{n^2 - 1} \left(A_k - \frac{2}{n+1} \frac{A_k g_k g_k^\top A_k}{g_k^\top A_k g_k} \right). \tag{12.85}$$

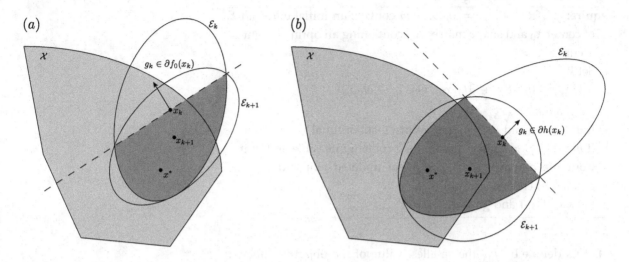

Figure 12.10 Ellipsoid algorithm: (a) if x_k is feasible, we cut a half-space containing points with objective value worse than $f_0(x_k)$; (b) if x_k is infeasible, we cut a half-space containing infeasible points.

Moreover, it can be proved that the volume of $\mathcal{E}_{k+1}$ is strictly smaller than the volume of $\mathcal{E}_k$, precisely

$$\text{vol}\,\mathcal{E}_{k+1} < e^{-\frac{1}{2n}}\,\text{vol}\,\mathcal{E}_k. \tag{12.86}$$

Observe that, since $x^* \in \mathcal{E}_k$, then if x_k is feasible it holds that

$$
\begin{aligned}
f_0(x^*) &\geq f(x_k) + g_k^\top (x^* - x_k) \\
&\geq f(x_k) + \inf_{x \in \mathcal{E}_k} g_k^\top (x - x_k)
\end{aligned}
$$

[see Example 9.14] $= f(x_k) - \sqrt{g_k^\top A_k g_k}.$

Hence, if x_k is feasible and $\sqrt{g_k^\top A_k g_k} \leq \epsilon$, we can stop the algorithm and conclude that ϵ-suboptimality on the objective function has been reached, since $f(x_k) - f_0(x^*) \leq \epsilon$. If instead x_k is infeasible, then it holds that, for all $x \in \mathcal{E}_k$,

$$
\begin{aligned}
h(x) &\geq h(x_k) + g_k^\top (x - x_k) \\
&\geq h(x_k) + \inf_{x \in \mathcal{E}_k} g_k^\top (x - x_k)
\end{aligned}
$$

[see Example 9.14] $= h(x_k) - \sqrt{g_k^\top A_k g_k}.$

Hence, if $h(x_k) - \sqrt{g_k^\top A_k g_k} > 0$ then $h(x) > 0$ for all $x \in \mathcal{E}_k$ and the problem has no feasible solution. We next summarize the scheme for the ellipsoid algorithm (Algorithm 18), and then state a result on the rate of convergence of this algorithm (in the case of a feasible problem).

Algorithm 18 Ellipsoid algorithm.

Require: $f_i : \mathbb{R}^n \to \mathbb{R}$, $i = 0, 1, \ldots, m$ convex; an initial ellipsoid $\mathcal{E}_0$ of center x_0 and shape matrix A_0, containing an optimal point x^*; accuracy $\epsilon > 0$.

1: Set $k = 0$
2: If $h(x_k) \leq 0$, let $g_k \in \partial f_0(x_k)$, else $g_k \in \partial h(x_k)$
3: Evaluate $\eta_k = \sqrt{g_k^\top A_k g_k}$
4: If $h(x_k) \leq 0$ and $\eta_k \leq \epsilon$, then return ϵ-suboptimal x_k and quit
5: If $h(x_k) - \eta_k > 0$, then declare the problem infeasible and quit
6: Compute parameters x_{k+1}, A_{k+1} of updated ellipsoid $\mathcal{E}_{k+1}$ according to (12.84), (12.85)
7: Let $k \leftarrow k + 1$ and go to 2.

Let us denote by $f_{0,k}^*$ the smallest value of the objective function obtained from 0 to k in the feasible iterations of the algorithm, that is

$$f_{0,k}^* = \min_{i=0,\ldots,k} \{f_0(x_i) : x_i \in \mathcal{X}\},$$

and let by convention $f_{0,k}^* = +\infty$ if none of the k iterations was feasible. Let further $B_\rho(x)$ denote the Euclidean ball in $\mathbb{R}^n$ of center x and radius ρ. The following proposition holds.

Proposition 12.3 (Convergence of the ellipsoid algorithm) *Consider problem (12.80) and assume that p^* is finite and that it is attained at an optimal solution x^*. Assume further that:*

1. *f_0 is Lipschitz continuous on $\mathbb{R}^n$, with Lipschitz constant G;*

2. *$\mathcal{E}_0$ is a ball of center x_0 and radius R, which contains x^*;*

3. *for given $\epsilon > 0$ there exists $\alpha \in (0, 1]$ and $x \in \mathcal{X}$ such that*

$$B_{\alpha\epsilon/G}(x) \subseteq \mathcal{E}_0 \cap \mathcal{X} \cap B_{\epsilon/G}(x^*).$$

Then it holds for the ellipsoid algorithm that

$$k > 2n^2 \ln \frac{GR}{\alpha\epsilon} \quad \Rightarrow \quad f_{0,k}^* \leq p^* + \epsilon.$$

Proof Observe that the cuts in the ellipsoid algorithm preserve at each iteration i either a half-space containing the whole feasible set (if x_i is infeasible), or a half-space of points whose objective value is no larger than the current one (if x_i is feasible). Therefore,

$$\mathcal{E}_0 \cap \mathcal{X} \cap \{x : f_0(x) \leq f_{0,k}^*\} \subseteq \mathcal{E}_k.$$

Suppose now that $f_{0,k}^* > p^* + \epsilon$. Then clearly the set $\{x : f_0(x) \leq p^* + \epsilon\}$ is contained in the set $\{x : f_0(x) \leq f_{0,k}^*\}$, hence

$$\mathcal{E}_0 \cap \mathcal{X} \cap \{x : f_0(x) \leq p^* + \epsilon\} \subseteq \mathcal{E}_0 \cap \mathcal{X} \cap \{x : f_0(x) \leq f_{0,k}^*\} \subseteq \mathcal{E}_k.$$
$$(12.87)$$

Since f_0 has Lipschitz constant G (that is $|f_0(x) - f_0(y)| \leq G\|x - y\|_2$ for all x, y), it follows that

$$\|x - x^*\|_2 \leq \frac{\epsilon}{G} \quad \Rightarrow \quad f_0(x) \leq p^* + \epsilon,$$

which means that the set $B_{\epsilon/G}(x^*) = \{x : \|x - x^*\|_2 \leq \epsilon/G\}$ is contained in the set $\{x : f_0(x) \leq p^* + \epsilon\}$. Hence, we obtain from (12.87) that

$$\mathcal{E}_0 \cap \mathcal{X} \cap B_{\epsilon/G}(x^*) \subseteq \mathcal{E}_k,$$

which implies that

$$\text{vol}(\mathcal{E}_0 \cap \mathcal{X} \cap B_{\epsilon/G}(x^*)) \leq \text{vol}(\mathcal{E}_k). \qquad (12.88)$$

Applying (12.86) recursively we also have that

$$\text{vol}(\mathcal{E}_k) \leq e^{-k/(2n)}\text{vol}(\mathcal{E}_0) = e^{-k/(2n)}\gamma_n R^n, \qquad (12.89)$$

where γ_n here denotes the volume of the Euclidean ball of unit radius in $\mathbb{R}^n$. Now, from Assumption 3 it follows that $\text{vol}(\mathcal{E}_0 \cap \mathcal{X} \cap B_{\epsilon/G}(x^*)) \geq \text{vol}(B_{\alpha\epsilon/G}(x^*)) = \alpha^n \gamma_n(\epsilon/G)^n$, for some $\alpha \in (0, 1]$, hence, chaining this inequality with (12.88) and (12.89), we obtain

$$\alpha^n \gamma_n(\epsilon/G)^n \leq e^{-k/(2n)}\gamma_n R^n,$$

from which we conclude that it must be that

$$k \leq 2n^2 \ln \frac{GR}{\alpha\epsilon}.$$

Recapitulating,

$$f_{0,k}^* > p^* + \epsilon \quad \Rightarrow \quad k \leq 2n^2 \ln \frac{GR}{\alpha\epsilon},$$

from which it follows that

$$k > 2n^2 \ln \frac{GR}{\alpha\epsilon} \quad \Rightarrow \quad f_{0,k}^* \leq p^* + \epsilon.$$

$$\square$$

Notice that Assumption 3 in Proposition 12.3 implies that the feasible set has a nonempty interior, since it must contain a full-dimensional ball of positive radius $\alpha\epsilon/G$.

12.5   Coordinate descent methods

Coordinate descent methods, or more generally block-coordinate descent methods, apply to problems where each variable (or block of variables) is *independently* constrained.[12] Specifically, we consider a special case of a generic minimization problem (12.61):

$$\min_{x=(x_1,\ldots,x_\nu)} f_0(x) \ : \ x_i \in \mathcal{X}_i, \ i = 1,\ldots,\nu. \qquad (12.90)$$

In words, the variable x can be decomposed into ν blocks $x_1,\ldots,x_\nu$, and each block x_i is independently constrained to belong to the set $\mathcal{X}_i$. Coordinate descent methods are based on iteratively minimizing with respect to one block, with all the other blocks being fixed. If

$$x^{(k)} = (x_1^{(k)}, \ldots, x_\nu^{(k)})$$

denotes the value of the decision variable at iteration k, partial minimization problems of the form

$$\min_{x_i \in \mathcal{X}_i} f_0(x_1^{(k)}, \ldots, x_{i-1}^{(k)}, x_i, x_{i+1}^{(k)}, \ldots, x_\nu^{(k)}), \qquad (12.91)$$

are solved. Different methods ensue, based on how exactly we form the next iterate. Other variants, not described here, arise with different line search strategies.

12.5.1   The Jacobi method

In the Jacobi method, we solve all the partial minimization problems (12.91), and then update all the blocks *simultaneously*. That is, for every $i = 1,\ldots,\nu$, we set

$$x_i^{(k+1)} = \arg\min_{x_i \in \mathcal{X}_i} f_0(x_1^{(k)}, \ldots, x_{i-1}^{(k)}, x_i, x_{i+1}^{(k)}, \ldots, x_\nu^{(k)}).$$

The scheme is depicted in Figure 12.11.

Convergence of such a method to the optimal solution of (12.90) is not guaranteed in general. A classical result guarantees existence and convergence to the optimal solution under certain hypotheses of contractivity. Namely, a map $F : \mathcal{X} \to \mathbb{R}^n$ is said to be *contractive* (with respect to some norm $\|\cdot\|$), if for some $\theta \in [0,1)$ it holds that

$$\|F(x) - F(y)\| \le \theta \|x - y\|, \quad \forall x, y \in \mathcal{X}.$$

The following convergence result holds.[13]

Theorem 12.3 *Let f_0 be continuously differentiable, and let the map*

$$F(x) = x - \gamma \nabla f_0(x)$$

[12] An example is the so-called box-constrained QP:

$$\min_x q(x) \ : \ l \le x \le u,$$

where q is a convex quadratic function, and $l, u \in \mathbb{R}^n$ impose upper and lower bounds on the variables.

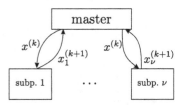

Figure 12.11 At iteration k, the Jacobi method solves ν sub-problems, where, in each sub-problem i, f_0 is minimized with respect to $x_i \in \mathcal{X}_i$, with all other variables being fixed at $x_j^{(k)}, j \ne i$.

[13] This, as well as many other results on distributed optimization methods, may be found in the book by D. Bertsekas and J. Tsitsiklis, *Parallel and Distributed Computation: Numerical Methods*, Athena Scientific, 1997.

be contractive for some $\gamma > 0$, and with respect to the following norm:

$$\|x\| = \max_{i=1,\dots,\nu} \|x_i\|_2 / w_i,$$

where w_i are positive scalars. Then problem (12.90) has a unique solution x^*, and the sequence of points $x^{(k)}$ generated by the Jacobi algorithm converges to x^* at a geometric rate.

12.5.2   Coordinate minimization method

The standard (block) coordinate minimization method (BCM), also known as the Gauss–Seidel method, works similarly to the Jacobi method, but in this algorithm the variable blocks are updated *sequentially*, according to the recursion

$$x_i^{(k+1)} = \arg\min_{x_i \in \mathcal{X}_i} f_0(x_1^{(k+1)}, \dots, x_{i-1}^{(k+1)}, x_i, x_{i+1}^{(k)}, \dots, x_\nu^{(k)}), \quad (12.92)$$

for $i = 1, \dots, \nu$. The logical scheme of this method is illustrated in Figure 12.12.

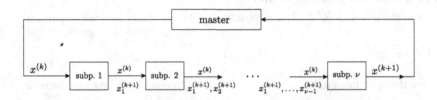

Figure 12.12 At iteration k, the block-coordinate minimization method solves sequentially ν sub-problems, in which variable blocks are successively updated according to (12.92).

Under the same assumptions as in Proposition 12.3, the sequence of points $x^{(k)}$ generated by the BCM algorithm converges to x^* at a geometric rate. Moreover, the following convergence result can be stated for the BCM algorithm for convex differentiable objective.

Theorem 12.4 *Assume f_0 is convex and continuously differentiable on $\mathcal{X}$. Moreover, let f_0 be strictly convex in x_i, when the other variable blocks x_j, $j \neq i$ are held constant. If the sequence $\{x^{(k)}\}$ generated by the BCM algorithm is well defined, then every limit point of $\{x^{(k)}\}$ converges to an optimal solution of problem (12.90).*

The sequential block-coordinate descent method may fail to converge, in general, for non-smooth objectives, even under convexity assumptions. An important exception, however, arises when f_0 is a composite function which can be written as the sum of a convex and differentiable function ϕ and a separable convex (but possibly non-smooth) term, that is

$$f_0(x) = \phi(x) + \sum_{i=1}^{\nu} \psi_i(x_i), \quad (12.93)$$

where ϕ is convex and differentiable, and ψ_i, $i = 1, \ldots, v$ are convex.[14] The following theorem holds.[15]

Theorem 12.5 *Let $x^{(0)} \in \mathcal{X}$ be an initial point for the BCM, and assume that the level set $S_0 = \{x : f_0(x) \leq f_0(x^{(0)})\} \in \operatorname{int} \operatorname{dom} f_0$ is compact. Assume further that f_0 has the form (12.93), where ϕ is convex and differentiable on S_0, and ψ_i, $i = 1, \ldots, v$, are convex. Then, every limit point of the sequence $\{x^{(k)}\}$ generated by the BCM converges to an optimal solution of problem (12.90).*

In particular, Theorem 12.5 is applicable to various ℓ_1-norm regularized problems, such as the LASSO, for which convergence of sequential coordinate descent methods is thus guaranteed.

12.5.3 *Power iteration and block coordinate descent*

As mentioned in Section 7.1, power iteration refers to a class of methods that can be applied to specific eigenvalue and singular value problems of linear algebra. Such methods are actually connected to coordinate descent. Consider for example the problem of finding a rank-one approximation to a given matrix[16] $A \in \mathbb{R}^{m,n}$. The problem can be expressed as

$$\min_{x,y} \|A - xy^\top\|_F^2, \tag{12.94}$$

and the stationarity conditions[17] for this problem are

$$x\|y\|_2^2 = A^\top y, \quad y\|x\|_2^2 = Ax.$$

In terms of the normalized vectors $u = x/\|x\|_2$, $v = y/\|y\|_2$, and with $\sigma \doteq \|x\|_2\|y\|_2$, we write the above as

$$Au = \sigma v, \quad A^\top v = \sigma u.$$

The stationarity conditions express that u, v are, respectively, the normalized left and right singular vectors of the matrix A, corresponding to the singular value σ. Hence, we can solve the above problem via SVD.

We now interpret this result in the context of block-coordinate descent. Rewriting problem (12.94) as

$$\min_{u,v,\sigma} \|A - \sigma uv^\top\|_F^2 \; : \; \|u\|_2 = \|v\|_2 = 1, \; \sigma \geq 0,$$

and solving for σ, we obtain that, at optimum, $\sigma = v^\top A u$, if the latter quantity is non-negative, and zero otherwise. Plugging this value back, we reduce the problem to

$$\max_{u,v} v^\top A u \; : \; \|u\|_2 = \|v\|_2 = 1.$$

[14] Notice that this setup includes the possibility of convex independent constraints on the variables of the form $x_i \in \mathcal{X}_i$, since functions ψ_i may include a term given by the indicator function of the set $\mathcal{X}_i$.

[15] See Theorem 4.1 in P. Tseng, Convergence of a block coordinate descent method for non-differentiable minimization, *J. Optimization Theory and Applications*, 2001.

[16] More details on rank approximation problems are given in Section 5.3.1.

[17] A point z is stationary for the problem
$$\min_z f(z)$$
with f differentiable if the gradient is zero at z: $\nabla f(z) = 0$; see Section 12.2.2.2.

We can apply block-coordinate descent to this problem, with the two blocks $u \in \mathbb{R}^m$, $v \in \mathbb{R}^n$. The Jacobi algorithm then becomes[18]

$$
\begin{aligned}
u_{k+1} &= \arg\max_{u\,:\,\|u\|_2=1} v_k^\top A u = \frac{A^\top v_k}{\|A^\top v_k\|_2}, \\
v_{k+1} &= \arg\max_{v\,:\,\|v\|_2=1} v^\top A u_{k+1} = \frac{A u_k}{\|A u_k\|_2}.
\end{aligned}
$$

[18] The Gauss–Seidel algorithm would have u_{k+1} instead of u_k in the v-update.

Provided the block coordinate descent converges to a stationary point, at such a point we have

$$
\sigma u = A^\top v, \quad \sigma v = A u, \quad \sigma = \|A^\top v\|_2 = \|A u\|_2.
$$

The Jacobi algorithm thus reduces to the power iteration method for computing the largest singular value of the matrix A, and associated vectors u, v.

12.5.4   Thresholded power iteration for sparse PCA

As mentioned in Section 13.5.2, sometimes it is desirable to solve a variant of the low-rank approximation problem of the form:

$$
\min_{x,y} \|A - xy^\top\|_\mathrm{F}^2 \;:\; \mathrm{card}(x) \le k, \;\; \mathrm{card}(y) \le h,
$$

where $A \in \mathbb{R}^{m,n}$, and $k \le m$, $h \le n$ control the cardinality (number of nonzero elements) in x, y. The problem is hard for all but small values of m, n. A heuristic that is efficient in practice consists of using the power iteration method with a modification: at each iteration, we reset to zero all but the largest k variables in x and the largest h variables in y. The recursion takes the form

$$
\begin{aligned}
u(k+1) &= T_h\!\left(\frac{Av(k)}{\|Av(k)\|_2} \right), \\
v(k+1) &= T_k\!\left(\frac{A^\top \overline{u}(k+1)}{\|A^\top \overline{u}(k+1)\|_2} \right),
\end{aligned}
$$

where $T_k(z)$ is the threshold operator, which zeros out all the elements in the input vector z, except the ones having the k largest magnitudes.

12.6   Decentralized optimization methods

In this section, we outline two techniques that can be used to solve certain classes of optimization problems in a *decentralized* way. By decentralized we mean, loosely speaking, that the actual numerical optimization process is executed, either physically or conceptually, on several computational units, instead of on a single, centralized, unit. For example, a large-scale problem that could be too large to be solved on a single processor may be split into smaller sub-problems that are solved independently on a grid of processors. Such a process typically requires also some form of coordination among the processing units, that can happen either on a peer-to-peer basis or require a central coordination unit, usually called the *master* unit. The simplest structure of an optimization problem that leads to a fully decentralized solution is that of a so-called *separable* problem, that is a problem of the form

$$p^* = \min_{x_1,\ldots,x_v} \sum_{i=1}^{v} f_{0,i}(x_i)$$
$$\text{s.t.:} \quad x_i \in \mathcal{X}_i, \qquad i = 1,\ldots,v,$$

where $x_i \in \mathbb{R}^{n_i}$, $i = 1,\ldots,v$, are the blocks of the overall variable $x = (x_1,\ldots,x_v)$, with $n_1 + \cdots + n_v = n$. Here, since the objective function is the sum of terms $f_{0,i}$ that depend only on x_i (the i-th block of the variable x), and since the x_is are constrained independently in the sets $\mathcal{X}_i$, it is straightforward to see that $p^* = \sum_{i=1}^{v} f_{0,i}^*$, where $f_{0,i}^*$, $i = 1,\ldots,v$, are the optimal values of the sub-problems

$$f_{0,i}^* = \min_{x_i} \quad f_{0,i}(x_i)$$
$$\text{s.t.:} \quad x_i \in \mathcal{X}_i.$$

The situation becomes more interesting, but also less obvious to solve, when some coupling is present, either in the objective function and/or in the constraints. For example, in a problem of the form

$$p^* = \min_{x_1,\ldots,x_v} \quad f_0(x_1,\ldots,x_v) \tag{12.95}$$
$$\text{s.t.:} \quad x_i \in \mathcal{X}_i, \qquad i = 1,\ldots,v$$

the variables x_i are constrained in independent sets $\mathcal{X}_i$ (we say that the feasible set of the problem is the Cartesian product of the $\mathcal{X}_i$s, that is $\mathcal{X} = \mathcal{X}_1 \times \cdots \times \mathcal{X}_v$), but are coupled in the objective function f_0, that now depends on the whole vector $x \in \mathcal{X}$. Similarly, in a problem of the form

$$p^* = \min_{x_1,\ldots,x_v} \quad \sum_{i=1}^{v} f_{0,i}(x_i) \tag{12.96}$$
$$\text{s.t.:} \quad x \in \mathcal{X},$$

the objective function is separable, but the variables x_i are coupled by the constraint $x \in \mathcal{X}$ (where $\mathcal{X}$ is in general *not* the Cartesian product of the $\mathcal{X}_i$s).

In the next section we discuss two classical methods for decentralized optimization, based on the idea of *decomposition*.

12.6.1 Decomposition methods

The idea of decomposition methods is to decompose an original problem into smaller sub-problems that can be independently solved by local processors, which are coordinated by a master process by means of some suitable exchange of messages. In their basic form, decomposition methods deal with problems of the form (12.96) with a separable objective but with some coupling in the constraints, where the coupling is typically in the form of a budget of resources constraint on the variables. Decomposition methods are typically classified into *primal* or *dual* methods. In primal methods, the master problem manages resources directly, by assigning individual budgets to the sub-problems; in dual methods the master problem manages resources indirectly, by assigning resource prices to the sub-problems, which then have to decide the amount of resources to be used depending on the price. These two types of decomposition method are described in the next sections.

12.6.1.1 Dual decomposition. Consider a problem of the form

$$p^* = \min_{x_1, \ldots, x_v} \quad \sum_{i=1}^{v} f_{0,i}(x_i) \tag{12.97}$$
$$\text{s.t.:} \quad x_i \in \mathcal{X}_i, \qquad i = 1, \ldots, v,$$
$$\sum_{i=1}^{v} h_i(x_i) \leq c.$$

In this problem, $h_i(x_i)$ represents the resource consumption relative to the decision x_i, and the last constraint has the interpretation of a budget limit on resource consumption. This last constraint is the one coupling the variables, since were it absent the problem would be separable. Let us consider the (partial) Lagrangian of this problem, obtained by "dualizing" the coupling constraint:

$$\mathcal{L}(x, \lambda) = \sum_{i=1}^{v} f_{0,i}(x_i) + \lambda^{\top} \left(\sum_{i=1}^{v} h_i(x_i) - c \right).$$

Then the dual function

$$g(\lambda) = \min_{x \in \mathcal{X}_1 \times \cdots \times \mathcal{X}_v} \mathcal{L}(x, \lambda)$$
$$= \sum_{i=1}^{v} \min_{x_i \in \mathcal{X}_i} \left(f_{0,i}(x_i) + \lambda^{\top} h_i(x_i) \right) - \lambda^{\top} c$$

is separable, that is, for given λ, the value of $g(\lambda)$ and the corresponding minimizers $x_i^*(\lambda)$ can be computed independently on v processors, each of which solves the minimization problem

$$g_i(\lambda) = \min_{x_i \in \mathcal{X}_i} f_{0,i}(x_i) + \lambda^{\top} h_i(x_i), \tag{12.98}$$

and returns the value $g_i(\lambda)$ and the minimizer $x_i^*(\lambda)$ to the master process. The role of the master process here is to solve the dual problem, that is to maximize $g(\lambda)$ over $\lambda \geq 0$:

$$d^* = \max_{\lambda \geq 0} \sum_{i=1}^{\nu} g_i(\lambda) - \lambda^\top c. \qquad (12.99)$$

This latter maximization can be performed by means of a projected gradient or subgradient method. We recall from Section 8.5.9 that a subgradient of $g(\lambda)$ is given by

$$\eta(\lambda) = \sum_{i=1}^{\nu} h_i(x_i^*(\lambda)) - c. \qquad (12.100)$$

Moreover, if $\mathcal{X}_i$ are compact and nonempty and the minimizers $x_i^*(\lambda)$ are unique for all λ, then $g(\lambda)$ is differentiable, and the above $\eta(\lambda)$ provides the gradient of g at λ. In any case, the master process can update the value of λ via a projected gradient or subgradient step

$$\lambda \leftarrow [\lambda + s \cdot \eta(\lambda)]_+, \qquad (12.101)$$

where s is a suitably chosen stepsize (found, e.g., via a backtracking line search when g is differentiable, or according to the rules of subgradient methods otherwise). The conceptual scheme of the dual decomposition method is depicted in Figure 12.13. The interpretation is that the master process updates the *prices* of resources, represented by the dual variable λ, and then the parallel processes find optimal allocations, for the given prices. This method will converge to the optimum of the dual problem (12.99). However, if strong duality holds, and the minimizers $x_i^*(\lambda)$ are unique for all λ, then the method also provides the primal optimal variables, since $x_i^*(\lambda) \to x_i^*$.

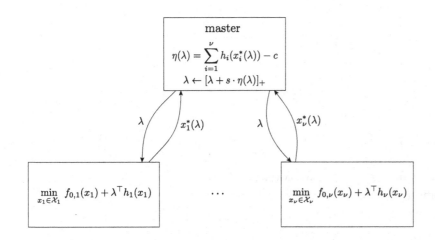

Figure 12.13 At each iteration of a dual decomposition method, processors $i = 1, \dots, \nu$ solve problems (12.98) for the given λ and return the minimizers $x_i^*(\lambda)$ and values $h_i(x_i^*(\lambda))$ to the master process. The master process aims at maximizing $g(\lambda)$, thus it builds a subgradient of g at λ according to (12.100) and updates λ according to (12.101).

Dual decomposition with coupling variables.   Dual decomposition can also be applied to problems where coupling is due to variables in the objective function, rather than to constraints. Consider for instance a problem of the form

$$p^* = \min_{x_1, x_2, y} \quad f_{0,1}(x_1, y) + f_{0,2}(x_2, y) \tag{12.102}$$
$$\text{s.t.:} \quad x_1 \in \mathcal{X}_1,\ x_2 \in \mathcal{X}_2,\ y \in \mathcal{Y},$$

where the decision variable is $x = (x_1, x_2, y)$. This problem would be separable, except for the presence of the coupling variable y in the objective function. A standard approach in this case is to introduce slack variables y_1, y_2 and an artificial equality constraint $y_1 = y_2$, so as to rewrite the problem as

$$p^* = \min_{x_1, x_2, y_1, y_2} \quad f_{0,1}(x_1, y_1) + f_{0,2}(x_2, y_2)$$
$$\text{s.t.:} \quad x_1 \in \mathcal{X}_1,\ y_1 \in \mathcal{Y},\ x_2 \in \mathcal{X}_2,\ y_2 \in \mathcal{Y},$$
$$y_1 = y_2.$$

Then, we "dualize" the constraint $y_1 = y_2$ by introducing a Lagrange multiplier λ, thus obtaining the Lagrangian

$$\mathcal{L}(x_1, x_2, y_1, y_2, \lambda) = f_{0,1}(x_1, y_1) + f_{0,2}(x_2, y_2) + \lambda^\top (y_1 - y_2).$$

The dual function

$$g(\lambda) = \min_{x_1 \in \mathcal{X}_1, x_2 \in \mathcal{X}_2, y_1, y_2 \in \mathcal{Y}} \mathcal{L}(x_1, x_2, y_1, y_2, \lambda)$$
$$= \min_{x_1 \in \mathcal{X}_1, y_1 \in \mathcal{Y}} f_{0,1}(x_1, y_1) + \lambda^\top y_1 + \min_{x_2 \in \mathcal{X}_2, y_2 \in \mathcal{Y}} f_{0,2}(x_2, y_2) - \lambda^\top y_2$$

is thus separable in the two groups of variables (x_1, y_1) and (x_2, y_2). A dual decomposition scheme is thus applicable, whereby, for given λ, the subprocesses compute in parallel and independently the two components of the dual function, and return the values of the minimizers $x_1^*(\lambda)$, $x_2^*(\lambda)$, $y_1^*(\lambda)$, $y_2^*(\lambda)$ to the master process. The master process in turn aims at maximizing $g(\lambda)$ by updating the current value of λ according to a gradient or subgradient step

$$\lambda \leftarrow \lambda + s \cdot \eta(\lambda),$$

where $\eta(\lambda)$ is a subgradient of g at λ (or the gradient, if $g(\lambda)$ is differentiable), which is given by

$$\eta(\lambda) = y_1^*(\lambda) - y_2^*(\lambda).$$

12.6.1.2 Primal decomposition.   Consider again problem (12.97), and introduce slack variables t_i, $i = 1, \ldots, \nu$. Then, the problem can be rewritten equivalently in the form

$$p^* = \min_{x_1, t_1, \ldots, x_v, t_v} \quad \sum_{i=1}^{v} f_{0,i}(x_i)$$
$$\text{s.t.:} \quad x_i \in \mathcal{X}_i, \qquad i = 1, \ldots, v,$$
$$\sum_{i=1}^{v} t_i \le c,$$
$$h_i(x_i) \le t_i, \quad i = 1, \ldots, v.$$

If we fix the value of $t = (t_1, \ldots, t_v)$ to some feasible values such that $\sum_{i=1}^{v} t_i \le c$, then this problem becomes separable, with optimal value $p^*(t) = p_1^*(t_1) + \cdots + p_v^*(t_v)$, where the values $p_i^*(t_i)$ are computed by solving the following decoupled sub-problems: for $i = 1, \ldots, v,$

$$p_i^*(t_i) = \min_{x_i} \quad f_{0,i}(x_i) \tag{12.103}$$
$$\text{s.t.:} \quad x_i \in \mathcal{X}_i,$$
$$h_i(x_i) \le t_i.$$

The fixed values of t_i have the interpretation of a direct assignment of resources. A master process should then manipulate t_i, $i = 1, \ldots, v$, so as to minimize $p^*(t)$, that is solve

$$\min_{t_1, \ldots, t_v} \quad \sum_{i=1}^{v} p_i^*(t_i)$$
$$\text{s.t.:} \quad \sum_{i=1}^{v} t_i \le c.$$

If $f_{0,i}$, $i = 1, \ldots, v$ are convex functions, then the discussion in Section 8.5.8 shows that also p_i^* are convex functions. Moreover, a subgradient of p_i^* at t_i is given by $-\lambda_{t_i}$, where λ_{t_i} is the optimal dual variable (Lagrange multiplier) relative to problem (12.103). Subgradient information can then be used by the master process for updating the resource assignments t_i, according to a projected subgradient step

$$t \leftarrow \left[\begin{bmatrix} t_1 \\ \vdots \\ t_v \end{bmatrix} + s \cdot \begin{bmatrix} \lambda_{t_1} \\ \vdots \\ \lambda_{t_v} \end{bmatrix} \right]_{\mathcal{T}}, \tag{12.104}$$

where s is a stepsize, and $[\cdot]_{\mathcal{T}}$ corresponds to projection onto the feasible set $\mathcal{T}$ of t, which in this case is $\mathcal{T} = \{t : \sum_i t_i \le c\}$. This primal decomposition scheme is illustrated in Figure 12.14.

Primal decomposition with coupling variables. Primal decomposition can also be applied to problems where coupling is due to variables in the objective function, rather than to constraints. Consider again

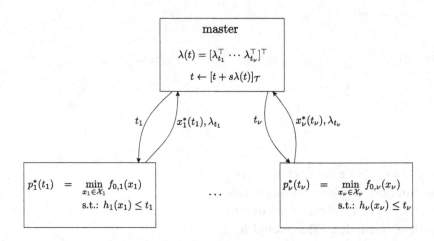

Figure 12.14 At each iteration of a primal decomposition method, processors $i = 1, \ldots, \nu$ solve problems (12.103) with the assigned resource budgets t_i and return the minimizers $x_i^*(t_i)$ as well as the corresponding dual variables λ_{t_i} to the master process. The master process aims at maximizing $p^*(t)$, thus it builds a subgradient of p^* at t and updates t according to (12.104).

a problem of the form (12.102), in which the decision variable is $x = (x_1, x_2, y)$, where y is the "complicating" variable that couples the problem. For fixed y, however, the problem decouples into two sub-problems

$$p_1^*(y) = \min_{x_1 \in \mathcal{X}_1} f_{0,1}(x_1, y), \qquad (12.105)$$

$$p_2^*(y) = \min_{x_2 \in \mathcal{X}_2} f_{0,2}(x_2, y), \qquad (12.106)$$

and the optimal value of (12.102) is given by

$$p^* = \min_{y \in \mathcal{Y}} p_1^*(y) + p_2^*(y). \qquad (12.107)$$

If $f_{0,i}(x_i, y)$ are jointly convex in (x_i, y), then, by the the partial minimizations in (12.105), (12.106), $p_i^*(y)$ are also convex. In this case, a primal decomposition approach may work as follows: a master process (which aims at solving (12.107)) sends a value $y \in \mathcal{Y}$ to the sub-problems (12.105), (12.106), which return to the master minimizers $x_1^*(y)$, $x_2^*(y)$, as well as subgradients $\eta_1(y)$, $\eta_2(y)$ of p_1^*, p_2^* at y, respectively. The master updates the y value according to a projected subgradient rule $y \leftarrow [y - s(\eta_1(y) + \eta_2(y))]_{\mathcal{Y}}$, and the process is iterated.

12.6.1.3 *Hierarchical decompositions.*

In the previous sections we outlined only a few cases and possibilities for decomposition structures. Actually, a variety of different situations may be dealt with via a combination of slack variables, fictitious equality constraints, partial dualizations, etc. Also, decomposition can be performed hierarchically over multiple levels, with a top level master and sub-problems that are themselves decomposed in sub-masters and sub-sub-problems, etc. As an example, consider a problem of the form

$$p^* = \min_{x_i, y} \quad \sum_{i=1}^{\nu} f_{0,i}(x_i, y)$$
$$\text{s.t.:} \quad x_i \in \mathcal{X}_i, \qquad i = 1, \ldots, \nu$$
$$\sum_{i=1}^{\nu} h_i(x_i) \le c,$$
$$y \in \mathcal{Y},$$

in which both coupling variables y and coupling constraints are present. In this case, one could first introduce a primal decomposition with respect to the complicating variable y, and then a dual decomposition with respect to the coupling constraint. This would lead to a three-level decomposition, with a primary master problem, a secondary dual master problem, and then the sub-problems. Alternatively, one could first introduce a dual decomposition, and then a primal one.

Example 12.1 (*Price/demand balancing in an electricity market*) Consider a schematic model of an electricity market, composed by an independent system operator (ISO), an aggregator, and a set of n electricity consumers, see Figure 12.15.

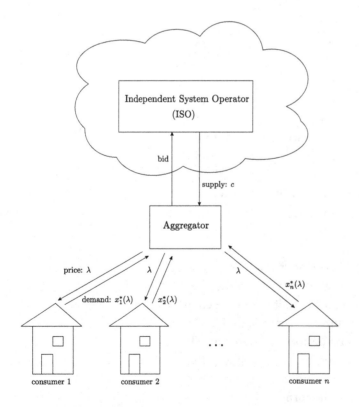

Figure 12.15 Consumer aggregation in an electricity market.

In a given time slot, the aggregator bids for c power units (say, Megawatts) from the ISO. The aggregator then commits to adjusting the aggregate power consumption within its service area, such that it closely follows the set-point c of the system operator. The aggregator should thus co-ordinate the electricity consumption of the n consumers within its service area, by allocating energy rationing responsibilities among individual consumers, in such a way that the overall power consumption is c, and a measure of users' aggregate "negative utility" is minimized. Denoting by x_i, $i = 1,\ldots,n$ the power allocated for the i-th consumer, the problem can be formulated as

$$\min_{x \in \mathbb{R}^n} \sum_{i=1}^{n} U_i(x_i) \quad \text{s.t.:} \quad \sum_{i=1}^{n} x_i = c, \tag{12.108}$$

where U_i is a function measuring the "negative utility" incurred by the i-th consumer, if x_i units of power were delivered to it. For example, we may consider a convex cost function of the type

$$U_i(x_i) = \alpha_i |x_i - \bar{x}_i|_i^{\beta},$$

where $\alpha_i > 0$, $\beta_i \geq 1$ are parameters, and $\bar{x}_i$ is the power allocation ideally desired by the i-th customer. We next show that a dual decomposition approach on problem (12.108) provides a market-based mechanism whereby consumers iteratively adjust their demands so as to reach the optimal value of (12.108). The Lagrangian of problem (12.108) is

$$\mathcal{L}(x,\lambda) \;=\; \sum_{i=1}^{n} U_i(x_i) + \lambda \left(\sum_{i=1}^{n} x_i - c \right)$$
$$=\; \sum_{i=1}^{n} (U_i(x_i) + \lambda x_i) - \lambda c.$$

Here, λ has the interpretation of price for unit of power that the aggregator proposes to consumers. The dual function $g(\lambda)$ can then be computed in a decoupled way as

$$g(\lambda) \;=\; \sum_{i=1}^{n} g_i(\lambda) - \lambda c,$$
$$g_i(\lambda) \;=\; \min_{x_i} U_i(x_i) + \lambda x_i,$$

where the minimizer $x_i^*(\lambda)$ of the problem $\min_{x_i} U_i(x_i) + \lambda x_i$ represents the response to price λ of the i-th customer, who seeks to optimally trade-off negative utility with cost. A subgradient of the dual function g at λ is given by

$$\eta(\lambda) = \sum_{i=1}^{n} x_i^*(\lambda) - c,$$

hence $g(\lambda)$ can be iteratively maximized via subgradient iterations

$$\lambda_{k+1} = \lambda_k + s_k \left(\sum_{i=1}^{n} x_i^*(\lambda_k) - c \right).$$

These iterations can be interpreted as a dynamic process known in economics as *tatonnement*: at stage k of the bargaining process the aggregator posts a price λ_k. Consumers respond with the quantities $x_i^*(\lambda_k)$

that they would be willing to consume at that price. In the next iteration $k+1$, the aggregator adjusts the price in order to better approximate the supply–demand equality constraint: if there is excess demand (i.e., $\sum_{i=1}^{n} x_i^*(\lambda) > c$) then the price increases, whereas if there is low demand (i.e., $\sum_{i=1}^{n} x_i^*(\lambda) < c$) then the price decreases. Under suitable hypotheses (e.g., strong duality holds, and all minimizers $x_i^*(\lambda)$ exist and are unique), this negotiation process converges to the optimal clearing price λ^* and to the corresponding optimal demands $x_i^*(\lambda^*)$, $i = 1, \ldots, n$.

12.7   Exercises

Exercise 12.1 (Successive projections for linear inequalities)   Consider a system of linear inequalities $Ax \leq b$, with $A \in \mathbb{R}^{m,n}$, where $a_i^\top$, $i = 1, \ldots, m$, denote the rows of A, which are assumed, without loss of generality, to be nonzero. Each inequality $a_i^\top x \leq b_i$ can be normalized by dividing both terms by $\|a_i\|_2$, hence we shall further assume without loss of generality that $\|a_i\|_2 = 1$, $i = 1, \ldots, m$.

Consider now the case when the polyhedron described by these inequalities, $\mathcal{P} \doteq \{x : Ax \leq b\}$ is nonempty, that is, there exists at least a point $\bar{x} \in \mathcal{P}$. In order to find a feasible point (i.e., a point in $\mathcal{P}$), we propose the following simple algorithm. Let k denote the iteration number and initialize the algorithm with any initial point $x_k = x_0$ at $k = 0$. If $a_i^\top x_k \leq b_i$ holds for all $i = 1, \ldots, m$, then we have found the desired point, hence we return x_k, and finish. If instead there exists i_k such that $a_{i_k}^\top x_k > b_{i_k}$, then we set $s_k \doteq a_{i_k}^\top x_k - b_{i_k}$, we update[19] the current point as

$$x_{k+1} = x_k - s_k a_{i_k},$$

and we iterate the whole process.

[19] This algorithm is a version of the so-called Agmon–Motzkin–Shoenberg *relaxation method* for linear inequalities, which dates back to 1953.

1. Give a simple geometric interpretation of this algorithm.

2. Prove that this algorithm either finds a feasible solution in a finite number of iterations, or it produces a sequence of solutions $\{x_k\}$ that converges asymptotically (i.e., for $k \to \infty$) to a feasible solution (if one exists).

3. The problem of finding a feasible solution for linear inequalities can be also put in relation with the minimization of the nonsmooth function $f_0(x) = \max_{i=1,\ldots,m}(a_i^\top x_k - b_i)$. Develop a subgradient-type algorithm for this version of the problem, discuss hypotheses that need be assumed to guarantee convergence, and clarify the relations and similarities with the previous algorithm.

Exercise 12.2 (Conditional gradient method) Consider a constrained minimization problem

$$p^* = \min_{x \in \mathcal{X}} f_0(x), \qquad (12.109)$$

where f_0 is convex and smooth and $\mathcal{X} \subseteq \mathbb{R}^n$ is convex and compact. Clearly, a projected gradient or proximal gradient algorithm could be applied to this problem, if the projection onto $\mathcal{X}$ is easy to compute. When this is not the case, the following alternative algorithm has been proposed.[20] Initialize the iterations with some $x_0 \in \mathcal{X}$, and set $k = 0$. Determine the gradient $g_k \doteq \nabla f_0(x_k)$ and solve

$$z_k = \arg\min_{x \in \mathcal{X}} g_k^T x.$$

Then update the current point as

$$x_{k+1} = (1 - \gamma_k) x_k + \gamma_k z_k,$$

where $\gamma_k \in [0, 1]$, and, in particular, we choose

$$\gamma_k = \frac{2}{k+2}, \quad k = 0, 1, \dots$$

Assume that f_0 has a Lipschitz continuous gradient with Lipschitz constant[21] L, and that $\|x - y\|_2 \le R$ for every $x, y \in \mathcal{X}$. In this exercise, you shall prove that

$$\delta_k \doteq f_0(x_k) - p^* \le \frac{2LR^2}{k+2}, \quad k = 1, 2, \dots \qquad (12.110)$$

1. Using the inequality

$$f_0(x) - f_0(x_k) \le \nabla f_0(x_k)^T (x - x_k) + \frac{L}{2} \|x - x_k\|_2^2,$$

which holds for any convex f_0 with Lipschitz continuous gradient,[22] prove that

$$f_0(x_{k+1}) \le f_0(x_k) + \gamma_k \nabla f_0(x_k)^T (z_k - x_k) + \gamma_k^2 \frac{LR^2}{2}.$$

Hint: write the inequality condition above, for $x = x_{k+1}$.

2. Show that the following recursion holds for δ_k:

$$\delta_{k+1} \le (1 - \gamma_k)\delta_k + \gamma_k^2 C, \quad k = 0, 1, \dots,$$

for $C \doteq \frac{LR^2}{2}$. *Hint:* use the optimality condition for z_k, and the convexity inequality $f_0(x^*) \ge f_0(x_k) + \nabla f_0(x_k)^T (x^* - x_k)$.

3. Prove by induction on k the desired result (12.110).

[20] Versions of this algorithm are known as the Franke–Wolfe algorithm, which was developed in 1956 for quadratic f_0, or as the Levitin–Polyak *conditional gradient algorithm* (1966).

[21] As defined in Section 12.1.1.

[22] See Lemma 12.1.

Exercise 12.3 (Bisection method) The bisection method applies to one-dimensional convex problems[23] of the form

$$\min_x f(x) \ : \ x_l \leq x \leq x_u,$$

where $x_l < x_u$ are both finite, and $f : \mathbb{R} \to \mathbb{R}$ is convex. The algorithm is initialized with the upper and lower bounds on x: $\underline{x} = x_l$, $\overline{x} = x_u$, and the initial x is set as the midpoint

$$x = \frac{\underline{x} + \overline{x}}{2}.$$

Then the algorithm updates the bounds as follows: a subgradient g of f at x is evaluated; if $g < 0$, we set $\underline{x} = x$; otherwise,[24] we set $\overline{x} = x$. Then the midpoint x is recomputed, and the process is iterated until convergence.

[24] Actually, if $g = 0$ then the algorithm may stop and return x as an optimal solution.

1. Show that the bisection method locates a solution x^* within accuracy ϵ in at most $\log_2(x_u - x_l)/\epsilon - 1$ steps.

2. Propose a variant of the bisection method for solving the unconstrained problem $\min_x f(x)$, for convex f.

3. Write a code to solve the problem with the specific class of functions $f : \mathbb{R} \to \mathbb{R}$, with values

$$f(x) = \sum_{i=1}^n \max_{1 \leq j \leq m} \left(\frac{1}{2} A_{ij} x^2 + B_{ij} x + C_{ij} \right),$$

where A, B, C are given $n \times m$ matrices, with every element of A non-negative.

Exercise 12.4 (KKT conditions) Consider the optimization problem[25]

[25] Problem due to Suvrit Sra (2013).

$$\min_{x \in \mathbb{R}^n} \ \sum_{i=1}^n \left(\tfrac{1}{2} d_i x_i^2 + r_i x_i \right)$$
$$\text{s.t.:} \quad a^\top x = 1, \ x_i \in [-1,1], \ i = 1, \ldots, n,$$

where $a \neq 0$ and $d > 0$.

1. Verify if strong duality holds for this problem, and write down the KKT optimality conditions.

2. Use the KKT conditions and/or the Lagrangian to come up with the fastest algorithm you can to solve this optimization problem.

3. Analyze the running time complexity of your algorithm. Does the empirical performance of your method agree with your analysis?

Exercise 12.5 (Sparse Gaussian graphical models) We consider the following problem in a symmetric $n \times n$ matrix variable X

$$\max_{X} \log \det X - \text{trace}(SX) - \lambda \|X\|_1 \: : \: X \succ 0,$$

where $S \succeq 0$ is a (given) empirical covariance matrix, $\|X\|_1$ denotes the sum of the absolute values of the elements of the positive definite matrix X, and $\lambda > 0$ encourages the sparsity in the solution X. The problem arises when fitting a multivariate Gaussian graphical model to data.[26] The ℓ_1-norm penalty encourages the random variables in the model to become conditionally independent.

[26] See Section 13.5.5.

1. Show that the dual of the problem takes the form

$$\min_{U} \: -\log \det(S + U) \: : \: |U_{ij}| \leq \lambda.$$

2. We employ a block-coordinate descent method to solve the dual. Show that if we optimize over one column and row of U at a time, we obtain a sub-problem of the form

$$\min_{x} \: x^\top Q x \: : \: \|x - x_0\|_\infty \leq 1,$$

where $Q \succeq 0$ and $x_0 \in \mathbb{R}^{n-1}$ are given. Make sure to provide the expression of Q, x_0 as functions of the initial data, and the index of the row/column that is to be updated.

3. Show how you can solve the constrained QP problem above using the following methods. Make sure to state precisely the algorithm's steps.

   - Coordinate descent.
   - Dual coordinate ascent.
   - Projected subgradient.
   - Projected subgradient method for the dual.
   - Interior-point method (any flavor will do).

   Compare the performance (e.g., theoretical complexity, running time/convergence time on synthetic data) of these methods.

4. Solve the problem (using block-coordinate descent with five updates of each row/column, each step requiring the solution of the QP above) for a data file of your choice. Experiment with different values of λ, report on the graphical model obtained.

Exercise 12.6 (Polynomial fitting with derivative bounds)

In Section 13.2, we examined the problem of fitting a polynomial of degree d through m data points $(u_i, y_i) \in \mathbb{R}^2$, $i = 1, \ldots, m$. Without loss of generality, we assume that the input satisfies $|u_i| \leq 1$, $i = 1, \ldots, m$. We parameterize a polynomial of degree d via its coefficients:

$$p_w(u) = w_0 + w_1 u + \cdots + w_d u^d,$$

where $w \in \mathbb{R}^{d+1}$. The problem can be written as

$$\min_w \|\Phi^\top w - y\|_2^2,$$

where the matrix Φ has columns $\phi_i = (1, u_i, \ldots, u_i^d)$, $i = 1, \ldots, m$. As detailed in Section 13.2.3, in practice it is desirable to encourage polynomials that are not too rapidly varying over the interval of interest. To that end, we modify the above problem as follows:

$$\min_w \|\Phi^\top w - y\|_2^2 + \lambda b(w), \tag{12.111}$$

where $\lambda > 0$ is a regularization parameter, and $b(w)$ is a bound on the size of the derivative of the polynomial over $[-1, 1]$:

$$b(w) = \max_{u\,:\,|u|\leq 1} \left| \frac{d}{du} p_w(u) \right|.$$

1. Is the penalty function b convex? Is it a norm?

2. Explain how to compute a subgradient of b at a point w.

3. Use your result to code a subgradient method for solving problem (12.111).

Exercise 12.7 (Methods for LASSO) Consider the LASSO problem, discussed in Section 9.6.2:

$$\min_x \frac{1}{2}\|Ax - y\|_2^2 + \lambda\|x\|_1,$$

Compare the following algorithms. Try to write your code in a way that minimizes computational requirements; you may find the result in Exercise 9.4 useful.

1. A coordinate-descent method.

2. A subgradient method, as in Section 12.4.1.

3. A fast first-order algorithm, as in Section 12.3.4.

Exercise 12.8 (Non-negative terms that sum to one) Let $x_i, i = 1, \ldots,$ n, be given real numbers, which we assume without loss of generality to be ordered as $x_1 \leq x_2 \leq \cdots \leq x_n$, and consider the scalar equation in variable ν that we encountered in Section 12.3.3.3:

$$f(\nu) = 1, \quad \text{where } f(\nu) \doteq \sum_{i=1}^{n} \max(x_i - \nu, 0).$$

1. Show that f is continuous and strictly decreasing for $\nu \leq x_n$.

2. Show that a solution ν^* to this equation exists, it is unique, and it must belong to the interval $[x_1 - 1/n, x_n]$.

3. This scalar equation could be easily solved for ν using, e.g., the bisection method. Describe a simpler, "closed-form" method for finding the optimal ν.

Exercise 12.9 (Eliminating linear equality constraints) We consider a problem with linear equality constraints

$$\min_x f_0(x) \ : \ Ax = b,$$

where $A \in \mathbb{R}^{m,n}$, with A full row rank: rank $A = m \leq n$, and where we assume that the objective function f_0 is decomposable, that is

$$f_0(x) = \sum_{i=1}^{n} h_i(x_i),$$

with each h_i a convex, twice differentiable function. This problem can be addressed via different approaches, as detailed in Section 12.2.6.

1. Use the constraint elimination approach of Section 12.2.6.1, and consider the function $\tilde{f}_0$ defined in Eq. (12.33). Express the Hessian of $\tilde{f}_0$ in terms of that of f_0.

2. Compare the computational effort[27] required to solve the problem using the Newton method via the constraint elimination technique, versus using the feasible update Newton method of Section 12.2.6.3, assuming that $m \ll n$.

[27] See the related Exercise 7.4.

III
Applications

13

Learning from data

If the facts don't fit the theory, change the facts.

Albert Einstein

THIS CHAPTER PROVIDES A GUIDED TOUR of some representative problems arising in machine learning, viewed through the lenses of optimization, and without any claim of exhaustivity. We first explore so-called supervised learning problems, where the basic issue is to fit some model to given response data. Then we turn to unsupervised learning problems, where the issue is to build a model for the data, without a particular response in mind.

Throughout this chapter, we will denote by $X = [x_1, \ldots, x_m] \in \mathbb{R}^{n,m}$ a generic matrix of data points, where $x_i \in \mathbb{R}^n$, $i = 1, \ldots, m$, the i-th data point, is also referred to as an *example*. We refer to a particular dimension of a generic data point x as a *feature*. Common features may include: the frequencies (or some other numerical score) of given words in a dictionary;[1] Boolean variables that determine the presence or absence of a specific feature, such as whether a specific actor played in the movie that the data point represents; or numerical values such as blood pressure, temperature, prices, etc.

[1] See the bag-of-words representation of text in Example 2.1.

13.1 *Overview of supervised learning*

In supervised learning, we seek to build a model of an unknown function $x \to y(x)$, where $x \in \mathbb{R}^n$ is the input vector and $y(x) \in \mathbb{R}$ is the corresponding output. We are given a set of observations, or *examples*, that is, a number of input–output pairs (x_i, y_i), $i = 1, \ldots, m$. We are to use these examples to learn a model of the function, $x \to \hat{y}_w(x)$, where w is a vector of model parameters. Once the model is obtained, we may perform a prediction: if x is a new input point, for which we have not observed the output, we set our prediction to be $\hat{y} = \hat{y}_w(x)$. If the response to be predicted is an arbitrary real number, we may posit a linear dependence of the output on some model parameters. Our model then takes the form $\hat{y}_w \doteq w^\top x$, where $w \in \mathbb{R}^n$ contains the model coefficients. More generally, we may assume $\hat{y}_w(x) = w^\top \phi(x)$, where $x \to \phi(x)$ is a given nonlinear mapping; setting $\phi(x) = (1, x)$ allows one to recover affine models.

Example 13.1 (*Predicting demand for inventory management*) A large store needs to predict the demand for a specific item from customers in a certain geographical area. It posits that the logarithm of the demand is an arbitrary real number, and that it depends linearly on a certain number of inputs (features): time of year, type of item, number of items sold the day before, etc. The problem is to predict tomorrow's demand, based on the observation of input–demand pairs over a recent past.

Sometimes the response to be predicted is binary, say $y(x) \in \{-1, 1\}$ for every input x; it is then referred to as a *label*. Then we may posit a sign-linear dependence, of the form $\hat{y}_w(x) = \text{sgn}(w^\top x + b)$, where $w \in \mathbb{R}^n$, $b \in \mathbb{R}$ contain the model coefficients. Again, we can use a more complex model that involves a given nonlinear mapping. Other types of response arising in classification, not discussed here, include categorical responses, or more complex objects such as trees.

Example 13.2 (*Classification of medical outcomes*) Binary classification problems arise often in a medical decision context, typically when trying to predict medical outcomes based on clinical measurements. The "Wisconsin Diagnostic Breast Cancer" (WDBC) data set is a publicly available database containing data related to $m = 569$ patients. For each patient, we have an input vector $x_i \in \mathbb{R}^{30}$ that is composed of 30 numerical fields obtained from a digitized image of a fine needle aspirate (FNA) of breast mass. The fields describe characteristics of the cell nuclei present in the image, such as the radius, texture, perimeter, area, smoothness, compactness, concavity, symmetry, and fractal dimension of the cell nucleus. The label associated with each input point is $+1$ if the diagnosis was malignant, and -1 if it was benign. The goal is to train a classifier that can then be used for diagnostic purposes (malignant/benign diagnosis) on a new patient. If we denote by $x \in \mathbb{R}^{30}$ the various measurements for the new patient, we can make a diagnosis according to the label predicted by the classifier on the new datum x.

Example 13.3 (*Binary classification of credit card applications*) A credit card company receives thousands of applications for new cards. Each application contains information about an applicant: age, marital status, annual salary, outstanding debts, etc. Problem: to decide whether an application should be approved, or to classify applications into two categories, approved and not approved.

Example 13.4 (*Other classification examples*) Binary classification problems arise in many situations. In recommendation systems, say for movies,[2] we have information about a given user's movie preferences; the goal of classification is to determine whether a given new movie will be liked by the user. In spam filtering, we may have a set of emails which we know are spam, and another set of emails that are known to be legitimate. In time-series prediction, we may try to determine whether a future value (such as a stock price) will increase or decrease with respect to its current value, based on the observation of past values.

[2] See, e.g., Section 11.4.1.4.

The learning (or *training*) problem is to find the "best" model coefficient vector w, such that $\hat{y}_w(x_i) \simeq y_i$ for $i = 1, \ldots, m$. We thus seek to minimize, over the model variable w, a certain measure of the mismatch between the vector of predictions $\hat{y}$ and the observed response y. In practice, we seek to make a good prediction on a yet unseen "test" point x. The key issue of out-of-sample performance is further discussed later.

Another key issue in supervised learning is to attach, to the predicted output $\hat{y}$, some guarantee of reliability; in other words, to quantify the probability of *misclassification* of the new datum.[3] This important aspect is the subject of statistical learning theory,[4] which is out of the scope of this book and will not be discussed further.

[3] In the medical setting of Example 13.2, this corresponds to a diagnosis error.
[4] See for instance the book by T. Hastie, R. Tibshirani, and J. Friedman, *The Elements of Statistical Learning*, Springer, 2008.

We first provide a detailed but basic example illustrating how fundamental concepts, such as regularization and out-of-sample performance, arise.

13.2   *Least-squares prediction via a polynomial model*

Consider a data set, as shown in Figure 13.1, of input–output pairs (x_i, y_i), with $x_i \in \mathbb{R}$, $i = 1, \ldots, m$. Our goal is to predict the value of y corresponding to some yet unseen value of the input x.

13.2.1   *Models*

Linear models.   We start with a basic linear model for the data, where we assume that, for a generic value of the input x:

$$y(x) = w_1 + x w_2 + e(x) = w^\top \phi(x) + e(x),$$

where w_1, w_2 are some weights, $\phi(x) \doteq (1, x)$, and $e(x)$ is an error term. To find the weight vector w, we use a least-squares approach,

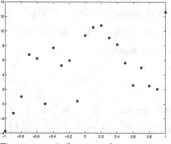

Figure 13.1 A data set of input–output pairs (x_i, y_i), $i = 1, \ldots, 20$.

which leads to the problem of minimizing the so-called *training error*

$$\min_{w} \frac{1}{m} \sum_{i=1}^{m} (y_i - \phi_i^\top w)^2,$$

where we have defined $\phi_i \doteq \phi(x_i) = (1, x_i)$, $i = 1, \dots, m$. The above problem can be expressed as a least-squares one:

$$\min_{w} \|\Phi^\top w - y\|_2^2, \tag{13.1}$$

where the vector $y \doteq (y_1, \dots, y_m)$ is the *response*, and the matrix $\Phi \in \mathbb{R}^{2,m}$ is the data matrix, with columns ϕ_i, $i = 1, \dots, m$. Here, our features include a constant feature set to 1 (the first row of Φ), and a second feature (the second row of Φ) that is simply the input x. The problem's objective involves the so-called *squared loss* function

$$\mathcal{L}(z, y) = \|z - y\|_2^2.$$

The process of using the available data to fit a specific model is often referred to as *training* stage, and problem (13.1) is called the training (or learning) problem.

Once problem (13.1) is solved, we can use the vector w for finding a predicted value of the output $\hat{y}(x)$ for an arbitrary input x: we set

$$\hat{y}(x) = \phi(x)^\top w = w_1 + w_2 x.$$

The rationale for our choice is that, since w made the error on the available data, $e = \Phi^\top w - y$, small, we expect that for the new input x, the prediction rule above is accurate. In practice, no guarantees can be obtained as to the accuracy of the prediction, unless some other assumptions are made about the way the data is generated.

Polynomial models. Clearly, if the data is close to linear in the input x, our prediction algorithm will perform well. If not, we can try a more elaborate model, for example a quadratic one:

$$y(x) = w_1 + x w_2 + x^2 w_3 + e(x) = w^\top \phi(x) + e(x),$$

where now $w \in \mathbb{R}^3$, and $\phi(x) = (1, x, x^2)$. Again, the problem of fitting w can be solved via LS:

$$\min_{w} \|\Phi^\top w - y\|_2^2,$$

where as before $y = (y_1, \dots, y_m)$, but now the data matrix $\Phi \in \mathbb{R}^{3,m}$ has columns $\phi_i = (1, x_i, x_i^2)$, $i = 1, \dots, m$. More generally, we can use a polynomial model of order k (see Figure 13.2):

$$y(x) = w_1 + x w_2 + x^2 w_3 + \dots + x^k w_{k+1} + e(x).$$

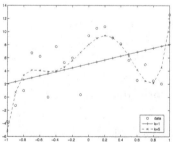

Figure 13.2 Two different polynomial models, one linear and the other of degree 7.

13.2.2   Evaluating performance

How good is our prediction? How can we compare our different polynomial models? In order to evaluate the performance of our approach, we can use a *leave-one-out* approach. Instead of using the entire data set (x_i, y_i), $i = 1, \ldots, m$, to fit our model, we use all but one data point, say we omit the j-th data point, (x_j, y_j). This data point will be our "test" (or, out-of-sample) point with which we will evaluate the performance. Thus, we separate our initial data set into two parts: one is the training set, containing all the data points except the j-th, and the other is the test set, which contains only the j-th data point. The leave-one-out method can be trivially extended to a larger number of points left out; yet other approaches use random partitions of the data into training and test sets, the latter typically containing 30% of the data. These different techniques all come under the same umbrella of *cross-validation*.

We proceed by solving the least-squares problem (13.1), with the last column of X and last element of vector y removed. We then compare the corresponding prediction with the actual value on the test data, and form the prediction error $(\hat{y}(x_j) - y_j)^2$. This out-of-sample error is potentially very different from the "in-sample" one, which would be measured by the optimal value of problem (13.1).

We can repeat the above process for all the values of the index j of the sample that we leave out, and compute the average test error:

$$\frac{1}{m} \sum_{j=1}^{m} (\hat{y}(x_j) - y_j)^2.$$

For each j, the j-th data point is not seen by the prediction algorithm. Hence, there is no reason to expect that the test error will be zero, or even small. If we plot the average test and training errors versus the order of the polynomial model, as in Figure 13.3, we notice that, as the order k grows, the training error decreases, while the test set error typically first decreases or remains constant, and then increases. This phenomenon is referred to as *over-fitting*. Models with higher complexity fit the training data better, but a model that is too complex (such as a high-degree polynomial) does not behave well on unseen data. Hopefully, there is a "sweet point" where the test set error is minimal.

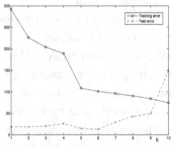

Figure 13.3 Training and test set error versus order of polynomial model.

13.2.3   Regularization and sparsity

Regularization. Assume that we have settled on a specific order k for our polynomial model. Polynomials of given degree k are not all created equal; some vary very wildly, with very large values of the derivative. If the new input point is not precisely known, wild variations in the polynomial's value may lead to a substantial classification test error.

In fact, the "complexity" of a polynomial model depends not only on the degree but also on the *size* of the coefficients. More formally it is possible to show[5] that for a polynomial $p_w(x) \doteq w_1 + w_2 x + \cdots + w_{k+1} x^k$, we have

[5] See Exercise 2.8.

$$\forall x \in [-1,1] \; : \; |\frac{\mathrm{d}}{\mathrm{d}x} p_w(x)| \le k^{3/2} \|w\|_2. \tag{13.2}$$

This means that, if a bound for data that is known *a priori* (e.g., $|x| \le 1$), then we can control the size of the derivative of our polynomial model of given order by making the Euclidean norm of the coefficient vector w small.

We are led to a modification of the optimization problem (13.1), where we add a penalty term involving $\|w\|_2$ to the loss function. For example, we can replace (13.1) with

$$\min_w \; \|\Phi^\top w - y\|_2^2 + \lambda \|w\|_2^2,$$

where $\lambda \ge 0$ is a regularization parameter. The above is a regularized LS problem, which can be solved by linear algebra methods, as discussed in Section 6.7.3.

The choice of a good parameter λ follows the same pattern as that of the order of the model. In fact, both parameters (regularization parameter λ and model order k) have to be simultaneously chosen by searching the corresponding two-dimensional space. In many cases, the extra degree of freedom does allow one to improve the test set error, as evidenced by Figure 13.4.

The squared Euclidean norm penalty is not the only way to control the size of the derivative. In fact, we can use many variants of the bound (13.2); for example:

$$\forall x \in [-1,1] \; : \; |\frac{d}{dx} p_w(u)| \le k \|w\|_1,$$

which corresponds to the penalized problem

$$\min_w \; \|\Phi^\top w - y\|_2^2 + \lambda \|w\|_1. \tag{13.3}$$

The above is a LASSO problem, which has been extensively discussed in Section 9.6.2.

The use of the squared Euclidean norm in the training problem has been historically preferred, as it makes the problem amenable to direct linear algebra methods. In contrast, the LASSO problem

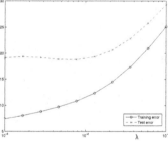

Figure 13.4 Training and test set error versus regularization parameter λ, for an order $k = 10$ fixed.

cannot be solved so transparently. With the advent of convex optimization methods, LASSO and related variants become computationally competitive with respect to regularized least squares. Using non-Euclidean norm penalties leads to tractable problems and offers useful alternatives, as seen next.

Sparsity.   In some cases, it is desirable to obtain a sparse model, that is, one for which the coefficient vector w has many zeros. As discussed in Example 9.11, the use of the ℓ_1-norm in the penalty term will encourage sparsity of the optimal w. For example, solving the LASSO problem (13.3) for $\lambda = 0.1$ produces the sparse polynomial

$$p_w(x) = 7.4950 + 10.7504x - 6.9644x^2 - 45.0750x^3 + 42.9250x^5 - 5.4516x^8 - 0.4539x^{15} + 9.2869x^{20},$$

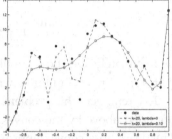

Figure 13.5 A sparse polynomial model, with $k = 20$ and $\lambda = 0.1$ used in the LASSO problem (13.3).

see Figure 13.5. Thus, the use of ℓ_1-norm penalty leads us to find a sparse polynomial that does not vary too wildly over the data span $[-1, 1]$.

13.3   Binary classification

Consider a binary classification problem, where we are given a number m of data points x_i and associated labels $y_i \in \{-1, 1\}$, $i = 1, \ldots, m$. Our goal is to predict the label $\hat{y}$ of a new, yet unseen, data point x.

13.3.1   Support vector machines

In Section 13.2, we had to predict a real-valued variable, and for that we initially used an affine function $x \to w^\top \phi(x)$, with $\phi(x) = (1, x)$. When predicting a label $y \in \{-1, 1\}$, the simplest modification is arguably the following prediction rule:[6]

$$\hat{y} = \text{sgn}(x^\top w + b).$$

[6] In this section, we single out the bias term b in the affine function, in order to simplify geometric interpretations seen later.

At first, we may try to find $w \in \mathbb{R}^n$, $b \in \mathbb{R}$ such that the average number of errors on the training set is minimized. We can formalize the latter with the 0/1 function

$$E(\alpha) \doteq \begin{cases} 1 & \text{if } \alpha < 0, \\ 0 & \text{otherwise.} \end{cases}$$

The classification rule $x \to \text{sgn}(w^\top x + b)$ produces an error on a given input–output pair (x, y) if and only if $y(w^\top x + b) < 0$. Hence, the average number of errors we observe on the training set for a given w is:

$$\frac{1}{m} \sum_{i=1}^{m} E(y_i(w^\top x_i + b)).$$

This leads to a training problem of the form

$$\min_{w,b} \frac{1}{m} \sum_{i=1}^{m} E(y_i(x_i^\top w + b)). \tag{13.4}$$

Unfortunately, the above problem is not convex and algorithms may be trapped in local minima, which is not desirable.[7]

To circumvent this problem, we notice that the function E above is bounded above by the convex "hinge" function $\alpha \to \max(0, 1 - \alpha)$. This leads to a convex approximation, in fact, an upper bound, on the training problem (13.4):

$$\min_{w,b} \frac{1}{m} \sum_{i=1}^{m} \max(0, 1 - y_i(x_i^\top w + b)). \tag{13.5}$$

This objective function, referred to as the *hinge loss*, is convex, and polyhedral; hence the above problem can be solved as a linear program. One possible LP formulation is

$$\min_{w,b,e} \frac{1}{m} \sum_{i=1}^{m} e_i \ : \ e \ge 0, \ e_i \ge 1 - y_i(x_i^\top w + b), \ i = 1, \ldots, m. \tag{13.6}$$

Once the model (that is, the pair (w, b)) is obtained, we can predict the label of a new data point $x \in \mathbb{R}^n$ via the classification rule $\hat{y} = \text{sgn}(w^\top x + b)$. The formulation in (13.5) is the basic building block in the so-called support vector machines (SVM) models. This name derives from a geometric interpretation, discussed in Section 13.3.3.

13.3.2 *Regularization and sparsity*

As in the example of Section 13.2, we may be interested in controlling the complexity of our model. One way to do so is to control the variation of the linear function $x \to x^\top w + b$, when x spans the observed input data. We note that the gradient of this function is simply w, hence controlling the size of w will achieve our goal.

If we assume that the data points are all contained in a Euclidean sphere of radius R centered at the origin, then we can use the ℓ_2 norm for regularization, since

$$\max_{x, x' \, : \, \|x\|_2 \le R, \ \|x'\|_2 \le R} |w^\top (x - x')| \le 2R\|w\|_2.$$

The corresponding learning problem (13.5) then becomes

$$\min_{w,b} \frac{1}{m} \sum_{i=1}^{m} \max(0, 1 - y_i(x_i^\top w + b)) + \lambda \|w\|_2^2, \qquad (13.7)$$

where $\lambda \geq 0$ is a regularization parameter, the choice of which is usually done via leave-one-out or similar methods.

If instead we assume that the data points are all contained in a box of size R, $\{x : \|x\|_\infty \leq R\}$, then the corresponding bound

$$\max_{x,x' \|x\|_\infty \leq R, \|x'\|_\infty \leq R} |w^\top (x - x')| \leq 2R\|w\|_1$$

leads to a modification to problem (13.5) as follows:

$$\min_{w,b} \frac{1}{m} \sum_{i=1}^{m} \max(0, 1 - y_i(x_i^\top w + b)) + \lambda \|w\|_1, \qquad (13.8)$$

where $\lambda \geq 0$ is again a regularization parameter.

The ℓ_1-norm in the above problem encourages sparsity of the optimal w. For such a vector, the scalar product $w^\top x$ involves only a few elements of x. Hence the technique allows one to identify a few key features (elements of x) that are instrumental in allowing a good prediction.

13.3.3 Geometric interpretations

The basic SVM model (13.5) and its regularized variants have a number of geometric interpretations.

Linearly separable data. We start by examining the case when the optimal value of problem (13.5) is zero, implying that a zero training error can be achieved. This means that there exists a vector $w \in \mathbb{R}^n$ and a scalar b such that

$$y_i(x_i^\top w + b) \geq 0, \quad i = 1, \dots, m. \qquad (13.9)$$

Geometrically, this means that the data is *linearly separable* by the hyperplane $\{x : w^\top x + b = 0\}$, in the sense that points x_i with label $y_i = 1$ are on one side of the hyperplane, while those with label $y_i = -1$ are on the other side, see Figure 13.6. The decision rule $\hat{y} = \mathrm{sgn}(w^\top x + b)$ corresponds to identifying which side of the hyperplane the new point x falls in.

Assume next that the data is *strictly linearly separable*, that is, that conditions (13.9) are satisfied with strict inequality signs. This is the case if and only if there exists a positive number $\beta > 0$ such that

$$y_i(x_i^\top w + b) \geq \beta, \quad i = 1, \dots, m.$$

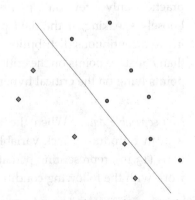

Figure 13.6 A data set that is strictly linearly separable.

In this case, we can divide both sides of the previous inequality by β, then normalize the model parameters (w, b), and obtain that strict linear separability is equivalent to the existence of (w, b) such that

$$y_i(x_i^\top w + b) \geq 1, \quad i = 1, \ldots, m. \tag{13.10}$$

The two "critical" hyperplanes $\mathcal{H}_\pm \doteq \{x : w^\top x + b = \pm 1\}$ delineate a slab inside which no data point lies. This slab separates the positively and negatively labelled data points, see Figure 13.7.

Maximum margin classifier. We observe that, in the previously considered case of strictly linearly separable data, there actually exist an infinite number of possible choices for (w, b) that satisfy the conditions (13.10), i.e., an infinite number of slabs may separate the data sets. A reasonable possibility for overcoming this non-uniqueness issue is to seek for the separating slab which maximizes the *separation margin*, i.e., the distance between the two "critical" hyperplanes $\mathcal{H}_\pm \doteq \{x : w^\top x + b = \pm 1\}$. Arguably, a hyperplane that maximizes the separation margin should have good properties: a test point x that will be close to the positively (or negatively) labelled data will have more chances to fall on the corresponding side of the decision hyperplane.

It may be easily verified that the distance between the two critical hyperplanes $\mathcal{H}_+$, $\mathcal{H}_-$ is given by $2/\|w\|_2$. To maximize the margin we may thus minimize $\|w\|_2$, while satisfying the labelling constraints. This results in the optimization problem

$$\min_{w,b} \|w\|_2 \, : \, y_i(x_i^\top w + b) \geq 1, \quad i = 1, \ldots, m, \tag{13.11}$$

which is often referred to as the *maximum margin SVM*, see Figure 13.8. The name of *support vectors* derives from the fact that, in practice, only a few data points lie on the critical boundaries $\mathcal{H}_\pm$. Loosely speaking, if the data points are drawn at random according to a continuous distribution, then the probability of having more than n feature points on the critical hyperplanes is zero. These special points lying on the critical hyperplanes are called *support vectors*.

Non-separable data. When the data is not separable (Figure 13.9), we can introduce "slack variables" in the strict separability conditions (13.10), representing penalties on the constraint violations. We work with the following conditions on (w, b) and a new error vector e:

$$e \geq 0, \quad y_i(x_i^\top w + b) \geq 1 - e_i, \quad i = 1, \ldots, m.$$

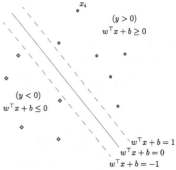

Figure 13.7 A strictly linearly separable data set and a separating slab.

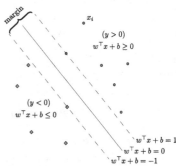

Figure 13.8 Maximum margin classifier.

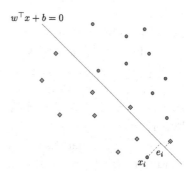

Figure 13.9 Approximate separation of a non-separable data set.

Ideally we would like to minimize the *number* of nonzero elements in e, that is, its cardinality. This would lead to a non-convex problem. However, we can approximate the problem and use the ℓ_1 norm instead of the cardinality function. Since $e \geq 0$, the ℓ_1 norm of e reduces to the sum of its elements. In this way, we obtain precisely the basic SVM problem previously presented in (13.6).

Classification with soft margin.   In the non-separable case, we then need to find a trade-off between the margin (distance between the two critical hyperplanes) and the classification error on the training set. This leads to a problem of the form (13.7), which we write using slack variables, as

$$\min_{w,b,e} \frac{1}{m} \sum_{i=1}^{m} e_i + \lambda \|w\|_2^2 \;:\; e \geq 0, \;\; e_i \geq 1 - y_i(x_i^\top w + b), \;\; i = 1, \ldots, m.$$
(13.12)

The objective function in (13.12) expresses the tradeoff between the width of the classification margin (which we like to be large) and the sum of all violation terms (which we like to be small). For these reasons this approach is referred to as classification with soft margin.

By construction, problem (13.12) is always feasible. If (w^*, b^*, e^*) is an optimal solution, and $e^* = 0$, then (w^*, b^*) is also optimal for the maximum-margin problem (13.11). On the other hand, if $e^* \geq 0$, $e^* \neq 0$, then the maximum margin problem is unfeasible, and the data is not separable. In this case, for all i such that e_i^* is nonzero, the corresponding constraint in (13.12) holds with equality, for otherwise we could reduce e_i^*, which would contradict optimality. Hence, e_i^* represents the normalized distance from the data point x_i to the hyperplane $y_i(w^\top x + b) = 1$, see Figure 13.10. For $e_i^* \in (0, 1)$, the corresponding data point x_i is inside the margin band, but still on the correct side of the decision boundary; for $e_i^* > 1$ the feature is instead misclassified.

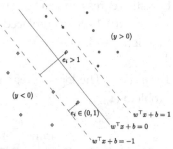

Figure 13.10 Classification with soft margin.

13.3.4   Robustness

We can interpret problems similar to the penalized variants (13.7) and (13.8) in terms of robustness. Assume that for every $i = 1, \ldots, m$, the data point x_i is only known to belong to a given sphere S_i of center $\hat{x}_i$, and radius ρ. First, we seek conditions for which the data is separable, irrespective of its precise location within the spheres; then we will try to maximize the level of robustness, that is, the radius of the spheres.

For a given model parameter vector (w, b), the conditions

$$y_i(x_i^\top w + b) \geq 0, \;\; i = 1, \ldots, m$$

hold for every choice of x_i in S_i, $i = 1, \ldots, m$, if and only if

$$y_i(\hat{x}_i^\top w + b) \geq \rho\|w\|_2, \quad i = 1, \ldots, m.$$

Due to the homogeneity in (w, b) of the above conditions, we can always assume $\rho\|w\|_2 = 1$. Maximizing ρ is then the same as minimizing $\|w\|_2$, and we are led to the problem (13.11), with x_i replaced with the centers $\hat{x}_i$. The geometrical interpretation is that we are separating spheres instead of points.

A similar interpretation holds for the ℓ_1-norm; in that case the assumption is that the data is unknown, but bounded within hypercubes of the form $\{x : \|x - \hat{x}\|_\infty \leq \rho\}$.

13.3.5   Logistic regression

Logistic regression is a variant of the SVM where the hinge function $\alpha \to \max(0, 1 - \alpha)$ is replaced with a smooth convex one, the logistic function $\alpha \to \ln(1 + e^{-\alpha})$.

The model.   The logistic regression counterpart to the SVM learning problem (13.5) is

$$\min_{w,b} \sum_{i=1}^{m} \ln(1 + e^{-y_i(x_i^\top w + b)}). \tag{13.13}$$

Of course, in practice, penalized variants are used. The above is a convex problem, since the log-sum-exp function in the objective is convex.[8] One advantage of logistic regression models is that they can be naturally extended to multi-class problems, where the labels are not binary but assume a finite number of values. We do not discuss these extensions here.

[8] See Example 2.14.

Probabilistic interpretation.   The main advantage of logistic regression over SVMs is that it actually corresponds to a probabilistic model. That model allows one to attach some level of probability for a new data point to belong to the positive or negative class. This is in contrast with SVMs, which provide a "yes or no" answer to the classification of a data point. The logistic model is as follows. The probability for a point x to have label $y \in \{-1, 1\}$ is postulated to be of the form

$$\pi_{w,b}(x, y) = \frac{1}{1 + e^{-y(w^\top x + b)}}.$$

The above makes sense geometrically: if a point is far away from the decision hyperplane defined by $w^\top x + b = 0$, well inside the region corresponding to the positively-labelled points, then the quantity $w^\top x + b$ is high, and the probability above is close to one. Conversely, if $w^\top x + b$ is highly negative, the point x is well inside the other side, and the probability above goes to zero.

To fit the model, we can use a maximum-likelihood approach. The likelihood is the probability that the model above would ascribe to the event that each data point x_i receives the observed label y_i, that is:

$$\prod_{i:y_i=+1} \pi_{w,b}(x_i, 1) \prod_{i:y_i=-1} \pi_{w,b}(x_i, -1).$$

Taking the negative logarithm and minimizing with respect to the model parameters (w, b) leads to problem (13.13).

Once a new data point is observed, we predict its label by computing the probabilities $\pi_{w,b}(x, \pm 1)$; our predicted label is the one that corresponds to the highest of the two values. This is the same as setting $\hat{y} = \operatorname{sgn}(w^\top x + b)$, as in SVM, with the added advantage that we can attach a probability level to our prediction.

13.3.6 Fisher discrimination

Fisher discrimination is yet another approach to binary classification. As we shall see, this problem can be solved via linear algebra methods. Fisher discrimination amounts to finding a direction $u \in \mathbb{R}^n$ in data space such that the projections of the positively and negatively labelled data along u are as far apart as possible, where the distance between the projected data set is measured relative to variability within each data class.

It is convenient to define two matrices corresponding to the positive and negative classes, respectively:

$$A_+ = [x_i]_{i:y_i=+1} \in \mathbb{R}^{n,m_+}, \quad B = [x_i]_{i:y_i=-1} \in \mathbb{R}^{n,m_-},$$

where $m_\pm = \operatorname{card}\{i : y_i = \pm 1\}$ denotes the size of each class. Similarly, we let $\bar{a}_+, \bar{a}_-$ denote the centroids of the two classes:

$$\bar{a}_\pm = \frac{1}{m_\pm} A_\pm \mathbf{1}.$$

Finally, let the centered data matrices be

$$\tilde{A}_\pm = A_\pm \left(I_{m_+} - \frac{1}{m_+} \mathbf{1}\mathbf{1}^\top \right).$$

The squared distance between the two centroids, along the direction u, is given by $(u^\top (\bar{a}_+ - \bar{a}_-))^2$. We take as a measure of discrimination this distance, normalized by the mean-square variation of the data.

The mean-square variation of the data set $A = [a_1, \ldots, a_p]$ around its average $\bar{a}$, along the direction u, is given by

$$
\begin{aligned}
s^2 &= \frac{1}{p} \sum_{i=1}^{p} (u^\top (a_i - \bar{a}))^2 = \frac{1}{p} \sum_{i=1}^{p} u^\top \tilde{a}_i \tilde{a}_i^\top u \\
&= \frac{1}{p} u^\top \left(\sum_{i=1}^{p} \tilde{a}_i \tilde{a}_i \right)^\top u \\
&= \frac{1}{p} u^\top \tilde{A} \tilde{A}^\top u.
\end{aligned}
$$

The discrimination criterion, to be maximized, is thus given by the normalized squared distance between the centroids, along the direction u:

$$
f_0(u) = \frac{u^\top (\bar{a}_+ - \bar{a}_-)(\bar{a}_+ - \bar{a}_-)^\top u}{u^\top \left(\frac{1}{m_+} \tilde{A}_+ \tilde{A}_+^\top + \frac{1}{m_-} \tilde{A}_- \tilde{A}_-^\top \right) u} = \frac{(u^\top c)^2}{u^\top M u},
$$

where $c \doteq \bar{a}_+ - \bar{a}_-$, and

$$
M \doteq \frac{1}{m_+} \tilde{A}_+ \tilde{A}_+^\top + \frac{1}{m_-} \tilde{A}_- \tilde{A}_-^\top.
$$

The problem then amounts to maximizing the ratio of two quadratic forms:

$$
\max_{u \neq 0} \frac{(u^\top c)^2}{u^\top M u}.
$$

By homogeneity, we can always rescale u so that $u^\top c = 1$, leading to the minimization problem

$$
\min_{u} u^\top M u \; : \; u^\top c = 1.
$$

The symmetric matrix M is positive semidefinite, hence the above is a convex quadratic program. Let us assume that $M \succ 0$ for simplicity. To solve the above convex problem, we use the Lagrangian

$$
\mathcal{L}(u, \mu) = \frac{1}{2} u^\top M u + \mu (1 - u^\top c).
$$

Optimal pairs (u, μ) are characterized by the conditions[9]

$$
0 = \nabla_u \mathcal{L}(u, \mu) = M u - \mu c, \; u^\top c = 1.
$$

[9] See Section 8.5.

Since M is positive definite, we obtain $u = \mu M^{-1} c$. Using the constraint $u^\top c = 1$, we obtain

$$
u = \frac{1}{c^\top M^{-1} c} M^{-1} c.
$$

The left panel in Figure 13.11 presents two data clouds in $\mathbb{R}^2$, and the arrow shows the optimal Fisher discriminating direction u. The histograms of the data projected along u are well separated, as shown in the right panel in the figure.

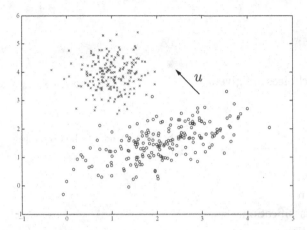 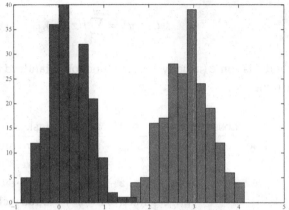

Figure 13.11 Fisher data discrimination.

13.4 A generic supervised learning problem

In this section, we consider a generic supervised learning problem of the form

$$\min_{w} \mathcal{L}(\Phi^{\top} w, y) + \lambda p(w), \qquad (13.14)$$

where $w \in \mathbb{R}^n$ contains the model parameters; p is a penalty function (usually a norm or a squared norm); λ is a regularization parameter; and the function $\mathcal{L}$, referred to as the *loss* function, serves to measure the mismatch between the predicted value and the response y. In the above, the matrix Φ contains some transformed data points; precisely, each column ϕ_i is of the form $\phi_i = \phi(x_i)$, where ϕ is some (possibly nonlinear) mapping, chosen by the user.[10]

The above class of problems is far from covering all supervised learning methods, but it does include least-squares regression, SVM, logistic regression, and more.

The prediction rule associated with the learning problem depends on the task and the model. In regression, the output to be predicted is a real number, and the prediction rule takes the form $\hat{y} = \phi(x)^{\top} w$. In binary classification, we use a sign-linear function, $\hat{y} = \text{sgn}(\phi(x)^{\top} w)$. We cover all these cases with the notation $\hat{y} = S(\phi(x)^{\top} w)$, where S is either the identity or the sign function.

13.4.1 Loss functions

The form of the loss function $\mathcal{L}$ depends on the task (regression or classification), that is, on the nature of the output to be predicted, as well as on other factors, such as prior assumptions on the data. It is

[10] See an illustration of this in Section 13.2.

common to assume that the loss is decomposable as a sum:

$$\mathcal{L}(z,y) = \sum_{i=1}^{m} l(z_i, y_i),$$

where l is some (usually convex) function. Standard choices are given in Table 13.1.

Loss function	$l(z,y)$	task
Euclidean squared	$\|z-y\|_2^2$	regression
ℓ_1	$\|z-y\|_1$	regression
ℓ_∞	$\|z-y\|_\infty$	regression
hinge	$\max(0, 1-yz)$	binary classification
logistic	$\ln(1 + e^{-yz})$	classification

Table 13.1 Standard choices of loss functions.

The specific choice of the loss function depends on the task, on assumptions about the data, and also on practical considerations such as available software.

13.4.2 *Penalty and constraint functions*

Various penalty functions can be used, depending again on the task and other considerations. The squared ℓ_2-norm has been historically preferred, since the resulting optimization problem (13.14) inherits nice properties such as smoothness (when the loss function itself is smooth).

Choosing a penalty. If we believe that the optimal w is sparse, that is, only a few features should dominate in the prediction of the output, an ℓ_1-norm penalty can be used to encourage sparsity. If the vector w is formed of blocks $w^{(i)}$, $i = 1, \ldots, p$, and we believe that many blocks should be zero vectors, we can use the composite norm

$$\sum_{i=1}^{p} \|w^{(i)}\|_2.$$

Indeed, the above is nothing else than the ℓ_1-norm of the vector of norms $(\|w^{(1)}\|_2, \ldots, \|w^{(p)}\|_2)$, the sparsity of which we seek to promote.

One consideration of practical importance: it is often desirable to design the penalty so that the learning problem (13.14) has a unique solution. To this end, a small squared ℓ_2-norm penalty is often added, in order to ensure strong convexity of the objective.

Constrained versions. Note that the penalty form (13.14) covers, as special cases, constrained variants of the form

$$\min_w \mathcal{L}(\Phi^\top w, y) \; : \; w \in \mathcal{C},$$

where $\mathcal{C}$ is a (usually convex) set, such as a Euclidean ball. The above can be formulated as the penalized problem, with an appropriate choice of p:

$$p(x) = \begin{cases} 1 & \text{if } w \in \mathcal{C}, \\ 0 & \text{otherwise.} \end{cases}$$

One may wonder about the practical difference between using a given penalty, its square, or the corresponding constrained version.[11] To illustrate this discussion, let us consider LASSO and two variants:

[11] See Exercise 13.5.

$$p^*_{\text{lasso}} \; \doteq \; \min_w \|\Phi^\top w - y\|_2^2 + \lambda \|w\|_1,$$

$$p^*_{\text{sqrt}} \; \doteq \; \min_w \|\Phi^\top w - y\|_2 + \mu \|w\|_1,$$

$$p^*_{\text{constr}} \; \doteq \; \min_w \|\Phi^\top w - y\|_2 \; : \; \|w\|_1 \le \alpha.$$

In the above, $\lambda \ge 0$, $\mu \ge 0$, and $\alpha \ge 0$ are regularization parameters. For a fixed triplet (λ, μ, α), the corresponding solutions of the above problem usually differ. However, when the parameters λ, μ, α span the non-negative real line, the path of solutions generated is the same for each problem.

13.4.3 Kernel methods

Kernel methods allow fast computations in the context of nonlinear rules, provided the learning problem involves a squared Euclidean penalty:

$$\min_w \mathcal{L}(\Phi^\top w, y) + \lambda \|w\|_2^2.$$

We observe that the first term, involving the loss, depends on w only via $\Phi^\top w$. Due to the fundamental theorem of linear algebra (see Section 3.2.4), any vector $w \in \mathbb{R}^n$ can be decomposed as $w = \Phi v + r$, where $\Phi^\top r = 0$. Then the squared Euclidean norm of w can be written as $\|\Phi v\|_2^2 + \|r\|_2^2$. Let us now express the above problem in terms of the new variables v, r:

$$\min_{v,r} \mathcal{L}(\Phi^\top \Phi v, y) + \lambda \left(\|\Phi v\|_2^2 + \|r\|_2^2 \right).$$

Clearly, the optimal value of r is zero.[12] Geometrically, this means that the optimal w lies in the span of the transformed data, that is, at optimum, $w = \Phi v$ for some v. The learning problem becomes

[12] This is true for the Euclidean norm penalty; in general, for example, with the ℓ_1-norm, this result does not hold.

$$\min_v \mathcal{L}(Kv, y) + \lambda v^\top K v, \tag{13.15}$$

where $K = \Phi^\top \Phi \in \mathbb{R}^{m,m}$ is the so-called *kernel matrix*. Thus, the above problem depends only on the scalar products $K_{ij} = \phi(x_i)^\top \phi(x_j)$, with $1 \le i, j \le m$.

A similar property holds for the prediction rule: if x is a new data point, the prediction rule takes the form $\hat{y} = S(\phi(x)^\top w) = S(\phi(x)^\top \Phi v)$. This expression also only involves scalar products, now between the transformed new point $\phi(x)$ and the transformed data $\phi(x_i)$.

Computationally, this means that everything depends on our ability to quickly evaluate scalar products between any pair of transformed points $\phi(x), \phi(x')$. Consider for example a two-dimensional problem case when $n = 2$, and when we seek a *quadratic* classification rule, of the form[13]

$$\hat{y} = w_1 + \sqrt{2}w_2 x_1 + \sqrt{2}w_3 x_2 + w_4 x_1^2 + \sqrt{2}w_5 x_1 x_2 + w_6 x_2^2.$$

This rule corresponds to the nonlinear mapping $x \to w^\top \phi(x)$, where

$$\phi(x) = (1, \sqrt{2}x_1, \sqrt{2}x_2, x_1^2, \sqrt{2}x_1 x_2, x_2^2).$$

We observe that, for any pair of points $x, x' \in \mathbb{R}^n$:

$$\phi(x)^\top \phi(x') = (1 + x^\top x')^2.$$

More generally, with a classification rule that is a polynomial of order d in x, we can form a mapping $\phi(x) \in \mathbb{R}^p$, with[14] $p = O(n^d)$, such that, for any pair of points $x, x' \in \mathbb{R}^n$:

$$\phi(x)^\top \phi(x') = (1 + x^\top x')^d.$$

Computing the above scalar product would *a priori* take an exponential (in n, the number of features) time, precisely $O(p) = O(n^d)$. The above formula allows a dramatic reduction of the computational complexity, to $O(n)$.

In order to compute the scalar products needed in the solution to the optimization problem (13.15), as well as in the prediction rule, we only need to form $O(m^2)$ scalar products. Since the complexity of solving the QP (13.15) is $O(m^3)$, the kernel method has a total complexity of $O(m^3 + nm^2)$, independent of the degree d. The bottom line: we can solve a learning problem having a Euclidean norm penalty with a polynomial model with as little effort as solving one with a linear model.

The kernel idea can be extended to non-polynomial models. Rather subtly, we can even bypass the explicit definition of the nonlinear mapping ϕ: it suffices to decide on the values of the kernel function

$$k(x, x') \doteq \phi(x)^\top \phi(x').$$

[13] The factor $\sqrt{2}$ is immaterial, as we can always rescale the elements of w accordingly.

[14] Here $O(q)$ refers to some linear function of q.

Of course, not every function $(x, x') \to k(x, x')$ defines a valid kernel: it has to be positive semidefinite, that is, in the form above for some mapping $x \to \phi(x)$. In practice, an added condition is that the kernel function values can be computed with a relatively low effort. Popular choices of valid kernel are given in Table 13.2.

Name	$k(x, x')$	parameters
linear	$x^\top x'$	none
polynomial	$(1 + x^\top x')^d$	d, the order
Gaussian	$e^{\frac{1}{2\sigma^2}\|x - x'\|_2^2}$	σ, the width

Table 13.2 Popular kernels, with associated parameters.

Example 13.5 (*Breast cancer classification*) Return to the breast cancer example 13.2, with the Wisconsin data set. In order to visualize a plot of the data and of the soft-margin separator, we reduced the dimensionality of the data from 30 to 2 dimensions, using the PCA approach (see Section 5.3.2). Notice that this may not be a good idea in practice, since data sets that are separable in the original space may not be separable when projected down to a lower-dimensional space. Here, however, this drastic reduction of dimensionality is performed only for the purpose of showing two-dimensional plots of the data.

Using the projected two-dimensional data points, we trained an SVM with Gaussian kernel (with $\sigma = 1$), by solving problem (13.15), with $\lambda = 1$. The results are shown in Figure 13.12, where $+$ denotes benign points, $*$ denotes malignant points, circles mark the support vectors, and the solid line represents the separation surface.

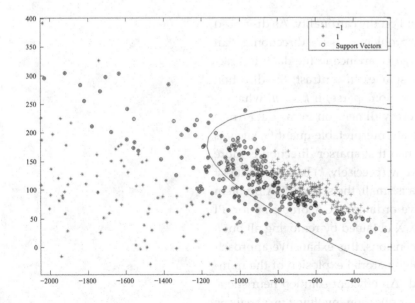

Figure 13.12 SVM classification on a projected two-dimensional breast cancer data set.

13.5 Unsupervised learning

In unsupervised learning, the data points $x_i \in \mathbb{R}^n$, $i = 1, \ldots, m$, do not come with assigned labels or responses. The task is to learn some information or structure about the data.

13.5.1 Principal component analysis (PCA)

In principal component analysis, the objective is to discover the most important, or informative, directions in a data set, that is the directions along which the data varies the most. PCA has been described in detail in Section 5.3.2. Numerically, the solution of the PCA problem boils down to computing the SVD of the data matrix. We next describe some variants on the basic PCA problem.

13.5.2 Sparse PCA

The PCA method has been widely applied in a variety of contexts, and many variants have been developed. A recent one, referred to as sparse PCA, is useful as it improves the interpretability of the resulting principal directions. In sparse PCA, a constraint is added to problem (5.18), in order to limit the number of nonzero elements in the decision variable:

$$\min_{z \in \mathbb{R}^n} \quad z^\top (\tilde{X}\tilde{X}^\top)z \qquad (13.16)$$
$$\text{s.t.:} \quad \|z\|_2 = 1, \ \ \text{card}(z) \leq k,$$

where $k \leq n$ is a user-defined bound on the cardinality. As discussed in Remark 5.3, the original PCA problem seeks a direction z that "explains," or accounts for, the largest variance in the data; in practice, that direction is typically not sparse. In contrast, the direction provided by sparse PCA is by definition sparse, if $k \ll n$, which is the case of interest in practice. If every dimension in the data (rows in the data matrix) corresponds to an interpretable quantity, such as price, temperature, volume, etc., then that sparser direction is more interpretable, as it involves only a few (precisely, k) basic quantities.

When the dimension of the data is small, the sparse PCA problem is easy to solve: it suffices to solve ordinary PCA problems for all the sub-matrices of the data matrix $\tilde{X}$ obtained by removing all but k dimensions (rows). For larger dimensions, this exhaustive approach becomes impossible due to the combinatorial explosion of the number of sub-matrices thus generated. An efficient solution heuristic is described in Section 12.5.4, and an application involving text analysis is given in Section 13.7.

Example 13.6 (*Sparse PCA of market data*) Coming back to Example 5.3, we observe that the direction of maximal variance, given as the first column of the matrix U, u_1, is a dense vector, with only one component (the fifth) close to zero:

$$u_1 = \begin{bmatrix} -0.4143 & -0.4671 & -0.4075 & -0.5199 & -0.0019 & -0.4169 \end{bmatrix}.$$

The corresponding largest singular value is $\sigma_1 = 1.0765$, while the ratio of variances is $\eta_1 = \sigma_1^2/(\sigma_1^2 + \cdots + \sigma_6^2) = 67.766\%$. This means that about 68% of the variance is contained in a specific non-trivial combination of almost all the indices, with weights given by the elements of u_1.

Now let us find a low-cardinality direction along which the variance is large, by solving the sparse PCA problem (13.16) with $k = 2$. We can search exhaustively for all combinations of two dimensions chosen in the original space of six dimensions. That is, we solve PCA problems for all the possible matrices $\tilde{X}_2$ obtained by removing all but two rows of the original matrix $\tilde{X}$. We find that the two indices corresponding to rows 4 and 5, namely the indices PAC and BOT, explain the most variance among all possible pairs. The corresponding largest variance of $\tilde{\sigma}_1 = 0.6826$ can be compared with the original largest variance of $\sigma_1 = 1.0765$. In terms of variance explained, the two indices have a ratio with respect to the total variance of $\tilde{\sigma}_1^2/(\sigma_1^2 + \cdots + \sigma_6^2) = 0.2725\%$. Thus, the two indices alone capture about 28% of the total market variance.

Example 13.7 (*Sparse PCA for topic discovery*) Sparse PCA can be used for discovering main topics in a large collection of documents. The text collection from *The New York Times* (NYT) in the UCI Machine Learning Repository contains $300,000$ articles and has a dictionary of $102,660$ unique words, resulting in a file of size 1 GB. The given data has no class labels, and for copyright reasons no filenames or other document-level metadata, such as article section information, are mentioned. The text has been transcribed into numerical form via a "bag-of-words" approach.

1st PC	2nd PC	3rd PC	4th PC	5th PC
million	point	official	president	school
percent	play	government	campaign	program
business	team	united_states	bush	children
company	season	u_s	administration	student
market	game	attack		
companies				

Table 13.3 Terms corresponding to the nonzeros in the top five sparse principal components for the NYT data.

The nonzero features (terms) corresponding to the top five sparse principal components are shown in Table 13.3. Each component represents a well-defined topic. Indeed, the first principal component for *The New York Times* is about business, the second one about sports, the third about US, the fourth about politics and the fifth about education. Even though no metadata were available, the sparse PCA still unambiguously identifies and perfectly corresponds to the topics used by *The New York Times* itself to classify articles on its own website.

The sparse PCA concept can be extended to finding a low-rank approximation to a matrix, with constraints on the cardinality of the vectors involved in the approximation. For example, the cardinality-constrained rank-one approximation problem is of the form

$$\min_{p,q} \|X - pq^\top\|_{\mathrm{F}} \; : \; p \in \mathbb{R}^n, \; q \in \mathbb{R}^m \, \mathrm{card}(p) \leq k, \; \mathrm{card}(q) \leq h,$$

where $X \in \mathbb{R}^{n,m}$, and $k \leq n, h \leq m$ are user-chosen integers used to constrain the number of nonzero elements in the vectors p, q. As with sparse PCA, the above can be solved with SVD when n, m are small. For larger sizes, the problem becomes very hard, but efficient heuristics, based on power iteration, are available, as seen in Section 12.5.4.

13.5.3 Non-negative matrix factorization

Non-negative matrix factorization (NNMF) refers to an attempt to approximate a given non-negative $n \times m$ data matrix X with non-negative low-rank components, as a low-rank product of the form $PQ^\top$, where both P, Q are non-negative matrices, with a low number of columns $k \ll \min(n, m)$. Hence, it is an apparently simple modification to the low-rank matrix factorization problem discussed in Section 5.3.2.4.

To illustrate NNMF, let us start with an ordinary rank-one approximation of a data matrix X:

$$\min_{p,q} \|X - pq^\top\|_{\mathrm{F}} \; : \; p \in \mathbb{R}^n, \; q \in \mathbb{R}^m.$$

If the objective value is small, and $X \approx pq^\top$ for some p, q, we may interpret p as a typical data point, q as a typical feature. If X is non-negative, there could be a rank-one approximation $X \approx pq^\top$, with some elements of p, q non-negative. In that case, it is hard to interpret the result; for example, we cannot interpret p as a typical data point, if it has negative components.

The NNMF approach applies to non-negative data matrices. If the target rank is $k = 1$, the NNMF problem can be formulated as

$$\min_{p,q} \|\tilde{X} - pq^\top\|_{\mathrm{F}} \; : \; p \in \mathbb{R}^n, \; p \geq 0, \; q \in \mathbb{R}^m, \; q \geq 0.$$

The interpretation of the result is that, if $X \approx pq^\top$ with p, q non-negative vectors, then each column of X is proportional to a single vector q, with different weights given by the vector p. Hence every data point follows a single "profile" q, up to a non-negative scaling factor specific to that data point. More generally, the NNMF problem is expressed as

$$\min_{P,Q} \|\tilde{X} - PQ^\top\|_F \ : \ P \in \mathbb{R}^{n,k}, \ P \geq 0, \ Q \in \mathbb{R}^{m,k}, \ Q \geq 0.$$

Here the interpretation is that the data points follow a linear combination of k basic profiles given by the columns of Q. The problem is non-convex, and hard to solve exactly. A heuristic based on block-coordinate descent[15] is often used, where the minimization is carried out with respect to P, Q alternately. Each sub-problem (of minimizing with respect to P or Q) is a convex problem, in fact a convex quadratic program.

[15] See Section 12.5.3.

Example 13.8 (*NNMF for topic discovery in text documents*) Using a bag-of-words approach (see Example 2.1), we can represent a collection of m text documents as a term-by-document occurrence matrix X, with X_{ij} the number of occurrences of term i in document j. For such a matrix, it is natural to use NNMF. A rank-one non-negative approximation of the form $X \approx pq^\top$ with $p \geq 0, q \geq 0$, expresses the fact that each document i (column of X) is approximately of the form $p_i q$. That is, under the rank-one approximation, all the documents are similar to a single document. Here, the non-negativity of p, q is crucial in order to interpret $p_i q$ as a document. A rank-k approximation would express the fact that all the documents are "mixes" of k basic documents.

Example 13.9 (*NNMF for student score matrix*) Consider a matrix of binary scores for involving four exams and five students:

$$X = \begin{bmatrix} 0 & 0 & 1 & 0 & 1 \\ 0 & 1 & 0 & 1 & 1 \\ 1 & 1 & 1 & 0 & 1 \\ 0 & 1 & 0 & 1 & 1 \end{bmatrix},$$

where X_{ij} is 1 if student j passed exam i, 0 otherwise. It turns out that we can write $X = PQ^\top$, with both P, Q having three columns only:

$$P = \begin{bmatrix} 0 & 1/2 & 0 \\ 1/2 & 0 & 0 \\ 0 & 1/2 & 1 \\ 1/2 & 0 & 0 \end{bmatrix}, \ Q = \begin{bmatrix} 1 & 0 & 0 \\ 1 & 0 & 2 \\ 0 & 2 & 0 \\ 0 & 0 & 2 \\ 0 & 2 & 2 \end{bmatrix}.$$

We can interpret the three columns of P, Q as representing three "skills" required to successfully pass exams. The matrix P provides a picture of which skills are needed for which exam; the matrix Q shows which skills are mastered by each student. For example, taking the first exam and the first student, we have, from P, that item 1 requires skill 2, but, from Q, we see that student 1 only masters skill 1, therefore exam 1 is failed by student 1. In fact, student 1's only success is over exam 3, since all other exams require either skill 2 or 3.

13.5.4 Robust PCA

A low-rank matrix approximation and the closely related PCA can be interpreted as expressing a given data matrix X as a sum $X = Y + Z$, where Y is low-rank, and Z is small.

Robust PCA[16] is another variation on the PCA theme, in which we seek to decompose a given $n \times m$ matrix X as a sum of a low-rank component and a *sparse* component. Formally:

$$\min_Z \, \mathrm{rank}(Y) + \lambda \, \mathrm{card}(Z) \; : \; Y, Z \in \mathbb{R}^{n,m}, \; X = Y + Z,$$

where $\mathrm{card}(Z)$ is the number of nonzero elements in the $n \times m$ matrix variable Z, and λ is a parameter that allows us to tradeoff a low rank of one component Y of X against the sparsity of the other component, Z. Such problems arise when some measured data is believed to have essentially a low-rank structure, but that there are "spikes" (or, outliers) in it, represented by the sparse component.

The robust PCA problem is hard to solve in general. As mentioned in Section 11.4.1.4, one can use the nuclear norm of a matrix[17] as a convex proxy of the rank, and the ℓ_1-norm to control the cardinality. This leads to a convex approximation to the robust PCA problem above:

$$\min_Z \, \|Y\|_* + \lambda \|Z\|_1 \; : \; Y, Z \in \mathbb{R}^{n,m}, \; X = Y + Z,$$

where $\|Z\|_1$ is the sum of the absolute elements in matrix Z, and $\|Y\|_*$ is the nuclear norm of Y (i.e., the sum of its singular values). The above is an SDP,[18] which is solvable as such for moderate problem sizes; for larger-scale instances it may be solved by specialized algorithms.

Example 13.10 (*Background modeling in video*)[19] Video data is a natural candidate for low-rank modeling, due to the correlation between frames. One of the most basic algorithmic tasks in video surveillance is to estimate a good model for the background variations in a scene. This task is complicated by the presence of foreground objects: in busy scenes, every frame may contain some anomaly. The background model needs to be flexible enough to accommodate changes in the scene, for example due to varying illumination. In such situations, it is natural to model the background variations as approximately low rank. Foreground objects,

---

[16] See E. J. Candès, X. Li, Y. Ma, and J. Wright, Robust principal component analysis?, *J. ACM*, 2011.

[17] See Section 5.2.2

[18] See Exercise 11.3.

[19] This example is based on: Robust Principal Component Analysis, by M. Balandat, W. Krichene, C. P. and K. K. Lam; EE 227A project report, UC Berkeley, May 2012.

Figure 13.13 A set of video frames. Top: the original set. Middle: the low-rank components, showing background only. Bottom: the sparse components, showing the foreground only.

such as cars or pedestrians, generally occupy only a fraction of the image pixels and hence can be treated as sparse errors.

We consider five frames from an original video, as shown at the top of Figure 13.13, which contains a passerby. The resolution of each frame is 176×144 pixels. We first separate them into three channels (RGB). For each channel, we stack each frame as a column of our matrix $X \in \mathbb{R}^{25344,5}$, and decompose X into a low-rank component Y and sparse component Z, via robust PCA. Then we combine the three channels again to form two sets of images, one with low-rank components and the other with sparse components. As shown in Figure 13.13, the low-rank component correctly recovers the background, while the sparse component correctly identifies the moving person.

13.5.5 Sparse graphical Gaussian models

The purpose of sparse graphical Gaussian modeling is to discover some hidden graph structure in a data set. The basic assumption is that the data has been generated by some multivariate Gaussian distribution, and the task is to learn the parameters of that distribution, encouraging models that exhibit a lot of conditional independence. To make these terms clearer, some background is in order.

Fitting a Gaussian distribution to data.   The density of a Gaussian distribution in $\mathbb{R}^n$ can be parameterized as

$$p(x) = \frac{1}{(2\pi \det S)^{n/2}} e^{-\frac{1}{2}(x-\mu)^{\top} S^{-1}(x-\mu)},$$

where the parameters are $\mu \in \mathbb{R}^n$ and S, a symmetric, positive definite $n \times n$ matrix. These parameters have a natural interpretation: μ is the mean of the distribution (its expected value), and S is its covariance matrix. Indeed, it can be readily verified that, by integration:

$$\mathbb{E}(x) = \int xp(x)\mathrm{d}x = \mu,$$

$$\mathbb{E}(x-\mu)(x-\mu)^\top = \int (x-\mu)(x-\mu)^\top p(x)\mathrm{d}x = S.$$

If a data set $X = [x_1, \ldots, x_m]$ is available, we may use the concept of maximum-likelihood in order to fit the Gaussian distribution, that is, find estimated values $\hat{\mu}, \hat{S}$ for the parameters μ, S. The idea is to maximize the value of the product of the densities $p(x_i)$, $i = 1, \ldots, m$. This leads to the problem

$$\max_{\mu, S} \, -\frac{1}{2} \sum_{i=1}^m (x_i - \mu)^\top S^{-1}(x_i - \mu) + \log \det S \; : \; S \succeq 0.$$

Solving for μ results in the estimate $\hat{\mu} = \mu^{\text{sample}}$, with μ^{sample} the *sample mean*:

$$\mu^{\text{sample}} \doteq \frac{1}{m} \sum_{i=1}^m x_i,$$

while the covariance matrix estimate is given as the solution to the optimization problem

$$\hat{S} = \arg\max_S \, \text{trace}(S^{-1}S^{\text{sample}}) - \log \det S \; : \; S \succ 0, \qquad (13.17)$$

where S^{sample} is the *sample covariance*

$$S^{\text{sample}} \doteq \frac{1}{m} \sum_{i=1}^m (x_i - \mu^{\text{sample}})(x_i - \mu^{\text{sample}})^\top.$$

Problem (13.17) is not convex as stated, but it becomes convex upon the change of variable $S \to P = S^{-1}$. We obtain

$$\max_P \, \text{trace}\, PS^{\text{sample}} + \log \det P \; : \; P \succ 0, \qquad (13.18)$$

which is convex in the matrix P. Assuming that the sample covariance matrix is positive definite, we can readily solve the above problem by looking at the optimality conditions, leading to $\hat{P} = (S^{\text{sample}})^{-1}$. We obtain that the maximum-likelihood estimate is just $\hat{S} = S^{\text{sample}}$. To summarize, the maximum-likelihood approach, in this case, simply sets the estimated mean and covariance matrix to their sample counterparts.

Conditional independence. Assume that the covariance matrix S is invertible. The inverse matrix $P \doteq S^{-1}$ is called the precision matrix, and its elements have a nice interpretation in terms of *conditional independence*. We say that a pair of variables (x_k, x_l) are conditionally independent if, when we fix all the other variables in x, the two random variables x_k and x_l are independent. This means that the resulting (conditional) probability density can be factored as a product of two scalar distributions, one depending on x_k and one on x_l. Now assume that the precision matrix has $P_{kl} = 0$ for some pair (k, l). The density function p given above satisfies

$$-2\ln p(x) = (x - \mu)^\top S^{-1}(x - \mu) = \sum_{i,j} P_{ij}(x_i - \mu_i)(x_j - \mu_j).$$

Clearly, when $P_{kl} = 0$, there is no term in the above quadratic function that contains both x_k and x_l. Thus, we can write, when fixing all the variables in x except x_k, x_l:

$$-2\ln p(x) = p_k(x_k) + p_l(x_l).$$

In words: p factors as a product of two functions, one that depends on x_k only and the other that depends on x_l only.

Example 13.11 Assume that a Gaussian distribution in $\mathbb{R}^3$ has zero mean, and that its 3×3 precision matrix P has $P_{12} = 0$. Let us check that x_1, x_2 are conditionally independent. We fix the variable x_3, and observe that

$$
\begin{aligned}
-2\ln p(x) &= P_{11}x_1^2 + P_{22}x_2^2 + 2P_{12}x_1x_2 + 2P_{13}x_1x_3 + 2P_{23}x_2x_3 + P_{33}x_3^2 \\
&= (P_{11}x_1^2 + 2P_{13}x_1x_3) + (P_{22}x_2^2 + 2P_{23}x_2x_3) + \text{constant}.
\end{aligned}
$$

As claimed, when x_3 is fixed, the density is written as a product of two terms, each involving x_1 or x_2 only.

Conditional independence is a natural concept when trying to understand relationships between random variables, based on observations. The more well-known concept of covariance may not be always helpful in practice. In many real-world data sets, all the variables are correlated with each other, and the covariance matrix is dense; there is no structure that becomes apparent by simply looking at the covariance matrix. In contrast, many variables can be conditionally independent; the corresponding precision matrix is sparse, in contrast with its inverse, the correlation matrix.

This motivates us to discover a conditional independence structure, for example in the form of a graph between the different random variables $x_1, \ldots, x_n$, which has no edge linking any conditionally independent pair (x_i, x_j). The sparser the graph, the easier it will be to read.

The following example illustrates how the precision matrix may be useful in revealing a structure that is otherwise hidden in the covariance matrix.

Example 13.12 (*A case with conditional independence*) Assume that we observe three variables x_1, x_2, x_3 in a large population, corresponding to shoe size, presence or absence of grey hair, and age, respectively. Assume that we compute a good estimate of the covariance matrix. Not surprisingly, that covariance matrix is dense, which means that all these variables are correlated. In particular, as age increases, shoe size also increases, as well as the occurrence of grey hair.

However, the precision matrix satisfies $P_{12} = 0$, which reveals that shoe size and presence or absence of grey hair are actually independent of each other, *for a given age*. This makes sense: if we were to bin the population in groups of similar age, we would find that for a given age group there is no statistical relationship between shoe size and grey hair.

Sparse precision matrix. In order to encourage conditional independence in the model estimated by maximum-likelihood, we modify the convex problem (13.18) with a penalty involving the ℓ_1-norm of P. This leads to the following variant:

$$\max_{P} \; \text{trace}\, PS^{\text{sample}} + \log \det P - \lambda \|P\|_1 \; : \; P \succ 0, \qquad (13.19)$$

where λ is a parameter that allows one to tradeoff the "goodness of fit" of our model, and the sparsity of the precision matrix.

In contrast with the original problem (13.18), the penalized problem above cannot be solved analytically. However, it is convex, and amenable to efficient algorithms.

Example 13.13 (*A graph of interest rates data*)

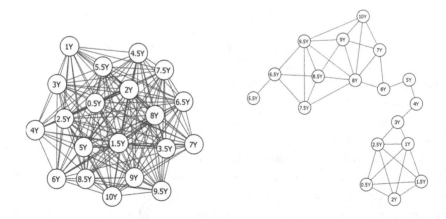

Figure 13.14 A sparse graphical model for the interest rate data.

In Figure 13.14, we represent the dependence structure of interest rates (sampled over a year) inferred from the inverse covariance matrix. Each node represents a specific interest rate maturity; edges between nodes are shown if the corresponding coefficient in the inverse covariance matrix is nonzero, that is, if they are conditionally dependent. We compare the solution to problem (13.19) for $\lambda = 0$ (which is then simply the inverse sample covariance matrix) and $\lambda = 1$. In the sparse solution the rates appear clearly clustered by maturity.

Example 13.14 (*A graph of Senators*) Return to the Senate voting data of Example 2.15. Figure 13.15 shows the sparse graph obtained from that data. We have solved problem (13.19) with $\lambda = 0.1$. The figure shows no edge between two nodes (Senators) whenever the corresponding element in the optimal sparse inverse covariance P is zero.

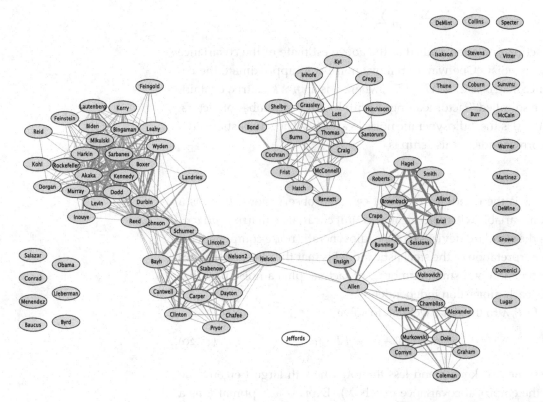

Figure 13.15 A sparse graphical model for the Senate voting data.

13.6   Exercises

Exercise 13.1 (SVD for text analysis) Assume you are given a data set in the form of an $n \times m$ term-by-document matrix X corresponding to a large collection of news articles. Precisely, the (i, j) entry in X is the frequency of the word i in the document j. We would like to visualize this data set on a two-dimensional plot. Explain how you would do the following (describe your steps carefully in terms of the

SVD of an appropriately centered version of X).

1. Plot the different news sources as points in word space, with maximal variance of the points.

2. Plot the different words as points in news-source space, with maximal variance of the points.

Exercise 13.2 (Learning a factor model) We are given a data matrix $X = [x^{(1)}, \ldots, x^{(m)}]$, with $x^{(i)} \in \mathbb{R}^n$, $i = 1, \ldots, m$. We assume that the data is centered: $x^{(1)} + \cdots + x^{(m)} = 0$. An (empirical) estimate of the covariance matrix is[20]

$$\Sigma = \frac{1}{m} \sum_{i=1}^{m} x^{(i)} x^{(i)\top}.$$

In practice, one often finds that the above estimate of the covariance matrix is noisy. One way to remove noise is to approximate the covariance matrix as $\Sigma \approx \lambda I + FF^\top$, where F is an $n \times k$ matrix, containing the so-called "factor loadings," with $k \ll n$ the number of factors, and $\lambda \geq 0$ is the "idiosyncratic noise" variance. The stochastic model that corresponds to this setup is

$$x = Ff + \sigma e,$$

where x is the (random) vector of centered observations, (f, e) is a random variable with zero mean and unit covariance matrix, and $\sigma = \sqrt{\lambda}$ is the standard deviation of the idiosyncratic noise component σe. The interpretation of the stochastic model is that the observations are a combination of a small number k of factors, plus a noise part that affects each dimension independently.

To fit F, λ to the data, we seek to solve

$$\min_{F, \lambda \geq 0} \|\Sigma - \lambda I - FF^\top\|_F. \qquad (13.20)$$

1. Assume λ is known and less than λ_k (the k-th largest eigenvalue of the empirical covariance matrix Σ). Express an optimal F as a function of λ, which we denote by $F(\lambda)$. In other words: you are asked to solve for F, with fixed λ.

2. Show that the error $E(\lambda) = \|\Sigma - \lambda I - F(\lambda)F(\lambda)^\top\|_F$, with $F(\lambda)$ the matrix you found in the previous part, can be written as

$$E(\lambda)^2 = \sum_{i=k+1}^{p} (\lambda_i - \lambda)^2.$$

Find a closed-form expression for the optimal λ that minimizes the error, and summarize your solution to the estimation problem (13.20).

[20] See Example 4.2.

3. Assume that we wish to estimate the risk (as measured by variance) involved in a specific direction in data space. Recall from Example 4.2 that, given a unit-norm n-vector w, the variance along the direction w is $w^\top \Sigma w$. Show that the rank-k approximation to Σ results in an underestimate of the directional risk, as compared with using Σ. How about the approximation based on the factor model above? Discuss.

Exercise 13.3 (Movement prediction for a time-series) We have a historical data set containing the values of a time-series $r(1), \ldots, r(T)$. Our goal is to predict if the time-series is going up or down. The basic idea is to use a prediction based on the sign of the output of an auto-regressive model that uses n past data values (here, n is fixed). That is, the prediction at time t of the sign of the value $r(t+1) - r(t)$ is of the form

$$\hat{y}_{w,b}(t) = \mathrm{sgn}\left(w_1 r(t) + \cdots + w_n r(t-n+1) + b\right),$$

In the above, $w \in \mathbb{R}^n$ is our classifier coefficient, b is a bias term, and $n \ll T$ determines how far back into the past we use the data to make the prediction.

1. As a first attempt, we would like to solve the problem

$$\min_{w,b} \sum_{t=n}^{T-1} (\hat{y}_{w,b}(t) - y(t))^2,$$

where $y(t) = \mathrm{sgn}(r(t+1) - r(t))$. In other words, we are trying to match, in a least-squares sense, the prediction made by the classifier on the training set, with the observed truth. Can we solve the above with convex optimization? If not, why?

2. Explain how you would set up the problem and train a classifier using convex optimization. Make sure to define precisely the learning procedure, the variables in the resulting optimization problem, and how you would find the optimal variables to make a prediction.

Exercise 13.4 (A variant of PCA) Return to the variant of PCA examined in Exercise 11.2. Using a (possibly synthetic) data set of your choice, compare the classical PCA and the variant examined here, especially in terms of its sensitivity to outliers. Make sure to establish an evaluation protocol that is as rigorous as possible. Discuss your results.

Exercise 13.5 (Squared vs. non-squared penalties) We consider the problems

$$
P(\lambda) \ : \ p(\lambda) \ \doteq \ \min_x \ f(x) + \lambda \|x\|,
$$

$$
Q(\mu) \ : \ q(\mu) \ \doteq \ \min_x \ f(x) + \frac{1}{2}\mu\|x\|^2,
$$

where f is a convex function, $\|\cdot\|$ is an arbitrary vector norm, and $\lambda > 0$, $\mu > 0$ are parameters. Assume that for every choice of these parameters, the corresponding problems have a unique solution.

In general, the solutions for the above problems for fixed λ and μ do not coincide. This exercise shows that we can scan the solutions to the first problem, and get the set of solutions to the second, and vice versa.

1. Show that both p, q are concave functions, and $\tilde{q}$ with values $\tilde{q}(\mu) = q(1/\mu)$ is convex, on the domain $\mathbb{R}_+$.

2. Show that

$$
p(\lambda) = \min_{\mu > 0} \ q(\mu) + \frac{\lambda^2}{2\mu}, \quad q(\mu) = \max_{\lambda > 0} \ p(\lambda) - \frac{\lambda^2}{2\mu}.
$$

For the second expression, you may assume that $\operatorname{dom} f$ has a nonempty interior.

3. Deduce from the first part that the paths of solutions coincide. That is, if we solve the first problem for every $\lambda > 0$, for any $\mu > 0$ the optimal point we thus find will be optimal for the second problem; and vice versa. It will be convenient to denote by $x^*(\lambda)$ (resp. $z^*(\mu)$) the (unique) solution to $P(\lambda)$ (resp. $Q(\mu)$).

4. State and prove a similar result concerning a third function

$$
r(\kappa) \ : \ r(\kappa) \doteq \min_x \ f(x) \ : \ \|x\| \le \kappa.
$$

5. What can you say if we remove the uniqueness assumption?

Exercise 13.6 (Cardinality-penalized least squares) We consider the problem

$$
\phi(k) \doteq \min_w \ \|X^\top w - y\|_2^2 + \rho^2\|w\|_2^2 + \lambda\operatorname{card}(w),
$$

where $X \in \mathbb{R}^{n,m}$, $y \in \mathbb{R}^m$, $\rho > 0$ is a regularization parameter, and $\lambda \ge 0$ allows us to control the cardinality (number of nonzeros) in the solution. This in turn allows better interpretability of the results. The above problem is hard to solve in general. In this exercise, we denote by $a_i^\top$, $i = 1, \ldots, n$ the i-th row of X, which corresponds to a particular "feature" (that is, dimension of the variable w).

1. First assume that no cardinality penalty is present, that is, $\lambda = 0$. Show that

$$\phi(0) = y^\top \left(I + \frac{1}{\rho^2} \sum_{i=1}^{n} a_i a_i^\top \right)^{-1} y.$$

2. Now consider the case $\lambda > 0$. Show that

$$\phi(\lambda) = \min_{u \in \{0,1\}^n} y^\top \left(I_m + \frac{1}{\rho^2} \sum_{i=1}^{n} u_i a_i a_i^\top \right)^{-1} y + \lambda \sum_{i=1}^{n} u_i.$$

3. A natural relaxation to the problem obtains upon replacing the constraints $u \in \{0,1\}^n$ with interval ones: $u \in [0,1]^n$. Show that the resulting lower bound $\phi(\lambda) \geq \underline{\phi}(\lambda)$ is the optimal value of the convex problem

$$\underline{\phi}(\lambda) = \max_{v} 2y^\top v - v^\top v - \sum_{i=1}^{n} \left(\frac{(a_i^\top v)^2}{\rho^2} - \lambda \right)_+ .$$

   How would you recover a suboptimal sparsity pattern from a solution v^* to the above problem?

4. Express the above problem as an SOCP.

5. Form a dual to the SOCP, and show that it can be reduced to the expression

$$\underline{\phi}(\lambda) = \|X^\top w - y\|_2^2 + 2\lambda \sum_{i=1}^{n} B\left(\frac{\rho x_i}{\sqrt{\lambda}} \right),$$

   where B is the (convex) *reverse Hüber function*: for $\xi \in \mathbb{R}$,

$$B(\xi) \doteq \frac{1}{2} \min_{0 \leq z \leq 1} \left(z + \frac{\xi^2}{z} \right) = \begin{cases} |\xi| & \text{if } |\xi| \leq 1, \\ \dfrac{\xi^2 + 1}{2} & \text{otherwise.} \end{cases}$$

   Again, how would you recover a suboptimal sparsity pattern from a solution w^* to the above problem?

6. A classical way to handle cardinality penalties is to replace them with the ℓ_1-norm. How does the above approach compare with the ℓ_1-norm relaxation one? Discuss.

14
Computational finance

OPTIMIZATION PLAYS an increasingly important role in computational finance. Since the seminal work of H. Markowitz, who won the Nobel prize in Economics (with M. Miller and W. Sharpe, in 1990) for his contributions related to the formalization of a financial trade-off decision problem into a convex optimization one, researchers and professionals in the finance field strove to devise suitable models for supporting tactical or strategic investment decisions. In this chapter we give a (necessarily limited) overview of some financial applications where convex optimization models provide a key instrument for enabling efficient solutions.

14.1 *Single-period portfolio optimization*

A basic financial portfolio model has been introduced in Example 2.6, Example 4.3, and Example 8.19. In this setting, the vector $r \in \mathbb{R}^n$ contains the random rate of returns of n assets over one fixed period of time of duration Δ (say, one day, one week, or one month), and $x \in \mathbb{R}^n$ is a vector describing the dollar value of an investor's wealth allocated in each of the n assets. In this section, we assume that the investor holds an initial portfolio $x(0) \in \mathbb{R}^n$ (so that the investor's initial total wealth is $w(0) \doteq \sum_{i=1}^n x_i(0) = \mathbf{1}^\top x(0)$), and that he wishes to perform transactions on the market so as to update the portfolio composition. The vector of transacted amounts is denoted by $u \in \mathbb{R}^n$ (where $u_i > 0$ if we increase the investment in the i-th asset, $u_i < 0$ if we decrease it, and $u_i = 0$ if we leave it unchanged), so that the updated portfolio composition, after transactions, is

$$x = x(0) + u.$$

We further assume that the portfolio is *self financing*, i.e., that the total wealth is conserved (no new cash is injected or retrieved), which is expressed by the *budget conservation* condition $\mathbf{1}^\top x = \mathbf{1}^\top x(0)$, that is

$$\mathbf{1}^\top u = 0.$$

Also, we consider without loss of generality that the initial wealth is of one unit (i.e., $w(0) = 1$). Sometimes, the investor is allowed to hold a virtually negative amount of an asset (i.e., it is admissible that some entries of x are negative). This financial artifice is called *short-selling*, and it is implemented in practice by borrowing the asset from the bank or broker, and selling it. If this feature is not allowed by the broker (or we do not want to use it), we impose on the problem the "no short-selling" constraint $x \geq 0$, that is

$$x(0) + u \geq 0 \quad \text{(no short-selling)}.$$

14.1.1 *Mean-variance optimization*

In a classical approach, named after the Nobel laureate economist H. Markowitz, one assumes that the expected value of the return vector $\hat{r} = \mathbb{E}\{r\}$, as well as the returns covariance matrix

$$\Sigma = \mathbb{E}\{(r - \hat{r})(r - \hat{r})^\top\},$$

are known. The return of the portfolio mix at the end of the period is then described by the random variable

$$\varrho(x) = r^\top x,$$

having expected value

$$\mathbb{E}\{\varrho(x)\} = \hat{r}^\top x$$

and variance

$$\text{var}\{\varrho(x)\} = x^\top \Sigma x.$$

In the Markowitz approach one postulates that the investor likes portfolios providing high expected return, while he is averse to *risk*, as measured by the variance of the portfolio (often called the *volatility*, in financial contexts). An optimal portfolio may then be obtained either by maximizing the expected value of ϱ subject to an upper limit on the admissible risk, or by minimizing risk subject to a lower bound on expected return. Letting μ denote the minimum desired level of expected return, and $\bar{\sigma}^2$ the maximum admissible level of risk, these two problems are expressed, respectively, as follows:

$$\rho(\bar{\sigma}) = \max_u \quad \hat{r}^\top x$$
$$\text{s.t.:} \quad x^\top \Sigma x \leq \bar{\sigma}^2,$$
$$x \in \mathcal{X}$$

and

$$\bar{\sigma}^2(\mu) = \min_u \quad x^\top \Sigma x$$
$$\text{s.t.:} \quad \hat{r}^\top x \geq \mu,$$
$$x \in \mathcal{X},$$

where we defined

$$\mathcal{X} \doteq \{x \in \mathbb{R}^n : x = x(0) + u, \ \mathbf{1}^\top x = w(0), \ \mathsf{ns} \cdot x \geq 0\},$$

where ns is a flag, which is assigned to the value 1 if short-selling is not allowed, and it is assigned to 0 otherwise. Since $\Sigma \succeq 0$, both problem formulations result in convex programs of quadratic type. In practice, these problems are usually solved for many increasing values of μ or $\bar{\sigma}$, so as to obtain a curve of optimal values in the risk/return plane, called the *efficient frontier*. Any portfolio x such that $(x^\top \Sigma x, \hat{r}^\top x)$ lies on this curve is called an *efficient portfolio*.

Example 14.1 (*Allocation over exchange-traded funds*) As a numerical example, consider the problem of allocating a portfolio among the following $n = 7$ assets:

1. SPDR Dow Jones Industrial ETF, ticker DIA;
2. iShares Dow Jones Transportation ETF, ticker IYT;
3. iShares Dow Jones Utilities ETF, ticker IDU;
4. First Trust Nasdaq-100 Ex-Tech EFT, ticker QQXT;
5. SPDR Euro Stoxx 50 ETF, ticker FEZ;
6. iShares Barclays 20+ Yr Treas. Bond ETF, ticker TLT;
7. iShares iBoxx USD High Yid Corp Bond ETF, ticker HYG.

Using 300 past observed data of daily returns until April 16, 2013, we estimated[1] an expected return vector and covariance matrix for these assets, resulting in

$$\hat{r} = \begin{bmatrix} 5.996 & 4.584 & 6.202 & 7.374 & 3.397 & 1.667 & 3.798 \end{bmatrix} \times 10^{-4},$$

$$\Sigma = \begin{bmatrix} 0.5177 & 0.596 & 0.2712 & 0.5516 & 0.9104 & -0.3859 & 0.2032 \\ 0.596 & 1.22 & 0.3602 & 0.7671 & 1.095 & -0.4363 & 0.2469 \\ 0.2712 & 0.3602 & 0.3602 & 0.2866 & 0.4754 & -0.1721 & 0.1048 \\ 0.5516 & 0.7671 & 0.2866 & 0.8499 & 1.073 & -0.4363 & 0.2303 \\ 0.9104 & 1.095 & 0.4754 & 1.073 & 2.563 & -0.8142 & 0.4063 \\ -0.3859 & -0.4363 & -0.1721 & -0.4363 & -0.8142 & 0.7479 & -0.1681 \\ 0.2032 & 0.2469 & 0.1048 & 0.2303 & 0.4063 & -0.1681 & 0.1478 \end{bmatrix} \times 10^{-4}.$$

[1] Estimating expected returns and the covariance matrix is a very delicate step in portfolio optimization. In this example, we just computed $\hat{r}$ and Σ, respectively, as the empirical mean and the empirical covariance matrix of the returns over the considered look-back period of 300 days.

We considered $N = 20$ equi-spaced values of μ in the interval $[3.54, 7.37] \times 10^{-4}$ and, for each value of μ, we solved problem (14.1) with ns = 1 (no short-selling allowed), thus obtaining a discrete-point approximation of the efficient frontier, as shown in Figure 14.1. For instance, the efficient portfolio number 13 has the following composition:

$$x = \begin{bmatrix} 0.0145 & 0.0014 & 54.79 & 31.68 & 0.0008 & 13.51 & 0.0067 \end{bmatrix} \times 0.01.$$

This portfolio is essentially composed of 54.8% of IDU, 31.7% of QQXT, and 13.5% of TLT, and provides an expected return of 0.0596% (daily), with a standard deviation of 0.494%.

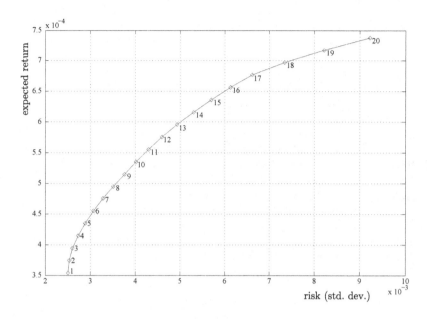

Figure 14.1 Efficient frontier.

14.1.2   *Portfolio constraints and transaction costs*

If required, further design constraints can be enforced on the portfolio composition, as discussed next.

14.1.2.1   Sector bounds. One can impose upper bounds on some elements of x, in order to limit the investment exposure in any individual asset. We can also require that the total sum invested in a given sector (corresponding to, say, the first k assets) does not exceed a fraction α of the total sum invested. This can be modeled by adding to the problem the following constraint:

$$\sum_{i=1}^{k} x_i \leq \alpha \mathbf{1}^\top x.$$

14.1.2.2   Diversification. We can also impose some diversification constraints. For example, we may impose that no group of k (with $k < n$) assets contains more than a fraction η of the total invested amount. We write this as

$$s_k(x) \doteq \sum_{i=1}^{k} x_{[i]} \leq \eta \mathbf{1}^\top x,$$

where $x_{[i]}$ is the i-th largest component of x, so that $s_k(x)$ is the sum of the k largest elements in x, see Example 9.10. Using the representation in (9.16), the diversification constraint can be expressed as follows: there exist a scalar t and an n-vector s such that

$$kt + \mathbf{1}^\top s \leq \eta \mathbf{1}^\top x, \quad s \geq 0, \quad s \geq x - t\mathbf{1}.$$

14.1.2.3 Transaction costs. We can take into account the presence of transaction costs in our optimization model. In particular, we may easily include a proportional cost structure of the form

$$\phi(u) = c\|u\|_1,$$

where $c \geq 0$ is the unit cost of transaction. If transaction costs are present, the budget conservation condition must take into account money lost for transactions: $\mathbf{1}^\top x(0) - \phi(u) = \mathbf{1}^\top x$, that is

$$\mathbf{1}^\top u + \phi(u) = 0.$$

This equality constraint is non-convex. However, if $\hat{r} \geq 0$, and we are considering problem (14.1), we can equivalently relax it into inequality form (see Example 8.19), so that problem (14.1) becomes

$$
\begin{aligned}
\rho(\bar{\sigma}) = \max_u \quad & \hat{r}^\top x && (14.1)\\
\text{s.t.:} \quad & x^\top \Sigma x \leq \bar{\sigma}^2, \\
& \mathbf{1}^\top u + c\|u\|_1 \leq 0, \\
& x = x(0) + u, \\
& \mathsf{ns} \cdot x \geq 0.
\end{aligned}
$$

14.1.3 Sharpe ratio optimization

The Sharpe ratio[2] (SR) is a measure of reward to risk, which quantifies the amount of expected return (in excess of the risk-free rate) per unit risk. The SR of a portfolio x is defined as

$$SR(x) = \frac{\mathbb{E}\{\varrho\} - r_f}{\sqrt{\mathrm{var}\{\varrho\}}} = \frac{\hat{r}^\top x - r_f}{\sqrt{x^\top \Sigma x}},$$

where $r_f \geq 0$ is the return of a *risk-free* asset (for instance, the return one gets from money deposited on a bank savings account), and it is assumed, as done previously, that $w(0) = \mathbf{1}^\top x(0) = 1$. The higher the SR value, the better the investment, from a reward/risk perspective. Assuming that the risk-free asset is not included among the assets considered in the portfolio, and under the condition that $\hat{r}^\top x > r_f$ and $\Sigma \succ 0$, the portfolio that maximizes the SR corresponds, geometrically, to the point of tangency to the efficient frontier of the

[2] After William Sharpe, Nobel laureate (1990).

line passing from the $(0, r_f)$ point (the so-called capital allocation line (CAL)), see Figure 14.2.

Under the above assumptions, finding a Sharpe-optimal portfolio amounts to solving the following problem

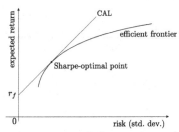

Figure 14.2 Capital allocation line and Sharpe-optimal point.

$$\max_{u} \quad \frac{\hat{r}^\top x - r_f}{\sqrt{x^\top \Sigma x}} \tag{14.2}$$
$$\text{s.t.:} \quad \hat{r}^\top x > r_f,$$
$$x \in \mathcal{X}.$$

The problem in this form is non-convex. However, we can elaborate it as follows: multiply the numerator and denominator of $SR(x)$ by a slack variable $\gamma > 0$, and let

$$\tilde{x} \doteq \gamma x = \gamma x(0) + \tilde{u}; \quad \tilde{u} \doteq \gamma u.$$

Then, since maximizing $SR(x)$ is equivalent to minimizing $1/SR(x)$ (due to the fact that both numerator and denominator are positive), we rewrite the problem as

$$\min_{\tilde{u}, \gamma > 0} \quad \frac{\sqrt{\tilde{x}^\top \Sigma \tilde{x}}}{\hat{r}^\top \tilde{x} - \gamma r_f}$$
$$\text{s.t.:} \quad \hat{r}^\top \tilde{x} > \gamma r_f,$$
$$\mathbf{1}^\top \tilde{u} = 0,$$
$$\tilde{x} = \gamma x(0) + \tilde{u},$$
$$\mathsf{ns} \cdot \tilde{x} \geq 0.$$

Notice that this problem is homogeneous in the variables $(\tilde{u}, \gamma)$, meaning that if $(\tilde{u}, \gamma)$ is a solution, then also $\alpha(\tilde{u}, \gamma)$ is a solution, for any $\alpha > 0$. We resolve this homogeneity by normalizing the denominator of the objective to be equal to one, that is, we impose that $\hat{r}^\top \tilde{x} - \gamma r_f = 1$. With this normalization, the problem becomes

$$\min_{\tilde{u}, \gamma > 0} \quad \sqrt{\tilde{x}^\top \Sigma \tilde{x}}$$
$$\text{s.t.:} \quad \mathbf{1}^\top \tilde{u} = 0,$$
$$\tilde{x} = \gamma x(0) + \tilde{u},$$
$$\hat{r}^\top \tilde{x} - \gamma r_f = 1,$$
$$\mathsf{ns} \cdot \tilde{x} \geq 0.$$

Since $\Sigma \succ 0$, we can factorize it as $\Sigma = \Sigma^{1/2} \Sigma^{1/2}$ (see Section 4.4.4), whence $\sqrt{\tilde{x}^\top \Sigma \tilde{x}} = \|\Sigma^{1/2} \tilde{x}\|_2$. The SR optimization problem can then be cast in the form of the following SOCP:

$$\min_{\tilde{u},\gamma>0,t} \quad t \tag{14.3}$$

$$\text{s.t.:} \quad \|\Sigma^{1/2}\tilde{x}\|_2 \le t,$$
$$\mathbf{1}^\top \tilde{u} = 0,$$
$$\tilde{x} = \gamma x(0) + \tilde{u},$$
$$\hat{r}^\top \tilde{x} - \gamma r_f = 1,$$
$$\text{ns} \cdot \tilde{x} \ge 0.$$

Once this problem is solved for $\tilde{u}, \gamma$, we recover the original decision variable u simply by $u = \gamma^{-1}\tilde{u}$.

For a numerical example, solving problem (14.3) on the data of Example 14.1, assuming a risk-free rate $r_f = 1.5 \times 10^{-4}$, we obtain a Sharpe-optimal portfolio composition of

$$x = [2.71 \quad 0.00 \quad 30.95 \quad 14.08 \quad 0.00 \quad 23.82 \quad 28.44]^\top \times 0.01,$$

with expected return 0.0460×0.01 and standard deviation 0.3122×0.01. This portfolio lies on the return/risk efficient frontier, at the point of tangency with the CAL line, as shown in Figure 14.3.

14.1.4 Value-at-Risk (VaR) optimization

Under the assumption that the return vector r has normal distribution, with known mean $\hat{r}$ and covariance Σ, the so-called *value-at-risk* of portfolio x, at level $\alpha \in (0, 1)$, is defined as (see Example 10.4)

$$\begin{aligned} \text{VaR}_\alpha(x) &= -\sup_\zeta : \text{Prob}\{\varrho(x) \le \zeta\} \le \alpha \\ &= -\inf_\zeta : \text{Prob}\{\varrho(x) \le \zeta\} \ge \alpha \\ &= -F_\varrho^{-1}(\alpha), \end{aligned}$$

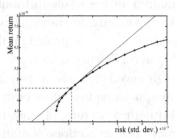

Figure 14.3 Sharpe-optimal portfolio (black circle) for $r_f = 1.5 \times 10^{-4}$.

where $F_\varrho^{-1}(\alpha)$ denotes the inverse cumulative distribution function of the normal random variable $\varrho(x)$ representing the portfolio return. The portfolio VaR is a popular measure of investment risk, often preferred to the classical variance measure. The meaning of VaR_α is that the probability that the investment results in a *loss*[3] higher than VaR_α is no larger than α. Equivalently, the investor is guaranteed to receive a return higher than $-\text{VaR}_\alpha$, with probability no smaller than $1 - \alpha$.

It has been shown in Example 10.4 that, for given $\alpha \in (0, 0.5)$,

$$\text{VaR}_\alpha(x) \le \gamma \quad \Leftrightarrow \quad = \Phi^{-1}(1 - \alpha)\|\Sigma^{1/2}x\|_2 \le \gamma + \hat{r}^\top x, \tag{14.4}$$

[3] We mean by *loss*, ℓ, the negative return, i.e., $\ell = -\varrho$. Thus, in our notation, VaR_α is typically a positive value, describing the maximum loss (up to probability $1 - \alpha$) that the investor may face.

where Φ is the standard normal cumulative distribution function. An optimal portfolio yielding maximal expected return, under a guaranteed upper bound γ on the VaR_α, is determined by solving the following SOCP:

$$\rho(\gamma) = \max_u \quad \hat{r}^\top x \tag{14.5}$$
$$\text{s.t.:} \quad \Phi^{-1}(1-\alpha)\|\Sigma^{1/2}x\|_2 \leq \hat{r}^\top x + \gamma,$$
$$x \in \mathcal{X}.$$

14.2 Robust portfolio optimization

The previous portfolio optimization models are based on the assumption that the *moments* (e.g., the expected value and covariance matrix), or the whole distribution, of the random return vector are known exactly. In practice, however, these quantities are not known, and need to be estimated on the basis of past historical data, possibly in conjunction with expert knowledge. Hence they are subject to high uncertainty due to estimation errors and/or wrong *a priori* assumptions on the nature of the random process describing the market returns. In turn, optimization problems that use "nominal" data may result in portfolio allocation decisions that are far from optimal in practice, i.e., optimal portfolios are known to be quite sensitive to the input data. A possibility for overcoming these issues consists of taking to some extent into account *a priori* the presence of uncertainty in these data, and to seek for "robust" portfolios, that perform well in the worst case against the uncertainty.

14.2.1 Robust mean-variance optimization

In a robust approach to mean-variance portfolio optimization, we assume that $\hat{r}$ and Σ are uncertain, and we model this uncertainty by considering a membership set for these parameters, i.e., we assume that $(\hat{r}, \Sigma)$ belongs to some given bounded uncertainty set $\mathcal{U}$. For a given portfolio x, the worst-case portfolio variance is given by

$$\sigma_{\text{wc}}^2 = \sup_{(\hat{r},\Sigma)\in\mathcal{U}} x^\top \Sigma x.$$

A robust version of the minimum variance portfolio design problem (14.1) would then be

$$\bar{\sigma}_{\text{wc}}^2(\mu) = \min_u \quad \sup_{(\hat{r},\Sigma)\in\mathcal{U}} x^\top \Sigma x \tag{14.6}$$
$$\text{s.t.:} \quad \inf_{(\hat{r},\Sigma)\in\mathcal{U}} \hat{r}^\top x \geq \mu,$$
$$x \in \mathcal{X}.$$

This problem can be solved efficiently and exactly only in some special cases, where the uncertainty set $\mathcal{U}$ is "simple" enough. For instance, if $\mathcal{U} = \{(\hat{r}, \Sigma) : \hat{r} \in \mathcal{U}_r, \Sigma \in \mathcal{U}_\Sigma\}$, where $\mathcal{U}_r, \mathcal{U}_\Sigma$ are interval sets

$$\mathcal{U}_r = \{\hat{r} : r_{\min} \le \hat{r} \le r_{\max}\},$$

$$\mathcal{U}_\Sigma = \{\Sigma : \Sigma_{\min} \le \Sigma \le \Sigma_{\max}, \Sigma \succeq 0\},$$

then, under the further assumptions that $x \ge 0$ and $\Sigma_{\max} \succeq 0$, problem (14.6) is equivalent to

$$\bar{\sigma}^2_{\mathrm{wc}}(\mu) = \min_u \quad x^\top \Sigma_{\max} x \qquad (14.7)$$
$$\text{s.t.:} \quad \hat{r}_{\min}^\top x \ge \mu,$$
$$\mathbf{1}^\top u = 0,$$
$$x = x(0) + u,$$
$$x \ge 0.$$

If the constraint $x \ge 0$ is not imposed, then problem (14.6) can still be solved efficiently, via an ad-hoc *saddle-point* algorithm, whose description is out of the scope of this presentation.[4]

[4] The interested reader may refer to the paper of Tütüncü and Koenig, Robust asset allocation, *Annals of Operations Research*, 2004.

Scenario approach.   For generic uncertainty sets $\mathcal{U}_r, \mathcal{U}_\Sigma$ and problems for which one cannot find an efficient robust reformulation, one can resort to the approximate scenario approach to robustness, as described in Section 10.3.4. In the present setting, one should assume a probability distribution on $\mathcal{U}_r, \mathcal{U}_\Sigma$, and collect N iid samples $(\hat{r}^{(i)}, \Sigma^{(i)})$, $i = 1, \ldots, N$, of the parameters. Then, we solve the scenario problem

$$\min_{x,t} \quad t \qquad (14.8)$$
$$\text{s.t.:} \quad x^\top \Sigma^{(i)} x \le t, \quad i = 1, \ldots, N,$$
$$\hat{r}^{(i)\top} x \ge \mu, \quad i = 1, \ldots, N,$$
$$x \in \mathcal{X}.$$

Solving this scenario problem with a value of N compatible with the prescription in (10.30) will provide a portfolio with a desired level of probabilistic robustness. The advantage of this approach is that it works for any generic structure of the uncertainty, and it does not require lifting the class of optimization problems one needs to solve, i.e., the "robustified" problem (14.8) remains a convex quadratic optimization problem.

14.2.2   Robust value-at-risk optimization

We have seen in Section 14.1.4 that, under the assumption that the return vector r has normal distribution with known mean $\hat{r}$ and covariance Σ, the value-at-risk of portfolio x at level $\alpha \in (0, 0.5)$ is given by Eq. (14.4). In practice, however, we may have two sources of uncertainty that could prevent the direct use of this formula: first, the actual distribution of the returns may not be exactly normal (distributional uncertainty) and, second, the values of the parameters $\hat{r}$, Σ may be known imprecisely (parameter uncertainty). We next discuss how we can tackle a VaR portfolio optimization problem under these two types of uncertainty.

14.2.2.1 Robust VaR under distributional uncertainty. We first suppose that the moments $\hat{r}$, Σ of the return distribution are known exactly, but the distribution itself is unknown. We then consider the class $\mathcal{P}$ of all possible probability distributions having mean $\hat{r}$ and covariance Σ. The worst-case probability of having a loss larger than some level ζ is

$$\text{Prob}_{\text{wc}}\{-\varrho(x) \geq \zeta\} \doteq \sup_{\mathcal{P}} \text{Prob}\{-\varrho(x) \geq \zeta\}, \qquad (14.9)$$

where the sup is computed with respect to all distributions in the class $\mathcal{P}$. The worst-case α-VaR is then defined as

$$\text{wc-VaR}_\alpha(x) \doteq \sup \zeta : \text{Prob}_{\text{wc}}\{-\varrho(x) \geq \zeta\} \leq \alpha,$$

and it clearly holds that

$$\text{wc-VaR}_\alpha(x) \leq \gamma \quad \Leftrightarrow \quad \text{Prob}_{\text{wc}}\{-\varrho(x) \geq \gamma\} \leq \alpha. \qquad (14.10)$$

Now, the Chebyshev–Cantelli inequality prescribes that, for any random variable z with finite variance, and $t > 0$, it holds that

$$\sup \text{Prob}\{z - \mathbb{E}\{z\} \geq t\} = \frac{\text{var}\{z\}}{\text{var}\{z\} + t^2},$$

where the sup is performed with respect to all distributions having the same mean and covariance. Therefore, applying the Chebyshev–Cantelli result to (14.9), and considering that the random variable $\varrho(x)$ has expected value $\hat{r}^\top x$ and variance $\|\Sigma^{1/2}x\|_2^2$, we have that

$$\begin{aligned} \text{Prob}_{\text{wc}}\{-\varrho(x) \geq \zeta\} &= \text{Prob}_{\text{wc}}\{-\varrho(x) + \hat{r}^\top x \geq \zeta + \hat{r}^\top x\} \\ &= \frac{\|\Sigma^{1/2}x\|_2^2}{\|\Sigma^{1/2}x\|_2^2 + (\zeta + \hat{r}^\top x)^2}. \end{aligned}$$

We hence have from (14.10) that, for $\alpha \in (0, 0.5)$,

$$\text{wc-VaR}_\alpha(x) \leq \gamma \quad \Leftrightarrow \quad \kappa(\alpha)\|\Sigma^{1/2}x\|_2 \leq \hat{r}^\top x + \gamma,$$

where

$$\kappa(\alpha) \doteq \sqrt{\frac{1-\alpha}{\alpha}}.$$

The distributionally-robust counterpart of the VaR optimization problem (14.5) is thus simply obtained by subsitituting in this program the coefficient $\Phi^{-1}(1-\alpha)$ of the SOC constraint with $\kappa(\alpha)$.

14.2.2.2 *Robust VaR under moment uncertainty.*

If, in addition to uncertainty in the distribution, we also have uncertainty in the value of the parameters $\hat{r}, \Sigma$, we can determine robust VaR portfolios via the following approach.[5] Assume that the uncertainty sets for $\hat{r}, \Sigma$ are interval ones:

$$
\begin{aligned}
\mathcal{U}_r &= \{\hat{r} : r_{\min} \le \hat{r} \le r_{\max}\}, \\
\mathcal{U}_\Sigma &= \{\Sigma : \Sigma_{\min} \le \Sigma \le \Sigma_{\max}\},
\end{aligned}
$$

[5] This approach is described in detail in the paper by El Ghaoui, Oks and Oustry, Worst-case value-at-risk and robust portfolio optimization: a conic programming approach, *Operations Research*, 2003.

and that there exists a $\Sigma \in \mathcal{U}_\Sigma$ such that $\Sigma \succ 0$. Then, defining wc-VaR$_\alpha(x)$ as the sup of VaR$_\alpha(x)$ over all $\hat{r} \in \mathcal{U}_r$, $\Sigma \in \mathcal{U}_\Sigma$, and over all distributions having mean $\hat{r}$ and covariance Σ, it can be proved that wc-VaR$_\alpha(x) \le \gamma$ holds if and only if there exist symmetric matrices $\Lambda_+, \Lambda_- \in S_+^n$, vectors $\lambda_+, \lambda_- \in \mathbb{R}^n$, and a scalar $v \in \mathbb{R}$, such that

$$
\begin{aligned}
&\operatorname{trace}(\Lambda_+ \Sigma_{\max}) - \operatorname{trace}(\Lambda_- \Sigma_{\min}) + \kappa^2(\alpha)v + \lambda_+^\top r_{\max} - \lambda_-^\top r_{\min} \le \gamma, \\
&\begin{bmatrix} \Lambda_+ - \Lambda_- & x/2 \\ x^\top/2 & v \end{bmatrix} \succeq 0, \\
&\lambda_+ \ge 0, \; \lambda_- \ge 0, \; \Lambda_+ \succeq 0, \; \Lambda_- \succeq 0, \\
&x = \lambda_- - \lambda_+.
\end{aligned}
$$

$$(14.11)$$

The robust counterpart of the VaR optimization problem (14.5), under distributional and moment uncertainty can therefore be stated explicitly in the form of the following SDP:

$$
\begin{aligned}
\max_{u,\Lambda_+,\Lambda_-,\lambda_+,\lambda_-,v} \quad & \hat{r}^\top x \\
\text{s.t.:} \quad & (14.11), \\
& x \in \mathcal{X}.
\end{aligned}
$$

14.3 *Multi-period portfolio allocation*

One problem with the investment allocation models proposed in the previous sections is that they consider a single investment period, and are hence focused at optimizing some portfolio performance at the end of this period. In practice, however, an investor, at the end of the period, faces again the problem of reallocating his wealth over the next period, and so on. Of course, he can then solve again the single period allocation problem, and iterate the strategy indefinitely. However, decisions based on a period-by-period strategy may be "myopic" for long-term objectives. That is, if the investor knows in advance that his investment targets are set over $T \geq 1$ forward periods, he might be better off by considering from the outset the multi-period nature of his investment problem, so as to obtain more far-sighted strategies.

In the next sections we shall discuss two *data-driven* techniques to optimize investment decisions over multiple periods. The first technique, that we shall call *open-loop* strategy, aims at determining at decision time $k = 0$ the whole sequence of future portfolio adjustments over the considered horizon from 0 to final time T. The second technique, which we shall call *closed-loop* strategy, is an improvement of the first strategy that adds flexibility to the future decisions, by allowing them to be functions (so-called *policies*) of the observed returns, thus reducing the impact of future uncertainty. Both approaches are data driven, that is they are based on the availability of a number N of simulated scenarios of the return streams.

We start with some notation and preliminaries: denote by $a_1, \ldots,$ a_n, a collection of assets, and by $p_i(k)$ the market price of a_i at time $k\Delta$, where k is a signed integer, and Δ is a fixed period of time. The simple return of an investment in asset i over the k-th period, from $(k-1)\Delta$ to $k\Delta$, is

$$r_i(k) \doteq \frac{p_i(k) - p_i(k-1)}{p_i(k-1)}, \; i = 1, \ldots, n; \; k = 1, 2, \ldots,$$

and the corresponding *gain* is defined as

$$g_i(k) \doteq 1 + r_i(k), \; i = 1, \ldots, n; \; k = 1, 2, \ldots$$

We denote by $r(k) \doteq [r_1(k) \; \cdots \; r_n(k)]^\top$ the vector of assets' returns over the k-th period, and by $g(k)$ the corresponding vector of gains. The notation $G(k) = \mathrm{diag}\,(g(k))$ indicates a diagonal matrix having the elements of $g(k)$ in the diagonal. The return and gain vectors are assumed to be random quantities. We let $T \geq 1$ denote the number of forward periods over which the allocation decisions need be taken, and we assume that there is available a *scenario-generating oracle* which is capable of generating a desired number N of independent and identically distributed (iid) samples of the forward return streams $\{r(1), \ldots, r(T)\}$.

14.3.1 *Portfolio dynamics*

We consider a decision problem over T periods (or *stages*), where at each period we have the opportunity of rebalancing our portfolio allocation, with the objective of obtaining a minimum level of a suitable cost function (to be discussed later) at the final stage, while guaranteeing satisfaction of portfolio constraints at each stage. The k-th decision period starts at time $k-1$ and ends at time k, see Figure 14.4.

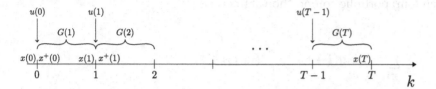

Figure 14.4 Investment decision horizon and periods.

We denote by $x_i(k)$ the euro value of the portion of the investor's total wealth invested in security a_i at time k. The portfolio at time k is the vector

$$x(k) \doteq \left[\begin{array}{ccc} x_1(k) & \cdots & x_n(k) \end{array} \right]^\top.$$

The investor's total wealth at time k is

$$w(k) \doteq \sum_{i=1}^{n} x(k) = \mathbf{1}^\top x(k).$$

Let $x(0)$ be the given initial portfolio composition at time $k = 0$ (for example, one may assume that $x(0)$ is all zeros, except for one entry representing the initial available amount of cash). At $k = 0$, we have the opportunity of conducting transactions on the market and therefore adjusting the portfolio by increasing or decreasing the amount invested in each asset. Just after transactions, the adjusted portfolio is $x^+(0) = x(0) + u(0)$, where $u_i(0) > 0$ if we increase the position on the i-th asset, $u_i(0) < 0$ if we decrease it, and $u_i(0) = 0$ if we leave it unchanged. Suppose now that the portfolio is held fixed for the first period of time Δ. At the end of this first period, the portfolio composition is

$$x(1) = G(1)x^+(0) = G(1)x(0) + G(1)u(0),$$

where $G(1) = \mathrm{diag}\,(g_1(1),\dots,g_n(1))$ is a diagonal matrix of the asset gains over the period from time 0 to time 1. At time $k = 1$, we perform again an adjustment $u(1)$ of the portfolio: $x^+(1) = x(1) + u(1)$, and then hold the updated portfolio for another period of duration Δ. At time $k = 2$ the portfolio composition is hence

$$x(2) = G(2)x^+(1) = G(2)x(1) + G(2)u(1).$$

Proceeding in this way for $k = 0,1,2,\dots$, we determine the iterative dynamic equations of the portfolio composition at the end of period $(k+1)$:

$$x(k+1) = G(k+1)x(k) + G(k+1)u(k), \quad k = 0,\dots,T-1 \quad (14.12)$$

as well as the equations for portfolio composition just after the $(k + 1)$-th transaction (see Figure 14.4)

$$x^+(k) = x(k) + u(k). \tag{14.13}$$

From (14.12) it results that the (random) portfolio composition at time $k = 1, \dots, T$ is

$$x(k) = \Phi(1, k)x(0) + \sum_{j=1}^{k} \Phi(j, k)u(j - 1), \tag{14.14}$$

where we defined $\Phi(v, k)$, $v \leq k$, as the *compounded gain* matrix from the beginning of period v to the end of period k:

$$\Phi(v, k) \doteq G(k)G(k - 1) \cdots G(v), \quad \Phi(k, k) \doteq G(k).$$

The portfolio expression can be rewritten compactly as

$$x(k) = \Phi(1, k)x(0) + \Omega_k u,$$

where

$$u \doteq \begin{bmatrix} u(0)^\top & \cdots & u(T - 2)^\top & u(T - 1)^\top \end{bmatrix}^\top,$$
$$\Omega_k \doteq \begin{bmatrix} \Phi(1, k) & \cdots & \Phi(k - 1, k) & \Phi(k, k) \mid 0 & \cdots & 0 \end{bmatrix}.$$

We thus have for the total wealth

$$w(k) = \mathbf{1}^\top x(k) = \phi(1, k)^\top x(0) + \omega_k^\top u,$$

where

$$\phi(v, k)^\top \doteq \mathbf{1}^\top \Phi(v, k),$$
$$\omega_k^\top \doteq \mathbf{1}^\top \Omega_k = [\phi(1, k)^\top \ \cdots \ \phi(k - 1, k)^\top \ \phi(k, k)^\top \mid 0 \ \cdots \ 0].$$

We consider the portfolio to be self-financing, that is

$$\sum_{i=1}^{n} u_i(k) = 0, \quad k = 0, \dots, T - 1,$$

and we include generic linear constraints in the model by imposing that the updated portfolios $x^+(k)$ lie within a given polytope $\mathcal{X}(k)$. The cumulative gross return of the investment over the whole horizon is

$$\varrho(u) \doteq \frac{w(T)}{w(0)} = \frac{\mathbf{1}^\top x(T)}{\mathbf{1}^\top x(0)} = \frac{\phi(1, T)^\top x(0)}{\mathbf{1}^\top x(0)} + \frac{1}{\mathbf{1}^\top x(0)} \omega_T^\top u.$$

We see that $\varrho(u)$ is an affine function of the decision variables u, with a random vector ω_T of coefficients that depends on the random gains over the T periods.

14.3.2 Optimal open-loop strategy

Assume now that N iid samples (scenarios) $\{G^{(i)}(k), k = 1, \ldots, T\}$, $i = 1, \ldots, N$, of the period gains are available from a scenario generating oracle. These samples produce in turn N scenarios for each of the Ω_k matrices, $k = 1, \ldots, T$, and hence of the ω_k and $\phi(1, k)$ vectors. We denote such scenarios by $\Omega_k^{(i)}$, $\omega_k^{(i)}$, $\phi^{(i)}(1, k)$, $i = 1, \ldots, N$, and by $x^{(i)}(k)$, $w^{(i)}(k)$, $\varrho^{(i)}(u)$, respectively, for the portfolio composition at time k, the total wealth at time k, and the cumulative final return, under the i-th scenario.

Using the sampled scenarios, we can construct several possible empirical risk measures, to be used as objectives to be minimized. Commonly employed objectives are for instance the following ones: let γ denote a given level of return at the final stage (often, but not necessarily, γ is set as the average of the end-of-horizon returns, i.e., $\gamma = \frac{1}{N} \sum_{i=1}^{N} \varrho^{(i)}$):

$$J_1 \doteq \frac{1}{N} \sum_{i=1}^{N} |\gamma - \varrho^{(i)}|, \tag{14.15}$$

$$J_2 \doteq \frac{1}{N} \sum_{i=1}^{N} (\gamma - \varrho^{(i)})^2, \tag{14.16}$$

$$J_{p1} \doteq \frac{1}{N} \sum_{i=1}^{N} \max(0, \gamma - \varrho^{(i)}), \tag{14.17}$$

$$J_{p2} \doteq \frac{1}{N} \sum_{i=1}^{N} \left(\max(0, \gamma - \varrho^{(i)}) \right)^2. \tag{14.18}$$

These objectives express the empirical average of a measure of deviation of the return from γ. In particular, (14.15) is proportional to the ℓ_1 norm of the deviations of the returns $\varrho^{(i)}$ from γ; (14.16) is proportional to the ℓ_2 norm of these deviations, while (14.17), (14.18) are asymmetric measures (sometimes referred to as *lower partial moments*): J_{p1} measures the empirical average of the return values $\varrho^{(i)}$ falling below level γ, while J_{p2} measures the average of the squares of the same deviations. The choice of the first- or second-order cost measures depends on the level of *risk aversion* of the investor, the higher degree in the measure reflecting a higher levels of risk aversion, due to the fact that large residuals are squared, and hence weight more on the cost.

Our open-loop multi-stage allocation strategy is determined by finding the portfolio adjustments $u = (u(0), \ldots, u(T-1))$ that minimize one of the described cost measures, subject to given portfolio composition constraints at each period of the investment horizon.

That is, we solve a problem of the form

$$J^*(\gamma) = \min_u \quad J(u)$$

$$\text{s.t.:} \quad x^{(i)+}(k) \in \mathcal{X}(k), \quad k = 0, \dots, T-1; \; i = 1, \dots, N,$$

$$\mathbf{1}^\top u(k) = 0, \quad k = 0, \dots, T-1,$$

where $J(u)$ is one of the mentioned costs, and where $x^{(i)+}(k)$ is given by (14.13), (14.14), under the i-th sampled scenario. These optimal allocations may be determined in a numerically efficient way by solving either a linear programming or a convex quadratic programming problem.

14.3.3   Closed-loop allocation with affine policies

The open-loop strategy discussed in the previous section may be suboptimal in an actual implementation, since all adjustment decisions $u(0), \dots, u(T-1)$ are computed at time $k = 0$. While the first decision $u(0)$ must be immediately implemented (here-and-now variable), the future decisions may actually wait-and-see the actual outcomes of the returns in the forward periods, and hence benefit from the uncertainty reduction that comes from these observations. For example, at time $k \geq 1$, when we need to implement $u(k)$, we have *observed* a realization of the asset returns over the periods from 1 to k. Hence, we would like to exploit this information, by considering *conditional* allocation decisions $u(k)$, that may react to the returns observed over the previous periods. This means that, instead of focusing on fixed decisions $u(k)$, we may wish to determine suitable *policies* that prescribe what the actual decision should be, in dependence of the observed returns from 1 up to k. In determining the structure of the decision policy one should evaluate a tradeoff between generality and numerical viability of the ensuing optimization problems. It has been observed in several research works[6] that linear or affine policies do provide an effective tradeoff, by allowing reactive policies to be efficiently computed via convex optimization techniques. In this section, we follow this route, and consider decisions prescribed by affine policies of the following form:

$$u(k) = \bar{u}(k) + \Theta(k)\,(g(k) - \bar{g}(k)), \quad k = 1, \dots, T-1, \qquad (14.19)$$

and $u(0) = \bar{u}(0)$, where $\bar{u}(k) \in \mathbb{R}^n$, $k = 0, \dots, T-1$ are "nominal" allocation decision variables, $g(k)$ is the vector of gains over the k-th period, $\bar{g}(k)$ is a given estimate of the expected value of $g(k)$, and $\Theta(k) \in \mathbb{R}^{n,n}$, $k = 1, \dots, T-1$, are the policy "reaction matrices,"

[6] See, e.g., G. Calafiore, Multi-period portfolio optimization with linear control policies, *Automatica*, 2008.

whose role is to adjust the nominal allocation with a term proportional to the deviation of the gain $g(k)$ from its expected value. Since the budget conservation constraint $\mathbf{1}^\top u(k) = 0$ must hold for any realization of the gains, we shall impose the restrictions

$$\mathbf{1}^\top \bar{u}(k) = 0, \quad \mathbf{1}^\top \Theta(k) = 0, \quad k = 0, 1, \ldots, T - 1.$$

14.3.3.1 Portfolio dynamics under affine policies. Applying the adjustment policy (14.19) to the portfolio dynamics equations (14.12), (14.13), we have

$$
\begin{aligned}
x^+(k) &= x(k) + \bar{u}(k) + \Theta(k)\left(g(k) - \bar{g}(k)\right), & (14.20) \\
x(k+1) &= G(k+1)x^+(k), \quad k = 0, 1 \ldots, T - 1, & (14.21)
\end{aligned}
$$

with $\Theta(0) \doteq 0$. From repeated application of (14.20), (14.21) we obtain the expression for the portfolio composition at a generic instant $k = 1, \ldots, T$:

$$x(k) = \Phi(1, k)x(0) + \Omega_k \bar{u} + \sum_{t=1}^{k} \Phi(t, k)\Theta(t - 1)\tilde{g}(t - 1), \quad (14.22)$$

where $\Theta(0) = 0$, and

$$
\begin{aligned}
\bar{u} &\doteq \begin{bmatrix} \bar{u}(0)^\top & \cdots & \bar{u}(T-2)^\top & \bar{u}(T-1)^\top \end{bmatrix}^\top, \\
\tilde{g}(k) &\doteq g(k) - \bar{g}(k), \quad k = 1, \ldots, T.
\end{aligned}
$$

A key observation is that $x(k)$ is an affine function of the decision variables $\bar{u}(k)$ and $\Theta(k)$, $k = 1, \ldots, T - 1$. The cumulative gross return of the investment over the whole horizon is then

$$
\begin{aligned}
\varrho(\bar{u}, \Theta) &= \frac{w(T)}{w(0)} = \frac{\mathbf{1}^\top x(T)}{\mathbf{1}^\top x(0)} \\
&= \frac{1}{\mathbf{1}^\top x(0)} \left(\phi(1, T)^\top x(0) + \omega_T^\top \bar{u} + \sum_{t=1}^{T} \Phi(t, T)\Theta(t - 1)\tilde{g}(t - 1) \right),
\end{aligned}
$$

which is again affine in the variables $\bar{u}$ and

$$\Theta \doteq [\Theta(1) \cdots \Theta(T - 1)].$$

14.3.3.2 Optimal strategy with affine policies. Given N iid samples (scenarios) of the period gains $\{G(k), k = 1, \ldots, T\}$, generated by a scenario generating oracle, we can determine optimal policies that minimize an objective of the form (14.15)–(14.18) by solving the convex optimization problem:

$$
\begin{aligned}
J_d^*(\gamma) = \min_{\bar{u}, \Theta} \quad & J(u, \Theta) \\
\text{s.t.:} \quad & x^{(i)+}(k) \in \mathcal{X}(k), \quad k = 0, \ldots, T - 1; \ i = 1, \ldots, N, \\
& \mathbf{1}^\top \bar{u}(k) = 0, \quad k = 0, \ldots, T - 1, \\
& \mathbf{1}^\top \Theta(k) = 0, \quad k = 1, \ldots, T - 1,
\end{aligned}
$$

where $x^{(i)+}(k)$ is given by (14.20), with $x(k)$ as in (14.22), under the i-th sampled scenario.

14.4 Sparse index tracking

We consider again the problem of replicating (tracking) an index, introduced in Example 9.13. This problem can be cast as the constrained least-squares program

$$
\begin{aligned}
\min_{x \in \mathbb{R}^n} \quad & \|Rx - y\|_2^2 \\
\text{s.t.:} \quad & \mathbf{1}^\top x = 1, \\
& x \geq 0,
\end{aligned}
$$

where $R \in \mathbb{R}^{T,n}$ is a matrix containing in column i, $i = 1, \ldots, n$, the historical returns of the component asset i over $T > n$ time periods, and $y \in \mathbb{R}^T$ is a vector containing the returns of a reference index over the same periods. The objective is to find proportions x of the component assets so as to match as closely as possible the return stream of the reference index. Typically, solution of this problem yields a mixing vector x that contains in different proportions *all* of the n component assets. In practice, however, n may be large, and the user may be interested in replicating the index using a *small* subset of the component assets, since fewer assets are easier to manage and involve less transaction cost. Therefore, the user may be willing to trade some tracking accuracy for "sparsity" of the solution x. We have seen in previous chapters that a usual approach for promoting sparsity in the solution consists of adding an ℓ_1-norm regularization term to the objective, that is to consider an objective of the form $\|Rx - y\|_2^2 + \lambda\|x\|_1$. Unfortunately, this approach is not bound to work in our context, since the decision variable is constrained in the standard simplex $\mathcal{X} = \{x : \mathbf{1}^\top x = 1, \ x \geq 0\}$, hence all feasible x have constant ℓ_1 norm, equal to one. We next describe an effective alternative approach for obtaining sparse solutions in problems where the variable is constrained in the simplex $\mathcal{X}$.

Consider first the ideal problem we would like to solve:

$$
p^* = \min_{x \in \mathcal{X}} \quad f(x) + \lambda\|x\|_0, \tag{14.23}
$$

where $f(x) \doteq \|Rx - y\|_2^2$, $\|x\|_0$ is the cardinality of x, and $\lambda \geq 0$ is a tradeoff parameter. Since we know this problem is non-convex and hard to solve, we seek an efficiently computable relaxation. To this end, we observe that

$$
\|x\|_1 = \sum_{i=1}^n |x_i| \leq \|x\|_0 \max_{i=1,\ldots,n} |x_i| \leq \|x\|_0 \|x\|_\infty.
$$

For $x \in \mathcal{X}$, the left-hand side of the above expression is one, hence we have that

$$x \in \mathcal{X} \quad \Rightarrow \quad \|x\|_0 \geq \frac{1}{\|x\|_\infty} = \frac{1}{\max\limits_{i=1,\ldots,n} x_i}.$$

Therefore, for $x \in \mathcal{X}$,

$$f(x) + \lambda \|x\|_0 \geq f(x) + \lambda \frac{1}{\max\limits_{i=1,\ldots,n} x_i}.$$

Thus, solving the problem

$$p_\infty^* = \min_{x \in \mathcal{X}} \; f(x) + \lambda \frac{1}{\max\limits_{i=1,\ldots,n} x_i}, \qquad (14.24)$$

we would find a lower bound $p_\infty^* \leq p^*$ of the original optimal objective value. Further, denoting by x_∞^* an optimal solution of (14.24), we immediately have that

$$f(x_\infty^*) + \lambda \|x_\infty^*\|_0 \geq p^* \geq p_\infty^*,$$

where the first inequality follows from (14.23). This relation permits us, once a solution x_∞^* is obtained, to check *a posteriori* its level of suboptimality, since it specifies that the "true" optimal level p^* must be contained in the interval $[p_\infty^*, \; f(x_\infty^*) + \lambda \|x_\infty^*\|_0]$.

Still, finding a solution to problem (14.24) is not immediately obvious, since this problem is not yet convex in x (the function $1/\max_i x_i$ is concave over $\mathcal{X}$). However, we can reason as follows:[7]

$$\begin{aligned} p_\infty^* &= \min_{x \in \mathcal{X}} f(x) + \min_i \frac{\lambda}{x_i} \\ &= \min_i \min_{x \in \mathcal{X}} f(x) + \frac{\lambda}{x_i} \\ &= \min_i \min_{x \in \mathcal{X}, t \geq 0} f(x) + t, \quad \text{s.t.: } tx_i \geq \lambda. \end{aligned}$$

[7] See Pilanci, El Ghaoui, and Chandrasekaran, Recovery of sparse probability measures via convex programming, *Proc. Conference on Neural Information Processing Systems*, 2012.

Further, the hyperbolic constraint $tx_i \geq \lambda$ can be expressed as an SOC constraint, using (10.2):

$$tx_i \geq \lambda, \; x_i \geq 0, \; t \geq 0 \quad \Leftrightarrow \quad \left\| \begin{array}{c} 2\sqrt{\lambda} \\ x_i - t \end{array} \right\|_2 \leq x_i + t.$$

The above reasoning proves that we can obtain a solution to the non-convex problem (14.24) by solving a series of n convex problems:

$$\begin{aligned} p_\infty^* = \min_{i=1,\ldots,n} \; &\min_{x,t} \; f(x) + t \\ \text{s.t.:} \quad &x \in \mathcal{X}, \\ &t \geq 0, \\ &\left\| \begin{array}{c} 2\sqrt{\lambda} \\ x_i - t \end{array} \right\|_2 \leq x_i + t. \end{aligned}$$

For a numerical example, consider again the data used in Example 9.13 concerning tracking of the MSCI WORLD index using $n = 5$ component indices. Solving (14.24) with $\lambda = 0.1$ we obtained a portfolio with only two nonzero components:

$$x_\infty^* = [95.19 \; 0 \; 4.81 \; 0 \; 0]^\top \times 0.01,$$

with a corresponding tracking error $\|Rx_\infty^* - y\|_2^2 = 0.0206$.

Remark 14.1 The approach to sparsity discussed here can clearly be applied to all other portfolio allocation problems presented in this chapter, besides the index tracking problem, whenever the decision variable is constrained in the standard simplex. For instance, it can be applied to find sparse portfolios in the mean–variance optimization problems described in Section 14.1.1, whenever a no-shortselling condition is imposed. Instead, when the portfolio variable is not constrained to be nonnegative (i.e., shortselling *is* allowed), then a standard relaxation using the ℓ_1 norm could be used to promote sparsity in the portfolio composition.

14.5   Exercises

Exercise 14.1 (Portfolio optimization problems) We consider a single-period optimization problem involving n assets, and a decision vector $x \in \mathbb{R}^n$ which contains our position in each asset. Determine which of the following objectives or constraints can be modeled using convex optimization.

1. The level of risk (measured by portfolio variance) is equal to a given target t (the covariance matrix is assumed to be known).

2. The level of risk (measured by portfolio variance) is below a given target t.

3. The Sharpe ratio (defined as the ratio of portfolio return to portfolio standard deviation) is above a target $t \geq 0$. Here both the expected return vector and the covariance matrix are assumed to be known.

4. Assuming that the return vector follows a known Gaussian distribution, ensure that the probability of the portfolio return being less than a target t is less than 3%.

5. Assume that the return vector $r \in \mathbb{R}^n$ can take three values $r^{(i)}$, $i = 1, 2, 3$. Enforce the following constraint: the smallest portfolio return under the three scenarios is above a target level t.

6. Under similar assumptions as in part 5: the average of the smallest two portfolio returns is above a target level t. *Hint:* use new variables $s_i = x^\top r^{(i)}$, $i = 1, 2, 3$, and consider the function $s \to s_{[2]} + s_{[3]}$, where for $k = 1, 2, 3$, $s_{[k]}$ denotes the k-th largest element in s.

7. The transaction cost (under a linear transaction cost model, and with initial position $x_{\text{init}} = 0$) is below a certain target.

8. The number of transactions from the initial position $x_{\text{init}} = 0$ to the optimal position x is below a certain target.

9. The absolute value of the difference between the expected portfolio return and a target return t is less than a given small number ϵ (here, the expected return vector $\hat{r}$ is assumed to be known).

10. The expected portfolio return is either above a certain value t_{up}, *or* below another value t_{low}.

Exercise 14.2 (Median risk) We consider a single-period portfolio optimization problem with n assets. We use past samples, consisting of single-period return vectors $r_1, \ldots, r_N$, where $r_t \in \mathbb{R}^n$ contains the returns of the assets from period $t - 1$ to period t. We denote by $\hat{r} \doteq (1/N)(r_1 + \cdots + r_N)$ the vector of sample averages; it is an estimate of the expected return, based on the past samples.

As a measure of risk, we use the following quantity. Denote by $\rho_t(x)$ the return at time t (if we had held the position x at that time). Our risk measure is

$$\mathcal{R}_1(x) \doteq \frac{1}{N} \sum_{t=1}^N |\rho_t(x) - \hat{\rho}(x)|,$$

where $\hat{\rho}(x)$ is the portfolio's sample average return.

1. Show that $\mathcal{R}_1(x) = \|R^\top x\|_1$, with R an $n \times N$ matrix that you will determine. Is the risk measure $\mathcal{R}_1$ convex?

2. Show how to minimize the risk measure $\mathcal{R}_1$, subject to the condition that the sample average of the portfolio return is greater than a target μ, using linear programming. Make sure to put the problem in standard form, and define precisely the variables and constraints.

3. Comment on the qualitative difference between the resulting portfolio and one that would use the more classical, variance-based risk measure, given by

$$\mathcal{R}_2(x) \doteq \frac{1}{N} \sum_{t=1}^N (\rho_t(x) - \hat{\rho}(x))^2.$$

Exercise 14.3 (Portfolio optimization with factor models – 1)

1. Consider the following portfolio optimization problem:

$$p^* = \min_{x} \quad x^\top \Sigma x$$
$$\text{s.t.:} \quad \hat{r}^\top x \geq \mu,$$

where $\hat{r} \in \mathbb{R}^n$ is the expected return vector, $\Sigma \in \mathbb{S}^n$, $\Sigma \succeq 0$ is the return covariance matrix, and μ is a target level of expected portfolio return. Assume that the random return vector r follows a simplified factor model of the form

$$r = F(f + \hat{f}), \quad \hat{r} \doteq F\hat{f},$$

where $F \in \mathbb{R}^{n,k}$, $k \ll n$, is a factor loading matrix, $\hat{f} \in \mathbb{R}^k$ is given, and $f \in \mathbb{R}^k$ is such that $\mathbb{E}\{f\} = 0$ and $\mathbb{E}\{ff^\top\} = I$. The above optimization problem is a convex quadratic problem that involves n decision variables. Explain how to cast this problem into an equivalent form that involves only k decision variables. Interpret the reduced problem geometrically. Find a closed-form solution to the problem.

2. Consider the following variation on the previous problem:

$$p^* = \min_{x} \quad x^\top \Sigma x - \gamma \hat{r}^\top x$$
$$\text{s.t.:} \quad x \geq 0,$$

where $\gamma > 0$ is a tradeoff parameter that weights the relevance in the objective of the risk term and of the return term. Due to the presence of the constraint $x \geq 0$, this problem does not admit, in general, a closed-form solution.

Assume that r is specified according to a factor model of the form

$$r = F(f + \hat{f}) + e,$$

where F, f, and $\hat{f}$ are as in the previous point, and e is an idiosyncratic noise term, which is uncorrelated with f (i.e., $\mathbb{E}\{fe^\top\} = 0$) and such that $\mathbb{E}\{e\} = 0$ and $\mathbb{E}\{ee^\top\} = D^2 \doteq \{d_1^2, \ldots, d_n^2\} \succ 0$. Suppose we wish to solve the problem using a logarithmic barrier method of the type discussed in Section 12.3.1. Explain how to exploit the factor structure of the returns to improve the numerical performance of the algorithm. *Hint:* with the addition of suitable slack variables, the Hessian of the objective (plus barrier) can be made diagonal.

Exercise 14.4 (Portfolio optimization with factor models – 2) Consider again the problem and setup in point 2 of Exercise 14.3. Let $z \doteq F^\top x$, and verify that the probem can be rewritten as

$$p^* = \min_{x \geq 0, z} \quad x^\top D^2 x + z^\top z - \gamma \hat{r}^\top x$$
$$\text{s.t.:} \quad F^\top x = z.$$

Consider the Lagrangian

$$\mathcal{L}(x, z, \lambda) = x^\top D^2 x + z^\top z - \gamma \hat{r}^\top x + \lambda^\top (z - F^\top x)$$

and the dual function

$$g(\lambda) \doteq \min_{x \geq 0, z} \mathcal{L}(x, z, \lambda).$$

Strong duality holds, since the primal problem is convex and strictly feasible, thus $p^* = d^* = \max_\lambda g(\lambda)$.

1. Find a closed-form expression for the dual function $g(\lambda)$.

2. Express the primal optimal solution x^* in terms of the dual optimal variable λ^*.

3. Determine a subgradient of $-g(\lambda)$.

Exercise 14.5 (Kelly's betting strategy) A gambler has a starting capital W_0 and repeatedly bets his whole available capital on a game where with probability $p \in [0, 1]$ he wins the stake, and with probability $1 - p$ he loses it. His wealth W_k after k bets is a random variable:

$$W_k = \begin{cases} 2^k W_0 & \text{with probability } p^k, \\ 0 & \text{with probability } 1 - p^k. \end{cases}$$

1. Determine the expected wealth of the gambler after k bets. Determine the probability with which the gambler eventually runs broke at some k.

2. The results of the previous point should have convinced you that the described one is a ruinous gambling strategy. Suppose now that the gambler gets more cautious, and decides to bet, at each step, only a fraction x of his capital. Denoting by w and ℓ the (random) number of times where the gambler wins and loses a bet, respectively, we have that his wealth at time k is given by

$$W_k = (1 + x)^w (1 - x)^\ell W_0,$$

where $x \in [0, 1]$ is the betting fraction, and $w + \ell = k$. Define the exponential rate of growth of the gambler's capital as

$$G = \lim_{k \to \infty} \frac{1}{k} \log_2 \frac{W_k}{W_0}.$$

(a) Determine an expression for the exponential rate of growth G as a function of x. Is this function concave?

(b) Find the value of $x \in [0, 1]$ that maximizes the exponential rate of growth G. Betting according to this optimal fraction is known as the optimal Kelly's gambling strategy.[8]

3. Consider a more general situation, in which an investor can invest a fraction of his capital on an investment opportunity that may have different payoffs, with different probabilities. Specifically, if $W_0 x$ dollars are invested, then the wealth after the outcome of the investment is $W = (1 + rx)W_0$, where r denotes the return of the investment, which is assumed to be a discrete random variable taking values $r_1, \ldots, r_m$ with respective probabilities $p_1, \ldots, p_m$ ($p_i \geq 0$, $r_i \geq -1$, for $i = 1, \ldots, m$, and $\sum_i p_i = 1$).

The exponential rate of growth G introduced in point 2 of this exercise is nothing but the expected value of the log-gain of the investment, that is

$$G = \mathbb{E}\{\log(W/W_0)\} = \mathbb{E}\{\log(1 + rx)\}.$$

The particular case considered in point 2 corresponds to taking $m = 2$ (two possible investment outcomes), with $r_1 = 1$, $r_2 = -1$, $p_1 = p$, $p_2 = 1 - p$.

(a) Find an explicit expression for G as a function of $x \in [0, 1]$.

(b) Devise a simple computational scheme for finding the optimal investment fraction x that maximizes G.

Exercise 14.6 (Multi-period investments) We consider a multi-stage, single-asset investment decision problem over n periods. For any given time period $i = 1, \ldots, n$, we denote by y_i the predicted return, σ_i the associated variance, and u_i the dollar position invested. Assuming our initial position is $u_0 = w$, the investment problem is

$$\phi(w) \doteq \max_u \sum_{i=1}^{n+1} \left(y_i u_i - \lambda \sigma_i^2 u_i^2 - c|u_i - u_{i-1}| \right) : u_0 = w, \; u_{n+1} = 0,$$

where the first term represents profit, the second, risk, and the third, approximate transaction costs. Here, $c > 0$ is the unit transaction cost and $\lambda > 0$ a risk-return tradeoff parameter. (We assume $\lambda = 1$ without loss of generality.)

1. Find a dual for this problem.

2. Show that ϕ is concave, and find a subgradient of $-\phi$ at w. If ϕ is differentiable at w, what is its gradient at w?

3. What is the sensitivity issue of ϕ with respect to the initial position w? Precisely, provide a tight upper bound on $|\phi(w+\epsilon) - \phi(w)|$ for arbitrary $\epsilon > 0$, and with y, σ, c fixed. You may assume ϕ is differentiable for any $u \in [w, w+\epsilon]$.

Exercise 14.7 (Personal finance problem) Consider the following personal finance problem. You are to be paid for a consulting job, for a total of $C = \$30,000$, over the next six months. You plan to use this payment to cover some past credit card debt, which amounts to $D = \$7000$. The credit card's APR (annual interest rate) is $r_1 = 15.95\%$. You have the following items to consider:

- At the beginning of each month, you can transfer any portion of the credit card debt to another card with a lower APR of $r_2 = 2.9\%$. This transaction costs $r_3 = 0.2\%$ of the total amount transferred. You cannot borrow any more from either of the credit cards; only transfer of debt from card 1 to 2 is allowed.

- The employer allows you to choose the schedule of payments: you can distribute the payments over a maximum of six months. For liquidity reasons, the employer limits any month's pay to $4/3 \times (C/6)$.

- You are paid a base salary of $B = \$70,000$ per annum. You cannot use the base salary to pay off the credit card debt; however, it affects how much tax you pay (see next).

- The first three months are the last three months of the current fiscal year and the last three months are the first three months of the next fiscal year. So if you choose to be paid a lot in the current fiscal year (first three months of consulting), the tax costs are high; they are lower if you choose to distribute the payments over several periods. The precise tax due depends on your gross annual total income G, which is your base salary, plus any extra income. The marginal tax rate schedule is given in Table 14.1.

- The risk-free rate (interest rate from savings) is zero.

- Time line of events: all events occur at the beginning of each month, i.e. at the beginning of each month, you are paid the chosen amount, and immediately you decide how much of each credit card to pay off, and transfer any debt from card 1 to card 2. Any outstanding debt accumulates interest at the end of the current month.

- Your objective is to maximize the total wealth at the end of the two fiscal years while paying off all credit card debt.

Table 14.1 Marginal tax rate schedule.

Total gross income G	Marginal tax rate	Total tax
$\$0 \leq G \leq \$80,000$	10%	$10\% \times G$
$\$80,000 \leq G$	28%	$28\% \times G$ plus $\$8000 = 10\% \times \$80,000$

1. Formulate the decision-making problem as an optimization problem. Make sure to define the variables and constraints precisely. To describe the tax, use the following constraint:

$$T_i = 0.1 \min(G_i, \alpha) + 0.28 \max(G_i - \alpha, 0), \qquad (14.25)$$

where T_i is the total tax paid, G_i is the total gross income in years $i = 1, 2$, and $\alpha = 80,000$ is the tax threshold parameter.

2. Is the problem a linear program? Explain.

3. Under what conditions on α and G_i can the tax constraint (14.25) be replaced by the following set of constraints? Is it the case for our problem? Can you replace (14.25) by (14.26) in your problem? Explain.

$$
\begin{aligned}
T_i &= 0.1 d_{1,i} + 0.28 d_{2,i}, \qquad (14.26) \\
d_{2,i} &\geq G_i - \alpha, \\
d_{2,i} &\geq 0, \\
d_{1,i} &\geq G_i - d_{2,i}, \\
d_{1,i} &\geq d_{2,i} - \alpha.
\end{aligned}
$$

4. Is the new problem formulation, with (14.26), convex? Justify your answer.

5. Solve the problem using your favorite solver. Write down the optimal schedules for receiving payments and paying off/transferring credit card debt, and the optimal total wealth at the end of two years. What is your total wealth W?

6. Compute an optimal W for $\alpha \in [70k, 90k]$ and plot α vs. W in this range. Can you explain the plot?

Exercise 14.8 (Transaction costs and market impact) We consider the following portfolio optimization problem:

$$\max_x \hat{r}^\top x - \lambda x^\top C x - c \cdot T(x - x^0) : x \geq 0, \quad x \in \mathcal{X}, \qquad (14.27)$$

where C is the empirical covariance matrix, $\lambda > 0$ is a risk parameter, and $\hat{r}$ is the time-average return for each asset for the given period. Here, the constraint set $\mathcal{X}$ is determined by the following conditions.

- No shorting is allowed.

- There is a budget constraint $x_1 + \cdots + x_n = 1$.

In the above, the function T represents transaction costs and market impact, $c \geq 0$ is a parameter that controls the size of these costs, while $x^0 \in \mathbb{R}^n$ is the vector of initial positions. The function T has the form

$$T(x) = \sum_{i=1}^{n} B_M(x),$$

where the function B_M is piece-wise linear for small x, and quadratic for large x; that way we seek to capture the fact that transaction costs are dominant for smaller trades, while market impact kicks in for larger ones. Precisely, we define B_M to be the so-called "reverse Hüber" function with cut-off parameter M: for a scalar z, the function value is

$$B_M(z) \doteq \begin{cases} |z| & \text{if } |z| \leq M, \\ \dfrac{z^2 + M^2}{2M} & \text{otherwise.} \end{cases}$$

The scalar $M > 0$ describes where the transition from a linearly shaped to a quadratically shaped penalty takes place.

1. Show that B_M can be expressed as the solution to an optimization problem:

   $$B_M(z) = \min_{v,w} v + w + \frac{w^2}{2M} \; : \; |z| \leq v + w, \; v \leq M, \; w \geq 0.$$

   Explain why the above representation proves that B_M is convex.

2. Show that, for given $x \in \mathbb{R}^n$:

   $$T(x) = \min_{w,v} \mathbf{1}^\top (v + w) + \frac{1}{2M} w^\top w \; : \; \begin{array}{l} v \leq M\mathbf{1}, \; w \geq 0, \\ |x - x^0| \leq v + w, \end{array}$$

   where, in the above, v, w are now n-dimensional vector variables, $\mathbf{1}$ is the vector of ones, and the inequalities are component-wise.

3. Formulate the optimization problem (14.27) in convex format. Does the problem fall into one of the known categories (LP, QP, SOCP, etc.)?

15
Control problems

DYNAMICAL SYSTEMS ARE physical systems evolving in time. Typically, these systems are modeled mathematically by a system of ordinary differential equations, in which command and disturbance signals act as *inputs* that, together with pre-existing *initial conditions*, generate the evolution in time of internal variables (the *states*) and of *output* signals. These kinds of model are ubiquitous in engineering, and are used to describe, for instance, the behavior of an airplane, the functioning of a combustion engine, the dynamics of a robot manipulator, or the trajectory of a missile.

Broadly speaking, the problem of *control* of a dynamical system amounts to determining suitable input signals so as to make the system behave in a desired way, e.g., to follow a desired output trajectory, to be resilient to disturbances, etc. Even an elementary treatment of control of dynamical systems would require an entire textbook. Here, we simply focus on very few specific aspects related to a restricted class of dynamical systems, namely finite-dimensional, linear, and time-invariant systems.

We start our discussion by introducing continuous-time models and their discrete-time counterparts. For discrete-time models, we highlight the connections between the input–output behavior over a finite horizon, and static linear maps described by systems of linear equations. We shall show how certain optimization problems arise naturally in this context, and discuss their interpretation in a control setting. This first part of the chapter thus deals with the so-called optimization-based control (in which control inputs are directly obtained by solving an optimization problem over a finite horizon), and with its iterative implementation over a sliding horizon, which yields the so-called model predictive control (MPC) paradigm. In a brief second part of the chapter we discuss instead a more classical control approach, based on infinite-horizon concepts such as stability, where optimization (SDP in particular) comes into the picture indirectly, as a tool for stability analysis or for the design of stabilizing feedback controllers.

15.1   Continuous and discrete time models

15.1.1   Continuous-time LTI systems

We start by considering the following simple example.

Example 15.1 (*A cart on a rail*) Consider a cart of mass m, moving along a horizontal rail, where it is subject to viscous damping (a damping which is proportional to the velocity) of coefficient β, see Figure 15.1. We let $p(t)$ denote the position of the center of mass of the cart, and $u(t)$ the force applied to the center of mass, where t is the (continuous) time.

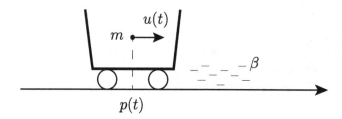

Figure 15.1 Cart on a rail.

The Newton dynamic equilibrium law then prescribes that

$$u(t) - \beta \dot{p}(t) = m\ddot{p}(t),$$

which is the second-order differential equation governing the dynamics of this system. If we introduce variables (states) $x_1(t) = p(t)$, $x_2(t) = \dot{p}(t)$, we may rewrite the Newton equation in the form of a system of two coupled differential equation of first order:

$$
\begin{aligned}
\dot{x}_1(t) &= x_2(t), \\
\dot{x}_2(t) &= \alpha x_2(t) + bu(t),
\end{aligned}
$$

where we defined

$$\alpha \doteq -\frac{\beta}{m}, \quad b \doteq \frac{1}{m}.$$

The system can then be recast in compact matrix form as

$$\dot{x}(t) = A_c x(t) + B_c u(t), \tag{15.1}$$

where

$$A_c = \begin{bmatrix} 0 & 1 \\ 0 & \alpha \end{bmatrix}, \quad B_c = \begin{bmatrix} 0 \\ b \end{bmatrix}.$$

With this system we can also associate an output equation, representing a signal y that is of some particular interest, e.g., the position of the cart itself:

$$y(t) = Cx(t), \quad C = [1 \ 0]. \tag{15.2}$$

Equations (15.1) and (15.2) are actually representative of a quite interesting class of dynamical systems, namely the so-called (strictly proper, finite dimensional) continuous-time linear time-invariant (LTI) systems. Such an LTI system is defined by the triplet of matrices (A, B, C), by means of the state equations

$$\dot{x} \ = \ Ax + Bu, \tag{15.3}$$
$$y \ = \ Cx, \tag{15.4}$$

where $x \in \mathbb{R}^n$ is the state vector, $u \in \mathbb{R}^m$ is the input vector, and $y \in \mathbb{R}^p$ is the output vector (x, u, y are all functions of time $t \in \mathbb{R}$, but we omit the explicit dependence on t for brevity). Given the value of the state at some instant t_0, and given the input $u(t)$ for $t \geq t_0$, there exists an explicit formula (usually known as Lagrange's formula) for expressing the evolution in time of the state of system (15.3):

$$x(t) = e^{A(t-t_0)}x(t_0) + \int_{t_0}^{t} e^{A(t-\tau)} Bu(\tau)\mathrm{d}\tau, \quad t \geq t_0. \tag{15.5}$$

15.1.2   *Discrete-time LTI systems*

For the analysis of certain dynamic phenomena, such as population dynamics or economics, it may be more appropriate to describe the system at *discrete* intervals of time. For example, if we want to model the economy of a country, we consider the relevant quantities entering the model (such as, for instance, the gross domestic product, or the unemployment rate) at some discrete intervals of time Δ (say, $\Delta =$ one week, or one month) rather than at "continuous" time instants. The "discrete" time is usually denoted by a signed integer variable $k = \dots, -1, 0, 1, 2, \dots$, representing time instants $t = k\Delta$. The class of (strictly proper, finite dimensional) discrete-time linear time-invariant (LTI) systems is represented by a system of first-order difference equations of the form

$$x(k+1) \ = \ Ax(k) + Bu(k), \tag{15.6}$$
$$y(k) \ = \ Cx(k). \tag{15.7}$$

Given the state of the system at time k_0, and given the input $u(k)$ for $k \geq k_0$, it can be readily verified by recursive application of Eq. (15.6) that we have

$$x(k) = A^{k-k_0}x(k_0) + \sum_{i=k_0}^{k-1} A^{k-i-1}Bu(i), \quad k \geq k_0. \tag{15.8}$$

15.1.2.1 Discretization. It happens very often in practice that a continuous-time dynamical systems must be analyzed and controlled by means of digital devices, such DSP and computers. Digital devices are inherently discrete-time objects, that can interact with the external world only at discrete instants $k\Delta$, where Δ is a small interval of time representing the I/O clock rate. It is therefore common to "convert" a continuous-time system of the form (15.3), (15.4) into its discrete-time version, by "taking snapshots" of the system at time instants $t = k\Delta$, where Δ is referred to as the *sampling interval*, assuming that the input signal $u(t)$ remains constant between two successive sampling instants, that is

$$u(t) = u(k\Delta), \quad \forall t \in [k\Delta, (k+1)\Delta).$$

This discrete-time conversion can be done as follows: given the state of system (15.3) at time $t = k\Delta$ (which we denote by $x(k) = x(k\Delta)$), we can use equation (15.5) to compute the value of the state at instant $(k+1)\Delta$:

$$
\begin{aligned}
x(k+1) &= e^{A\Delta}x(k) + \int_{k\Delta}^{(k+1)\Delta} e^{A((k+1)\Delta - \tau)} Bu(\tau)d\tau \\
&= e^{A\Delta}x(k) + \int_{k\Delta}^{(k+1)\Delta} e^{A((k+1)\Delta - \tau)} Bd\tau\, u(k) \\
&= e^{A\Delta}x(k) + \int_{0}^{\Delta} e^{A\tau} Bd\tau\, u(k).
\end{aligned}
$$

This means that the sampled version of the continuous-time system (15.3), (15.4) evolves according to a discrete-time recursion of the form (15.6), (15.7) with

$$
\begin{aligned}
x(k+1) &= A_\Delta x(k) + B_\Delta u(k), \\
y(k) &= Cx(k),
\end{aligned}
$$

where

$$A_\Delta = e^{A\Delta}, \quad B_\Delta = \int_0^\Delta e^{A\tau} Bd\tau. \tag{15.9}$$

Example 15.2 Consider the continuous-time model of a cart on a rail in Example 15.1, and assume the numerical values $m = 1$ kg, $\beta = 0.1$ Ns/m. By discretizing the system (15.1) with sample time $\Delta = 0.1$ s, we obtain a discrete-time system

$$
\begin{aligned}
x(k+1) &= Ax(k) + Bu(k), \tag{15.10} \\
y(k) &= Cx(k), \tag{15.11}
\end{aligned}
$$

where A and B are computed according to (15.9), as follows. We first observe by direct computation that

$$A_c^2 = A_c A_c = \begin{bmatrix} 0 & \alpha \\ 0 & \alpha^2 \end{bmatrix}, \qquad A_c^3 = A_c^2 A_c = \begin{bmatrix} 0 & \alpha^2 \\ 0 & \alpha^3 \end{bmatrix}, \dots,$$

$$A_c^k = A_c^{k-1} A_c = \begin{bmatrix} 0 & \alpha^{k-1} \\ 0 & \alpha^k \end{bmatrix},$$

hence

$$
\begin{aligned}
e^{A_c t} &= I + \sum_{k=1}^{\infty} \frac{1}{k!} A_c^k t^k = I + \sum_{k=1}^{\infty} \frac{1}{k!} \begin{bmatrix} 0 & \alpha^{k-1} \\ 0 & \alpha^k \end{bmatrix} t^k \\
&= \begin{bmatrix} 1 & \sum_{k=1}^{\infty} \frac{1}{k!} \alpha^{k-1} t^k \\ 0 & 1 + \sum_{k=1}^{\infty} \frac{1}{k!} \alpha^k t^k \end{bmatrix} \\
&= \begin{bmatrix} 1 & \frac{1}{\alpha}(e^{\alpha t} - 1) \\ 0 & e^{\alpha t} \end{bmatrix},
\end{aligned}
$$

and

$$A = e^{A_c \Delta} = \begin{bmatrix} 1 & \frac{1}{\alpha}(e^{\alpha \Delta} - 1) \\ 0 & e^{\alpha \Delta} \end{bmatrix} = \begin{bmatrix} 1 & 0.0995017 \\ 0 & 0.9900498 \end{bmatrix}.$$

Similarly, for B we have

$$
\begin{aligned}
B &= \int_0^{\Delta} e^{A_c \tau} B_c d\tau = \int_0^{\Delta} \begin{bmatrix} 1 & \frac{1}{\alpha}(e^{\alpha \tau} - 1) \\ 0 & e^{\alpha \tau} \end{bmatrix} \begin{bmatrix} 0 \\ b \end{bmatrix} d\tau \\
&= b \int_0^{\Delta} \begin{bmatrix} \frac{1}{\alpha}(e^{\alpha \tau} - 1) \\ e^{\alpha \tau} \end{bmatrix} d\tau \\
&= \frac{b}{\alpha} \begin{bmatrix} \frac{1}{\alpha}(e^{\alpha \Delta} - 1) - \Delta \\ e^{\alpha \Delta} - 1 \end{bmatrix} = \begin{bmatrix} 0.0049834 \\ 0.0995016 \end{bmatrix}.
\end{aligned}
$$

15.2   Optimization-based control synthesis

15.2.1   Synthesis of the control command for state targeting

In this section we concentrate on a generic discrete-time system of the form (15.6), (15.7), with a scalar input signal $u(k)$ and scalar output signal $y(k)$ (such systems are called SISO, which stands for single-input single-output). This model may come from an originally discrete-time representation of some system, or as a result of discretization of an originally continuous-time system. We address the following control problem: given an initial state $x(0) = x_0$, a target state x_T, and a target integer instant $T > 0$, determine a sequence of control actions $u(k)$, $k = 0, \dots, T-1$ to be assigned to the system such that the state of the system at time T is equal to x_T. In other words, we seek the sequence of control commands that drive the system's state to the target, at the desired instant.

This problem can be formalized by using Eq. (15.8), which yields

$$
x(T) = A^T x_0 + \sum_{i=0}^{T-1} A^{T-i-1} B u(i)
$$

$$
= A^T x_0 + [A^{T-1}B \ A^{T-2}B \ \cdots \ AB \ B] \begin{bmatrix} u(0) \\ u(1) \\ \vdots \\ u(T-2) \\ u(T-1) \end{bmatrix}
$$

$$
= A^T x_0 + R_T \mu_T,
$$

where we defined the *T-reachability matrix* $R_T \doteq [A^{T-1}B \ \cdots \ B]$, and the vector μ_T containing the sequence of control actions. The assigned control problem is thus equivalent to finding a vector $\mu_T \in \mathbb{R}^T$ such that

$$
R_T \mu_T = \xi_T, \quad \xi_T \doteq x_T - A^T x_0, \tag{15.12}
$$

where $R_T \in \mathbb{R}^{n,T}$ is the T-reachability matrix of the system. Equation (15.12) is quite interesting: it says that the state of a discrete-time LTI system at time T is a linear function of the sequence of input commands from time 0 to $T - 1$, see Figure 15.2.

Determining an input sequence that drives the systems to the desired state amounts to solving for μ_T a system of linear equations of the form (15.12). We assume that $T \geq n$ and that R_T is full row rank (i.e., that the system is fully T-reachable, in the language of control engineering). Then, Eq. (15.12) always admits a solution and, if $T > n$, it actually admits an infinite number of possible solutions. This means that there are infinitely many possible input sequences that permit us to reach the desired target state. Among all possibilities it seems natural to choose an input sequence such that the target is reached with "minimum effort," where the "effort" may be measured by some norm of the vector μ_T containing the input sequence.

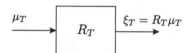

Figure 15.2 Linear map from command sequence to state.

15.2.1.1 Minimum-energy control. For example, the energy of the input signal over the finite time interval $\{0, \ldots, T - 1\}$ is defined as

$$
\text{energy of } u(k) = \sum_{k=0}^{T-1} |u(k)|^2 = \|\mu_T\|_2^2.
$$

Finding the minimum-energy command sequence that reaches the desired target amounts therefore to finding the minimum ℓ_2-norm solution of a system of linear equations (see Section 6.3.2):

$$
\min_{\mu_T} \ \|\mu_T\|_2^2
$$

$$\text{s.t.:} \quad R_T \mu_T = \xi_T,$$

which has the explicit solution (under the hypothesis of full reachability)

$$\mu_T^* = R_T^\dagger \xi_T = R_T^\top (R_T R_T^\top)^{-1} \xi_T, \tag{15.13}$$

where the matrix $G_T \doteq R_T R_T^\top$ is known as the T-reachability *Gramian* of the system.

Another question related to the minimum-energy control problem is the determination of the set of all possible states that can be reached at time T using input sequences of unit energy. That is, we seek to describe the set

$$\mathcal{E}_T = \{x = R_T \mu_T + A^\top x_0 : \|\mu_T\|_2 \leq 1\}.$$

Besides a translation of a constant bias term $A^\top x_0$, the above set is just the image set of the unit ball $\{\mu_T : \|\mu_T\|_2 \leq 1\}$ under the linear map described by the matrix R_T. Therefore, from Lemma 6.4, we have that the reachable set with unit-energy input is the ellipsoid with center $c = A^\top x_0$ and shape matrix $G_T = R_T R_T^\top$, that is

$$\mathcal{E}_T = \{x : (x - c)^\top G_T^{-1} (x - c) \leq 1\}.$$

15.2.1.2 Minimum-fuel control.

An alternative approach for determining the command sequence would be to minimize the ℓ_1 norm of the command sequence, instead of the ℓ_2 norm. The ℓ_1 norm of μ_T is proportional to the consumption of "fuel" needed to produce the input commands. For example, in aerospace applications, control inputs are typically forces produced by thrusters that spray compressed gas, hence $\|\mu_T\|_1$ would be proportional to the total quantity of gas necessary for control actuation. The problem now becomes

$$\min_{\mu_T} \quad \|\mu_T\|_1 \tag{15.14}$$
$$\text{s.t.:} \quad R_T \mu_T = \xi_T,$$

which can be recast as an equivalent LP

$$\min_{\mu_T, s} \quad \sum_{k=0}^{T-1} s_k$$
$$\text{s.t.:} \quad |u(k)| \leq s_k, \quad k = 0, \ldots, T-1,$$
$$\qquad R_T \mu_T = \xi_T.$$

Example 15.3 Consider the discretized model of a cart on a rail of Example 15.2. Given initial conditions $x(0) = [0 \ 0]^\top$, we seek minimum-energy and minimum-fuel input sequences that bring the cart into position $p(t) = 1$, with zero velocity, at time $t = 10$ s. Since the sampling time is $\Delta = 0.1$ s, we have that the final integer target time is

$T = 100$, hence our input sequence is composed of 100 unknown values: $\mu_T = [u(0), \dots, u(T-1)]^\top$. The minimum-energy control sequence is found by applying Eq. (15.13), which yields the results shown in Figure 15.3. The minimum-fuel control sequence is instead found by solving the LP in (15.14), which yields the results shown in Figure 15.4. We observe that the shape of input signals obtained from the minimum-energy approach and from the minimum-fuel one are very different qualitatively. In particular, the minimum-fuel solution is *sparse*, meaning in this case that the control action is zero everywhere except for the initial and the final instants (in the control lingo, this is called a *bang-bang* control sequence).

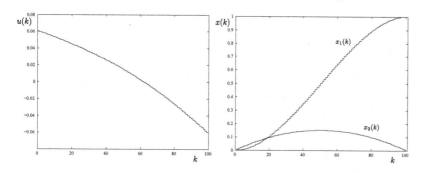

Figure 15.3 Minimum-energy control signal (left) and resulting state trajectory (right).

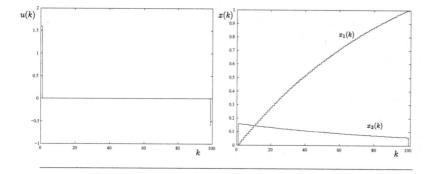

Figure 15.4 Minimum-fuel control signal (left) and resulting state trajectory (right).

15.2.2 Synthesis of the control command for trajectory tracking

In Section 15.2.1 we discussed the problem of reaching a target state, without worrying about the trajectory travelled by the states between the initial and the target instants. Here we study instead the problem of finding a control sequence such that the output $y(k)$ of a discrete-time LTI system tracks as closely as possible an assigned reference trajectory $y_{\text{ref}}(k)$, over a given finite time horizon $k \in \{1, \dots, T\}$. We next assume that the system is SISO, and that $x(0) = 0$, and $y_{\text{ref}}(0) = 0$.

From Eq. (15.8), with $x_0 = 0$, we have that

$$x(k) = A^{k-1}Bu(0) + \cdots + ABu(k-2) + Bu(k-1).$$

Then, considering the output equation $y(k) = Cx(k)$, we obtain

$$y(k) = CA^{k-1}Bu(0) + \cdots + CABu(k-2) + CBu(k-1), \quad k = 1, \ldots, T.$$

Rewriting these latter equations in matrix format, we have

$$\begin{bmatrix} y(1) \\ y(2) \\ \vdots \\ y(T) \end{bmatrix} = \begin{bmatrix} CB & 0 & \cdots & 0 \\ CAB & CB & \cdots & 0 \\ \vdots & \vdots & \ddots & \vdots \\ CA^{T-1}B & \cdots & CAB & CB \end{bmatrix} \begin{bmatrix} u(0) \\ u(1) \\ \vdots \\ u(T-1) \end{bmatrix},$$

that is, defining the output sequence $\mathcal{Y}_T$ and the "transition matrix" $\Phi_T \in \mathbb{R}^{T,T}$ in the obvious way,

$$\mathcal{Y}_T = \Phi_T \mu_T. \tag{15.15}$$

The matrix Φ_T has a particular structure (constant values along the diagonals) known as Toeplitz structure; moreover, Φ_T is lower triangular, therefore it is invertible whenever $CB \neq 0$. Further, the coefficients appearing in the first column of Φ_T are known as the "impulse response" of the system, since they represent the sequence of output values obtained from the system when the input signal is a discrete impulse, that is $u(0) = 1$ and $u(k) = 0$ for all $k \geq 1$. We observe therefore that, on a fixed and finite time horizon, the input–output behavior of a discrete-time LTI system is described by the linear map (15.15), which is defined via the transition matrix Φ_T. Now, if the desired output sequence is assigned to some reference sequence $y_{\text{ref}}(k)$ for $k \in \{1, \ldots, T\}$, we may determine the input sequence that produces the target output sequence by solving (15.15) for μ_T (such a solution exists and it is unique whenever $BC \neq 0$). Defining $\mathcal{Y}_{\text{ref}} = [y_{\text{ref}}(1) \cdots y_{\text{ref}}(T)]^\top$, this is also equivalent to finding μ_T such that

$$\min_{\mu_T} \|\Phi_T \mu_T - \mathcal{Y}_{\text{ref}}\|_2^2. \tag{15.16}$$

This latter least-squares formulation has the advantage of providing an input sequence also when Eq. (15.15) happens to be singular. In this case (singular Φ_T) we determine an input sequence μ_T such that the ensuing output $\mathcal{Y}_T$ is "close," albeit not identical, to $\mathcal{Y}_{\text{ref}}$, in the least-squares sense.

Problem (15.16) translates in mathematical language to a control design "specification," namely, to find the control input such that the *tracking error* with respect to the given reference output $y_{\text{ref}}(k)$ is minimized. However, this is hardly the only specification for a control system. Notice for instance that all the focus in problem (15.16) is on the tracking error, while we are neglecting completely the *command activity*, that is the behavior of the input sequence μ_T. In reality, one must take into account, for instance, limits on the maximum amplitude of the input, slew rate constraints, or at least keep under control the energy content of the input signal, or its fuel consumption. For example, even when Φ_T is invertible, it is good practice to consider the following regularized version of the tracking problem, in which we introduced a penalty term proportional to the energy of the input signal:

$$\min_{\mu_T} \; \|\Phi_T \mu_T - \mathcal{Y}_{\text{ref}}\|_2^2 + \gamma \|\mu_T\|_2^2. \qquad (15.17)$$

Such a formulation allows for a tradeoff between tracking accuracy and input energy: typically a small tracking error requires a high-energy input signal. If the computed input (say, obtained for $\gamma = 0$) has too high energy, we may try to solve the problem again with a higher $\gamma > 0$, etc., until a satisfactory tradeoff is found.

Of course, many variations on the above theme are possible. For instance, one may wish to penalize the ℓ_1 norm of the input, resulting in the regularized problem

$$\min_{\mu_T} \; \|\Phi_T \mu_T - \mathcal{Y}_{\text{ref}}\|_2^2 + \gamma \|\mu_T\|_1,$$

or one may need to add to (15.17) explicit constraints on the amplitude of the command signal:

$$\min_{\mu_T} \quad \|\Phi_T \mu_T - \mathcal{Y}_{\text{ref}}\|_2^2 + \gamma \|\mu_T\|_2^2$$
$$\text{s.t.:} \quad |\mu_T| \leq u_{\max},$$

which results in a convex QP. Also, constraints on the instantaneous rate of variation of the signal (slew rate) can be handled easily, by observing that

$$\begin{bmatrix} u(1) - u(0) \\ u(2) - u(1) \\ \vdots \\ u(T-1) - u(T-2) \end{bmatrix} = D\mu_T, \quad D \doteq \begin{bmatrix} -1 & 1 & 0 & \cdots & 0 \\ 0 & -1 & 1 & \cdots & 0 \\ \vdots & \vdots & \ddots & \ddots & \vdots \\ 0 & \cdots & 0 & -1 & 1 \end{bmatrix}.$$

If we denote by $s_{\max}$ the limit on the input slew rate, a regularized problem with both maximum amplitude and slew rate constraints on the input signal can be posed in the form of the following convex QP:

$$\min_{\mu_T} \quad \|\Phi_T \mu_T - \mathcal{Y}_{\text{ref}}\|_2^2 + \gamma \|\mu_T\|_2^2$$
$$\text{s.t.:} \quad |\mu_T| \leq u_{\max}, \; |D\mu_T| \leq s_{\max}.$$

Example 15.4 Consider again the discretized model of a cart on a rail of Example 15.2. Given initial conditions $x(0) = [0 \; 0]^\top$ and the following reference output trajectory:

$$y_{\text{ref}}(k) = \sin(\omega k \Delta), \quad \omega = \frac{2\pi}{10}, \quad k = 1, \ldots, 100,$$

we seek an input signal such that the output of system (15.10) follows y_{ref} as closely as possible, over the interval $k \in \{1, \ldots, 100\}$. To this end, we consider the regularized LS formulation in (15.17). We first solved the problem with $\gamma = 0$. In this case, since $CB = 0.005 \neq 0$, exact output tracking can (in principle) be achieved. The results are shown in Figure 15.5: we observe that indeed the tracking error is numerically zero (notice the 10^{-15} scale in the $e(k)$ graph on the right panel). However, this is achieved at the expense of an input signal $u(k)$ that is oscillating strongly and with high amplitude, see left panel in Figure 15.5. One reason for this behavior is that although Φ_T is invertible whenever $CB \neq 0$, its condition number rapidly increases with T (in our problem, the condition number of Φ_T is approximately 10^6).

Figure 15.5 Input signal $u(k)$ (left) and tracking error $e(k) = y(k) - y_{\text{ref}}(k)$ (right), for $\gamma = 0$.

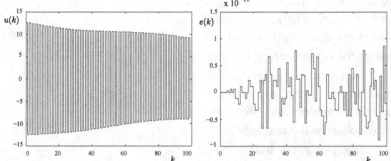

Allowing for a nonzero penalization of the input energy decreases the condition number of the LS coefficient matrix (see Section 6.7.3.1) and greatly improves the behavior of the input signal. For example, assuming $\gamma = 0.1$ we obtain the results shown in Figure 15.6. We observe that the tracking error is now higher than before (but perhaps still acceptable, depending on the specific application), while the input signal has smaller amplitude and it is substantially smoother than in the case with $\gamma = 0$.

Figure 15.6 Input signal $u(k)$ (left) and tracking error $e(k) = y(k) - y_{\text{ref}}(k)$ (right), for $\gamma = 0.1$.

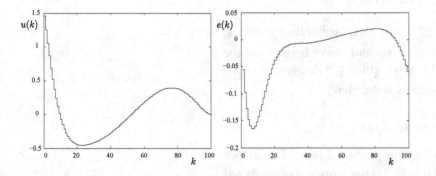

15.2.3 Model predictive control

Model predictive control (MPC) is an effective and widely employed optimization-based technology for the control of dynamical systems. Here, we briefly illustrate the MPC idea on LTI discrete-time systems. With the control design approach discussed in the previous sections, the *whole* control sequence $u(0), \ldots, u(T-1)$ over the fixed time horizon of length T is computed beforehand (i.e., at time 0), and then applied to the system. This may pose some problems in practice, since the actual system to which the computed input is applied may change over time due to disturbances. If the time horizon T is large, the system's state at later instants, due to disturbances, may be substantially different from what we predicted beforehand at time 0, when we computed the control sequence. This is indeed a general problem when one has to plan decisions (control inputs, in this case) over multiple time periods. Intuition suggests that it would be better to compute the sequence at time $k = 0$, call it $u^{(0)} = (u_{0|0}, \ldots, u_{T-1|0})$, and apply to the system at $k = 0$ only the first of these controls, $u(0) = u_{0|0}$. Then, we "wait and see" how the actual system reacts to this first input (that is, we measure the actual state $x(1)$), and recompute a new sequence (call it $u^{(1)} = (u_{0|1}, \ldots, u_{T-1|1})$) for a one-step-forward shifted horizon. Hence, at time $k = 1$ we apply to the system the first value $u(1) = u_{0|1}$ of this sequence, and then iterate the whole process indefinitely (i.e., we observe the outcome $x(2)$, compute a new sequence $u^{(2)} = (u_{0|2}, \ldots, u_{T-1|2})$, apply $u(2) = u_{0|2}$, etc.). This idea is called a *sliding-horizon* implementation of an MPC control law.

In a basic form, MPC computes the predicted control sequence at time k by solving an optimization problem of the following type:

$$\min_{u^{(k)}} \quad J = \sum_{j=0}^{T-1} \left(x_{j|k}^{\top} Q x_{j|k} + u_{j|k}^{\top} R u_{j|k} \right) + x_{T|k}^{\top} S x_{T|k} \qquad (15.18)$$

$$\text{s.t.:} \quad u_{\text{lb}}^{(j)} \leq u_{j|k} \leq u_{\text{ub}}^{(j)}, \quad j = 0, \ldots, T-1,$$
$$F^{(j)} x_{j|k} \leq g^{(j)}, \quad j = 1, \ldots, T,$$

where $Q \in \mathbb{S}_+^n$, $S \in \mathbb{S}_+^n$, $R \in \mathbb{S}_{++}^q$ are given weighting matrices, $u_{\text{ub}}^{(j)}, u_{\text{lb}}^{(j)} \in \mathbb{R}^q$ are vectors of given upper and lower bounds on the predicted control input, and $F^{(j)} \in \mathbb{R}^{m,n}$, $g^{(j)} \in \mathbb{R}^m$ describe possible polyhedral constraints on the system's state. Here,

$$u^{(k)} \doteq (u_{0|k}, \ldots, u_{T|k})$$

is the decision variable, containing the predicted control sequence computed at time k, and $x_{j|k}, j = 0, \ldots, T$, is the sequence of predicted states, which satisfy the system's dynamic equations

$$x_{i+1|k} = Ax_{i|k} + Bu_{i|k}, \quad i = 0, \ldots, T-1, \qquad (15.19)$$

with initial condition $x_{0|k} = x(k)$, where $x(k)$ is the actual system's state, observed at time k. Thus, at each $k = 0, 1, \ldots$, we observe $x(k)$, then solve problem (15.18) to find the optimal predicted sequence $u^{(k)} = (u_{0|k}, \ldots, u_{T|k})$, and finally apply to the actual system the control input $u(k) = u_{0|k}$, and iterate. The objective in (15.18) quantifies a tradeoff between energy in the states and in the control signal (the terms $x_{j|k}^\top Q x_{j|k}$, $u_{j|k}^\top R u_{j|k}$ are weighted ℓ_2 norms of the state and control vectors, respectively), plus a similar penalty term $x_{T|k}^\top S x_{T|k}$ on the terminal state. The purpose of this control strategy is to drive (regulate) the system's state asymptotically to zero. One specific feature of MPC is that it permits us to directly take into account constraints on the control signal (e.g., upper and lower bounds), as well as constraints on the allowable state trajectory.

Notice that applying equation (15.19) recursively for $i = 0$ to j, we obtain

$$x_{j|k} = A^j x(k) + R_j u^{(k)}, \quad R_j = [A^{j-1}B \;\; \cdots \;\; AB \;\; B \;\; 0 \;\; \cdots \;\; 0].$$

This shows that $x_{j|k}$ is a linear function of the vector of decision variables $u^{(k)}$. Substituting this expression into the objective and constraints of problem (15.18), we easily see that this problem is a convex quadratic program in the $u^{(k)}$ variable.

15.3   *Optimization for analysis and controller design*

In this section we briefly discuss some situations in which optimization techniques can be used in order to analyze some system properties (such as stability), or to design a suitable controller for a system. These techniques usually focus on asymptotic properties of the system and, contrary to the ones discussed in the previous sections, are *indirect*, in the sense that optimization is not used for directly determining the control sequence; instead, it is used to determine the values of some parameters (e.g., controller gains) related to ideal devices (controllers) that should be connected to the system (for example in a feedback configuration) in order to control it. There exists a really extensive literature on the topic of optimization methods (especially SDP-based) for control systems analysis and design.[1] Here, we just give the flavor of this type of results by considering a couple of very basic problems related to stability and feedback stabilization of LTI systems.

[1] See, e.g., S. Boyd, L. El Ghaoui, E. Feron, and V. Balakrishnan, *Linear Matrix Inequalities in System and Control Theory*, SIAM, 1994; R. E. Skelton, T. Iwasaki, and K. Grigoriadis, *A Unified Algebraic Approach to Linear Control Design*, 1998; or C. Scherer, S. Weiland, *Linear Matrix Inequalities in Control*, 2004, available online.

15.3.1   Continuous-time Lyapunov stability analysis

The continuous-time LTI system

$$\dot{x}(t) = Ax(t) + Bu(t) \qquad (15.20)$$

is said to be (asymptotically) stable if, for $u(t) = 0$, it holds that $\lim_{t \to \infty} x(t) = 0$, for all initial conditions $x(0) = x_0$. In words, the *free response* of the system goes asymptotically to zero, independent of the initial conditions. It is well known from Lyapunov stability theory that a necessary and sufficient condition for stability of this system is the existence of a matrix $P \succ 0$ such that

$$A^\top P + PA \prec 0, \qquad (15.21)$$

which is a (strict) LMI condition on the matrix variable P. Existence of such a matrix P provides a *certificate* of stability, in the form of a quadratic Lyapunov function $V(x) = x^\top P x$. Notice that both conditions $P \succ 0$ and (15.21) are homogeneous. This means that if some $\bar{P}$ satisfies the conditions

$$\bar{P} \succ 0, \quad A^\top \bar{P} + \bar{P}A = -\bar{Q}, \ \bar{Q} \succ 0$$

then these conditions are satisfied also for any $P = \alpha \bar{P}$, with $\alpha > 0$. In particular, if we choose

$$\alpha = \max\{\lambda_{\min}^{-1}(\bar{P}), \lambda_{\min}^{-1}(\bar{Q})\},$$

then

$$P = \alpha \bar{P} \succeq I, \quad A^\top P + PA \preceq -I.$$

Therefore, stability is equivalent to satisfaction of the following two non-strict LMIs in P:

$$P \succeq I, \quad A^\top P + PA \preceq -I.$$

The problem of finding a Lyapunov stability certificate P can then be posed as an SDP of the form

$$\begin{aligned}
\min_{P,\nu} \quad & \nu \\
\text{s.t.:} \quad & A^\top P + PA \preceq -I, \\
& I \preceq P \preceq \nu I,
\end{aligned}$$

where the last LMI constraints imply that the condition number of P is $\leq \nu$, hence we seek a Lyapunov matrix with minimal condition number.

   An equivalent approach can alternatively be formulated in terms of the inverse Lyapunov matrix $W = P^{-1}$. Indeed, by multiplying

Eq. (15.21) on the left and on the right by P^{-1} we obtain the alternative stability condition $\exists W \succ 0$ such that $WA^\top + AW \prec 0$, which, by following the same previous reasoning on homogeneity, is equivalent to

$$\exists W \succ I: \quad WA^\top + AW \prec -I. \tag{15.22}$$

An optimal (minimum condition number) inverse Lyapunov certificate can be found by solving the following SDP problem:

$$\begin{aligned} \min_{W,\nu} \quad & \nu \\ \text{s.t.:} \quad & WA^\top + AW \preceq -I, \\ & I \preceq W \preceq \nu I. \end{aligned}$$

15.3.2 Stabilizing state-feedback design

The previous approach can be extended from stability analysis to design of a feedback stabilizing control law. Assume that the control input takes the following *state-feedback* form:

$$u(t) = Kx(t), \tag{15.23}$$

where $K \in \mathbb{R}^{m,n}$ is the state-feedback *gain* matrix (or controller), to be designed so as to make the controlled system stable. Plugging (15.23) into the state equation (15.20), we obtain the controlled system

$$\dot{x} = (A + BK)x.$$

This system is stable if and only if conditions (15.22) hold for the closed-loop system matrix $A_{cc} = A + BK$, that is if and only if

$$\exists W \succ I: \quad WA^\top + AW + (KW)^\top B^\top + B(KW) \prec -I.$$

Introducing a new variable $Y \doteq KW$, these conditions are rewritten as

$$\exists W \succ I: \quad WA^\top + AW + Y^\top B^\top + BY \prec -I,$$

which are LMI conditions in the variables W and Y. We can then find a stabilizing feedback gain by first solving a convex optimization problem under the previous LMI constraint, and then recovering K as $K = YW^{-1}$. For instance, the objective of the optimization problem can be chosen as a tradeoff of the condition number of W and the norm of Y:

$$\begin{aligned} \min_{W,Y,\nu} \quad & \nu + \eta \|Y\|_2 \\ \text{s.t.:} \quad & WA^\top + AW + Y^\top B^\top + BY \prec -I, \\ & I \preceq W \preceq \nu I, \end{aligned}$$

where $\eta \geq 0$ is a tradeoff parameter.

15.3.2.1 Robust feedback design. The described approach for feedback stabilization based on LMIs is quite flexible, as it can be easily extended to the situation where the system to be stabilized is *uncertain*. Here, we only discuss the simplest case of uncertainty affecting the system, namely the so-called *scenario*, or polytopic, uncertainty. Consider the system

$$\dot{x}(t) = A(t)x(t) + B(t)u(t), \tag{15.24}$$

where now the systems matrices A, B can vary inside a polytope of given matrices (A_i, B_i), $i = 1, \ldots, N$, that is

$$(A(t), B(t)) \in \text{co}\{(A_1, B_1), \ldots, (A_N, B_N)\}.$$

The matrix pairs (A_i, B_i) are the vertices of the polytope, and they can be interpreted as scenarios, or different possible realizations of an uncertain plant, as measured under different operating conditions, epochs, etc. It can be proved that a *sufficient* condition for stability of the uncertain system (15.24) is the existence of a common quadratic Lyapunov function for all the vertex systems, that is

$$\exists W \succ I : \quad WA_i^\top + A_i W \prec -I, \quad i = 1, \ldots, N.$$

Based on this sufficient condition, we can for instance design a feedback control law of the form (15.23) that robustly stabilizes the uncertain system. The controlled system is described by

$$\dot{x}(t) = (A(t) + B(t)K)x(t),$$

and it is stable if

$$\exists W \succ I : \quad W(A_i + B_i K)^\top + (A_i + B_i K)W \prec -I, \quad i = 1, \ldots, N.$$

Introducing the variable $Y = KW$, the feedback gain matrix K of a stabilizing controller can be found by solving the following problem:

$$\begin{aligned}
\min_{W,Y,\nu} \quad & \nu + \eta\|Y\|_2 \tag{15.25}\\
\text{s.t.:} \quad & WA_i^\top + A_i W + Y^\top B_i^\top + B_i Y \prec -I, \quad i = 1, \ldots, N,\\
& I \preceq W \preceq \nu I,
\end{aligned}$$

and then recovering K as $K = YW^{-1}$.

Example 15.5 (*Robust stabilization of an inverted pendulum*) Consider the simplified model of an inverted pendulum depicted in Figure 15.7, constituted by a point mass m on the tip of a rigid rod of length ℓ, where $\theta(t)$ denotes the angular position of the rod w.r.t. the vertical axis, $u(t)$ is the input torque applied at the joint, and g is the gravity acceleration.

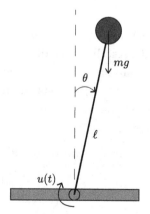

Figure 15.7 An inverted pendulum.

The Newton equilibrium equations for this system prescribe that

$$m\ell^2\ddot\theta(t) = mg\ell \sin\theta(t) + u(t).$$

The dynamics of the system are thus described by a second-order *non-linear* differential equation. However, as far as the angle $\theta(t)$ remains small, we may approximate the nonlinear equation by a linear differential equation, obtained by assuming that $\sin\theta(t) \simeq \theta(t)$, for $|\theta(t)| \simeq 0$. The approximated dynamic equation is

$$m\ell^2\ddot\theta(t) = mg\ell\theta(t) + u(t),$$

hence, introducing state variables $x_1(t) = \theta(t)$, $x_2(t) = \dot\theta(t)$, the system is approximately described by a continuous-time LTI system

$$\dot x = Ax + Bu, \quad A = \begin{bmatrix} 0 & 1 \\ \alpha & 0 \end{bmatrix}, \; B = \begin{bmatrix} 0 \\ \beta \end{bmatrix},$$

where $\alpha \doteq \frac{g}{\ell}$, $\beta \doteq \frac{1}{m\ell^2}$. This system is unstable, independently of the (positive) numerical values of the parameters. Suppose we are given the following numerical values:

$$g = 9.8\,\mathrm{m/s^2},\; m = 0.5\,\mathrm{kg},\; \ell \in [0.9, 1.1]\,\mathrm{m},$$

that is, we are uncertain about the length of the rod. We want to design a state-feedback control law $u(t) = Kx(t)$ which is robustly stabilizing for this system. To this end, we observe that the system matrices may vary along the segment defined by the two vertex systems

$$A_1 = \begin{bmatrix} 0 & 1 \\ 10.8889 & 0 \end{bmatrix}, \; B_1 = \begin{bmatrix} 0 \\ 2.4691 \end{bmatrix};$$

$$A_2 = \begin{bmatrix} 0 & 1 \\ 8.9091 & 0 \end{bmatrix}, \; B_2 = \begin{bmatrix} 0 \\ 1.6529 \end{bmatrix}.$$

We can thus apply the sufficient condition for stabilizability of polytopic systems, and find the control gain K by solving the convex optimization problem in (15.25), with $N = 2$ and $\eta = 1$. Numerical solution via CVX yielded the robustly stabilizing controller $K = -[6.7289, \; 2.3151]$, together with the Lyapunov certificate

$$P = W^{-1} = \begin{bmatrix} 0.8796 & 0.2399 \\ 0.2399 & 0.5218 \end{bmatrix}.$$

Remark 15.1 *Analysis in the frequency domain.* The analysis and design approach outlined in the previous sections is based on a *time-domain* representation of the system. An alternative classical approach exists in which the system is analyzed in a transformed domain, usually called the *frequency domain*. The passage from time domain to frequency domain is obtained by means of a special mapping that transforms a time-domain signal into a function of a complex variable. This mapping, called the (unilateral) *Laplace transform,*[2] associates with a time-domain

[2] The use of the Laplace transform (or of the z-transform, for discrete-time systems) in the analysis and design of control systems is a classic topic, covered in any introductory textbook on linear systems, see, e.g., J. P. Hespanha, *Linear Systems Theory*, Princeton, 2009; W. J. Rough, *Linear System Theory*, Prentice-Hall, 1993; or the more advanced treatments in T. Kailath, *Linear Systems*, Prentice-Hall, 1980; F. Callier, C. A. Desoer, *Linear System Theory*, Springer, 2012, C.-T. Chen, *Linear System Theory and Design*, Oxford, 1999.

signal $w(t) \in \mathbb{C}^n$ the function

$$W(s) \doteq \int_0^\infty w(t)e^{-st}dt,$$

where $s \in \mathbb{C}$, and $W(s)$ is well defined for all s such that the above integral converges to a finite value. The restriction of $W(s)$ to purely imaginary values of s is denoted by

$$\hat{W}(\omega) \doteq W(s)|_{s=j\omega} = \int_0^\infty w(t)e^{-j\omega t}dt,$$

and it coincides with the Fourier transform of $w(t)$, when $w(t)$ is a *causal* signal.[3] If w is real, then $\hat{W}(-\omega)^\top = \hat{W}(\omega)^\star$, where * denotes the transpose conjugate. It follows that, for a real signal, the squared norm $\|\hat{W}(\omega)\|_2^2 = \hat{W}(\omega)^\star\hat{W}(\omega) = \hat{W}(-\omega)^\top\hat{W}(\omega)$ is centrally symmetric with respect to ω.

For a continuous-time LTI system

$$\dot{x}(t) = Ax(t) + Bu(t), \quad x(0) = 0, \quad (15.26)$$
$$y(t) = Cx(t) + Du(t), \quad (15.27)$$

where $x(t) \in \mathbb{R}^n$ is the state, $u(t) \in \mathbb{R}^p$ is the input, $y(t) \in \mathbb{R}^m$ is the output, and the dimensions of the system's matrices (A, B, C, D) are defined accordingly. The *transfer matrix* of the system is defined as

$$H(s) \doteq C(sI - A)^{-1}B + D.$$

The transfer matrix provides a simple description of some aspects of the system behaviour in the Lapace domain, since in this domain the input–output relation is simply described as the product

$$Y(s) = H(s)U(s),$$

where $U(s)$, $Y(s)$ are the Laplace transforms of $u(t)$, $y(t)$, respectively. Each entry $H_{ij}(s)$ of the matrix $H(s)$ is the Laplace transform of a real signal $h_{ij}(t)$ which represents the i-th entry $y_i(t)$ of the output vector when the j-th input $u_j(t)$ is an impulse (a Dirac delta function), and all other inputs $u_k(t)$, $k \neq j$, are zero. Also, if the input signal admits a Fourier transform $\hat{U}(\omega)$, and the system is stable, then the Fourier transform of the output is

$$\hat{Y}(\omega) = H(j\omega)\hat{U}(\omega),$$

and $\hat{H}(\omega) \doteq H(j\omega)$, for $\omega \geq 0$, gives the so-called *frequency response* of the system.

[3] A causal signal is one which is identically zero for $t < 0$. Signals used in control are usually assumed to be causal, since the instant $t = 0$ is conventionally assumed as the time at which the system is "switched on."

15.3.3 *Discrete-time Lyapunov stability and stabilization*

Similarly to the continuous-time case, a discrete-time LTI system

$$x(k+1) = Ax(k) + Bu(k) \quad (15.28)$$

is said to be (asymptotically) stable if its free response goes asymptotically to zero, independently of the initial conditions. A necessary and sufficient condition for stability of (15.28) is the existence of a matrix $P \succ 0$ such that

$$A^\top P A - P \prec 0, \tag{15.29}$$

or, equivalently, by homogeneity,

$$\exists P \succeq I : \quad A^\top P A - P \preceq -I.$$

Also, defining $W = P^{-1}$ and multiplying Eq. (15.29) on the left and on the right by W, we obtain the conditions

$$\exists W \succ 0 : \quad W - W A^\top W^{-1} A W \succ 0,$$

which can be de-homogenized as

$$\exists W \succeq I : \quad W - W A^\top W^{-1} A W \succeq I.$$

Recalling the Schur complement rule, the above conditions can be written as a single LMI condition in W:

$$\begin{bmatrix} W - I & W A^\top \\ A W & W \end{bmatrix} \succeq 0. \tag{15.30}$$

This latter formulation is particularly useful for feedback control design: suppose we want to design a state-feedback law $u(k) = Kx(k)$ for system (15.28). Then, the controlled system is

$$x(k+1) = (A + BK)x(k),$$

which is stable if and only if (by applying (15.30))

$$\begin{bmatrix} W - I & W(A + BK)^\top \\ (A + BK)W & W \end{bmatrix} \succeq 0.$$

Defining a new variable $Y = KW$, this becomes an LMI condition on W and Y. A stabilizing control gain K can thus be obtained, for instance, by solving the following optimization problem in variables W, Y:

$$\begin{aligned} \min_{W,Y,\nu} \quad & \nu + \eta \|Y\|_2 \\ \text{s.t.:} \quad & \begin{bmatrix} W - I & W A^\top + Y^\top B^\top \\ A W + B Y & W \end{bmatrix} \succeq 0, \tag{15.31} \\ & W \preceq \nu I, \end{aligned}$$

and then recovering K as $K = YW^{-1}$. This approach can also be easily extended to robust stabilization of uncertain polytopic systems, following a route similar to the one previously outlined for continuous-time systems.

15.4 Exercises

Exercise 15.1 (Stability and eigenvalues) Prove that the continuous-time LTI system (15.20) is asymptotically stable (or stable, for short) if and only if all the eigenvalues of the A matrix, $\lambda_i(A)$, $i = 1, \ldots, n$, have (strictly) negative real parts.

Prove that the discrete-time LTI system (15.28) is stable if and only if all the eigenvalues of the A matrix, $\lambda_i(A)$, $i = 1, \ldots, n$, have moduli (strictly) smaller than one.

Hint: use the expression $x(t) = e^{At}x_0$ for the free response of the continuous-time system, and the expression $x(k) = A^k x_0$ for the free response of the discrete-time system. You may derive your proof under the assumption that A is diagonalizable.

Exercise 15.2 (Signal norms) A continuous-time *signal* $w(t)$ is a function mapping time $t \in \mathbb{R}$ to values $w(t)$ in either $\mathbb{C}^m$ or $\mathbb{R}^m$. The *energy* content of a signal $w(t)$ is defined as

$$E(w) \doteq \|w\|_2^2 = \int_{-\infty}^{\infty} \|w(t)\|_2^2 dt,$$

where $\|w\|_2$ is the 2-norm of the signal. The class of finite-energy signal contains signals for which the above 2-norm is finite.

Periodic signals typically have infinite energy. For a signal with period T, we define its *power* content as

$$P(w) \doteq \frac{1}{T} \int_{t_0}^{t_0+T} \|w(t)\|_2^2 dt.$$

1. Evaluate the energy of the harmonic signal $w(t) = v e^{j\omega t}$, $v \in \mathbb{R}^m$, and of the causal exponential signal $w(t) = v e^{at}$, for $a < 0$, $t \geq 0$ ($w(t) = 0$ for $t < 0$).

2. Evaluate the power of the harmonic signal $w(t) = v e^{j\omega t}$ and of the sinusoidal signal $w(t) = v \sin(\omega t)$.

Exercise 15.3 (Energy upper bound on the system's state evolution) Consider a continuous-time LTI system $\dot{x}(t) = Ax(t)$, $t \geq 0$, with no input (such a system is said to be *autonomous*), and output $y(t) = Cx$. We wish to evaluate the energy contained in the system's output, as measured by the index

$$J(x_0) \doteq \int_0^{\infty} y(t)^\top y(t) dt = \int_0^{\infty} x(t)^\top Q x(t) dt,$$

where $Q \doteq C^\top C \succeq 0$.

1. Show that if the system is stable, then $J(x_0) < \infty$, for any given x_0.

2. Show that if the system is stable and there exists a matrix $P \succeq 0$ such that

$$A^\top P + PA + Q \preceq 0,$$

then it holds that $J(x_0) \leq x_0^\top P x_0$. *Hint:* consider the quadratic form $V(x(t)) = x(t)^\top P x(t)$, and evaluate its derivative with respect to time.

3. Explain how to compute a minimal upper bound on the state energy, for the given initial conditions.

Exercise 15.4 (System gain) The *gain* of a system is the maximum energy amplification from the input signal to output. Any input signal $u(t)$ having finite energy is mapped by a stable system to an output signal $y(t)$ which also has finite energy. Parseval's identity relates the energy of a signal $w(t)$ in the time domain to the energy of the same signal in the Fourier domain (see Remark 15.1), that is

$$E(w) \doteq \|w\|_2^2 = \int_{-\infty}^{\infty} \|w(t)\|_2^2 dt = \frac{1}{2\pi} \int_{-\infty}^{\infty} \|\hat{W}(\omega)\|_2^2 d\omega \doteq \|\hat{W}\|_2^2.$$

The *energy gain* of system (15.26) defined as

$$\text{energy gain} \doteq \sup_{u(t):\|u\|_2 < \infty, u \neq 0} \frac{\|y\|_2^2}{\|u\|_2^2}.$$

1. Using the above information, prove that, for a stable system,

$$\text{energy gain} \leq \sup_{\omega \geq 0} \|H(j\omega)\|_2^2,$$

where $\|H(j\omega)\|_2$ is the spectral norm of the transfer matrix of system (15.26), evaluated at $s = j\omega$. The (square-root of the) energy gain of the system is also known as the $\mathcal{H}_\infty$-norm, and it is denoted by $\|H\|_\infty$.

Hint: use Parseval's identity and then suitably bound a certain integral. Notice that equality actually holds in the previous formula, but you are not asked to prove this.

2. Assume that system (15.26) is stable, $x(0) = 0$, and $D = 0$. Prove that if there exists $P \succeq 0$ such that

$$\begin{bmatrix} A^\top P + PA + C^\top C & PB \\ B^\top P & -\gamma^2 I \end{bmatrix} \preceq 0 \qquad (15.32)$$

then it holds that

$$\|H\|_\infty \leq \gamma.$$

Devise a computational scheme that provides you with the lowest possible upper bound γ^* on the energy gain of the system.

Hint: define a quadratic function $V(x) = x^\top P x$, and observe that the derivative in time of V, along the trajectories of system (15.26), is

$$\frac{dV(x)}{dt} = x^\top P \dot{x} + \dot{x}^\top P x.$$

Then show that the LMI condition (15.32) is equivalent to the condition that

$$\frac{dV(x)}{dt} + \|y\|^2 - \gamma^2 \|u\|^2 \leq 0, \quad \forall x, u \text{ satisfying (15.26)},$$

and that this implies in turn that $\|H\|_\infty \leq \gamma$.

Exercise 15.5 (Extended superstable matrices) A matrix $A \in \mathbb{R}^{n,n}$ is said to be continuous-time *extended superstable*[4] (which we denote by $A \in E_c$) if there exists $d \in \mathbb{R}^n$ such that

$$\sum_{j \neq i} |a_{ij}| d_j < -a_{ii} d_i, \ d_i > 0, \quad i = 1, \ldots, n.$$

[4] See B. T. Polyak, Extended superstability in control theory, *Automation and Remote Control*, 2004.

Similarly, a matrix $A \in \mathbb{R}^{n,n}$ is said to be discrete-time extended superstable (which we denote by $A \in E_d$) if there exists $d \in \mathbb{R}^n$ such that

$$\sum_{j=1}^{n} |a_{ij}| d_j < d_i, \ d_i > 0, \quad i = 1, \ldots, n.$$

If $A \in E_c$, then all its eigenvalues have real parts smaller than zero, hence the corresponding continuous-time LTI system $\dot{x} = Ax$ is stable. Similarly, if $A \in E_d$, then all its eigenvalues have moduli smaller than one, hence the corresponding discrete-time LTI system $x(k+1) = Ax(k)$ is stable. Extended superstability thus provides a *sufficient* condition for stability, which has the advantage of being checkable via feasibility of a set of linear inequalities.

1. Given a continuous-time system $\dot{x} = Ax + Bu$, with $x \in \mathbb{R}^n$, $u \in \mathbb{R}^m$, describe your approach for efficiently designing a state-feedback control law of the form $u = -Kx$, such that the controlled system is extended superstable.

2. Given a discrete-time system $x(k+1) = Ax(k) + Bu(k)$, assume that matrix A is affected by interval uncertainty, that is

$$a_{ij} = \hat{a}_{ij} + \delta_{ij}, \quad i, j = 1, \ldots, n,$$

where $\hat{a}_{ij}$ is the given nominal entry, and δ_{ij} is an uncertainty term, which is only known to be bounded in amplitude as $|\delta_{ij}| \leq \rho r_{ij}$, for given $r_{ij} \geq 0$. Define the radius of extended superstability as the largest value ρ^* of $\rho \geq 0$ such that A is extended superstable for all the admissible uncertainties. Describe a computational approach for determining such a ρ^*.

16
Engineering design

ONE AREA WHERE recent advances in convex modeling and optimization has had a tremendous impact in the last few decades is
certainly that of engineering design. With the advent of reliable optimization techniques (first, linear programming and, later, quadratic
and conic programming) engineers started to look back at various
analysis and design problems, discovering that they could be formulated, and hence efficiently solved, via convex models. Entire fields,
such as automatic control and circuit analysis, were revolutionized
in the 1990s by the introduction of convex programming methodologies (SDP, in particular). Today, convex models are routinely used
to solve relevant problems in, to mention just a few, structural mechanics, identification, process control, filter design, macromodeling
of electrical circuits, logistics and management, network design, etc.
In this chapter, we detail a selection of these applications.

16.1 Digital filter design

A single-input single-output digital filter is a dynamical system, with
scalar input signal $u(t)$ and scalar output signal $y(t)$, where t stands
for the (discrete) time variable. A finite-impulse response (FIR) filter
is a particular type of filter, which has the form

$$y(t) = \sum_{i=0}^{n-1} h_i u(t-i), \quad t \in \mathbb{Z},$$

where $h_0, \ldots, h_{n-1}$ is called the *impulse response* of the filter. The
name derives from the fact that the time response of the filter to
the discrete-time impulse

$$u(t) = \begin{cases} 1 & \text{if } t = 0, \\ 0 & \text{otherwise,} \end{cases}$$

is precisely the finitely supported signal

$$y(t) = \begin{cases} h_t & \text{if } 0 \le t \le n-1, \\ 0 & \text{otherwise.} \end{cases}$$

The discrete-time Fourier transform of the impulse response is a complex-valued function $H : \mathbb{R} \to \mathbb{C}$ with values

$$H(\omega) = \sum_{t=0}^{n-1} h_t e^{-j\omega t}, \quad \omega \in [-\pi, \pi].$$

This function is important, since it dictates how the filter responds to periodic signals. Precisely, if the input is a complex exponential $u(t) = e^{j\omega_0 t}$, then the output will be the scaled complex exponential $y(t) = H(\omega_0)e^{j\omega_0 t}$. Since $H(\omega)$ is 2π-periodic and $H^*(\omega) = H(-\omega)$, we can restrict the analysis to the normalized frequency interval $[0, \pi]$. A simple example of a FIR filter is a moving average filter: a moving average filter of length 2 is of the form

$$y(t) = \frac{1}{2}(u(t) + u(t-1)), \quad t \in \mathbb{R}.$$

16.1.1   Linear-phase FIR filters

A special class of FIR filters is the so-called Type I *linear-phase* filters. Type I linear phase FIR filters have the characteristic of having an odd number of terms (taps): $n = 2N + 1$, with an impulse response that is symmetric around the midpoint:

$$h_t = h_{n-1-t}, \quad t = 0, \dots, n-1.$$

The qualifier "linear phase" comes from the fact that, for such filters, the frequency response has the form

$$H(\omega) = e^{-j\omega N} \tilde{H}(\omega), \quad \omega \in [0, \pi],$$

where $\tilde{H}(\omega)$ is the *real-valued* function

$$\tilde{H}(\omega) = h_N + 2 \sum_{t=0}^{N-1} h_t \cos((N-t)\omega),$$

called the *amplitude response* of the filter. Notice that the actual phase of $H(\omega)$ may be discontinuous, since it is given by $-\text{sgn}(\tilde{H}(\omega))\omega N$, and the modulus of $H(\omega)$ is given by

$$|H(\omega)| = |\tilde{H}(\omega)|.$$

In some cases, it is more convenient to work with $H(\omega)$ expressed in terms of the "continuous linear phase" $\theta(\omega) = -\omega N$ and the amplitude response $\tilde{H}(\omega)$ (which is real, but can take on positive as well as negative values), rather than with the actual phase and modulus. This is, for instance, the case in our context, since $\tilde{H}(\omega)$ is a simple linear function of the design parameters h, i.e., we can write

$$\tilde{H}(\omega) = a^\top(\omega)h, \quad h = \begin{bmatrix} h_0 \\ h_1 \\ \vdots \\ h_{N-1} \\ h_N \end{bmatrix}, \; a(\omega) = \begin{bmatrix} 2\cos\omega N \\ 2\cos\omega(N-1) \\ \vdots \\ 2\cos\omega \\ 1 \end{bmatrix}.$$

Notice further that, without loss of generality, we can choose the impulse response vector h to be such that $\tilde{H}(0) > 0$.

16.1.2 Low-pass FIR design specifications

A design problem involving FIR filters typically involves choosing the filter's impulse response h so as to achieve a desired shape in the amplitude response. For example, one may want to ensure that the filter rejects high-frequency signals, but passes low-frequency ones (low-pass filter). These requirements may be mapped to the filter's amplitude response as specified next, see also Figure 16.1 for a graphical illustration of the amplitude design constraints.

- Stop-band constraints:

$$-\delta_s \le \tilde{H}(\omega) \le \delta_s, \quad \omega \in [\Omega_s, \pi],$$

where Ω_s is a "stop-band" frequency, and $\delta_s > 0$ corresponds to an attenuation level we seek to achieve at high frequencies.

- Pass-band constraints:

$$1 - \delta_p \le \tilde{H}(\omega) \le 1 + \delta_p, \quad \omega \in [0, \Omega_p],$$

where Ω_p is a "pass-band" frequency bound, and $\delta_p > 0$ is a bound on the pass-band ripple at low frequencies.

Notice that, for δ_s, δ_p "small," the above constraints are approximately equivalent to the following constraints on the (base 10) logarithm of the modulus of $H(\omega)$:

$$\log|H(\omega)| \le \log \delta_s, \quad \omega \in [\Omega_s, \pi],$$
$$-\ln(10)\delta_p \le \log|H(\omega)| \le \ln(10)\delta_p, \quad \omega \in [0, \Omega_p].$$

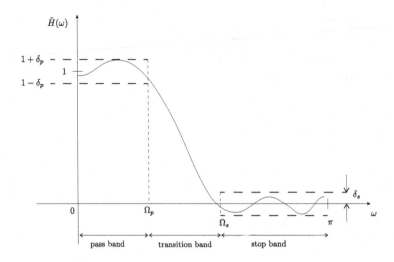

Figure 16.1 Template for low-pass FIR design.

16.1.2.1 *Frequency discretization.*

Notice that the previous design constraints involve a set of linear inequality constraints *at each frequency* ω in the respective intervals. Therefore, they actually involve an *infinite* number of linear constraints. To overcome this issue, we simply discretize the frequency intervals. Instead of enforcing the constraints for every frequency in an interval, we enforce them on a finite frequency grid inside the interval. We choose a finite set of frequencies ω_i, $i = 1, \ldots, N_s$ that belong to the high-frequency region $[\Omega_s, \pi]$, and approximate the stop-band constraint by a *finite* number of linear inequalities in h:

$$-\delta_s \le a^\top(\omega_i)h \le \delta_s, \quad i = 1, \ldots, N_s.$$

Likewise, we choose another set of frequencies ω_i, $i = N_s + 1, \ldots, N_s + N_p$ belonging to the low-frequency region $[0, \Omega_p]$, and approximate the pass-band ripple constraints by a finite number of linear inequalities in h:

$$1 - \delta_p \le a^\top(\omega_i)h \le 1 + \delta_p,, \quad i = N_s + 1, \ldots, N_s + N_p.$$

16.1.3 *FIR design via linear programming*

Under the previous assumptions, a low-pass FIR filter design problem can now be formulated in various ways as a linear program. For instance, one may fix the desired stop-band attenuation level $\delta_s > 0$ and find h and δ_p such that the pass-band ripple is minimized:

$$\min_{h \in \mathbb{R}^{N+1}, \delta_p \in \mathbb{R}} \quad \delta_p$$
$$\text{s.t.:} \quad -\delta_s \le a^\top(\omega_i)h \le \delta_s, \qquad i = 1, \ldots, N_s,$$
$$1 - \delta_p \le a^\top(\omega_i)h \le 1 + \delta_p, \quad i = N_s + 1, \ldots, N_s + N_p.$$

Alternatively, we can minimize the stop-band attenuation level δ_s subject to a given bound on the pass-band ripple $\delta_p > 0$, which results in the LP

$$\min_{h \in \mathbb{R}^{N+1}, \delta_s \in \mathbb{R}} \quad \delta_s$$

$$\text{s.t.:} \quad -\delta_s \le a^\top(\omega_i)h \le \delta_s, \qquad i = 1, \ldots, N_s,$$

$$1 - \delta_p \le a^\top(\omega_i)h \le 1 + \delta_p, \quad i = N_s + 1, \ldots, N_s + N_p.$$

Also, one may obtain designs that strike a tradeoff between the δ_p level and the δ_s level by solving the following LP for different values of the weighting parameter $\mu \ge 0$:

$$\min_{h \in \mathbb{R}^{N+1}, \delta_s \in \mathbb{R}, \delta_p \in \mathbb{R}} \quad \delta_s + \mu \delta_p$$

$$\text{s.t.:} \quad -\delta_s \le a^\top(\omega_i)h \le \delta_s, \qquad i = 1, \ldots, N_s,$$

$$1 - \delta_p \le a^\top(\omega_i)h \le 1 + \delta_p, \quad i = N_s + 1, \ldots, N_s + N_p.$$

16.1.4 A numerical example

We present here a numerical example of minimization of the stop-band attenuation level δ_s, subject to a given bound on the pass-band ripple δ_p. We choose parameters $N = 10$ (so the filter has $n = 2N + 1 = 21$ taps), $\Omega_p = 0.35\pi$, $\Omega_s = 0.5\pi$, $\delta_p = 0.02$ (i.e, -33.98 dB), and we discretized the frequency intervals with $N_s = N_p = 100$ linearly spaced points. Solving the corresponding LP we obtain an optimal stop-band attenuation $\delta_s = 0.0285$ (i.e., -30.9 dB). The amplitude response plot is shown in Figure 16.2, the corresponding logarithmic plot of $|H(\omega)|$ is shown in Figure 16.3, and the filter coefficients (impulse response) are shown in Figure 16.4.

16.1.5 Filter design by approximate reference matching

An alternative approach for designing general linear-phase FIR filters is based on the idea of providing a "reference" desired amplitude response and then trying to find filter parameters h so that the filter response is "as close as possible" to the reference response, over all frequencies.

16.1.5.1 *LS design.* Let $\tilde{H}_{\text{ref}}(\omega)$ be the given reference amplitude response, and consider a discretization ω_i, $i = 1, \ldots, M$ of the frequency interval $[0, \pi]$. Then a possible design approach is to look for filter coefficients h such that the following mismatch measure is minimized:

$$\min_{h} \quad \sum_{i=1}^{M} (\tilde{H}(\omega_i) - \tilde{H}_{\text{ref}}(\omega_i))^2.$$

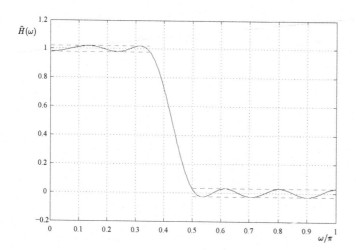

Figure 16.2 Amplitude response of FIR filter.

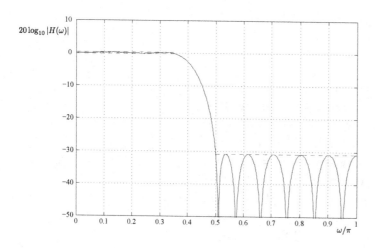

Figure 16.3 Modulus of the FIR filter as a function of (normalized) frequency.

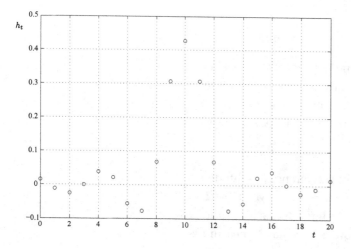

Figure 16.4 Impulse response of FIR filter.

Since $\tilde{H}(\omega) = a^{\top}(\omega)h$, this is clearly an LS problem:

$$\min_h \ \|Ah - b\|_2^2,$$

where

$$A = \begin{bmatrix} a^{\top}(\omega_1) \\ \vdots \\ a^{\top}(\omega_M) \end{bmatrix}, \quad b = \begin{bmatrix} \tilde{H}_{\mathrm{ref}}(\omega_1) \\ \vdots \\ \tilde{H}_{\mathrm{ref}}(\omega_M) \end{bmatrix}.$$

A variant of this approach is obtained by introducing weights in the mismatch error at different frequencies: one chooses a frequency weight profile $w_i \geq 0$, $i = 1, \ldots, M$, and solves the modified problem

$$\min_h \ \sum_{i=1}^{M} w_i^2 (\tilde{H}(\omega_i) - \tilde{H}_{\mathrm{ref}}(\omega_i))^2.$$

In this setting, a relatively high value of w_i implies that it is important to reduce the mismatch at frequency ω_i, while a relatively low value of w_i implies that the mismatch error at ω_i is less important. This approach leads to the weighted LS problem

$$\min_h \ \|W(Ah - b)\|_2^2, \quad W = \mathrm{diag}\,(w_1, \ldots, w_M).$$

As a numerical example, we considered a filter with $N = 10$, $M = 200$ linearly spaced discretized frequencies, and a reference response equal to one up to pass-band frequency $\Omega_p = 0.35\pi$ and equal to zero for $\omega > \Omega_s = 0.5\pi$. The reference amplitude linearly decreases from one to zero in the transition band. Using constant (unit) frequency weights, the result of the LS optimization problem yielded the amplitude response shown in Figure 16.5 and the corresponding modulus diagram shown in Figure 16.6.

16.1.5.2 Chebyshev design. A solution for the LS design is readily computed. However, such an approach does not permit us to control the pointwise maximum mismatch error between the desired (reference) response and the actual filter response. A Chebyshev-type design instead aims at minimizing the maximum weighted mismatch, that is

$$\min_h \ \max_{i=1,\ldots,M} \ w_i |\tilde{H}(\omega_i) - \tilde{H}_{\mathrm{ref}}(\omega_i)|.$$

This problem can be expressed as the following LP:

$$\begin{aligned}
\min_{h,\gamma} \quad & \gamma \\
\text{s.t.:} \quad & w_i(a^{\top}(\omega_i)h - \tilde{H}_{\mathrm{ref}}(\omega_i)) \leq \gamma, \quad i = 1, \ldots, M, \\
& w_i(a^{\top}(\omega_i)h - \tilde{H}_{\mathrm{ref}}(\omega_i)) \geq -\gamma, \quad i = 1, \ldots, M.
\end{aligned}$$

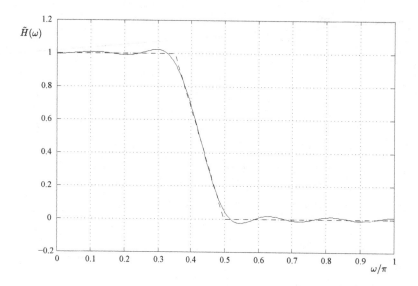

Figure 16.5 Amplitude response of the FIR filter obtained by LS matching. The dashed line is the reference response $\tilde{H}_{ref}(\omega)$.

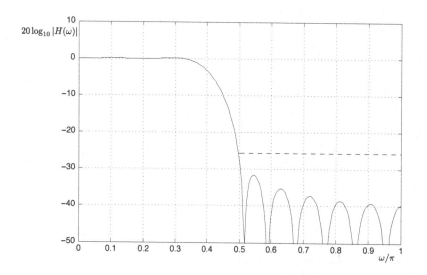

Figure 16.6 Modulus of the FIR filter obtained by LS matching. The dashed line shows the modulus level at the stop-band frequency $\Omega_s = 0.5$, which is equal to 0.0518 (i.e., -25.71 dB).

Application of this design approach to numerical data as in the previous example yielded the amplitude response shown in Figure 16.7 and the corresponding modulus diagram shown in Figure 16.8. The maximum absolute discrepancy between the desired and actual amplitude response resulted to be equal to $\gamma = 0.0313$ (i.e., -30.1 dB).

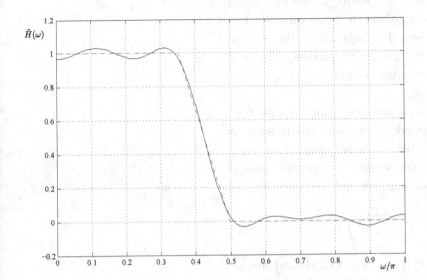

Figure 16.7 Amplitude response of the FIR filter obtained by Chebyshev matching. The dashed line is the reference response $\tilde{H}_{ref}(\omega)$.

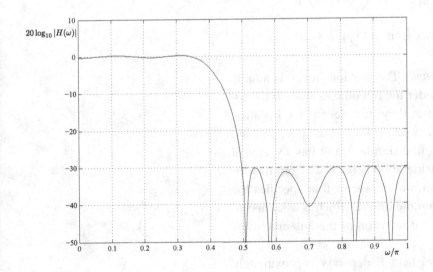

Figure 16.8 Modulus of the FIR filter obtained by Chebyshev matching. The dashed line shows the modulus level at the stop-band frequency $\Omega_s = 0.5$, which is equal to 0.0313 (i.e., -30.1 dB).

16.2   *Antenna array design*

In an antenna array, the outputs of several emitting antenna elements are linearly combined to produce a composite array output. The array output has a directional pattern that depends on the relative weights, or scale factors, used in the combining process. The goal of weight design is to choose the weights so as to achieve a desired directional pattern.

The basic element in a transmitting antenna is an isotropic harmonic oscillator, which emits a spherical, monochromatic wave at wavelength λ and frequency ω. The oscillator generates an electromagnetic field whose electrical component, at a certain point p located at distance d from the antenna, is given by

$$\frac{1}{d}\text{Re}\left(z \cdot \exp\left(j\left(\omega t - \frac{2\pi d}{\lambda} \right) \right) \right),$$

where $z \in \mathbb{C}$ is a design parameter that allows us to scale and change the phase of the electrical field. We refer to this complex number as the *weight* of the antenna element. We now place n such oscillators at the locations $p_k \in \mathbb{R}^3$, $k = 1, \ldots, n$. Each oscillator is associated with a complex weight $z_k \in \mathbb{C}$, $k = 1, \ldots, n$. Then the total electrical field received at a point $p \in \mathbb{R}^3$ is given by the following weighted sum:

$$E = \text{Re}\left(\exp(j\omega t) \cdot \sum_{k=1}^{n} \frac{1}{d_k} z_k \cdot \exp\left(\frac{-2\pi j d_k}{\lambda} \right) \right),$$

where $d_k = \|p - p_k\|_2$ is the distance from p to p_k, $k = 1, \ldots, n$.

Far-field approximation for linear arrays. The previous formula admits an approximated simplification under the hypotheses that *(a)* the oscillators form a linear array, that is they are placed, for instance, on an equidistant grid of points placed on the x-axis as $p_k = \ell k e_1$, $k = 1, \ldots, n$, with $e_1 = [1\,0\,0]$ (the first standard unit basis vector of $\mathbb{R}^3$), and *(b)* the point p under consideration is far away from the origin: $p = ru$, where $u \in \mathbb{R}^3$ is a unit-norm vector that specifies the direction, and r is the distance from the origin, which is assumed to be much larger than the geometric dimension of the antenna array, i.e., $r \gg n\ell$, see Figure 16.9.

For a linear array, the electrical field E depends approximately only on the angle φ between the array and the far away point under consideration. Indeed, we have

$$d_k = r\sqrt{1 + (k\ell/r)^2 + 2(k\ell/r)\cos\varphi} \simeq r + k\ell\cos\varphi, \quad \text{for } k\ell/r \text{ small,}$$

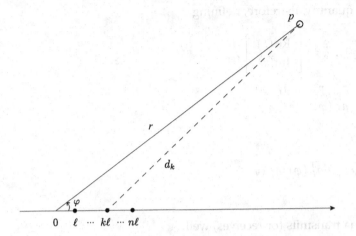

Figure 16.9 Linear antenna array.

and a good approximation to E is of the form

$$E \simeq \frac{1}{r}\mathrm{Re}\left(\exp(j\omega t - 2\pi j r/\lambda)\cdot D_z(\varphi)\right),$$

where the function $D_z : [0, 2\pi] \to \mathbb{C}$ is called the *diagram* of the antenna:

$$D_z(\varphi) \doteq \sum_{k=1}^{n} z_k \cdot \exp\left(\frac{-2\pi j k \ell \cos \varphi}{\lambda}\right). \tag{16.1}$$

We have used the subscript "z" to emphasize that the diagram depends on the choice of the complex weight vector $z = [z_1, \ldots, z_n]$.

An analogous result holds for a linear array of *receiver* antennas, see Figure 16.10: a harmonic plane wave with frequency ω and wavelength λ is incident from the direction φ and propagates across the array. The signal outputs are converted to baseband (complex numbers), weighted by the weights z_k, and summed to give again the linear array beam pattern $D_z(\varphi)$.

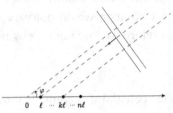

Figure 16.10 A linear array of receiver antennas.

16.2.1 Shaping the antenna diagram

The squared modulus of the antenna diagram, $|D_z(\varphi)|^2$, turns out to be proportional to the directional density of electromagnetic energy sent by the antenna. Hence, it is of interest to "shape" (by choice of the zs) the magnitude diagram $|D_z(\cdot)|$ in order to satisfy some directional requirements. Notice that $D_z(\varphi)$, for fixed φ, is a linear function of the real and imaginary parts of the weights z. In particular, from (16.1), we have that

$$D_z(\varphi) = a^\top(\varphi)z, \quad a^\top(\varphi) = [a_1(\varphi) \cdots a_n(\varphi)],$$
$$a_k(\varphi) = \exp\left(\frac{-2\pi j k \ell \cos \varphi}{\lambda}\right), \quad k = 1, \ldots, n.$$

Notice further that $D_z(\varphi)$ is a complex quantity, therefore, defining

$$a_R(\varphi) = \mathrm{Re}\,(a(\varphi)),\ a_I(\varphi) = \mathrm{Im}(a(\varphi)),\ \zeta = \begin{bmatrix} \mathrm{Re}\,(z) \\ \mathrm{Im}(z) \end{bmatrix},$$

$$C(\varphi) = \begin{bmatrix} a_R^\top(\varphi) & -a_I^\top(\varphi) \\ a_I^\top(\varphi) & a_R^\top(\varphi) \end{bmatrix},$$

we have

$$
\begin{aligned}
D_z(\varphi) &= [a_R^\top(\varphi)\ -a_I^\top(\varphi)]\zeta + j[a_I^\top(\varphi)\ a_R^\top(\varphi)]\zeta, \\
|D_z(\varphi)| &= \|C(\varphi)\zeta\|_2.
\end{aligned}
$$

A typical requirement is that the antenna transmits (or receives) well along a desired direction (on or near a given angle), and not for other angles. In this way, the energy sent is concentrated around a given "target" direction, say $\varphi_{\mathrm{target}} = 0°$, and small outside of that band. Another type of requirement involves the thermal noise power generated by the antenna.

Normalization.   First, we normalize the energy sent along the target direction. When multiplying all weights by a common nonzero complex number, we do not vary the directional distribution of energy; therefore we lose nothing by normalizing the weights so that

$$D_z(0) = 1.$$

This constraint is equivalent to two linear equality constraints in the real and imaginary parts of the decision variable $z \in \mathbb{C}^n$:

$$[a_R^\top(0)\ -a_I^\top(0)]\zeta = 1, \quad [a_I^\top(0)\ a_R^\top(0)]\zeta = 0.$$

Sidelobe level constraints.   We next define a "pass-band" $[-\phi,\ \phi]$, where $\phi > 0$ is given, inside which we wish the energy to be concentrated; the corresponding "stop-band" is the outside of this interval. To enforce the concentration of energy requirement, we require that

$$|D_z(\varphi)| \le \delta, \quad \forall \varphi : |\varphi| \ge \phi,$$

where δ is a desired attenuation level on the stop-band (this is sometimes referred to as the sidelobe level). The sidelobe level constraint actually entails an infinite number of constraints. One practical possibility for dealing with this continuous infinity of constraints is to simply discretize them, by imposing

$$|D_z(\varphi_i)| \le \delta, \quad i = 1, \dots, N,$$

where $\varphi_1, \ldots, \varphi_N$ are N regularly spaced discretized angles in the stop-band. This is a set of N SOC constraints in the real and imaginary parts of z:

$$\|C(\varphi_i)\zeta\|_2 \leq \delta, \quad i = 1, \ldots, N.$$

For example, as depicted in Figure 16.11, the magnitude diagram must pass through the point on the right at $\phi = 0°$, and otherwise be contained in the white area. In the stop-band (shaded area), the magnitude diagram must remain below the level δ, at least at the discretized points.

Thermal noise power constraint. It is often desirable also to control the thermal noise power generated by the emitting antennas. It turns out that this power is proportional to the squared Euclidean norm of the (complex) vector z, that is:

$$\text{thermal noise power} = \alpha \|z\|_2^2 = \sum_{i=1}^{n} |z_i|^2.$$

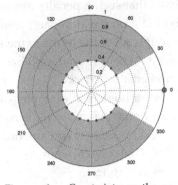

Figure 16.11 Constraints on the antenna diagram.

16.2.2   Least-squares design approach

A first, simplified, approach to tackle the antenna design problem by taking into account the tradeoff between the sidelobe level attenuation and thermal noise power is a least-squares one. Instead of imposing individual constraints on the stop-band levels at each discretized angle φ_i, we include in the objective a term that penalizes overall the sum of squared stop-band levels. That is, we consider the problem

$$\min_z \|z\|_2^2 + \mu \sum_{i=1}^{N} |D_z(\varphi_i)|^2, \quad \text{s.t.: } D_z(0) = 1,$$

where $\mu \geq 0$ acts as a tradeoff parameter. This is an equality-constrained LS problem, which can be written more explicitly in the variable ζ containing the real and imaginary parts of z, as follows:

$$\min_\zeta \|\zeta\|_2^2 + \mu \zeta^\top A \zeta, \quad \text{s.t.: } C(0)\zeta = \begin{bmatrix} 1 \\ 0 \end{bmatrix},$$

where we defined

$$A = \sum_{i=1}^{N} C^\top(\varphi_i) C(\varphi_i).$$

Since $A \succeq 0$, this problem is a convex QP. By factoring $A = F^\top F$, we can further write it in LS form as

$$\min_\zeta \left\| \begin{bmatrix} I \\ \sqrt{\mu} F \end{bmatrix} \zeta \right\|_2^2, \quad \text{s.t.: } C(0)\zeta = \begin{bmatrix} 1 \\ 0 \end{bmatrix}.$$

Note that such a penalty approach will not enforce *a priori* a desired bound δ on the stop-band level. We can only hope that, for μ large enough, all of the terms in the sum will go below the desired threshold δ. The achieved stop-band attenuation level can thus only be checked *a posteriori*.

For a numerical example, we set the following parameters: number of antennas $n = 16$, wavelength $\lambda = 8$, pass-band width $\phi = \pi/6$, distance between antennas $\ell = 1$, number of discretized angles $N = 100$, tradeoff parameter $\mu = 0.5$. The solution via CVX is shown in Figure 16.12.

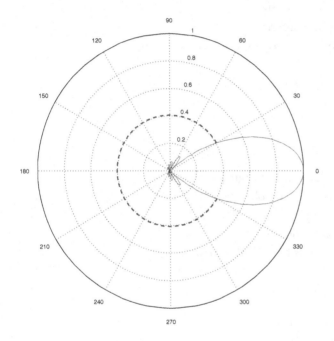

Figure 16.12 Antenna diagram from the LS solution.

The resulting (square-root of) thermal noise power is $\|z\|_2 = 0.5671$, and the maximum stop-band level is 0.4050 (highlighted with a dashed circle arc in Figure 16.12). Also, we may solve the problem repeatedly for increasing values of μ, and plot the corresponding tradeoff curve between the square-root of the thermal noise power $\|z\|_2$ and the stop-band attenuation level, as shown in Figure 16.13.

16.2.3 SOCP design approach

The actual antenna design problem, with explicit constraints on the stop-band levels at the discretized angles, can be cast directly as an SOCP. One possibility is to miminize the thermal noise power subject to sidelobe level constraints, which results in the explicit SOCP in the

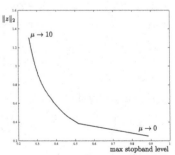

Figure 16.13 Tradeoff curve for values of μ in the interval $[0, 10]$.

vector ζ containing the real and imaginary parts of z:

$$\min_{\zeta\in\mathbb{R}^{2n},\gamma}\quad \gamma$$
$$\text{s.t.:}\quad C(0)\zeta = [1\ 0]^{\top},$$
$$\|C(\varphi_i)\zeta\|_2 \le \delta,\ \ i=1,\dots,N,$$
$$\|\zeta\|_2 \le \gamma.$$

Alternatively, one may minimize the sidelobe-level attenuation δ, subject to a given bound γ on the thermal noise power, or else one can solve for an optimal tradeoff between stop-band attenuation and noise, by considering the SOCP

$$\min_{\zeta\in\mathbb{R}^{2n},\gamma,\delta}\quad \delta + w\gamma$$
$$\text{s.t.:}\quad C(0)\zeta = [1\ 0]^{\top},$$
$$\|C(\varphi_i)\zeta\|_2 \le \delta,\ \ i=1,\dots,N,$$
$$\|\zeta\|_2 \le \gamma,$$

where $w \ge 0$ is a given tradeoff parameter.

As a numerical example, solution of problem (16.2.3), with $\delta = 0.35$, yielded $\|z\|_2 = 0.435$ and the polar diagram shown in Figure 16.14.

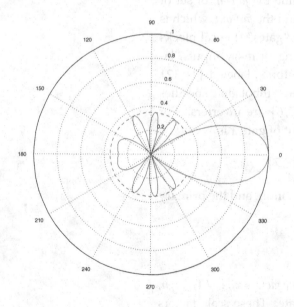

Figure 16.14 Antenna diagram resulting from problem (16.2.3), with $\delta = 0.35$.

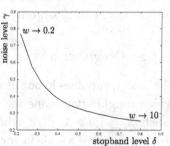

Figure 16.15 Tradeoff curve for values of w in the interval $[0.2, 10]$.

Also, we may construct a tradeoff curve by solving problem (16.2.3) repeatedly for increasing values of w. Figure 16.15 shows this curve for $w \in [0.2, 10]$.

16.3 Digital circuit design

We consider the problem of designing a digital circuit known as a *combinational logic block*. In such a circuit, the basic building block is a *gate*: a circuit that implements some simple Boolean function. A gate could be, for example, an *inverter*, which performs logical inversion, or a more complicated function such as a "not-AND" or NAND, which takes two Boolean inputs A, B to produce the inverse of (A and B). The basic idea is to design (size) the gates in such a way that the circuit operates fast, yet occupies a small area. There are other factors, such as power, involved in the design problem but we will not discuss this here. The design variables are scale factors that determine the size of each gate, as well as its basic electrical parameters. Together with the (here, fixed) topology of the circuit, these parameters in turn influence the speed of the circuit. In this section, we shall use geometric programming models to address the circuit design problem. GPs have a long history in circuit design.[1]

[1] The presentation given here, as well as the example, is taken from the paper by S. Boyd, S.-J. Kim, D. Patil, and M. Horowitz, Digital circuit optimization via geometric programming, *Operations Research*, 2005, which also contains many references to related literature.

16.3.1 Circuit topology

The combinational circuit consists of connected gates, with primary inputs and outputs. We assume that there are no loops in the corresponding graph. For each gate, we can define the *fan-in*, or set of predecessors of the gate in the circuit graph, and the *fan-out*, which is its set of successors. The circuit consists of n "gates" (logical blocks with a few inputs and one output) connecting primary inputs, labeled $\{8, 9, 10\}$ in Figure 16.16, to primary outputs, labeled $\{11, 12\}$ in the figure. Each gate is represented by a symbol that specifies its type; for example, the gates labelled $\{1, 3, 6\}$ are inverters. For example, for this circuit, the fan-in and fan-out of gate 4 is

$$\mathrm{FI}(4) = \{1, 2\}, \quad \mathrm{FO}(4) = \{6, 7\}.$$

By definition, the primary inputs have empty fan-in, and the primary outputs have empty fan-out.

16.3.2 Design variables

The design variables in our models are the *scale factors* x_i, $i = 1, \ldots, n$, which roughly determine the size of each gate. These scale factors satisfy $x_i \geq 1$, $i = 1, \ldots, n$, where $x_i = 1$ corresponds to a minimum-sized gate, while a scale factor $x_i = 16$ corresponds to the case when all the devices in the gate have 16 times the widths of those in the minimum-sized gate. The scale factors determine the size, and var-

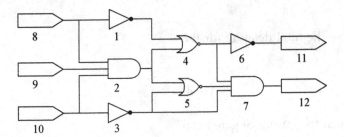

Figure 16.16 A digital circuit example.

ious electrical characteristics, such as resistance and conductance, of the gates. These relationship can be well approximated as follows.

- The *area* $A_i(x)$ of gate i is proportional to the scale factor x_i, i.e., $A_i(x) = a_i x_i$, for some $a_i > 0$.

- The *intrinsic capacitance* of gate i is of the form

$$C_i^{\text{intr}}(x) = C_i^{\text{intr}} x_i,$$

where C_i^{intr} are positive coefficients.

- The *load capacitance* of gate i is a linear function of the scale factors of the gates in the fan-out of gate i:

$$C_i(x) = \sum_{j \in \text{FO}(i)} C_j x_j,$$

where C_j are positive coefficients.

- Each gate has a *resistance* that is inversely proportional to the scale factor (the larger the gate, the more current can pass through it):

$$R_i(x) = r_i / x_i,$$

where r_i are positive coefficients.

- The *gate delay* is a measure of how fast the gate implements the logical operation it is supposed to perform; this delay can be approximated as

$$D_i(x) = 0.7 R_i(x)(C_i^{\text{intr}}(x) + C_i(x)).$$

We observe that all the above quantities are posynomial functions in the (positive) design vector x.

16.3.3   Design objective

A possible design objective is to minimize the total delay D for the circuit. We can express the total delay as

$$D = \max_{1 \leq i \leq n} T_i, \qquad (16.2)$$

where T_i represents the latest time at which the output of gate i can make a transition, assuming that the primary input signals transition at $t = 0$. That is, T_i is the maximum delay over all paths that start at primary input and end at gate i. We can express T_i via the recursion

$$T_i = \max_{j \in \mathrm{FI}(i)} (T_j + D_i). \qquad (16.3)$$

The operations involved in the computation of D involve only addition and point-wise maximum. Since each D_i is a posynomial in x, we can express the total delay as a generalized posynomial in x. For the circuit in Figure 16.16, the total delay D can be expressed as

$$
\begin{aligned}
T_i &= D_i, \quad i = 1, 2, 3; \\
T_4 &= \max(T_1, T_2) + D_4; \\
T_5 &= \max(T_2, T_3) + D_5; \\
T_6 &= T_4 + D_6; \\
T_7 &= \max(T_3, T_4, T_5) + D_7; \\
D &= \max(T_6, T_7).
\end{aligned}
$$

16.3.4   A circuit design problem

We now consider the problem of choosing the scale factors x_i to minimize the total delay, subject to an area constraint:

$$\min_x D(x) \quad \text{s.t.:} \ \ A_i(x) \leq A_{\max}, \ x_i \geq 1, \quad i = 1, \dots, n,$$

where $A_{\max}$ is an upper bound on the area of each gate. Since D is a generalized posynomial in x, the above can be cast as a GP. In order to find a compact and explicit GP representation, we may use the intermediate variables T_i, which we encountered in our definition of the delay, by observing that the equality relation (16.3) can be replaced by the inequality relation

$$T_i \geq \max_{j \in \mathrm{FI}(i)} (T_j + D_i), \qquad (16.4)$$

without changing the value of the optimization problem at optimum. The reason for this is due to a reasoning analogous to the one presented in Section 8.3.4.5: since the objective D in (16.2) is non-decreasing with respect to T_i, and since T_i increases if T_j, $j \in \mathrm{FI}(i)$, increase, the optimum will select all T_i as small as possible, hence equality will hold in (16.4). Further, this latter inequality is equivalent to

$$T_i \geq T_j + D_i, \quad \forall j \in \mathrm{FI}(i).$$

We then obtain the following explicit GP representation of the design problem:

$$
\begin{aligned}
\min_{x, T_i > 0, D} \quad & D \\
\text{s.t.:} \quad & A_i(x) \leq A_{\max}, \ x_i \geq 1, & i = 1, \dots, n; \\
& D \geq T_i, & i = 1, \dots, n; \\
& T_i \geq T_j + D_i(x), \ \forall j \in \mathrm{FI}(i), & i = 1, \dots, n.
\end{aligned}
$$

For a numerical example, we considered the circuit in Figure 16.16, with $A_{\max} = 16$ and the other data as shown in Table 16.1.

Numerical solution using CVX yields optimal scale factors

$$
x^* = \begin{bmatrix} 4.6375 \\ 2.0000 \\ 4.8084 \\ 1.6000 \\ 1.0000 \\ 1.0000 \\ 1.0000 \end{bmatrix},
$$

Gate	C_i	C_i^{intr}	r_i	a_i
1,3,6	3	3	0.48	3
2,7	4	6	0.48	8
4,5	5	6	0.48	10

Table 16.1 Data for the circuit in Figure 16.16.

corresponding to a minimal delay $D^* = 7.686$.

16.4   Aircraft design

It has recently been observed that several problems related to aircraft structural and operational design can be cast in the form of geometric programs, and hence solved efficiently via convex optimization. We here present a simple example of a wing design problems.[2]

Our task is to design a wing with total area S, wing span b, and wing aspect ratio $A = b^2/S$, so as to minimize the drag force

$$D = \frac{1}{2}\rho V^2 C_D S,$$

where V is the aircraft cruise speed (in m/s), ρ is the air density, and C_D is the drag coefficient, see Figure 16.17.

When an airplane is in steady flight, it must satisfy two basic equilibrium conditions, as shown in Figure 16.18: the lift force L must

[2] Our presentation is taken from the paper Geometric programming for aircraft design optimization, by W. Hoburg and P. Abbeel, in proc. *Structures, Structural Dynamics and Materials Conference*, 2012, to which the reader is referred for further details.

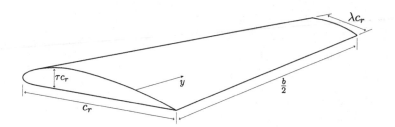

Figure 16.17 Single-taper wing geometry: c_r is the root chord, $b/2$ is the half wing span, λ is the chord tapering factor, and τ is the airfoil thickness to chord ratio.

counterbalance the weight W of the airplane, and the thrust T must balance the drag D, that is

$$L = W, \quad T = D.$$

Denoting by C_L the lift coefficient, the lift is given by

$$L = \frac{1}{2}\rho V^2 C_L S.$$

The drag coefficient is modeled as the sum of three terms: fuselage parasite drag, wing parasite drag, and induced drag, as specified next:

$$C_D = \frac{CDA_0}{S} + kC_f \frac{S_{\text{wet}}}{S} + \frac{C_L^2}{\pi A e}, \tag{16.5}$$

where CDA_0 is the fuselage drag area, k is a form factor that accounts for pressure drag, S_{wet} is the wetted area (i.e., the actual area of the whole surface of the wing, while S is the area of a 2D projection, or shadow, of the wing), C_L is the lift coefficient, and e is the Oswald efficiency factor. For a fully turbulent boundary layer, the skin friction coefficient C_f can be approximated as

$$C_f = \frac{0.0074}{\text{Re}^{0.2}},$$

where

$$\text{Re} = \frac{\rho V}{\mu}\sqrt{S/A}$$

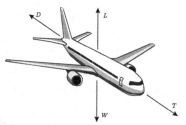

Figure 16.18 In steady flight conditions, the lift L equals the weight W, and the thrust T equals the drag D.

is the Reynolds number at the mean chord[3] $c = \sqrt{S/A}$, being μ the viscosity of air. The total aircraft weight W is modeled as the sum of a fixed weight W_0 and wing weight W_w, which is in turn given by

$$W_w = k_s S + k_l \frac{N_{\text{ult}} b^3 \sqrt{W_0 W}}{S\tau},$$

where k_s, k_l are suitable constants, N_{ult} is the ultimate load factor for structural sizing, and τ is the airfoil thickness to chord ratio (note that W_w is a function of W itself). The weight equations are coupled to the drag equations by the constraint that, in steady flight conditions,

[3] The mean chord is defined as the value c in the interval $[\lambda c_r, c_r]$ such that $S = bc$.

the lift force must equal the weight, that is $L = W$. Hence, it must hold that

$$\frac{1}{2}\rho V^2 C_L S = W,$$
$$W_0 + W_w = W.$$

Finally, the airplane must be capable of flying at a minimum speed $V_{\min}$ at the time of landing, without stalling. This requirement may be imposed via the constraint

$$\frac{1}{2}\rho V_{\min}^2 C_L^{\max} S \geq W,$$

where $C_L^{\max}$ is the maximum lift coefficient at landing (flaps down).

We must choose values of S, A, and V that minimize drag, subject to all the above relations. Constant parameters are given in Table 16.2. This problem appears as a difficult optimization problem, involving nonlinearly coupled variables. The key point, however, is that in this design problem we can relax the posynomial *equality* constraints into posynomial *inequality* constraints, without modifying the optimal objective value of the problem. For instance, we can equivalently replace the equality relation (16.5) by the inequality one

$$C_D \geq \frac{\text{CDA}_0}{S} + kC_f\frac{S_{\text{wet}}}{S} + \frac{C_L^2}{\pi Ae},$$

provided that C_D does not appear in any other monomial equality constraint, and that the objective and inequality constraints are all monotone increasing (or constant) in C_D. Under these conditions, if the equality relation (16.5) did not hold at the optimum, we could clearly decrease C_D until achieving equality, without increasing the objective or moving the solution outside the feasible set. Since these conditions are satisfied in our current setting, we can write our design problem into explicit GP format, as follows.

$$\min_{A,S,V,C_D,C_L,C_f,\text{Re},W,W_w} \quad \frac{1}{2}\rho V^2 C_D S$$

$$\text{s.t.:} \quad \frac{0.074}{C_f \text{Re}^{0.2}} = 1, \quad \frac{2W}{\rho V^2 C_L S} = 1,$$

$$\frac{2W}{\rho V_{\min}^2 C_L^{\max} S} \leq 1, \quad \frac{\rho V}{\mu \text{Re}}\sqrt{S/A} = 1,$$

$$\frac{\text{CDA}_0}{C_D S} + k\frac{C_f}{C_D}\frac{S_{\text{wet}}}{S} + \frac{C_L^2}{C_D \pi Ae} \leq 1,$$

$$k_s\frac{S}{W_w} + k_l\frac{N_{\text{ult}}A^{3/2}\sqrt{W_0 W S}}{W_w \tau} \leq 1,$$

$$\frac{W_0}{W} + \frac{W_w}{W} \leq 1.$$

Constant parameters are given in Table 16.2. Solving the GP yields the optimal design reported in Table 16.3.

Quantity	Value	Units	Description
CDA_0	0.0306	m^2	fuselage drag area
ρ	1.23	Kg/m^3	air density
μ	1.78×10^{-5}	Kg/ms	viscosity of air
S_{wet}/S	2.05		wetted area ratio
k	1.2		form factor
e	0.96		Oswald efficiency factor
W_0	4940	N	aircraft weight excluding wing
N_{ult}	2.5		ultimate load factor
τ	0.12		airfoil thickness to chord ratio
V_{min}	22	m/s	landing speed
C_L^{max}	2.0		max C_L, flaps down
k_s	45.42	N/m^2	
k_l	8.71×10^{-5}	m^{-1}	

Table 16.2 Constants in the aircraft design example.

Quantity	Value	Units	Description
A	12.7		wing aspect ratio
S	12.08	m^2	wing surface
V	38.55	m/s	cruise speed
C_D	0.0231		drag coefficient
C_L	0.6513		lift coefficient
C_f	0.0039		skin friction coefficient
Re	2.5978×10^6		Reynolds number
W	7189.1	N	total weight
W_w	2249.1	N	wing weight

Table 16.3 Optimal design for the aircraft example.

The GP model allows us to easily obtain globally optimal values of the design parameters. In a real design setting, however, the designer would want to consider a range of possible tradeoffs between competing objectives (e.g., increase the landing speed and/or the cruise speed). These tradeoffs can be readily explored numerically, by solving the above GP for a range of $V_{\min}$ and V values. Also, several other aspects can be included in the optimization model. For example, the model may account for the weight of fuel, by writing

$$W = (W_0 + W_w)(1 + \theta_{\text{fuel}}),$$

where θ_{fuel} is the fuel mass fraction. In turn, the fuel mass fraction is related to the aircraft reachable range by means of Brequet's range equation

$$R = \frac{h_{\text{fuel}}}{g} \eta_0 \frac{L}{D} \log(1 + \theta_{\text{fuel}}),$$

where it is assumed that the lift to drag ratio L/D remains constant, η_0 is the overall fuel power to thrust power efficiency factor, and h_{fuel} relates the fuel mass flow rate to the fuel power, according to the equation $P_{\text{fuel}} = \dot{m}_{\text{fuel}} h_{\text{fuel}}$. In steady flight conditions, it must hold that

$$TV \le \eta_0 P_{\text{fuel}}.$$

Breguet's equation is not directly amenable to posynomial format, but it can be well approximated by a Taylor series expansion, by writing

$$1 + \theta_{\text{fuel}} = \exp\left(\frac{gRD}{h_{\text{fuel}} \eta_0 L}\right),$$

and observing that the series expansion of the exponential function has posynomial structure. Hence, Breguet's equation can be approximately included in a GP model by imposing

$$z = \frac{gRD}{h_{\text{fuel}} \eta_0 L},$$
$$\theta_{\text{fuel}} \ge z + \frac{z^2}{2!} + \frac{z^3}{3!} + \cdots$$

16.5 Supply chain management

In this section we discuss a problem arising in production engineering, regarding control of an inventory level under demand uncertainty, over multiple discrete time periods.[4] The problem consists of making ordering, stocking, and storage decisions over a given time horizon composed of T stages, with the objective of minimizing cost and satisfying the demand. Cost is due to actual purchase cost, as

[4] The treatment presented here follows the setup and notation originally proposed in the paper by Bertsimas and Thiele, A robust optimization approach to supply chain management, *Operations Research*, 2006.

well as from costs due to holding and shortage. Considering one single good, and denoting by $x(k)$ the stock level of this good at time k, the basic evolution of the stock level in (discrete) time can be written as

$$x(k+1) = x(k) + u(k) - w(k), \quad k = 0, 1, \ldots, T-1,$$

where $x(0) = x_0$ is a given initial stock level, $u(k)$ is the stock ordered at time k, and $w(k)$ is the demand during the period from k to $k+1$. Assuming that we incur a unit holding cost h for storage, a unit shortage cost p, and that the unitary cost for buying the good is c, at each time k the stage cost is given by

$$cu(k) + \max(hx(k+1), -px(k+1)).$$

Considering further an upper bound M on the size of any order, we may write the T-stage inventory control problem as

$$\min_{u(0),\ldots,u(T-1)} \sum_{k=0}^{T-1} cu(k) + \max(hx(k+1), -px(k+1))$$
$$\text{s.t.:} \quad 0 \leq u(k) \leq M, \, k = 0, \ldots, T-1,$$

where

$$x(k) = x_0 + \sum_{i=0}^{k-1} (u(i) - w(i)), \quad k = 1, \ldots, T.$$

Introducing slack variables $y(0), \ldots, y(T-1)$, the problem may be cast in the form of the following linear program:

$$\min_{u(0),\ldots,u(T-1),y(0),\ldots,y(T-1)} \sum_{k=0}^{T-1} y(k)$$
$$\text{s.t.:} \quad cu(k) + hx(k+1) \leq y(k), \, k = 0, \ldots, T-1,$$
$$cu(k) - px(k+1) \leq y(k), \, k = 0, \ldots, T-1,$$
$$0 \leq u(k) \leq M, \, k = 0, \ldots, T-1.$$

Further, defining vectors $u = (u(0), \ldots, u(T-1))$, $y = (y(0), \ldots, y(T-1))$, $w = (w(0), \ldots, w(T-1))$, and $x = (x(1), \ldots, x(T))$, we rewrite the problem in compact notation as

$$\min_{u,y} \quad \mathbf{1}^\top y \tag{16.6}$$
$$\text{s.t.:} \quad cu + hx \leq y, \quad cu - px \leq y,$$
$$0 \leq u \leq M\mathbf{1},$$

where

$$x = x_0\mathbf{1} + Uu - Uw, \quad U \doteq \begin{bmatrix} 1 & 0 & 0 & \cdots & 0 \\ 1 & 1 & 0 & \cdots & 0 \\ \vdots & \vdots & \ddots & \ddots & \vdots \\ 1 & 1 & 1 & \cdots & 1 \end{bmatrix}.$$

In this setup, we assumed that the demand $w(k)$ is deterministically known over all forward periods $k = 0, \ldots, T - 1$, and that all decisions are set at time $k = 0$. In the next sections we discuss how to release these two assumptions.

16.5.1 Robustness against interval-uncertain demand

We here assume that the demand w is imprecisely known. Specifically, we consider the situation where the demand may fluctuate within ρ percent from a nominal expected demand $\hat{w} \geq 0$, that is

$$w \in W \doteq \{w : w_{\text{lb}} \leq w \leq w_{\text{up}}\}, \tag{16.7}$$

where

$$w_{\text{lb}} \doteq (1 - \rho/100)\hat{w}, \quad w_{\text{ub}} \doteq (1 + \rho/100)\hat{w}.$$

We then seek an ordering sequence u that minimizes the worst-case cost over all admissible demands, that is

$$\begin{aligned} \min_{u,y} \quad & \mathbf{1}^\top y \\ \text{s.t.:} \quad & cu + hx \leq y, \; \forall w \in W, \\ & cu - px \leq y, \; \forall w \in W, \\ & 0 \leq u \leq M\mathbf{1}. \end{aligned} \tag{16.8}$$

Since $x = x_0 \mathbf{1} + Uu - Uw$, where U has non-negative elements, and since each element of w is known to belong to an interval, the constraint $cu + hx \leq y, \; \forall w \in W$ is satisfied if and only if

$$cu + h\left(x_0 \mathbf{1} + Uu - Uw_{\text{lb}}\right) \leq y,$$

and the constraint $cu - px \leq y, \; \forall w \in W$ is satisfied if and only if

$$cu - p\left(x_0 \mathbf{1} + Uu - Uw_{\text{ub}}\right) \leq y.$$

The optimal worst-case decisions are thus obtained by solving the following LP:

$$\begin{aligned} \min_{u,y} \quad & \mathbf{1}^\top y \\ \text{s.t.:} \quad & cu + h\left(x_0 \mathbf{1} + Uu - Uw_{\text{lb}}\right) \leq y, \\ & cu - p\left(x_0 \mathbf{1} + Uu - Uw_{\text{ub}}\right) \leq y, \\ & 0 \leq u \leq M\mathbf{1}. \end{aligned} \tag{16.9}$$

16.5.2 Affine ordering policies under interval uncertainty

In the previous problem formulation, all forward ordering decisions $u(0), \ldots, u(T - 1)$ are computed a time $k = 0$, in an "open-loop" fashion. In practice, however, one would implement only the first decision $u(0)$ (the so-called "here and now" decision), and then wait and see what happens until the next decision time. At time $k = 1$, the actual realization of the uncertain demand is observed, hence this information is available at the time when decision $u(1)$ is to be taken. It is then intuitively clear that the new decision $u(1)$ could benefit from this information, and that the original approach that neglects it is thus sub-optimal. In general, the decision $u(k)$ to be taken at time k may benefit from the information on the *observed* demand from time 0 up to $k - 1$ (this demand is thus no longer "uncertain" at time k), which is to say that $u(k)$ is a generic function of the past demand, i.e., $u(k) = \varphi_k(w(0), \dots, w(k-1))$. In such a reactive, or "closed-loop" approach, the optimization problem would amount to finding the optimal functions φ_k (usually called *policies*) such that the worst-case cost is minimized. However, searching for fully generic functions makes the problem hard (the search should be conducted over an "infinite-dimensional" set of all possible functions φ_k). Therefore, a common approach is to fix a parameterization of φ_k, and then optimize over the finite-dimensional parameter set (see also Section 14.3.3 for an analogous approach that we described for multi-period optimization problems arising in a financial context). An effective approach is, for instance, to consider *affine* parameterizations, that is to consider, for $k = 1, \dots, T - 1$, policies φ_k of the form

$$u(k) = \varphi_k(w(0), \dots, w(k-1)) = \bar{u}(k) + \sum_{i=0}^{k-1} \alpha_{k,i}\left(w(i) - \hat{w}(i)\right),$$

where $\bar{u}(k)$ are the "nominal" decisions, $u(0) = \bar{u}(0)$, and $\alpha_{k,i}$ are parameters that permit us, for $k \geq 1$, to correct the nominal decisions proportionally to the deviation of the actual demand $w(i)$ from its nominal value $\hat{w}(i)$. In matrix notation, we thus write

$$u = \bar{u} + A(w - \hat{w}), \quad A \doteq \begin{bmatrix} 0 & 0 & \cdots & 0 \\ \alpha_{1,0} & 0 & \cdots & 0 \\ \alpha_{2,0} & \alpha_{2,1} & \cdots & 0 \\ \vdots & \vdots & \ddots & \vdots \\ \alpha_{T-1,0} & \cdots & \alpha_{T-1,T-2} & 0 \end{bmatrix},$$

where $\bar{u} = (\bar{u}(0), \dots, \bar{u}(T-1))$. In particular, under the interval uncertainty model (16.7), the demand fluctuation $\tilde{w} \doteq w - \hat{w}$ lies in the symmetric vector interval

$$-\bar{w} \leq \tilde{w} \leq \bar{w}; \quad \bar{w} \doteq \frac{\rho}{100}\hat{w} \geq 0.$$

Substituting the expression for u in problem (16.8), we obtain

$$\min_{\bar{u}, y, A} \quad \mathbf{1}^\top y \tag{16.10}$$

$$\text{s.t.:} \quad (c\bar{u} + cA\tilde{w}) + h(x_0\mathbf{1} + U(\bar{u} + A\tilde{w}) - Uw) \leq y, \; \forall w \in W,$$

$$(c\bar{u} + cA\tilde{w}) - p(x_0\mathbf{1} + U(\bar{u} + A\tilde{w}) - Uw) \leq y, \; \forall w \in W,$$

$$0 \leq \bar{u} + A\tilde{w} \leq M\mathbf{1}, \; \forall w \in W.$$

Observe that, if v is a vector, and $\tilde{w} \geq 0$, then

$$\max_{-\tilde{w} \leq \bar{w} \leq \tilde{w}} v^\top \bar{w} = |v|^\top \tilde{w},$$

$$\min_{-\tilde{w} \leq \bar{w} \leq \tilde{w}} v^\top \bar{w} = -|v|^\top \tilde{w},$$

where $|v|$ is a vector formed by the absolute values of the entries of v. Applying this rule row-wise to the constraints of problem (16.10), we obtain that the robust problem is equivalent to

$$\min_{\bar{u}, y, A} \quad \mathbf{1}^\top y$$

$$\text{s.t.:} \quad c\bar{u} + hU\bar{u} + hx_0\mathbf{1} - hU\tilde{w} + |cA + hUA - hU|\tilde{w} \leq y,$$

$$c\bar{u} - pU\bar{u} - px_0\mathbf{1} + pU\tilde{w} + |cA - pUA + pU|\tilde{w} \leq y,$$

$$\bar{u} + |A|\tilde{w} \leq M\mathbf{1},$$

$$\bar{u} - |A|\tilde{w} \geq 0.$$

Introducing three slack lower-triangular matrices Z_1, Z_2, Z_3, we can rewrite this problem in a format readily amenable to LP standard form, as follows:

$$\min_{\bar{u}, y, A, Z_1, Z_2, Z_3} \quad \mathbf{1}^\top y \tag{16.11}$$

$$\text{s.t.:} \quad c\bar{u} + hU\bar{u} + hx_0\mathbf{1} - hU\tilde{w} + Z_1\tilde{w} \leq y,$$

$$c\bar{u} - pU\bar{u} - px_0\mathbf{1} + pU\tilde{w} + Z_2\tilde{w} \leq y,$$

$$\bar{u} + Z_3\tilde{w} \leq M\mathbf{1},$$

$$\bar{u} - Z_3\tilde{w} \geq 0,$$

$$|cA + hUA - hU| \leq Z_1,$$

$$|cA - pUA + pU| \leq Z_2,$$

$$|A| \leq Z_3.$$

Once this problem is solved, we obtain the ordering policy and the uncertain stock level as

$$u = \bar{u} + A\tilde{w} \tag{16.12}$$

$$x = x_0\mathbf{1} + U(\bar{u} - \tilde{w}) + (UA - U)\tilde{w}.$$

We can then obtain upper and lower limits for the orders and stock level as follows:

$$
\begin{aligned}
u_{\text{lb}} &= \bar{u} - |A|\bar{w}, \\
u_{\text{ub}} &= \bar{u} + |A|\bar{w}, \\
x_{\text{lb}} &= x_0 \mathbf{1} + U(\bar{u} - \hat{w}) - |UA - U|\bar{w}, \\
x_{\text{ub}} &= x_0 \mathbf{1} + U(\bar{u} - \hat{w}) + |UA - U|\bar{w}.
\end{aligned}
$$

The actual order level $u(k)$ to be executed at time k will be computed online according to the policy (16.12), as time unfolds and the realized values of $\bar{w}(i)$, $i = 0, \ldots, k-1$ are observed. What we know *a priori* is that the value $u(k)$ will be contained in the interval $[u_{\text{lb}}(k), u_{\text{ub}}(k)]$, for $k = 0, \ldots, T-1$.

16.5.2.1 A numerical example. To illustrate these ideas, let us consider a decision horizon of $T = 12$ periods (e.g., months), with the following data: purchase cost $c = 5$, storage cost $h = 4$, shortage cost $p = 6$, upper bound on orders $M = 110$. Further, we assume that the nominal demand follows a sinusoidal path

$$
\hat{w}(k) \doteq 100 + 20 \sin\left(2\pi \frac{k}{T-1}\right), \quad k = 0, \ldots, T-1, \tag{16.13}
$$

and that the initial stock level is $x_0 = 100$. Under no uncertainty (demand level w equal to the nominal one $\hat{w}$), solving problem (16.6) yields an optimal cost of $5,721.54$, and the order and stock level profiles shown in Figure 16.19.

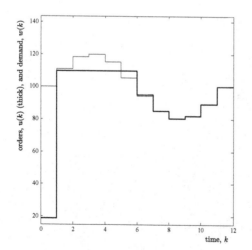
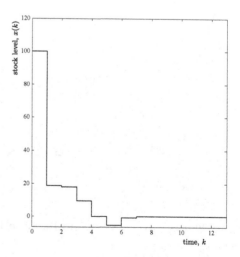

Considering next the presence of $\rho = 15\%$ uncertainty on the demand, we solved the robust "open-loop" problem (16.9), obtaining

Figure 16.19 Optimal orderings under nominal demand. Left panel: demand and ordering profiles. Right panel: stock level profile. Optimal nominal cost is $5,721.54$.

a worst-case cost of 11,932.50, and the order and stock level bounds profiles shown in Figure 16.20.

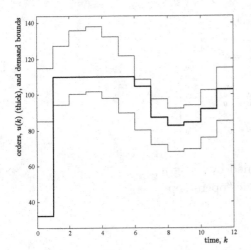
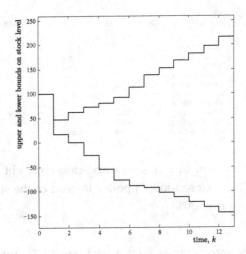

Under the same 15% uncertainty level, we solved the robust "closed-loop" problem (16.11), obtaining a worst-case cost of 8,328.64, and the order and stock level bounds profiles shown in Figure 16.21.

Figure 16.20 Optimal "open-loop" orderings under $\rho = 15\%$ uncertain demand. Left panel: demand bounds and ordering profile. Right panel: stock level bounds. Optimal worst-case cost is 11,932.50.

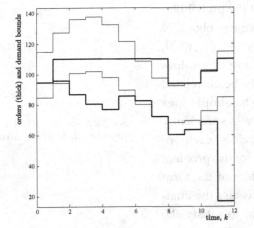
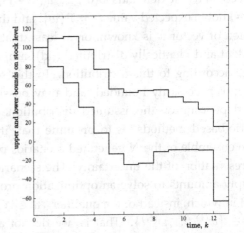

Figure 16.21 Optimal "closed-loop" orderings under $\rho = 15\%$ uncertain demand. Left panel: demand and ordering bounds profiles. Right panel: stock level bounds. Optimal worst-case cost is 8,328.64.

Also, the optimal parameters of the ordering policy (16.12) resulted to be

$$\bar{u}^\top = \begin{bmatrix} 94.69 & 103.0 & 98.36 & 95.26 & 93.49 & 97.83 & 96.26 & 91.25 & 77.33 & 78.68 & 85.2 & 63.4 \end{bmatrix},$$

$$A = \begin{bmatrix}
0 & 0 & 0 & 0 & 0 & 0 & 0 & 0 & 0 & 0 & 0 & 0 \\
0.4697 & 0 & 0 & 0 & 0 & 0 & 0 & 0 & 0 & 0 & 0 & 0 \\
0.2506 & 0.4743 & 0 & 0 & 0 & 0 & 0 & 0 & 0 & 0 & 0 & 0 \\
0.1332 & 0.2518 & 0.4828 & 0 & 0 & 0 & 0 & 0 & 0 & 0 & 0 & 0 \\
0.07069 & 0.1324 & 0.254 & 0.4868 & 0 & 0 & 0 & 0 & 0 & 0 & 0 & 0 \\
0.007817 & 0.01524 & 0.04958 & 0.1856 & 0.4395 & 0 & 0 & 0 & 0 & 0 & 0 & 0 \\
0.009959 & 0.0148 & 0.03499 & 0.0915 & 0.2258 & 0.4529 & 0 & 0 & 0 & 0 & 0 & 0 \\
0.02248 & 0.05136 & 0.08535 & 0.114 & 0.1627 & 0.2676 & 0.4913 & 0 & 0 & 0 & 0 & 0 \\
0.01838 & 0.03096 & 0.04785 & 0.06251 & 0.0879 & 0.1427 & 0.2594 & 0.5104 & 0 & 0 & 0 & 0 \\
0.009341 & 0.01578 & 0.02445 & 0.03202 & 0.04514 & 0.07356 & 0.1343 & 0.2649 & 0.5463 & 0 & 0 & 0 \\
0.004745 & 0.008067 & 0.01258 & 0.01654 & 0.02343 & 0.03848 & 0.07088 & 0.1412 & 0.2945 & 0.6839 & 0 & 0 \\
0.01105 & 0.02035 & 0.03276 & 0.04322 & 0.06086 & 0.09775 & 0.1717 & 0.3123 & 0.547 & 0.8799 & 1.372 & 0
\end{bmatrix}.$$

The improvement in terms of worst-case cost, obtained by using a "reactive" (or "closed-loop") policy instead of the static "open-loop" approach, is over 30%.

16.5.3 Scenario approach for generic stochastic uncertainty

When the uncertainty on the demand level $w(k)$ is stochastic, and cannot be effectively captured via the interval model previously discussed, we can follow a simple and effective, albeit approximate, approach based on sampled scenarios of the uncertainty, see Section 10.3.4. In such an approach, we assume that the demand vector w (which contains the demands $w(0), \ldots, w(T-1)$) is a random vector with a known expected value $\hat{w} = \mathbb{E}\{w\}$, and that the probability distribution of vector w is known, or, at least, that we can obtain N independent and identically distributed (iid) samples $w^{(1)}, \ldots, w^{(N)}$, generated according to this distribution. In this setting, the values $w(k)$ are not necessarily bounded, and may possibly be correlated in time, depending on the assumed distribution. The simple idea of scenario-based methods[5] is to presume that, if N is sufficiently large, the ensemble of the N generated scenarios provides a reasonable representation of the uncertainty. The scenario-robust problem then simply amounts to solve an optimization problem of the form of (16.10) in which, instead of the qualifier $\forall w \in W$, we use the qualifier $\forall w \in \{w^{(1)}, \ldots, w^{(N)}\}$. That is, we do not aim to satisfy the constraints *for all possible* realizations of the uncertainty, but only for the sampled scenario values. Clearly, the so-obtained solution will not be robust in a deterministic, worst-case sense, but rather in a relaxed, probabilistic one. We may attain a good robustness level by choosing N large enough: following Eq. (10.30), for given $\alpha \in (0, 1)$, if the number of scenarios is selected so that

[5] See G. Calafiore, Random convex programs, *SIAM J. Opt.*, 2010.

$$N \geq \frac{2}{1 - \alpha}(n + 10), \tag{16.14}$$

where n is the total number of decision variables in the optimization problem, then the scenario solution will be "probabilistic robust" to level α (usually, however, good results are obtained even for a smaller number of scenarios).

In the present context, we shall then solve a linear optimization problem of the form

$$\min_{\bar{u}, y, A} \mathbf{1}^\top y \tag{16.15}$$
$$\text{s.t.:} \quad cu^{(i)} + hx^{(i)} \leq y, \quad i = 1, \ldots, N,$$
$$cu^{(i)} - px^{(i)} \leq y, \quad i = 1, \ldots, N,$$
$$0 \leq u^{(i)} \leq M\mathbf{1}, \quad i = 1, \ldots, N,$$

where

$$u^{(i)} = \bar{u} + A(w^{(i)} - \hat{w}), \quad i = 1, \ldots, N,$$
$$x^{(i)} = x_0\mathbf{1} + Uu^{(i)} - Uw^{(i)}, \quad i = 1, \ldots, N.$$

As a numerical example, we consider again the data used in Section 16.5.2.1. In this case, problem (16.15) has

$$n = T + T + \frac{T(T-1)}{2} = 90$$

decision variables. Considering a robustness level $\alpha = 0.9$, rule (16.14) suggests using $N = 2,000$ scenarios in the problem. We further assume that the expected demand is given by (16.13), and that vector w has a normal distribution with covariance matrix

$$\Sigma = \text{diag}\left(\sigma_0^2, \ldots, \sigma_{T-1}^2\right),$$

with variances that increase with time (to model a situation in which the uncertainty on the demand level is higher for demands that are farther away in time) as

$$\sigma_k^2 = (1+k)\bar{\sigma}^2; \quad k = 0, \ldots, T-1, \quad \text{with } \bar{\sigma}^2 = 1.$$

Solving one instance of the scenario problem (16.15) yielded an optimal worst-case (over the considered scenarios) cost of 6,296.80, and the following optimal policy:

$$\bar{u}^\top = \begin{bmatrix} 24.4939 & 108.77 & 108.603 & 108.808 & 110.0 & 110.0 & 98.7917 & 83.7308 & 80.6833 & 81.5909 & 89.8084 & 94.6564 \end{bmatrix},$$

$$A = \begin{bmatrix}
0 & 0 & 0 & 0 & 0 & 0 & 0 & 0 & 0 & 0 & 0 & 0 \\
0.3774 & 0 & 0 & 0 & 0 & 0 & 0 & 0 & 0 & 0 & 0 & 0 \\
0.02567 & 0.2845 & 0 & 0 & 0 & 0 & 0 & 0 & 0 & 0 & 0 & 0 \\
-0.1184 & 0.2025 & 0.1344 & 0 & 0 & 0 & 0 & 0 & 0 & 0 & 0 & 0 \\
0 & 0 & 0 & 0 & 0 & 0 & 0 & 0 & 0 & 0 & 0 & 0 \\
0 & 0 & 0 & 0 & 0 & 0 & 0 & 0 & 0 & 0 & 0 & 0 \\
0.4868 & 0.03163 & 0.496 & 0.4467 & 0.4006 & 0.5332 & 0 & 0 & 0 & 0 & 0 & 0 \\
0.3849 & -0.1964 & 0.254 & 0.2024 & 0.4895 & 0.2111 & 0.6726 & 0 & 0 & 0 & 0 & 0 \\
0.135 & 0.072 & -0.1547 & 0.1603 & -0.03367 & 0.02405 & 0.04367 & 0.5894 & 0 & 0 & 0 & 0 \\
-0.2695 & 0.6738 & 0.1186 & -0.2117 & 0.08057 & 0.001069 & 0.1795 & 0.2256 & 0.6008 & 0 & 0 & 0 \\
0.2583 & 0.1248 & 0.04462 & 0.07893 & 0.05989 & 0.09069 & 0.007373 & 0.3262 & 0.2858 & 0.5088 & 0 & 0 \\
0.4861 & 0.6754 & 0.009505 & 0.3034 & 0.3853 & 0.08153 & 0.1048 & 0.4221 & 0.6108 & 0.3069 & 1.086 & 0
\end{bmatrix}.$$

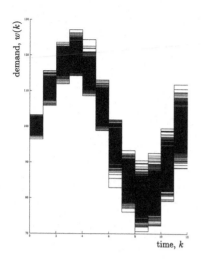
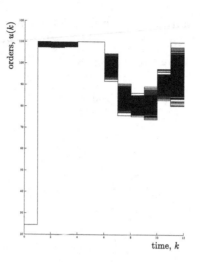
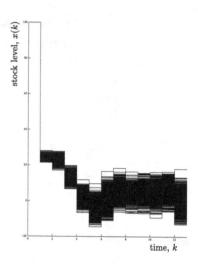

Figure 16.22 Monte Carlo simulation of the optimal scenario policy against $1,500$ randomly generated demand profiles. Left panel: random demands. Center panel: ordering profiles. Right panel: stock level profiles.

We can then test *a posteriori* the performance of this policy via a Monte Carlo method, by generating new scenarios (different from the ones used in the optimization), and simulating the behavior of the policy against these random demand profiles. Using $1,500$ new demand scenarios, we obtained the simulation shown in Figure 16.22.

The histogram of the costs resulting from each scenario is also shown in Figure 16.23: the maximum cost in this simulation was $6,190.22$, which happened to be below the scenario-worst-case cost of $6,296.80$.

16.6   Exercises

Exercise 16.1 (Network congestion control) A network of $n = 6$ peer-to-peer computers is shown in Figure 16.24. Each computer can upload or download data at a certain rate on the connection links shown in the figure. Let $b^+ \in \mathbb{R}^8$ be the vector containing the packet transmission rates on the links numbered in the figure, and let $b^- \in \mathbb{R}^8$ be the vector containing the packet transmission rates on the reverse links, where it must hold that $b^+ \geq 0$ and $b^- \geq 0$.

Define an arc–node incidence matrix for this network:

$$
A \doteq \begin{bmatrix}
1 & 0 & 1 & 1 & 0 & 0 & 0 & 0 \\
-1 & 1 & 0 & 0 & 0 & 0 & 0 & 0 \\
0 & 0 & 0 & -1 & 1 & 0 & 0 & 0 \\
0 & -1 & -1 & 0 & 0 & -1 & -1 & 0 \\
0 & 0 & 0 & 0 & -1 & 1 & 0 & 1 \\
0 & 0 & 0 & 0 & 0 & 0 & 1 & -1
\end{bmatrix},
$$

and let $A_+ \doteq \max(A, 0)$ (the positive part of A), $A_- \doteq \min(A, 0)$ (the

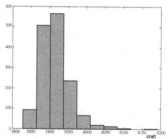

Figure 16.23 Histogram of costs obtained from Monte Carlo simulation of the optimal scenario policy against $1,500$ randomly generated demand profiles.

negative part of A). Then the total output (upload) rate at the nodes is given by $v_{\text{upl}} = A_+ b^+ - A_- b^-$, and the total input (download) rate at the nodes is given by $v_{\text{dwl}} = A_+ b^- - A_- b^+$. The net outflow at nodes is hence given by

$$v_{\text{net}} = v_{\text{upl}} - v_{\text{dwl}} = Ab^+ - Ab^-,$$

and the flow balance equations require that $[v_{\text{net}}]_i = f_i$, where $f_i = 0$ if computer i is not generating or sinking packets (it just passes on the received packets, i.e., it is acting as a relay station), $f_i > 0$ if computer i is generating packets, or $f_i < 0$ if it is sinking packets at an assigned rate f_i.

Each computer can download data at a maximum rate of $\bar{v}_{\text{dwl}} = 20$ Mbit/s and upload data at a maximum rate of $\bar{v}_{\text{upl}} = 10$ Mbit/s (these limits refer to the total download or upload rates of a computer, through all its connections). The level of congestion of each connection is defined as

$$c_j = \max(0, (b_j^+ + b_j^- - 4)), \quad j = 1, \ldots, 8.$$

Assume that node 1 must transmit packets to node 5 at a rate $f_1 = 9$ Mbit/s, and that node 2 must transmit packets to node 6 at a rate $f_2 = 8$ Mbit/s. Find the rate on all links such that the average congestion level of the network is minimized.

Exercise 16.2 (Design of a water reservoir) We need to design a water reservoir for water and energy storage, as depicted in Figure 16.25.

The concrete basement has a square cross-section of side length b_1 and height h_0, while the reservoir itself has a square cross-section of side length b_2 and height h. Some useful data is reported in Table 16.4.

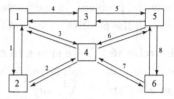

Figure 16.24 A small network.

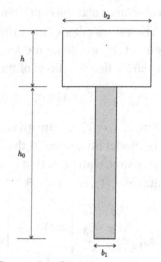

Figure 16.25 A water reservoir on a concrete basement.

Table 16.4 Data for reservoir problem.

Quantity	Value	Units	Description
g	9.8	m/s^2	gravity acceleration
E	30×10^9	N/m^2	basement long. elasticity modulus
ρ_w	10×10^3	N/m^3	specific weight of water
ρ_b	25×10^3	N/m^3	specific weight of basement
J	$b_1^4/12$	m^4	basement moment of inertia
N_{cr}	$\pi^2 J E / (2h_0)^2$	N	basement critical load limit

The critical load limit N_{cr} of the basement should withstand at least twice the weight of water. The structural specification $h_0/b_1^2 \leq 35$ should hold. The form factor of the reservoir should be such that $1 \leq b_2/h \leq 2$. The total height of the structure should be no larger than 30 m. The total weight of the structure (basement plus reservoir full of water) should not exceed 9.8×10^5 N. The problem is to find the dimensions b_1, b_2, h_0, h such that the potential energy P_w of the stored water is maximal (assume $P_w = (\rho_w h b_2^2) h_0$). Explain if and how the problem can be modeled as a convex optimization problem and, in the positive case, find the optimal design.

Exercise 16.3 (Wire sizing in circuit design) Interconnects in modern electronic chips can be modeled as conductive surface areas deposed on a substrate. A "wire" can thus be thought of as a sequence of rectangular segments, as shown in Figure 16.26.

We assume that the lengths of these segments are fixed, while the widths x_i need to be sized according to the criteria explained next. A common approach is to model the wire as the cascade connection of RC stages, where, for each stage, $S_i = 1/R_i$, C_i are, respectively, the conductance and the capacitance of the i-th segment, see Figure 16.27.

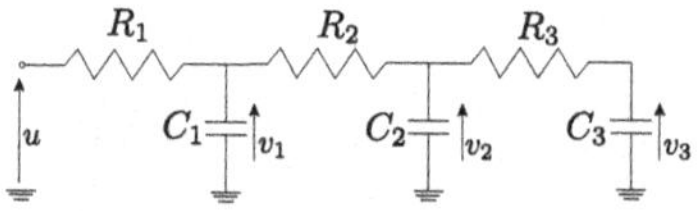

The values of S_i, C_i are proportional to the surface area of the wire segment, hence, since the lengths ℓ_i are assumed known and fixed, they are affine functions of the widths, i.e.,

$$S_i = S_i(x_i) = \sigma_i^{(0)} + \sigma_i x_i, \quad C_i = C_i(x_i) = c_i^{(0)} + c_i x_i,$$

where $\sigma_i^{(0)}, \sigma_i, c_i^{(0)}, c_i$ are given positive constants. For the three-segment wire model illustrated in the figures, one can write the following set of dynamic equations that describe the evolution in time of the node voltages $v_i(t)$, $i = 1, \ldots, 3$:

$$\begin{bmatrix} C_1 & C_2 & C_3 \\ 0 & C_2 & C_3 \\ 0 & 0 & C_3 \end{bmatrix} \dot{v}(t) = - \begin{bmatrix} S_1 & 0 & 0 \\ -S_2 & S_2 & 0 \\ 0 & -S_3 & S_3 \end{bmatrix} v(t) + \begin{bmatrix} S_1 \\ 0 \\ 0 \end{bmatrix} u(t).$$

These equations are actually expressed in a more useful form if we introduce a change of variables

$$v(t) = Qz(t), \quad Q = \begin{bmatrix} 1 & 0 & 0 \\ 1 & 1 & 0 \\ 1 & 1 & 1 \end{bmatrix},$$

from which we obtain

$$\mathcal{C}(x)\dot{z}(t) = -\mathcal{S}(x)z(t) + \begin{bmatrix} S_1 \\ 0 \\ 0 \end{bmatrix} u(t),$$

where

$$\mathcal{C}(x) \doteq \begin{bmatrix} C_1 + C_2 + C_3 & C_2 + C_3 & C_3 \\ C_2 + C_3 & C_2 + C_3 & C_3 \\ C_3 & C_3 & C_3 \end{bmatrix}, \quad \mathcal{S}(x) \doteq \mathrm{diag}\,(S_1, S_2, S_3).$$

Figure 16.26 A wire is represented as a sequence of rectangular surfaces on a substrate. Lengths ℓ_i are fixed, and the widths x_i of the segments are the decision variables. This example has three wire segments.

Figure 16.27 RC model of a three-segment wire.

Clearly, $\mathcal{C}(x)$, $\mathcal{S}(x)$ are symmetric matrices whose entries depend affinely on the decision variable $x = (x_1, x_2, x_3)$. Further, one may observe that $\mathcal{C}(x)$ is nonsingular whenever $x \geq 0$ (as is physically the case in our problem), hence the evolution of $z(t)$ is represented by (we next assume $u(t) = 0$, i.e., we consider only the free-response time evolution of the system)

$$\dot{z}(t) = -\mathcal{C}(x)^{-1}\mathcal{S}(x)z(t).$$

The *dominant time constant* of the circuit is defined as

$$\tau = \frac{1}{\lambda_{\min}(\mathcal{C}(x)^{-1}\mathcal{S}(x))},$$

and it provides a measure of the "speed" of the circuit (the smaller τ, the faster is the response of the circuit).

Describe a computationally efficient method for sizing the wire so as to minimize the total area occupied by the wire, while guaranteeing that the dominant time constant does not exceed an assigned level $\eta > 0$.

Index

Printed in the United States
By Bookmasters